Problem Solving and Data Analysis using Minitab

Problem Solving and Data Analysis using Minitab

A clear and easy guide to Six Sigma methodology

Rehman M. Khan

Chartered Chemical Engineer and Six Sigma Black Belt, Loughborough, UK

A John Wiley & Sons, Ltd., Publication

This edition first published 2013

Registered office
John Wiley & Sons Ltd, The Atrium, Southern Gate, Chichester, West Sussex, PO19 8SQ, United Kingdom

For details of our global editorial offices, for customer services and for information about how to apply for permission to reuse the copyright material in this book please see our website at www.wiley.com.

Library of Congress Cataloging-in-Publication Data

Khan, Rehman M.
Problem solving and data analysis using Minitab : a clear and easy guide to six sigma methodology / Rehman M. Khan.
pages cm
Includes index.
ISBN 978-1-118-30757-1 (hardback)
1. Mathematical statistics--Data processing. 2. Problem solving--Statistical methods. 3. Minitab.
4. Six sigma (Quality control standard) I. Title.
QA276.45.M56K43 2013
658.4'013028553--dc23

2012035286

A catalogue record for this book is available from the British Library.

ISBN: 978-1-118-30757-1

Typeset in 10/13 Frutiger LT Std by Laserwords Private Limited, Chennai, India

Contents

Acknowledgements

Firstly, I would like to thank God for giving me the ability and circumstance to write this book.

I would like to thank my wife, Mahwish, and my children, Iqra, Humzah and Raeesa, for being very patient with me and accepting that I would be spending all of my spare time outside of work on this project. Mahwish and Humzah (12 years old at that time) also helped me with the proof-reading and transferring the script from Power Points to Microsoft Word.

I would like to acknowledge my parents' effort to continually challenge me and my brother in our education. I think my dad would have been very proud at the release of this book and I know my mum is.

I would like to thank our family down in London, particularly, my mum and wife's parents who did not see us very much during the production of the manuscript but still continued to support us.

I want to thank all my family and friends around the world for their support. There are too many people to name so I will just say where you live, Loughborough Bedford, London, Preston, Birmingham, Sheffield, Pakistan, North America, Australia, Saudi Arabia. If I haven't mentioned where you live please feel free to associate yourself with one of the places I did mention.

I also want to mention Springfields Fuels Limited near Preston. I spent most of my working life there and they introduced me to Six Sigma. I also want to mention my current employer, British Gypsum, and thank them for my continued Six Sigma training. In particular, Gary Pilcher who supported my training and then the start of my training course within British Gypsum. There is a fantastic team at East Leake, I want to thank everyone there for their moral support, in particular the other members of the Melnik 6 (Greg Bere, Paul Brauer, Matt Carey, Lee Chaplin & Gary Parkins).

I want to thank the good people at Wiley for publishing this book. I also want to thank the good people at Minitab for letting me use their excellent software.

Some teachers you never forget. So I would also like to thank Mr Crowley who was my Mathematics teacher at Hayes Manor School. He took me through my 'O', 'AO' and 'A' Level Maths.

Finally, I would like to thank the people that have bought this book. I am hoping that this is the start of the journey for you as well as me.

CHAPTER 1
Introduction

Confucius said 'I hear and I forget. I see and I remember. I do and I understand'. This proverb sums up the spirit of this book as it is very practical and it involves the reader. Every chapter contains examples and exercises that will capture the reader and ensure the information is passed on in a memorable way.

This book is aimed at numerical professionals, students or academics who wish to learn and apply statistical techniques for problem solving, process improvement or data analysis without getting bogged down in theory. In fact anyone that wants to be data driven in their decision making should use this book to understand how to use Minitab.

The vehicle for the statistical techniques is Minitab version 16. However, most of it will also be applicable to version 15. The notes will indicate if a feature is in Minitab 16 only by stating 'M16 only'. At this point I could give you a list of the new features that are incorporated within Minitab 16 but that would be a bit pointless, so I will attempt to point out the important new features as we go through the modules. You will definitely need to have access to a copy of Minitab when going through the book and it would be advantageous to have a copy of Microsoft Excel as well.

Minitab has developed considerably between versions 15 and 16. One of the main enhancements is the Assistant which helps users select the appropriate test, enter the data and interpret the results. The Assistant is available for a number of the key test procedures. Users of older versions of Minitab and also Minitab 16 have the option of using the non-Assistant methods which are accessed via the traditional drop down menus. For convenience, within the book I will refer to this method as the Classic method.

I am a chartered Chemical Engineer and Six Sigma Black Belt. I have worked in the nuclear industry, cosmetics industry and in construction products manufacturing. My main role has always been process improvement and the projects that I have completed have won awards and saved millions of pounds. Whilst honing my problem solving and data analysis skills I recognised that there was a huge gap in the availability of appropriate training materials and yet there was a huge demand from numerical professionals to learn the skills. I was disappointed not to be able to find the right sort of books to help me learn and understand how to use Minitab. However, I was fortunate enough to have my employers put me through my Six Sigma training.

My aim is not to blind the reader with mathematical theory but to teach problem solving and data analysis through the use of statistical analysis in a very graphical and accessible way. The book uses example based learning that each reader can work through at their own pace. Each example is broken down into the very exact steps that must be followed in order to work through the complex analysis. After the examples there is usually an exercise so that the reader can be assured that they have understood the key learning points. However, even the answers do not leave the reader cold with just a single numerical solution. The exercise answers show graphic milestones that the reader must achieve in order to reach the endpoint of the analysis. The example and exercise data sets can be downloaded from the Wiley publishing website, www.wiley.com/go/six_sigma_methodology.

Problem Solving and Data Analysis using Minitab: A clear and easy guide to Six Sigma methodology, First Edition. Rehman M. Khan.

I have found that other books teaching this subject make themselves inaccessible to beginners either because complex statistical theory is put before practical learning or there is an attempt to cover everything that Minitab can do. The main strengths of this book are that it is a training course in book form and it teaches a much sought after skill set. It teaches the reader using a logical and stepwise methodology. The examples and exercises take the reader through key learning points and, because they are so very easy to follow, they build the reader's confidence. The book does not cover every possible topic. For instance it does not cover the handling of attribute data. This is due to space limitations and the fact that in a numerical world we come across this type of data set less and less. However, if there are procedures that you wish to learn with attribute data I suggest that you learn the equivalent test for continuous data and then use the help system to understand how to handle attribute data. Also, we will not be learning Six Sigma project methodology as it would probably double the page count.

I would like to introduce a metaphor for problem solving. I am standing in Field A but when I solve my problem I will be able to move over to Field B which is a better place. In between is a rather large wall, and this wall represents the problem. The wall has a gate and when I find the key, which can be represented by the answer, I can open the gate and step through. The problem is that I need to find the key and it can be anywhere in Field A. However, to help me find the key I have a metal detector. The thing that determines how well I use the metal detector is my own skill. Well, in the real world the metal detector is Minitab and to find the solution to numerical problems I must ensure that I can utilise this fantastic tool. This book's sole purpose is to increase your skill level with Minitab. The intention is to do this by giving you a bit of theory and then getting you to solve problems using Minitab. Luckily, Minitab will make our life easy by doing all the complicated maths. All we need to do is tell it what to do and then understand what it is telling us.

Let's have a look at what we are going to be learning throughout this book. If you are beginner then I strongly recommend that you start from the very beginning and at least go through Chapters 2–4 before you start skipping forward to topics of interest. If you are an experienced user of Minitab then feel free to launch into any of the topics that are of interest to you.

1. Introduction
   This is the wordiest chapter in the book and you're doing well to get through it. In the rest of the book we are going to keep you busy in front of your computer working on Minitab.

2. Minitab Navigation
   In this chapter you are going to get the feel of Minitab. We will start by discussing the Minitab windows and then the drop down menus. As discussion is not a very good way of learning we will then import some data and look at the data formatting system used within Minitab. For additional learning we will put together some simple graphs and we will learn to add additional content to those graphs. This has a twofold purpose in that the reader will experience some of the stand alone graphing tools and will become familiar with Minitab navigation. Later you will see that a lot of Minitab's statistical procedures produce their own graphs which are very useful for data analysis. We also need to learn how to use the report pad so that we can send reports to Microsoft Word and PowerPoint.

3. Basic Statistics
   This is the chapter where we learn the basic background statistics that we need to know to help us understand what is coming up in the more demanding chapters. If you are new to statistical methods you have probably only been introduced to descriptive statistics. In this chapter we learn one of the most important concepts in the course and that is the difference between descriptive

statistics and inferential statistics and how that links to the concept of the entire population and a sample set from that population.

4. Hypothesis Testing
   You could say that hypothesis testing is one of the foundational chapters of the statistical procedures that we are going to be using. Once we get to grips with the theory we will start using some of the test procedures within Minitab. We will start by looking at a single set of data and determining whether the mean of the population could be a particular value. We will then compare two data sets and attempt to infer conclusions about them.

5. Analysis of Variance
   The analysis of variance (ANOVA) procedure builds on the last chapter and lets us make comparisons between two or more groups. We start by looking at the one way ANOVA procedure and move onto the ANOVA General Linear Model (GLM). The GLM allows us to model multiple factors and multiple levels.

6. Measurement System Analysis
   This chapter teaches us procedures to help us understand whether our measurement systems are adequate and reliable. The measurement system can be an instrument, like a weigh scale, or it could be a person making a judgment about a product or process. Either way we want to know if the measurement systems are free from human failings and whether they can be relied upon.

7. Statistical Process Control
   Statistical process control (SPC) is about monitoring your process and looking for unusual occurrences by using control charts. Finding out the type of unusual behaviour and when it occurs helps us to identify the initiating events. We can then put controls in place that will eliminate the initiating events and thereby making the process more stable.

8. Process Capability
   We use process capability to measure our customer requirements against what the process is actually delivering. This is often done at the start of the project to measure the gaps and then again at the end of the project to demonstrate the improvement.

9. Correlation and Regression
   These two terms are often confused. In the world of Minitab correlation is about establishing whether two parameters have a linear relationship and checking the strength of that relationship. Regression goes further by trying to fit an equation to model the relationship. We will learn procedures for single and multiple factor regression.

10. Design of Experiment
    Don't worry. Design of experiment (DOE) is not always about being in a laboratory and conducting experiments. DOE is actually an optimised methodology for checking selected inputs and checking whether they have a statistically significant effect on a selected response. DOE tells us how strongly each factor is affecting the response and whether any of the interactions are important to the response.

    There are many types of DOE and we will concentrate on two of the more commonly used types. DOE comes with its own terminology so we will spend a bit of time getting to grips with the terminology before we get into DOE proper.

11. Help
    Minitab has a number of comprehensive help systems that can help you should you get stuck. This chapter will introduce the help systems and show you what is available so you will be able to quickly get the information that you need.

This book originally started out as a course that I ran within British Gypsum. I then decided that I could do with less time with my wife and family and it would be a good idea to convert the course into a book. One of the concepts that I introduced during the course was the Statistical Charter. The charter was a self imposed set of rules designed to protect the stakeholders, customers and no doubt our own reputations. The charter has the following functions:

- Stops us from turning fiction into fact.
- Warns stakeholders regarding the certainty of our conclusions.
- Sets limits so that we can safely execute the statistical procedures.

You will see that the charter manifests itself within the care that we take to set up the test procedures. I can give you guidance that, for example, we should try and aim for a power of between 80–90% but when it comes to using the tests within the real world you may choose to ignore the advice. However, most of the advice given here is derived from the Minitab White Papers and gives a robust methodology for conducting the procedures.

Statistics can have a bad reputation. It was Benjamin Disraeli who said 'There are lies, damn lies and statistics' and we don't want to perpetuate that reputation. Personally, I think a particular quote from the Toby McGuire Spiderman movie is more apt for the exponents of statistical analysis: 'With great power comes great responsibility'.

The statistical techniques taught within this book are most commonly found within Six Sigma. They are used for problem solving, data analysis and the reduction of variation within the DMAIC Framework. There are other business improvement frameworks such as World Class Manufacturing (WCM) and these tools and techniques sit equally well within the Focussed Improvement (FI) Pillar of WCM. However, the use of Six Sigma is far more widespread and it is more commonly associated with Process Improvement and Minitab.

The table below shows how the statistical toolset can be used within the DMAIC Framework. The tools used are not set in stone as the methodology allows the use of any tool which gets the job done.

For each of the chapters I would strongly recommend that you initially read the theory section in the chapter and then, when it comes to examples in the book, follow the instructions to work through each example on a PC with a copy of Minitab. When you are happy that you understand the methodology applied to solve the example problems then try a single exercise without using the book. After completing the exercise go back and review the solution shown within the book. If you are happy that you have successfully completed the exercise then complete all the remaining exercises in the same way. If not, review the notes and the solution to establish where you went wrong and then repeat the exercise. As a gift to the reader for reading this far into the introduction I will tell you that the exercises are usually easier than the examples.

Define	Measure	Analyse	Improve	Control
Used to set Measures of performance for project, Define gap between process outputs and customer requirements. Tools used: Project Charter, Critical To Tree, SIPOC.	Used to start understanding the process. Carry out Measurement Systems Analysis (MSA) on the Key Output Process Variables (KOPV) and captute the process baseline. Statistical Tools used: SPC, Process Capability, MSA. Tools used: Project Charter, Critical To Tree, SIPOC.	Used to define the Key Process Input Variables (KPIVs). Need to identify sources of variation and screen potential causes. MSA of KPIVs. Tools used: Process Mapping, Brainstorming, Fishbone Diagram, Cause and Effect Matrix, FMEA. Statistical Tools used: MSA, SPC, Hypothesis Testing, ANOVA, Correlation and Regression, DOE	Optimises settings of KPIVs to deliver the required output. Statistically vali- dates new operating conditions. Tools used: Updated FMEA, Cost/Benefit Analysis, Error Proofing Statistical Tools used: MSA, SPC, Hypothesis Testing, ANOVA, Correlation and Regression, DOE	Locks in and documents optimised parameters. Measures improvement and sustainability. Tools used: Updated FMEA, Control Plan, Procedures reviewed, engineering and operaating, Success Sheets/Best Practise Sheets, Project Report and Presentations. Statistical Tools used: SPC, Process Capability,

In case you are wondering, all of the data sets used are fictitious. They were designed to show particular aspects of the methodology being taught.

And finally, if you are in need of some inspiration go onto YouTube and search for the 'Minitab song', it keeps me motivated!

CHAPTER 2
Minitab Navigation

2.1 Windows

In this Module we want you to get the feel of Minitab. We are going to do this by first discussing the Minitab windows and then the dropdown menus. As discussion is not a very good way of learning we will then import some data and get you to start working with that data.

1. Open Minitab.

You will be presented with the screen below.

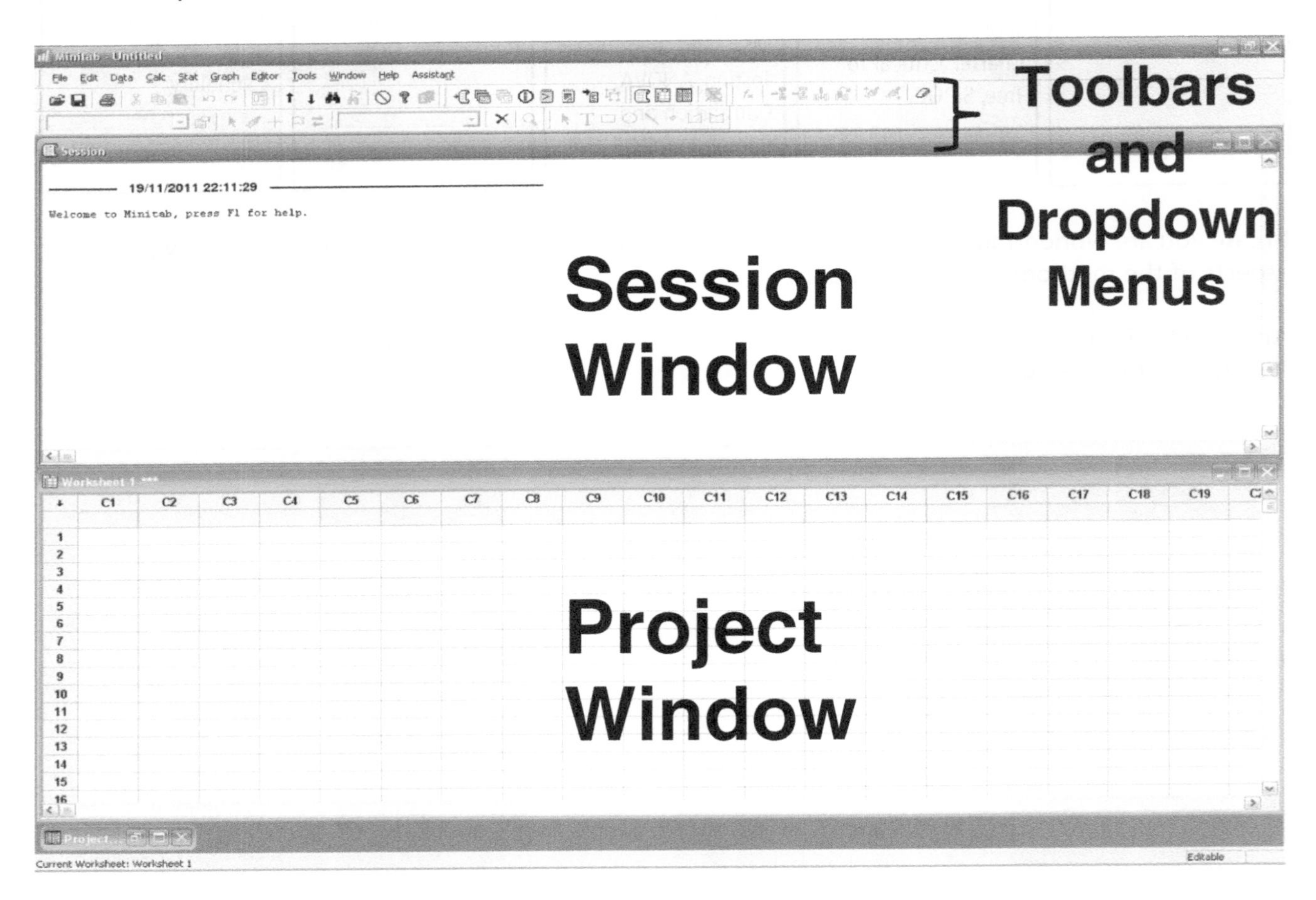

At the top of the screen there are the dropdown menus and toolbars, which work like most Microsoft applications. Below that we have two windows: the session window and the project window. The graphs window is not shown at start up but it is also commonly used.

Problem Solving and Data Analysis using Minitab: A clear and easy guide to Six Sigma methodology, First Edition. Rehman M. Khan.

The project window is mainly used to enter data into Minitab and is used in a similar way to Microsoft Excel. However, the main difference is that Minitab wants you to present data in columns and is quite strict about having the same data format within a column.

The session window is where the user is given numerical outputs. You will have guessed that the graphs window is where the user is given graphical outputs. Graphs can exist in their own windows but it's best to keep them minimised and only call them up when you need them from the graphs folder.

Now that we know what the windows do, let's learn to move between the windows.

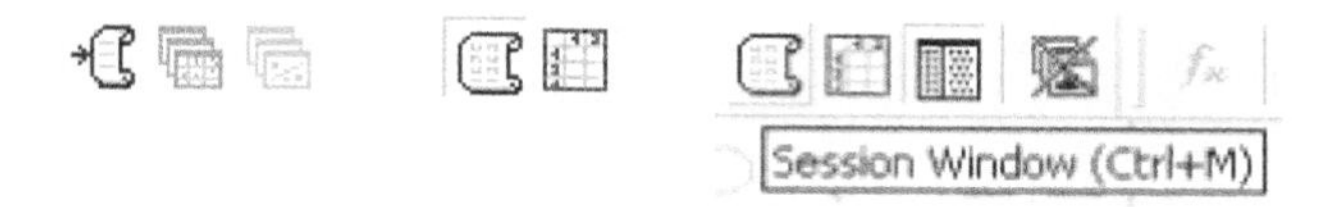

2. Locate the three folder icons in the tool bar as shown in the above, left figure.

3. Single click between the live icons a couple of times. Notice how the folder icons not only open the corresponding window but also a project manager window. The window icons, as shown in the central figure above will open the particular window and not the project manager. The window you select will cover the complete screen if the windows were previously maximised. Give that a go now and finish up on the project window.

4. If you forget what the icons are for, you can hover over them with the mouse and a pop-up box will open to tell you. As shown in the above, right figure.

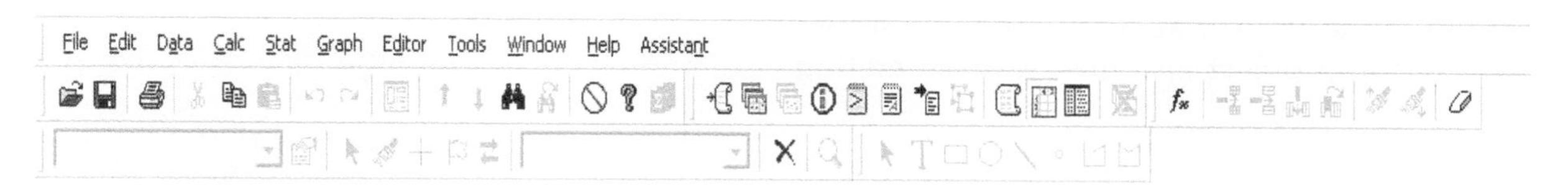

The figure above shows the dropdown menu and default toolbars for Minitab 16 (M16). The figure below the next paragraph is for Minitab 15 (M15). In fact all the dropdown menus are virtually the same between M15 and M16, with the exception of the Assistant, which is a new feature on M16. There are many new features within M16 but these are better explored as we go through the modules because the majority of them are for experienced users.

Next we will look at the dropdown menus. Later we will explore the key icons within the toolbar by working with some data which we will import.

2.2 Dropdown Menus

The File dropdown should be familiar as it is similar to most Microsoft programs.

Project and Worksheet Description allow the user to enter information about either item. This can be very useful when you refer back to the work in the future.

Open Worksheet and Other Files both allow the import of data from various common file types.

Query Database is a facility that allows the user to import information from a database.

There are options for printing presented here, however, from my experience printing for the purpose of producing reports is best done via the Report Pad.

File
New... Ctrl+N
Open Project... Ctrl+O
Save Project Ctrl+S
Save Project As...
Project Description...
Open Worksheet...
Save Current Worksheet
Save Current Worksheet As...
Worksheet Description...
Close Worksheet
Query Database (ODBC)...
Open Graph...
Other Files
Save Window As...
Print Worksheet... Ctrl+P
Print Setup...
Exit

The Edit tab is not used that often. This is because right clicking on an object also gives custom editing options.

Cut, Copy and Paste are more easily done with Ctrl+X, C and V, respectively.

Worksheet Links can be useful. They allow the user to link a section of a spreadsheet to the data within the project window. Depending on how the link is managed, changes to the spreadsheet will automatically be applied to the data within the project window.

The Edit Last Dialog facility is tremendously useful but I find it easier to use directly from the toolbar.

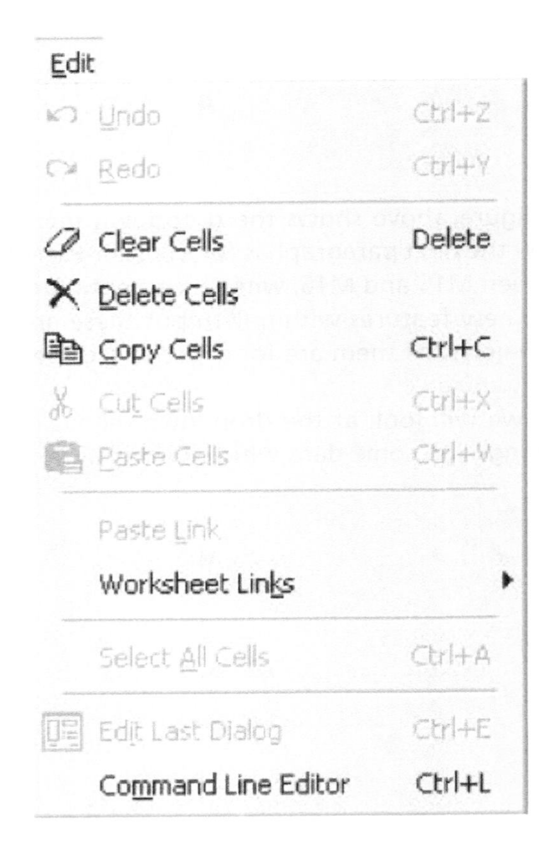

The top half of the toolset is used to arrange the data that is in the worksheet window. On occasion subsets may need to be moved from the same column to adjacent columns and vice versa. Copy contains predefined copying procedures that can be used as shortcuts.

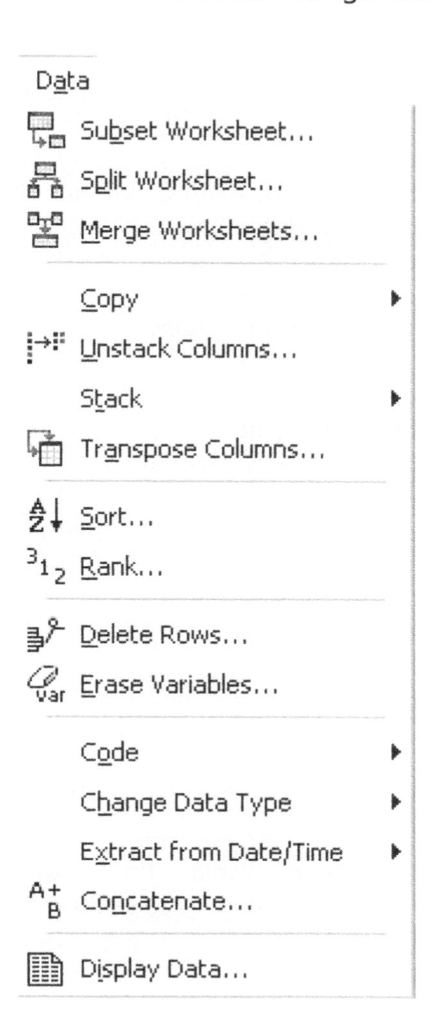

As in Excel, Transpose Columns will spin the selected data by 90°. We need to be wary of data types though.

Code works to replace selected items throughout a column and lets you convert data types in the process.

Change Data Type is useful in correcting the data type.

Concatenate is the same as the Excel function and adds text strings together.

Using Calc will be covered in the module as it is quite useful. Column and Row Statistics pretty much do as expected and give you an output in the session window.

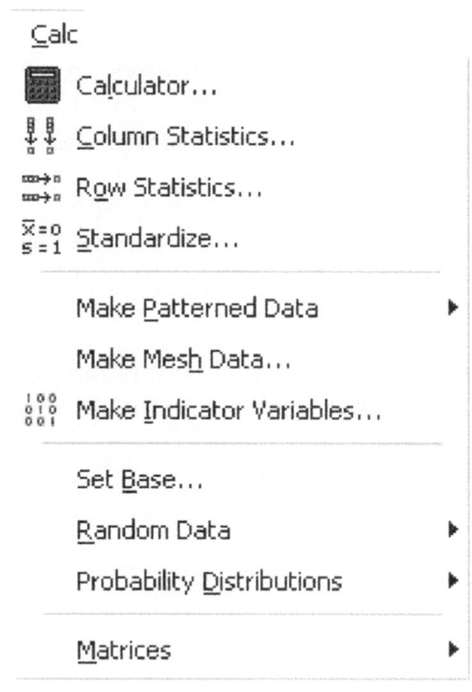

The central and bottom sections are used to generate different types of data distributions. Making mesh data can be quite fun in a nerdy sort of way.

The Stat menu is the heart of Minitab and contains the majority of the statistical tests.

This book will focus on a number of these menus and submenus. We will be going into Regression, ANOVA, DOE, Control Charts and Quality Tools submenus in later modules.

Basic Statistics will be the most frequently used submenu.

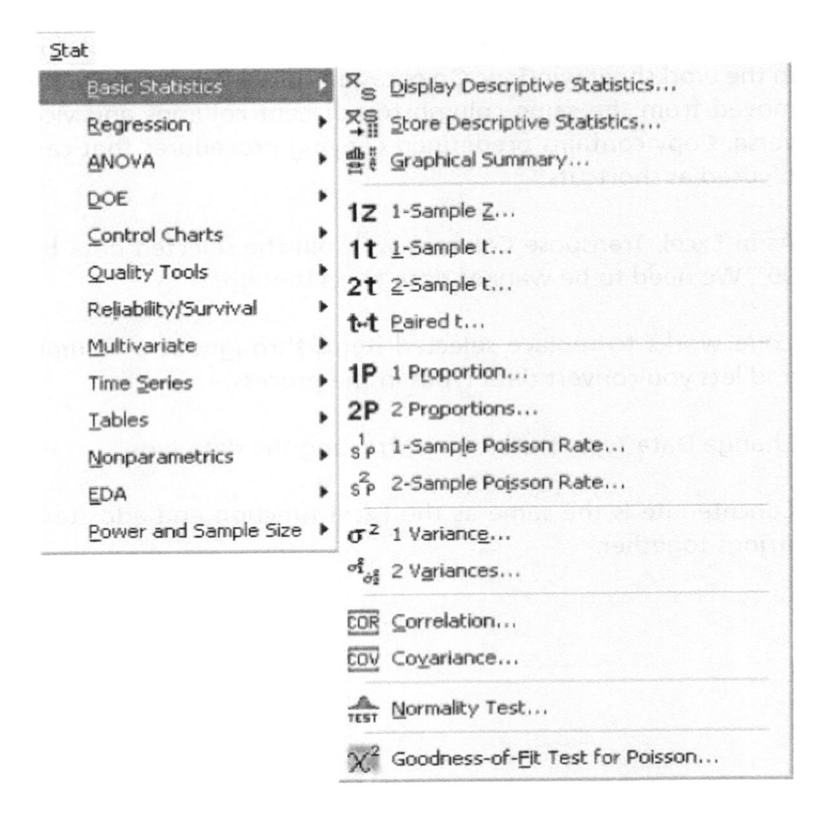

Using graphs to display data is usually a quick and easy way to see trends and should be done in the early stages of analysis.

The Graph menu in Minitab supplies a large variety of graphing tools which can provide a very different insight into the data than you would normally see with standard spreadsheet programs.

Minitab will generate a large number of these graphs when statistical tests are carried out.

We are going to look at a number of these graphs within this module.

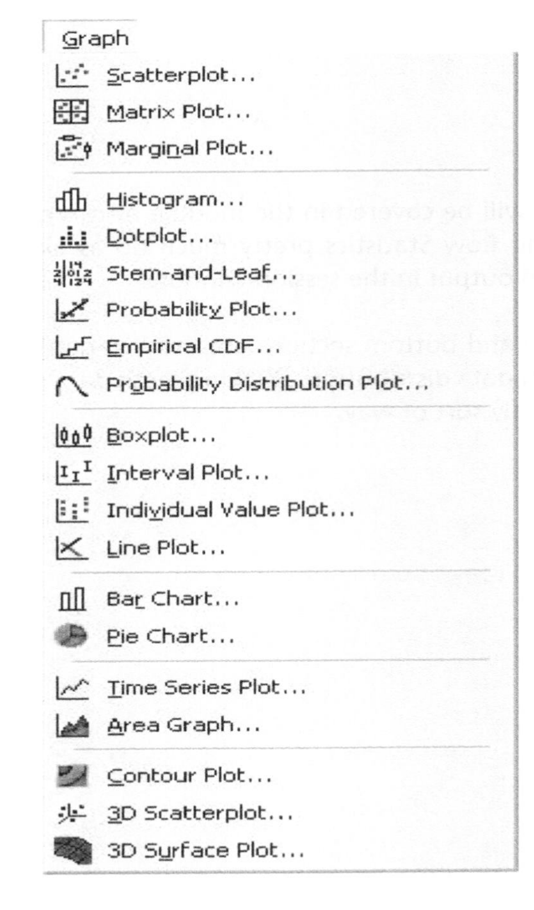

The Editor dropdown menu contains a number of standard editing functions. A number of these are available by right clicking on the live worksheet.

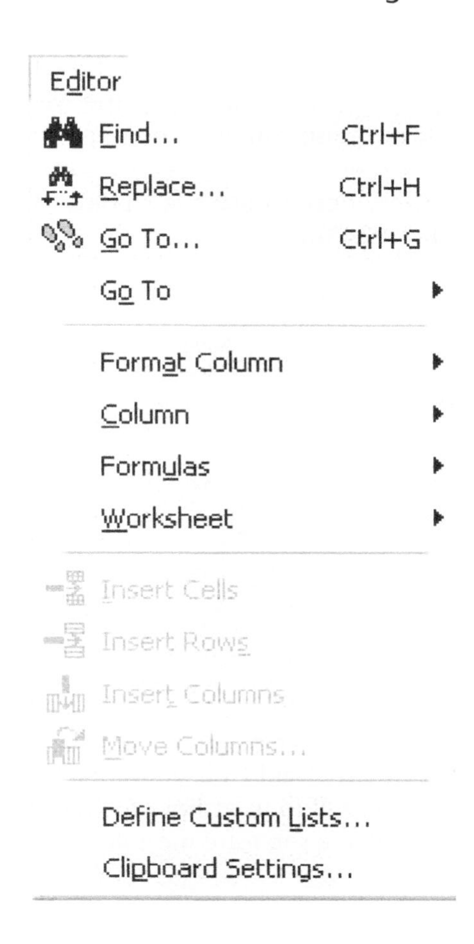

Column formatting is done here.

Define Custom Lists is quite interesting. It can be used to define a standard list that when set up can then be added into the worksheet by entering the starting element of the list and then pulling down on the handle of the active cell. This is particularly useful if you have a custom list that you regularly want to add to the project window.

The three applications listed at the top of the Tools menu can be launched from here.

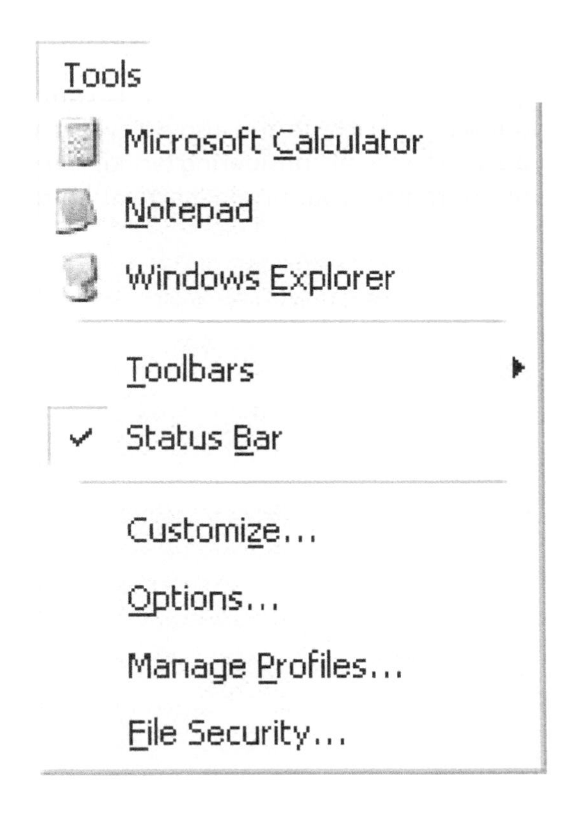

Toolbars can be used to set which toolbars are active. For example, the 3D Graph Editing tool bar can be activated from here.

Customize can be used to change a large number of features. It is best fully explored by the reader in their own time.

Amongst other things Options is used to set:

- the root directory for saving work,
- the background colour on all graphs,
- the output options on statistical tests.

File Security can be used to set password protection for our files.

The icons in Window are most useful when you are working on a large project whilst utilising all the windows. They can be used to quickly organise your windows and graphs.

We will show how to use the Update All Graphs Now function later in this module.

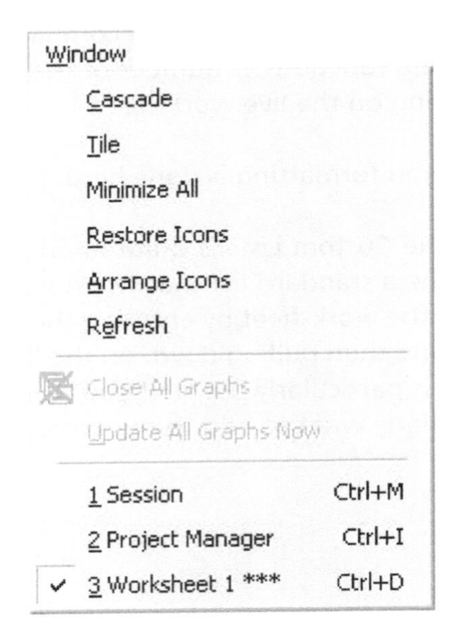

The Help menu should be your first option when faced with problems. We will look at how to use the Help system in much greater detail in a separate module.

Minitab actually comes supplied with a number of sample data sets which are utilised within the help systems in order to demonstrate statistical procedures.

Licensing is a M16 only option. It contains information about your Minitab License and also a clever feature to Move License. If you are mobile or considering working from home it would allow you to transfer your product activation to another location.

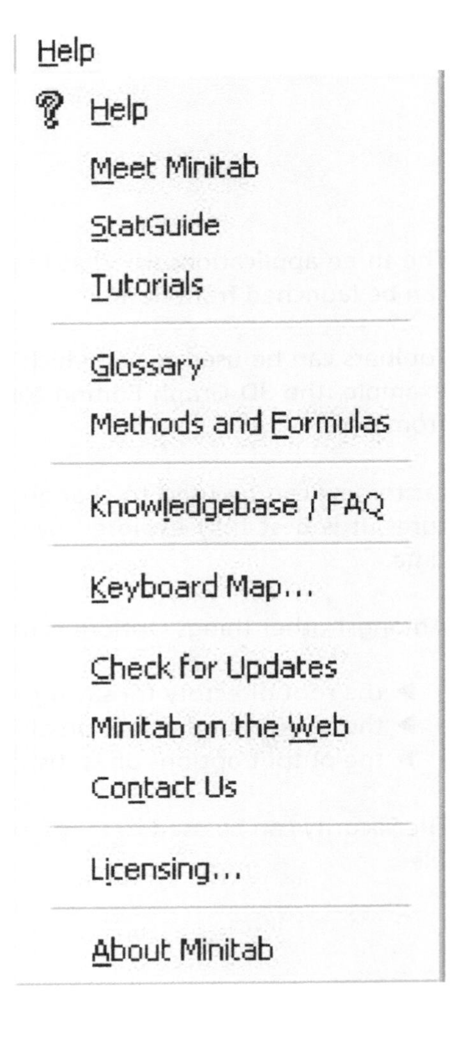

The Assistant dropdown is M16 only and is one of the key advances in the update. The Assistant will lead you through the tool selection and ensure that the pre-requisites are correct. It will also provide you with an interpretation of the results. It could make guide texts and training redundant!

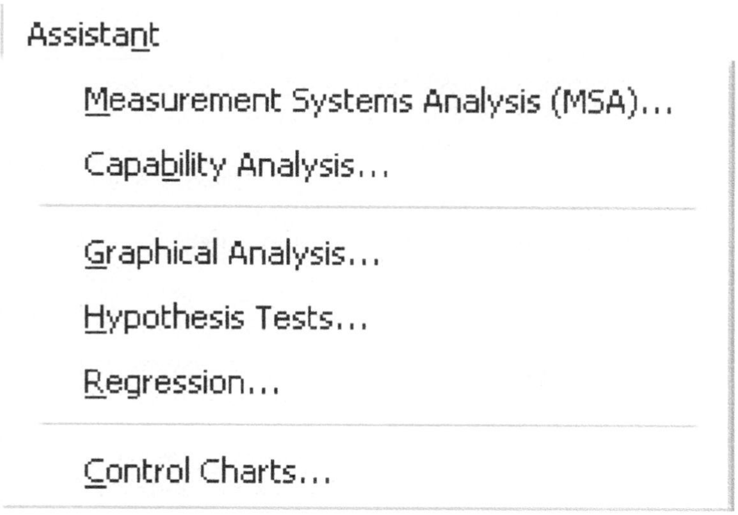

Assistant

- Measurement Systems Analysis (MSA)...
- Capability Analysis...
- Graphical Analysis...
- Hypothesis Tests...
- Regression...
- Control Charts...

2.3 Importing Data

We are now going to look at importing some data into Minitab. We will then carry out a number of exercises on that data. These will include checking the data format within the columns and, if we need to, we will change the data format. We will then manipulate the data by adding columns together and then produce a number of different types of graphs and add content to those graphs. Once we have a few graphs we will look at how we can export our data so that others can view it. We will finish the session by carrying out some Minitab customisation and also using the Assistant to produce some graphs.

The table below shows the data that we are going to work with. The data concerns a piece of equipment that requires both gas and electricity to run. The equipment has two zones. The Total Cost row relates to how much it costs to run the equipment. Person relates to the operator who was running the equipment and Stage refers to a particular event that we want to monitor; in this case it relates to when maintenance was conducted on the machine. Stage changes from condition Before to Stage1 so we can tell when the maintenance occurred.

The data currently lives within Microsoft Excel and it is formatted in rows not columns. Each row has a title cell and then 38 data cells.

1. Open File 02 Minitab Navigation.xls using Microsoft Excel.

2. Go to 'Data in rows' worksheet.

Minitab analyses data in columns. We need to change the arrangement of this data so that it is in columns. Minitab has a Transpose function but it is better to prepare the data in Excel and then transfer it to Minitab. This is due to wanting to maintain both the column headings and the mixed data types in the example data.

	A	B	C	D
1	Date	03/10/2012	03/10/2012	03/10/2012
2	Time	03:00:00	04:00:00	05:00:00
3	Total Cost	266.53577	266.53577	266.53577
4	Zone1 Electricity	129	129	130
5	Zone2 Electricity	120	120	121
6	Zone1 Gas	3222	3232.9	3222
7	Zone2 Gas	2927.1	2927.1	2938
8	Feedrate	17.2a	17.2	17.2
9	Stage	Before	Before	Before
10	Person	Chappo	Miff	Omer

3. In order to shift the table around so that the data is in columns first copy target cells A1 to AM10.

4. Go to worksheet 'Sheet2'.

5. Right click on cell A1 and use the Paste Special command from within the dropdown menu.

6. Select Transpose from the selection box that opens and then click OK.

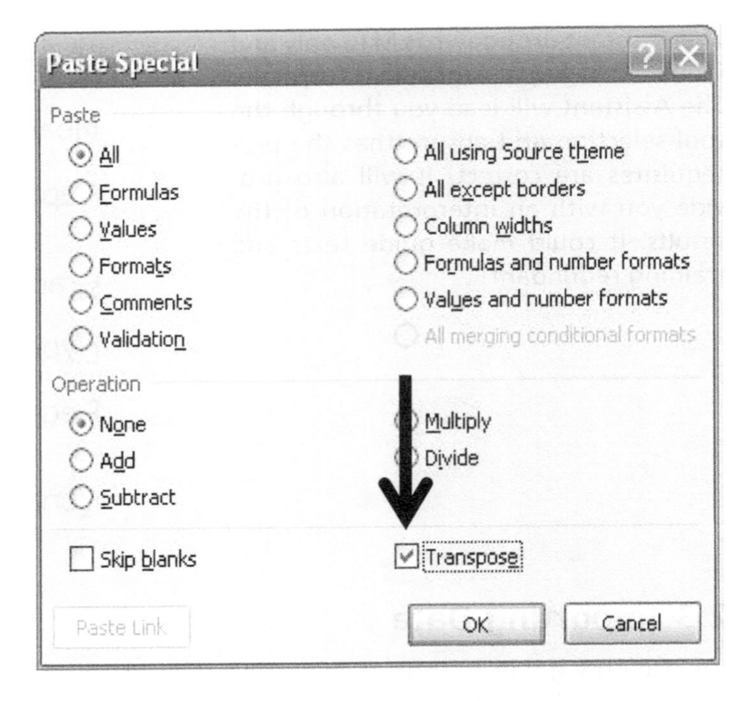

If you had trouble you can use the table that I have already transposed. It is in the 'Data in Columns' worksheet. There are two easy ways to transfer the data from an Excel worksheet into Minitab.

7. Copy cells A1 to J39 of the transposed cells by highlighting and then pressing Ctrl+C.

8. Then Paste the data, by pressing Ctrl+V, into Minitab with Cell A1 being copied into the title row directly under 'C1' and above Row1 in the shaded row. Minitab uses these rows for titles. Pasting under this row would cause the title row to be counted as data and this would upset the column formatting.

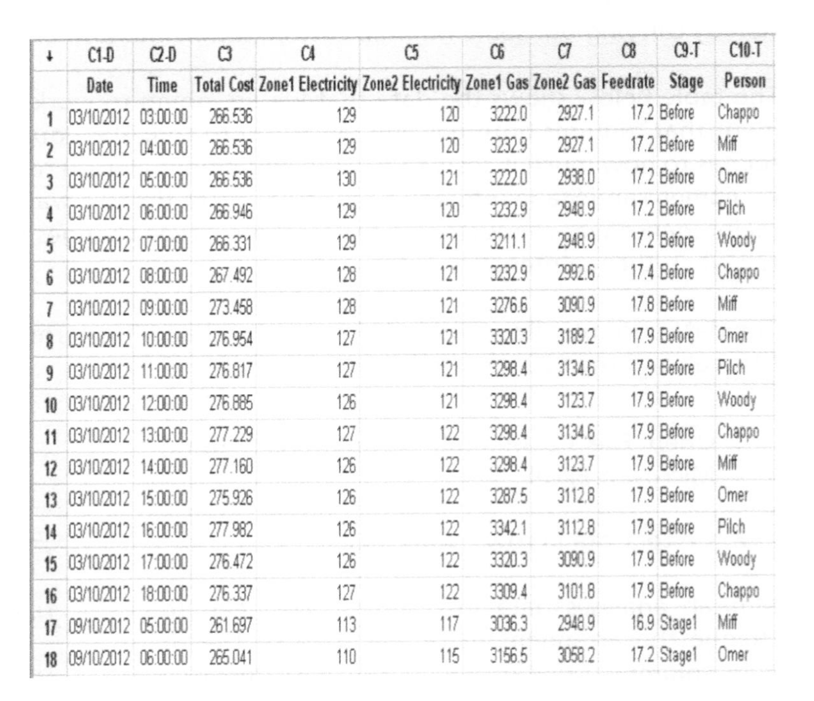

↓	C1-D	C2-D	C3	C4	C5	C6	C7	C8	C9-T	C10-T
	Date	Time	Total Cost	Zone1 Electricity	Zone2 Electricity	Zone1 Gas	Zone2 Gas	Feedrate	Stage	Person
1	03/10/2012	03:00:00	266.536	129	120	3222.0	2927.1	17.2	Before	Chappo
2	03/10/2012	04:00:00	266.536	129	120	3232.9	2927.1	17.2	Before	Miff
3	03/10/2012	05:00:00	266.536	130	121	3222.0	2938.0	17.2	Before	Omer
4	03/10/2012	06:00:00	266.946	129	120	3232.9	2948.9	17.2	Before	Pilch
5	03/10/2012	07:00:00	266.331	129	121	3211.1	2948.9	17.2	Before	Woody
6	03/10/2012	08:00:00	267.492	128	121	3232.9	2992.6	17.4	Before	Chappo
7	03/10/2012	09:00:00	273.458	128	121	3276.6	3090.9	17.8	Before	Miff
8	03/10/2012	10:00:00	276.954	127	121	3320.3	3189.2	17.9	Before	Omer
9	03/10/2012	11:00:00	276.817	127	121	3298.4	3134.6	17.9	Before	Pilch
10	03/10/2012	12:00:00	276.885	126	121	3298.4	3123.7	17.9	Before	Woody
11	03/10/2012	13:00:00	277.229	127	122	3298.4	3134.6	17.9	Before	Chappo
12	03/10/2012	14:00:00	277.160	126	122	3298.4	3123.7	17.9	Before	Miff
13	03/10/2012	15:00:00	275.926	126	122	3287.5	3112.8	17.9	Before	Omer
14	03/10/2012	16:00:00	277.982	126	122	3342.1	3112.8	17.9	Before	Pilch
15	03/10/2012	17:00:00	276.472	126	122	3320.3	3090.9	17.9	Before	Woody
16	03/10/2012	18:00:00	276.337	127	122	3309.4	3101.8	17.9	Before	Chappo
17	09/10/2012	05:00:00	261.697	113	117	3036.3	2948.9	16.9	Stage1	Miff
18	09/10/2012	06:00:00	265.041	110	115	3156.5	3058.2	17.2	Stage1	Omer

The alternate method of importing data into Minitab is shown below.

9. From within Minitab click File <<Open Worksheet.
10. Navigate to the folder that contains the data file. Ensure that the correct file type has been selected. Then double click on the file that you want to open.
11. Minitab will put the data from each Excel worksheet into a separate Minitab Worksheet. Notice how the formatting of the Data in Rows sheet is in disarray.
12. Close all the Minitab worksheets with the exception of the Data in Columns worksheet.

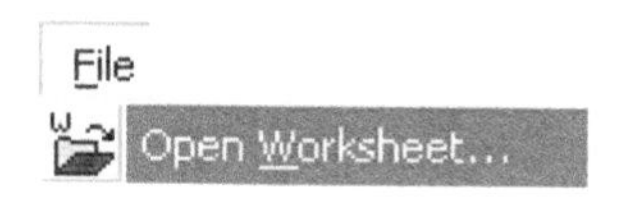

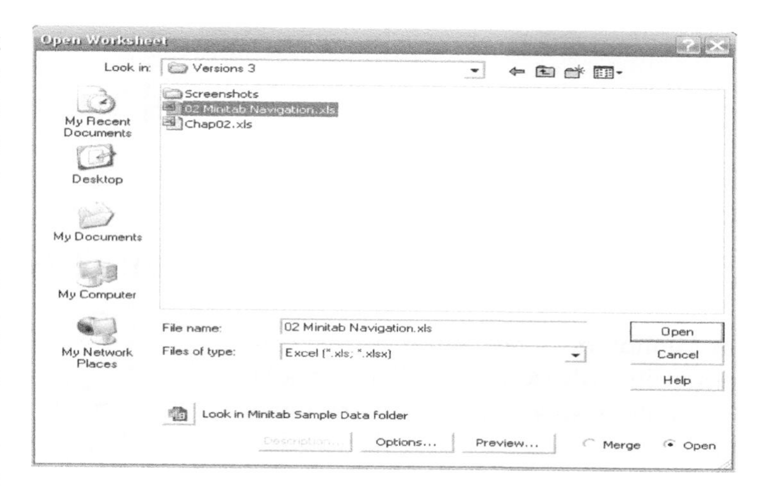

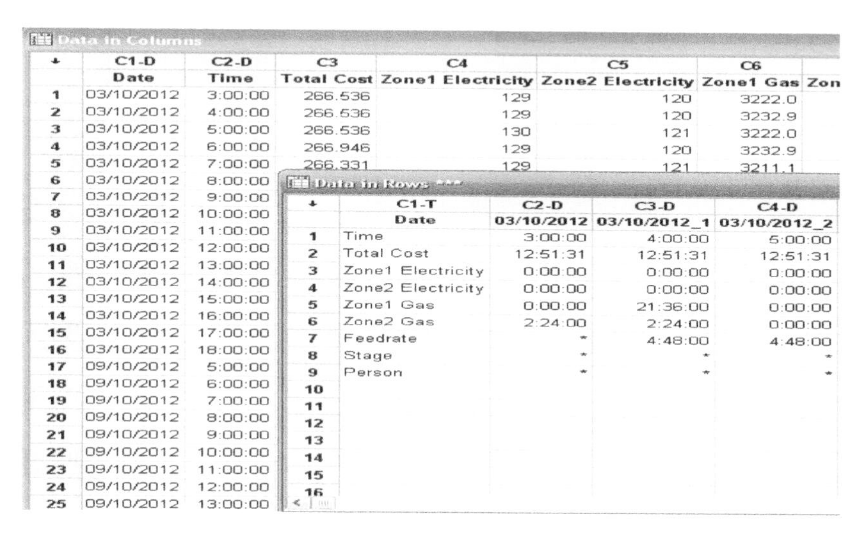

Data in Columns

↓	C1-D Date	C2-D Time	C3 Total Cost	C4 Zone1 Electricity	C5 Zone2 Electricity	C6 Zone1 Gas
1	03/10/2012	3:00:00	266.536	129	120	3222.0
2	03/10/2012	4:00:00	266.536	129	120	3232.9
3	03/10/2012	5:00:00	266.536	130	121	3222.0
4	03/10/2012	6:00:00	266.946	129	120	3232.9
5	03/10/2012	7:00:00	266.331	129	121	3211.1
6	03/10/2012	8:00:00				
7	03/10/2012	9:00:00				
8	03/10/2012	10:00:00				
9	03/10/2012	11:00:00				
10	03/10/2012	12:00:00				
11	03/10/2012	13:00:00				
12	03/10/2012	14:00:00				
13	03/10/2012	15:00:00				
14	03/10/2012	16:00:00				
15	03/10/2012	17:00:00				
16	03/10/2012	18:00:00				
17	09/10/2012	5:00:00				
18	09/10/2012	6:00:00				
19	09/10/2012	7:00:00				
20	09/10/2012	8:00:00				
21	09/10/2012	9:00:00				
22	09/10/2012	10:00:00				
23	09/10/2012	11:00:00				
24	09/10/2012	12:00:00				
25	09/10/2012	13:00:00				

Data in Rows ***

↓	C1-T Date	C2-D 03/10/2012	C3-D 03/10/2012_1	C4-D 03/10/2012_2
1	Time	3:00:00	4:00:00	5:00:00
2	Total Cost	12:51:31	12:51:31	12:51:31
3	Zone1 Electricity	0:00:00	0:00:00	0:00:00
4	Zone2 Electricity	0:00:00	0:00:00	0:00:00
5	Zone1 Gas	0:00:00	21:36:00	0:00:00
6	Zone2 Gas	2:24:00	2:24:00	0:00:00
7	Feedrate	*	4:48:00	4:48:00
8	Stage	*	*	*
9	Person	*	*	*

2.4 Column Formats

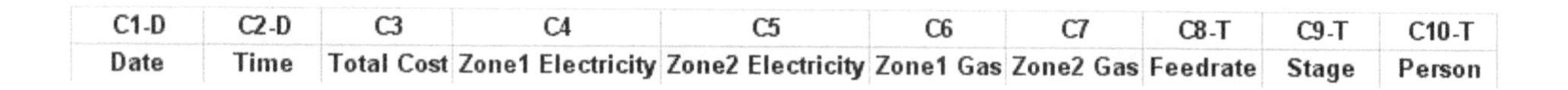

C1-D	C2-D	C3	C4	C5	C6	C7	C8-T	C9-T	C10-T
Date	Time	Total Cost	Zone1 Electricity	Zone2 Electricity	Zone1 Gas	Zone2 Gas	Feedrate	Stage	Person

Notice the column numbering system and column formatting as shown within the above figure.

Minitab uses:

- D for Date/Time,
- T for Text,
- A blank for numeric.

There is only one type of format allowed in each column.

Notice that column C8 is incorrectly formatted as text due to a typing error.

1. Click into the cell and remove the 'a'.

Notice that Minitab does not automatically correct the formatting. Also Minitab will not handle the data as numeric so we have to correct the formatting.

C7	C8-T	C9-T
Zone2 Gas	Feedrate	Stage
2927.1	17.2a	Before
2927.1	17.2	Before
2938.0	17.2	Before
2948.9	17.2	Before

C7	C8-T	C9-T
Zone2 Gas	Feedrate	Stage
2927.1	17.2	Before
2927.1	17.2	Before
2938.0	17.2	Before
2948.9	17.2	Before

2. Click Data <<Change Data Type <<Text to Numeric

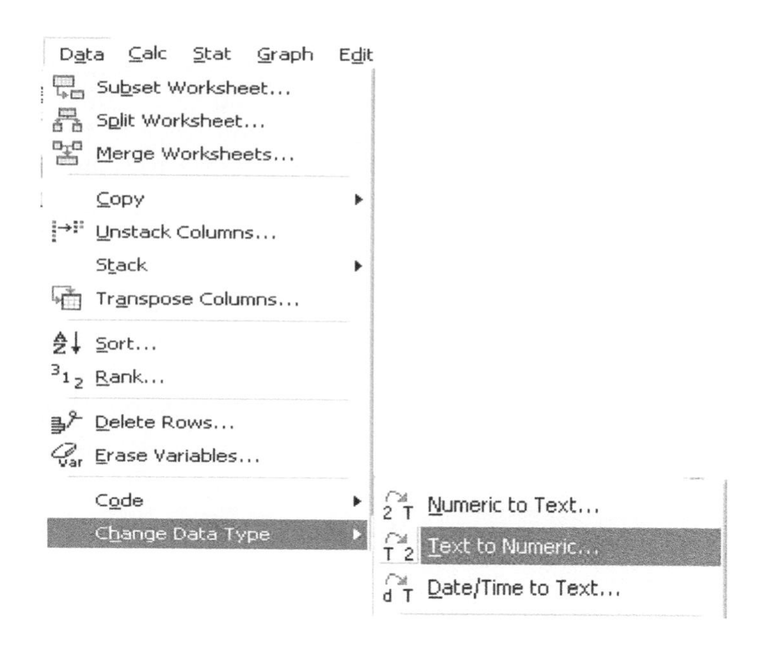

3. The Text to Numeric menu box opens. Minitab uses the top input box labelled Change Text Columns to ask which column needs changing. You can select Feedrate either by double clicking on it or by highlighting it with a single click and then pressing Select. The destination for our numeric column is going to be C8. We are going to overwrite our old text column by selecting C8. C8 is typed directly into the bottom input box. Then click OK.

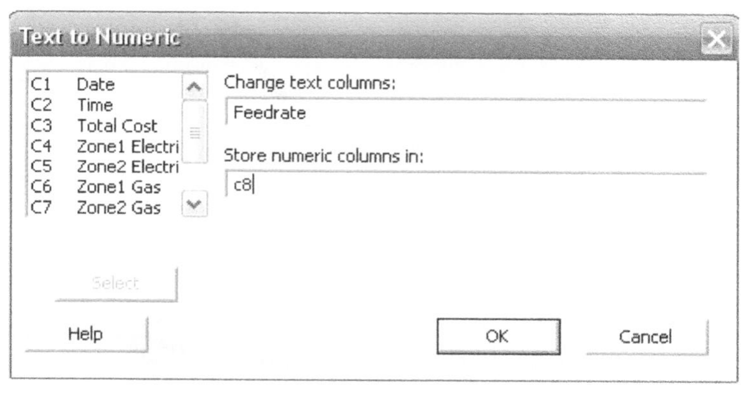

4. Click the Show Info icon. This gives us a summary of the data that is within the current worksheet. We can see that the format of column 8 has been changed to numeric. Additionally, we can see that column 9 has the title Stage. It contains 38 cells of data and none of the data is missing. The column format is text.

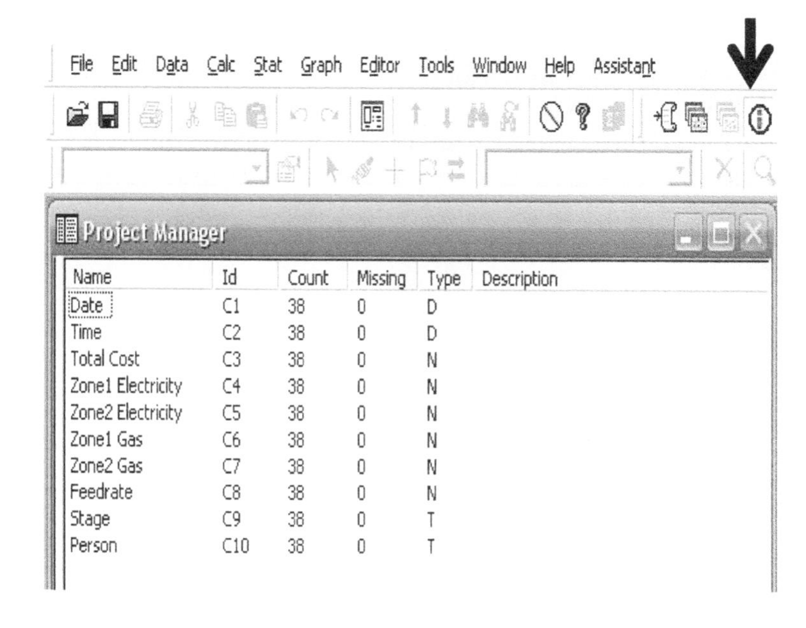

2.5 The Calculator

In the next part of the exercise we are going to use the Calculator to add the values in two columns together. To show a different way of adding columns together we are then going to use the Assign Formula function to add two different columns together. The figure below shows the layout of the calculator.

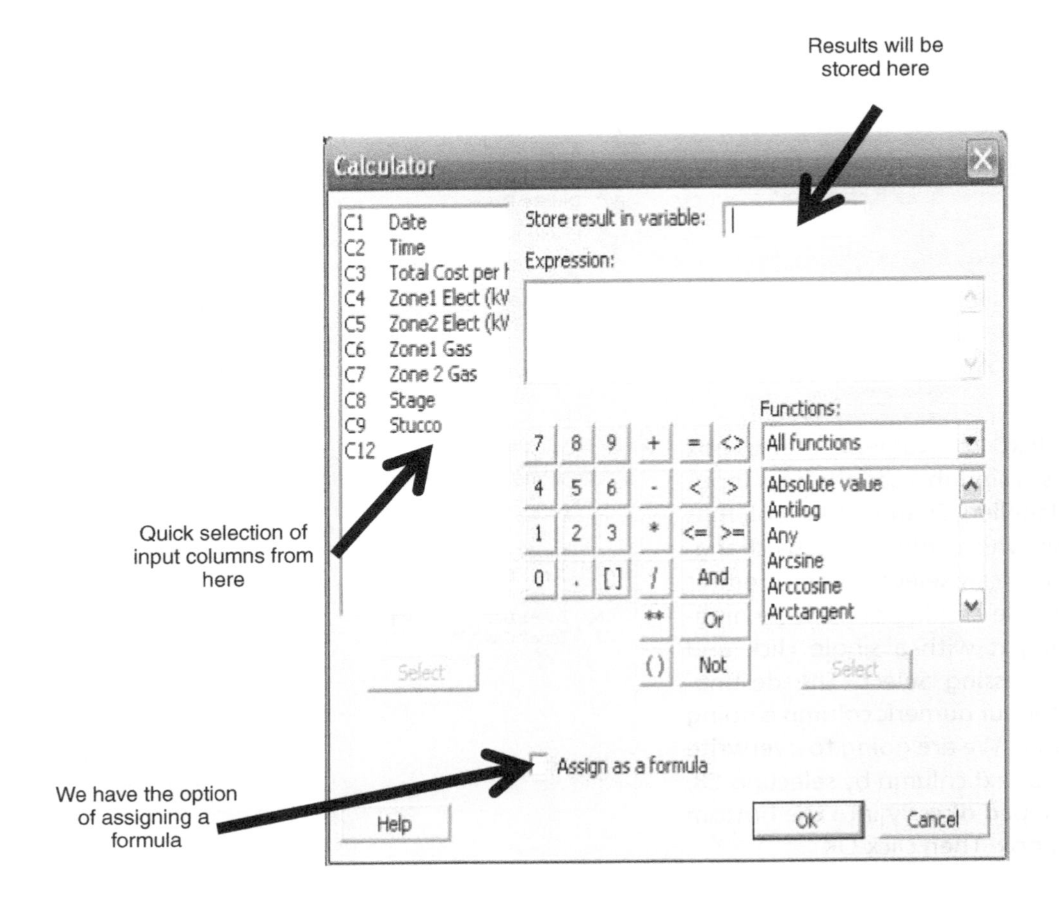

1. Click Calc <<Calculator.

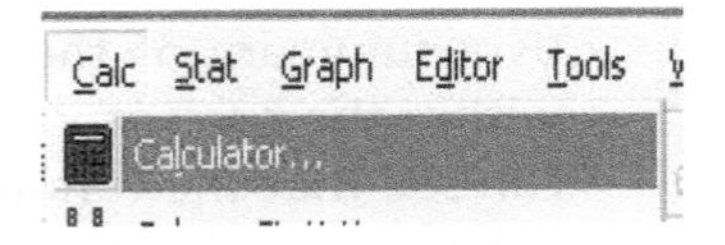

The calculator menu box opens. The calculator menu is used to mathematically manipulate columns of data. The output is stored in the current worksheet.

2. Input the same expression as shown.
3. Store results in C11.
4. Click OK.
5. C11 will now contain the result of adding together each of the rows in the stated columns within the expression. Initially the resultant column will not have a heading. Double click within the heading area of C11 and insert 'Total Elec' as the column heading.

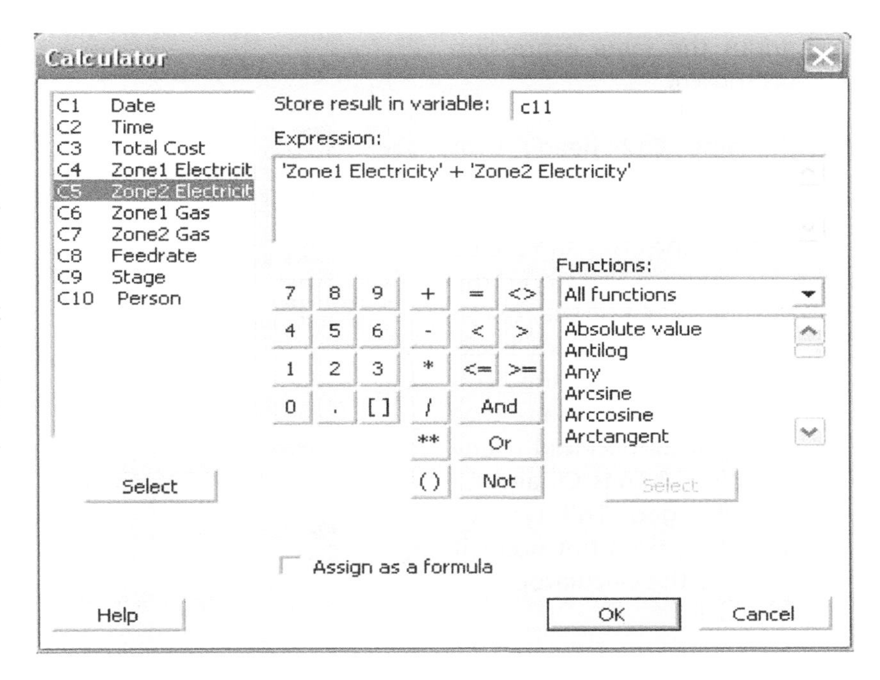

6. Move the cursor to C12 row 1.
7. Click on the Assign Formula button icon in the toolbar.
8. Notice the Formula box that opens. It is similar to the Calculator; however there is nowhere to assign the destination of the data as that is preselected by placing the active cell.

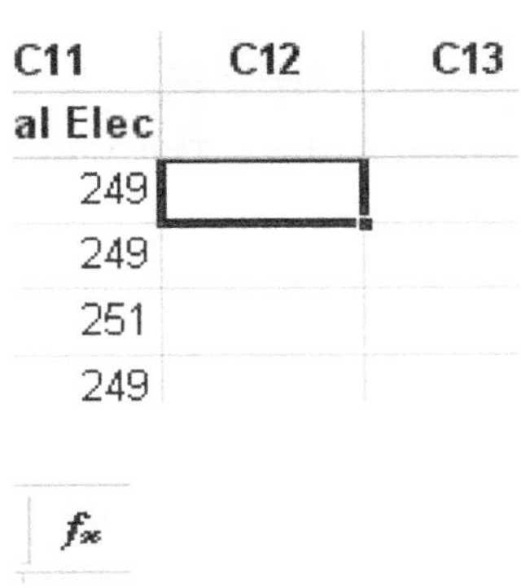

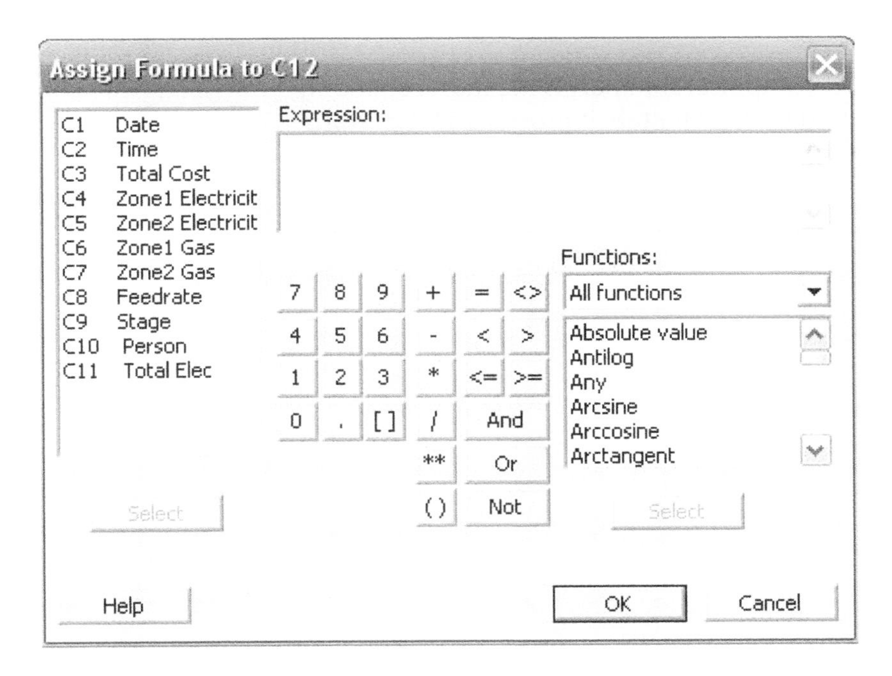

9. Input the same expression as shown.
10. Click 'OK'.
11. Rename C12 heading as 'Total Gas'. Minitab has added together each row in the Zone1 Gas and Zone2 Gas columns and stored the results in C12. The green cross with the black background, within the column heading, means that C12 will be updated if any cells in columns C6 and C7 are changed. This type of updating does not occur if you use the calculator.

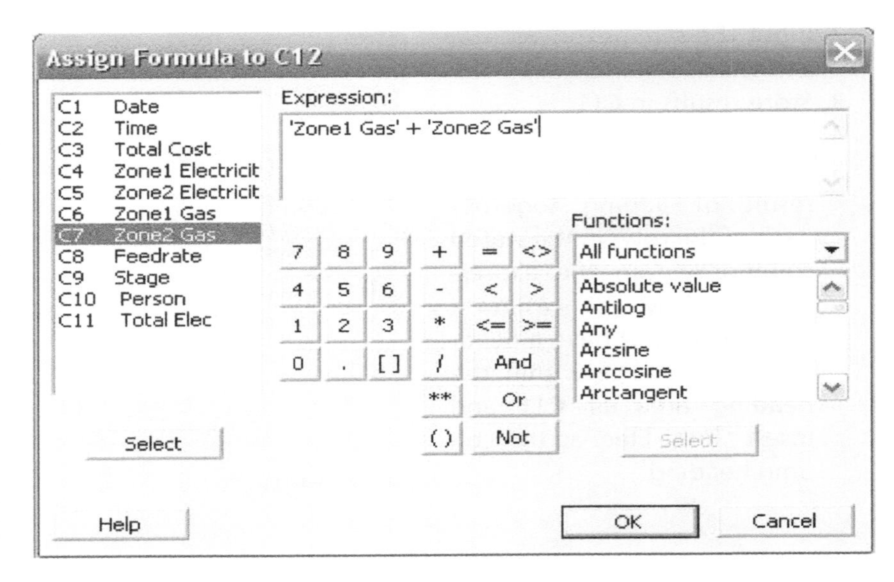

2.6 Basic Graphs

We are going to continue analysing our example data. This time we are going to carry out a batch of statistical analysis that Minitab calls the Graphical Summary. This battery of tests is an excellent first choice when examining data.

1. Click Stat <<Basic Statistics <<Graphical Summary
2. From the Graphical Summary menu box Enter Total Gas as the variable to be investigated.
3. Then click OK.

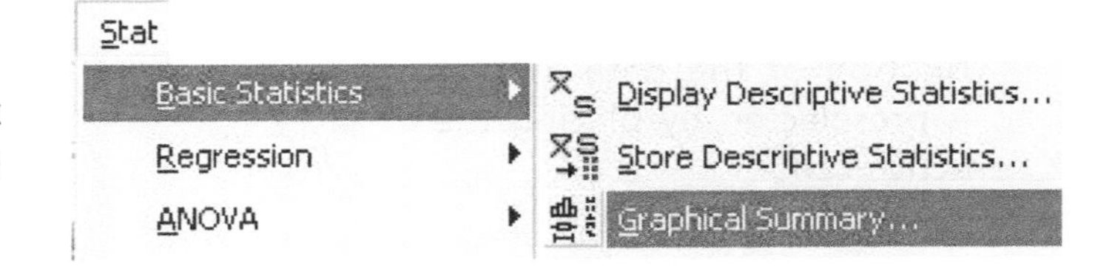

The Graphical Summary is the most commonly used tool in Minitab. It provides a lot of statistical information in one quick procedure.

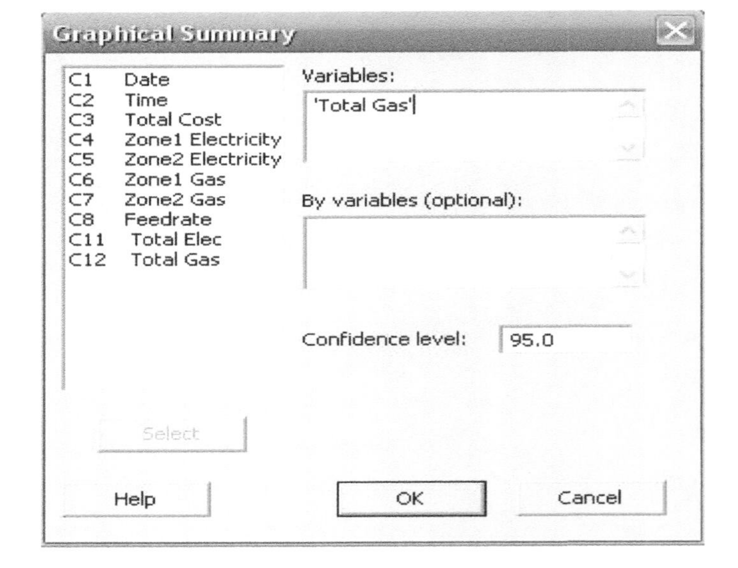

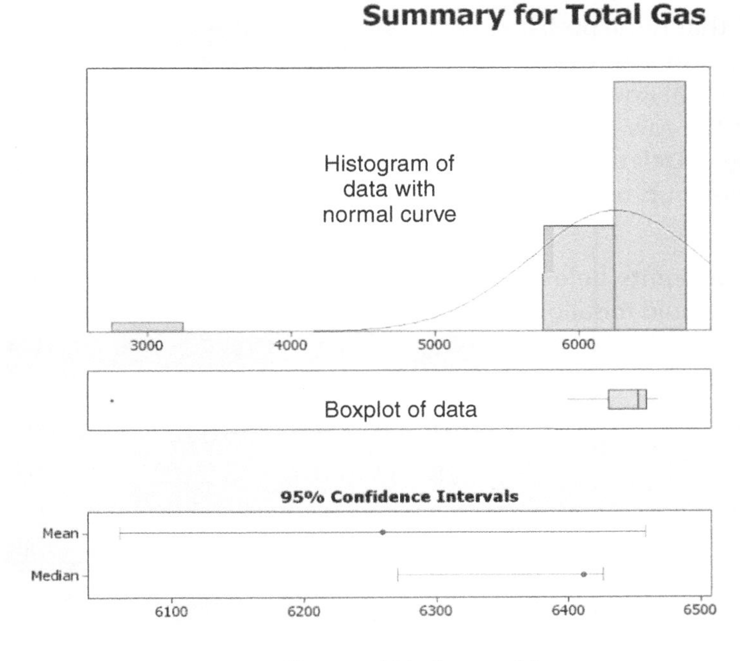

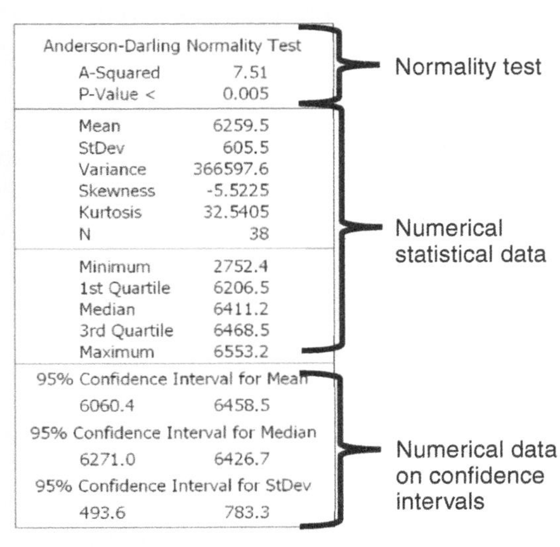

Anderson-Darling Normality Test	
A-Squared	7.51
P-Value <	0.005
Mean	6259.5
StDev	605.5
Variance	366597.6
Skewness	-5.5225
Kurtosis	32.5405
N	38
Minimum	2752.4
1st Quartile	6206.5
Median	6411.2
3rd Quartile	6468.5
Maximum	6553.2
95% Confidence Interval for Mean	
6060.4	6458.5
95% Confidence Interval for Median	
6271.0	6426.7
95% Confidence Interval for StDev	
493.6	783.3

Looking at the boxplot we see that there is an outlier. This is shown as an asterisk and is separated from the tails of the boxplot.

4. Hover the mouse pointer over the outlier in the boxplot to see the value of the point and the data row. We see that this data has come from Row 25 of the Total Gas Data.

In the real world we would investigate outliers thoroughly. For the example we are going to say it has been confirmed that this point is low because the machine suffered a stop during this data collection period.

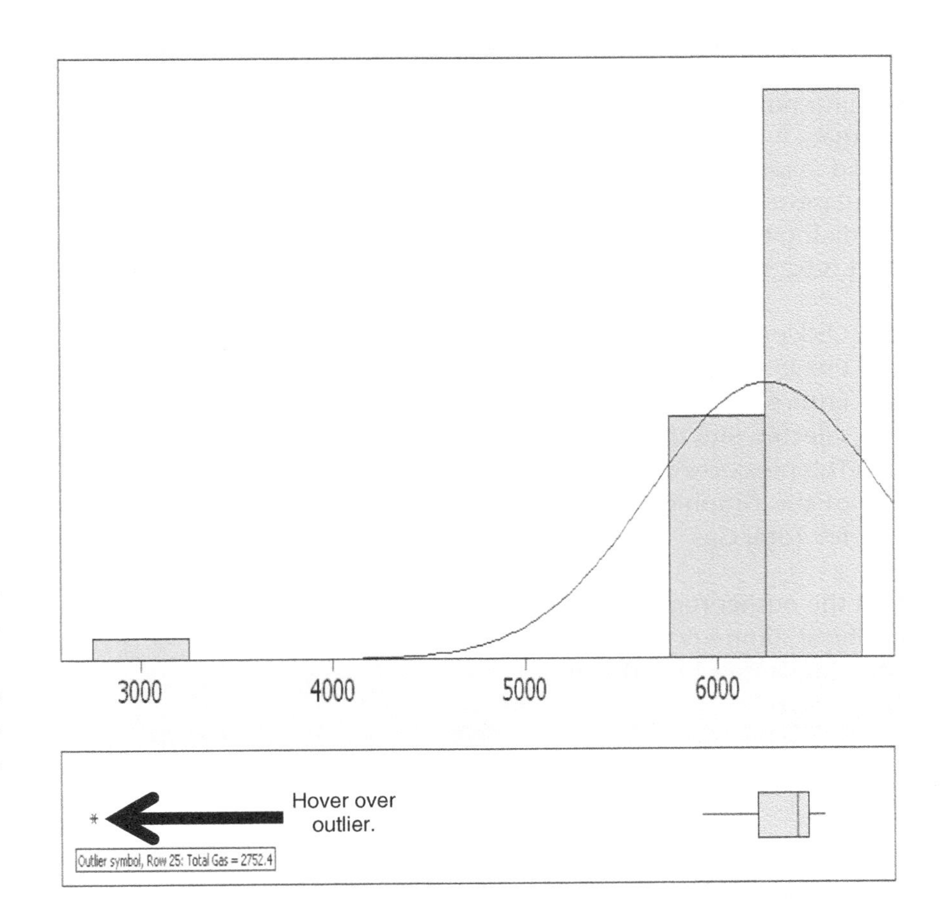

5. Click on the project window icon.
6. We see that it is not just the data in C12 that is the problem. Due to the stop the rest of the data in the row is also lower than expected. Click on the actual Row number for row 25. This will highlight all of the row.
7. Right click on the Row number and select Delete Cells. Row 25 is deleted. And all the rows move up one slot automatically.

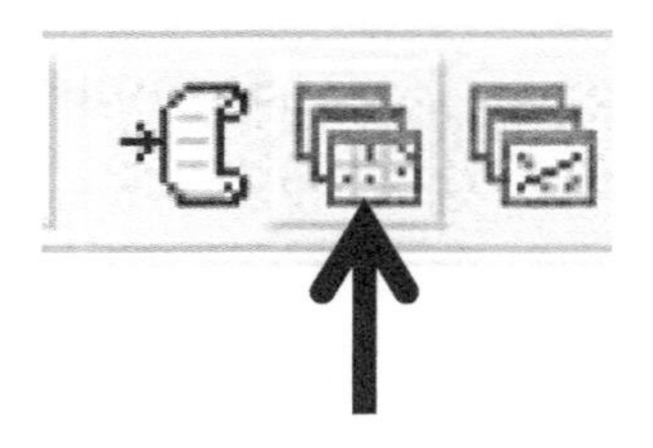

Note: if a single cell is deleted, all the data points below would still shift up one row. Beware that this would misalign adjacent rows within the data.

23	09/10/2012	11:00:00	273.407
24	09/10/2012	12:00:00	273.752
25	09/10/2012	13:00:00	190.236
26	09/10/2012	14:00:00	262.232
27	09/10/2012	15:00:00	269.088
28	09/10/2012	16:00:00	277.795

Undo Ctrl+Z
Redo Ctrl+Y
Clear Cells Delete
Delete Cells

We are now going to produce exactly the same Graphical Summary again to highlight a number of learning points. One is to show that instead of navigating through Stat <<Basic Statistics <<Graphical Summary we can just click on the Edit Last Dialog Box icon. This remembers the last dialog box we went to and takes us straight back to it. The second is that the dialog box remembers our last entries. This is useful most of the time but can occasionally be problematic as all sublevel settings will also be stored. The dialog box can be reset to initial settings by pressing F3 but don't do this now. The final learning point is to do with updating graphs. Follow the lesson through to find out about that.

8. Click on the Edit Last dialog box icon or press Ctrl+E.
9. Ensure 'Total Gas' is still selected and then click OK. This produces another copy of the Graphical Summary for Total Gas.

With the outlier removed a new graphical summary is shown for Total Gas. Notice how the graph has changed.

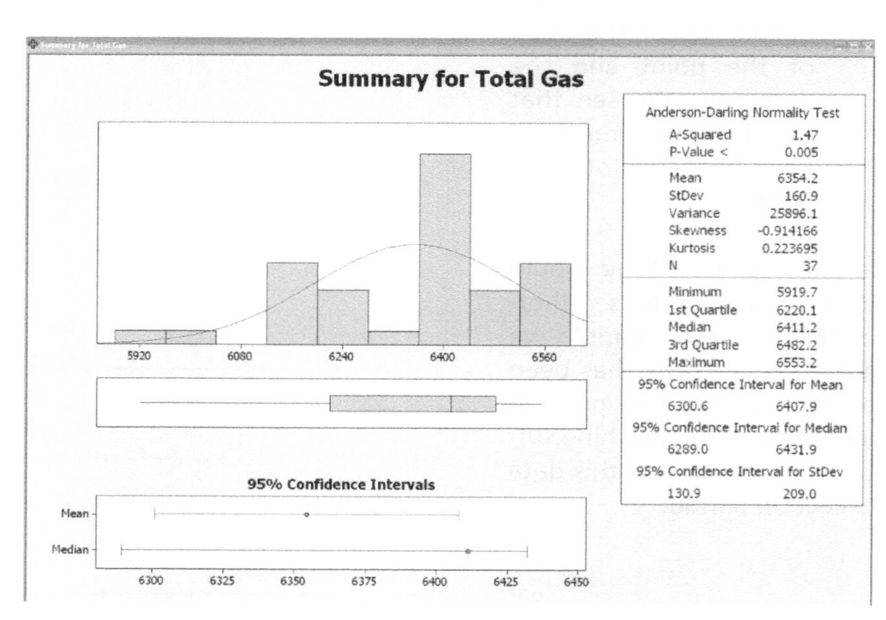

10. Click on the Show Graphs Folder icon. This shows that we have two copies of the graph.
11. Cycle between the two graphs a couple of times by double clicking on the top and bottom text.

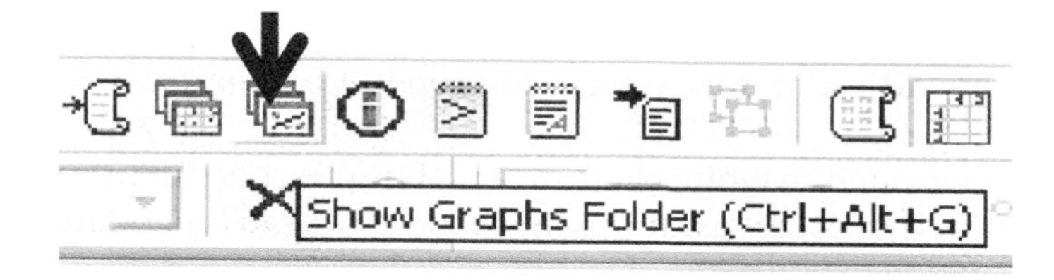

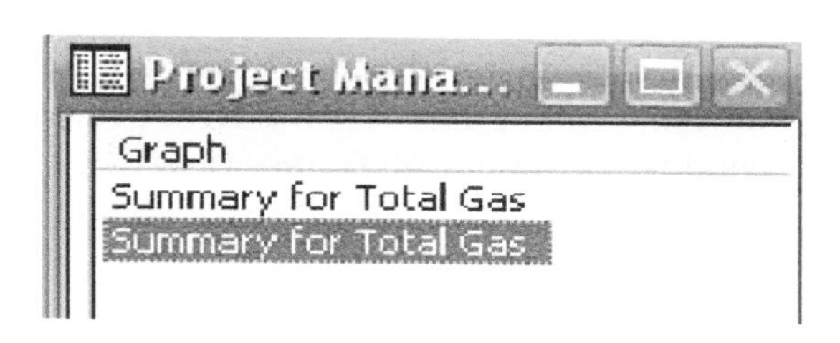

Notice that there is a difference in the type of cross in the top left corner of each graph.

The white diagonal cross on the original graph indicates that it is not current.

The green cross indicates that the graph is up to date with the data in the project window.

These charts cannot be updated. They must be redrawn.

12. Close the Graphical Summary that is not up to date.

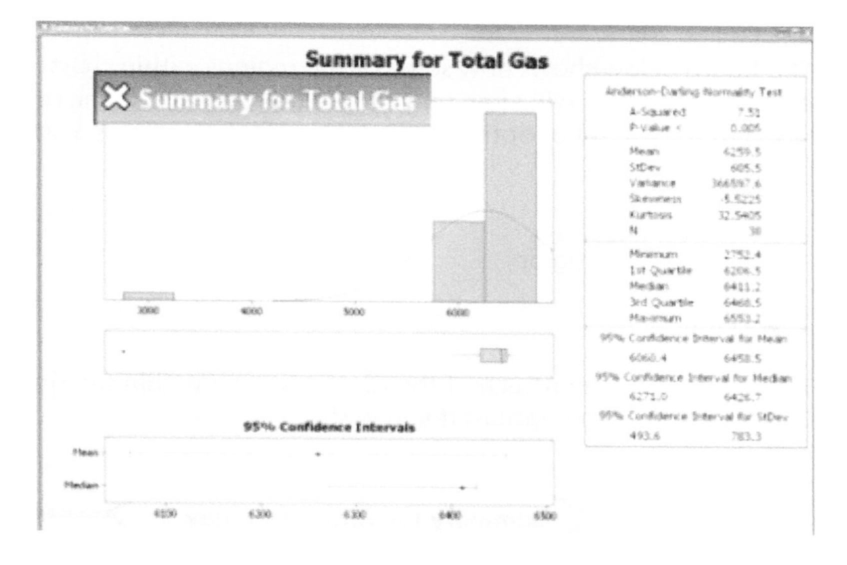

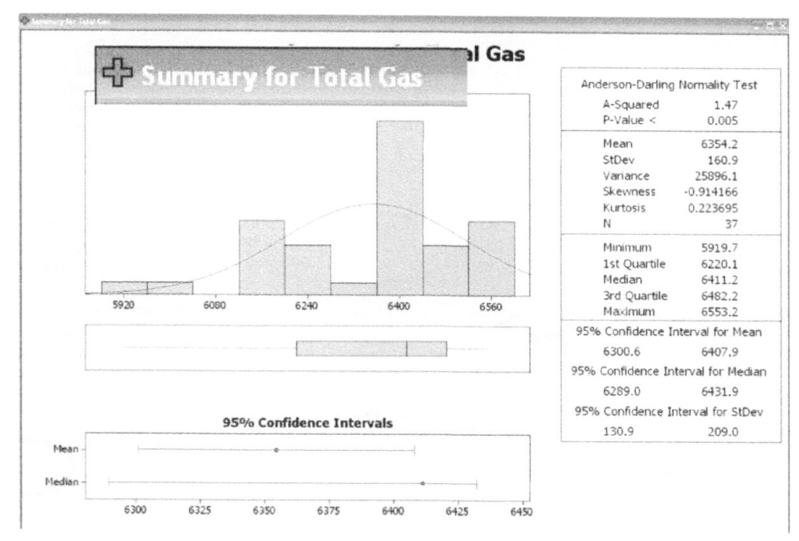

Some charts, usually Control Charts, can be updated and they have slightly different indicators in the top left corner.

The green cross in the white circle means that this chart is current. Additionally, if the source data changes the chart can be updated.

The yellow circle with the horizontal yellow bar means that the chart is not current. Right clicking on the chart and selecting 'Update Graph Automatically' will update it. An alternative way of doing this is to go Window <<Update All Graphs Now.

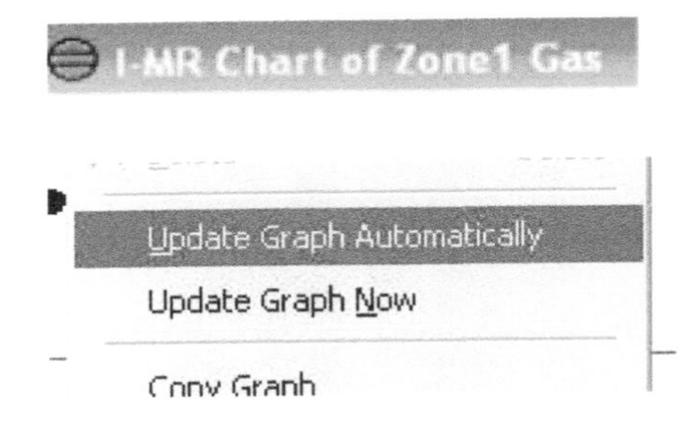

The figure below shows how some of the regions within charts can be edited. Double clicking on a specific region or item will open up an input box that allows the region or item to be edited. Each region or item has its own set or options for editing. The following is a list of some of the items that can be edited:

- Graph scales,
- Graph bars,
- Background region,
- Titles,
- Data.

Try double clicking on each of the regions shown within the rings or the items to open the options box. Change some of the options if you wish.

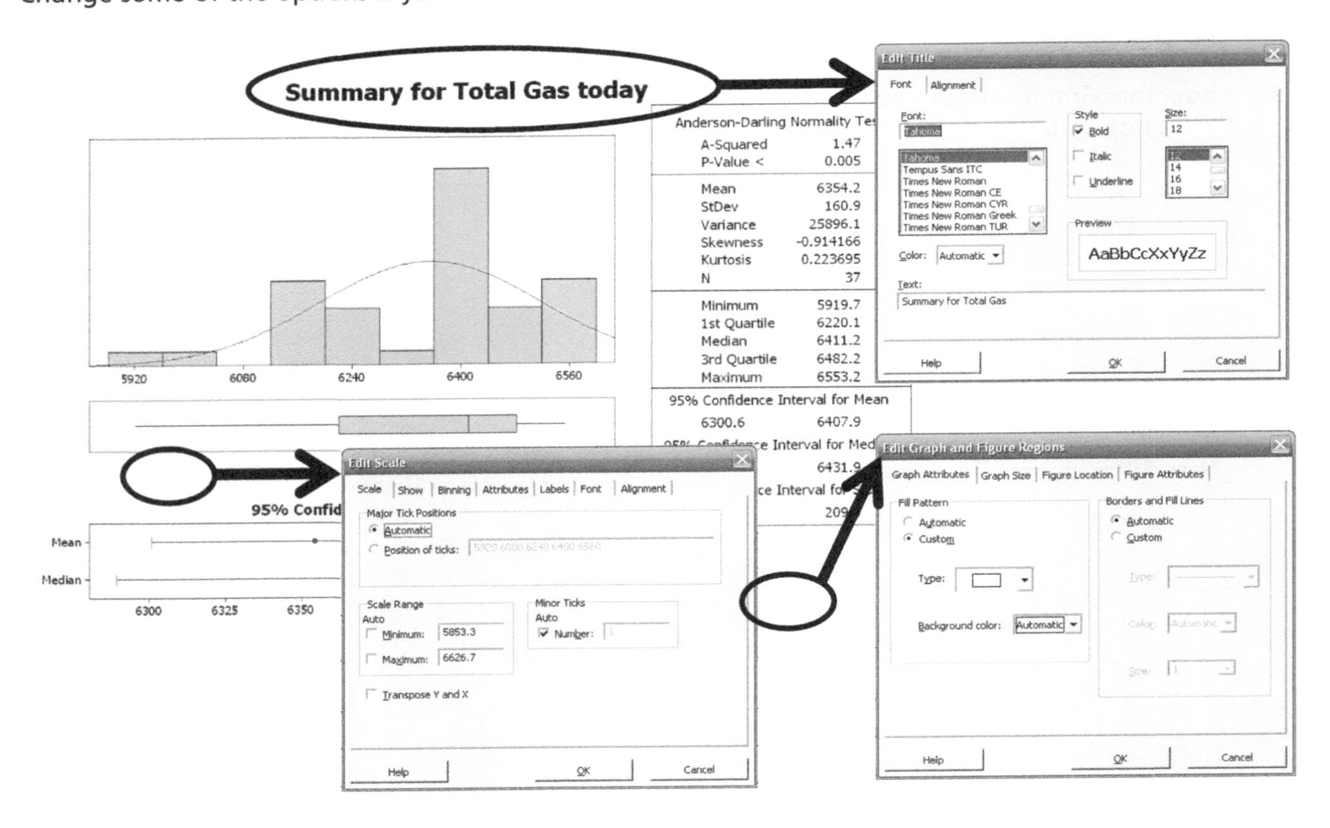

When you make changes to a chart they won't be carried through to the next chart that you produce. There is a way to make some global changes to charts. I like to do my bit for the environment by setting the background colour to my charts as white, thereby saving printer ink. I will show you how to access these global options.

13. Click on Tools <<Options
14. Double click on each of the following: Graphics, then Regions and then Graph.
15. Amongst the options that are available will be background colour. This can then be set to white. Then click OK.

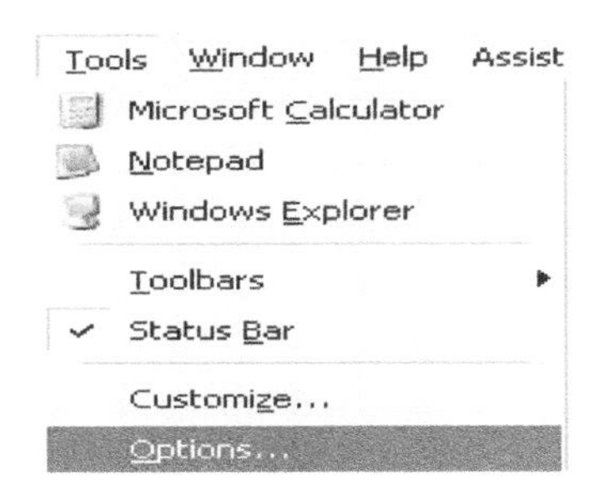

Note that if you want to change other global settings they can be accessed via this menu.

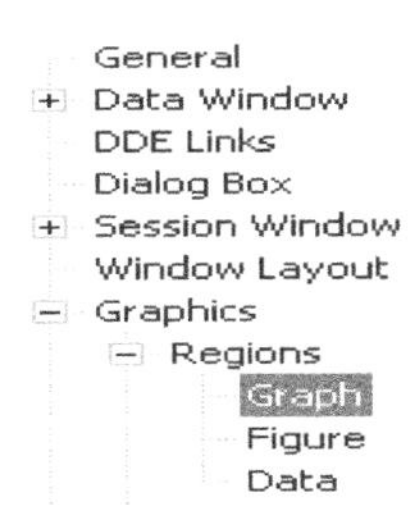

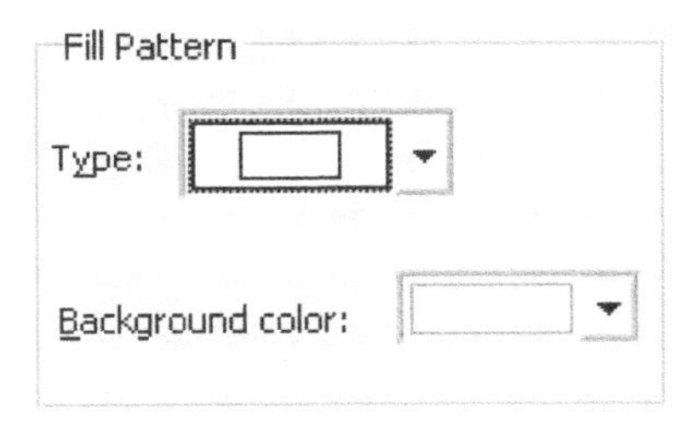

2.7 Adding Detail to Graphs

We are now going to learn how to add additional content to the graphs, such as reference lines and text.

1. Select the second of the Graphical Summaries that we produced.
2. Right click on the graph on the histogram graph region.
3. Go to <<Add <<Reference Lines.
4. We will show a reference line at 6300 on this graph as this indicates the target level for Total Gas. Enter 6300 into the lower cell. Then click OK.

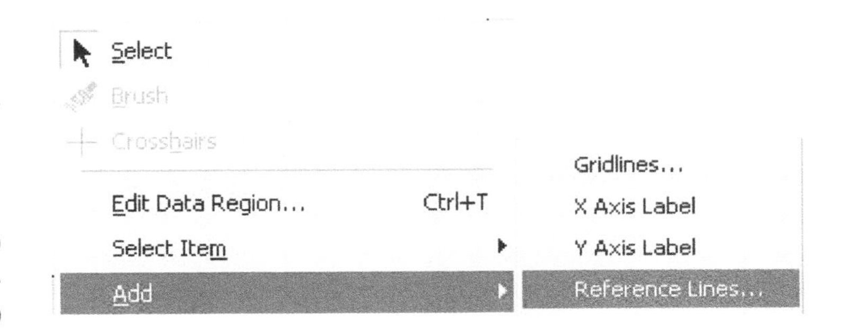

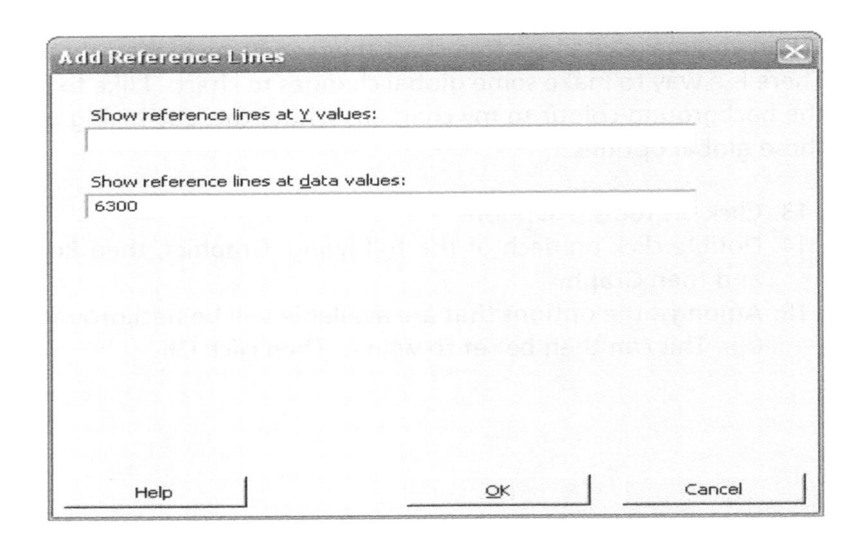

Within the toolbars there is a drawing toolbar. This is similar to other programs so the icons should be familiar.

1. Click on the Text icon to add text to the graph.
2. Drag an area on the histogram where you want the text to be placed. This can be adjusted later.
3. When you release the mouse button after dragging, a text box will open allowing you to enter the text shown. Then click OK.
4. In order to change the look of the text, double click on it. Go to the Font tab and change the text colour to red. There are also other text options available here.

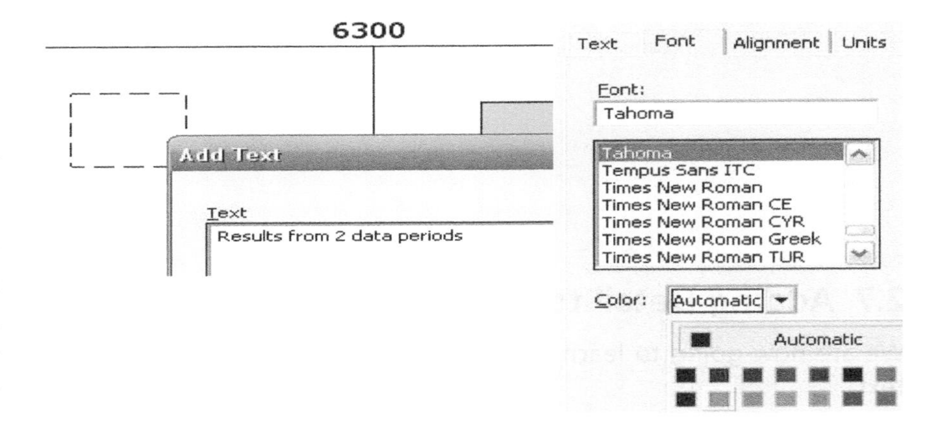

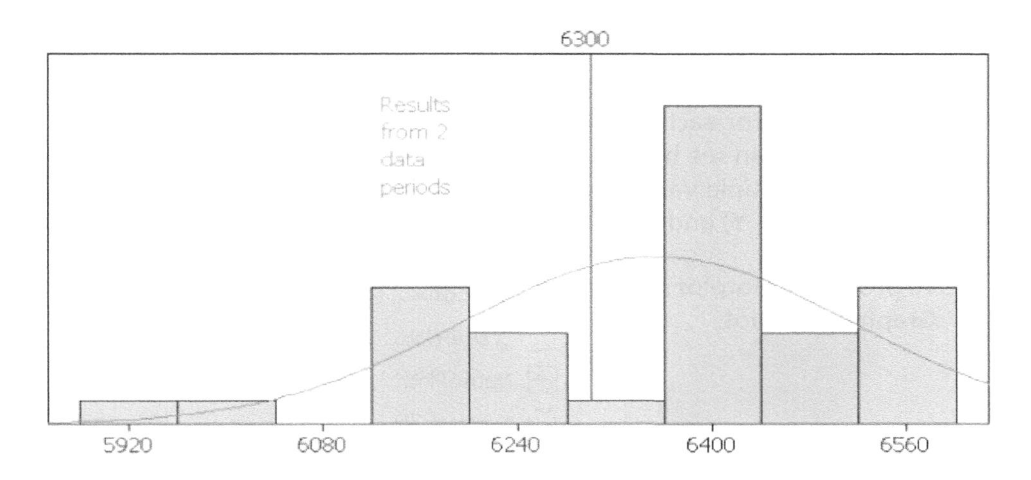

2.8 Saving Graphs

We can save our current graph as a separate file. We have a number of options such as setting the file type and we can even save it as a picture file.

1. Click File <<Save Graph As…

A normal save box will appear where you can select the destination and file type.

For the highest compatibility use .jpg and .tif file types.

Saving the graph as a .mgf file will then require Minitab to open the graph.

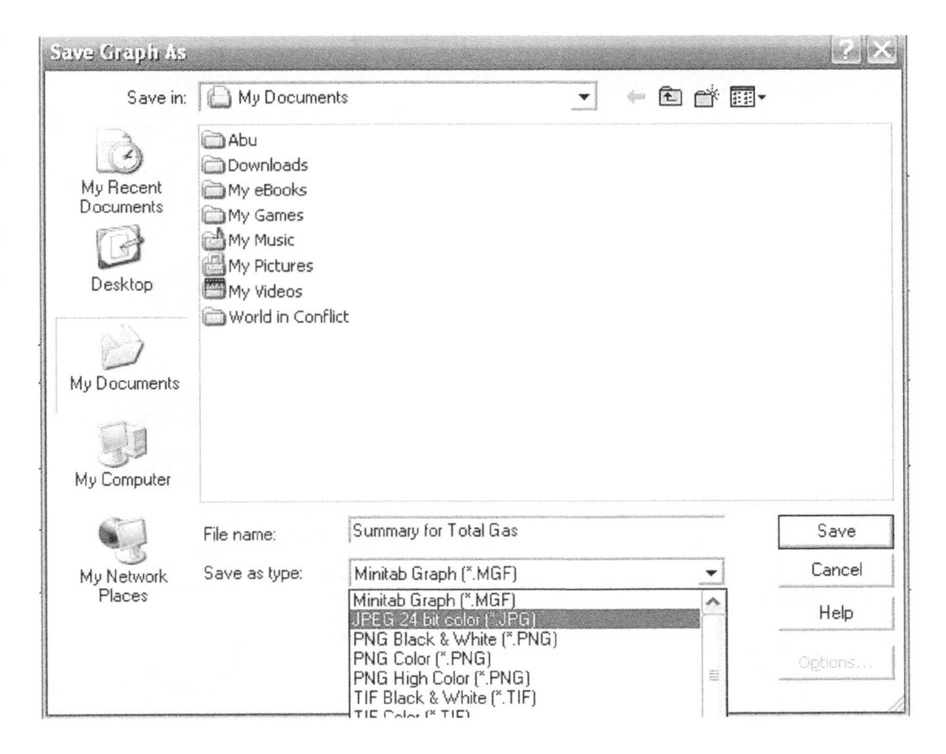

2.9 Dotplots

We will now look at producing Dotplots and Boxplots. Both graphs use almost the same options for displaying groups. For each graph that we produce we will look at displaying the inputs with different Groupings so you can see how the selection works. The first distinction made is whether the graph is to contain One or Multiple Variables (One Y or Multiple Ys). We will start with the simplest selection, which is one variable (One Y) and no groups. You will see a similar selection method for other types of graphs.

1. To produce a Dotplot select Graph <<Dotplot.

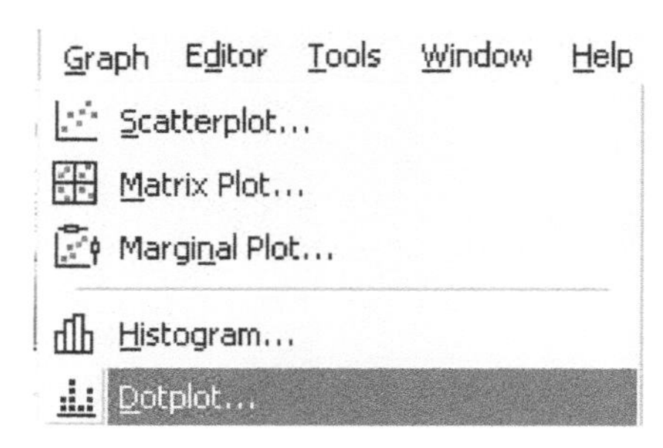

2. At the Dotplot menu select One Y <<Simple.
3. Then click OK.

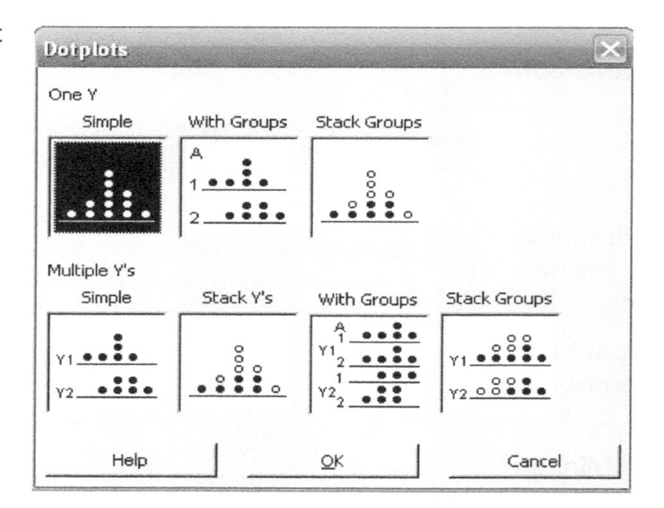

4. Select 'Total Elec' as the Graph variables.
5. Then click OK.

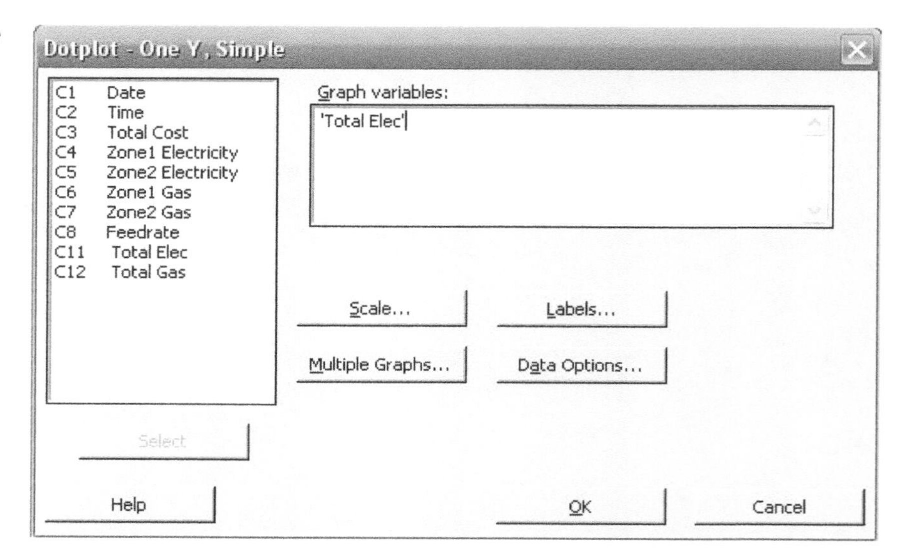

6. For this particular data set the dotplot shows that there are two distinct distributions within the Total Elec data. However, at this stage we cannot make conclusions about what is separating the data.

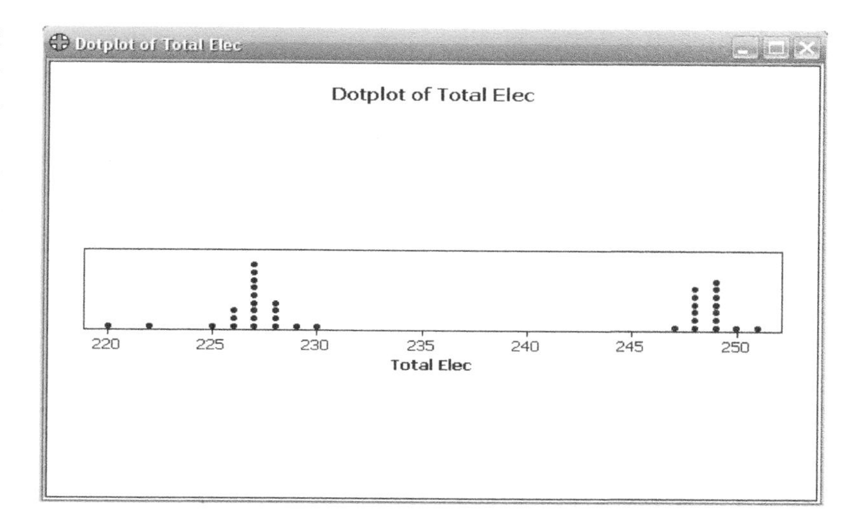

We will now produce a dotplot from a single variable (One Y). We will use the Stage column to separate the data into two groups. You might remember earlier we said that the Stage column contains categorical data and we were using it to tell us when maintenance occurred. This would help us tell if maintenance had any effect on the other variables.

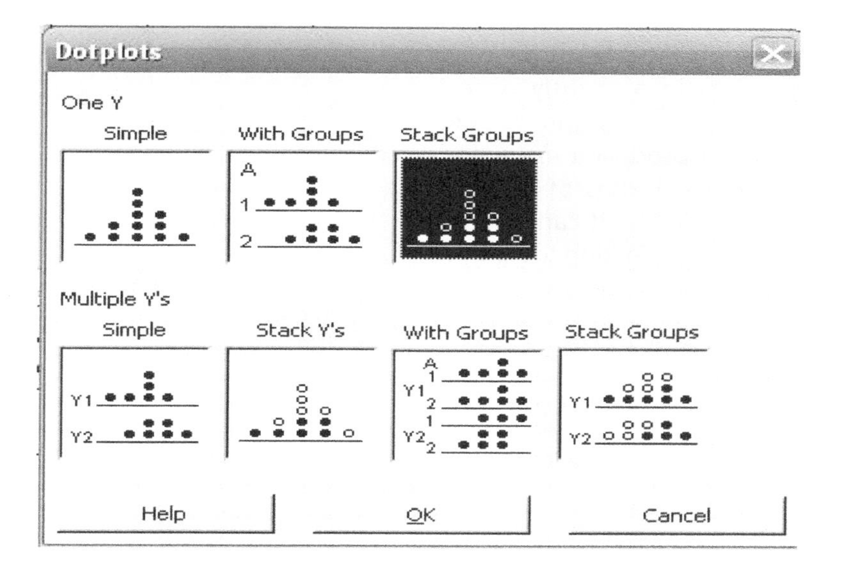

1. Select Graph <<Dotplot.
2. At the Dotplot menu select One Y <<Stack Groups. Click OK.
3. Select 'Total Elec' as the variable.
4. Select 'Stage' as the Categorical variable. Note that the tick box for Stack dots of last categorical variable is ticked.
5. Click OK.

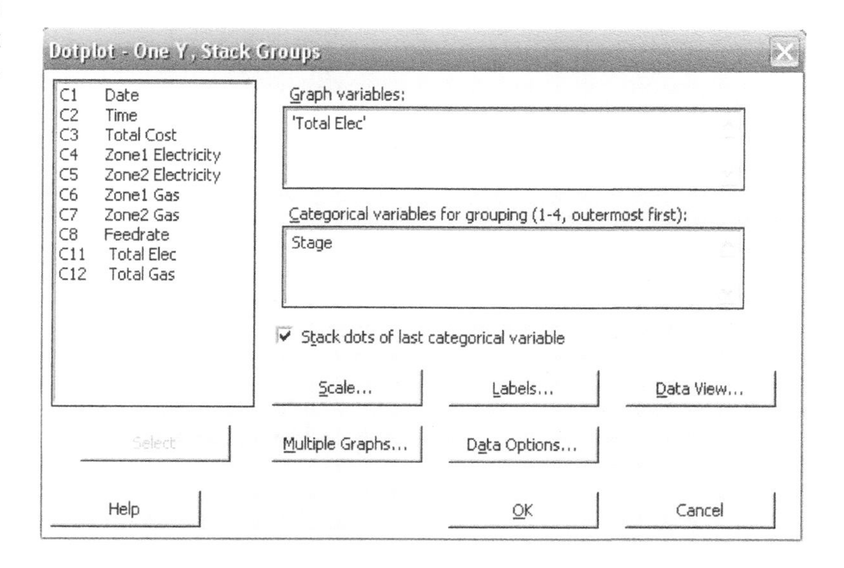

The following graph is displayed. It can be seen that the two stages of the data produce two separate distributions. We could conclude that our electricity usage has reduced after the maintenance activity.

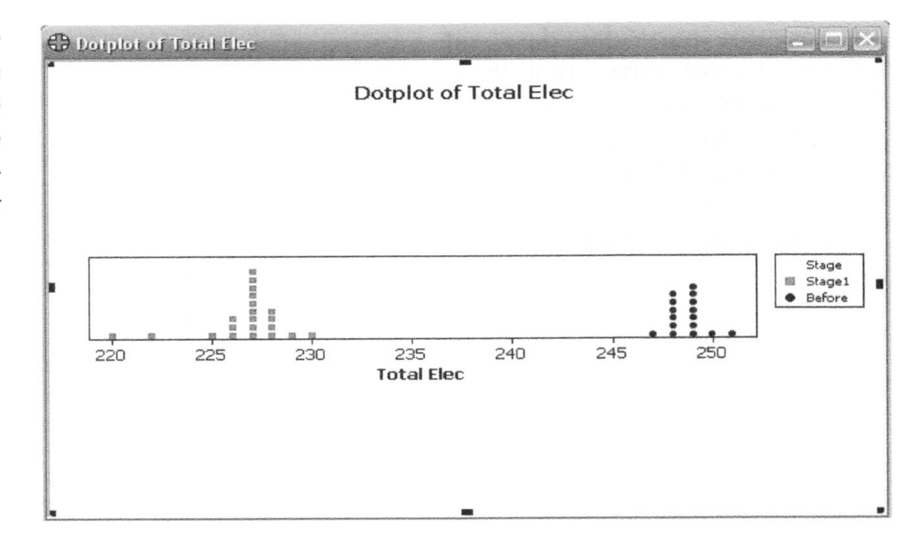

The graph (right) shows the same data in a slightly different way. This graph would have been produced if we had unticked Stack dots of last categorical variable. It can also be produced by selecting the graph type as One Y, with Groups.

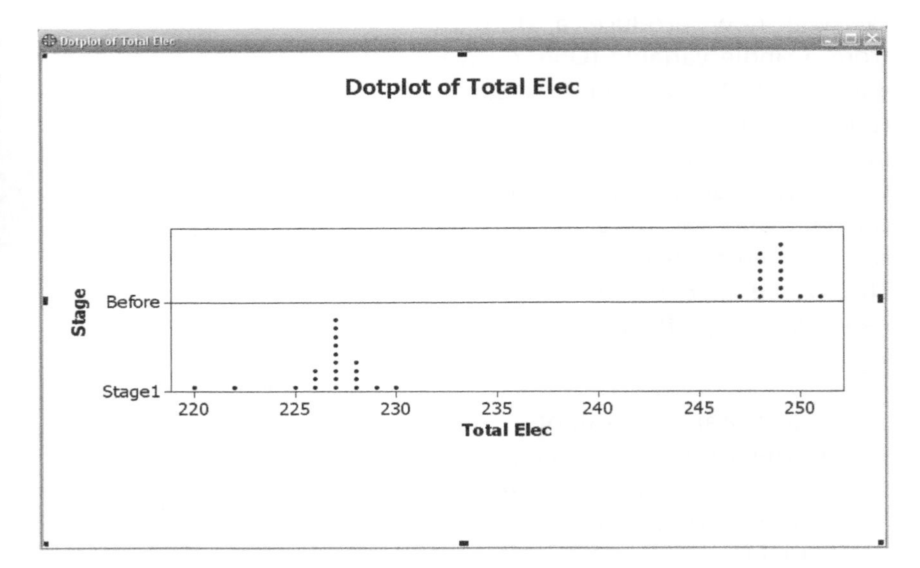

We will now break the rules that I just told you. We are going to display two variables using a One Y graph option.

6. Select Graph <<Dotplot.
7. At the Dotplot menu select One Y <<Stack Groups.

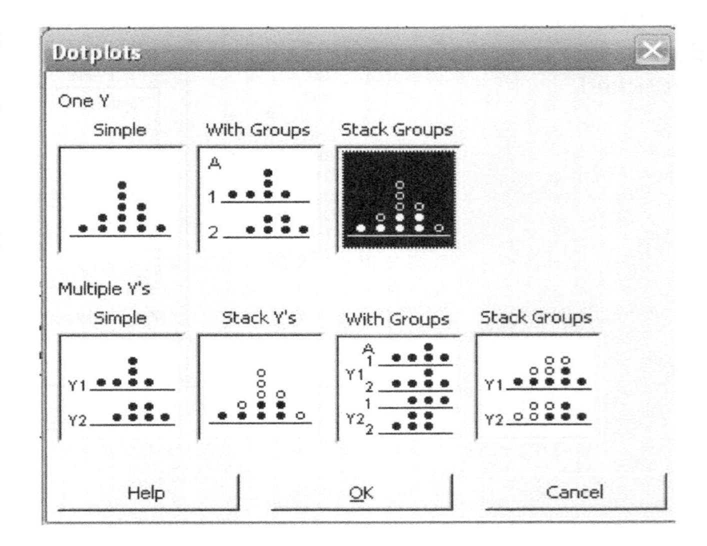

8. Select 'Total Elec' and then 'Total Gas' as the variables.
9. Select 'Stage' as the Categorical variable.
10. Click Multiple Graphs.

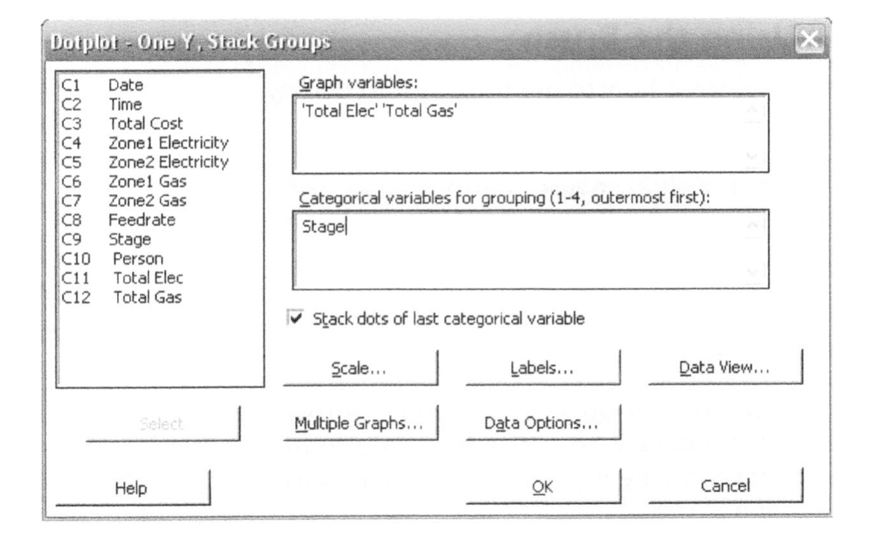

11. Select 'In separate Panels of the same graph.
12. 'Then click OK and OK again.

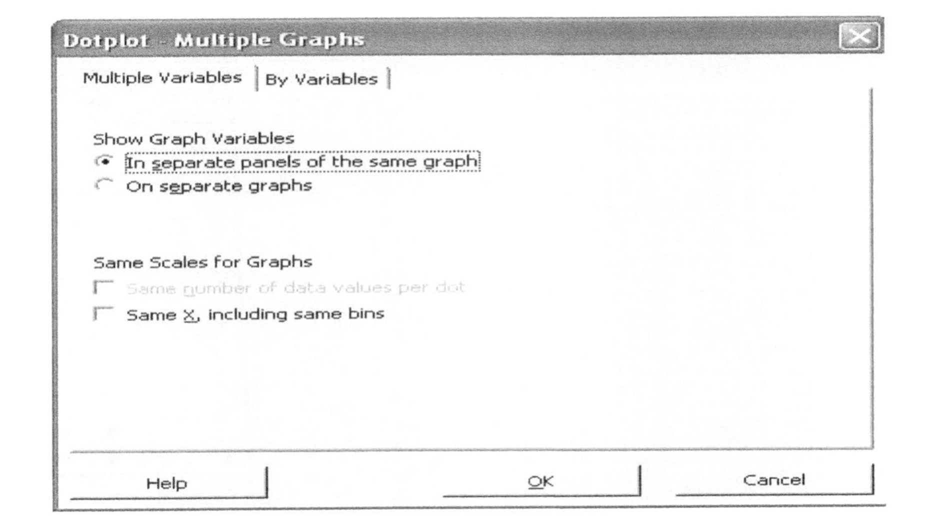

The following two graphs will be displayed together, with each graph within its own panel.

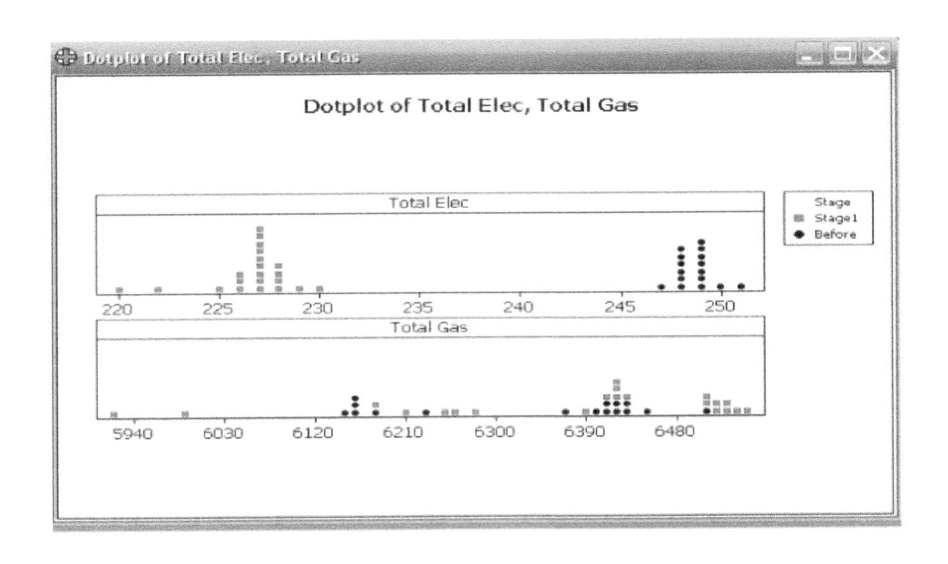

2.10 Using the Brush

We will now learn how to use the Brush to quickly obtain data pertaining to data points on graphs.

1. Click on the Brush Mode icon.
2. Using the left mouse button, drag the pointer over a selection of data points on the graph.

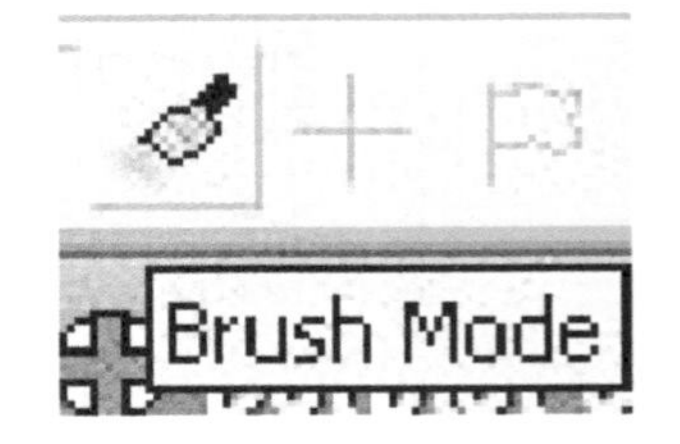

Notice how the pop up box tells you the row number of the data. Also notice that data from the same rows will be highlighted in other areas of the graph.

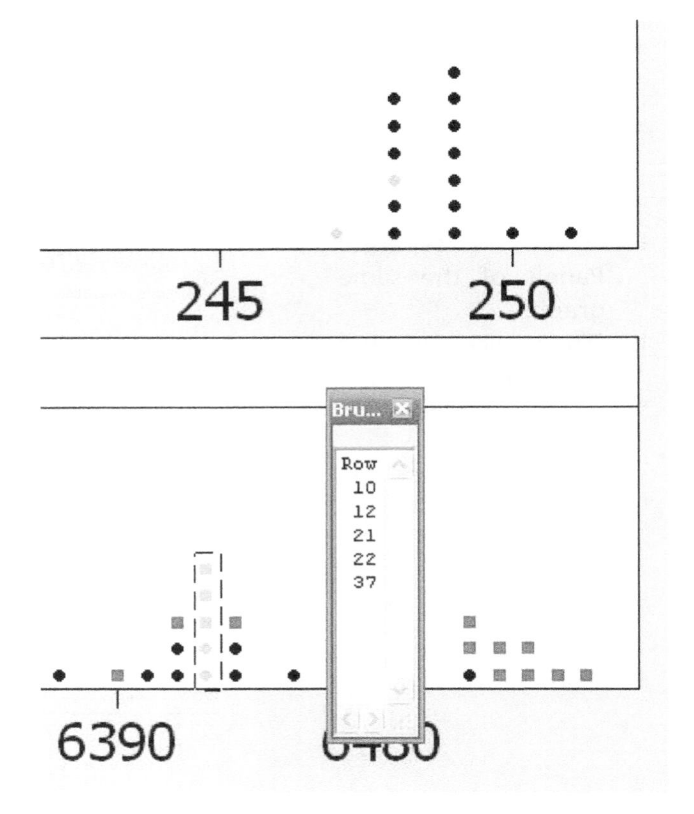

The Brush tool can also be used to display data from other data columns.

3. Ensure Brush Mode is selected.

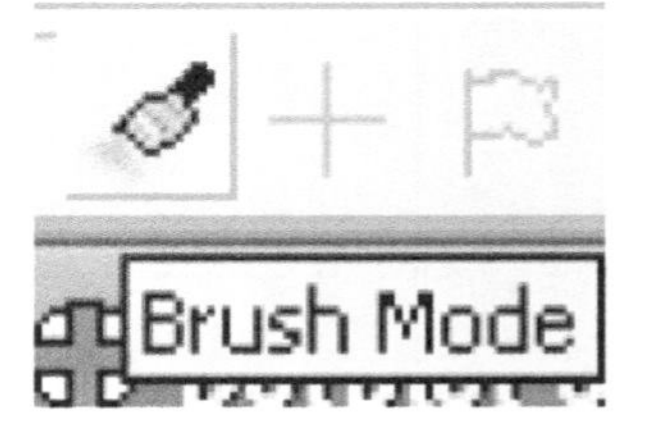

4. Select Editor <<Set ID Variables.

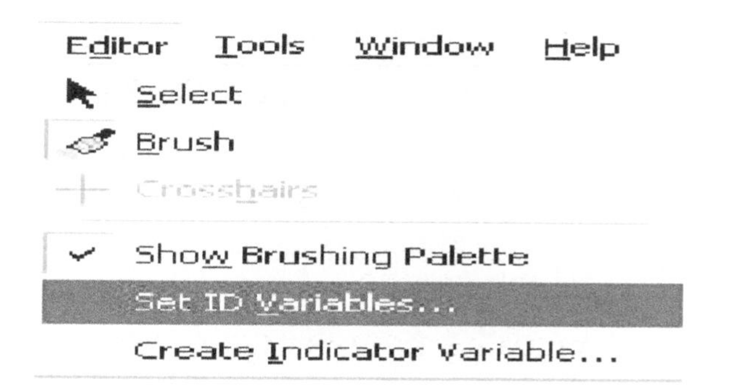

5. Select 'Date', 'Time' and 'Feedrate' as the required variables.

6. Then click OK.

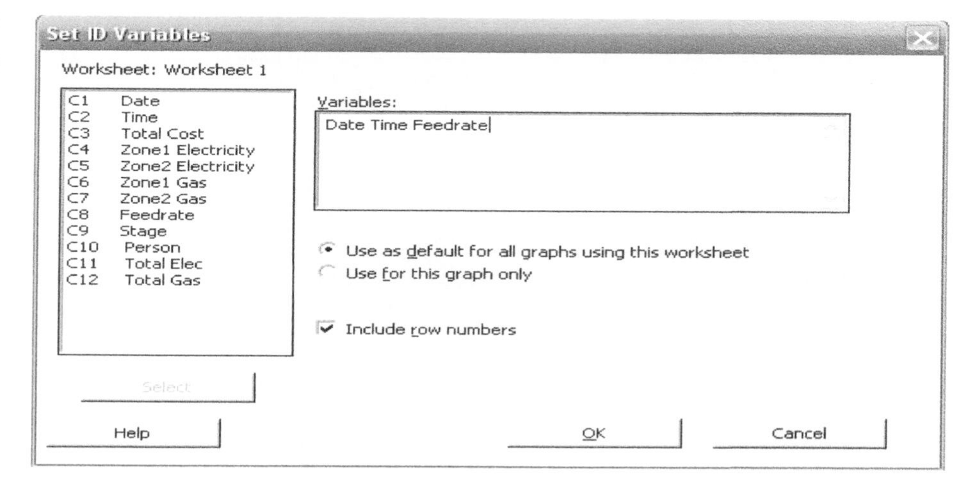

7. Again use the Brush to drag over a number of data points. Notice how the data from our selected variables is displayed.

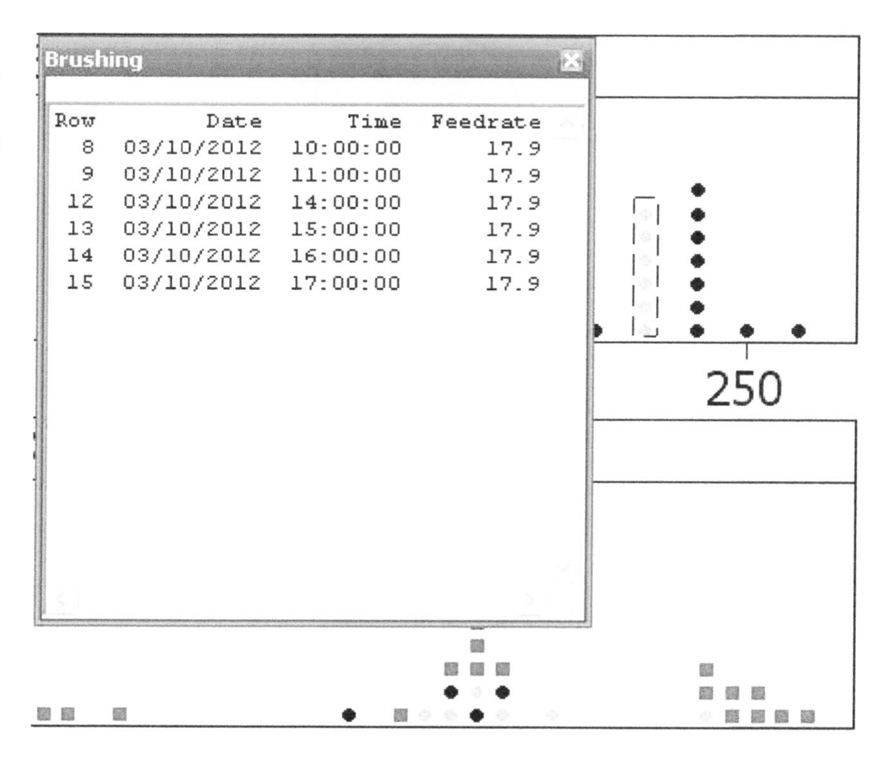

2.11 Boxplots

We will now look at Multiple Y graphs but this time we are going to produce boxplots.

1. Select Graph <<Boxplot.
2. At the Boxplot menu select Multiple Ys <<Simple

3. Click OK.

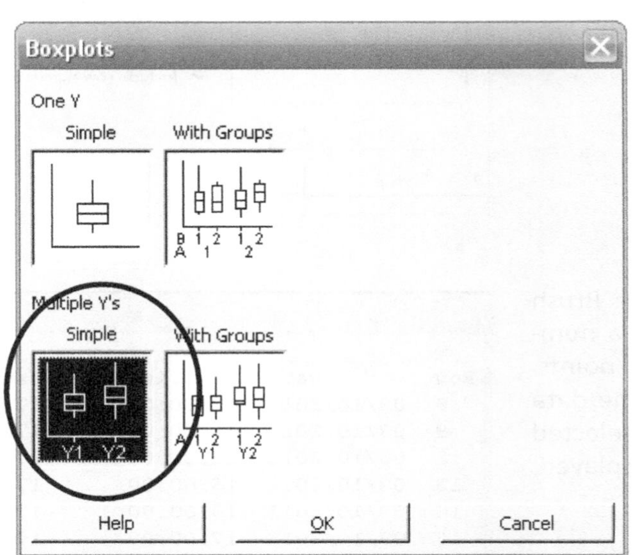

4. Select 'Zone 1 Gas' and then 'Zone 2 Gas' as the variables.
5. Then click OK.

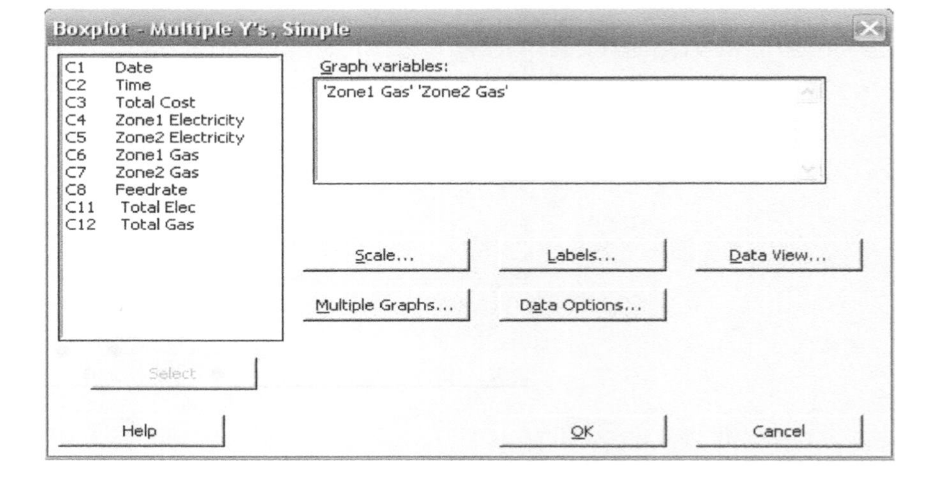

The boxplot is displayed. From the boxplot we can begin to conclude that Zone1 uses more gas than Zone2. This is something we could confirm using hypothesis testing. We will learn more about this in Chapter 4.

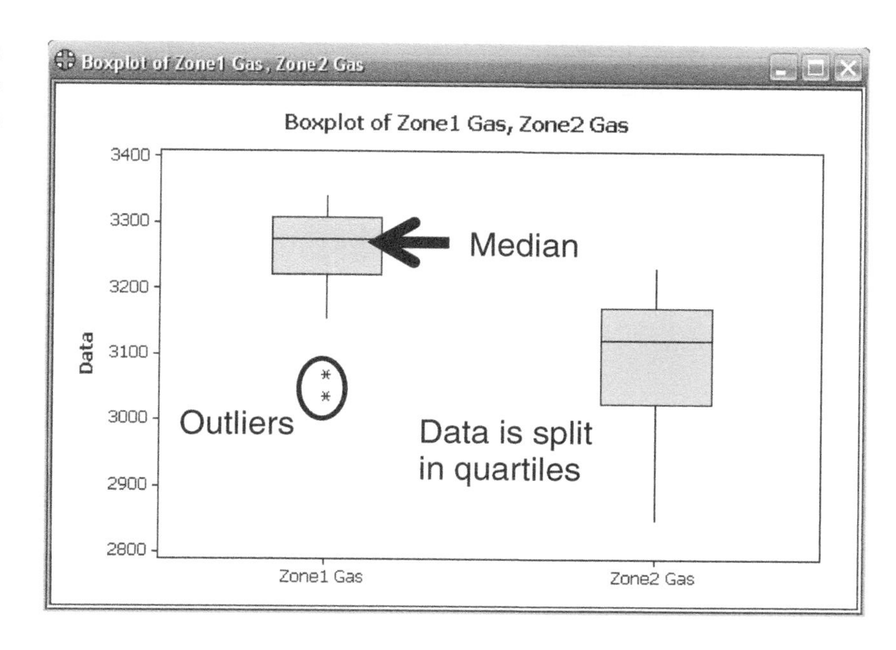

Notice the key features of the boxplot:

- The line in the box denotes the median.
- The data is split into quartiles between the tails and the two sections of the box.
- Outliers are represented by asterisks.

This is the last of the boxplots and we are going to select Multiple Ys, With Groups. Then we are going to add some additional content to the graph.

6. Select Graph <<Boxplot.

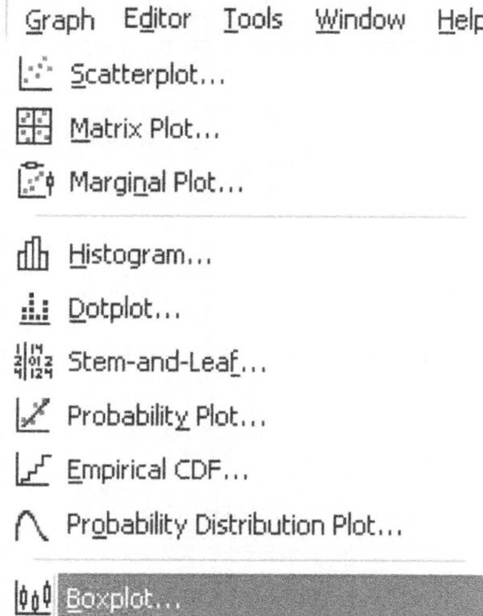

7. At the Boxplot menu select Multiple Ys <<With Groups.
8. Click OK.

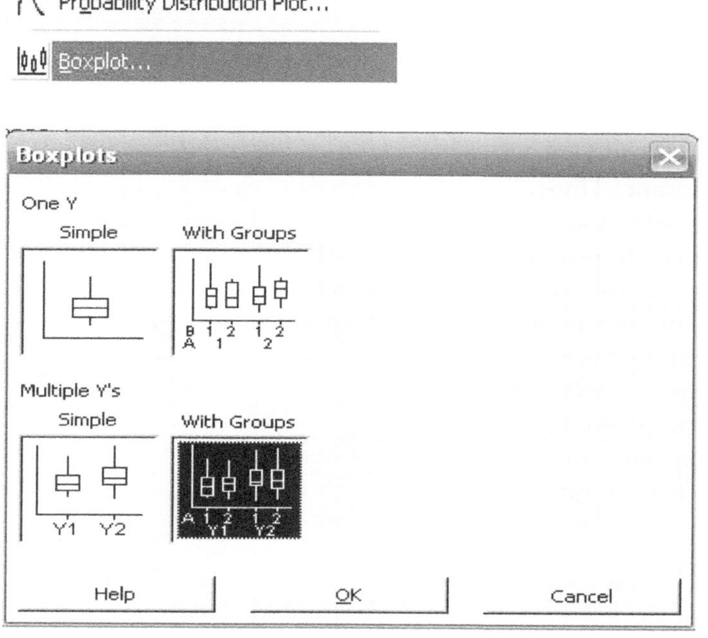

9. Select 'Zone1 Gas' and then 'Zone2 Gas' as the variables.
10. Select Stage as the Categorical Variable
11. Then click OK.

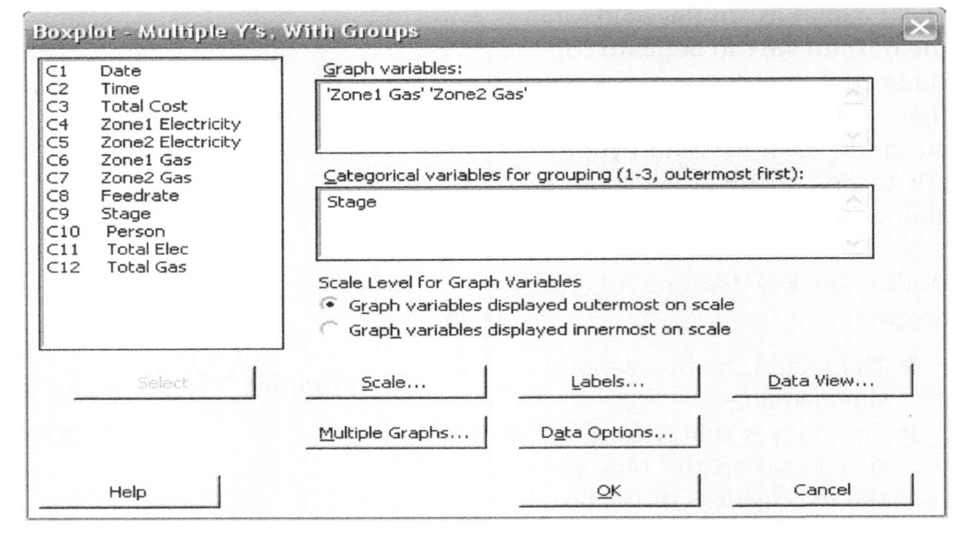

The boxplot is displayed. From the graph we could conclude that the maintenance activity has not impacted gas usage in the same way it did electricity usage.

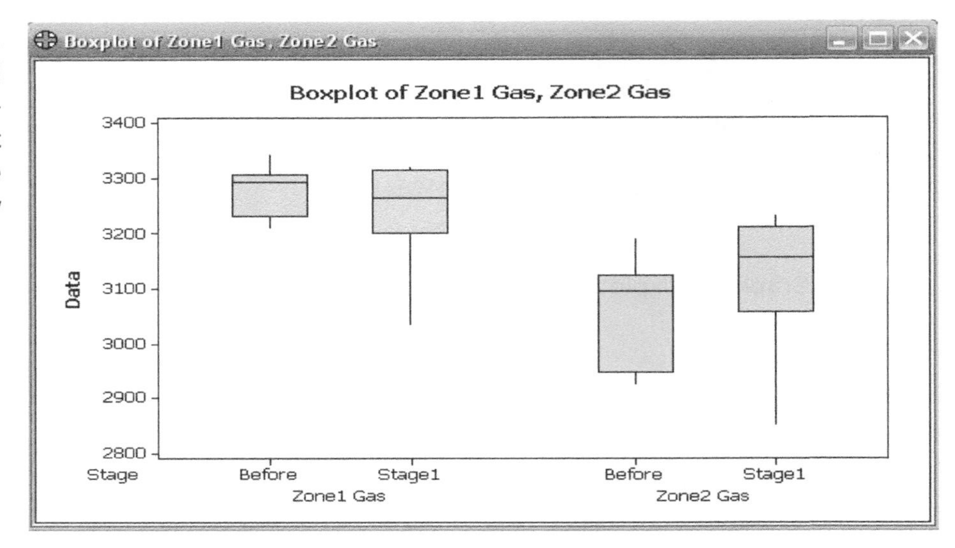

12. Right click on the graph panel.
13. Go to <<Add <<Reference Lines.
14. Enter 3100 into the upper cell to produce a horizontal reference line on the boxplot. Then Click OK. A value of 3100 has been selected to be the target for the individual zone gas usage.

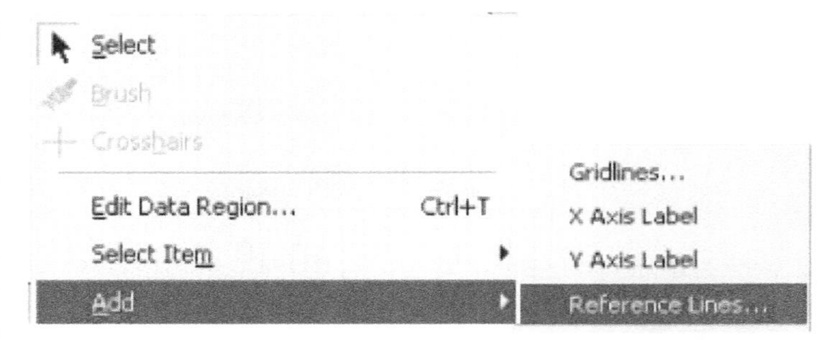

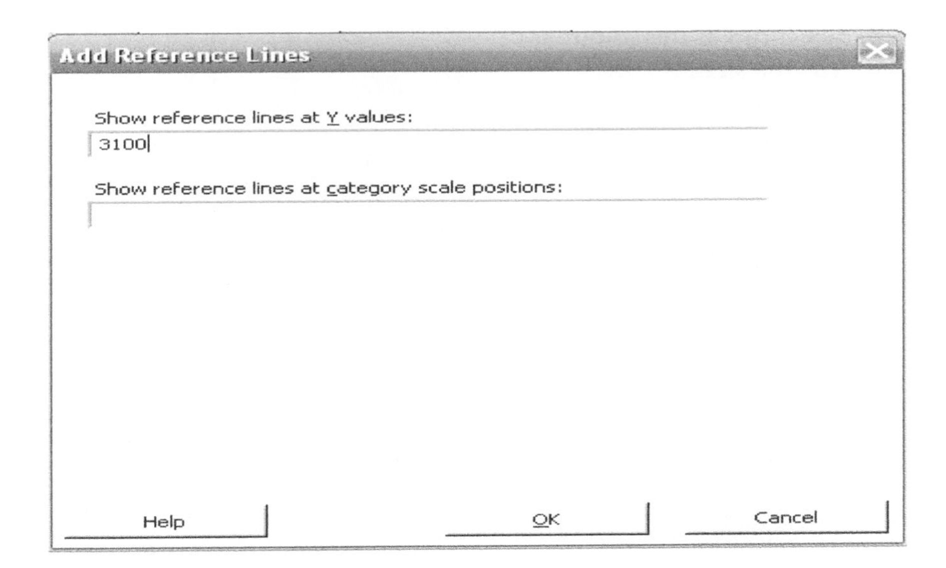

The following steps change the colour of the reference line to red and also its thickness.

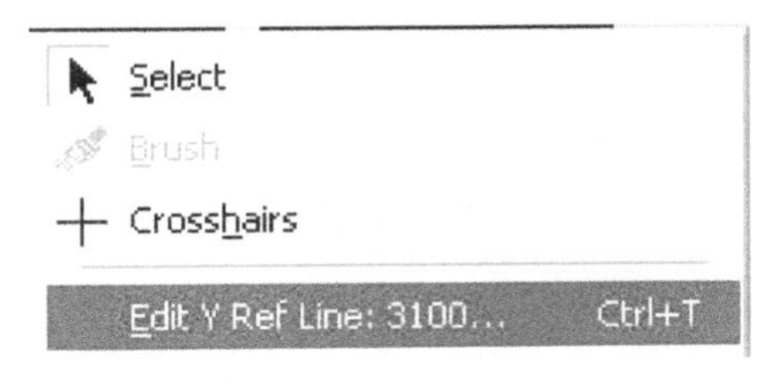

15. Right click on the reference line. Click on Edit Y Ref Line.
16. Alternatively, double click on the reference line.
17. On the Attributes tab change the radio button to Custom.

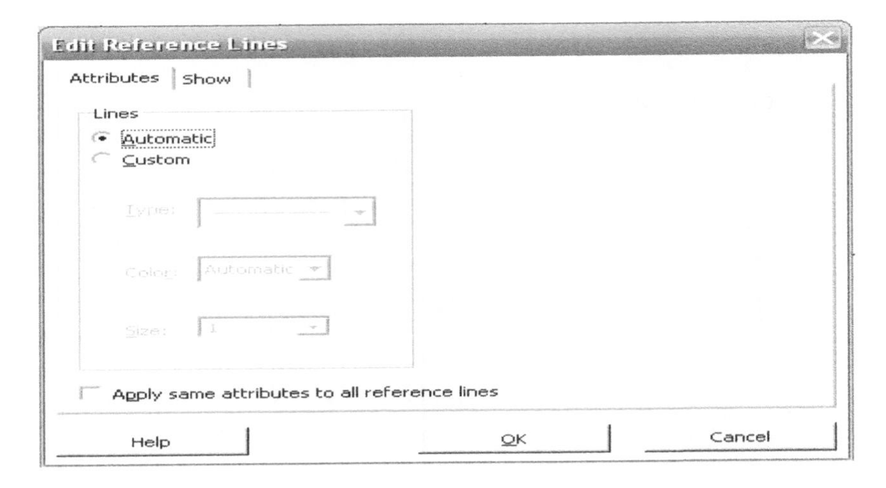

18. Select red as the colour and 5 as the size.
19. Click OK.

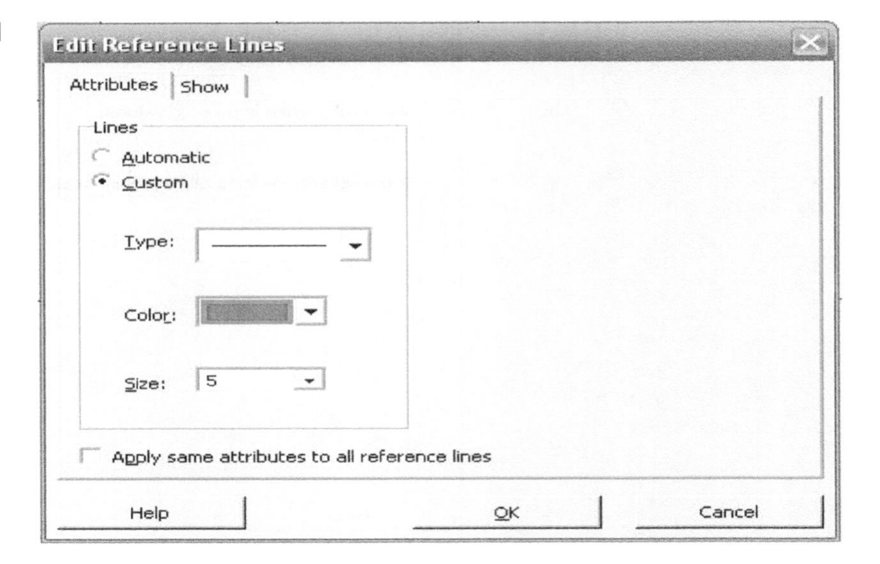

We will add the mean values of the data onto the graph together with a symbol which denotes the position of the mean.

20. Right click on the graph panel.
21. Click on Add <<Data Labels.

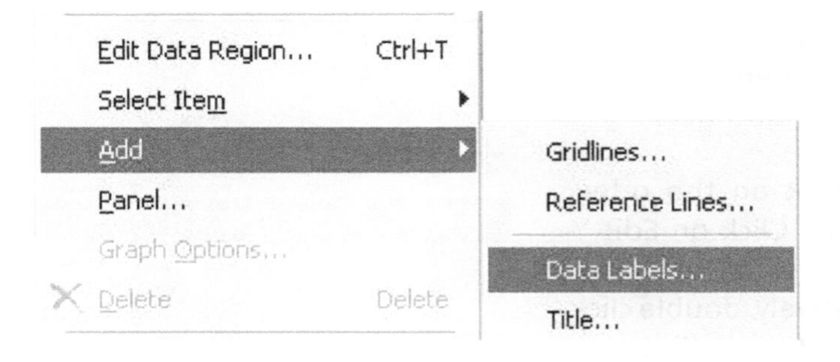

22. Click onto the Means tab.

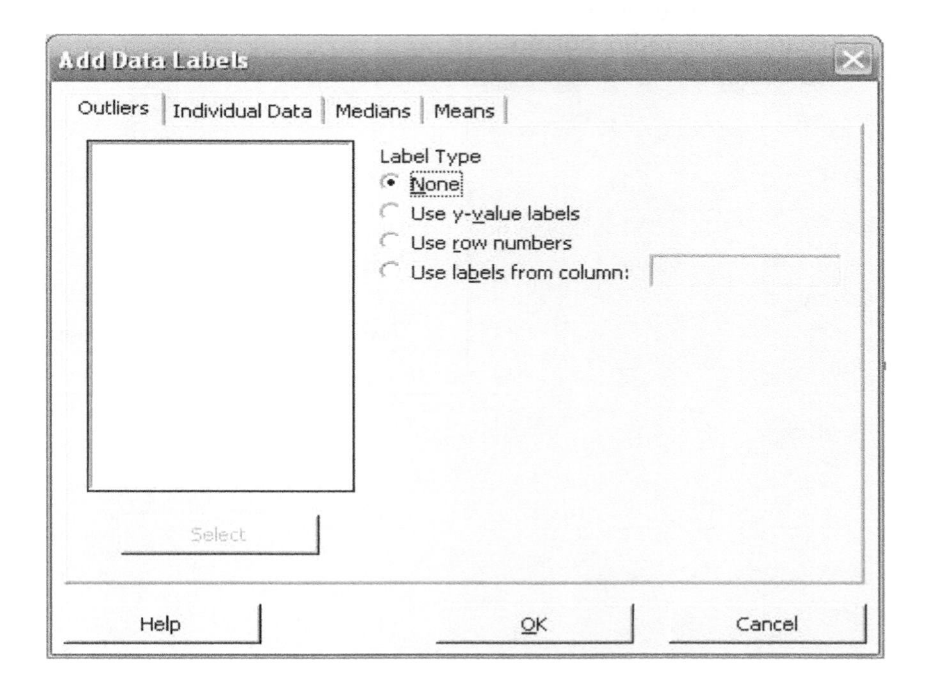

23. Click on 'Use y-value labels'.
24. Click OK.

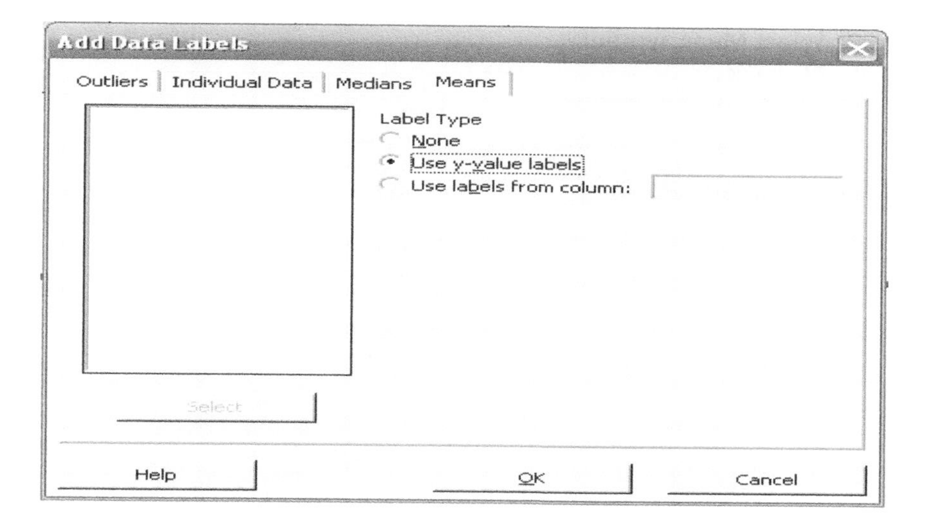

25. Right click onto the graph.
26. Select Add <<Data Display.

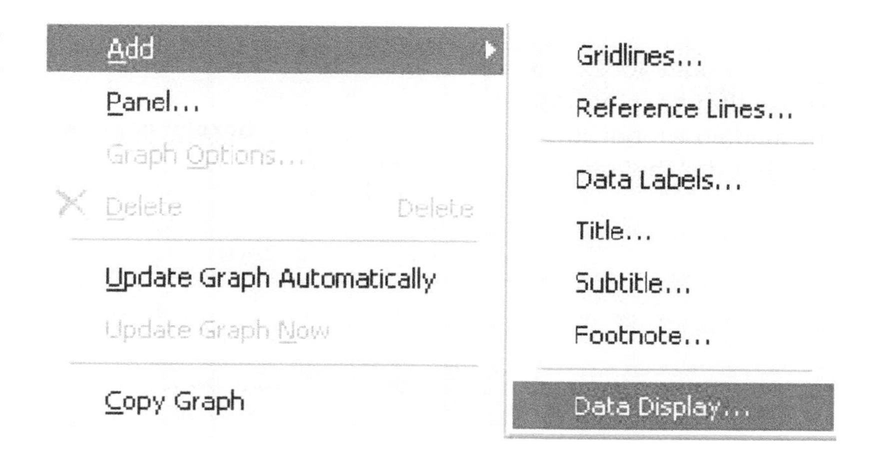

The Add Data Display menu box opens. It shows all the items that can be added onto the graph. The default items for this chart are already ticked.

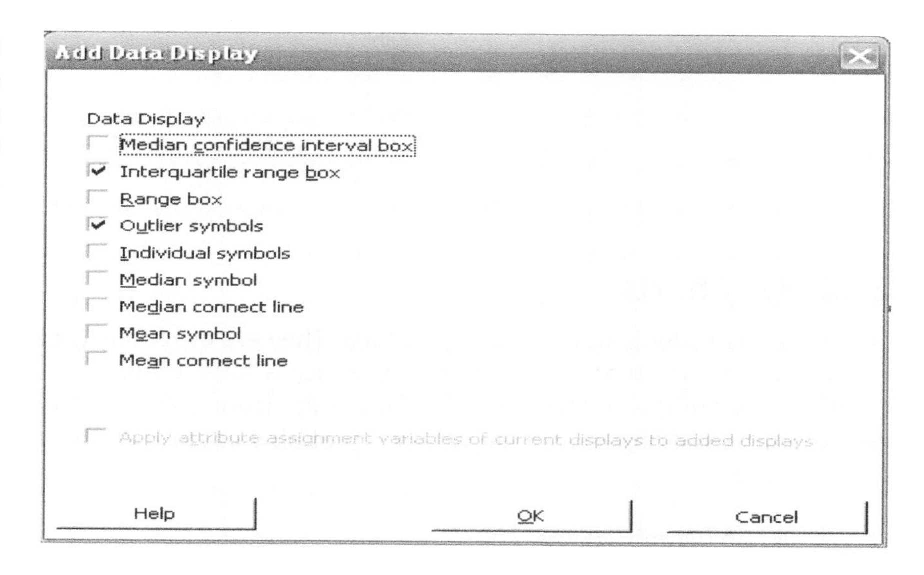

27. Select Mean symbol.
28. Click OK.

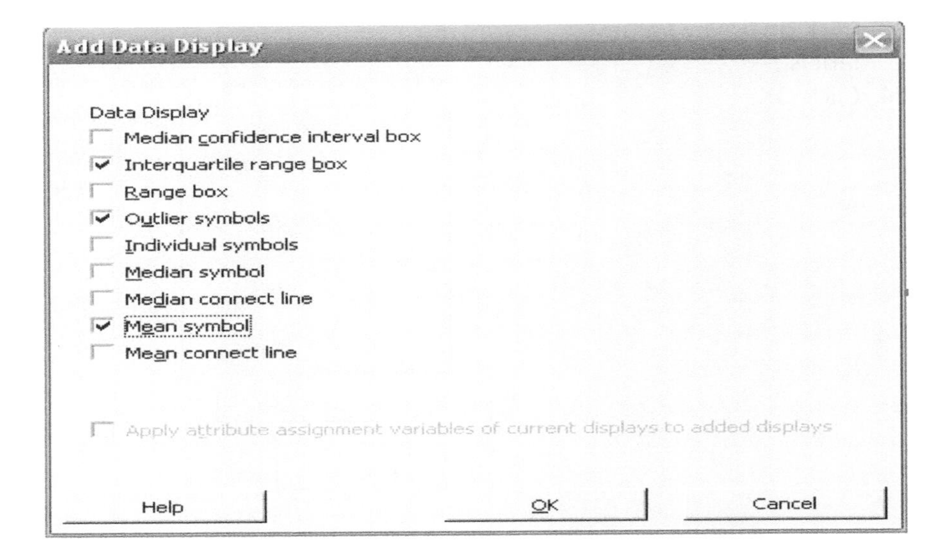

The graph should look like this. Notice that boxplots display the median symbol automatically but not the mean.

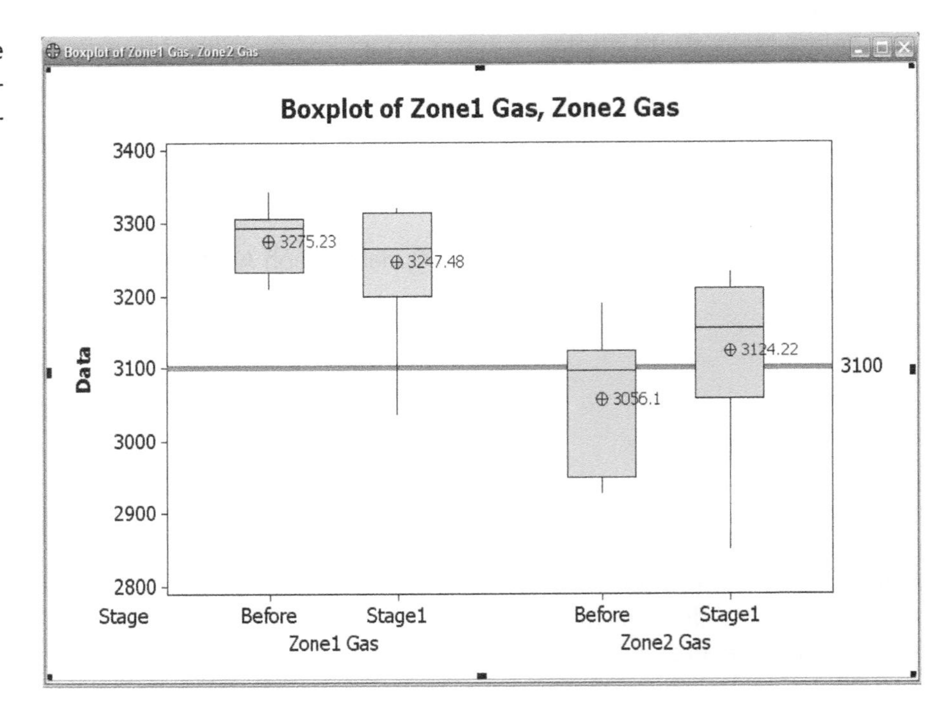

2.12 Bar Charts

Let's now have a look at Minitab's Bar Charts. They are especially useful when we start to use the inbuilt functions. Let's say that we wanted to produce a basic Pareto chart showing the standard deviation (I will abbreviate this to StDev) of total gas usage from each of the operators. Let's have a look how we would do this in Minitab and see why it is probably easier than producing the same graph in a spreadsheet program.

1. Click on Graph <<Bar Chart

2. Change the selector that appears to A function of a variable.

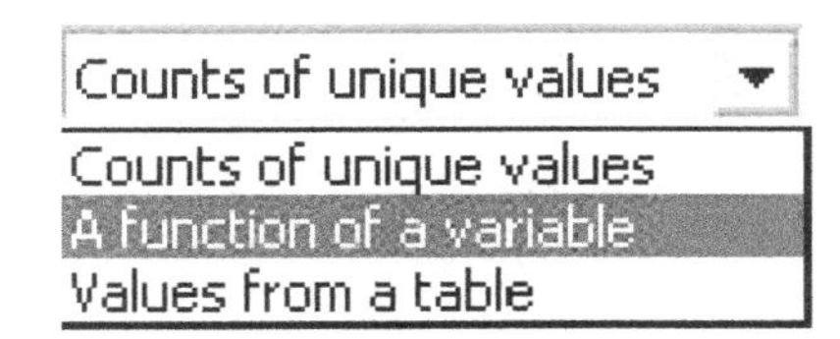

3. The bar graph that we will be producing will be a One Y, Simple.
4. Click OK.

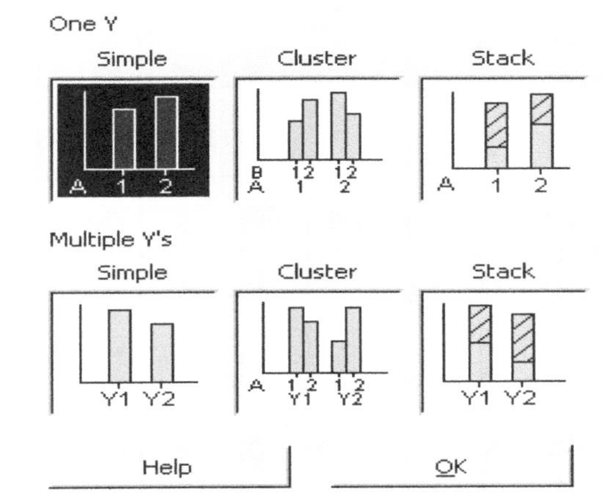

5. From the Function selector box select StDev.
6. Select Total Gas as the Graph variables.
7. Select the column heading Person for Categorical variable.
8. Then Click on chart options.

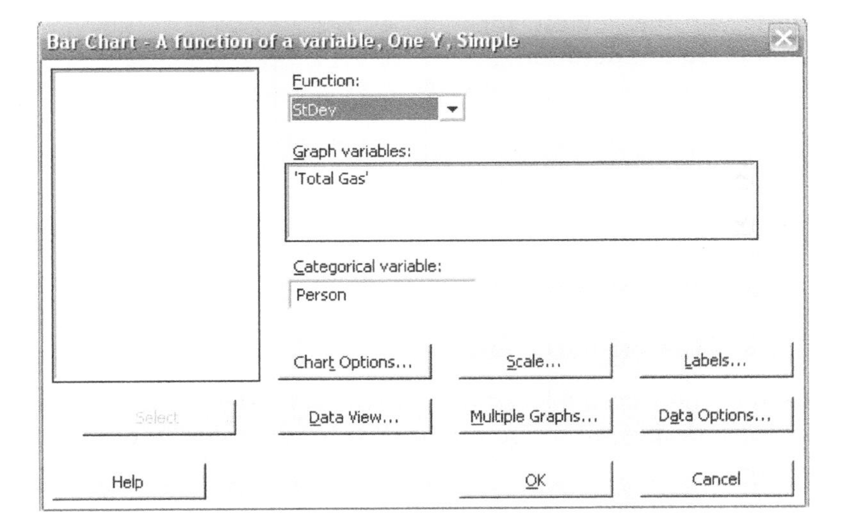

9. In order to make this a Pareto chart, where the value of the bars decreases as you read from left to right, we need to select Decreasing Y.

Order Main X Groups By
Default
Increasing Y
Decreasing Y

10. Click OK and OK again in order to produce the Pareto Chart.

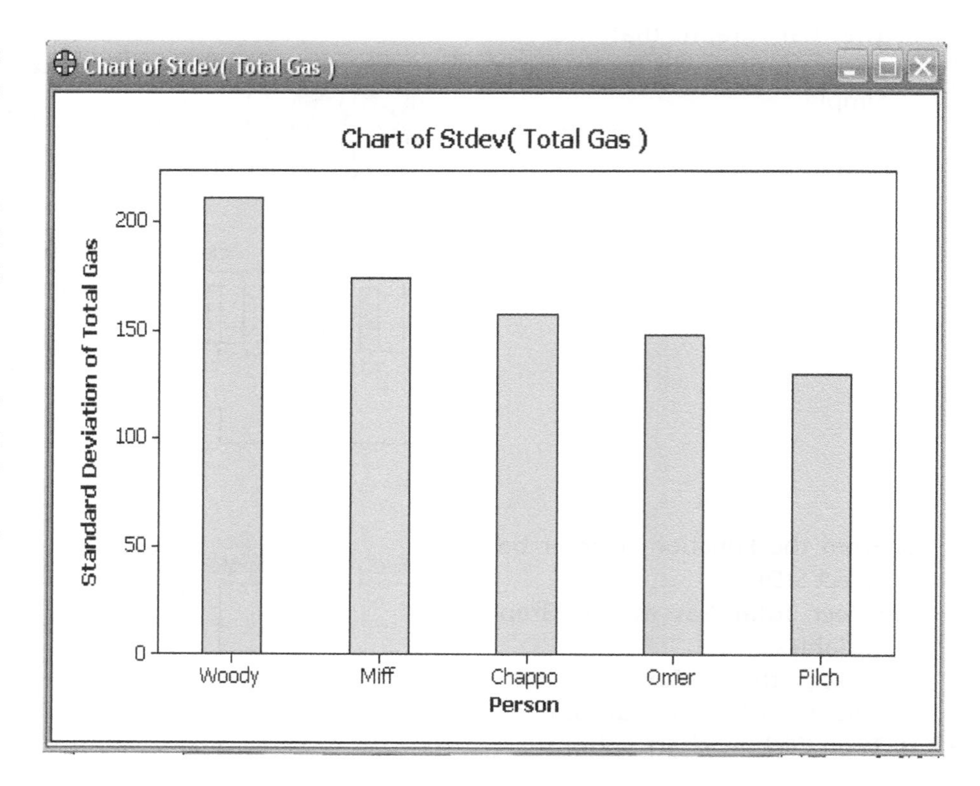

Note that Minitab has a specialised Pareto chart which is listed under Stat <<Quality Tools.

The Pareto chart of Total Gas standard deviation by operator is displayed. It can be seen that Woody has the highest StDev and Pilch had the lowest for this period of operation.

2.13 The Layout Tool

We have produced a lot of graphs within the same worksheet. Minitab gives a tool to display and print multiple graphs on the same sheet. It is called the Layout Tool. It is a very useful tool for the collation of graphs.

1. Click on the Show Graphs Folder icon.
2. Select any graph by double clicking it within the Project Manager. Then left click onto the graph within the Graph Window to select it.

3. Select Editor <<Layout Tool.

Note: if the graph is not selected the Layout Tool will not be selectable.

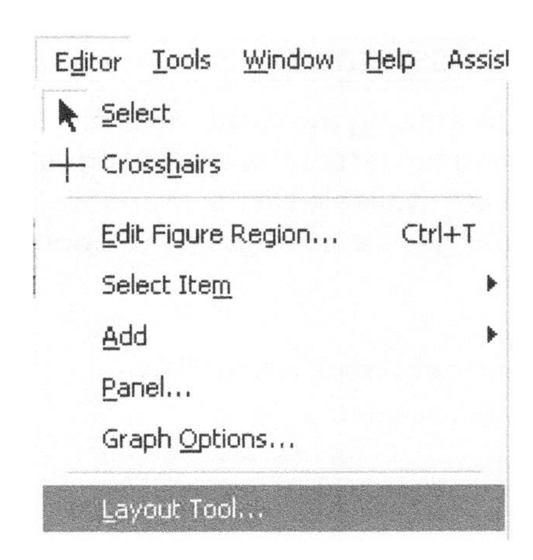

The layout tool will appear. It already contains the graph that you just selected.

In the top left is the grid selection for the layout.

Below that is a list of graphs that can be selected for the layout.

4. Select three other graphs for the Layout.
5. Then click on Finish.

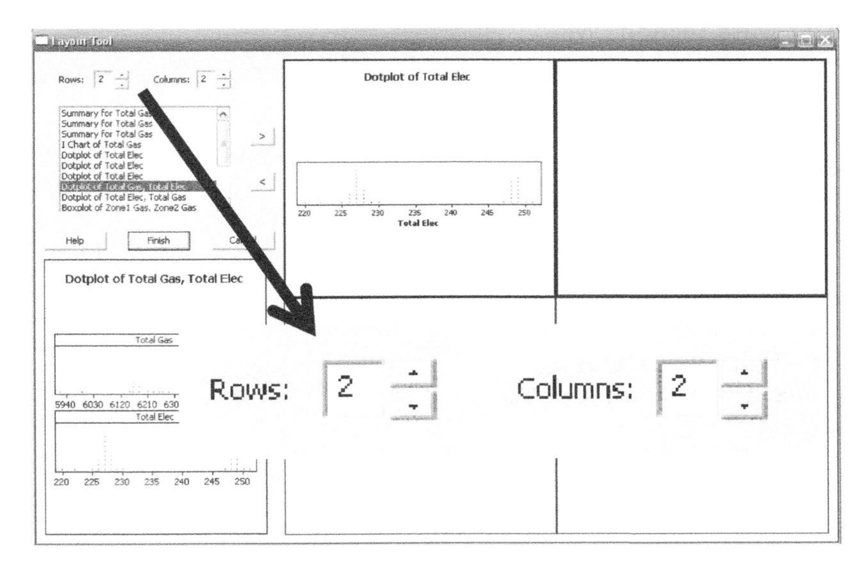

A new graph appears with our four selected graphs as panels. It should be noted that the graphs can still be edited but they cannot be updated.

The name of the graph appears as Layout in the Project Manager. It can be renamed by right clicking on Layout within the Project Manager.

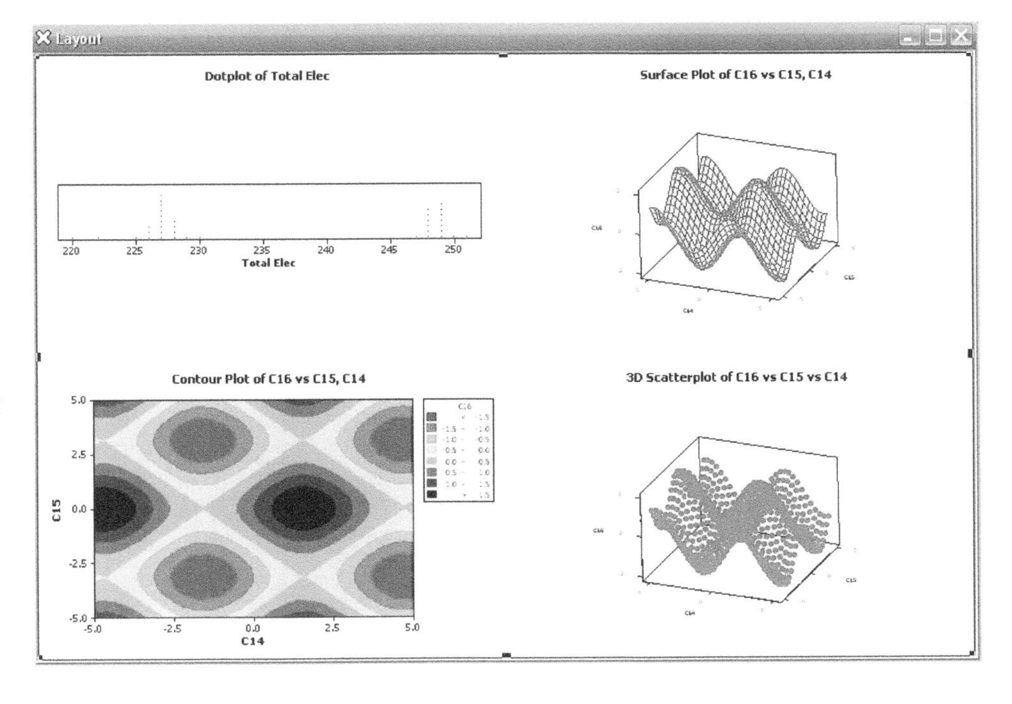

2.14 Producing Graphs with the Assistant

In this section we are going to have a quick look at using the Assistant to produce graphs. As stated earlier this is an M16 feature. If you have M15 don't feel too left out as all of the graphs and procedures accessed via the Assistant have their equivalents which are available via the dropdown menus. The Assistant offers a very graphical method of selection which also gives a lot of guidance about the selection.

1. Click on Assistant <<Graphical Analysis

Assistant
Measurement Systems Analysis (MSA)...
Capability Analysis...
Graphical Analysis...

The selector menu for Graphical Analysis opens. There are three sections and so far we have only looked at graphs within the Distribution of Data section. We will be covering Variables over Time and Relationships between Variables in later chapters. If you know the type of graph that you want, you can click on the graph type and open the data entry menu box straight away. Let's pretend we need help to decide.

2. Click on Help me choose within the Distribution of Data section.

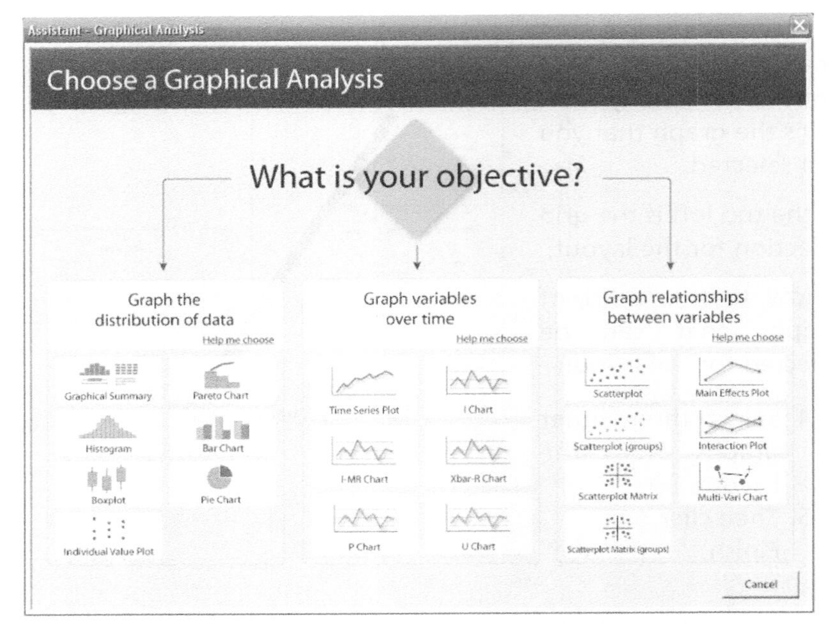

We are presented with a decision tree to help us choose the type of graph that we want. However, if we don't understand the terms being used within the decision tree the Assistant will give us further help by explaining the terms and taking us through the decision tree.

3. Click on Click to start above the Data type decision icon.

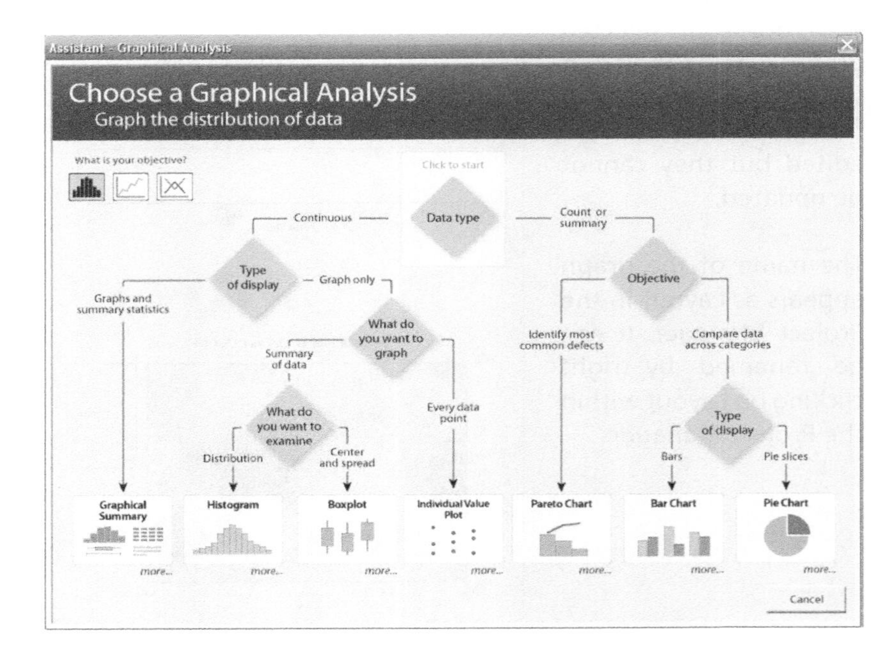

In the previous figure, the first decision that had to be made was whether the data that we were going to graph was Continuous Data or whether it was Count of Summary Data. The Assistant explains both choices within the decision. We must click Next below the appropriate choice.

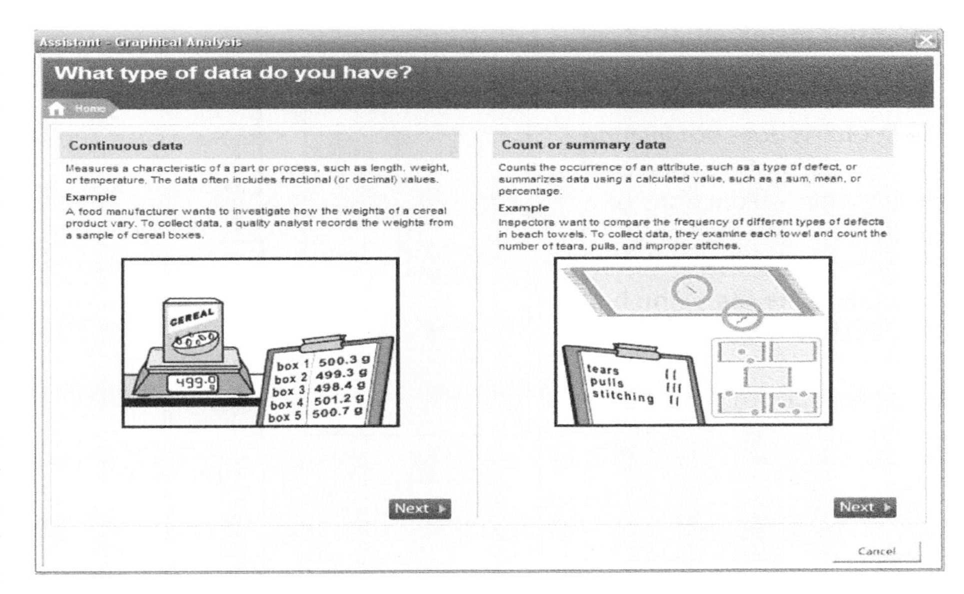

4. Follow the decisions through to get to Bar Chart. You will need to select Count/Summary Data, Compare data across categories and then Bars.
5. Note: even at this point we can press the cross icons to get additional information if we don't understand any of the terms or requirements. Press the icon to create the graph.

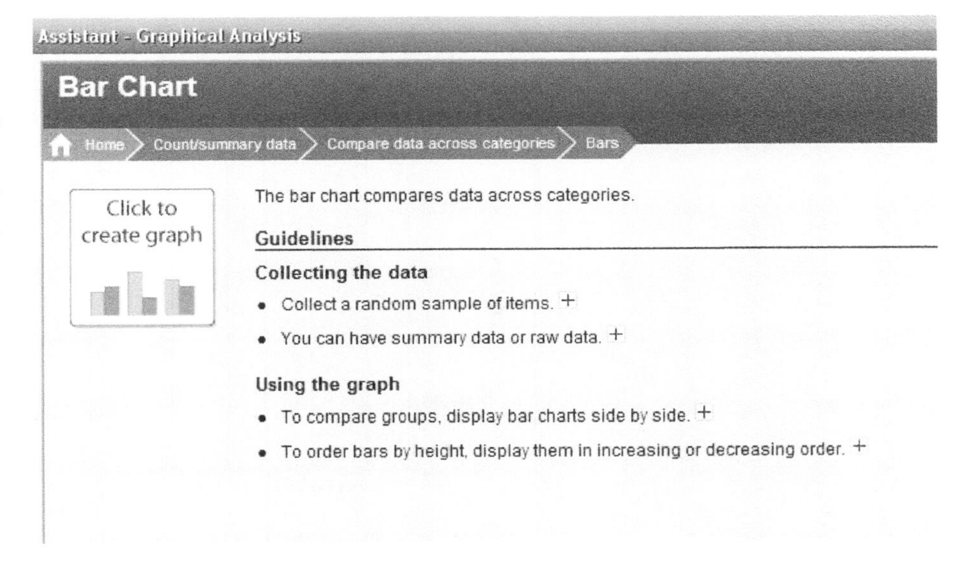

Using this menu we will produce a Pareto chart of Total Gas. Note that our selection in Bar Charts does not include the options that we used previously, that is Functions of a variable.

6. Complete the menu box as shown and click OK.

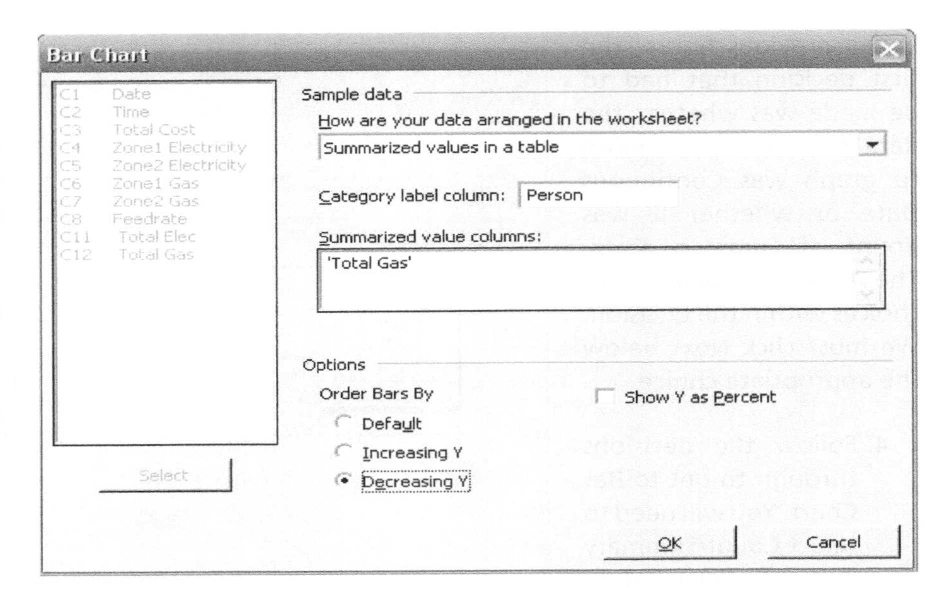

A Pareto chart of Total Gas is produced.

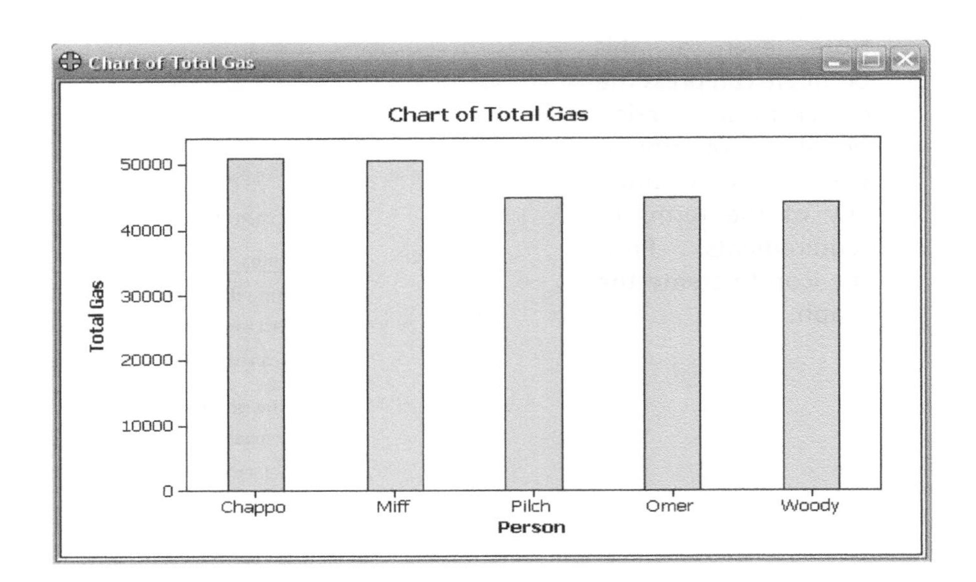

2.15 Producing Reports

In this section we are going to see how information can be exported to a Word Processing program such as Microsoft Word in order to produce reports and increase accessibility for others. You will need to have the appropriate program available on your PC.

1. Select any graph and right click on the graph panel.
2. Click on Append Graph to Report. We can then start selecting other items for the Report Pad. Data/calculations from the Session window can also be appended to the Report pad by right clicking.

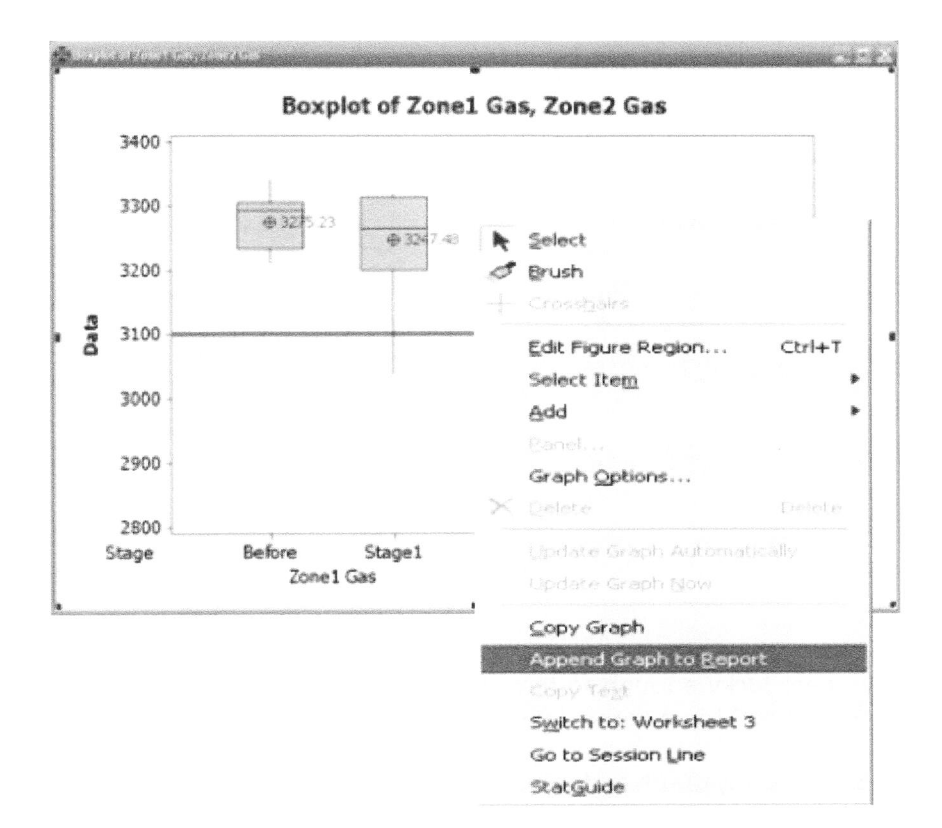

3. When you have finished sending items to the Report Pad click the Show Report Pad icon.

4. Right click onto the Report Pad folder icon and then select Copy to Word Processor or Save Report As...

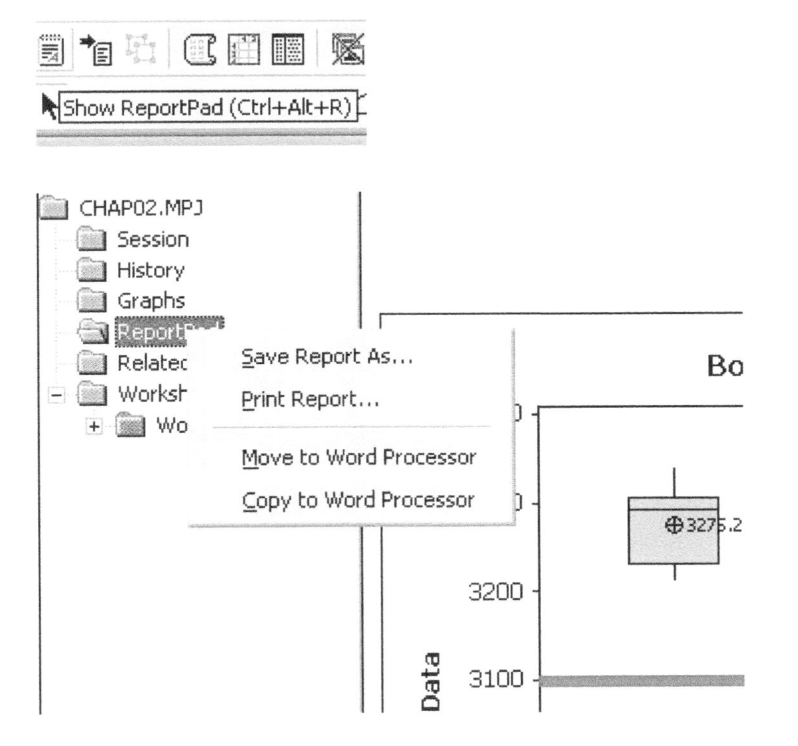

5. The contents of the Report Pad can now be copied into a word processer in .rtf format and subsequently converted to .doc format for Microsoft Word.

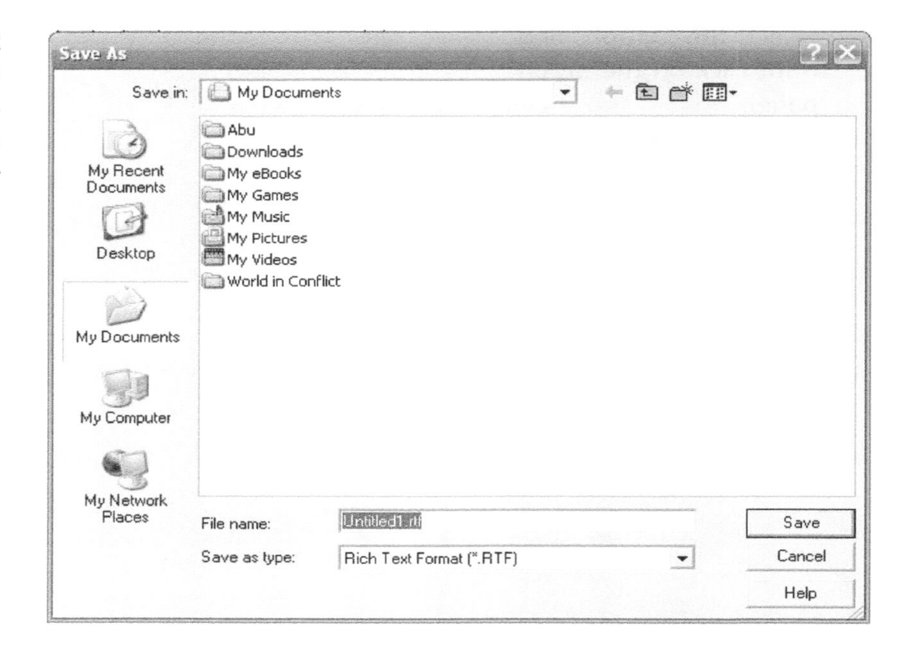

If you are lucky enough to have Minitab 16 you can export directly to Microsoft Word or PowerPoint. You will need to have the appropriate program available on your PC.

6. Select any graph and right click on the graph panel.
7. Click on Send Graph to Microsoft Word or PowerPoint. This action will export the graph or data immediately.

I have put some data into the Session Window and shown how that is exported as well. If you have not got Word or Powerpoint open, Minitab will open it for you.

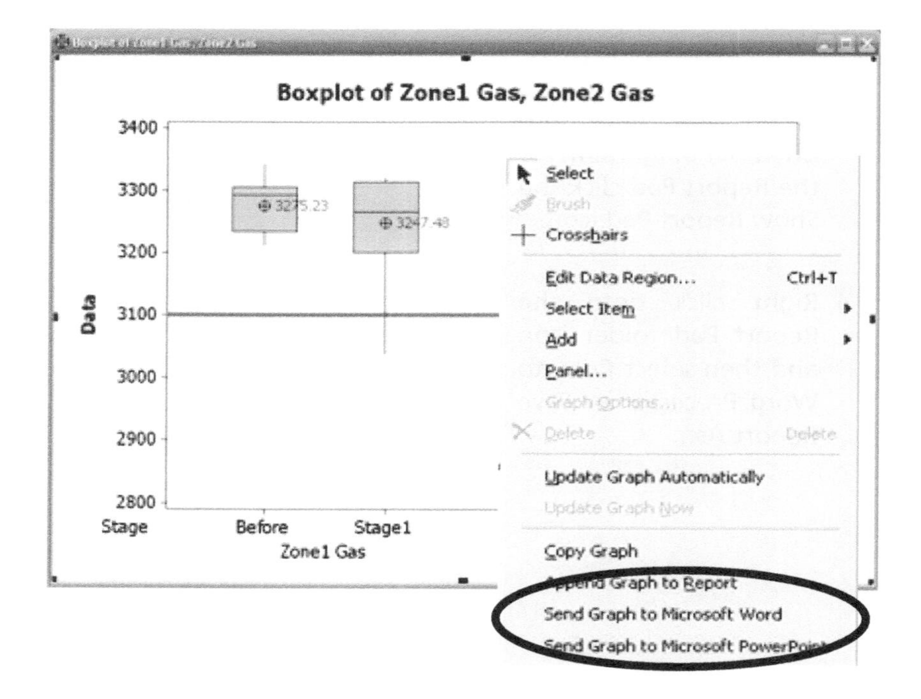

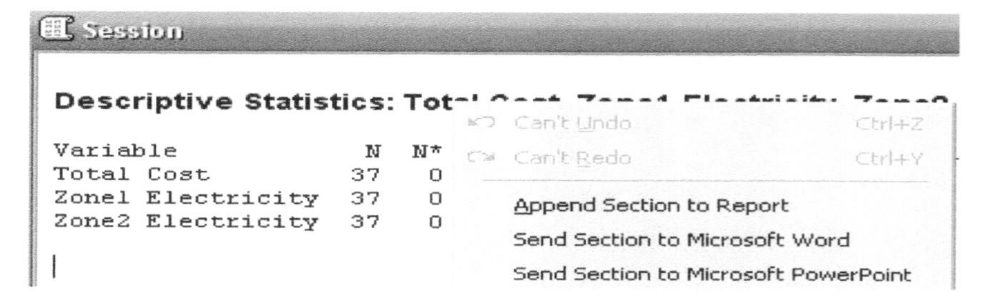

2.16 Creating a New Project/Worksheet Button

Finally, after every example and exercise you will need to start afresh and open a new project. I am not going to tell you how to save your work because that is the same as in any other work based program. However, to make things easier we will install the New Project/Worksheet button into our toolbars.

1. Click Tools <<Customize

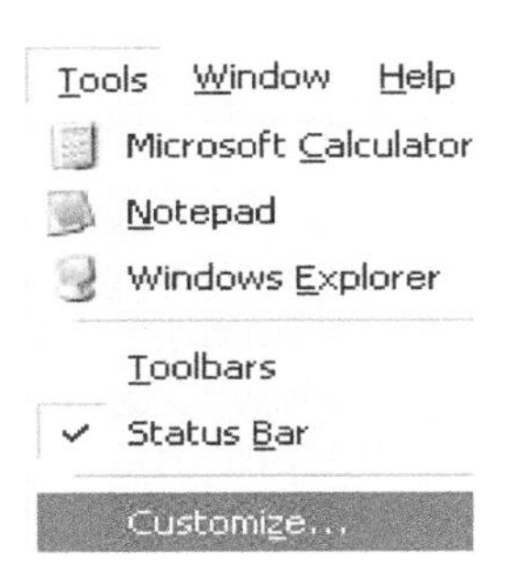

2. On the Commands tab ensure File is selected in Categories. Find the New icon and simply drag it into your toolbar. I have placed it next to the Open Project icon on my copy of Minitab.

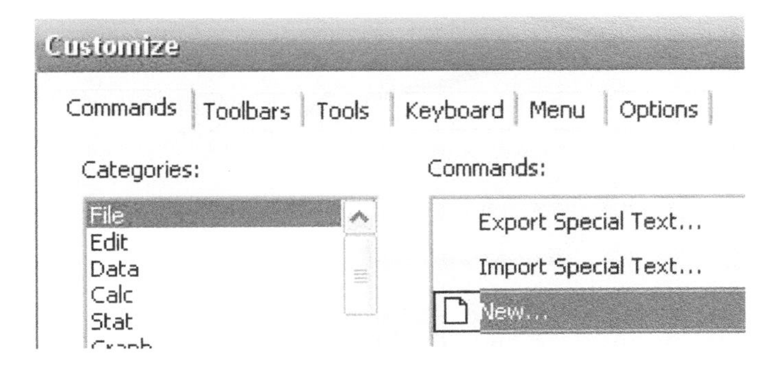

3. Click on the New icon and select Minitab Project. If you have not already saved your work you will be prompted to do so. After which you will begin afresh with a blank project.

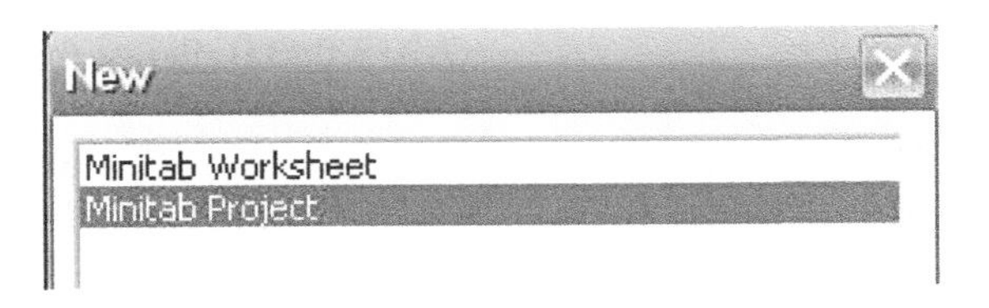

CHAPTER 3
Basic Statistics

3.1 Types of Data

When we start looking at any system we need to get a quick grasp of the data that the system is working with, as this dictates the tools and techniques that we will use to analyse the system. We may even decide that the quality of the data is too low to understand process behaviour and then the initial step would be to change how we monitor the process.

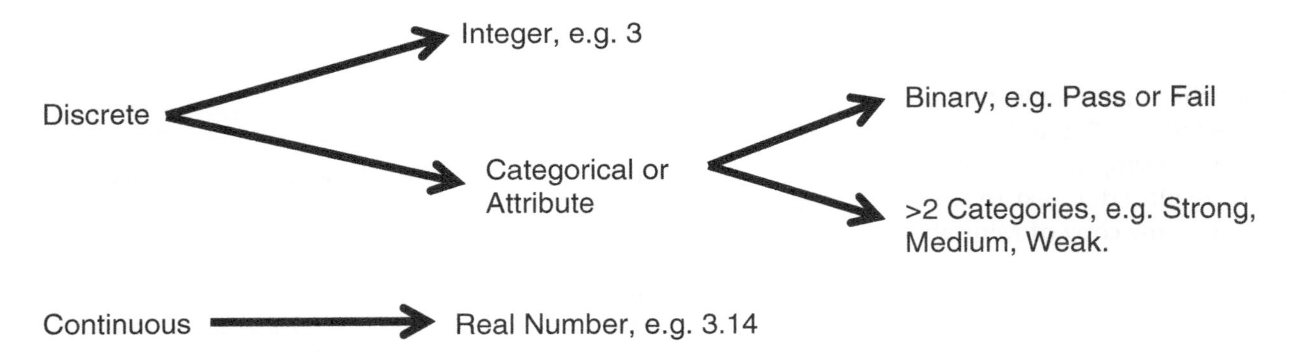

Continuous data gives us the most value and that is why most modern measuring systems will produce continuous data to the highest possible resolution. We will only focus on integers and real numbers for that reason.

3.2 Central Location

In order to demonstrate the next two concepts, Measures of Central Location and Measures of Dispersion, we are going to use the following continuous data set:

7.2, 3.0, 3.0, 9.1, 4.0, 5.0, 9.0, 6.3, 2.0

There are $N = 9$ numbers within our data set. The data set is our complete set of values so we call it our population. We say that we are using descriptive statistics because we have the values of the entire population within our calculations.

Mean, Median and Mode are three very common measures of central location that you will know already.

Mean: The mean, μ, is the result when all of the values are added together and then divided by the number of values within the data set, N.

Problem Solving and Data Analysis using Minitab: A clear and easy guide to Six Sigma methodology, First Edition. Rehman M. Khan.

$$\mu = \frac{7.2+3.0+3+9.1+4.0+5.0+9.0+6.3+2.0}{9} = 5.4$$

Equation (3.1)

Median: The midpoint in a string of sorted data. If we sort our data in ascending order we get:

2.0, 3.0, 3.0, 4.0, 5.0, 6.3, 7.2, 9.0, 9.1
↑

5.0 is the median as it is at the midpoint of the ordered data. If there is an even number of elements within the data set then the mean of the middle two numbers is taken to be the median.

Mode: The most frequently occurring value. The mode is 3.0 as it occurs twice within the data set and all the other numbers only occur once.

Depending upon which measure of central location that we decide to use, our value for central location could be 5.4, 5.0 or 3.0.

3.3 Dispersion

Moving onto Measures of Dispersion. The easiest of these is Range.

Range is the difference between the largest and the smallest of the elements within the data set.

Data Set = 7.2, 3.0, 3.0, 9.1, 4.0, 5.0, 9.0, 6.3, 2.0

Range = Max Value – Min Value = 9.1 – 2.0 = 7.1

Equation (3.2)

Let's have a look at how data is distributed about the mean, μ.

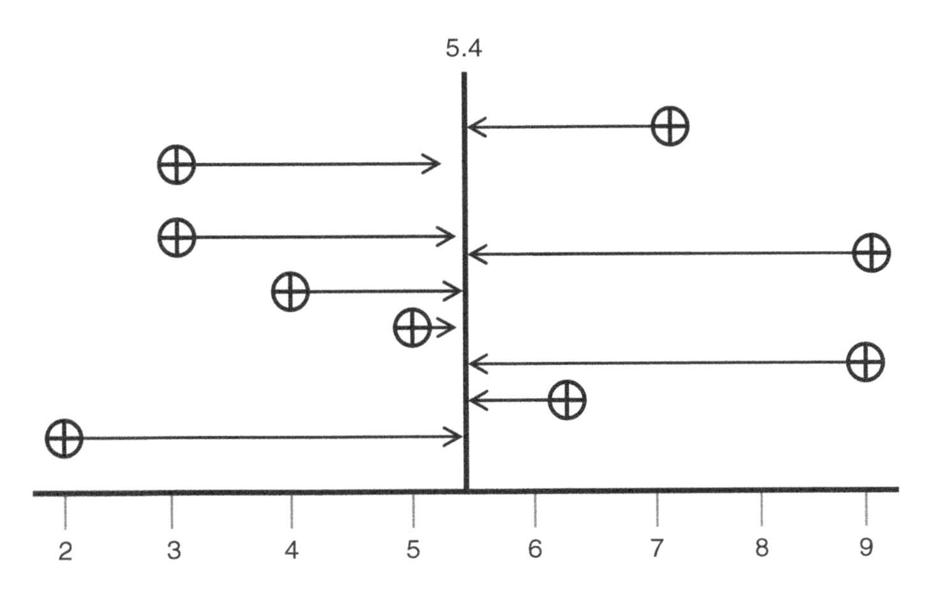

The deviation is the distance from each point to the mean. By definition the sum of the deviations is zero.

Deviation = $(X - \mu)$

Equation (3.3)

$$\text{Sum of Deviations Squared} = \sum_{x=1}^{N}(xi - \mu)^2$$

Equation (3.4)

The Sum of Deviations Squared may also be called the sum of squares. Note that we are using the sum of squares procedure on the deviations.

$$\text{Variance} = \sigma^2 = \frac{\sum_{x=1}^{N}(xi - \mu)^2}{N}$$

Equation (3.5)

$$\text{StDev.P} = \sigma = \sqrt{\frac{\sum_{x=1}^{N}(x_i - \mu)^2}{N}}$$

Equation (3.6)

Remember these are equations for population parameters. StDev.P stands for population standard deviation.

The table below shows how the StDev.P for our data set is derived manually.

Data Set	Mean	Deviation	Deviation Sqr	Sum of Sqrs of Deviations	Variance	StDev.P
7.2	5.4	1.8	3.24	55.9	6.21	2.49
3.0		−2.4	5.76			
3.0		−2.4	5.76			
9.1		3.7	13.69			
4.0		−1.4	1.96			
5.0		−0.4	0.16			
9.0		3.6	12.96			
6.3		0.9	0.81			
2.0		−3.4	11.56			

3.4 Descriptive Statistics

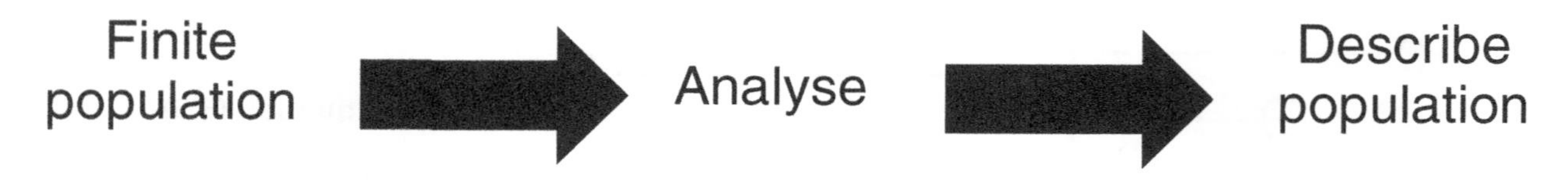

Descriptive statistics are for the case where we have data on a complete population. We analyse that complete data set which then gives us information to describe the population. Let's have a look at descriptive statistics within Minitab.

This is a short exercise on Descriptive Statistics. We are going to import some data into Minitab and then demonstrate how to use Descriptive Statistics. We will be using this data for the remainder of the chapter so you do not need to start a new project after every example/exercise.

1. Open the Excel file '03 Basic Stats.xls' and go to sheet 1.
2. Import the four columns into Minitab by copying and pasting.

Normal	Exercise1	Exercise2	Exercise3
98.70073	113.5166	5.643678	98.1995
106.4411	115.2099	4.300852	105.7378
97.85076	100.7614	142.7615	98.22182
96.77334	105.5682	309.3205	107.2309
94.64565	96.26837	100.9789	104.7541
99.17578	115.5223	61.55869	107.124
95.1274	110.5756	14.82981	100.3159
90.7505	103.2359	87.44837	105.6286
105.2022	114.7376	33.26984	102.1976
95.88524	104.1737	58.96926	97.89239
101.2302	111.5185	32.05817	99.14487
91.63122	109.199	16.68909	100.5925
104.4955	104.4998	101.8246	96.85037
102.3799	120.0891	176.4651	104.9549
111.0015	112.5191	29.50158	91.30796
95.94029	108.6777	52.95856	97.77616
95.26619	119.4542	4.219999	102.7522
92.98437	109.4609	5.005408	106.6444
101.771	112.2243	25.40558	104.91
96.20273	109.9257	3.977879	101.3001
96.69518	103.5963	5.718547	
94.52719	116.5084	20.52797	
99.69039	105.1016	53.0038	
93.04756	104.5246	18.73714	
98.93617	110.3089	193.551	
100.8563	117.8197	8.744367	

3. Check that the data has been correctly transferred into Minitab. All columns should be numeric.

C1	C2	C3	C4
Normal	Exercise1	Exercise2	Exercise3
98.701	113.517	5.644	98.199
106.441	115.210	4.301	105.738
97.851	100.761	142.761	98.222
96.773	105.568	309.320	107.231

4. Click on the column information button.

5. The data type for all four columns is numeric, as indicated by the N under Type.

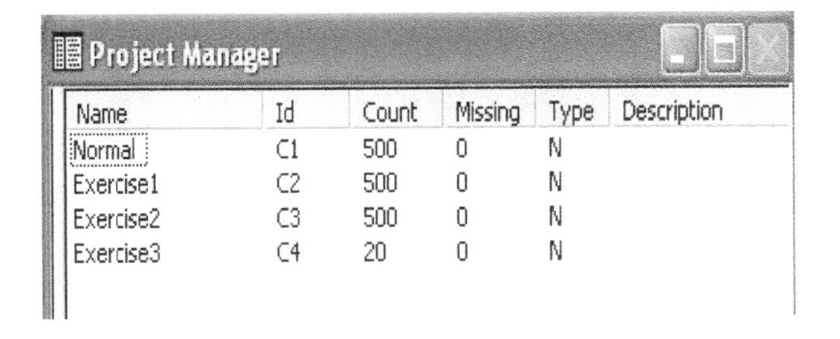
Project Manager

Name	Id	Count	Missing	Type	Description
Normal	C1	500	0	N	
Exercise1	C2	500	0	N	
Exercise2	C3	500	0	N	
Exercise3	C4	20	0	N	

6. Click Stat <<Basic Statistics <<Display Descriptive Statistics.

7. Complete the menu box as shown.

8. Click on Statistics

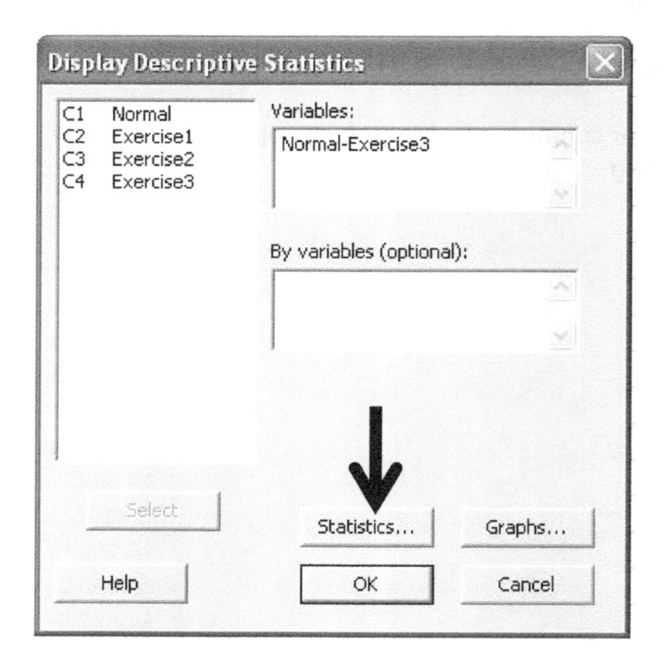

9. The menu box shows the range of statistics that are available. The default statistics are already ticked. Change the selection to the one shown.
10. Click OK and then click on Graphs.

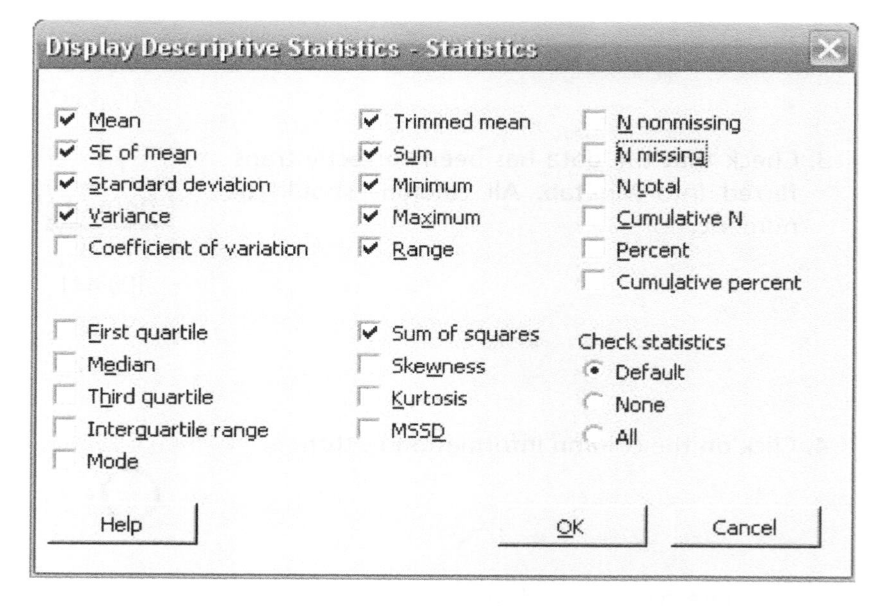

11. The menu box shows the four graphs that are available. On this occasion we are not going to produce a graph.
12. Then click OK and OK again.

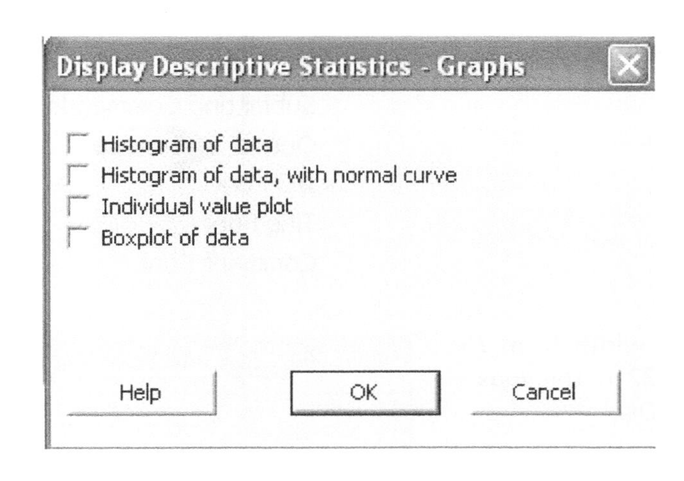

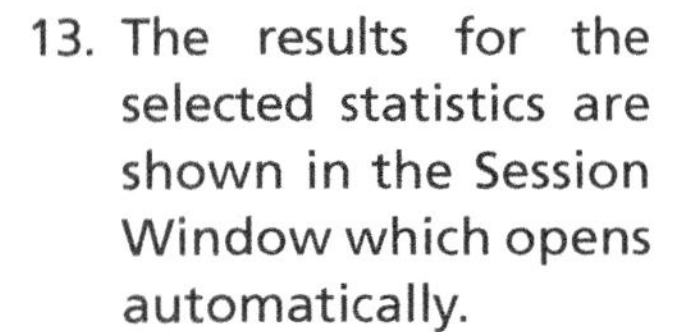

13. The results for the selected statistics are shown in the Session Window which opens automatically.

Descriptive Statistics: Normal, Exercise1, Exercise2, Exercise3

```
Variable     Mean  SE Mean  TrMean  StDev  Variance       Sum  Sum of Squares
Normal     100.35    0.218  100.34   4.87     23.72  50174.97      5046893.58
Exercise1  100.13    0.498  100.11  11.14    124.21  50065.26      5075040.66
Exercise2   53.39     2.34   47.10  52.41   2746.99  26694.25      2795914.76
Exercise3  101.68    0.951  101.94   4.25     18.07   2033.54       207106.81

Variable   Minimum  Maximum   Range
Normal       87.55   118.06   30.51
Exercise1    75.28   121.81   46.52
Exercise2     3.10   373.81  370.71
Exercise3    91.31   107.23   15.92
```

Notice that the data is too wide for the Session Window and Minitab has to start a new table. We can change the default session window width.

14. Click on Tools <<Options.

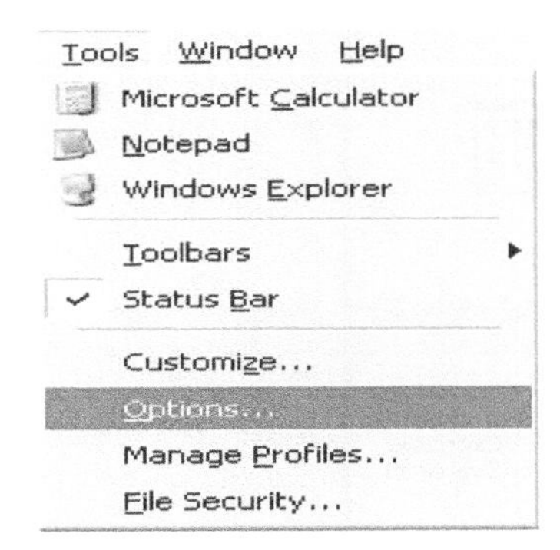

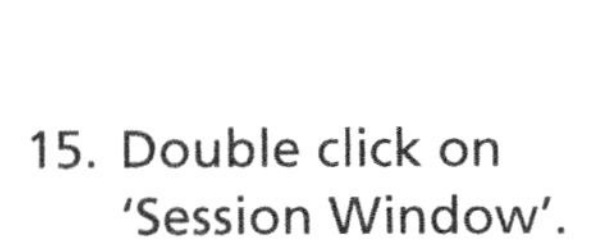

15. Double click on 'Session Window'.

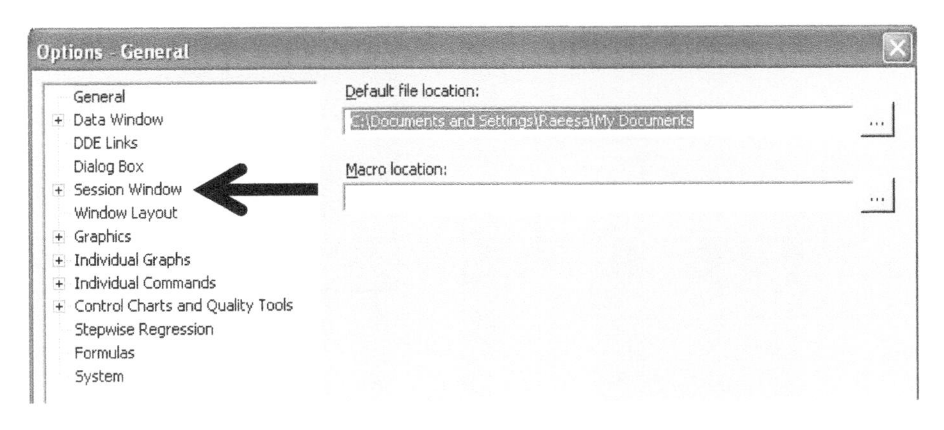

16. Then double click on 'Output'.

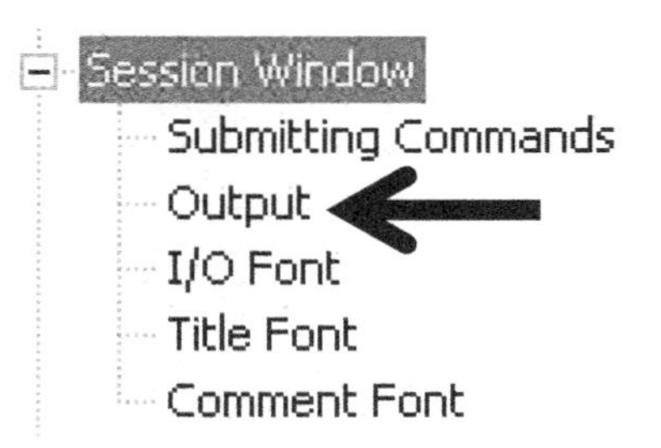

17. Change the Line width from 79 to 132, where 132 is the maximum. Then click OK.

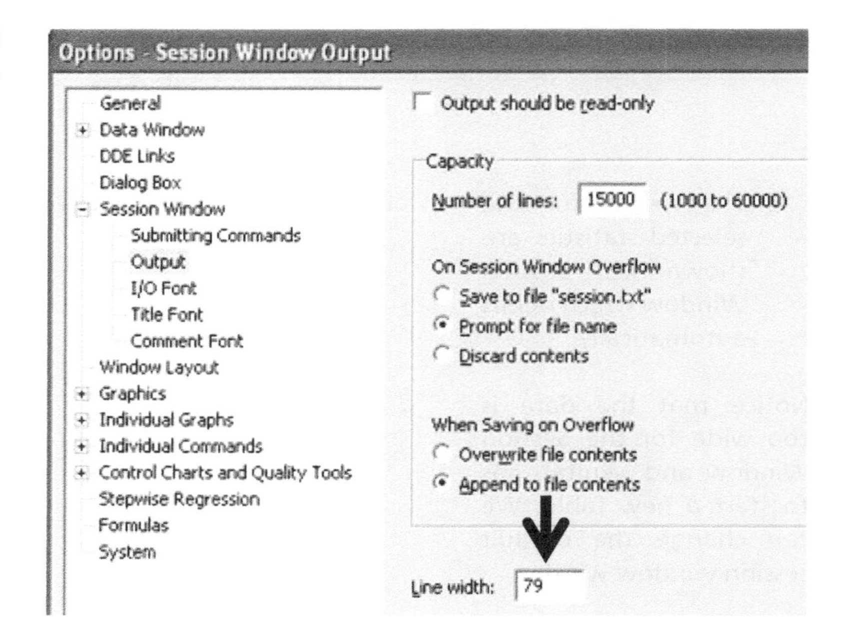

18. Click on the Edit Last Dialog icon.

19. Then click OK to display the same selection of Descriptive Statistics.

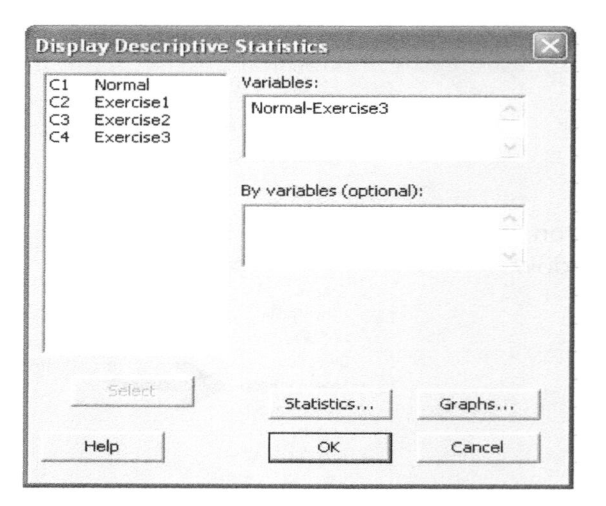

Notice that Minitab can now display all the data in a single table.

Descriptive Statistics: Normal, Exercise1, Exercise2, Exercise3

Variable	Mean	SE Mean	TrMean	StDev	Variance	Sum	Sum of Squares	Minimum	Maximum	Range
Normal	100.35	0.218	100.34	4.87	23.72	50174.97	5046893.58	87.55	118.06	30.51
Exercise1	100.13	0.498	100.11	11.14	124.21	50065.26	5075040.66	75.28	121.81	46.52
Exercise2	53.39	2.34	47.10	52.41	2746.99	26694.25	2795914.76	3.10	373.81	370.71
Exercise3	101.68	0.951	101.94	4.25	18.07	2033.54	207106.81	91.31	107.23	15.92

Now that we know how to use Descriptive Statistics within Minitab we will use this procedure on the data set that we used when learning about measures of dispersion and central location.

1. Open the Excel file '03 Basic Stats.xls' and go to sheet2.

	A
1	Data Set
2	7.2
3	3
4	3
5	9.1
6	4
7	5
8	9
9	6.3
10	2

2. Import the single columns into Minitab. Instead of opening a new project just put it into column 5 of the current worksheet.

C5
Data Set
7.2
3.0
3.0
9.1
4.0
5.0
9.0
6.3
2.0

3. Click on the Edit Last Dialog icon.
4. Press F3 to clear the previous input. Then select Data Set as the required Variables.

Variables:

'Data Set'

5. Click on Statistics.
6. Change the selection to that shown. This will calculate the items that we manually calculated for our Data Set.
7. Then click OK and OK again.

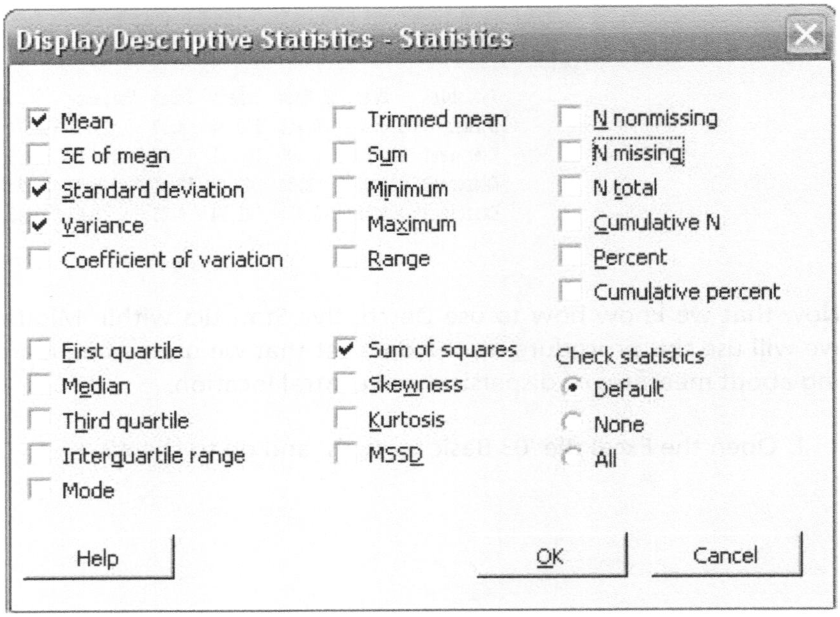

Display Descriptive Statistics - Statistics

- [x] Mean
- [ ] SE of mean
- [x] Standard deviation
- [x] Variance
- [ ] Coefficient of variation
- [ ] Trimmed mean
- [ ] Sum
- [ ] Minimum
- [ ] Maximum
- [ ] Range
- [ ] N nonmissing
- [ ] N missing
- [ ] N total
- [ ] Cumulative N
- [ ] Percent
- [ ] Cumulative percent
- [ ] First quartile
- [ ] Median
- [ ] Third quartile
- [ ] Interquartile range
- [ ] Mode
- [x] Sum of squares
- [ ] Skewness
- [ ] Kurtosis
- [ ] MSSD

Check statistics
- (•) Default
- ( ) None
- ( ) All

Help OK Cancel

8. Look at the results produced within the Session Window and compare those with the numbers that we calculated earlier.

Descriptive Statistics: Data Set

```
                                        Sum of
Variable    Mean  StDev  Variance      Squares
Data Set   5.400  2.643     6.987      318.340
```

Mean	Sum of Sqrs of Deviations	Variance	StDev.P
5.4	55.9	6.21	2.49

We see that the mean is the same; however, all the other terms that were calculated are different. It is worth trying to understand why this is. The sum of squares (SS) is different because initially we calculated SS for the deviations and now we have just calculated the SS of the data set, not the deviations. The variance and StDev are different because we initially calculated them as population parameters; however, Minitab has calculated them as Sample parameters. This leads us on nicely to look at the difference between population and sample statistics and also descriptive and inferential statistics.

3.5 Inferential Statistics

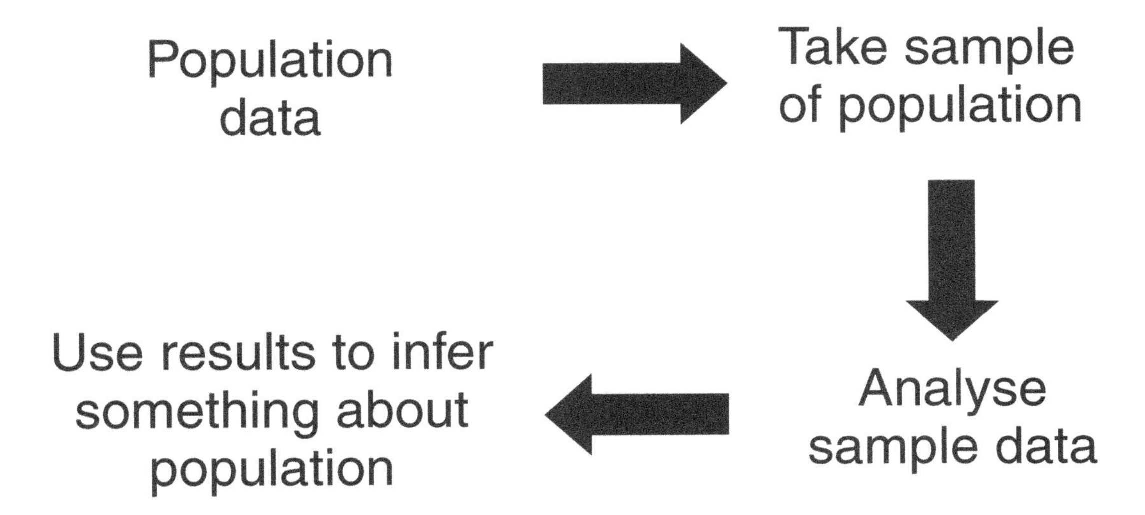

Imagine that we have a factory where we are monitoring parameters like weight, flow, concentration and so on. When we start to capture data we will probably not take the complete data set but only take a small fraction of it. We will then analyse our captured data and then we will want to infer something about the population. This is where sample or inferential statistics come in. For example, let's say that our factory is making 100 000 widgets every day. We take a sample of 500 widgets in the day. We weigh all the widgets and find that the average weight is 25 g. The average weight of the 500 samples widgets is 25 g; and that's descriptive statistics. We then say that all the widgets weighed by the factory have an average weight of 25 g; and that's inferential statistics because we have not checked the weight of the complete population.

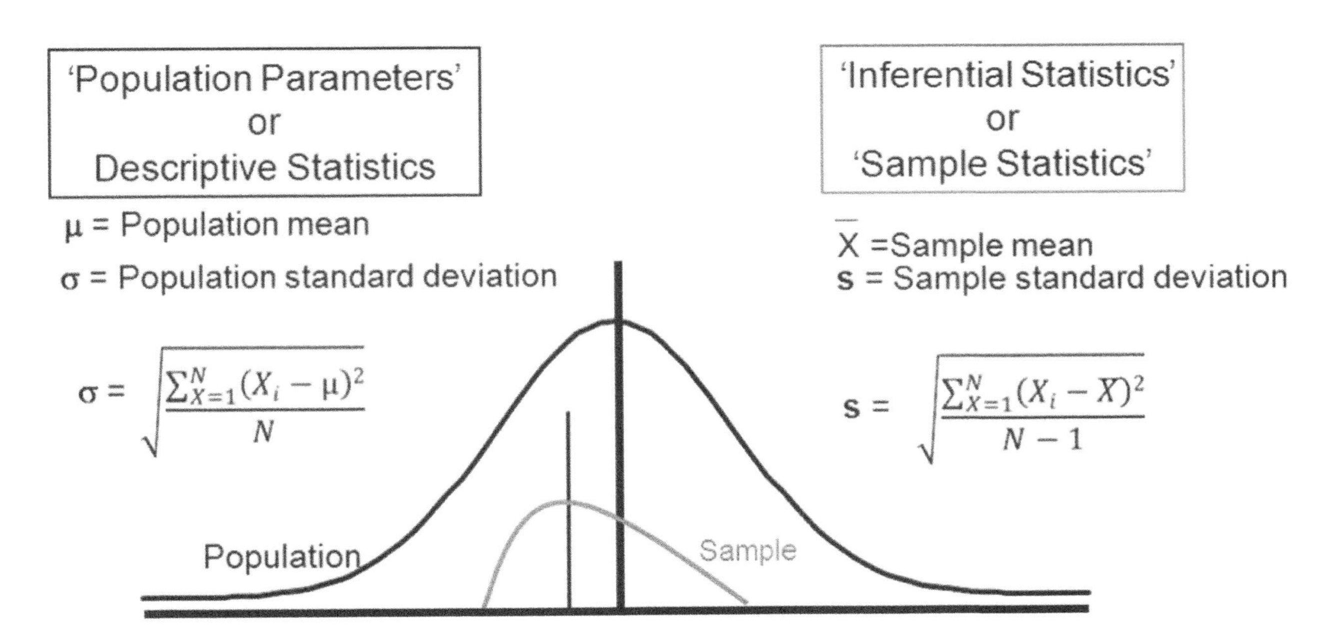

Usually it helps us to randomly select our sample from our population. However, when you use sample statistics to infer something about the population there is a chance that you will be wrong. Therefore, we use confidence intervals to let us know how likely we are to be correct.

3.6 Confidence Intervals

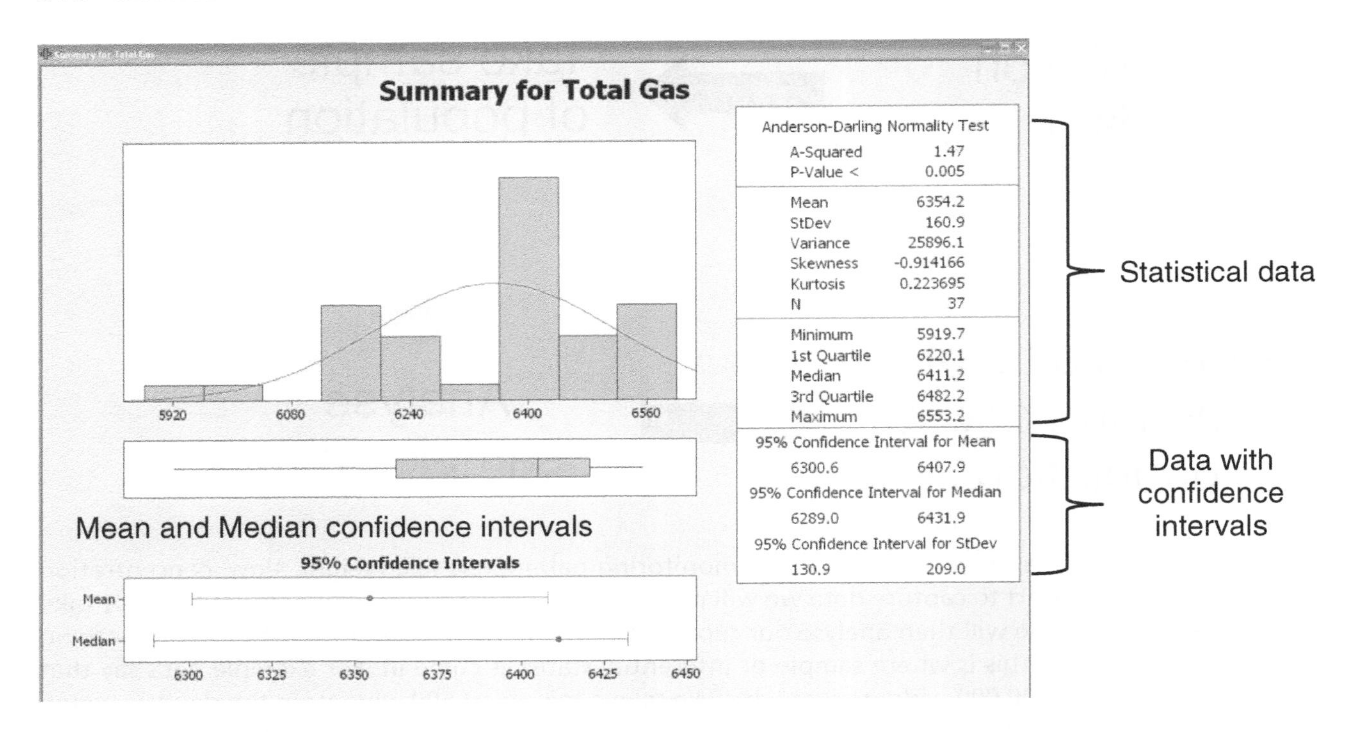

If you worked your way through Chapter 2 you will have seen confidence intervals within the Graphical Summaries that were produced. In the figure above, Minitab is showing us confidence intervals for the Mean, Median and Standard Deviation. This means that from the 37 data points provided Minitab is calculating that the population will have a Mean between 6300.6 and 6407.9. It is 95% confident that this estimate for the population is correct. The mean for the 37 data points is 6354.2.

$$\text{Confidence Interval for Mean} = \text{Mean} \pm \text{Confidence Factor} \times \text{Variability}$$

Equation (3.7)

Minitab will calculate all the required confidence intervals for you. It is useful to know how a confidence interval is put together so you can appreciate the factors that affect the size of the confidence interval, see Equation (3.7).

The Confidence Factor is based on our initial statement of how confident we want to be in our estimate. The default for Minitab is 95% (there will be a lot more on this in the next chapter) but the Confidence Factor is also based on sample size. The greater the sample size the lower the value of the Confidence Factor.

The Variability is based on the sample standard deviation divided by the number of samples square rooted.

Assuming we keep our confidence level at 95%, you can see that a large sample size and low standard deviation within the sample will make the confidence interval smaller.

Increasing the confidence level, to say 99%, and keeping everything else the same will result in a larger confidence interval.

Example of Confidence Intervals

At the S Todd Beans factory they fill beans into cans so that the average amount of beans per can is nearly 100 g. They decide to weigh a random sample of 20 cans, in order to determine what the average of all beans produced might be. Let's say that the data for 20 randomly selected cans is listed in the column called Exercise3 in the data that you previously imported into Minitab earlier this chapter.

What is a 95% confidence interval for the true mean fill using the data provided?

1. Click Stat <<Basic Statistics <<Graphical Summary.
2. Select Exercise3 as the variable, then press OK.

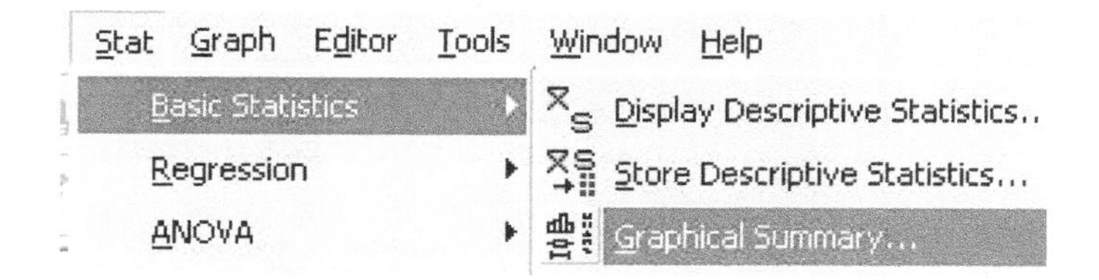

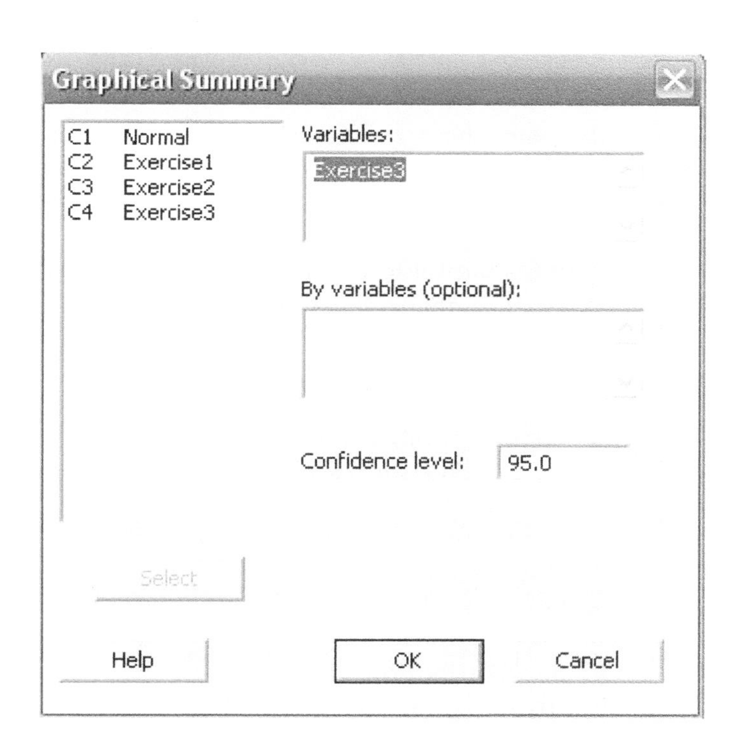

Summary for Exercise3

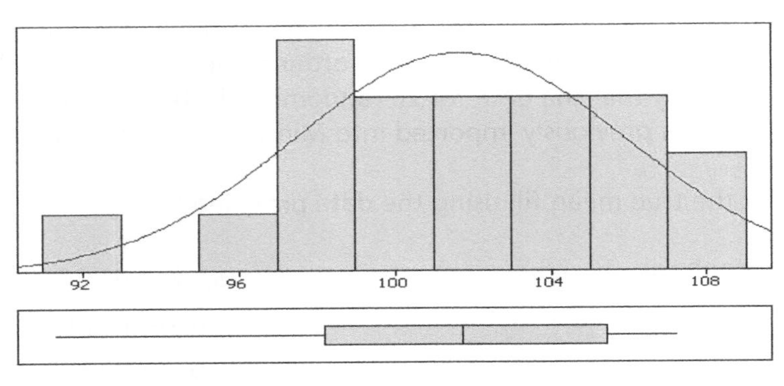

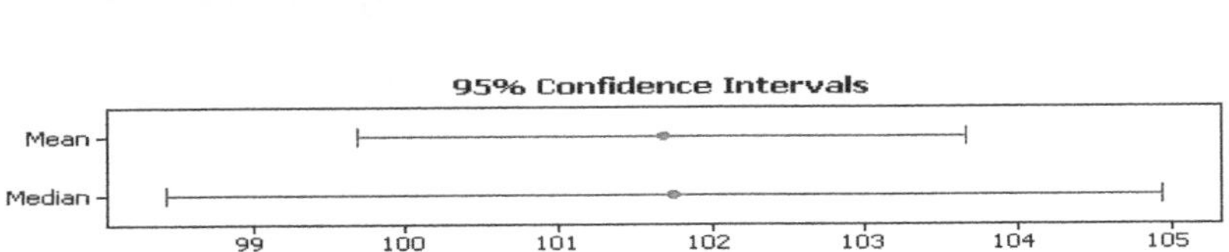

Anderson-Darling Normality Test	
A-Squared	0.44
P-Value	0.260
Mean	101.68
StDev	4.25
Variance	18.07
Skewness	-0.578802
Kurtosis	0.032257
N	20
Minimum	91.31
1st Quartile	98.21
Median	101.75
3rd Quartile	105.46
Maximum	107.23
95% Confidence Interval for Mean	
99.69	103.67
95% Confidence Interval for Median	
98.44	104.94
95% Confidence Interval for StDev	
3.23	6.21

The 20 data points have a mean of 101.68. The 95% confidence interval for the mean is between 99.69 and 103.67. This means that we can be 95% confident that the mean for all cans of beans will be between 99.69 and 103.67 g.

Before learning about inferential statistics the incorrect procedure would have been to sample 20 cans and then state that the mean of the population was 101.68 g.

3.7 Normal Distribution

One of the fundamentals within statistics is to consider the type of distribution that we get from our data. Distribution means the shape of the histogram we would get where the *y* axis is frequency.

The Normal Distribution (or Gaussian distribution) is particularly important because it forms the basis for a number of test procedures, it shows up in the natural world and manmade world and it is an outcome of the Central Limit Theorem. Another important point to remember is that the Normal Distribution is not the only type of distribution.

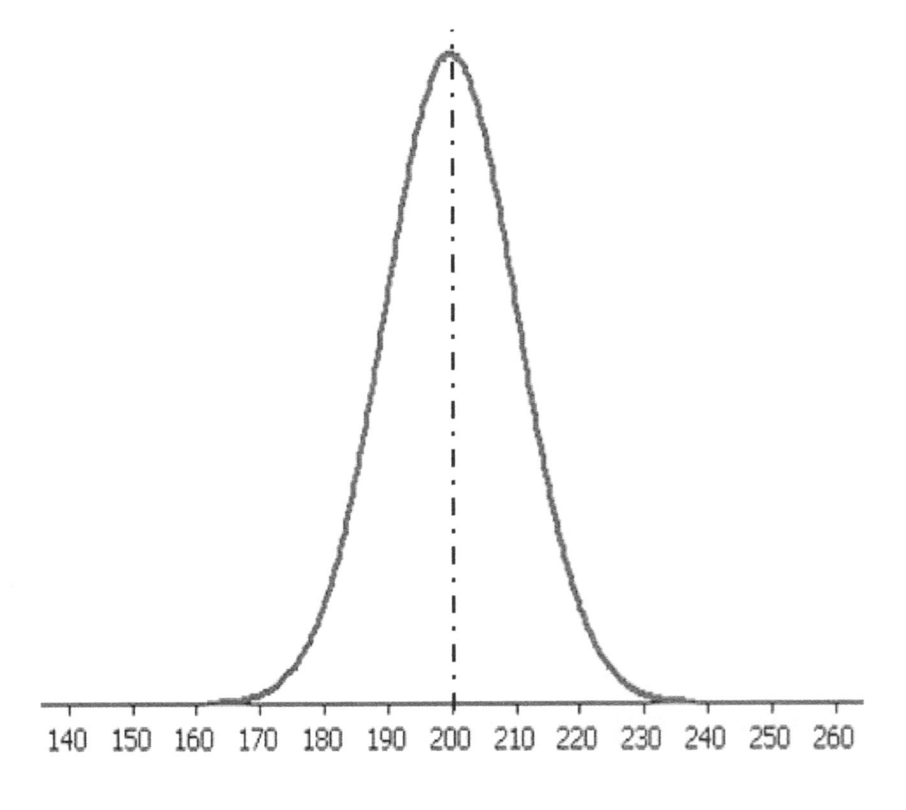

Slight departures from the normal distribution, like skewness and kurtosis, can give us important information about why our process is not preforming as expected.

The normal distribution shown in the figure has a mean of 200 and a standard deviation of 10.

Additionally, the normal distribution can be described by a number of generic rules.

- It is described by its mean and standard deviation.
- The tails extend to ± infinity.
- All observations will appear under the curve.
- The curve is symmetrical.
- The mean, mode and median are almost equal.

68.27% of the data is within one standard deviation of the mean.

95.45% of the data is within two standard deviations of the mean.

99.73% of the data is within three standard deviations of the mean.

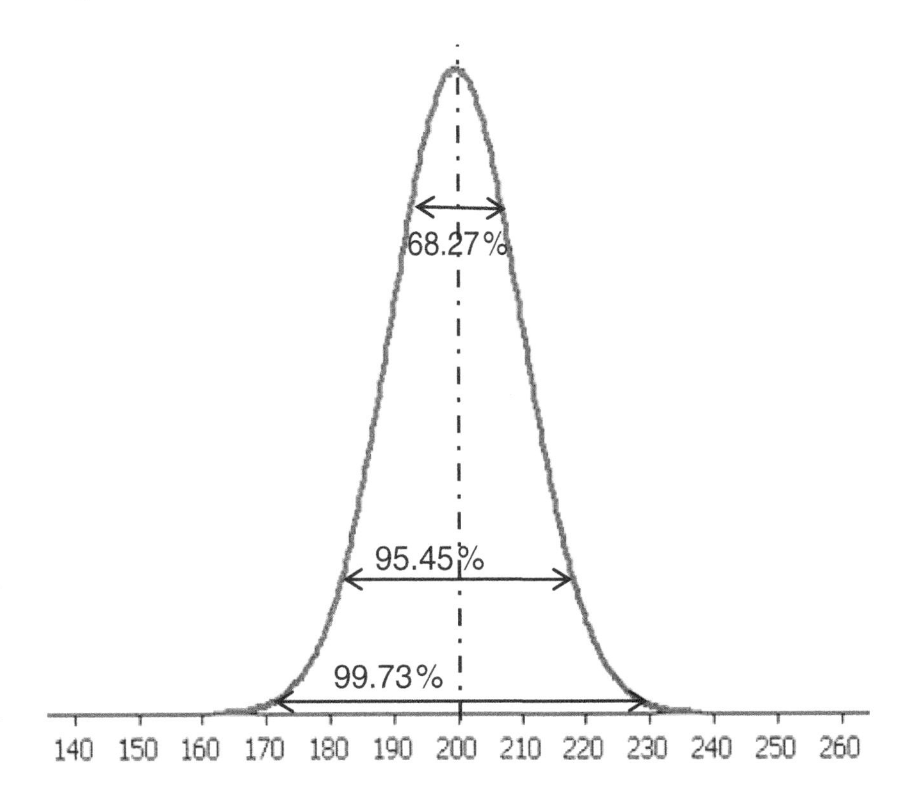

3.8 Deviations from Normality

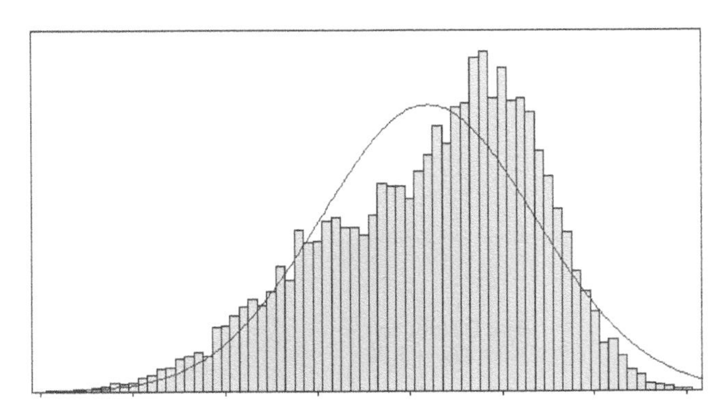

Left or Negative skewed data.
The mean will be to the left of the Median.

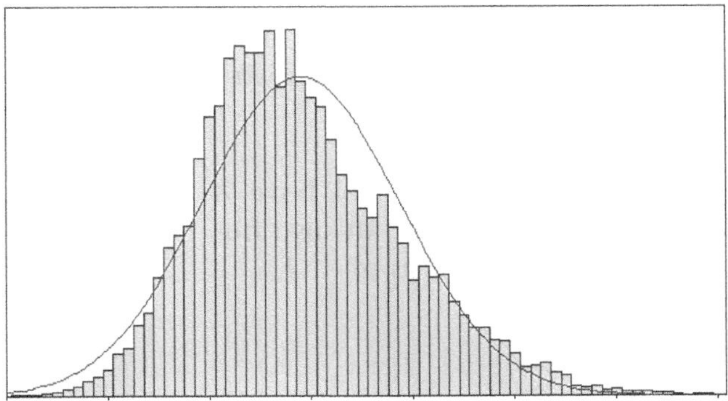

Right or Positive skewed data.
The mean will be to the right of the Median.

Above, two distributions are shown where there have been slight departures from normality, for whatever reason. One of the distributions is left skewed and the other right skewed. If we wanted to improve the process one of the first steps might be to find out what is causing the skewed distributions if a normal distribution was expected.

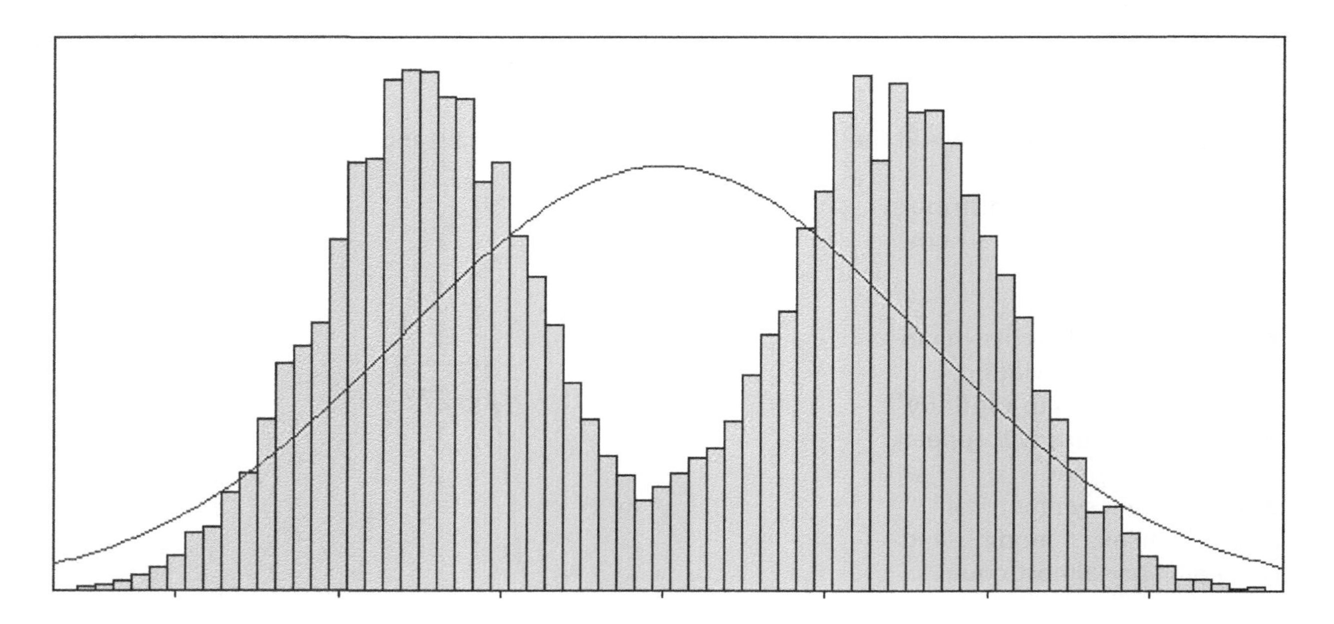

Another departure from normality that we sometimes see is the bimodal distribution. This is actually two normal distributions joined together. It could be that during the data collection a change has occurred that resulted in two different processes. The data should only contain one distribution. Sometimes this can be hard to see if the two separate distributions are close together.

Kurtosis is also a measure of how a distribution can deviate from normality.

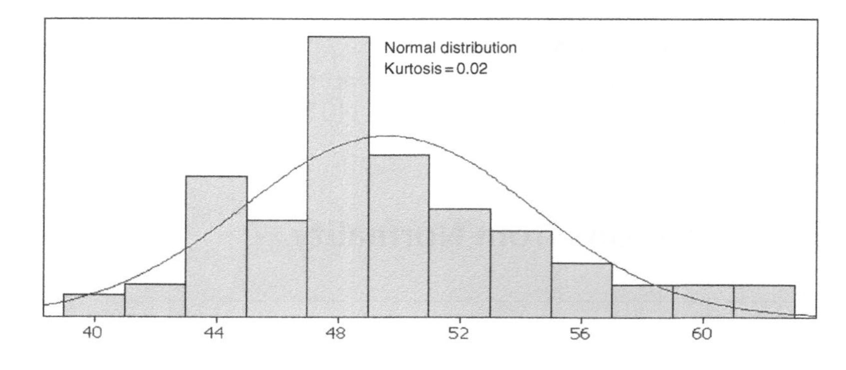

Kurtosis is a measure of how flat or peaked a distribution is. We previously said that a normal distribution has 68.27% of the population within 1 standard deviation of the mean. A flat distribution will have less than 68.0% and a peaked distribution will have more.

A normal distribution will have a kurtosis of 0. A totally flat distribution will have a kurtosis of –1.2.

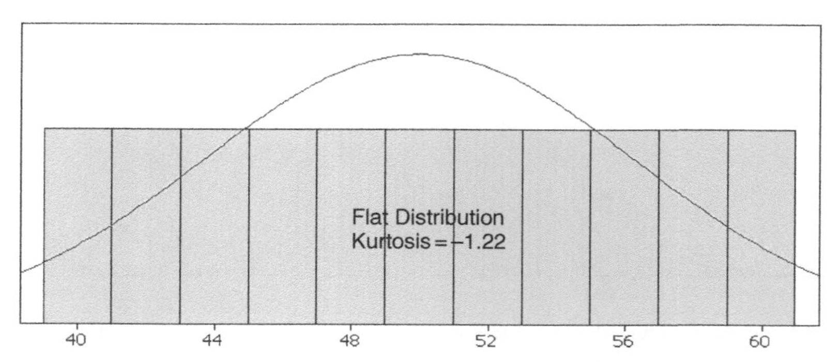

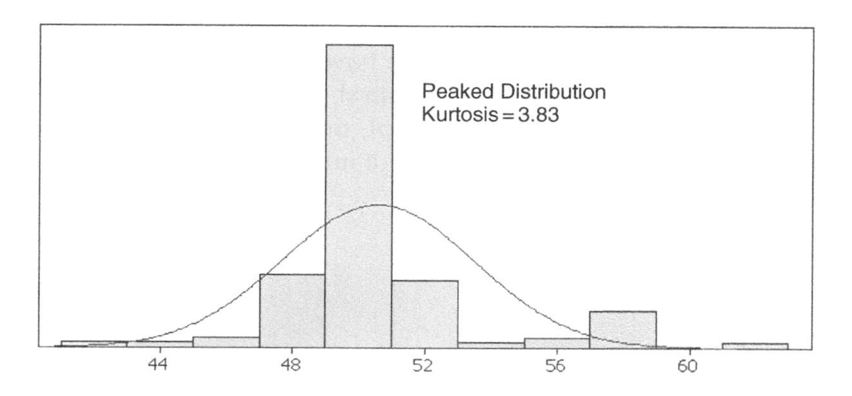

It is important for us to be able to recognise a normal distribution as it tells us how to advance with a number of test procedures. There are two main ways of doing this; one is to use a Normal Probability Plot and the other is to use the Anderson–Darling (AD) Normality Test. Both the Normal Probability Plot and the AD test result are given in Minitab's Normality Test Procedure. Additionally, the result of the AD test is given in the top left corner of the Graphical Summary.

The normal probability plot is a graphical technique. In order to establish if the distribution is normally distributed, check the straightness of the line in the plot by applying 'the Fat Pencil Test'. If you can cover all of the points on the line using a fat pencil then the distribution is deemed to be normal.

As stated earlier the results of the AD Normality Test are given on the Graphical Summary, which again highlights the usefulness of the Graphical Summary. If the P value from the Anderson–Darling test is greater than our significance level (usually 0.05 by default), then the data can be treated as normal.

There are some rules:

The Anderson–Darling test is not robust for small sample sizes. For samples less than 20, use 'the Fat Pencil' test. If a fat pencil can cover all of the points on the normal probability plot, the data can be treated as normal; but be wary of the results. **Remember always protect the customer.**

We are now going to run through a few normality tests and apply the rules that we have just learnt. Check that you have still got the data shown overleaf loaded in the project window. If not, open Excel file 03 Basic Stats.xls and import the data into Minitab.

Normal	Exercise1	Exercise2	Exercise3
98.70073	113.5166	5.643678	98.1995
106.4411	115.2099	4.300852	105.7378
97.85076	100.7614	142.7615	98.22182
96.77334	105.5682	309.3205	107.2309
94.64565	96.26837	100.9789	104.7541
99.17578	115.5223	61.55869	107.124
95.1274	110.5756	14.82981	100.3159
90.7505	103.2359	87.44837	105.6286
105.2022	114.7376	33.26984	102.1976
95.88524	104.1737	58.96926	97.89239
101.2302	111.5185	32.05817	99.14487
91.63122	109.199	16.68909	100.5925
104.4955	104.4998	101.8246	96.85037
102.3799	120.0891	176.4651	104.9549
111.0015	112.5191	29.50158	91.30796
95.94029	108.6777	52.95856	97.77616
95.26619	119.4542	4.219999	102.7522
92.98437	109.4609	5.005408	106.6444
101.771	112.2243	25.40558	104.91
96.20273	109.9257	3.977879	101.3001
96.69518	103.5963	5.718547	
94.52719	116.5084	20.52797	
99.69039	105.1016	53.0038	
93.04756	104.5246	18.73714	
98.93617	110.3089	193.551	
100.8563	117.8197	8.744367	

1. Click Stat <<Basic Statistics <<Normality Test.
2. Select Normal as the Variable to be tested.
3. Then click OK.

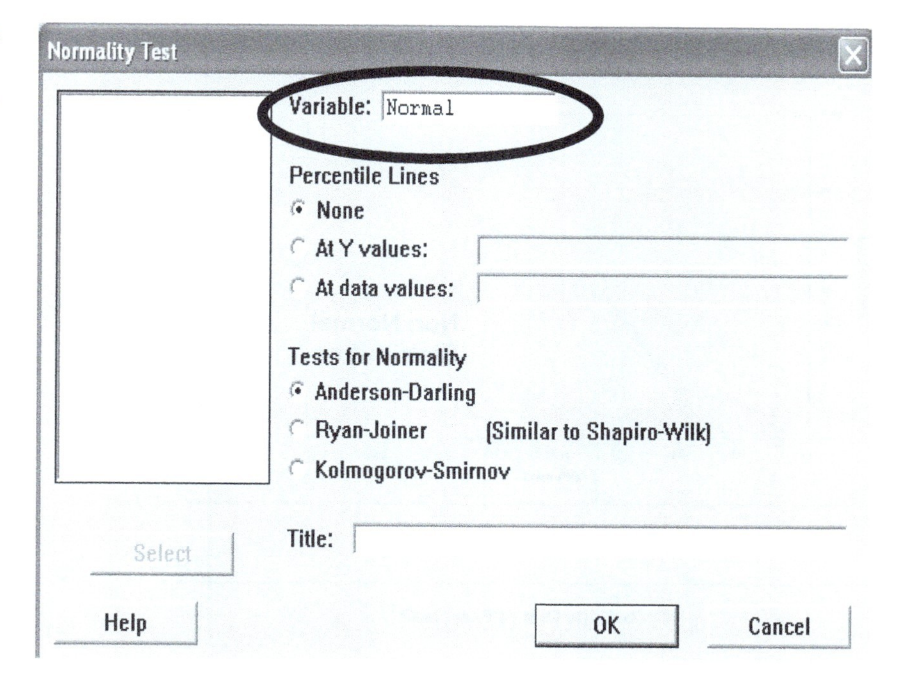

The data looks reasonably straight and we are happy that we can cover all the points with a 'Fat Pencil'. Therefore, we conclude that data within the column labelled Normal is likely to have come from a normally distributed population.

The AD *P* value is also greater than 0.05, also indicating normality. As an exercise test the other columns. Carry out the normality test for the remaining three columns, then produce a graphical summary for all four columns. Are they normally distributed? Does the assumption of normality carry any risk?

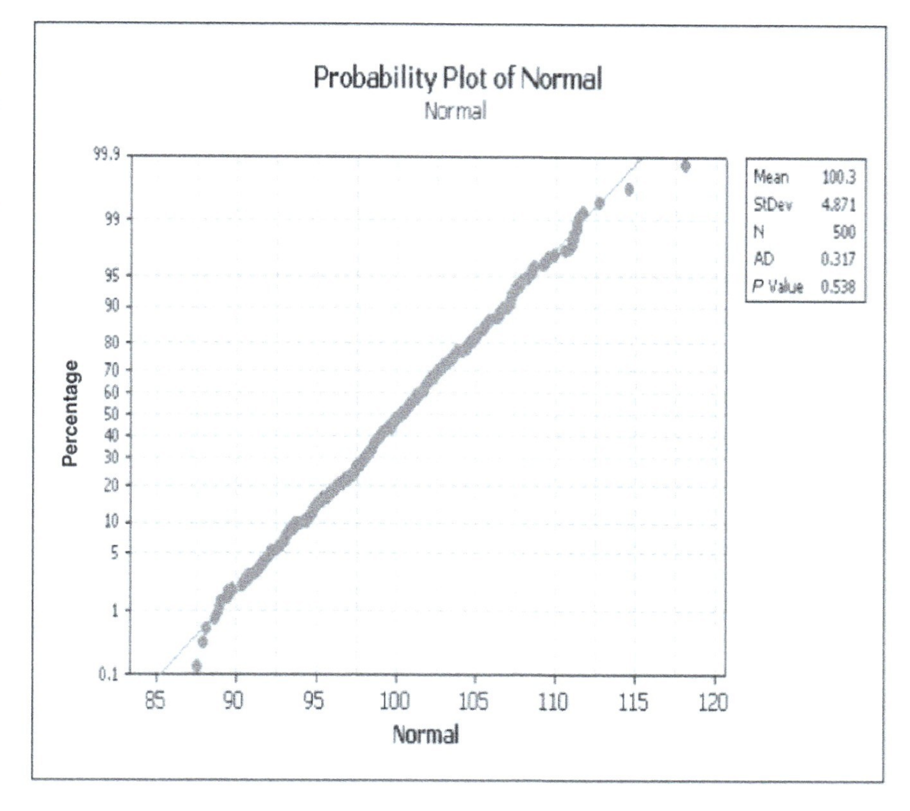

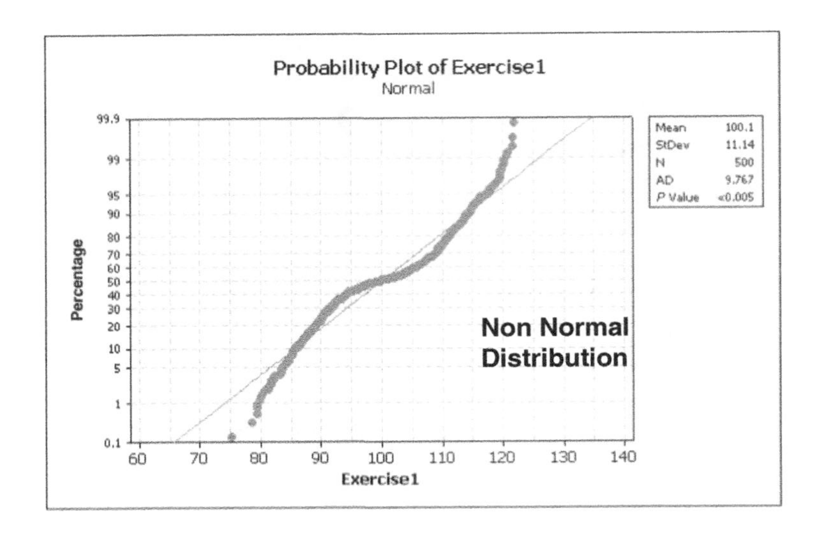
Probability Plot of Exercise1
Normal
Percentage
Exercise1
Mean 100.1
StDev 11.14
N 500
AD 9.767
P Value <0.005
Non Normal Distribution

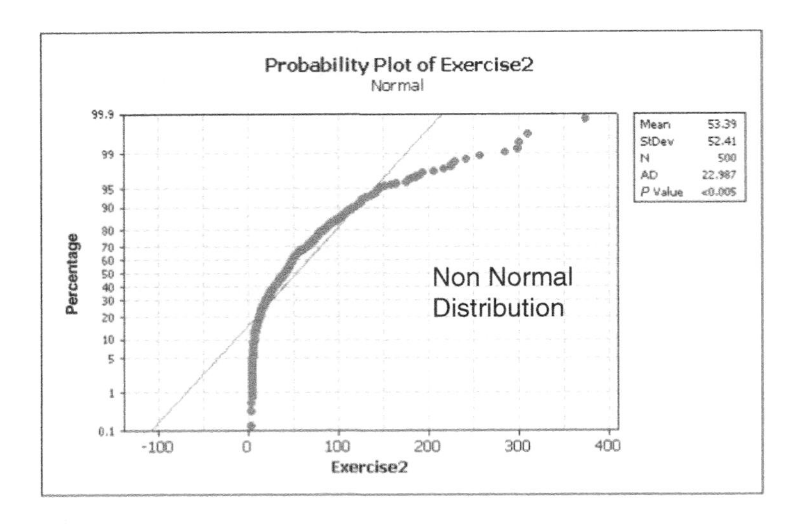
Probability Plot of Exercise2
Normal
Percentage
Exercise2
Mean 53.39
StDev 52.41
N 500
AD 22.987
P Value <0.005
Non Normal Distribution

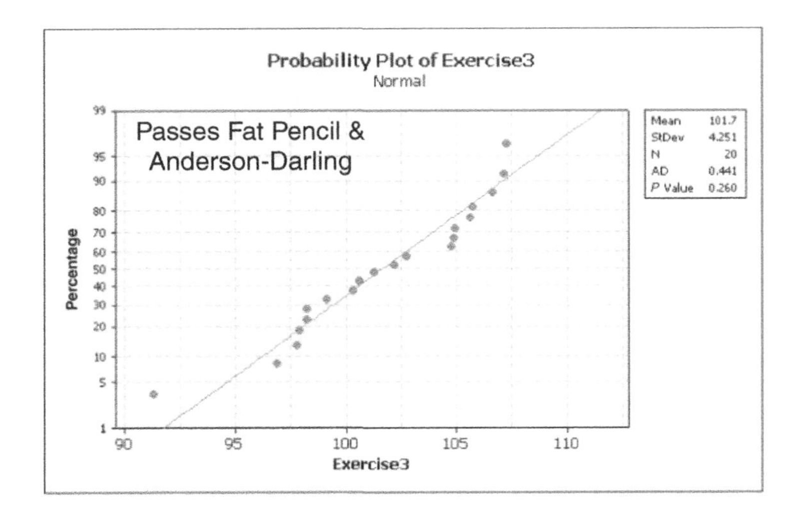
Probability Plot of Exercise3
Normal
Percentage
Exercise3
Mean 101.7
StDev 4.251
N 20
AD 0.441
P Value 0.260
Passes Fat Pencil &
Anderson-Darling

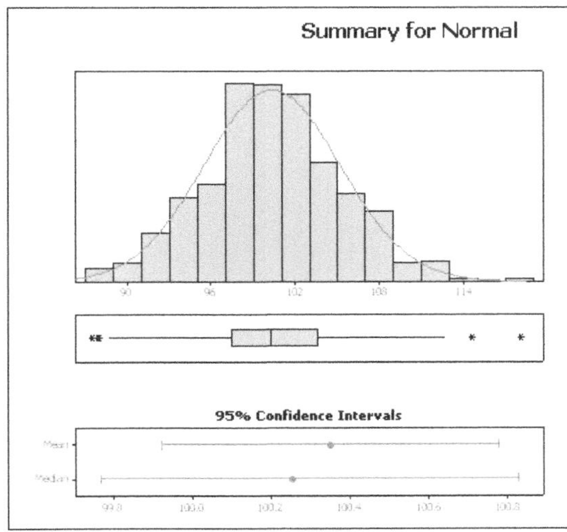
Summary for Normal
Anderson-Darling Normality Test
A-Squared 0.32
P-Value 0.538
Mean 100.35
StDev 4.87
Variance 23.72
Skewness 0.068502
Kurtosis 0.101748
N 500
Minimum 87.55
1st Quartile 97.41
Median 100.25
3rd Quartile 103.56
Maximum 118.06
95% Confidence Interval for Mean
99.92 100.78
95% Confidence Interval for Median
99.77 100.83
95% Confidence Interval for StDev
4.59 5.19
95% Confidence Intervals

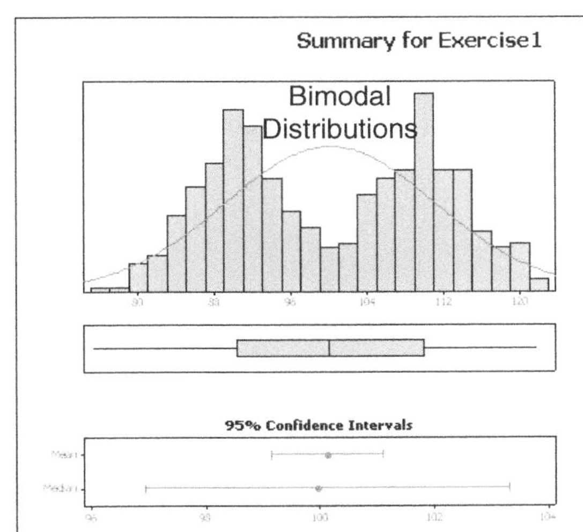
Summary for Exercise1
Bimodal
Distributions
Anderson-Darling Normality Test
A-Squared 9.77
P-Value < 0.005
Mean 100.13
StDev 11.14
Variance 124.21
Skewness 0.01678
Kurtosis -1.25565
N 500
Minimum 75.28
1st Quartile 90.35
Median 99.96
3rd Quartile 109.84
Maximum 121.81
95% Confidence Interval for Mean
99.15 101.11
95% Confidence Interval for Median
96.95 103.30
95% Confidence Interval for StDev
10.49 11.88
95% Confidence Intervals

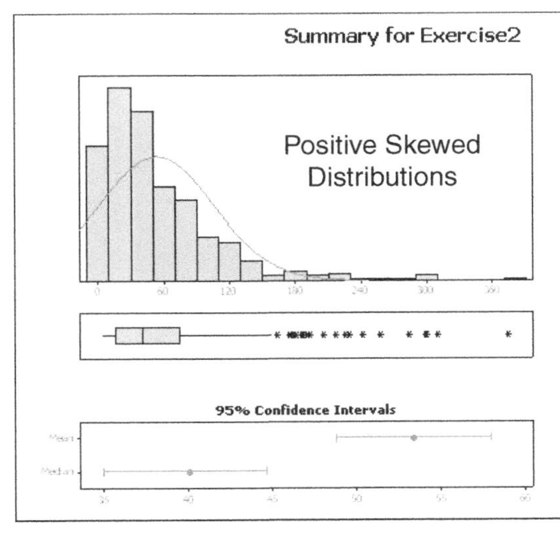
Summary for Exercise2
Positive Skewed
Distributions
Anderson-Darling Normality Test
A-Squared 22.99
P-Value < 0.005
Mean 53.388
StDev 52.412
Variance 2746.992
Skewness 2.17573
Kurtosis 6.80016
N 500
Minimum 3.100
1st Quartile 15.277
Median 40.076
3rd Quartile 73.533
Maximum 373.809
95% Confidence Interval for Mean
48.783 57.994
95% Confidence Interval for Median
34.972 44.656
95% Confidence Interval for StDev
49.352 55.879
95% Confidence Intervals

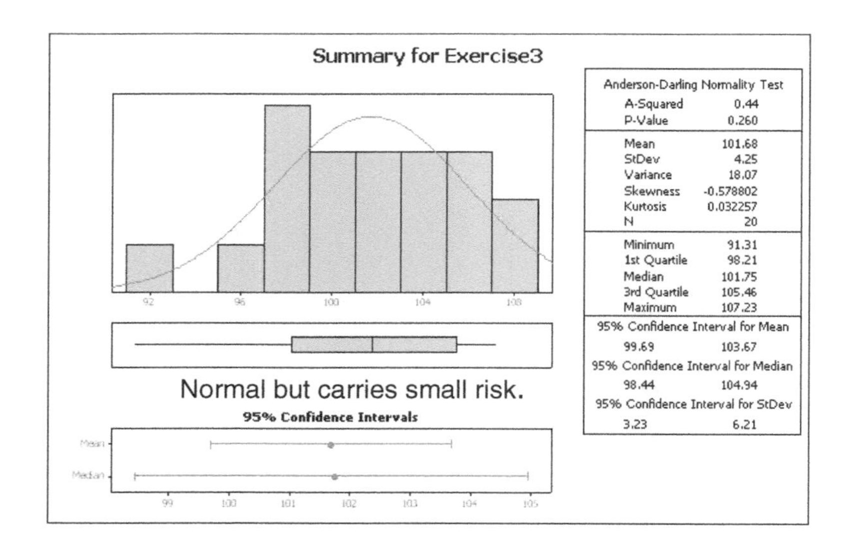

3.9 Central Limit Theorem

The Central Limit Theorem (CLT) is the foundation of Hypothesis Testing. The CLT states that, if you have a non normal distribution and you take the mean of a set number of samples and keep repeating this process to build up a distribution of means, the resultant distribution of means will be more normally distributed and have a narrower distribution than the parent.

$$S = \frac{\sigma}{\sqrt{N}}$$

Equation (3.8)

Where S = sample StDev, σ = population StDev and N = the number of samples taken each time. We are not going to be using the CLT directly but it is worth knowing of its existence.

CHAPTER 4
Hypothesis Testing

Hypothesis Testing is one of the foundational chapters of the statistical procedures that we are going to be using.

In the last chapter we learnt how we could take a sample from a population, analyse that sample, apply confidence intervals and then infer some properties of the population from our sample data. Hypothesis Testing builds on this by helping us make specific decisions about our population from the sample data.

Hypothesis Testing is vital because in the real world we should be much more interested in the population even though it is the sample data that helps us understand the population. It is essential in problem solving to know what the population is doing. Imagine claiming a major breakthrough in a project based on sample data without having made any checks on the population. What if the conclusions you made about the sample data did not apply to the population data. You would then have to explain to the Boss why the project was not delivering as you claimed.

Using the Hypothesis Testing procedure we are going to learn how to test a population is performing against a specific mean. Then we are going to start comparing variances and means for two populations in the same manner.

All the data files for this chapter are in file '04 Hypothesis Testing.xls'

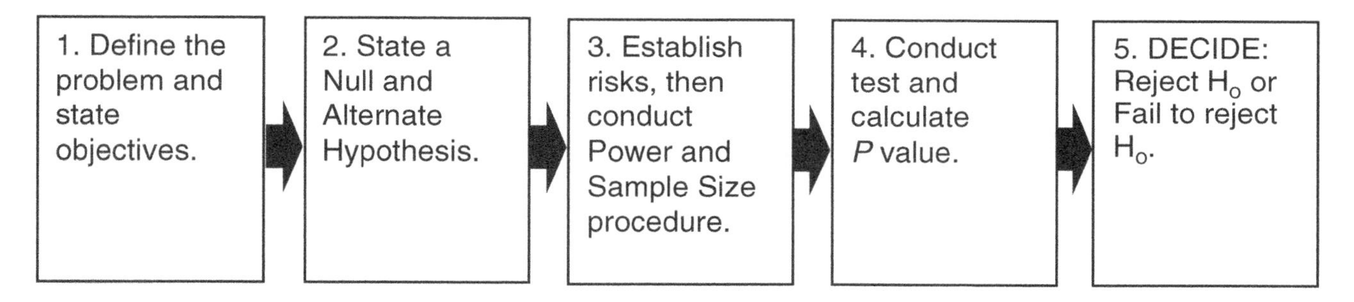
1. Define the problem and state objectives.
2. State a Null and Alternate Hypothesis.
3. Establish risks, then conduct Power and Sample Size procedure.
4. Conduct test and calculate *P* value.
5. DECIDE: Reject H_o or Fail to reject H_o.

Hypothesis Testing follows the procedure shown above. We will start by looking at the main steps in the procedure to build our knowledge before we go into the examples. When we start looking at the examples our journey within the procedure will be highlighted. After a couple of goes it will become second nature and we will stop using the flowchart. For Minitab 16 users the Assistant feature instantly does it all for you, including the power and sample size procedure, but it is still worth knowing the traditional way or Classic Method. We need to be aware that sometimes the Assistant will use a slightly different test procedure to the one used within the Classic Method and we need to be aware that our assumptions and test limitations may be slightly different in each case.

Problem Solving and Data Analysis using Minitab: A clear and easy guide to Six Sigma methodology, First Edition. Rehman M. Khan.

Steps 1 and 2 are fairly straightforward and will become second nature once practiced a few times.

It was stated in the last chapter that there is a risk of being wrong about the decision you make because the sample you took was not representative of the population. We will be developing the risk model in more detail by looking at a procedure called Power and Sample Size (P&SS) in Step 3.

Step 4 is the actual test procedure. This will vary slightly from test to test and you will even need to use a test within a test when comparing the means of two populations. Don't worry though, by then it will be second nature.

Finally, Step 5 is making the decision with regard to the result. You will learn the phrase 'If the P is low, the null must go.' You will use this mantra throughout your statistical career, so it is worth mentioning early within the text.

4.1 The Problem Statement

This is a statement of the problem that you are looking to solve and it will have only two possible outcomes. It will show that you have knowledge of the test procedure that you are going to apply and the type of data that you have. A succinct sentence is sufficient. Examples are:

- Is the mean of Group 1 equal to that of Group 2, or is it different?
- Could the variance of Group 1 be the same as that of Group 2, or is it different?
- Could the mean of this population be equal to 230, or is it different?

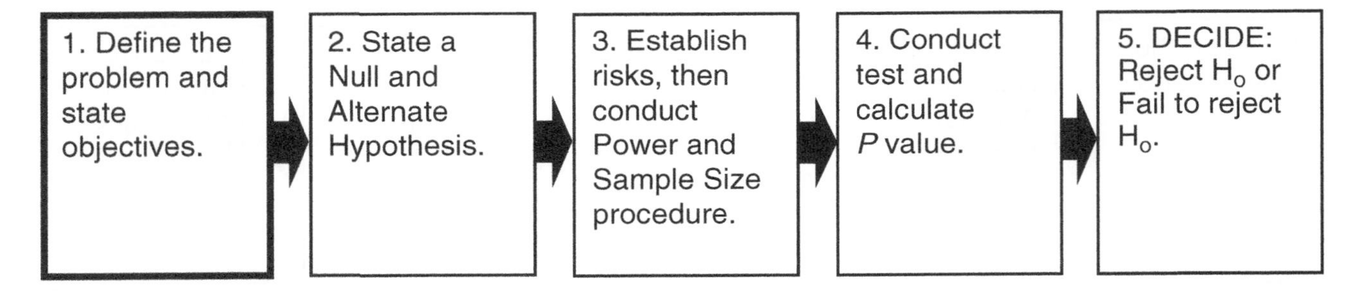
1. Define the problem and state objectives.
2. State a Null and Alternate Hypothesis.
3. Establish risks, then conduct Power and Sample Size procedure.
4. Conduct test and calculate P value.
5. DECIDE: Reject H_0 or Fail to reject H_0.

All the problem statements shown are for two sided tests. If we take the top example, we want to know if the means are the same or different. To be different, the mean of Group 2 has to be either greater than or less than that of Group 1. An example of a one sided test would be: is the mean of Group 1 less than or equal to that of Group 2, or is it greater than the mean of Group 2? The advantage of a one sided test is that it will have a greater power than the equivalent two sided test.

4.2 Null and Alternate Hypotheses

There are two possible outcomes to our Problem Statement and at this point we don't know which is correct. From our Problem Statement we must decide which of the outcomes is the Null Hypothesis and which is the Alternate Hypothesis. The Null Hypothesis (H_o) states that either there is no difference, or the parameters being investigated are equal. The Alternate Hypothesis (H_a) states that there is a difference, or the parameters being investigated are not equal. Previously, one of the examples of a problem statement was 'Is the mean of Group 1 equal to that of Group 2, or different?'

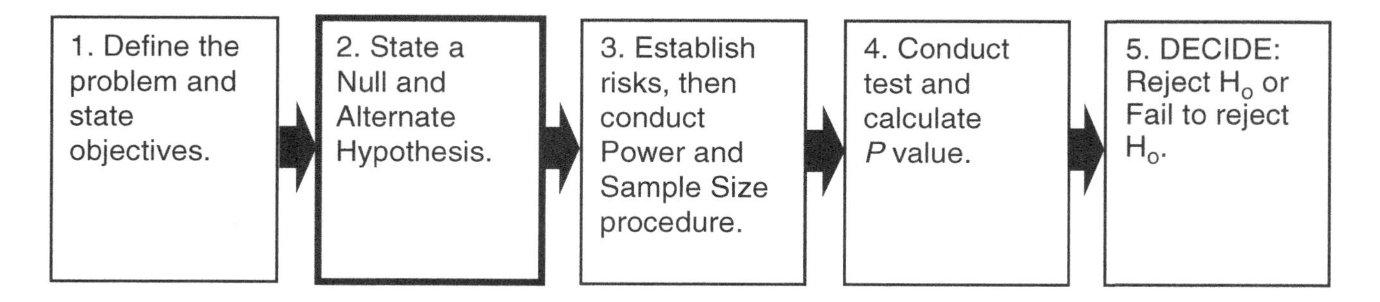
1. Define the problem and state objectives.
2. State a Null and Alternate Hypothesis.
3. Establish risks, then conduct Power and Sample Size procedure.
4. Conduct test and calculate P value.
5. DECIDE: Reject H0 or Fail to reject H0.

The Null Hypothesis would be the means of Group 1 and Group 2 are equal.

$$\text{Or} \quad H_o : \mu_{Group1} = \mu_{Group2}$$

The Alternate Hypothesis would be the means of Group 1 and Group 2 are not equal.

$$\text{Or} \quad H_a : \mu_{Group1} \neq \mu_{Group2}$$

Consider a one sided test with the following problem statement: 'Is the mean of Group 1 less than or equal to that of Group 2, or is it greater than that of Group 2?'

The Null Hypothesis would be the mean of Group 1 is less than or equal to that of Group 2.

$$\text{Or} \quad H_o : \mu_{Group1} \leq \mu_{Group2}$$

The Alternate Hypothesis would be the mean of Group 1 is greater than that of Group 2.

$$\text{Or} \quad H_a : \mu_{Group1} > \mu_{Group2}$$

4.3 Establishing the Risks

Let's say the two possible outcomes to our Problem Statement are: (1) no difference (aligned to H_o) and (2) difference (H_a). We need to assign a risk, α, of incorrectly saying there was a difference when there was none. α is called the significance and $1 - \alpha$ is called the confidence. The default value used in Minitab for α is 0.05 and this is where we get our 95% confidence as a default. There will be an option within the test menus for you to change this value. Put another way, the α risk is the risk of the instance when the data supports the rejection of the H_o, but it is wrong. This risk of a false positive or Type I error is also sometimes called the 'Producer's Risk'.

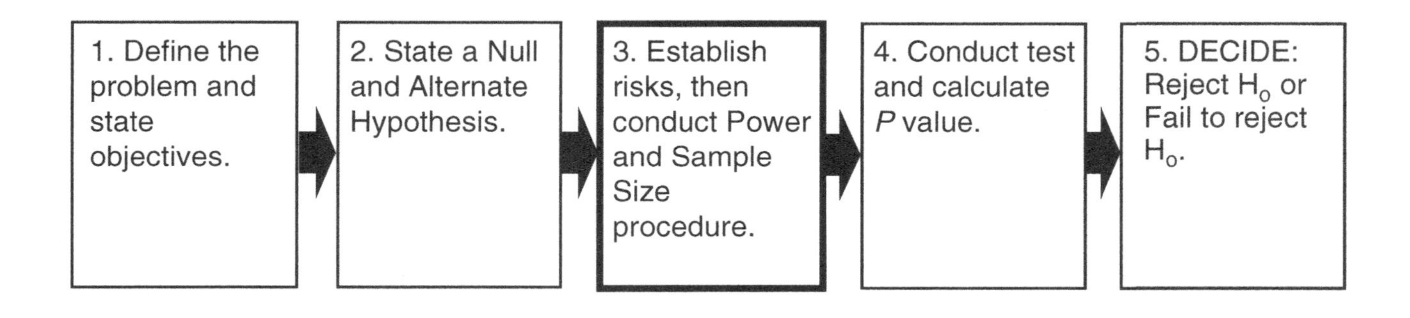
1. Define the problem and state objectives.
2. State a Null and Alternate Hypothesis.
3. Establish risks, then conduct Power and Sample Size procedure.
4. Conduct test and calculate P value.
5. DECIDE: Reject H0 or Fail to reject H0.

The β risk relates to the instance when there really is a difference. The β risk is the chance of the test incorrectly concluding that there was not a difference when there actually was a difference. The risk of a false negative or Type II error is also sometimes called the 'Consumer's Risk'. Statistically speaking, the data states that we should not reject the H_o, but in reality we should reject the Ho. The term 'Power' is given by 1 – β. Typically, we want a power of around 80–90% for our test procedures. The power of the test is the probability of correctly concluding there was a difference when in reality there was a difference.

The Significance, α risk, will be set within the test procedure. Changing the α risk will change the confidence intervals of the test statistic.

The β risk or Power is calculated prior to testing within the Power and Sample Size Procedure.

It is easier to see how the α and β risks are derived in the following matrix.

		The decision you make, or your opinion	
		There is **not** a difference	There is a difference
Reality, or the **Truth**	There is **not** a difference	Right Correct Decision Confidence = 1 – α	Wrong Type I Error a risk or Producers risk
	There is a difference	Wrong Type II Error β risk or Consumers risk	Right Correct Decision Power = 1 – β

Just to reinforce a couple of the terms from the matrix above:

- α risk: Risk of finding a difference when there really isn't one (false positive).
- β risk: Risk of not finding a difference when there really is one (false negative).

- Unless we test the complete population we will never know the truth. Therefore, we try and stack the odds in our favour by having high values of confidence and power.

The figure below shows the same matrix with the axis now referring to the null hypothesis versus our decision.

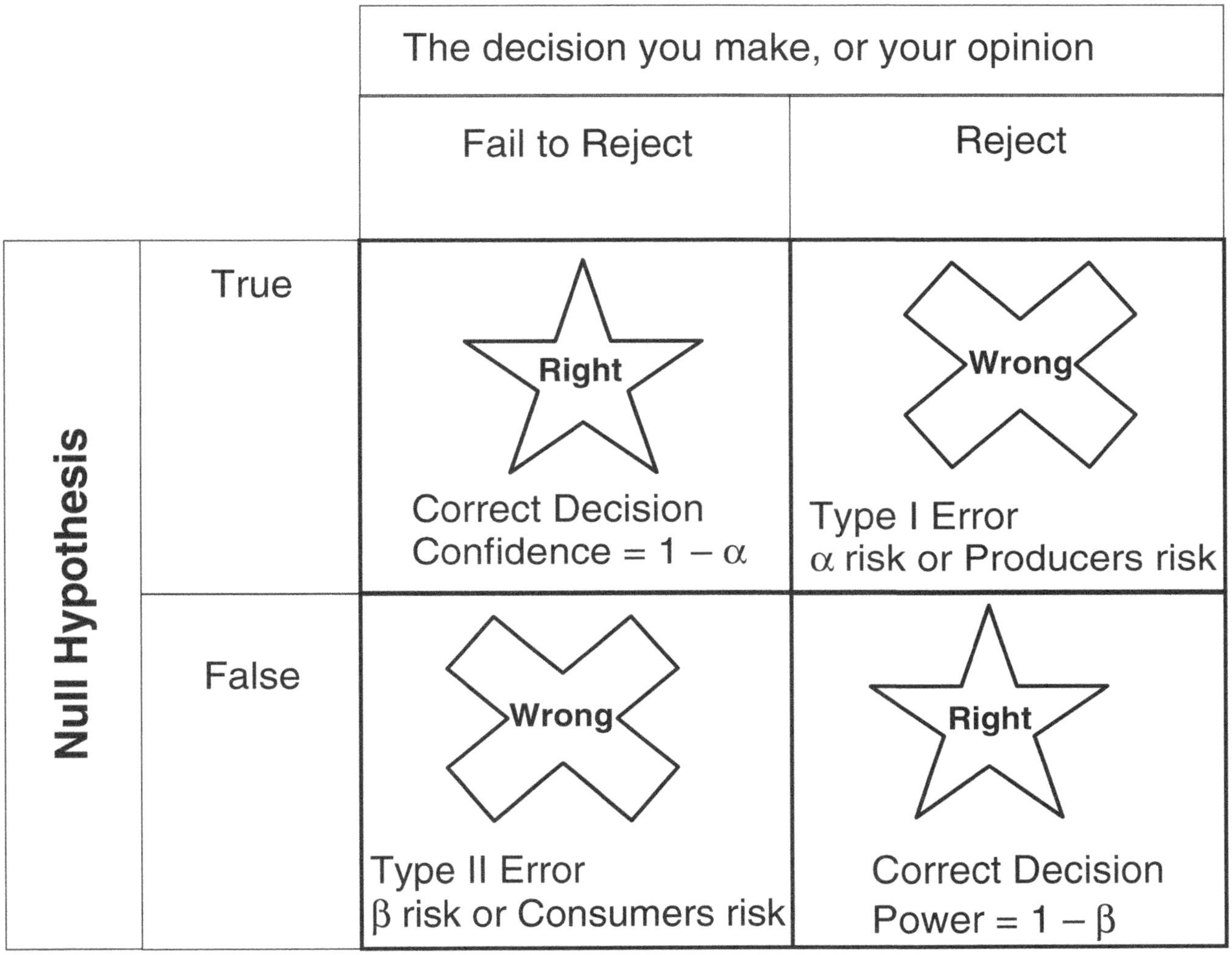

		The decision you make, or your opinion	
		Fail to Reject	Reject
Null Hypothesis	True	Right Correct Decision Confidence = 1 − α	Wrong Type I Error α risk or Producers risk
	False	Wrong Type II Error β risk or Consumers risk	Right Correct Decision Power = 1 − β

Remember the α and β risk are set by you. The default or recommended values need to be changed if the consequence of the risk is unacceptable. If injury was the consequence I would hope that a confidence level of 95% was too low.

We will now go into an example showing the Power and Sample Size Procedure. Prior to showing the test procedures for the first time there will be a generic table detailing what the procedure is used for and its assumptions and limitations.

4.4 Power and Sample Size

Procedure

- Power & Sample Size Testing for a 1 sample *t* test.

What's it used for?

- Given the standard deviation and another two out of the three following factors; the number of samples required, the power of the test or the detectable difference the procedure is used to calculate the third missing factor.

Assumptions and Limitations

- The standard deviation is required for this procedure. Use historical data to calculate the StDev. If that is not available it can be calculated from test data. Failing that an estimate will have to be made which would then be checked against the test data when that is available.

Example 1. Power and Sample Size for a 1 sample *t* test.

Prawn video games want to establish the mean time to finish the first level in one of their games. They will use a 1 sample *t* test but initially they want to calculate the power of the test.

At a significance level of 5% they want a power of 80% and want to be able to detect a difference of ±5 min. They have resources to conduct 14 independent tests. And they know that the historical standard deviation is 7 min.

No file required. Note: for the purposes of demonstrating P&SS we will not define the problem statement and will not list the null and alternate hypothesis for this problem.

1. Click Stat <<P&SS << 1-Sample *t*…

Stat
Power and Sample Size ▸
1t 1-Sample t...

2. Complete the menu box as shown. Note that 'Differences' refers to detectable difference. Also note that, as long as StDev is available, only two out of the other three factors are required and the third is calculated. In this example we will be calculating power so we leave it blank.
3. Then click OK.

Power and Sample Size for 1-Sample t

Specify values for any two of the following:

Sample sizes: 14

Differences: -5 5

Power values:

Standard deviation: 7

Options... Graph...

Help OK Cancel

Minitab gives us a graphical output and also has the same information reflected within the session window.

Power is on the *y* axis and the detectable difference is on the *x* axis. The curve shows how the power would change if we specified a different 'difference', that is a difference of 6 or –6 would give a power greater than 80%. It can also be seen that the power curve is two sided as it caters for both a positive difference and a negative difference.

Hovering the mouse over the red dot, which represents our point of interest, displays the actual power with a difference of ±5.

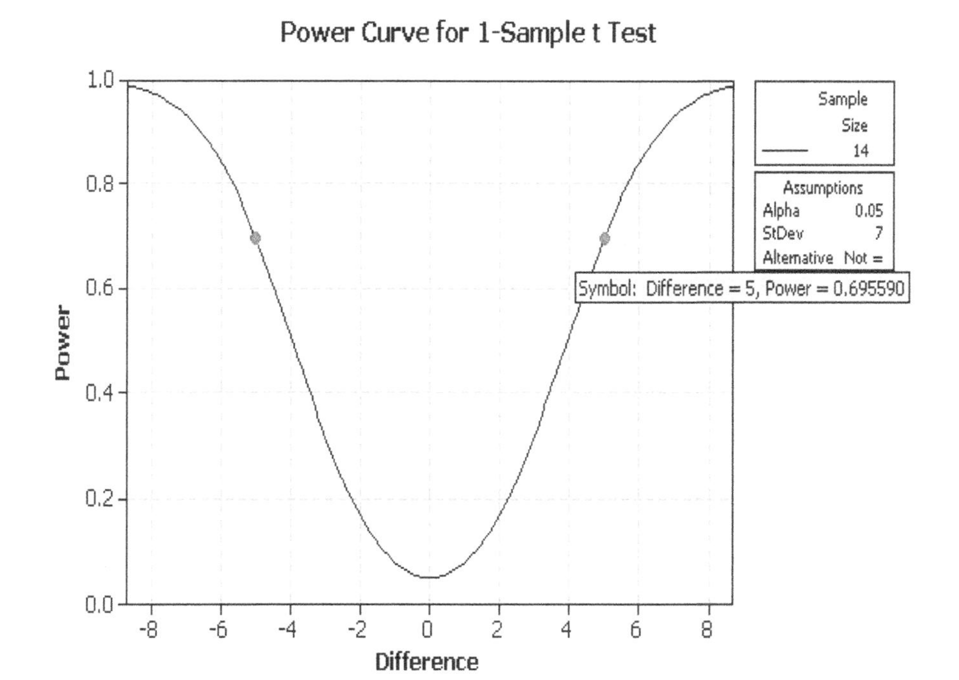

Example 1 continued. Power and Sample size for a 1 sample *t*-test.

Prawn video games realise that only a power of 70% can be achieved with the 14 tests but they don't have the resources to conduct more tests. Someone from the team then suggests that they are only really interested if the level is taking longer to complete. Therefore, they should look at using a one sided test.

They decide that they are only really interested in looking at a difference of +5 min.

Under the same conditions what power can they achieve now?

1. Click on the Edit Last icon. It should open the menu box for P&SS for 1 Sample *t*.
2. Complete the menu box as shown. Remember to change the selection for difference.
3. Then click Options.

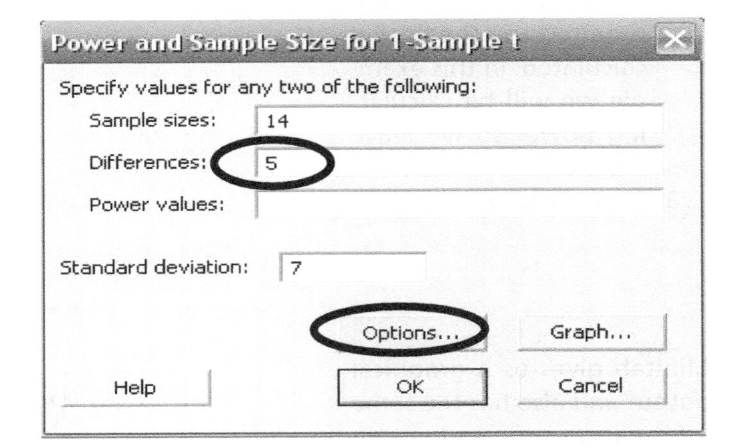

4. For the Alternative Hypothesis select Greater than. It is by using these radio buttons that we make our choice for a one sided test. The default selection is 'Not equal' (two sided).
5. Then click OK and OK again.

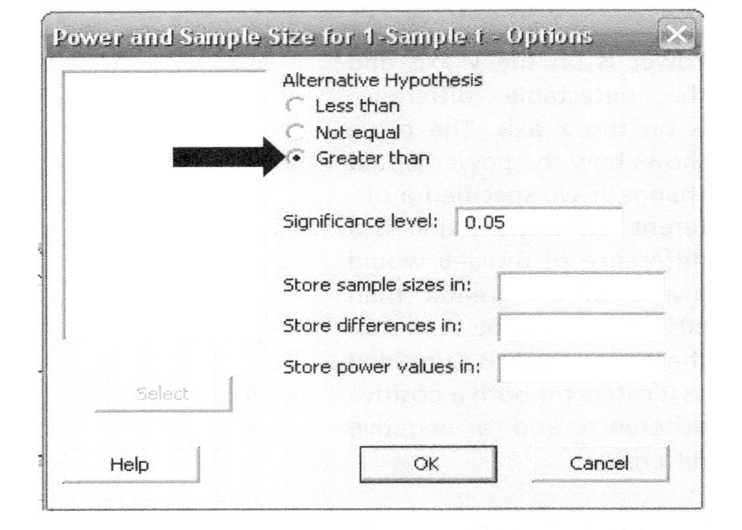

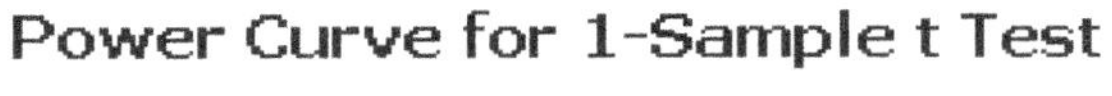

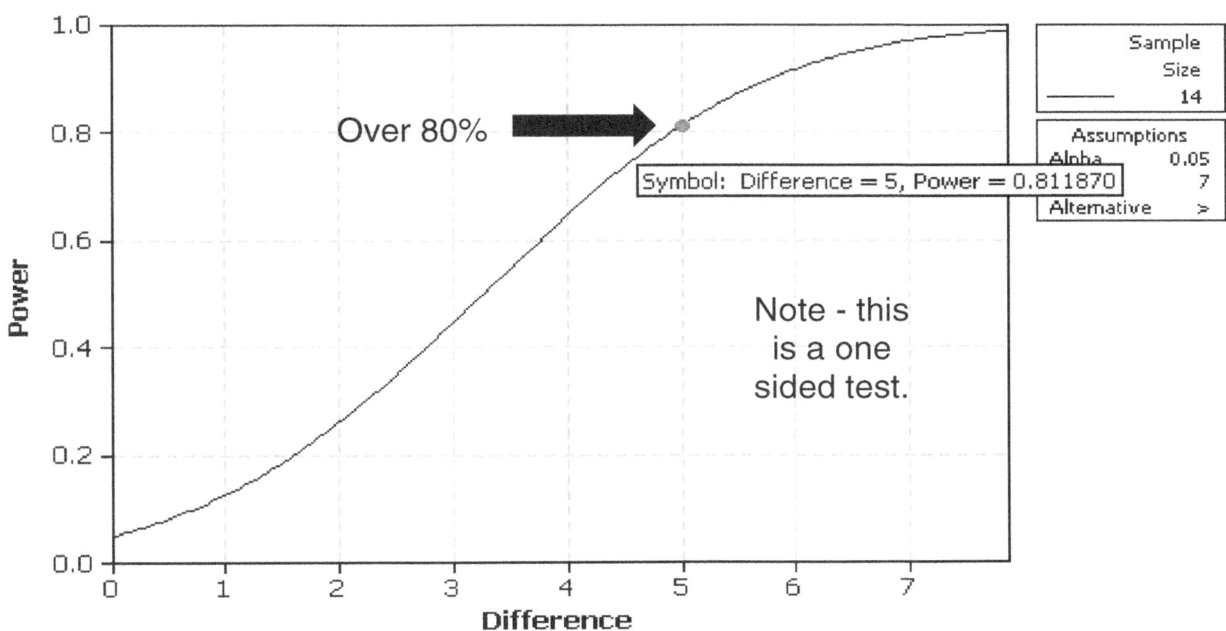

Prawn video games can achieve a power of over 80% using 14 tests with a one sided test. This means that if there is a difference there is an 80% chance of detecting it correctly. As we did not change the confidence of the test, that remains at the default level of 95%. Which means that if the null hypothesis is correct and there is no difference we will have a 95% chance of detecting it correctly.

The power of a test is as important as the significance and should be stated as a part of the results.

It is worth having an understanding of the parameters that affect the power.

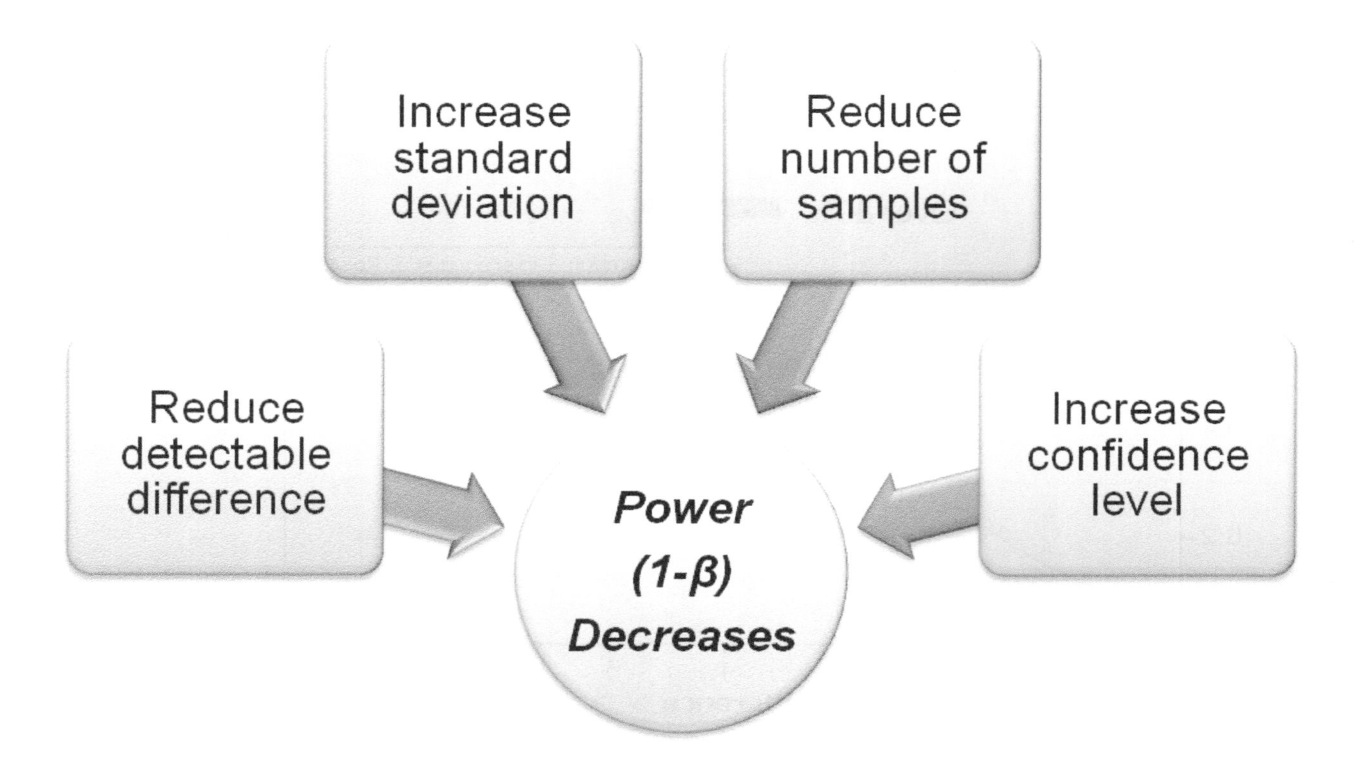
Increase standard deviation
Reduce number of samples
Reduce detectable difference
Increase confidence level
Power
(1-β)
Decreases

The interaction chart shows the effects on Power of changing a number of the key parameters available within test procedures.

4.5 Conducting the Test and Evaluating the Results

Now that we understand about power and sample size we are going to look at how to conduct the tests. We will learn:

- 1 Sample *t* Test for comparing the mean of a population to a target value.
- Paired *t* Test for comparing the means of two dependent populations.
- 2 Variances Test for comparing the levels of variation of two independent populations.
- 2 Sample *t* Test for comparing the means of two independent populations.

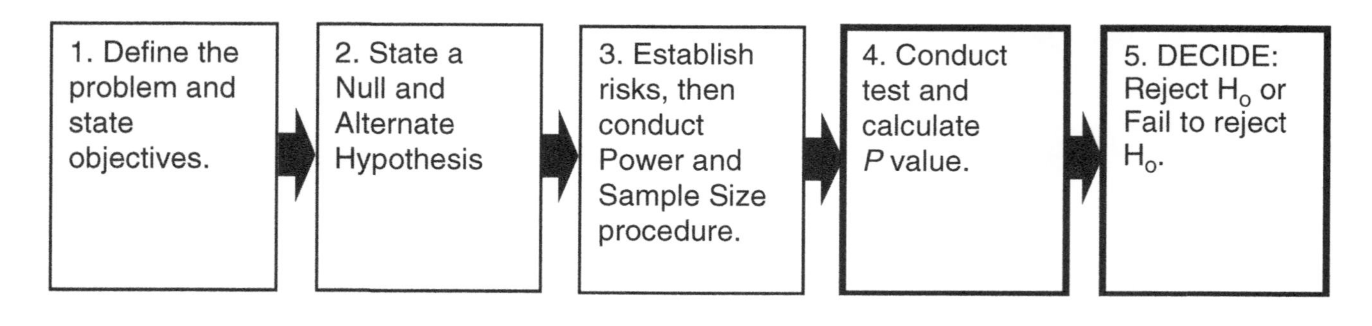
1. Define the problem and state objectives.
2. State a Null and Alternate Hypothesis
3. Establish risks, then conduct Power and Sample Size procedure.
4. Conduct test and calculate P value.
5. DECIDE: Reject H_o or Fail to reject H_o.

We will use examples to consider each of the tests using the standard procedure shown above. Then for M16 users we will show how the Assistant would tackle the example. There will of course also be some exercises along the way.

Prior to conducting the hypothesis tests it is always wise to take the opportunity to view the data using the Graphical Summary. It's good to get a feel for the numbers and the shape of the distribution and the impact that these will have on the accuracy of the test that you are about to conduct.

The result of the test will be shown in the session window. The result will take the form of a value again.

In statistical terms:

- If the *P* value is greater than our significance level (usually 0.05) we conclude that there is insufficient evidence to reject the null hypothesis.
- If the *P* value is less than our significance level (usually 0.05) we conclude that there is sufficient evidence to reject the null hypothesis and accept the alternate hypothesis.

I was taught 'If the P is low, the null must go' and this works well in the practical world.

Now that we know how to evaluate the test results let's go back to our example of the video game manufacturer.

4.6 One Sample *t* Test

Procedure	• 1 Sample *t* Test
What's it used for?	• This test is used to assess a population against a target mean.

Assumptions and Limitations	• The following assumptions must be true about our sample data
	  • The sample data must be randomly selected from the population.
	  • Data must be continuous and uni model (a single distribution)
	  • Sample data is likely to be from a normally distributed population
	• This test is fairly robust to non normal data as long as the sample size is greater than 20.

Example 2. 1 Sample *t*-test.

Prawn video games want to establish if the mean time to finish the first level from one of their games is greater than 200 min. They have already conducted the power and sample size procedure and were satisfied that the power of the 1 sample *t* test was going to be 80%. That is with a one sided test, sample size of 14, detectable difference of 5 min and standard deviation of 7 min.

The manufacturer conducts the 14 independent play tests and records the results. Establish if the mean completion time of the population is likely to be more than 200 min from the results of the 14 play tests?

Worksheet: Level Time.

Problem Statement

Establish if the mean completion time of the population is likely to be more than 200 min from the 14 play test results?

Null Hypothesis: Population Mean ≤ 200 min

Alternate Hypothesis: Population Mean > 200 min

We have already established α and β

Prior to conducting the test procedure we must ensure that the sample data meets the requirements for the test. The best way of doing this is to look at the Graphical Summary.

1. Open worksheet Level Times and transfer the data into Minitab.

C1
Level Comp time
198.7
204.0
204.6
209.2
207.0
206.9
206.3
207.0
204.8
214.9
200.6
208.5
217.7
205.0

2. Examine the Graphical Summary by clicking Stat <<Basic Statistics <<Graphical Summary.

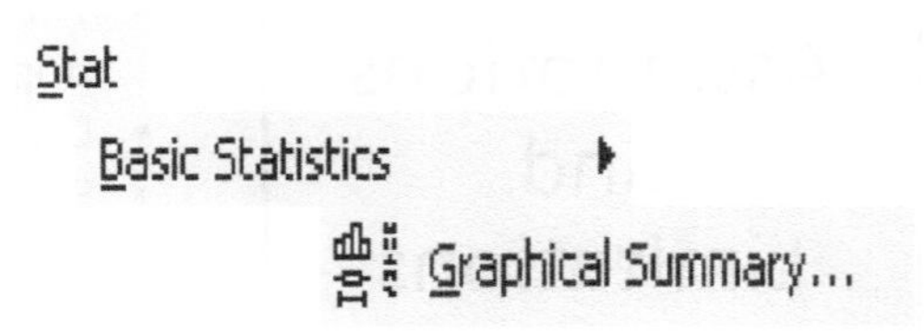

3. Select C1 and then click OK.

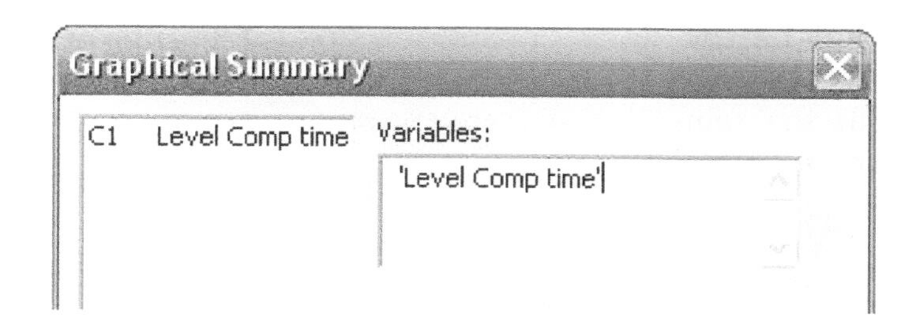

The mean for the sample is 206.8 and we can see the 95% CI for the mean.

The AD *P* value is 0.137 but we should be wary as the sample size is small so we will carry out a normality test.

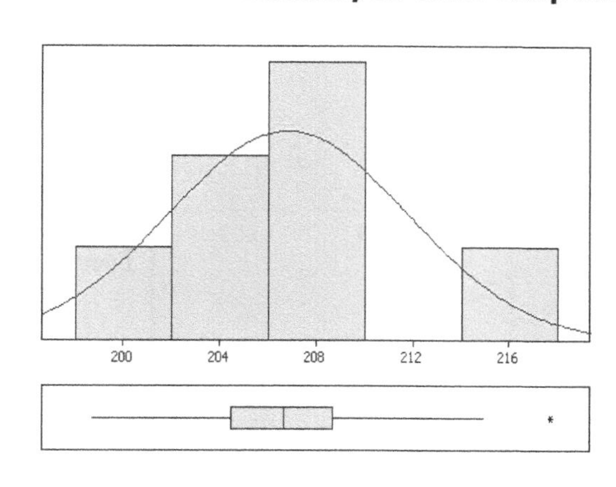

Anderson-Darling Normality Test	
A-Squared	0.54
P-Value	0.137
Mean	206.80
StDev	4.94
Variance	24.41
Skewness	0.76546
Kurtosis	1.15980
N	14
Minimum	198.70
1st Quartile	204.45
Median	206.60
3rd Quartile	208.68
Maximum	217.70
95% Confidence Interval for Mean	
203.95	209.65
95% Confidence Interval for Median	
204.57	208.54
95% Confidence Interval for StDev	
3.58	7.96

4. Click Stat <<Basic Statistics <<Normality Test.

Stat

Basic Statistics

TEST Normality Test...

5. Select C1 as the Variable.
6. Then click OK.

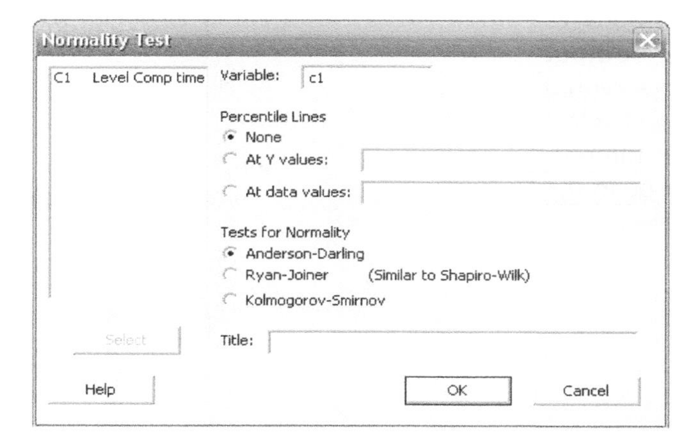

The dots can just be covered with a Fat Pencil so we will accept that the sample came from a normal distribution.

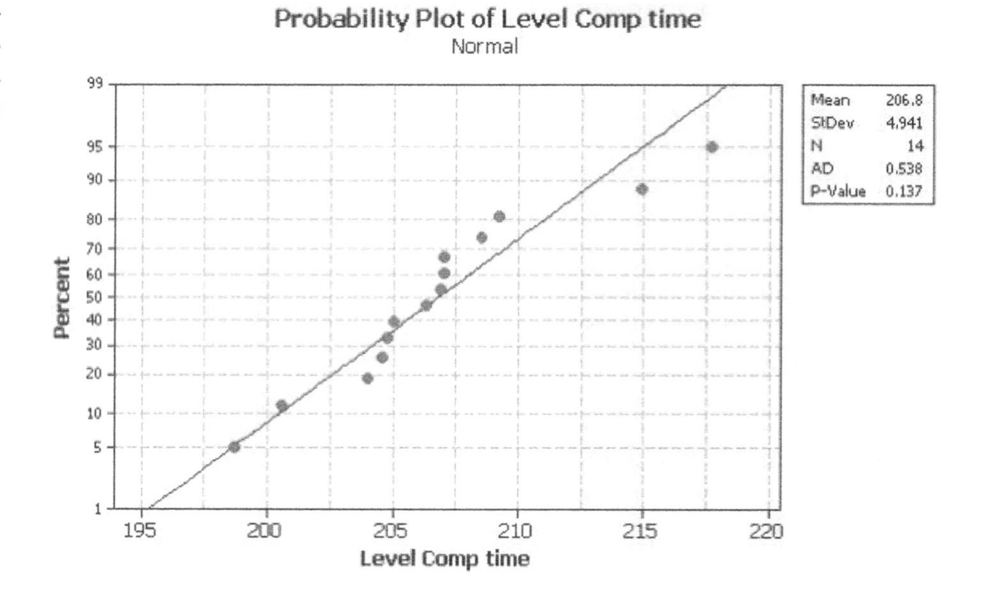

Finally we can now start the actual test procedure.

7. Click Stat <<Basic Statistics <<1-Sample t.

Stat
Basic Statistics
1t 1-Sample t...

8. Select C1 as the data column.
9. Enter 200 as the Hypothesized mean, this is the value that we are testing against, 200 min.
10. Then click Graphs to specify the output graph that we would like. If we don't select a graph then the output will only be in the session window.

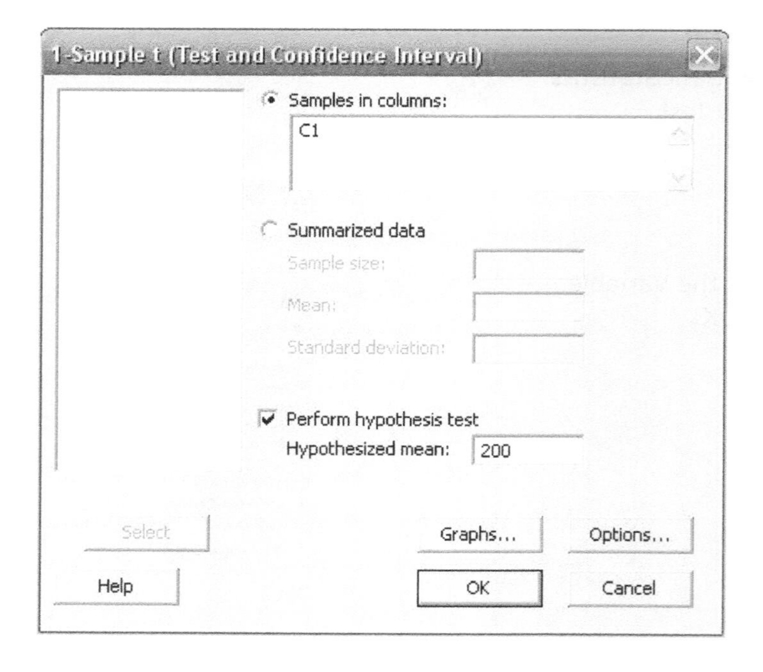

11. Select Boxplot of data.

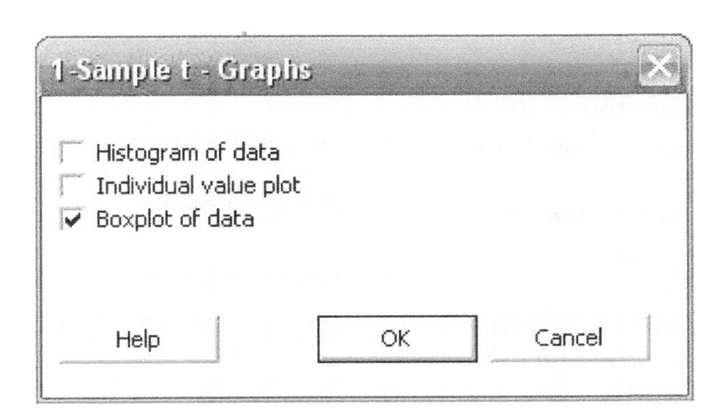

12. Then click OK.
13. Click Options.
14. Use the dropdown menu to select 'greater than' as the Alternative hypothesis. The default is always 'not equal'.
15. Then click OK and OK again.

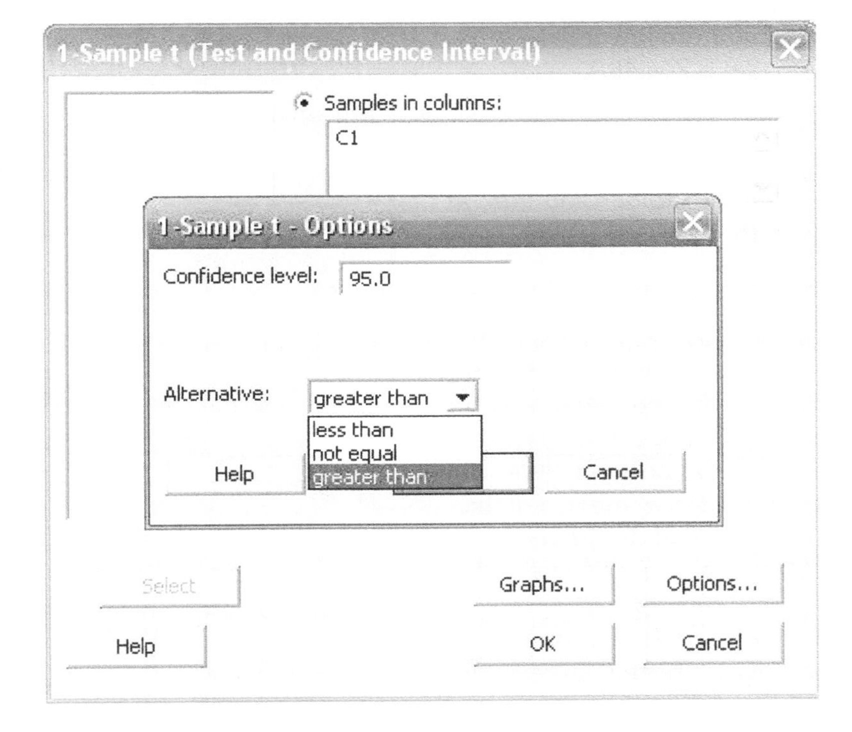

16. To have a look at the result of the test click on the Show Session Folder icon.

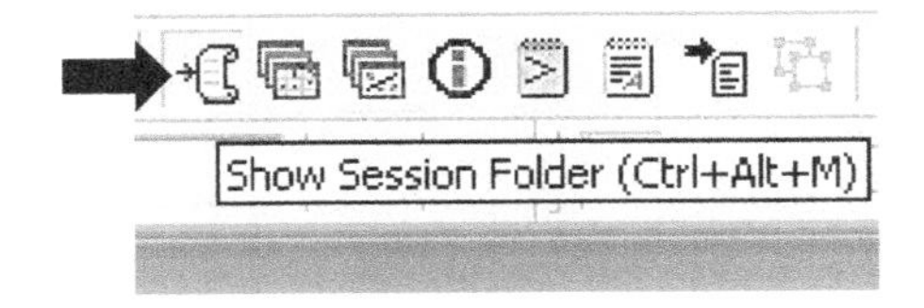

17. Locate our test result and look at the *P* value. If it is less than the α (0.05) then we can reject the null hypothesis. Our *P* value is 0.00 so we reject the null hypothesis that the mean completion time is 200 min or less (if the *P* is low the null must go!). Note that mean of the samples was 206.8 and the lower 95% interval is 204.46.

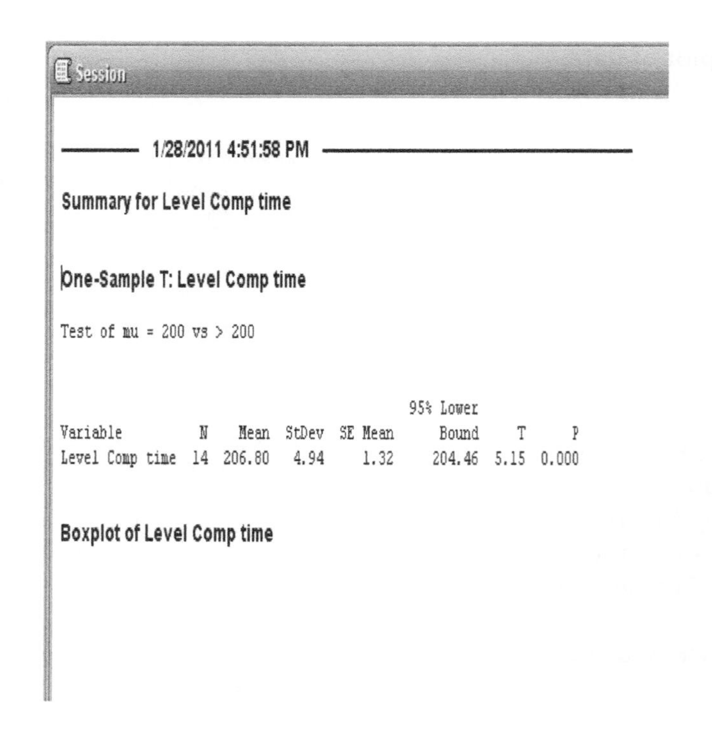

18. In order to see this result graphically go to the graph folder and look at the boxplot.

Graphically it can be see that H_o is not within the confidence interval for the mean. It is for this reason we have rejected the null hypothesis. We can state that the mean completion time will be greater than 200 min.

As we requested a one sided test, which tested whether the mean of the population was greater than 200 min, only a lower bound confidence interval is shown on the graph.

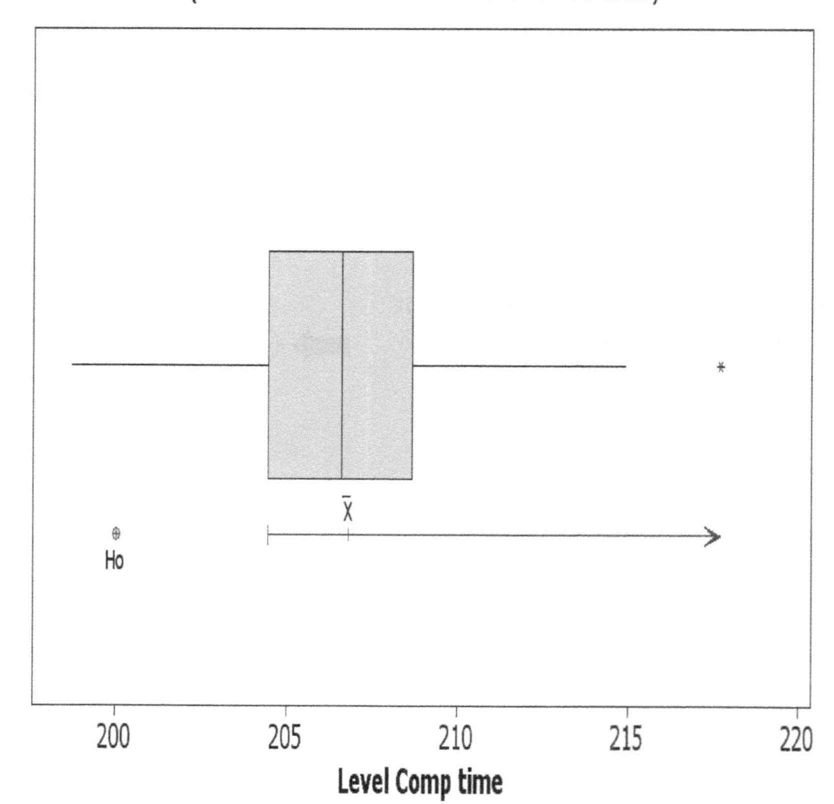

We are going to have a look at using the Assistant to repeat the 1 Sample *t* Test. As stated earlier the Assistant is an M16 feature. The Assistant offers a graphical selection for the appropriate hypothesis test. The Assistant will also give guidance on interpreting the result of the test and additional relevant information such as P&SS, normality and outliers within the data.

1. Click on Assistant<<Hypothesis Tests.

Assistant

Hypothesis Tests...

The tests within the Assistant are broken down into three groups: (1) tests relating to a single sample, (2) tests which have two samples and (3) the comparison of more than two samples.

2. Click on 'Help me choose' within the 'Compare one sample with a target' group.

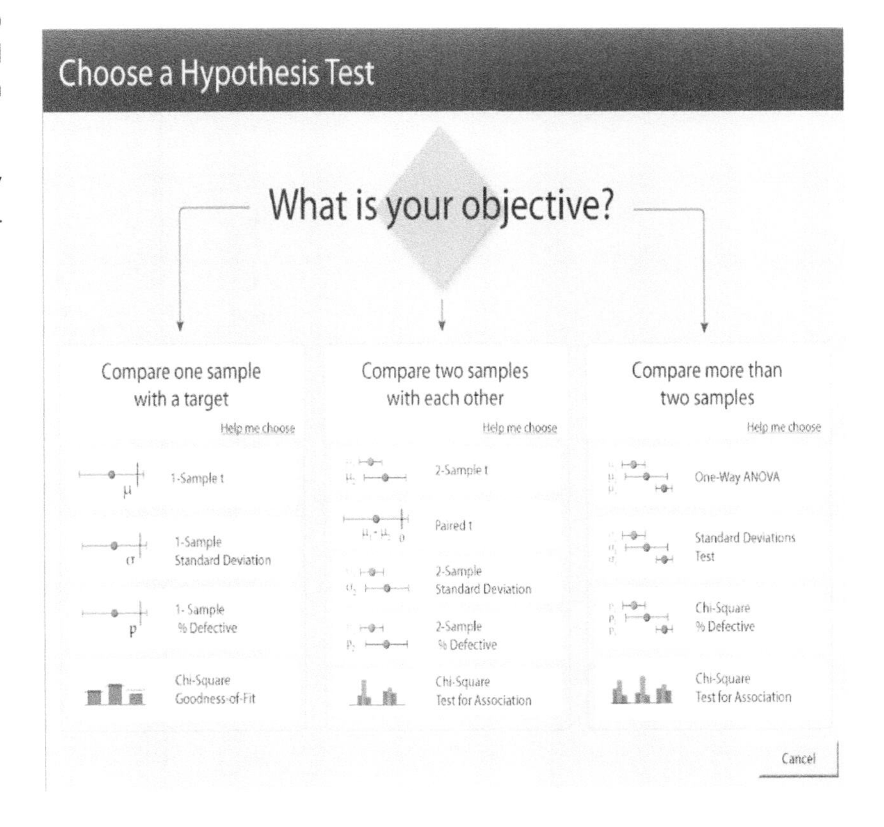

We are presented with a decision tree to help us choose the type of test that we want. We can click on the diamonds in order to get more information about our selection.

As a reminder, this book does not deal with Attribute data as Continuous data is far more common and useful.

3. Click on the button for the 1 Sample *t* Test.

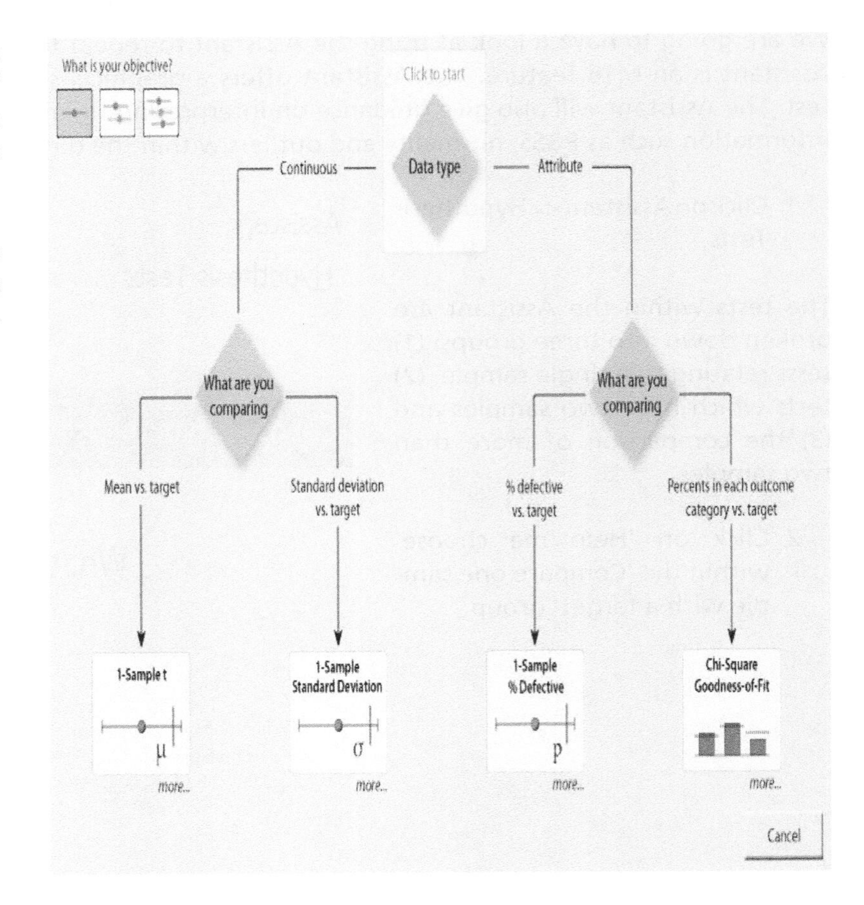

4. Complete the menu box as shown and click OK.

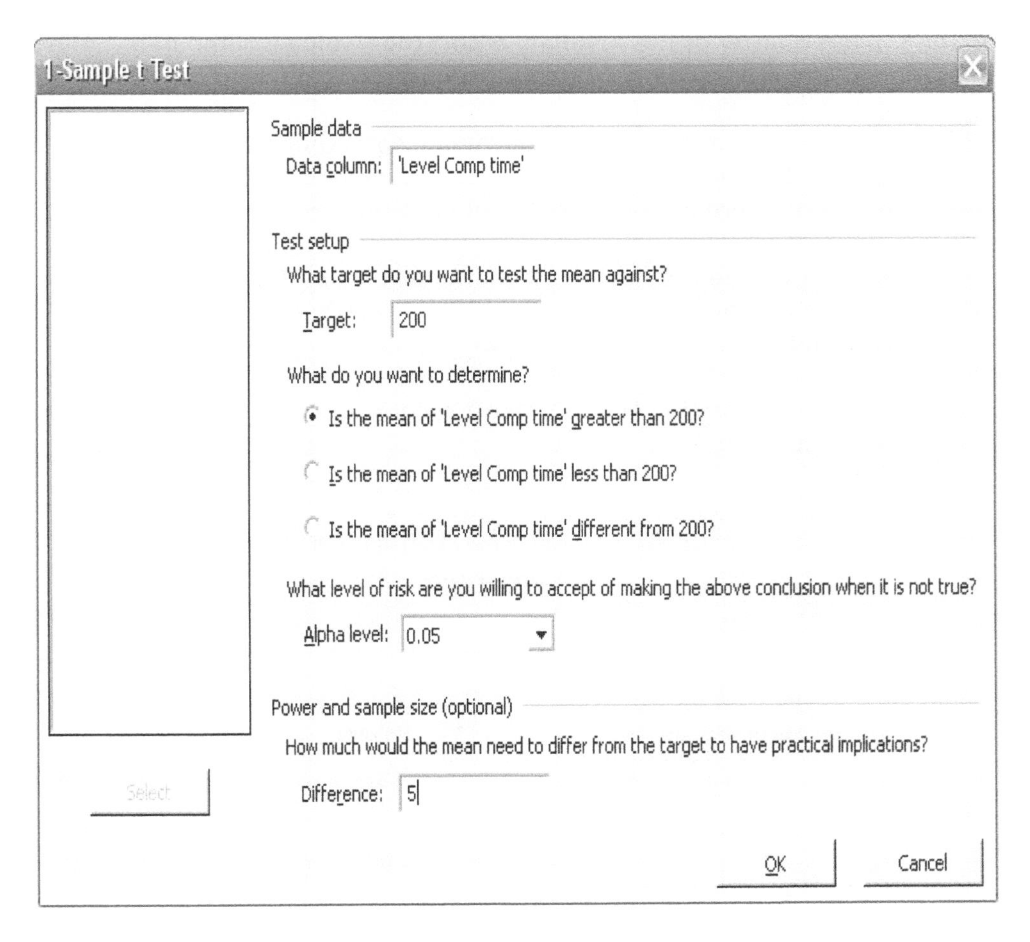

The menu box takes the same inputs as the previous menus for the 1 sample t test.

Note that there is not a selection menu for the type of graphical output that we will get. Also, the Power and Sample Size selection only takes in a single parameter. This is because StDev will be calculated from the sample data and so will sample size.

The Assistant gives three output pages: the Summary Report, the Diagnostic Report and the Report Card. For the purposes of navigation each of these pages can be handled as a graph.

On the top left quarter of the summary report the *P* value is compared to the significance level. There is an orange marker on a sliding scale indicating the actual *P* value.

Below this we see a dotplot of the actual sample data. Graphically we see that 200 is not greater than the 95% lower bound confidence level for the population mean, 204.46. If 200 had been greater than the 95% lower bound confidence level we would not have rejected the null hypothesis.

On the top right we see the statistics from the test. Note that the StDev of the sample data was 4.94. As we are conducting a one sided test Minitab uses a 90% confidence interval to describe the population mean. If we conducted a two sided test Minitab would use a 95% CI.

On the bottom right, within the comments section, you can read the conclusion of the test procedure and edit the text.

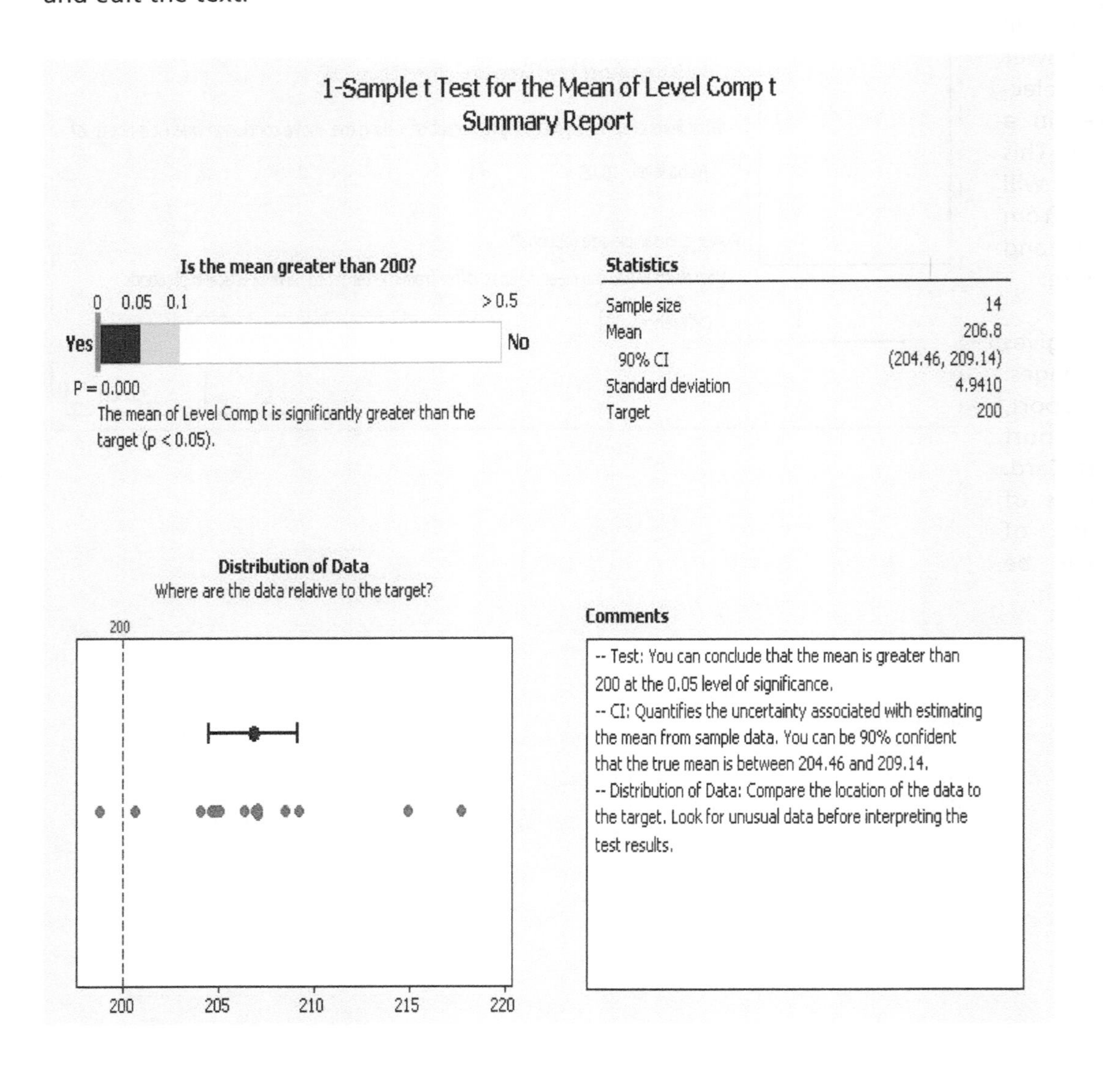

On the top of the Diagnostic Report we are presented with an SPC chart. The data point that is coloured red is deemed to be unusual and may have occurred under unusual circumstances. Minitab is warning us that these kinds of events could affect our conclusion. We will be covering SPC in detail within Chapter 7.

The bottom left and right sections of this panel deal with Power and Sample Size. If you relied only upon the Assistant you would get these results after you had potentially conducted test runs and had taken samples. That could be wasteful if you then found your power to be inadequate.

As we previously stated that a difference of five was of interest to us, we are told that our power is 97.3%. This is much higher than the menu driven test because originally we used a historical StDev of 7.0. Here, the Assistant is calculating the StDev from the samples and using a value of 4.94. To make the correct decision about which figure to quote and use always think about the application and think about protecting the stakeholders.

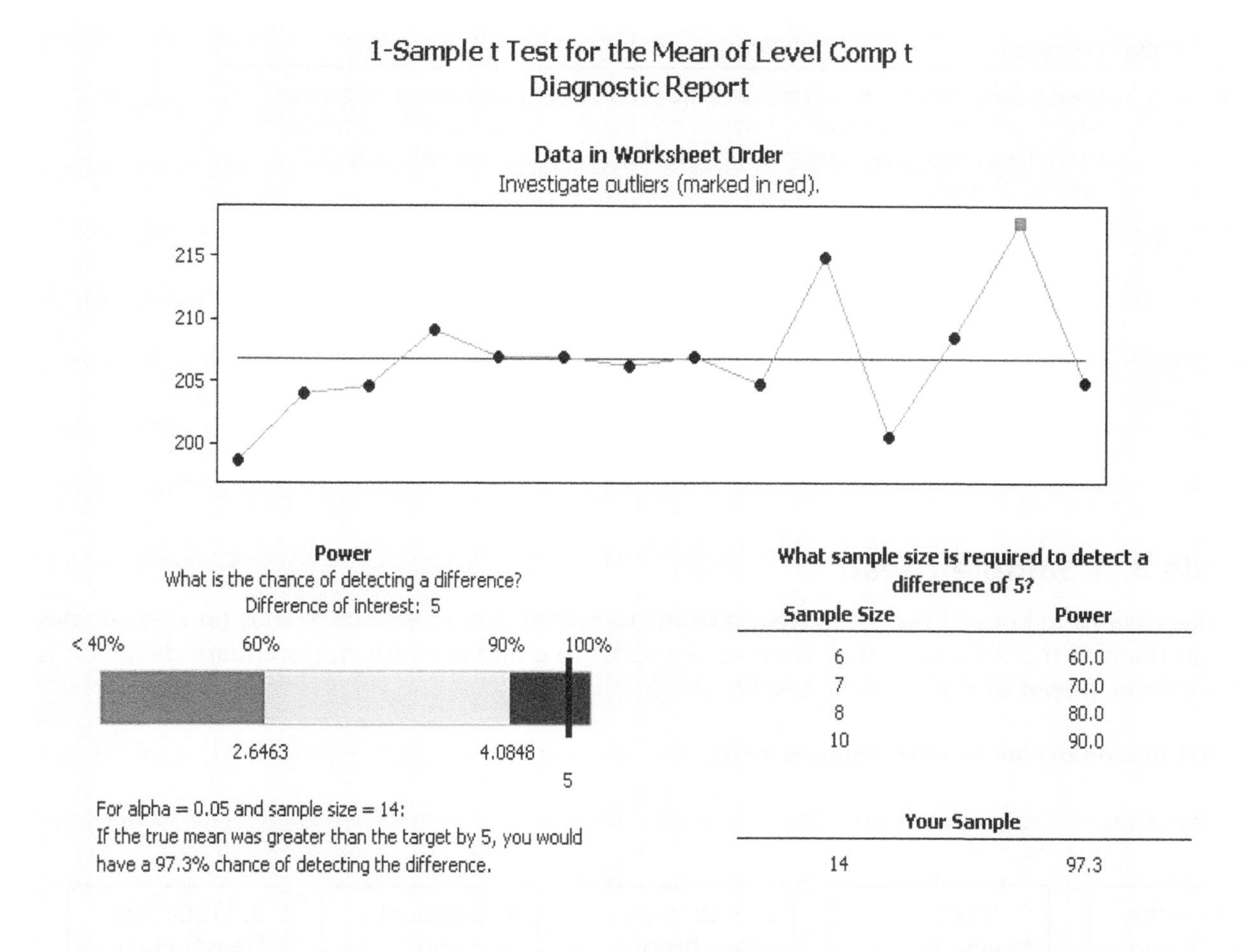

Finally, we have the Report Card. We are again told that we had one unusual data point. In the real world we would investigate if there were any unusual circumstances behind that point.

We get a warning due to our small sample size as it could lead to an inaccurate assessment of normality.

Finally, we get a satisfactory indicator relating to Power and Sample Size. This is because there is a sufficient difference between the mean and the target.

Let's move onto the next example. We are going to go through that from beginning to end using the hypothesis testing procedure. Initially we will use the classic menus. We will then analyse the data with the Assistant.

1-Sample t Test for the Mean of Level Comp t
Report Card

Check	Status	Description
Unusual Data	⚠	One data point (row 13) is unusual compared to the others. Because unusual data can have a strong influence on the results, you should try to identify the cause of its unusual nature. Correct any data entry or measurement errors. Consider removing data that are associated with special causes and repeating the analysis.
Normality	⚠	Because your sample size is less than 20, normality can be an issue. If the data are not normally distributed, the p-value may be inaccurate with small samples. In addition, unusual data can have a strong influence on the test results. Because normality cannot be reliably checked with small samples, you should use caution when interpreting the test results.
Sample Size	✓	The sample is sufficient to detect a difference between the mean and the target.

Example 3. 1 Sample *t* Test.

Parky Bakery wants to know if the mean weight of the buns that they bake is 99 g. They take 50 samples and weigh them. If the difference they want to detect is 1.0 g and the historical standard deviation is 1.5 g what is the power of the 1 sample *t* test?

Establish if the mean weight of the buns is 99 g?

Worksheet: Buns.

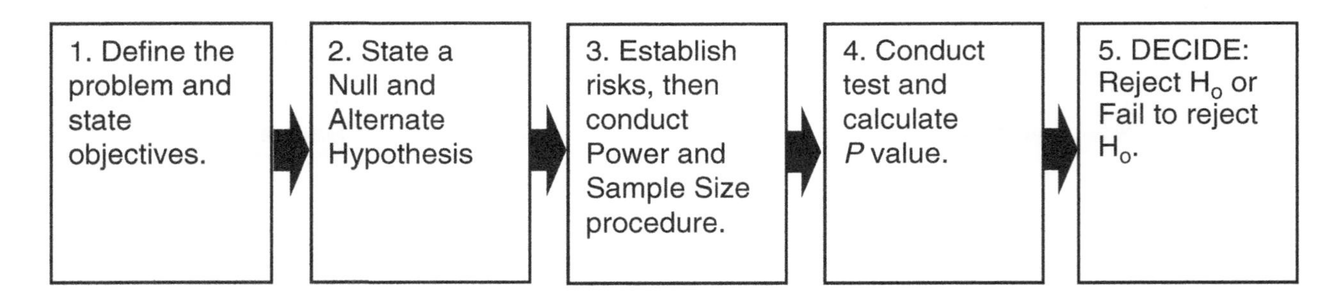
1. Define the problem and state objectives.
2. State a Null and Alternate Hypothesis
3. Establish risks, then conduct Power and Sample Size procedure.
4. Conduct test and calculate P value.
5. DECIDE: Reject H_0 or Fail to reject H_0.

The figure above shows the Hypothesis Testing procedure again. We will be using the steps to guide ourselves through the test procedure.

Problem Statement

Establish if the mean weight of the buns is 99 g.

(We can see that this is a two sided test.)

Null Hypothesis

Population Mean = 99 g

Alternate Hypothesis

Population Mean ≠ 99 g

Calculating the Power

1. Start a new project.
2. Click Stat <<P&SS << 1-Sample t.
3. Complete the menu box as shown.
4. Then click OK.

Stat

Power and Sample Size ▸

1t 1-Sample t...

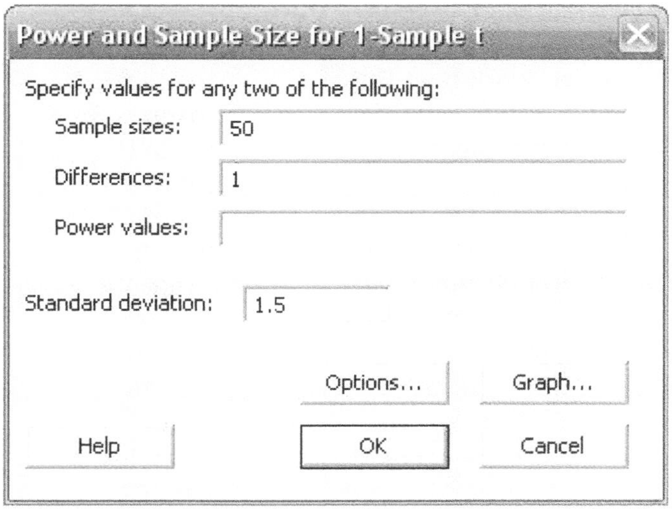

Power and Sample Size for 1-Sample t

Specify values for any two of the following:

Sample sizes: 50

Differences: 1

Power values:

Standard deviation: 1.5

Options... Graph...

Help OK Cancel

We have a power of 0.996. This means that if there is a difference there is a 99.6% chance of detecting it correctly. As we did not change the confidence of the test that remains at the default level of 95%. Which means that if the null hypothesis is correct and there is no difference we will have a 95% chance of detecting that correctly.

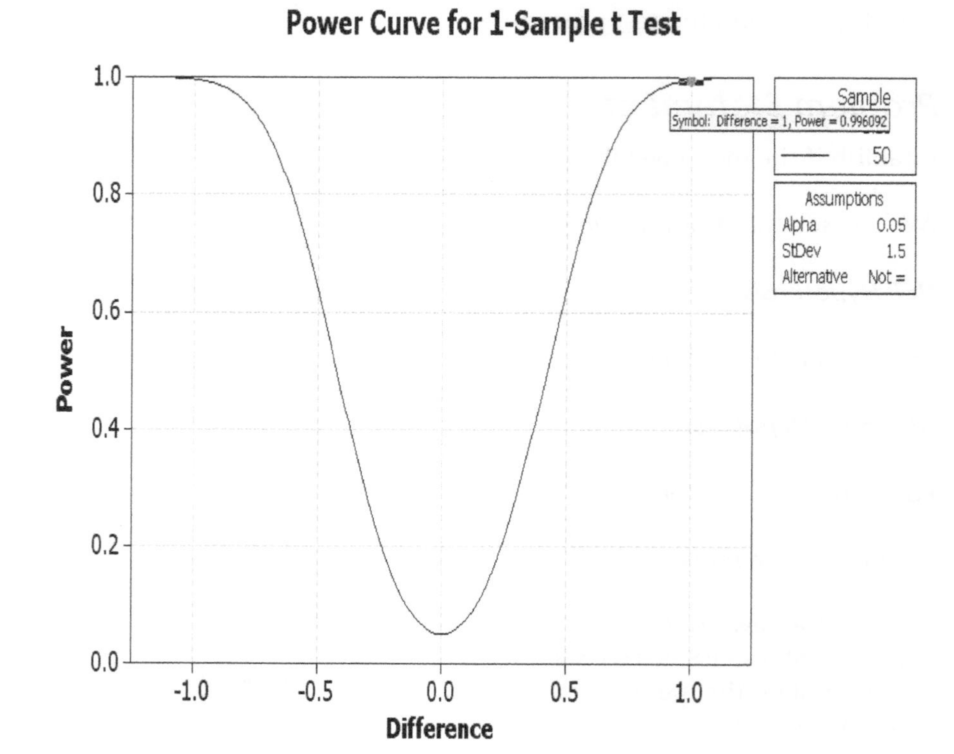

Calculating the Test Statistic

5. Open worksheet Buns and transfer the data into Minitab.
6. Examine the Graphical Summary by clicking Stat <<Basic Statistics <<Graphical Summary.
7. Select C1 and then click OK.

C1
Bun Weights
99.582
101.342
100.311
101.674
101.922
100.615
99.589
101.722
100.309
100.176
100.076
101.711
100.915
97.502

Stat

Basic Statistics ▸

Graphical Summary...

The mean for the samples is 100.38 and we can see the 95% CI for the mean.

The AD *P* value is 0.103 which indicates this data is from a normally distributed population

As the sample size is sufficiently large we do not need to use a separate normality test.

We can now conduct the test procedure.

Summary for Bun Weights

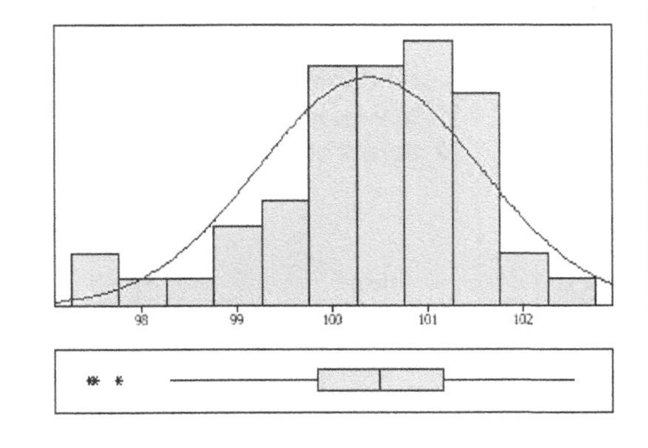

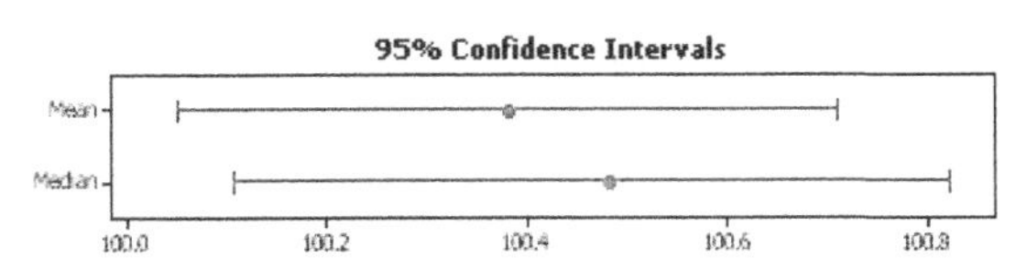

Anderson-Darling Normality Test	
A-Squared	0.62
P-Value	0.103
Mean	100.38
StDev	1.16
Variance	1.35
Skewness	-0.689614
Kurtosis	0.455762
N	50
Minimum	97.46
1st Quartile	99.83
Median	100.48
3rd Quartile	101.14
Maximum	102.51
95% Confidence Interval for Mean	
100.05	100.71
95% Confidence Interval for Median	
100.11	100.82
95% Confidence Interval for StDev	
0.97	1.45

8. Click Stat <<Basic Statistics <<1-Sample t.

Stat

Basic Statistics ▸

1t 1-Sample t...

9. Select C1 as the data column.
10. Enter 99 as the Hypothesized Mean.
11. Then click Graphs.

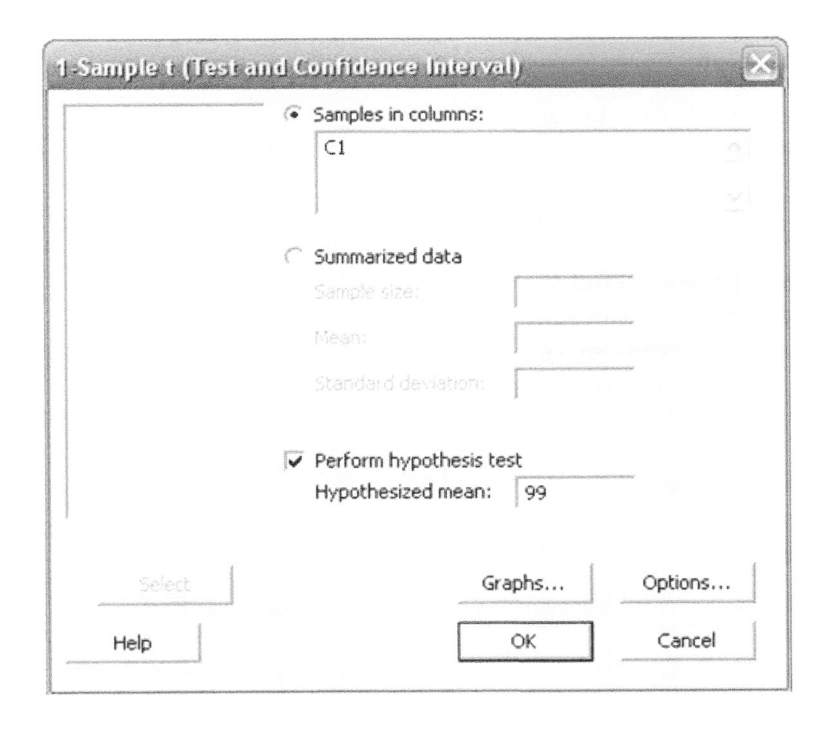

12. Select Boxplot of Data.
13. Then click OK and OK again.

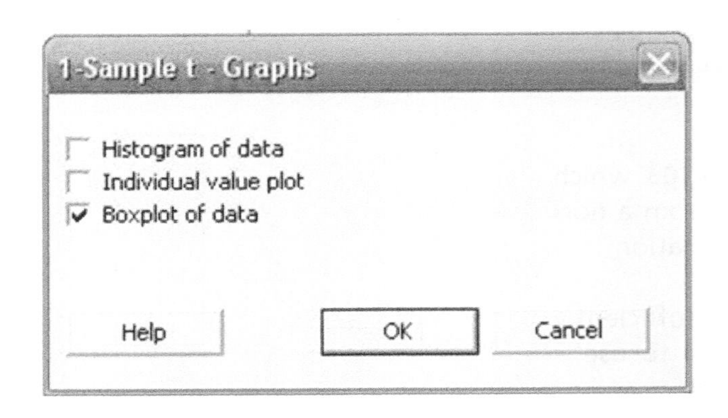

Decision Time

14. Click on the Show Session Folder icon.

15. Locate our test result and look at the P value. If it is less than the α (0.05) then we can reject the null hypothesis.
16. Go to the graph folder and look at the boxplot.

One-Sample T: Bun Weights

Test of mu = 99 vs not = 99

Variable	N	Mean	StDev	SE Mean	95% CI	T	P
Bun Weights	50	100.381	1.161	0.164	(100.051, 100.711)	8.41	0.000

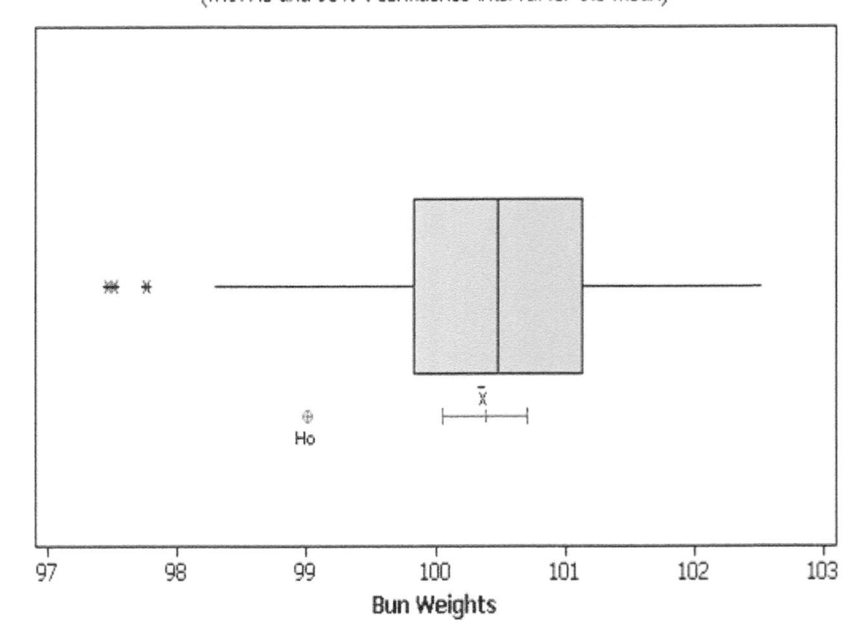

Graphically it can be see that H_o is not within the confidence interval for the mean.

Since we have rejected the null hypothesis we can state that the mean bun weight is not 99 g. We can say it is higher.

We will now run this example again using the Assistant.

1. Click on Assistant <<Hypothesis Tests.

Assistant

Hypothesis Tests...

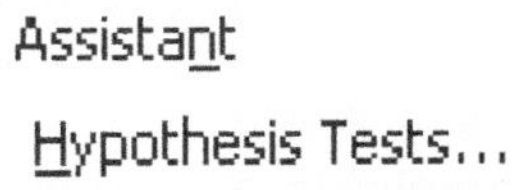

2. Click on 1-Sample t within the 'Compare one sample with a target' group.

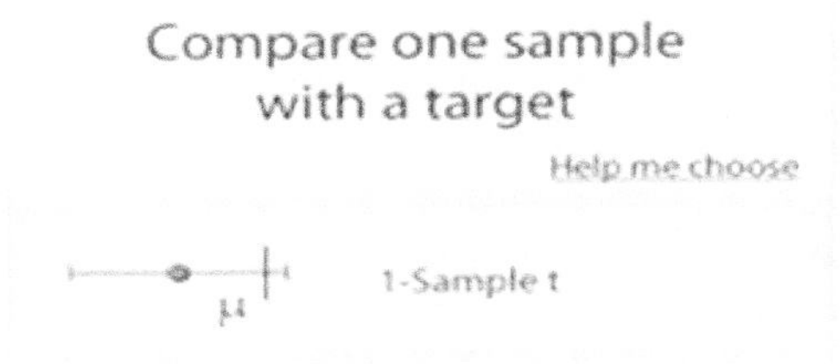

Compare one sample with a target

Help me choose

1-Sample t

3. Complete the menu box as shown and click OK. The menu box takes the same inputs as the previous menus for the 1 sample *t* test.

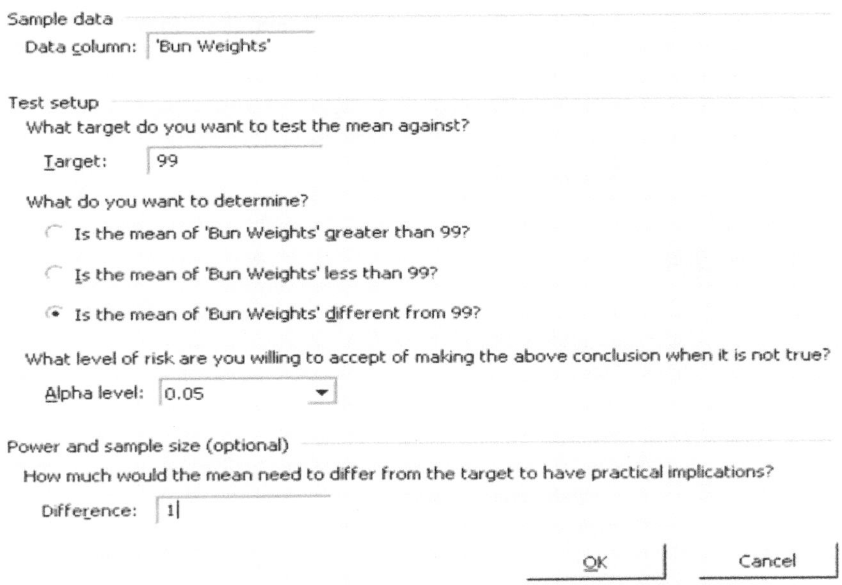

Sample data

Data column: 'Bun Weights'

Test setup

What target do you want to test the mean against?

Target: 99

What do you want to determine?

Is the mean of 'Bun Weights' greater than 99?

Is the mean of 'Bun Weights' less than 99?

Is the mean of 'Bun Weights' different from 99?

What level of risk are you willing to accept of making the above conclusion when it is not true?

Alpha level: 0.05

Power and sample size (optional)

How much would the mean need to differ from the target to have practical implications?

Difference: 1

OK Cancel

From the *P* value indicator in the top left quarter of the Summary Report we see that the *P* value = 0. Therefore the real mean will differ from our target value.

Below this we see the histogram of the actual sample data. Graphically we see that 99 is outside the 95% confidence interval for the population mean. You get a histogram rather than a dotplot when you have 20 or more data points.

On the top right we see the statistics from the test. Note that the StDev of the sample data was 1.16 compared to the 1.5 we used in the Classic Method.

On the bottom right, within the comments section, you can read the conclusion of the test procedure. The comments can be edited to add additional information.

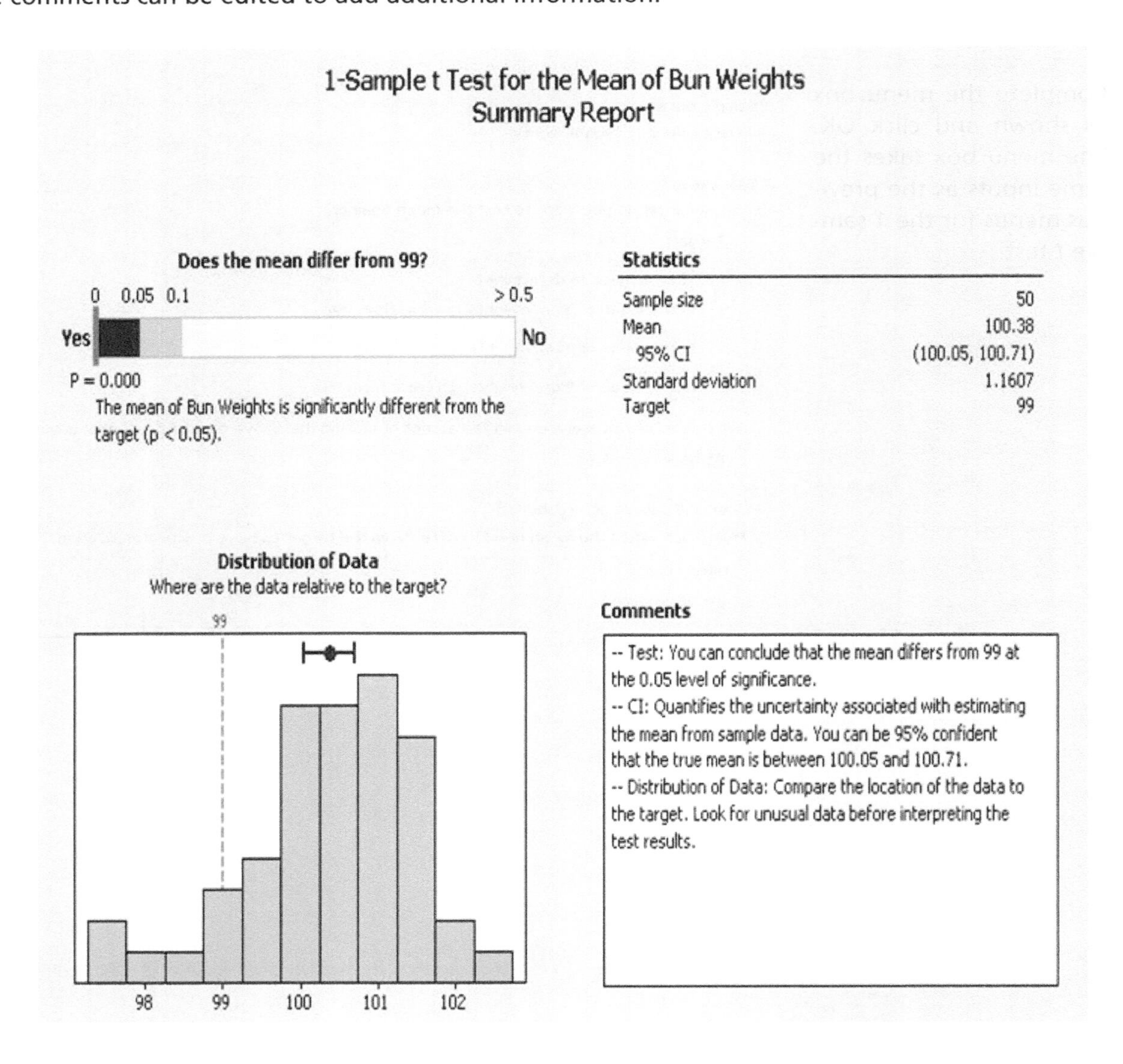

On the top of the Diagnostic Report we are presented with the SPC chart. There are three data points that Minitab deems to be outliers.

The power achieved using the Assistant is again higher than previously calculated. It is again due to the fact that a lower StDev was used in the calculation.

1-Sample t Test for the Mean of Bun Weights
Diagnostic Report

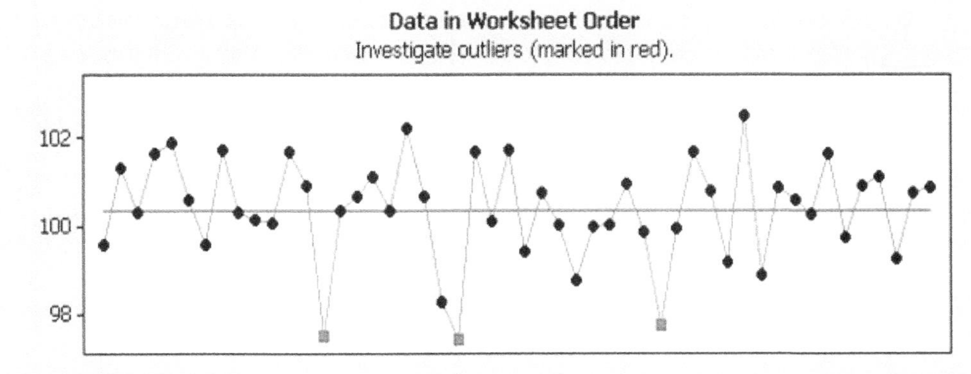

Power
What is the chance of detecting a difference?
Difference of interest: 1

< 40% 60% 90% 100%

0.37059 0.54283 1

For alpha = 0.05 and sample size = 50:
If the true mean differed from the target by 1 in either direction, you would have a 100.0% chance of detecting the difference.

What sample size is required to detect a difference of 1?

Sample Size	Power
9	60.0
11	70.0
13	80.0
17	90.0

Your Sample

50	100.0

Power is a function of the sample size and the standard deviation. If the power is not satisfactory, consider increasing the sample size.

Finally, we have the Report Card. We are again told that we had some unusual data points.

We don't get any further warnings.

1-Sample t Test for the Mean of Bun Weights
Report Card

Check	Status	Description
Unusual Data	⚠	Some of the data points are unusual compared to the others. Because unusual data can have a strong influence on the results, you should try to identify the cause of their unusual nature. These points are marked in red on the Diagnostic Report. You can hover over a point or use Minitab's brushing feature to identify the worksheet row. Correct any data entry or measurement errors. Consider removing data that are associated with special causes and repeating the analysis.
Normality	✓	Because your sample size is at least 20, normality is not an issue. The test is accurate with nonnormal data when the sample size is large enough.
Sample Size	✓	The sample is sufficient to detect a difference between the mean and the target.

It's time for an exercise. Complete the question using the classic method. Use the full procedure for hypothesis testing. Then if you have M16 use the Assistant and compare the answer.

Exercise 1. 1 Sample *t* test.

Bear is a keen cyclist and likes to buy all the latest equipment. He wants to buy the latest lights for his bike and therefore must show that his current lights are failing. He keeps a spreadsheet showing the recharge time for his lights. If the time becomes different to the recharge time stated by the manufacturer he will have the excuse he needs to buy new lights.

Bear has 75 data points. If the difference he wants to detect is 0.5 h and the historical standard deviation is 0.8 h what is the power of the 1 sample *t* test?

Use the graphical summary to check the sample data. Establish if the mean recharge time of the lights is different to 17 h? Display the comparison of the null hypothesis to the 95% CI for the population mean using a boxplot.

The data is in Worksheet: Lights.

Problem Statement

Establish if the mean recharge time of the lights is different to 17 h.

(As we are only checking to see if there is a difference a two sided test is required.)

Null Hypothesis

Population Mean = 17 h

Alternate Hypothesis

Population Mean ≠ 17 h

Based on the historical StDev the power for the test is very high. The default significance level will be used for the test.

Power and Sample Size

```
1-Sample t Test

Testing mean = null (versus not = null)
Calculating power for mean = null + difference
Alpha = 0.05  Assumed standard deviation = 0.8


                 Sample
Difference         Size     Power
       0.5           75  0.999639
```

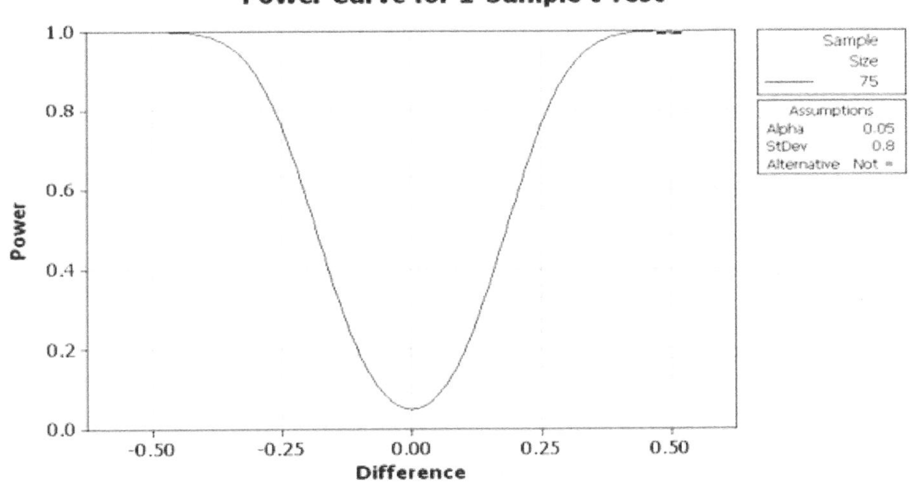

The graphical summary shows that the sample distribution is uni modal and normally distributed.

Summary for Lights

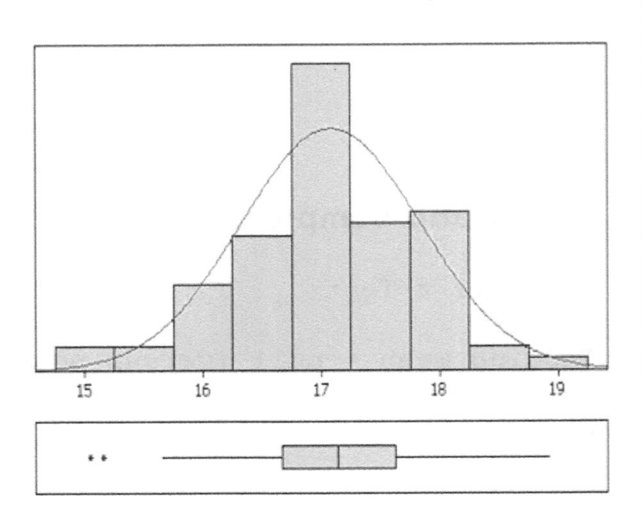

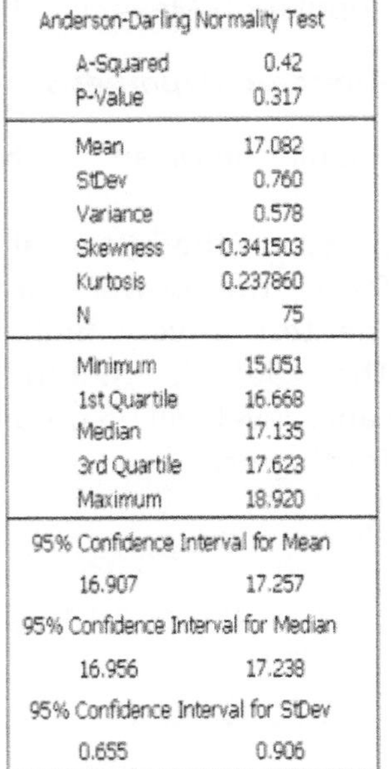

Anderson-Darling Normality Test	
A-Squared	0.42
P-Value	0.317
Mean	17.082
StDev	0.760
Variance	0.578
Skewness	-0.341503
Kurtosis	0.237860
N	75
Minimum	15.051
1st Quartile	16.668
Median	17.135
3rd Quartile	17.623
Maximum	18.920
95% Confidence Interval for Mean	
16.907	17.257
95% Confidence Interval for Median	
16.956	17.238
95% Confidence Interval for StDev	
0.655	0.906

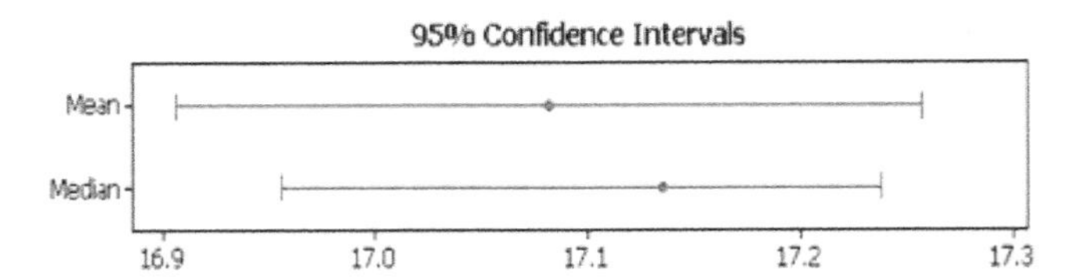

The boxplot shows that the estimated mean lies within the 95% CI for the population mean.

Boxplot of Lights

(with Ho and 95% t-confidence interval for the mean)

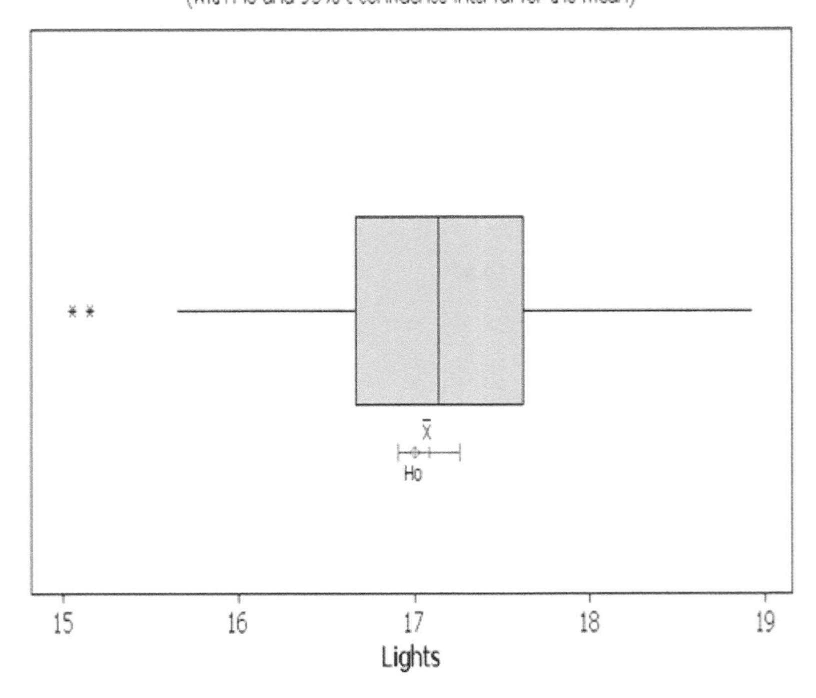

One-Sample T: Lights

```
Test of mu = 17 vs not = 17


Variable   N     Mean    StDev  SE Mean         95% CI               T      P
Lights    75  17.0819  0.7603   0.0878  (16.9070, 17.2568)  0.93  0.354
```

As the *P* value is above the alpha value we cannot reject the null hypothesis. We cannot say the mean recharge time of the lights is different to 17 h.

Bear didn't like the result and bought some new lights anyway.

Exercise 1. Using the Assistant.

The *P* value = 0.354. Therefore the target is not significantly different from the real mean.

Within the histogram we see how the target mean is within the 95% confidence interval for the population mean.

On the top right we see the statistics from the test. Note that the StDev of the sample data was 0.76 compared to the 0.8 we used in the Classic Method.

1-Sample t Test for the Mean of Lights
Summary Report

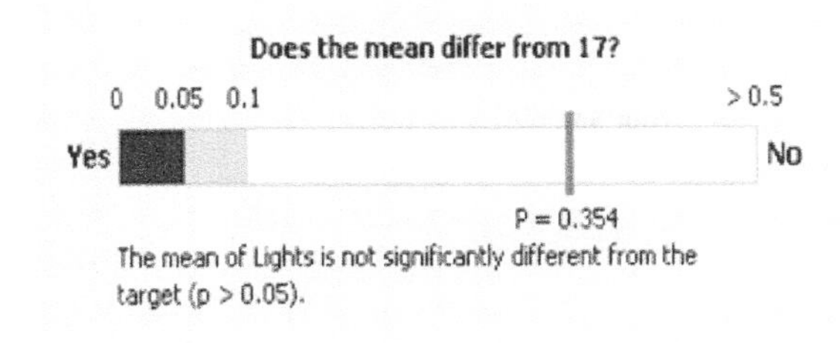

Statistics

Sample size	75
Mean	17.082
95% CI	(16.907, 17.257)
Standard deviation	0.76025
Target	17

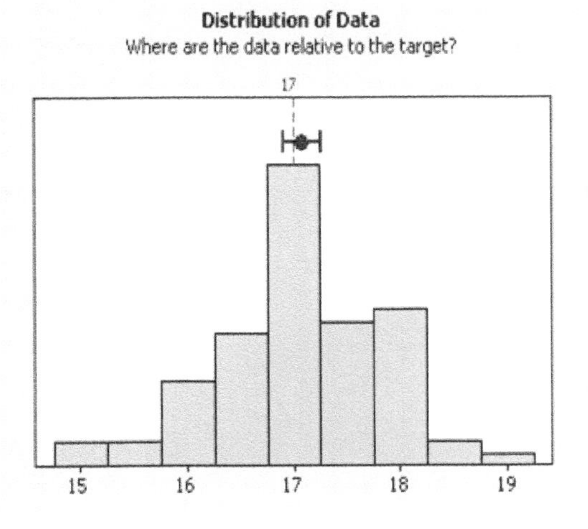

Comments

-- Test: There is not enough evidence to conclude that the mean differs from 17 at the 0.05 level of significance.
-- CI: Quantifies the uncertainty associated with estimating the mean from sample data. You can be 95% confident that the true mean is between 16.907 and 17.257.
-- Distribution of Data: Compare the location of the data to the target. Look for unusual data before interpreting the test results.

On the top of the Diagnostic Report we are presented with the SPC chart. There are two data points that Minitab deems to be outliers.

The power achieved using the Assistant is again higher than previously calculated. It is again due to the fact that a lower StDev was used in the calculation.

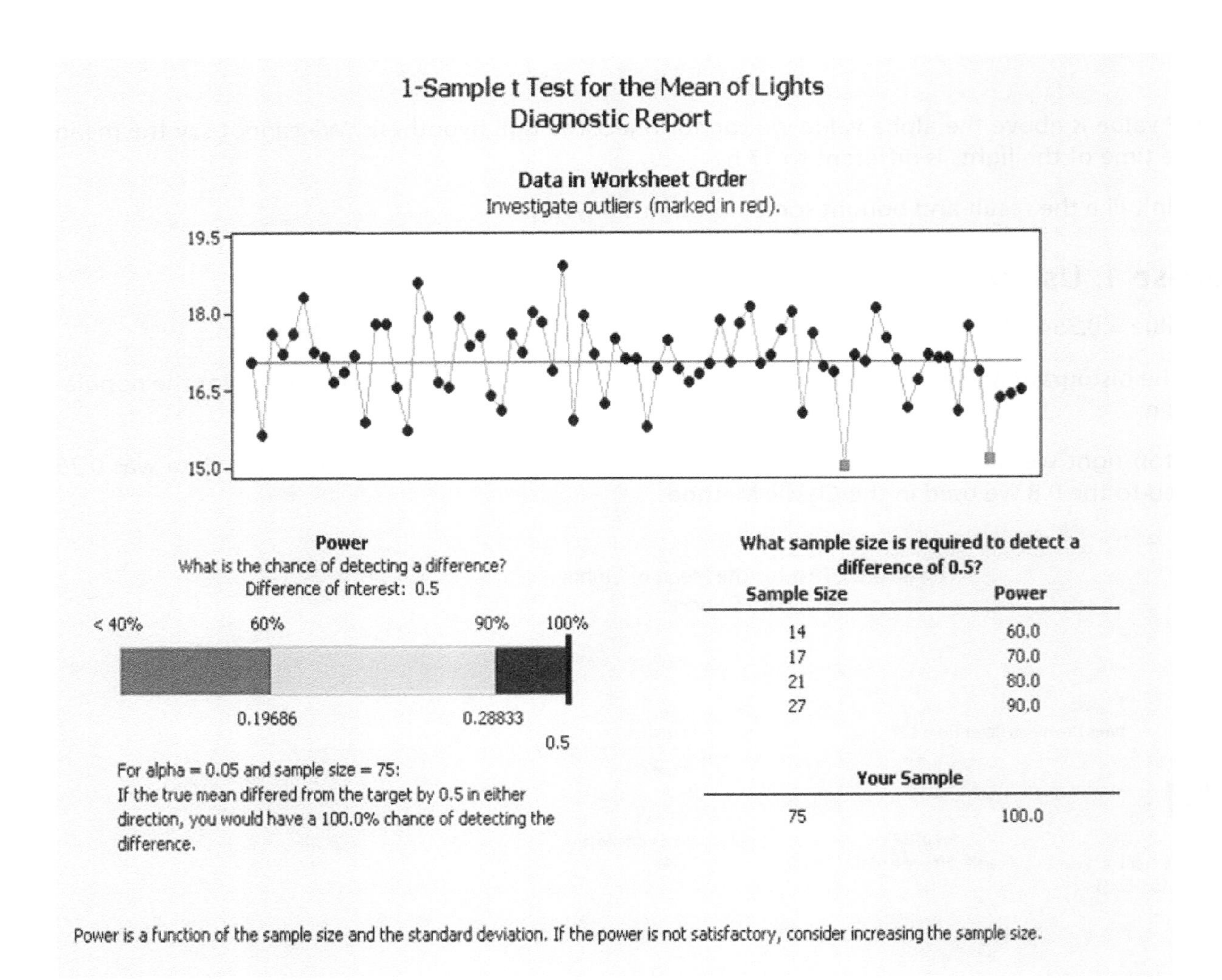

Finally, we have the Report Card. We are again told that we had some unusual data points.

We don't get any further warnings.

1-Sample t Test for the Mean of Lights
Report Card

Check	Status	Description
Unusual Data	⚠	Some of the data points are unusual compared to the others. Because unusual data can have a strong influence on the results, you should try to identify the cause of their unusual nature. These points are marked in red on the Diagnostic Report. You can hover over a point or use Minitab's brushing feature to identify the worksheet row. Correct any data entry or measurement errors. Consider removing data that are associated with special causes and repeating the analysis.
Normality	✓	Because your sample size is at least 20, normality is not an issue. The test is accurate with nonnormal data when the sample size is large enough.
Sample Size	✓	Although the test results are not significant, the power is adequate. Based on your sample size, standard deviation, and alpha, you have a 100.0% chance of detecting a difference of 0.5 between the mean and the target. Because the power is adequate, you can conclude that it is unlikely that there is a difference of 0.5 or larger.

4.7 Paired *t* Test

Procedure	• Paired *t* Test
What's it used for?	• A paired *t* test is used to test the mean difference between two sets of dependent samples. • They usually involve a human element that gives the dependency between the two sets of data.
Assumptions and Limitations	• The difference of the data sets should form a normally distributed population, the test is fairly robust to non normal data when the sample size is large (>20). • M15 users can use Power and Sample Size for the 1 Sample *t* test.

The Paired *t* Test

The Paired *t* Test is used to check the mean difference between two sets of dependent data. To help understand what is meant by dependent data think of using a set of people to test two types of running shoe. The test will be to see if 100 m times are different using each type of shoe. There are two obvious ways of arranging the test

1. Split the runners into two groups, one group uses shoe A and the other shoe B. Then compare the times.

2. Have all the runners run with shoe A and then repeat the run with shoe B. Look at the difference in time for each person.

For method 1, if the ability of the runners was to vary greatly between the groups the difference between the shoes might get lost within the difference of the groups. Method 2 seems a better choice because we can filter out the variation from our system by using the system twice and then looking at the difference.

In fact, the Paired *t* Test is the same as conducting a 1 Sample t Test on the difference of the two data sets and using 0 as the hypothesised mean.

We will learn more about the Paired *t* Test by going straight into the examples. We will use the same procedure as we have used previously for hypothesis testing.

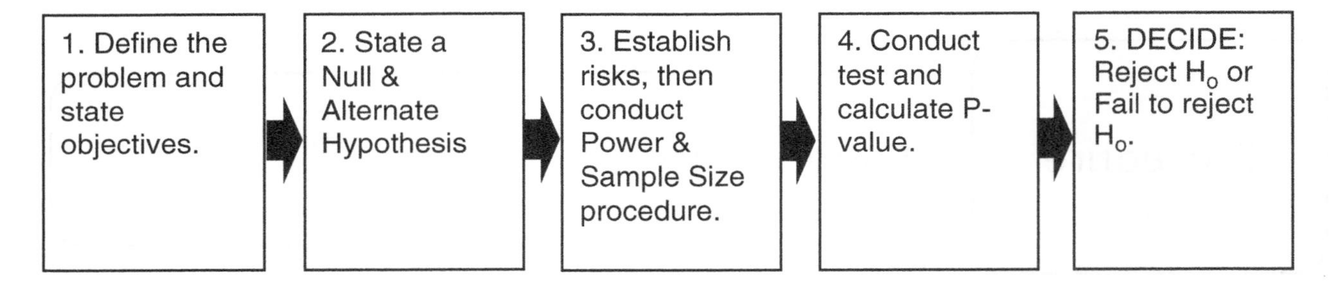
1. Define the problem and state objectives.
2. State a Null & Alternate Hypothesis
3. Establish risks, then conduct Power & Sample Size procedure.
4. Conduct test and calculate P-value.
5. DECIDE: Reject H_o or Fail to reject H_o.

Example 4. Paired *t* Test.

An archery equipment manufacturer has developed a new arrow release, the Sureshot2. They want to know if it performs differently to the Sureshot1. The company gets 50 archers of varying ability and measures their average score with both types of release. With a historical StDev of 5, what is the power of the test?

Is the Sureshot2 an improvement over the previous model by an average score of 2?

Worksheet: Archery.

Problem Statement

Establish if the Sureshot2 gives a better archery score than the Sureshot1.

Null Hypothesis

Score with Sureshot2 ≤ Score with Sureshot1.

Alternate Hypothesis

Score with Sureshot2 > Score with Sureshot1.

(If the Sureshot2 is better it will give a higher score.)

We will now establish the power.

1. Click Stat <<P&SS << Paired t (M15 users can use P&SS for the 1 Sample t test).

Stat
Power and Sample Size ▸ Paired t...

2. Complete the menu box as shown.
3. Click on Options.

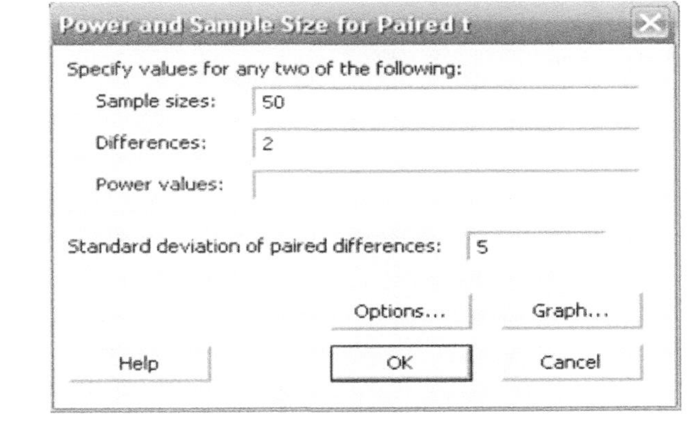

4. Select the radio button for setting the alternate hypothesis to Greater than. As we are using Greater than, this is a one sided test.
5. Click OK and OK again.

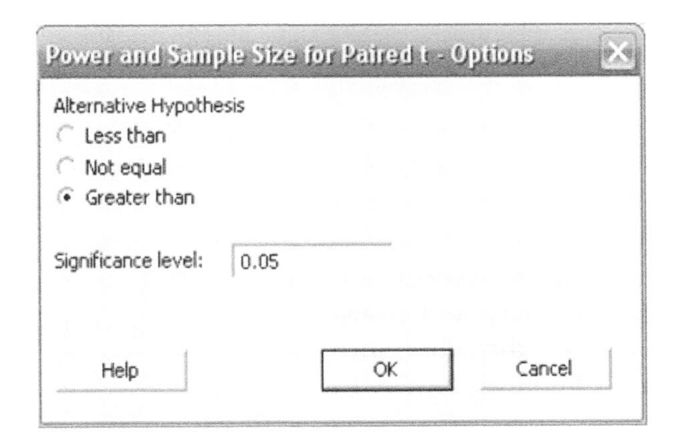

Two power curves are shown here; the one specifically for the Paired *t* Test and the one for the 1 Sample *t* Test. It can be seen that they are identical and both give the same result for a Power of 87.3%.

For our test, this means that if there is a difference we have an 87.3% chance of correctly stating that there was a difference.

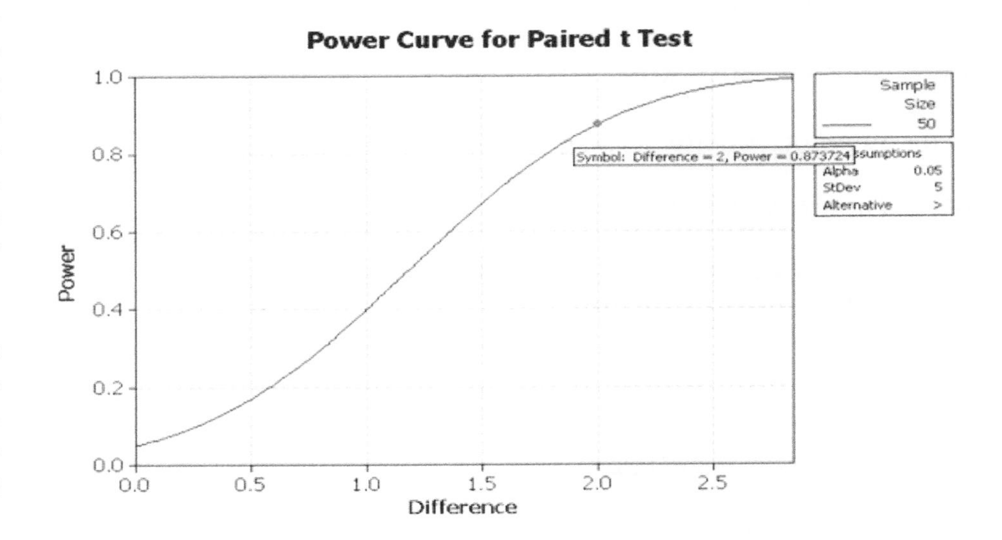

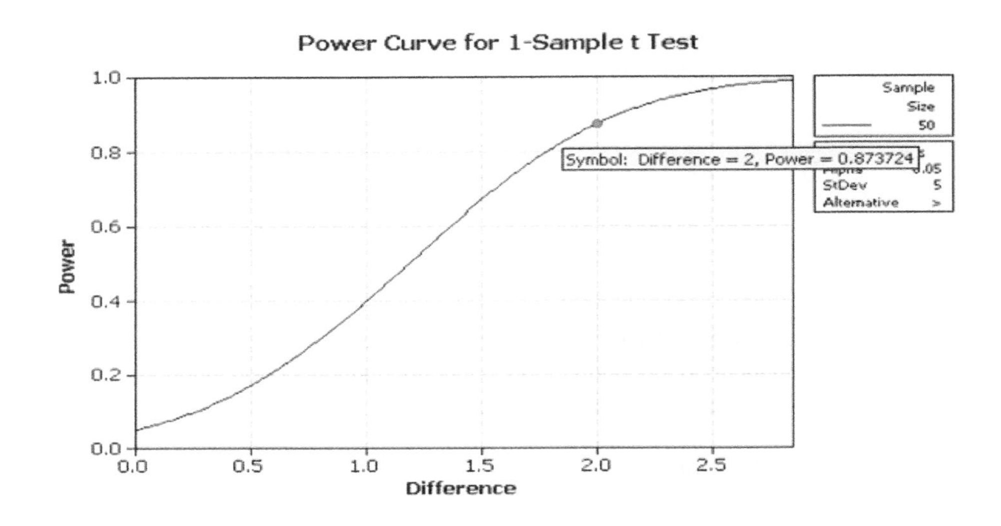

6. Go to Worksheet: Archery and transfer the data into Minitab.

↓	C1	C2
	Sureshot1	Sureshot2
1	58.19	59.66
2	49.06	46.22
3	46.21	52.55
4	49.33	48.17
5	54.14	50.52

7. Click Stat <<Basic Statistics <<Paired t

Stat

Basic Statistics ▸ t-t Paired t...

8. Select the columns as shown. Ensure the order is correctly entered.
9. Then click on Graphs.

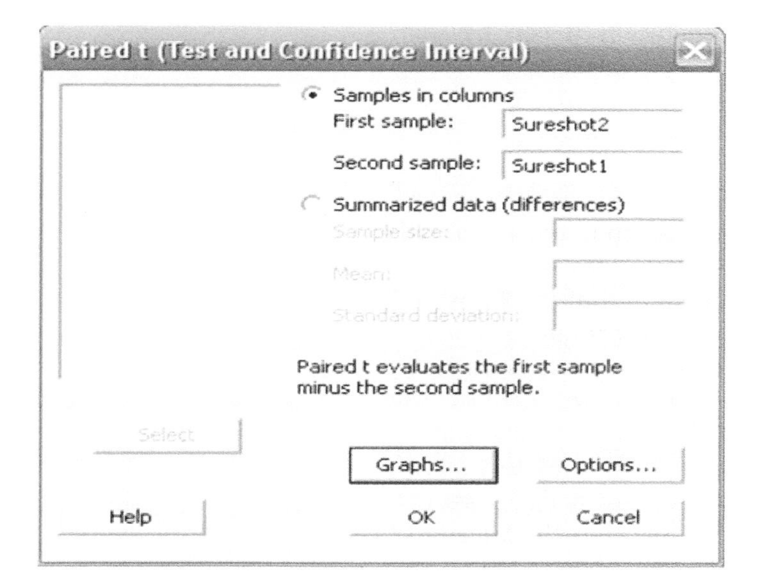

10. Select the Boxplot of data and then click on OK.
11. Click on Options.

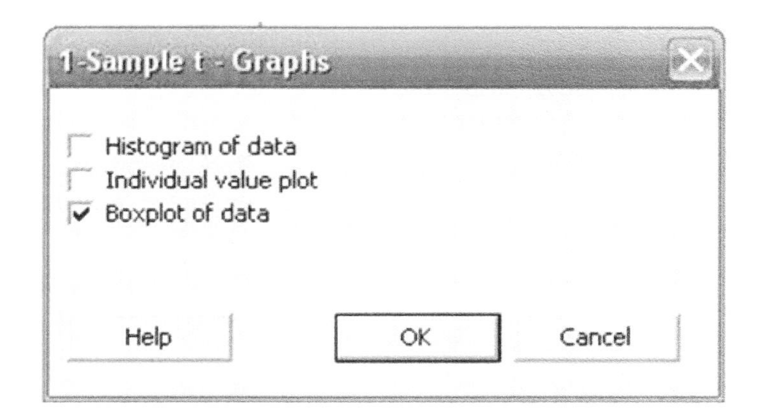

12. Change the Alternative drop down menu to 'greater than'.
13. Click on OK and OK again.

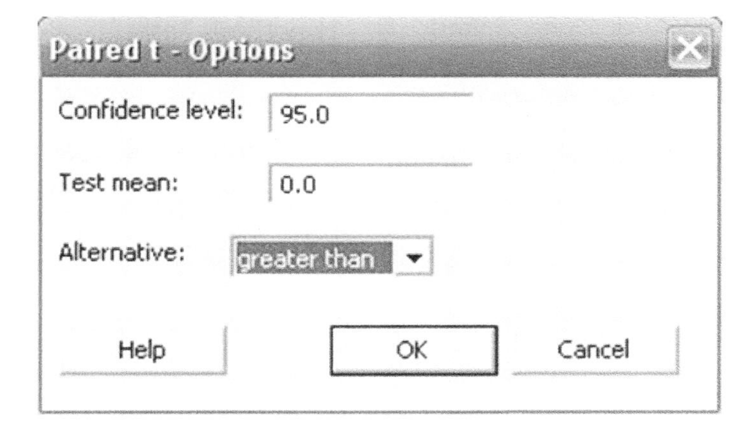

Graphically it can be seen that H_0 exceeds the lower end of the confidence interval for the mean.

From the session window we see that the mean difference was −1.924. This indicates that the Sureshot2 actually produced a lower score than the Sureshot1. However, we cannot say if it is statistically significant. With the test which we have just done we can only say that the Sureshot2 was not better than the Sureshot1 as the *P* value was 0.994.

We will now use the Assistant to conduct the same test (M16 only).

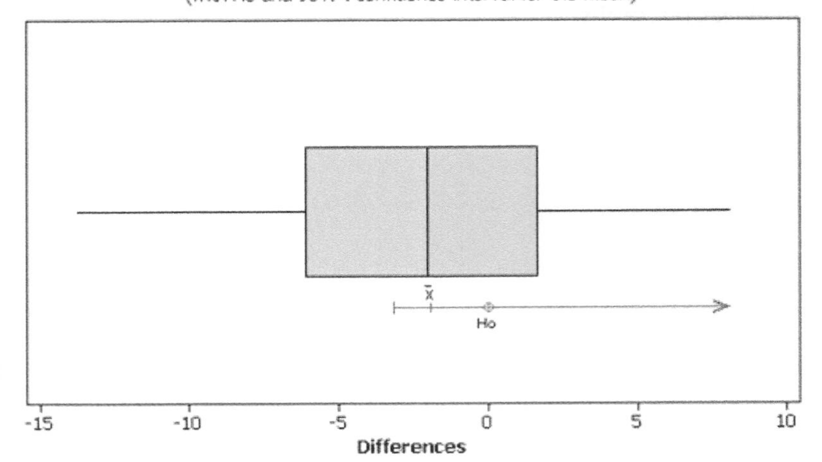

Paired T-Test and CI: Sureshot2, Sureshot1

```
Paired T for Sureshot2 - Sureshot1

             N     Mean  StDev  SE Mean
Sureshot2   50    48.89   7.75     1.10
Sureshot1   50    50.81   5.66     0.80
Difference  50   -1.924  5.192    0.734


95% lower bound for mean difference: -3.155
T-Test of mean difference = 0 (vs > 0): T-Value = -2.62  P-Value = 0.994
```

1. Click on Assistant <<Hypothesis Tests.
2. Click on Paired t within the 'Compare two samples with each other' group.
3. Complete the menu box as shown and click OK.

Assistant
Hypothesis Tests...

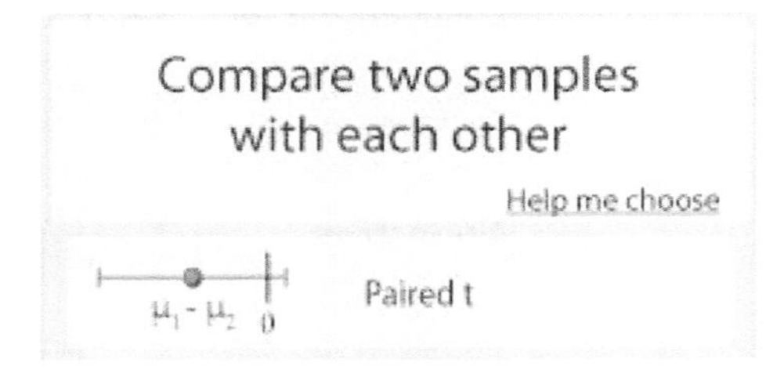

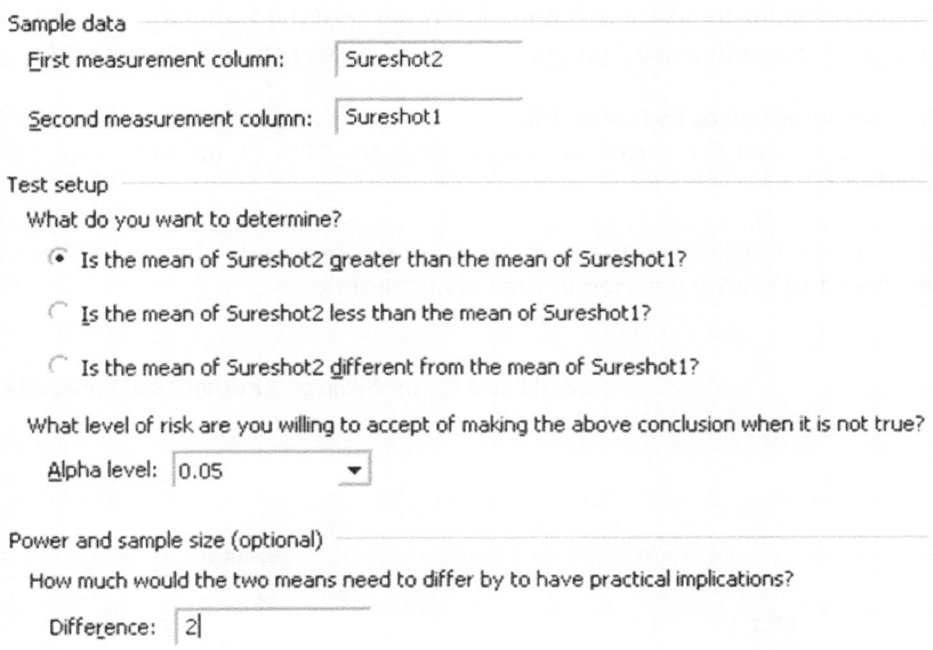

Sample data

First measurement column: Sureshot2

Second measurement column: Sureshot1

Test setup

What do you want to determine?

(•) Is the mean of Sureshot2 greater than the mean of Sureshot1?

( ) Is the mean of Sureshot2 less than the mean of Sureshot1?

( ) Is the mean of Sureshot2 different from the mean of Sureshot1?

What level of risk are you willing to accept of making the above conclusion when it is not true?

Alpha level: 0.05

Power and sample size (optional)

How much would the two means need to differ by to have practical implications?

Difference: 2

From the *P* value indicator in the top left quarter of the Summary Report we see that the *P* value = 0.994. Therefore we cannot say that the Sureshot2 is better than the Sureshot1.

The histogram gives us our first indication that the Sureshot2 was actually worse than the Surehot1. It can be seen that the zero is actually outside the 90% CI. (Minitab uses a 90% CI because it was a one sided test).

On the top right we see the statistics from the test. Note that the StDev of the sample data was 5.16 compared to the 5.0 we used in the Classic Method.

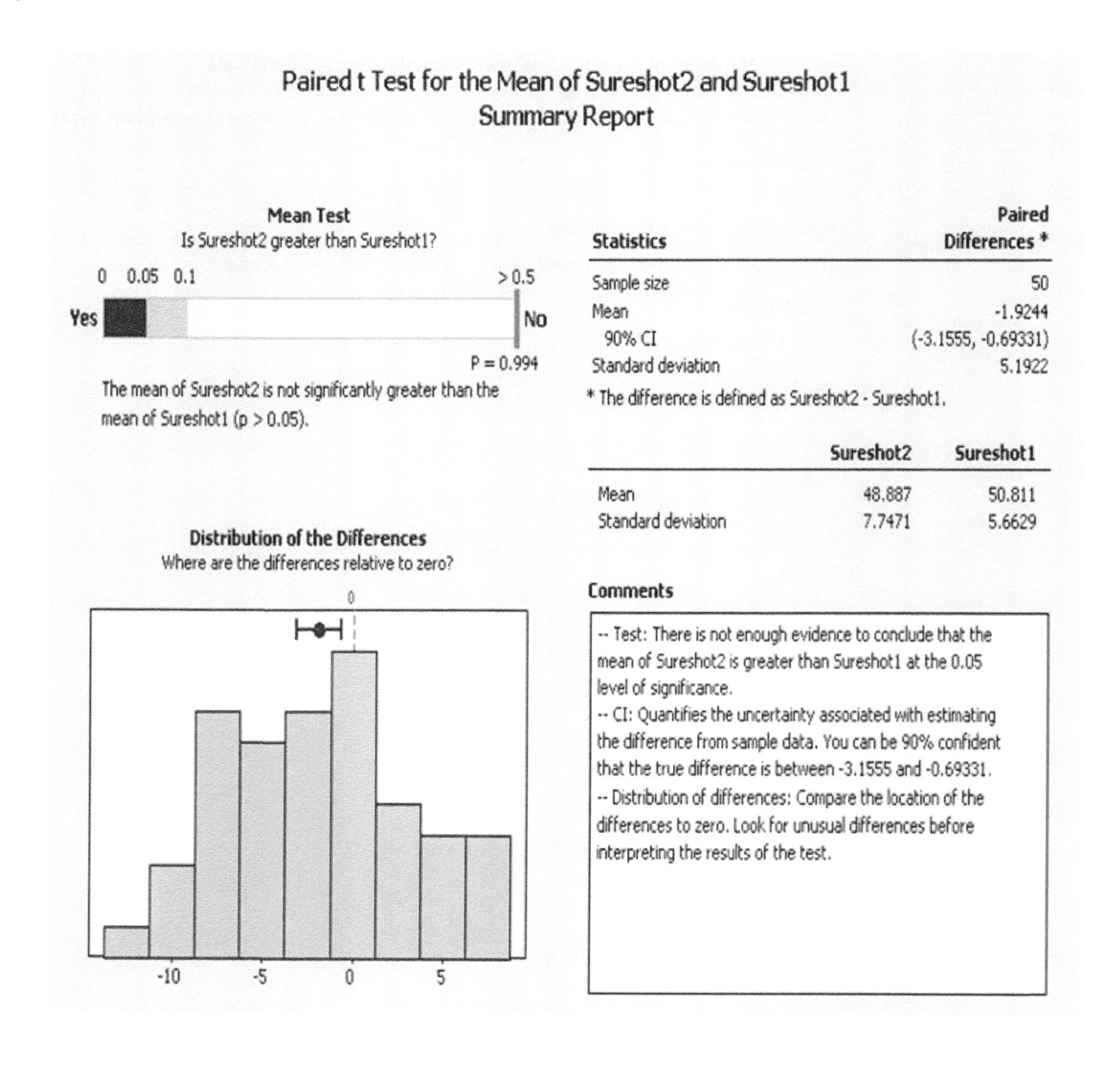

On the top of the Diagnostic Report we are presented with the SPC chart which is now based on the difference of the points. None of our points are deemed to be unusual.

This is the first time that the Assistant has calculated a lower power than the initial check. Again this is due to the StDev.

Paired t Test for the Mean of Sureshot2 and Sureshot1
Diagnostic Report

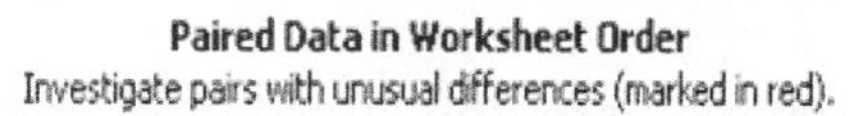

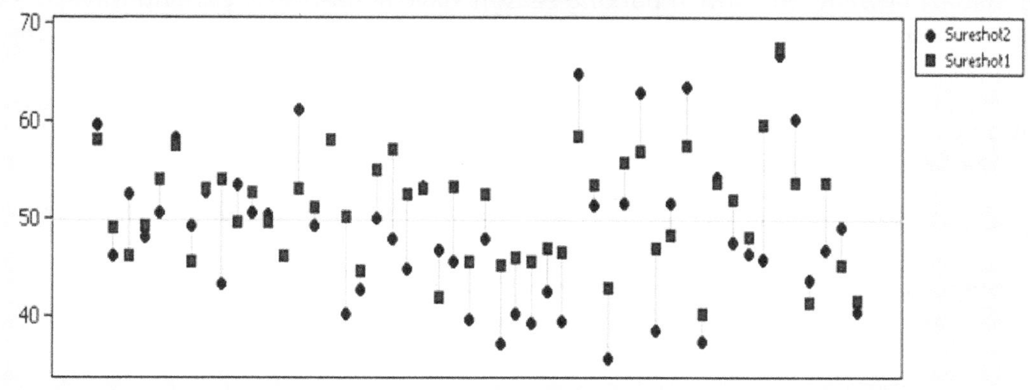

Power
What is the chance of detecting a difference?
Difference of interest: 2

< 40% | 60% | 90% | 100%

1.4134 | 2.1792

2

For alpha = 0.05 and sample size = 50:
If the true mean of Sureshot2 was 2 greater than Sureshot1,
you would have a 85.1% chance of detecting the difference.

What sample size is required to detect a difference of 2?

Sample size	Power
26	60.0
34	70.0
44	80.0
60	90.0

Your Sample

50	85.1

Power is a function of the sample size and the standard deviation. If the power is not satisfactory, consider increasing the sample size.

On the Report Card we are told that we did not have any unusual points within the data and that normality is not an issue.

We are given a warning that we may have the data sets the wrong way round, but we haven't.

Paired t Test for the Mean of Sureshot2 and Sureshot1
Report Card

Check	Status	Description
Unusual Data	✓	There are no unusual paired differences. Unusual data can have a strong influence on the results.
Normality	✓	Because your sample size is at least 20, normality is not an issue. The test is accurate with nonnormal data when the sample size is large enough.
Sample Size	i	Although you want to determine whether the mean of Sureshot2 is greater than Sureshot1, the data suggest that the mean of Sureshot2 is actually less than Sureshot1. It is unusual for the test result to be significant in the less than direction when the opposite comparison is chosen. Confirm that you selected the correct comparison in Minitab.

It's time for another exercise. Complete the exercise using the Classic Method. Use the full procedure for hypothesis testing. Then if you have M16 use the Assistant and compare the answer.

Exercise 2. Paired *t* test.

A motoring magazine wants to know if parking sensors have an effect on parking time. The magazine gets 50 drivers of varying ability and measures parking time (seconds) with and without sensors on the same car.

If they want a power >90% and the historical standard deviation is 8, what is the maximum difference that can be measured (as an integer)?

Do parking sensors change parking time?

Display the comparison of the data to the null hypothesis using a boxplot.

Worksheet: Parking Times.

Problem Statement

Establish if parking sensors change the parking time.

Null Hypothesis

Parking time With Sensors = Without Sensors

Alternate Hypothesis

Parking time With Sensors ≠ Without Sensors

(This is a two sided test.)

In order to find the values of difference to give us a power of >90% we must guess a couple of values. I have chosen 3 and 4. As long as I put a space between the two numbers they will both be displayed on the graph together.

A difference of 4 s gives us a power of 93.3%.

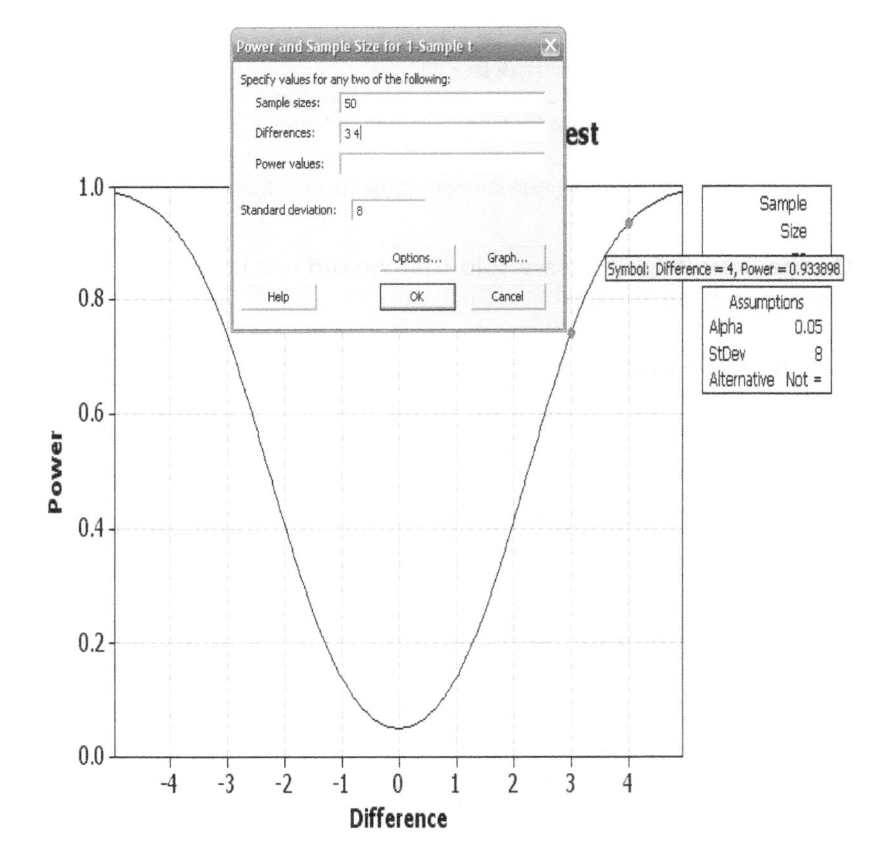

Graphically it can be seen that H_o sits within the confidence interval for the mean. This means that there is not a difference in parking times with or without sensors.

From the Session window we see that we have *P* value of 0.822. 'As the *P* is not low the null cannot go'. Meaning that we cannot reject the null hypothesis.

We will now use the Assistant to conduct the same test (M16 only).

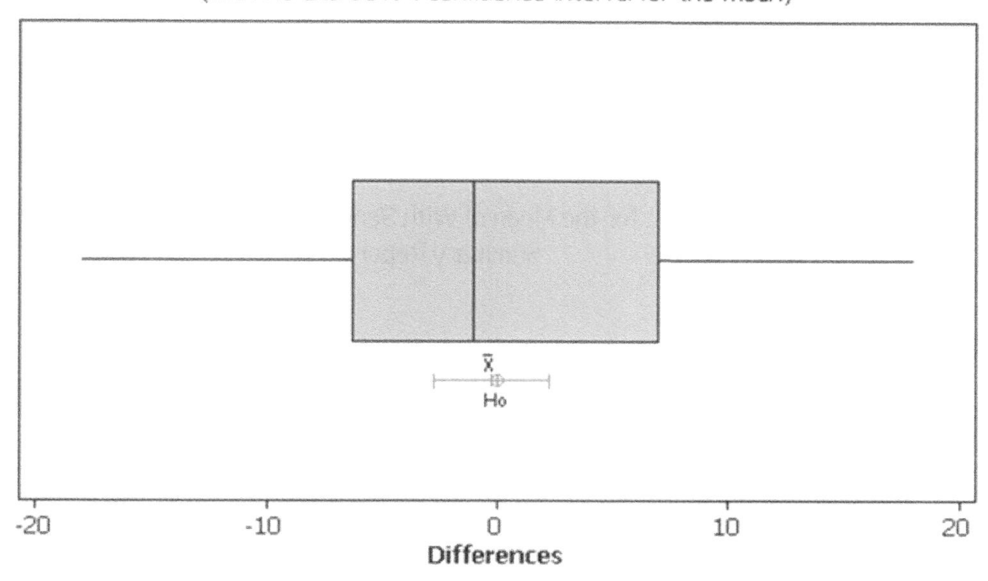

Paired T-Test and CI: With Sensors, Without Sensors

```
Paired T for With Sensors - Without Sensors

                  N    Mean  StDev  SE Mean
With Sensors     50  129.42  10.82     1.53
Without Sensors  50  129.70   6.31     0.89
Difference       50   -0.28   8.73     1.23

95% CI for mean difference: (-2.76, 2.20)
T-Test of mean difference = 0 (vs not = 0): T-Value = -0.23  P-Value = 0.822
```

From the *P* value indicator in the top left quarter of the summary report we see that the *P* value = 0.882. Again, this means that we cannot reject the null hypothesis.

The histogram shows us that the zero target clearly lies within the confidence interval.

On the top right we see the statistics from the test. Note that the StDev of the sample data was 8.7 compared to the 8.0 we used in the Classic Method.

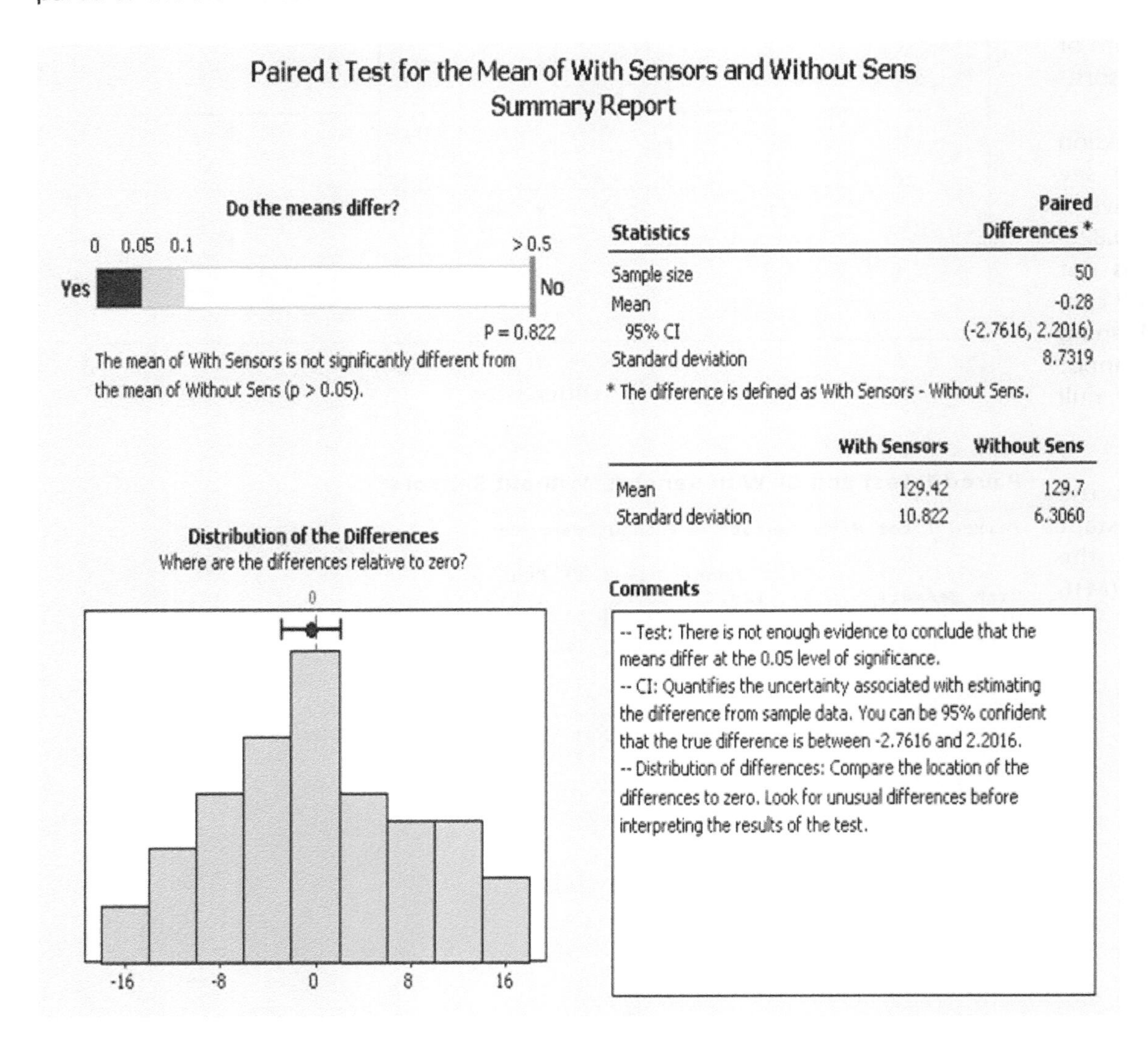

On the top of the Diagnostic Report we are presented with the SPC chart which is now based on the difference of the points. None of our points are deemed to be unusual.

The Assistant has calculated a lower power than the initial check. Again this is due to the StDev.

Paired t Test for the Mean of With Sensors and Without Sens
Diagnostic Report

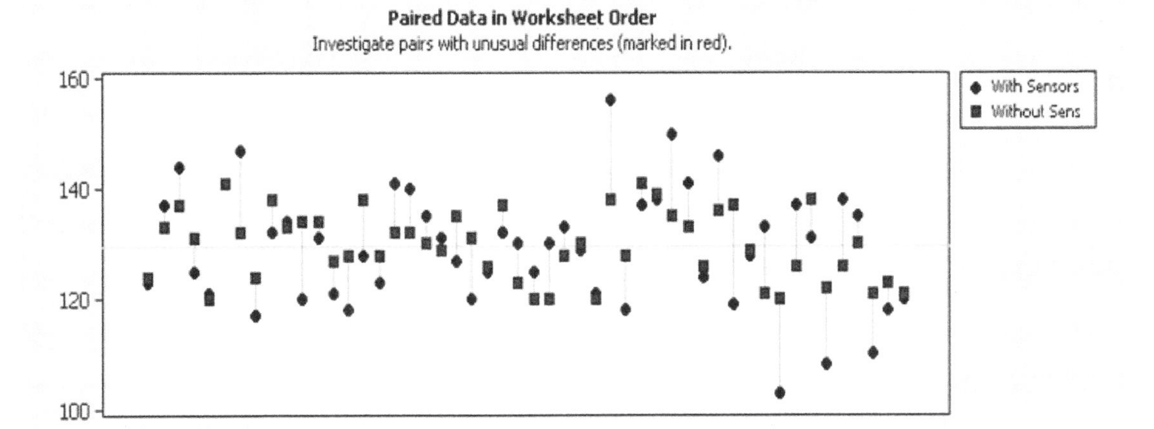

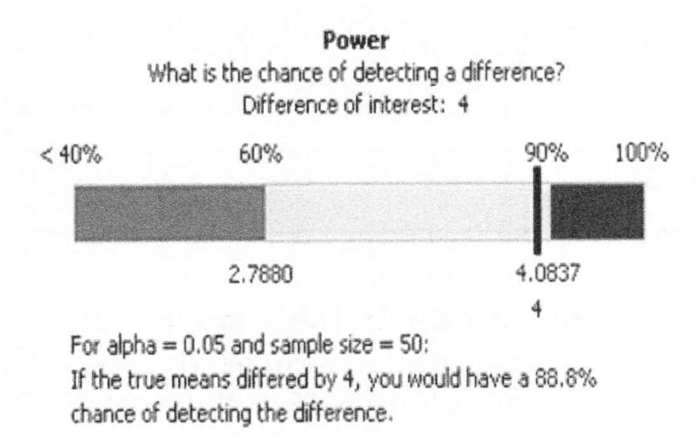

What sample size is required to detect a difference of 4?

Sample size	Power
26	60.0
32	70.0
40	80.0
53	90.0

Your Sample

50	88.8

Power is a function of the sample size and the standard deviation. If the power is not satisfactory, consider increasing the sample size.

On the report card we are told that we did not have any unusual points within the data and that normality is not an issue.

We are given a warning as the power is calculated to be 88.8% with the Assistant.

Paired t Test for the Mean of With Sensors and Without Sens
Report Card

Check	Status	Description
Unusual Data	✓	There are no unusual paired differences. Unusual data can have a strong influence on the results.
Normality	✓	Because your sample size is at least 20, normality is not an issue. The test is accurate with nonnormal data when the sample size is large enough.
Sample Size	⚠	Based on your sample size, standard deviation of the paired differences, and alpha, you only have a 88.8% chance of detecting a difference of 4 between the means. To have a 90% chance of detecting a difference of 4, you need to increase both sample sizes to 53. Some practitioners feel that an 80% chance of detection is sufficient. If this is your case, you can conclude that it is unlikely that there is a difference of 4 or larger.

We have looked at comparing a single population to a mean and we have looked at comparing the means of two dependent populations using the Paired *t* Test. We are now going to move on and look at comparing two independent populations. We will start by comparing the variance of the populations and then move on to comparing the means.

4.8 Two Variance Test

Procedure	• 2 Variance Test
What's it used for?	• The 2 Variance test is used to determine whether the variances or StDevs of two populations are different.
Assumptions and Limitations	• Power and Sample Size for this test was introduced in M16. • There are two procedures used within Minitab. These are the *F* test, for data that is likely to have come from a normally distributed population, and the Levene (Levene/Brown–Forsythe) Test, is for all continuous distributions. • The Assistant uses the Levene test for all distributions. • The recommendation for the Assistant is to have a sample size of at least 20 within the smallest group to ensure the test performs well. • Sample sizes can be unequal.

It is worth spending a bit of time and effort understanding the options available to different Minitab users for this test.

M15 users do not have the option of conducting Power and Sample Size for this test. M15 users have slightly different input menus for the 2 Variances Test. The options for the output graph are fixed but the graph contains the *P* values for the *F* test and Levene's test.

M16 users have the option of conducting Power and Sample Size and have the option of selecting one or more of four output graphs using the Classic Method during the test procedure. The *P* values for the *F* test and Levene's test are presented in the session window.

For both Classic Methods it is recommended the data be checked for normality using the Graphical Summary as M15 and M16 users are always presented with the results of both the *F* test and Levene's test. The user can then use the appropriate *P* value when making the decision regarding equal variances.

The Classic Method for checking two variances can be used in the Classic Method of the 2 Sample *t* Test, which we will be looking at later in this section.

M16 users also have the option of using the Assistant, which will always use Levene's test.

We will now go through the same example using the M16 Classic Method and using the Assistant. M15 users should be able to work through the latter half of the first example as the differences in the menus are minor.

Exercise 5. 2 Variances Test.

Scotty, the maintenance engineer, has repaired worn parts on a feeder system that regulates the amount of additives to a process. The control system on the feeder will try and maintain the same mean but there is always some variation from the feeder. Scotty understands the process metrics and he wants to know if he has changed the variation of the feeder with the work that he has done.

Before conducting the test procedure, help Scotty understand the power that he can expect from the test. He thinks that 75 data points before and after the work will be sufficient. Scotty feels that if the StDev decreased by a ratio (Before/After) of less than 0.75 he would be concerned.

Worksheet: Feeder.

Problem Statement

Establish if the StDev from the feeder increased after maintenance

Null Hypothesis $\frac{StDev(before)}{StDev(After)} = 1$

Alternate Hypothesis $\frac{StDev(before)}{StDev(After)} < 1$

We need to maintain a consistent system of how we arrange our data between the numerator and the denominator. We will put the before data as the numerator and the after data as the denominator. This means that an increase in post maintenance StDev from the feeder will result in ratio of less than 1. As we are only interested in a worsening StDev we can employ a one sided test to increase the power.

The P&SS Test for 2 Variances is available in M16 only.

1. Click Stat <<P&SS << 2 Variances.

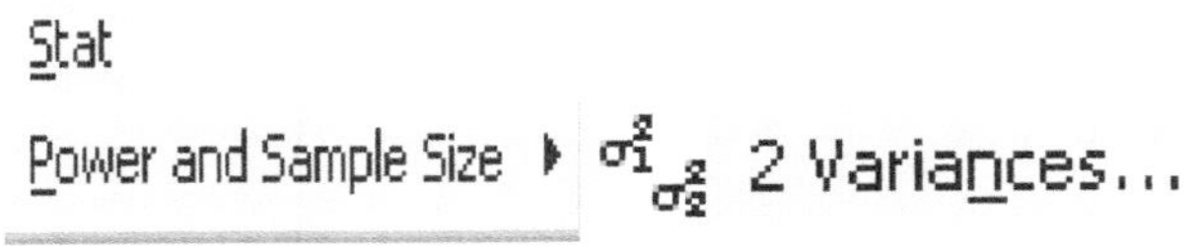

Stat
Power and Sample Size ▸ σ_1^2 / σ_2^2 2 Variances...

2. Complete the menu box as shown. Ensure the uppermost selection box is set to StDevs and not Variances.
3. Then click on Options to set the type of test.

Power and Sample Size for 2 Variances

Enter ratios of standard deviations

Specify values for any two of the following:

Sample sizes: 75

Ratios: 0.75

(StDev 1 / StDev 2)

Power values:

Options... Graph...

Help OK Cancel

4. Set the Alternate Hypothesis to Less Than. Note that the method has been set for Levene's Test.
5. Click OK and OK again.
6. Look at the graph that is produced. We see that we would have a power of 73% with Levene's test. We will repeat the Power and Sample Size procedure but select the F test.

Alternative Hypothesis

Less Than

Not equal

Greater Than

Significance level: 0.05

Method

Levene's Test

F Test

Help

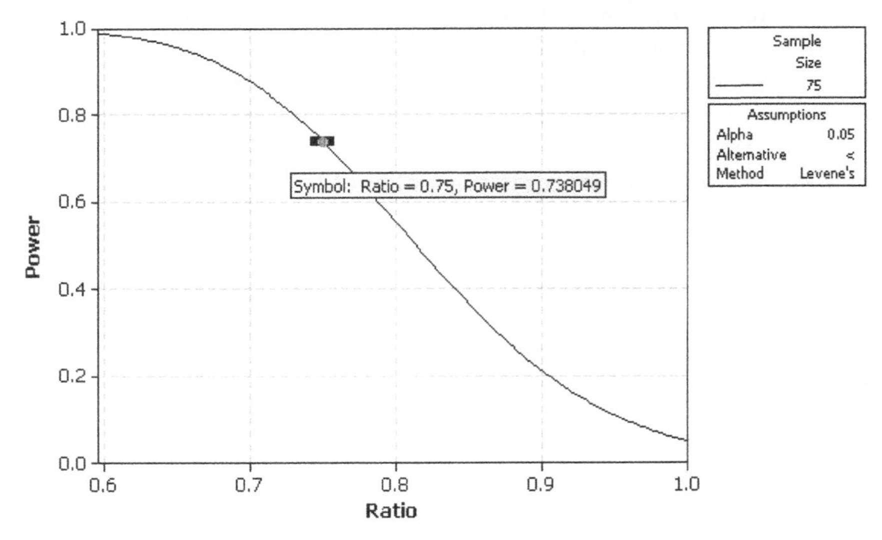

7. Click on the Edit Last icon.
8. Click on Options and set the Method to the F Test.
9. Click OK and OK again.

Method
Levene's Test
F Test

From the graph that was produced we see that we get a higher power if our data was applicable for use with the *F* test.

Usually we would conduct P&SS prior to conducting the test and collecting data. Therefore we would not know whether our data was likely to come from a normally distributed population and we would not know if the *F* test was applicable. However, after collecting the data it is worth checking due to the increased power of the *F* test.

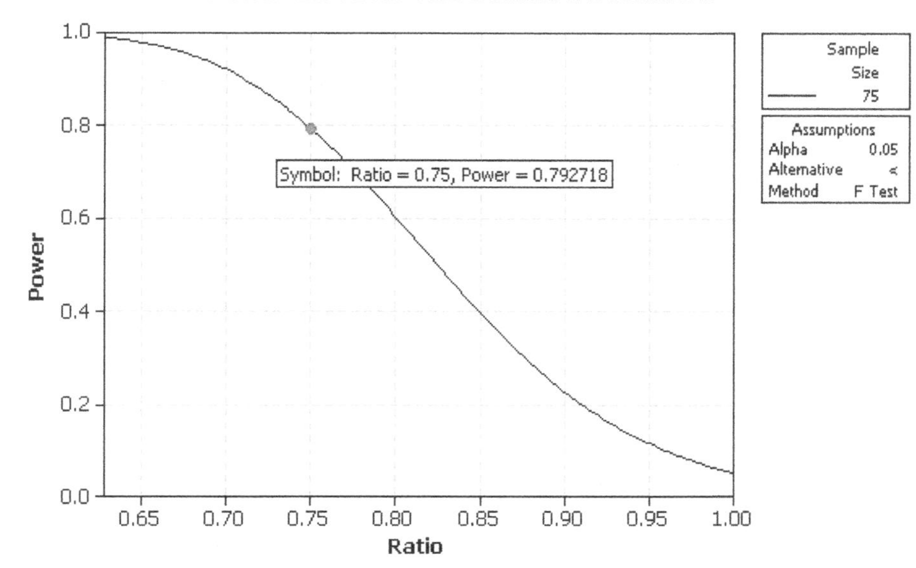

10. Open worksheet Feeder and transfer the data into Minitab. It's not relevant but notice that Scotty has provided more data than originally planned.

C1	C2
Before	After
1020	872
1121	1144
921	1158
951	865

11. Examine the Graphical Summary by selecting Stat <<Basic Statistics <<Graphical Summary. We want to establish if we can apply the F test or whether we have to use Levene's test.

12. Select both columns of data and then click OK.

Graphical Summary

C1 Before
C2 After

Variables:

Before After

It can be seen that both the *P* values are below 0.05 and therefore the samples are not likely to have come from normal distributions. This means that we need to use Levene's test results. Also, note the StDev's. The ratio of the Before to After StDev is 112.86/105.8 = 1.067

Summary for Before

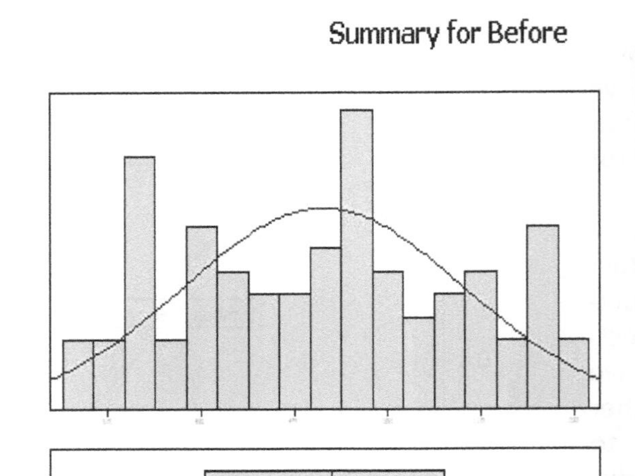

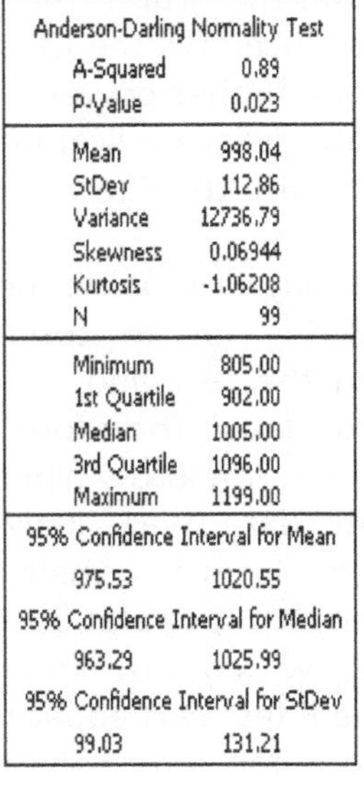

Anderson-Darling Normality Test	
A-Squared	0.89
P-Value	0.023
Mean	998.04
StDev	112.86
Variance	12736.79
Skewness	0.06944
Kurtosis	-1.06208
N	99
Minimum	805.00
1st Quartile	902.00
Median	1005.00
3rd Quartile	1096.00
Maximum	1199.00
95% Confidence Interval for Mean	
975.53	1020.55
95% Confidence Interval for Median	
963.29	1025.99
95% Confidence Interval for StDev	
99.03	131.21

95% Confidence Intervals

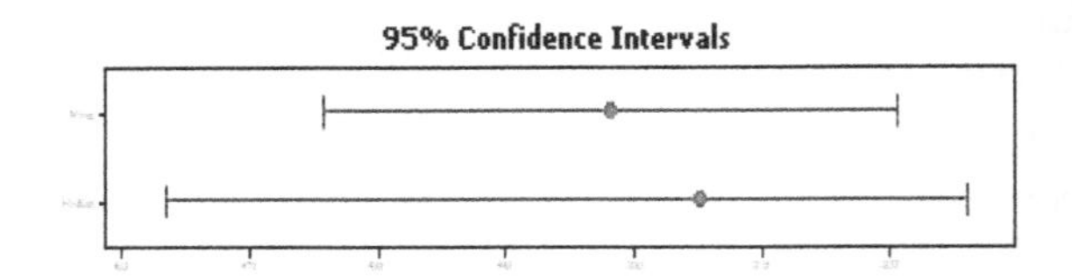

Summary for After

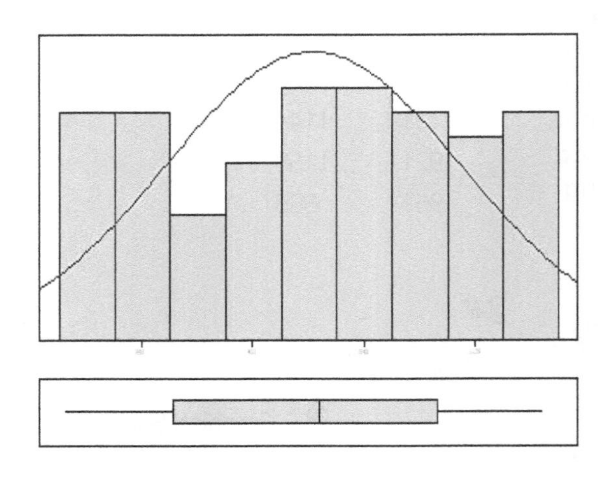

Anderson-Darling Normality Test	
A-Squared	1.10
P-Value	0.007
Mean	1002.9
StDev	105.8
Variance	11194.5
Skewness	-0.13769
Kurtosis	-1.21349
N	76
Minimum	825.0
1st Quartile	902.3
Median	1007.0
3rd Quartile	1092.5
Maximum	1167.0
95% Confidence Interval for Mean	
978.8	1027.1
95% Confidence Interval for Median	
972.8	1051.0
95% Confidence Interval for StDev	
91.2	125.9

95% Confidence Intervals

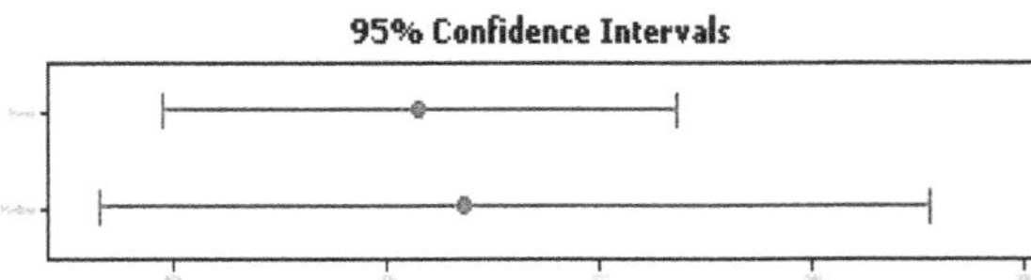

13. Click Stat <<Basic Statistics << 2 Variances.

Stat

Basic Statistics ▸ 2 Variances...

14. Complete the menu box as shown. Ensure the uppermost selection box is set to Samples in different columns. Click on Graphs.

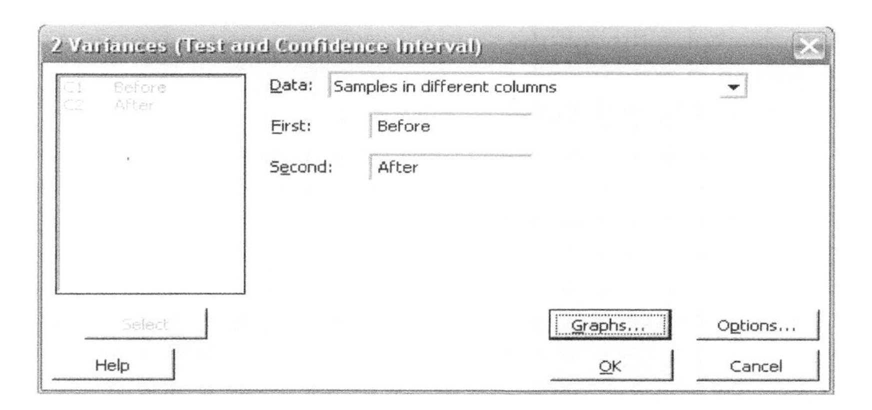

15. M16 allows the selection of any of the graphs shown. M15 will always display two plots on a single page which also displays the *P* values.

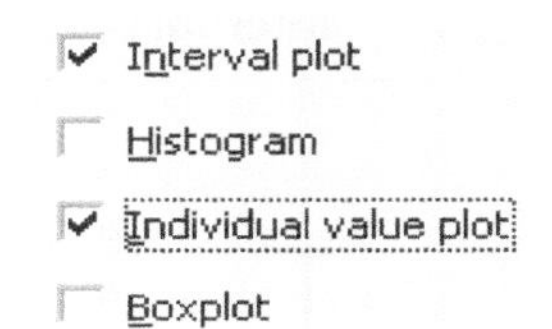

16. Click OK and then click onto Options. Change the Alternative hypothesis to less than. Click OK and OK again.

2 Variances - Options

Confidence level: 95.0

Hypothesized ratio: StDev 1 / StDev 2

Value: 1

Alternative: less than

Help OK Cancel

From the dotplot it can be seen that there is slightly less variation in the After group. It can also be seen that the variation is almost equal on either side of the mean symbol, although the before group appears to be slightly more loaded on the tails.

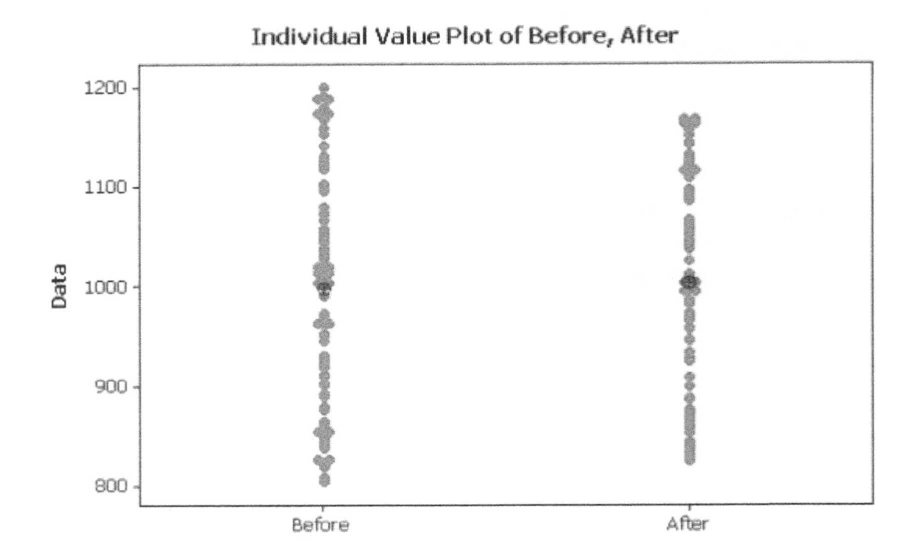

The interval plot shows the 95% CIs for the StDevs of both groups. Note that Minitab has swapped the order of the groups. The interval for the After group does appear to be lower.

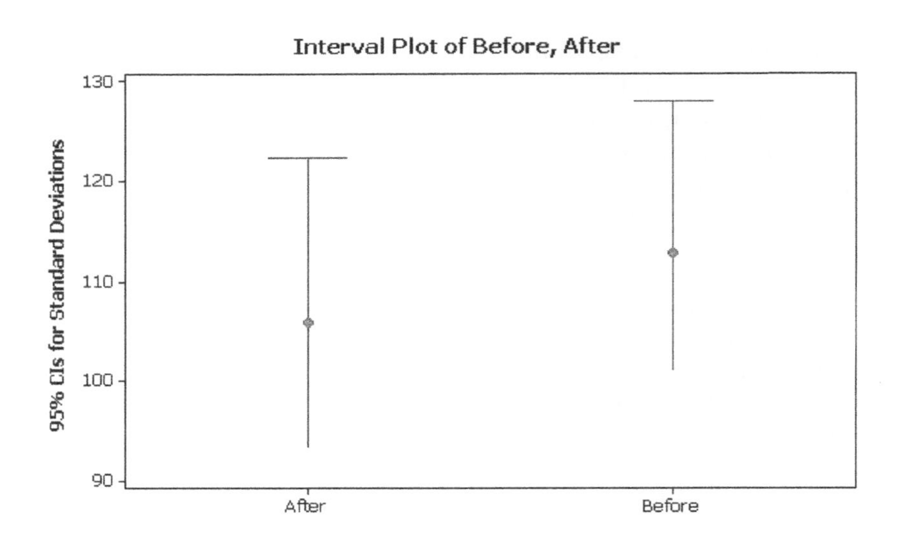

M15 users will be able to see the *P* values from both tests on the output graph. M16 users need to go to the session window.

As our data was not normally distributed we need to use the results of the Levene's test. The *P* value for Levene's test is 0.676 so we cannot reject the null hypothesis. This means that the StDev is not greater after the maintenance.

Scotty can be satisfied that his maintenance work has not adversely affected the feeder variation.

We will now conduct the same procedure using the Assistant.

Test and CI for Two Variances: Before, After

```
Method

Null hypothesis         Sigma(Before) / Sigma(After) = 1
Alternative hypothesis  Sigma(Before) / Sigma(After) < 1
Significance level      Alpha = 0.05


Statistics

Variable   N    StDev   Variance
Before    99  112.857  12736.794
After     76  105.804  11194.489

Ratio of standard deviations = 1.067
Ratio of variances = 1.138


95% One-Sided Confidence Intervals

               Upper Bound   Upper Bound
Distribution    for StDev   for Variance
of Data             Ratio          Ratio
Normal              1.273          1.621
Continuous          1.221          1.490


Tests

                                                Test
Method                            DF1  DF2  Statistic  P-Value
F Test (normal)                    98   75       1.14    0.720
Levene's Test (any continuous)      1  173       0.21    0.676
```

1. Click on Assistant <<Hypothesis Tests.
2. Click on 2-Sample Standard Deviation within the 'Compare two samples with each other' group.
3. Complete the menu box as shown and click OK. Ensure the Test Setup is done correctly. We want to test if the StDev of before is less than After. With a bit of simple maths we can work out that we need to check the power for a 25% difference.

Assistant

Hypothesis Tests...

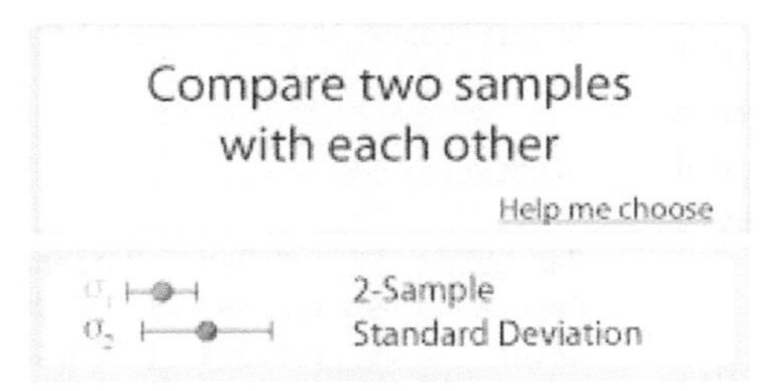

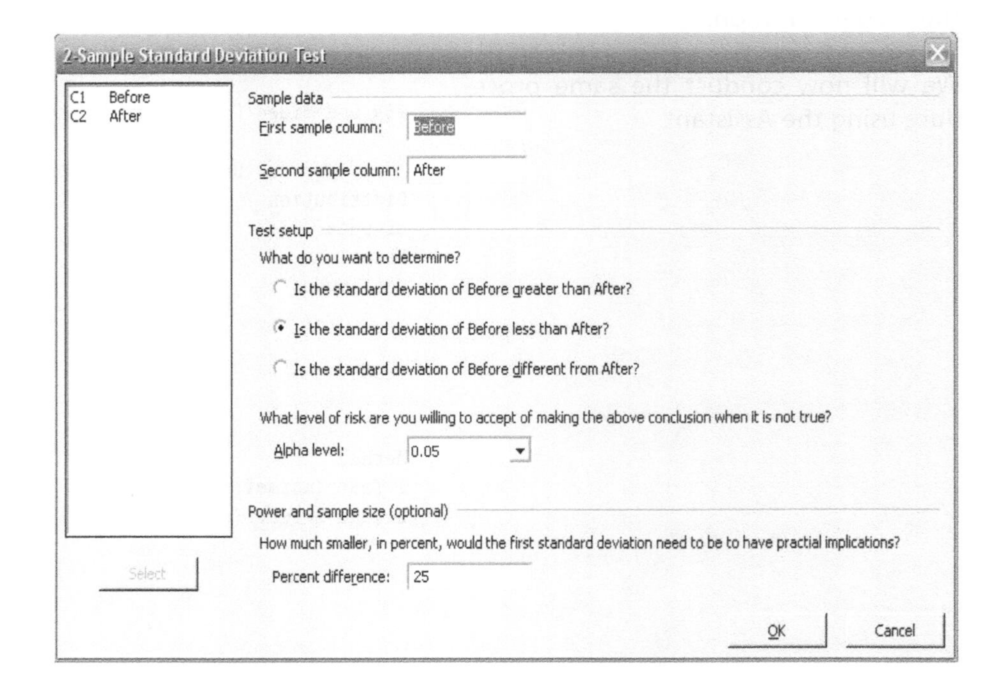

The test delivers a P value = 0.676, which is the same as the results for Levene's test we calculated using the Classic Method.

We are given an additional chart which shows the CIs for StDev. We can see that they clearly overlap indicating that there isn't a difference.

The histogram shows the spread of the data. Again we can see that After data actually has less spread.

In the comments we are told that we cannot reject the null hypothesis.

2-Sample Standard Deviation Test for Before and After
Summary Report

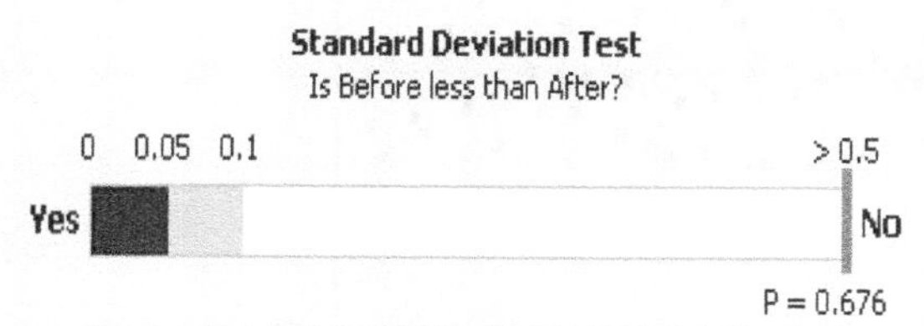

The standard deviation of Before is not significantly less than After (p > 0.05).

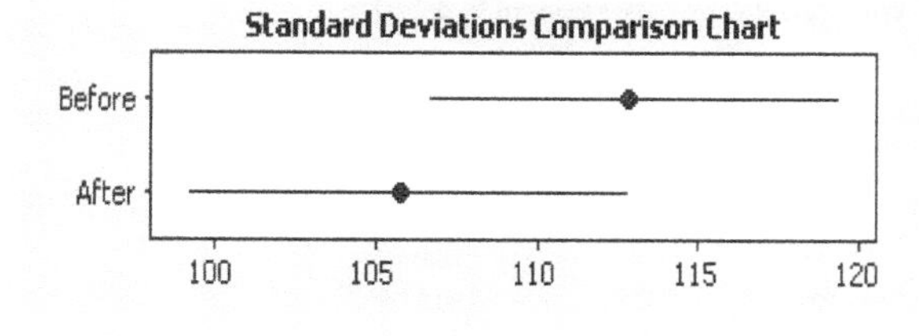

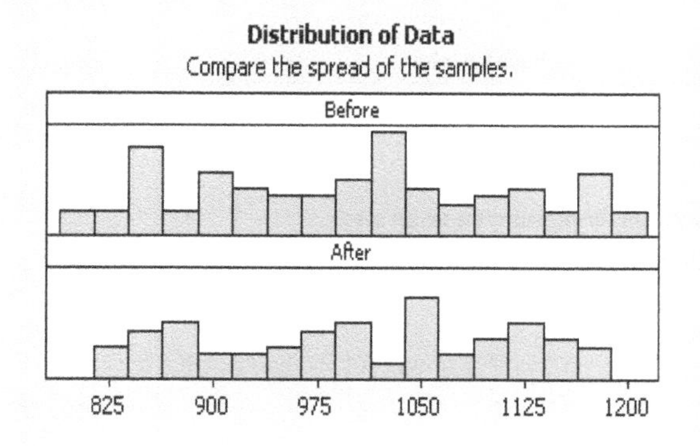

Statistics	Before	After
Sample size	99	76
Mean	998.04	1002.9
Standard deviation	112.86	105.80
90% CI	(104.8, 123.6)	(97.91, 116.9)

Comments

-- Test: There is not enough evidence to conclude that the standard deviation of Before is less than After at the 0.05 level of significance.
-- Comparison chart: Red intervals indicate that the standard deviations differ. Consider the size of the difference to determine if it has practical implications.
-- Distribution of Data: Compare the spread of the samples. Look for unusual data before interpreting the results of the test.

The SPC charts show that none of our points are deemed to be unusual because none of the points are marked in red.

The power reported for the test is 79.3% using the actual sample sizes of 99 and 76.

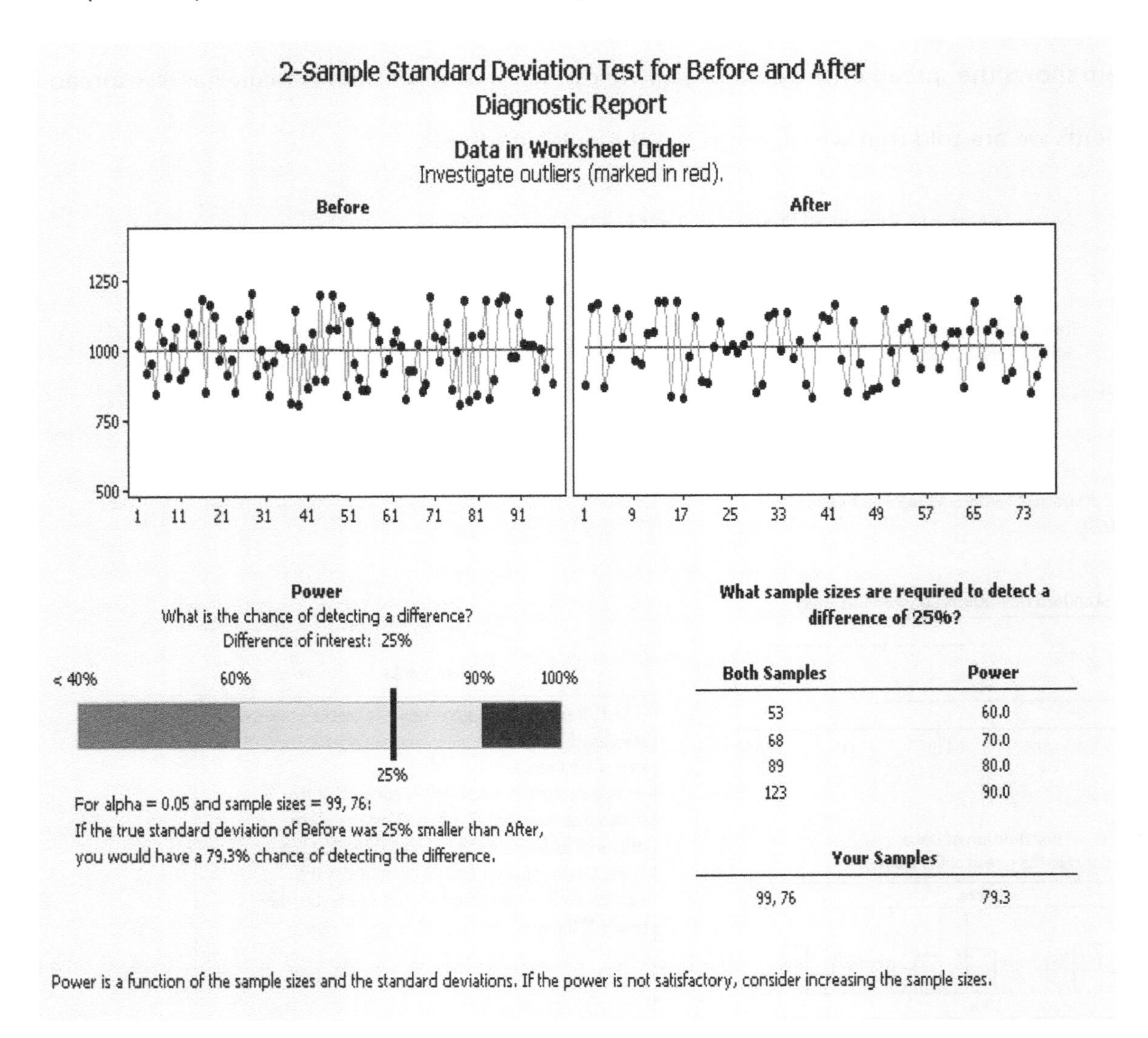

On the report card we are told that we did not have any unusual points within the data and that normality is not an issue.

We are given a warning regarding the power of the test. We would need both sample sizes to be 123 to obtain a power of 90% or 89 to get a power of 80%.

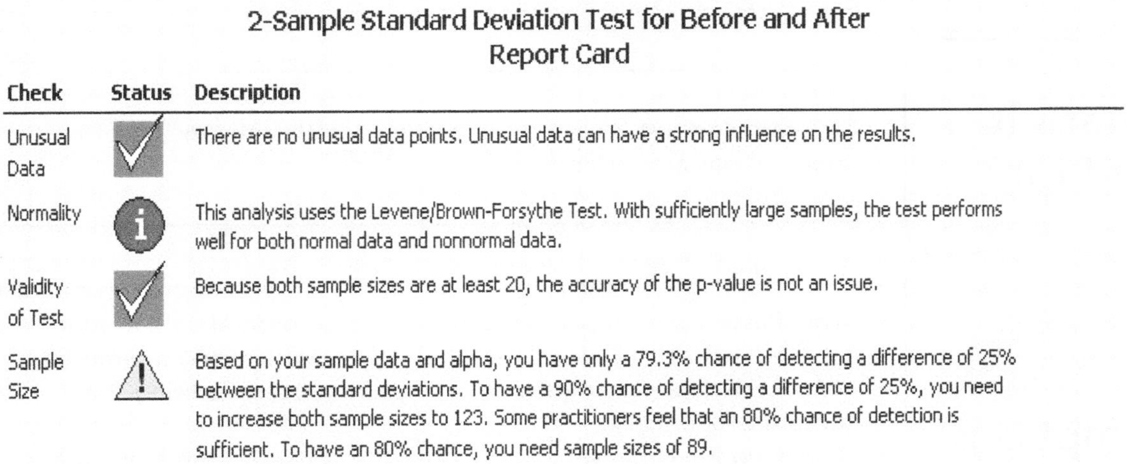

2-Sample Standard Deviation Test for Before and After
Report Card

Check	Status	Description
Unusual Data	✓	There are no unusual data points. Unusual data can have a strong influence on the results.
Normality	i	This analysis uses the Levene/Brown-Forsythe Test. With sufficiently large samples, the test performs well for both normal data and nonnormal data.
Validity of Test	✓	Because both sample sizes are at least 20, the accuracy of the p-value is not an issue.
Sample Size	⚠	Based on your sample data and alpha, you have only a 79.3% chance of detecting a difference of 25% between the standard deviations. To have a 90% chance of detecting a difference of 25%, you need to increase both sample sizes to 123. Some practitioners feel that an 80% chance of detection is sufficient. To have an 80% chance, you need sample sizes of 89.

The last procedure that we are going to look at in this chapter is the 2 Sample *t* Test. This is used to compare the mean of two populations. As such it is widely used to understand the effects of changes on processes.

We will go straight into looking at worked examples to learn the test procedure. When using the Classic Method we need to know if our two groups have equal variances prior to conducting the 2 Sample *t* Test. Again, M16 users can also use the Assistant for the 2 Sample *t* Test.

4.9 Two Sample *t* Test

Procedure	• 2 Sample *t* Test
What's it used for?	• A 2 sample *t* test is used to determine whether the means of two populations are different.
Assumptions and Limitations	• The Classic Method performs well with non normal data as long as the data is continuous, reasonably symmetrical and uni modal. We check these assumptions prior to the procedure using the graphical summary. • Checking for equal variances within the Classic Method increases test reliability. The Graphical Summaries can also be used to make conclusions as to whether the populations have equal variances. • The method used by the Assistant does not require the user to test for equal variances prior to the test. • For both methods sample size is important for test validation and power.

Example 6. 2 Sample *t* test.

Two CPU manufacturers both make chips that are rated at 2200 MHz. Overclocking is running the CPU at higher speed than its rated speed in order to get more performance from the computer. A sample of CPUs is taken from each manufacturer. Each chip is overclocked to run at its highest possible speed. The maximum speed and manufacturer for each chip is recorded.

The historical StDev is known to be 20 and 50 samples are to be tested from each manufacturer. If we want a power of 90% from the 2 sample *t* test what is the maximum difference that can be detected?

Using the data provided in Worksheet: CPU Speed, establish if there is a difference between the chips in terms of overclocking to the highest speed. Then establish which manufacturer's chips can be over-clocked to the highest speed.

Problem Statement

Establish which chips can be overclocked to the highest speeds.

Null Hypothesis

Pintel Speed = AMB Speed

Alternate Hypothesis

Pintel Speed ≠ AMB Speed

(This will be a two sided test. If we detect a difference we will use the graphical outputs to establish which chips are faster.)

1. Click Stat <<P&SS <<2-Sample t.

Stat

Power and Sample Size ▸ 2t 2-Sample t...

2. Complete the menu box as shown. There are a couple of points to note here. The sample size listed should be for the smallest group as this will give us a safety margin. The estimate for StDev should be for the pooled value.

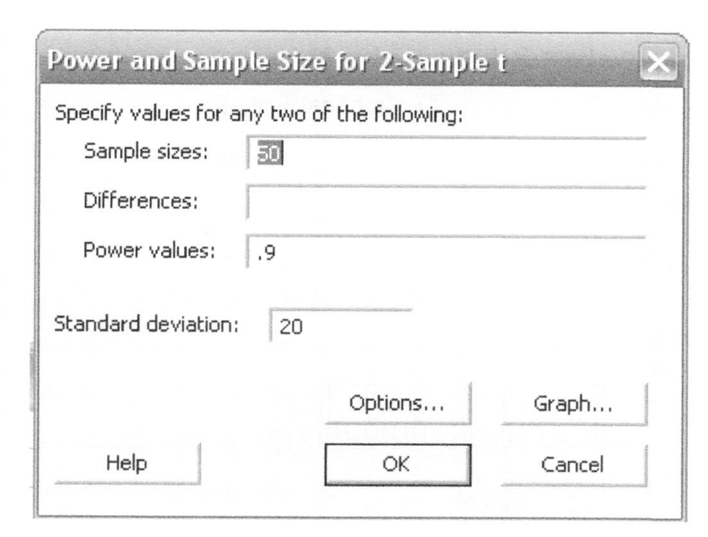

3. Then click OK.

It can be seen that under the test conditions stated we would be able to detect a difference of about 13 with a power of 90%.

This is acceptable so we can continue with the test procedure.

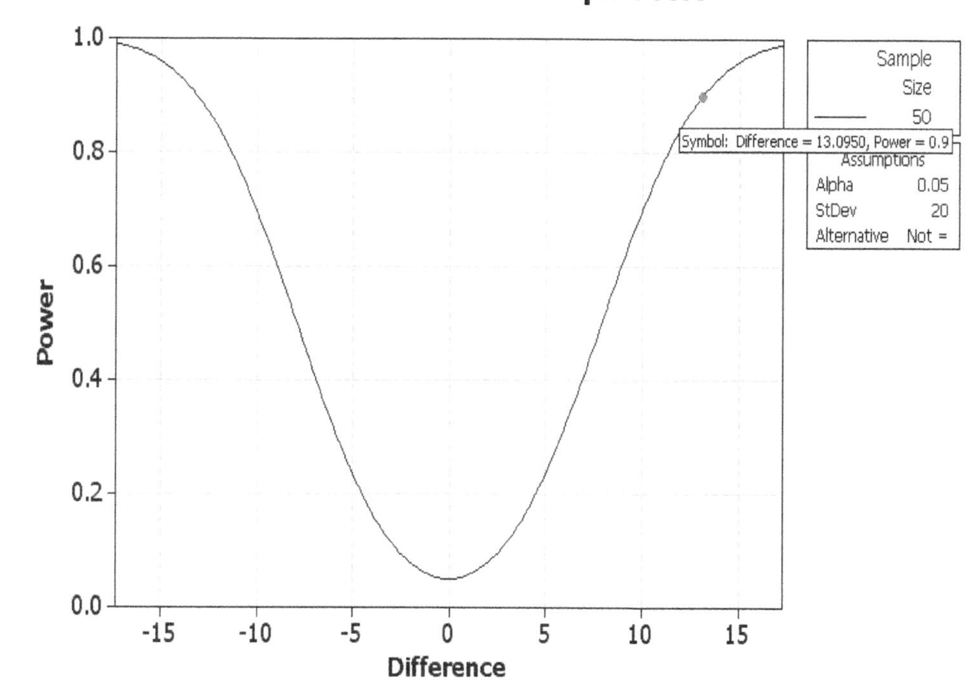

4. Open worksheet CPU Speed and transfer the data into Minitab.

C1	C2
Pintel	**AMB**
2362.15	2370.55
2318.35	2356.94
2397.74	2405.93
2351.03	2401.80

5. We need to check the underlying assumptions about the data by using the Graphical summary. It is also a good idea to get a feel for the data by checking the Graphical summary before the test procedure is started. Click Stat <<Basic Statistics <<Graphical Summary.

Stat
Basic Statistics
Graphical Summary...

6. Select Pintel and AMB. Then click OK.

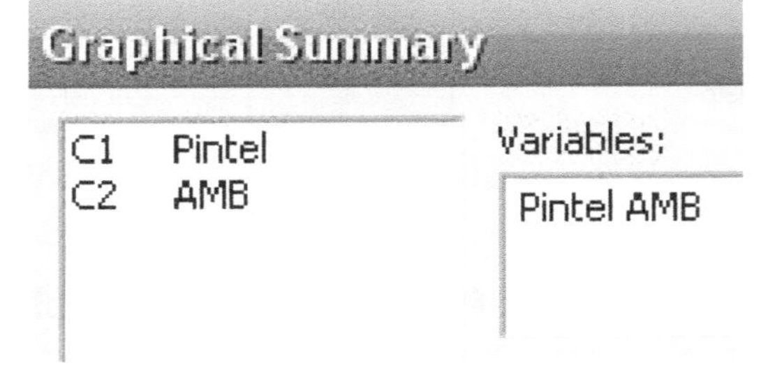

From the graphical summaries it can be seen that our data is reasonably symmetrical and unimodal. The *P* values indicate normality. The StDevs are 21.9 and 18.2 and the 95% CIs for StDevs overlap. Using our knowledge and experience we could assume equal variances with this data. However we could also use the 2 Variances Test, the session window results are shown below.

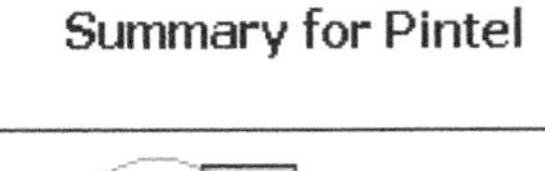

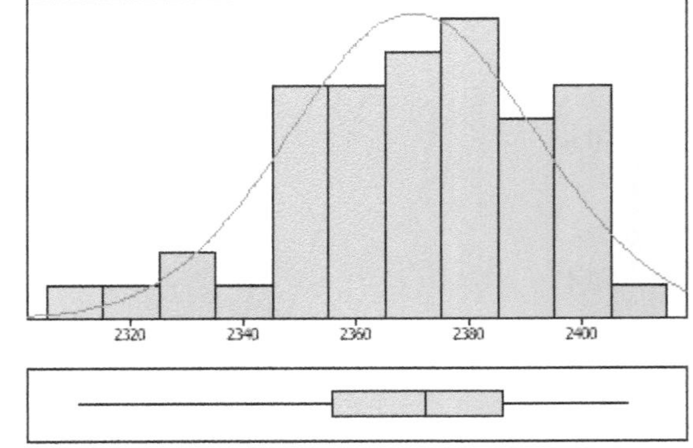

Anderson-Darling Normality Test	
A-Squared	0.31
P-Value	0.533
Mean	2369.9
StDev	21.9
Variance	478.5
Skewness	-0.536866
Kurtosis	0.197266
N	50
Minimum	2310.7
1st Quartile	2355.6
Median	2372.2
3rd Quartile	2385.7
Maximum	2407.8
95% Confidence Interval for Mean	
2363.7	2376.1
95% Confidence Interval for Median	
2362.9	2378.4
95% Confidence Interval for StDev	
18.3	27.3

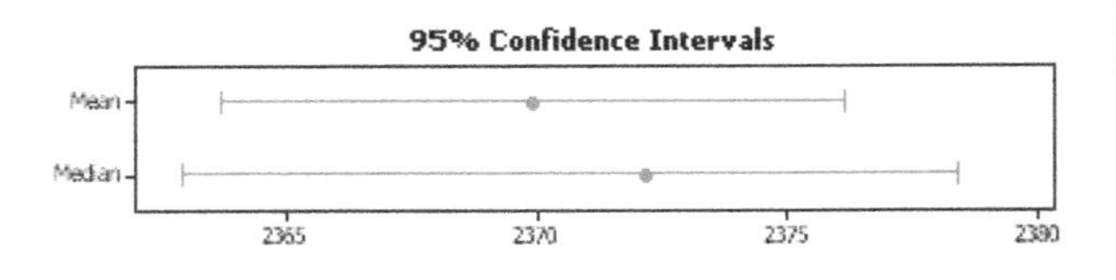

Summary for AMB

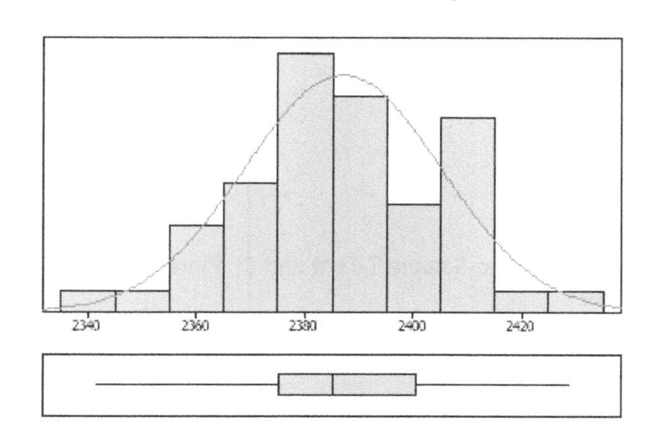

95% Confidence Intervals

Mean

Median

2382 2384 2386 2388 2390 2392

Anderson-Darling Normality Test	
A-Squared	0.22
P-Value	0.812
Mean	2387.0
StDev	18.2
Variance	330.6
Skewness	-0.148454
Kurtosis	0.071037
N	50
Minimum	2341.5
1st Quartile	2375.4
Median	2385.3
3rd Quartile	2400.7
Maximum	2428.6
95% Confidence Interval for Mean	
2381.8	2392.2
95% Confidence Interval for Median	
2382.0	2392.0
95% Confidence Interval for StDev	
15.2	22.7

Method	DF1	DF2	Test Statistic	P-Value
F Test (normal)	49	49	1.45	0.199
Levene's Test (any continuous)	1	98	1.60	0.209

Now that we know the assumptions are valid regarding the sample data and we can assume equal variances let's conduct the actual test.

Stat

Basic Statistics ▸

2t 2-Sample t...

7. Click Stat <<Basic Statistics << 2-Sample t.

8. Complete the menu box as shown. Make sure Assume equal variances is ticked.

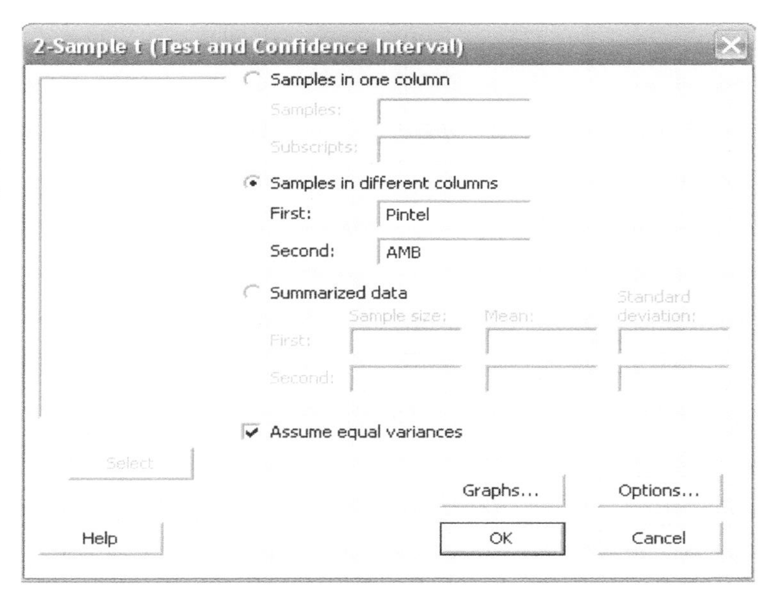

9. Then click on Graphs and select Boxplots.
10. Click OK and OK again.

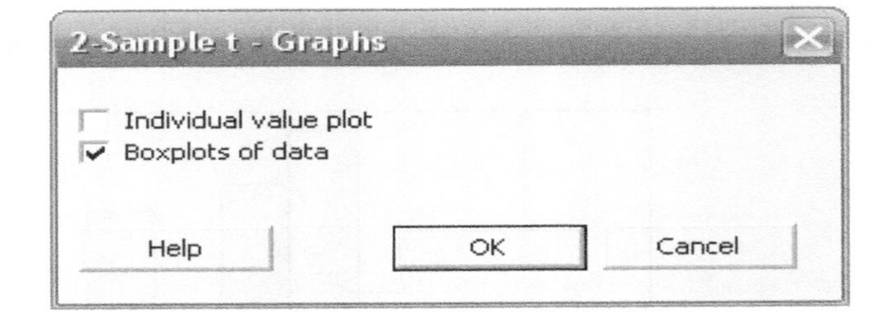

From the test results in the session window we can see that the *P* value is zero and the mean speed for AMB chips is faster. Therefore, we can say AMB chips can be overclocked to higher speeds than Pintel chips. The estimate for the difference was actually 17.

The boxplot confirms that the AMB chips were faster than Pintel. However, we would not know from the boxplot alone if the difference was significant.

We will now conduct the equivalent procedure with the Assistant.

Two-Sample T-Test and CI: Pintel, AMB

```
Two-sample T for Pintel vs AMB

         N    Mean  StDev  SE Mean
Pintel  50  2369.9   21.9      3.1
AMB     50  2387.0   18.2      2.6


Difference = mu (Pintel) - mu (AMB)
Estimate for difference:  -17.09
95% CI for difference:  (-25.07, -9.11)
T-Test of difference = 0 (vs not =): T-Value = -4.25  P-Value = 0.000  DF = 98
Both use Pooled StDev = 20.1144
```

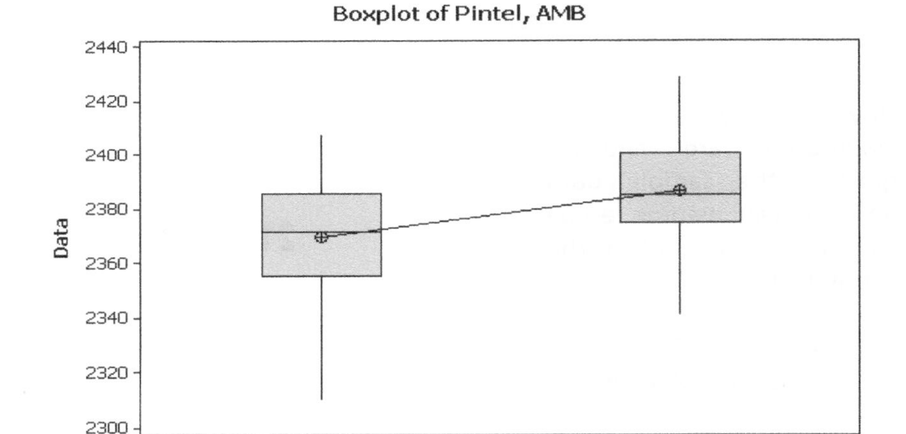

1. Click on Assistant <<Hypothesis Tests.
2. Click on 2-Sample t within the 'Compare two samples with each other' group.
3. For Test Setup we want to check if the mean of Pintel is different to AMB. For P&SS we will enter a Difference of 13. Complete the menu box as shown and click OK.

Assistant

Hypothesis Tests...

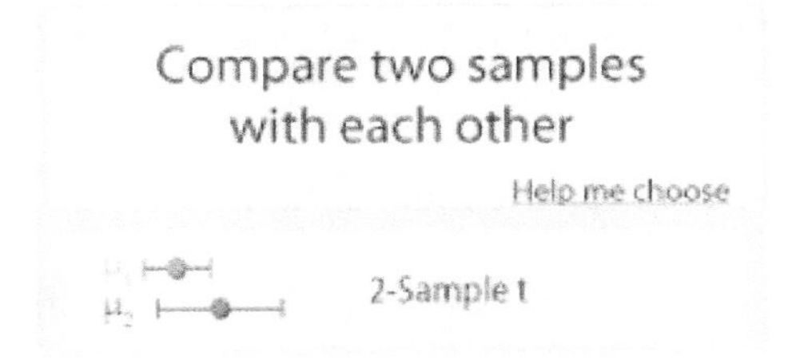

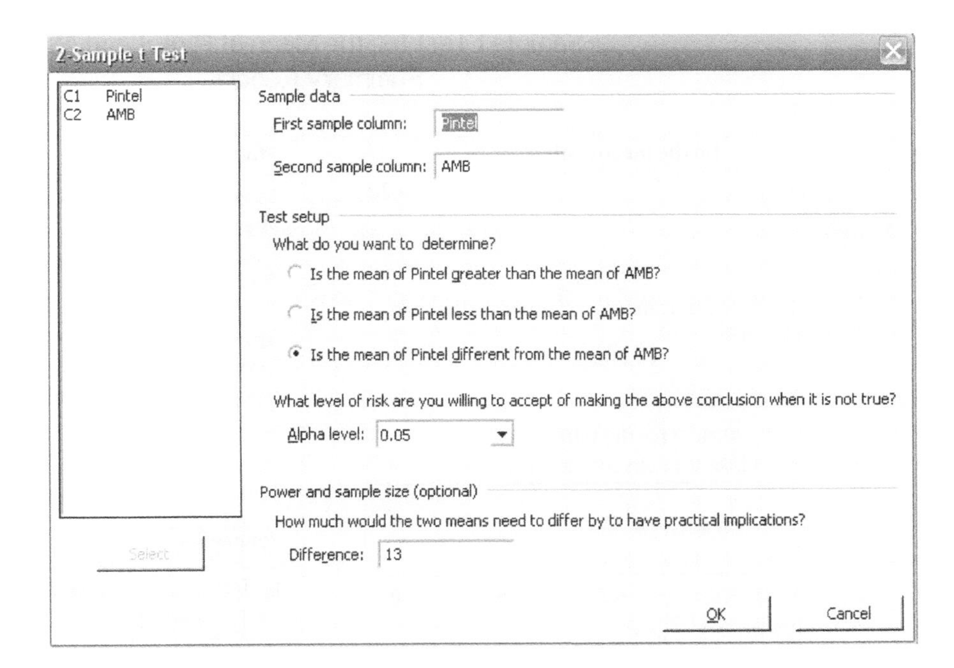

The test delivers a P value = 0.0, telling us that the populations have different means.

We are given an additional chart which shows the 95% CI for the difference. If the interval had gone through the zero point we would have had to conclude that there was not a difference between the means of the populations. The difference is negative only due to the order that we entered the columns into the input menu box. Under Statistics in the top left we are told that 'The difference is defined as Pintel–AMB'.

Within the histograms we see the 95% I for the means. Again they do not overlap.

In the comments we are told that we can conclude that the means differ at the 5% significance level.

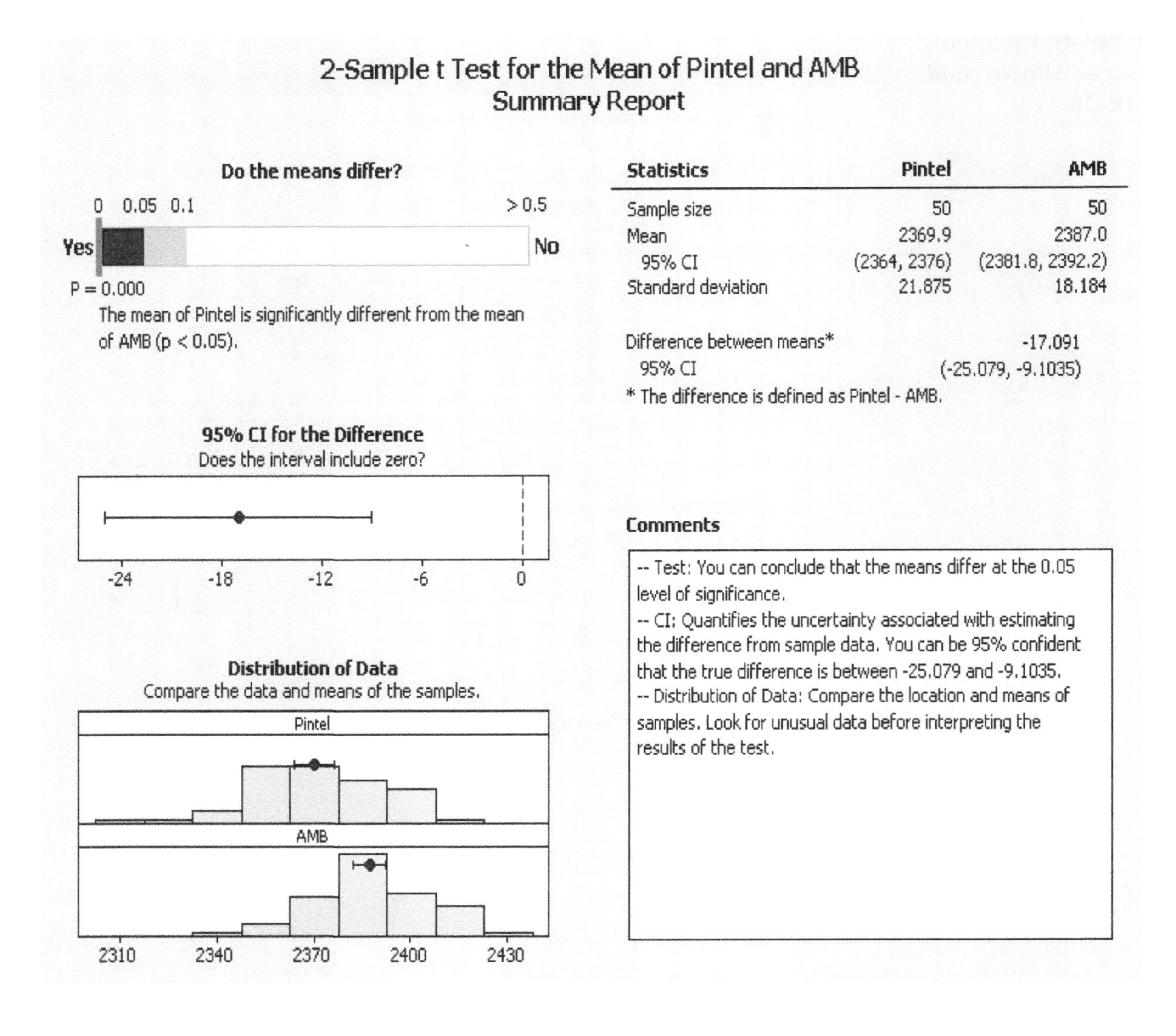

The SPC charts show that none of our points are deemed to be unusual because none of the points are marked in red.

The power reported using the pooled StDev for the sample data was 89.2% in order to be able to detect a difference of 13.

2-Sample t Test for the Mean of Pintel and AMB
Diagnostic Report

Data in Worksheet Order
Investigate outliers (marked in red).

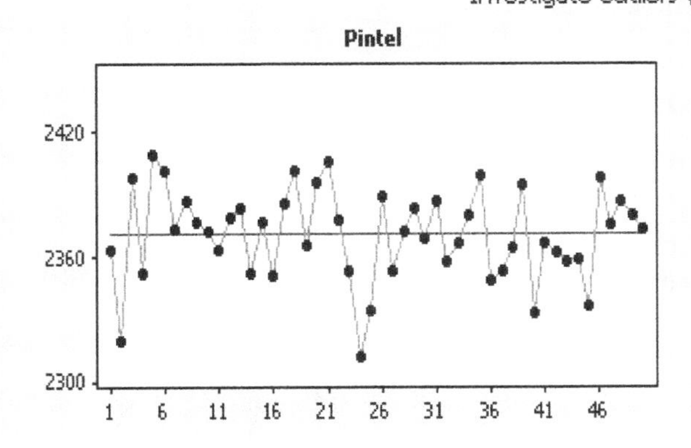

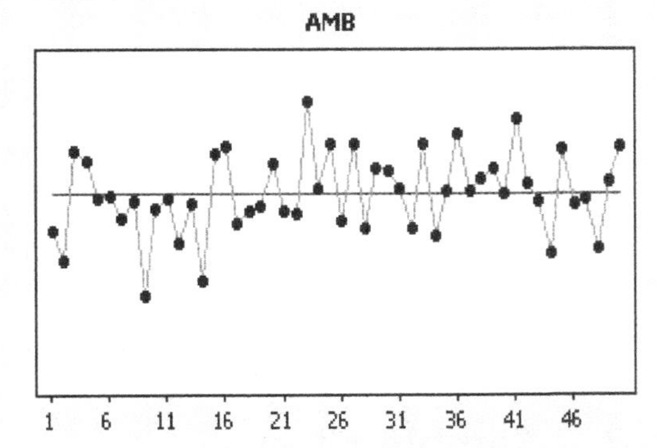

Power
What is the chance of detecting a difference?
Difference of interest: 13

< 40% 60% 90% 100%

8.9950 13.174
13

For alpha = 0.05 and sample sizes = 50:
If the true means differed by 13, you would have a 89.2% chance of detecting the difference.

What sample sizes are required to detect a difference of 13?

Both Samples	Power
25	60.0
31	70.0
39	80.0
52	90.0

Your Samples	
50, 50	89.2

Power is a function of the sample sizes and the standard deviations. If the power is not satisfactory, consider increasing the sample sizes.

On the Report Card we are told that we did not have any unusual points within the data and that normality is not an issue and we had sufficient data points.

We are given information stating that the test method used within the Assistant does not require equal variances to be confirmed.

2-Sample t Test for the Mean of Pintel and AMB
Report Card

Check	Status	Description
Unusual Data	✓	There are no unusual data points. Unusual data can have a strong influence on the results.
Normality	✓	Because both sample sizes are at least 15, normality is not an issue. The test is accurate with nonnormal data when the sample sizes are large enough.
Sample Size	✓	The sample is sufficient to detect a difference between the means.
Equal Variance	i	The 2-sample t used by Minitab's Assistant does not assume or require that the two samples have equal variances. Research shows that the test performs well with unequal variances, even when the sample sizes are not equal.

Example 7. 2 Sample *t* Test.

A city has two driving test centres. They both conduct a large number of tests every month and the pass rate is recorded for each of the sites every month.

It has already been established that to detect a difference of 2% with the pooled standard deviation of 3%, then 49 samples are required to achieve a power of 90%.

Use the data provided to establish if the mean pass rates from the two centres are different. Use the Classic Method and the Assistant to carry out the analysis.

Worksheet: Test Centre.

1. Open worksheet Test Centre and transfer the data into Minitab.

C1	C2
SiteA	SiteB
86.27	82.60
88.90	85.32
87.38	85.98
87.92	83.32

2. We need to check the underlying assumptions about the data by using the Graphical summary. Click Stat <<Basic Statistics <<Graphical Summary.

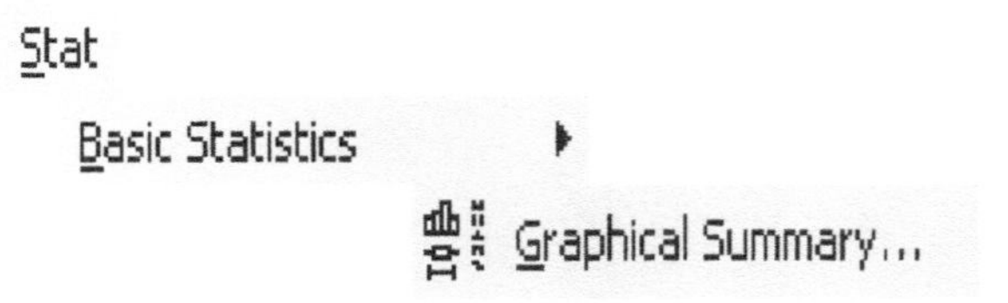

3. Select SiteA and SiteB. Then click OK.

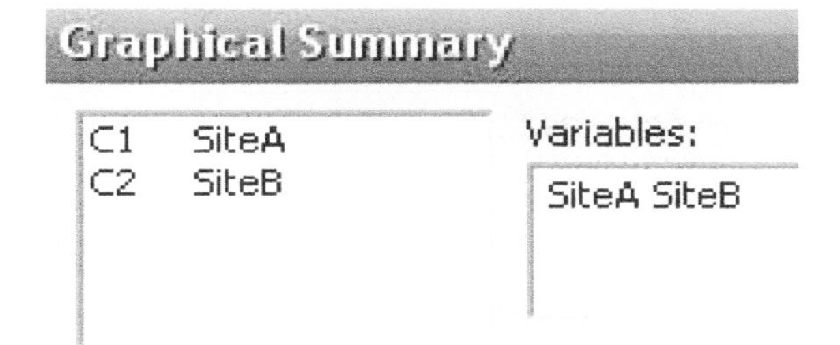

The distribution for SiteA is normal, contains continuous data and is uni modal. Unfortunately, SiteB has a bi modal distribution. This means that the test results will be unreliable and we should not proceed. We could take this as an opportunity to explore why there is a bi modal distribution and see if this can increase our understanding of how SiteB is working. The problem solver within you should feel some excitement at this opportunity. Let's see how the Assistant gets on with this problem.

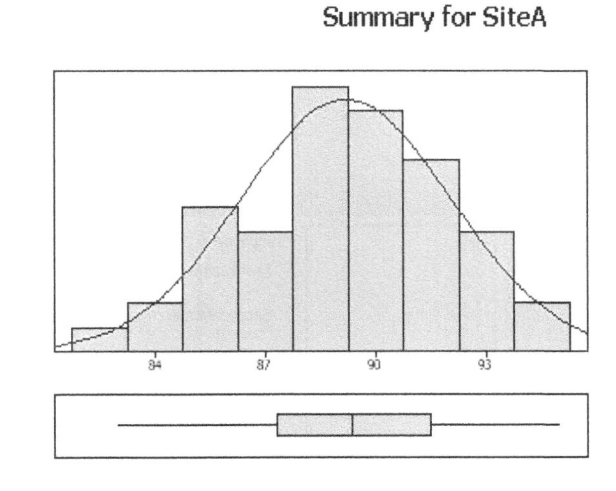

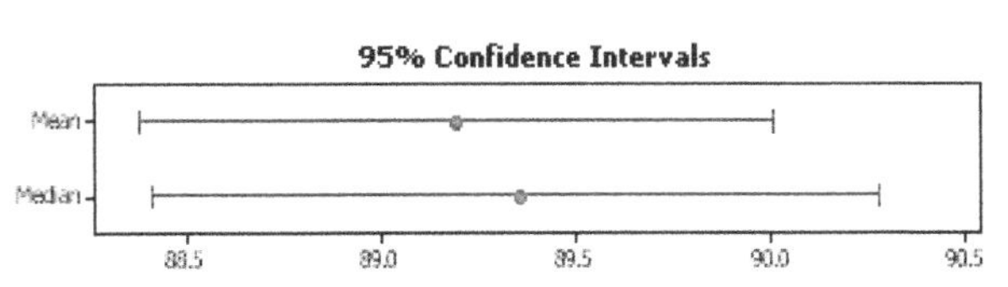

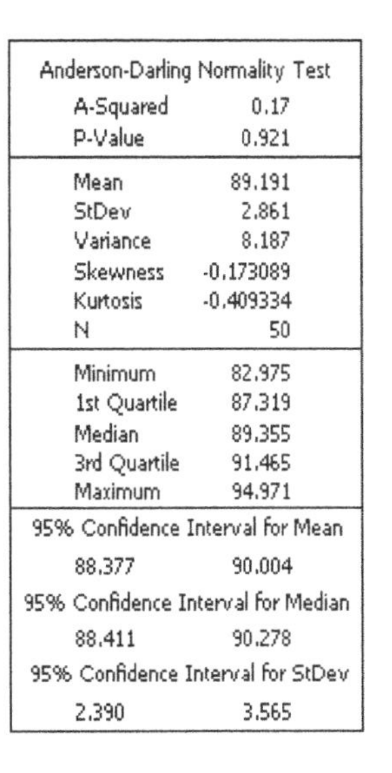

Anderson-Darling Normality Test	
A-Squared	0.17
P-Value	0.921
Mean	89.191
StDev	2.861
Variance	8.187
Skewness	-0.173089
Kurtosis	-0.409334
N	50
Minimum	82.975
1st Quartile	87.319
Median	89.355
3rd Quartile	91.465
Maximum	94.971
95% Confidence Interval for Mean	
88.377	90.004
95% Confidence Interval for Median	
88.411	90.278
95% Confidence Interval for StDev	
2.390	3.565

Summary for SiteB

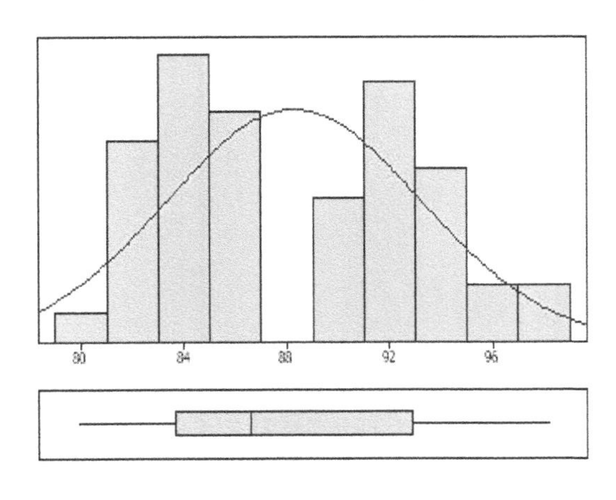

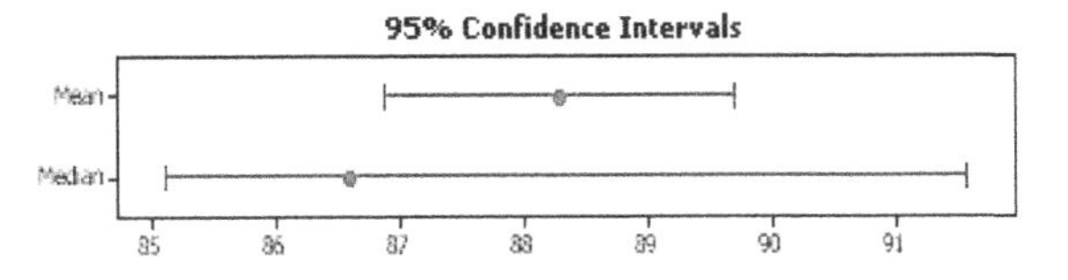

Anderson-Darling Normality Test	
A-Squared	1.57
P-Value <	0.005
Mean	88.282
StDev	4.952
Variance	24.520
Skewness	0.25958
Kurtosis	-1.26046
N	50
Minimum	79.948
1st Quartile	83.667
Median	86.585
3rd Quartile	92.860
Maximum	98.203
95% Confidence Interval for Mean	
86.875	89.690
95% Confidence Interval for Median	
85.108	91.572
95% Confidence Interval for StDev	
4.136	6.171

1. Click on Assistant <<Hypothesis Tests.
2. Click on 2-Sample t within the 'Compare two samples with each other' group.
3. Complete the menu box as shown and click OK. For P&SS we have entered a Difference of 3.

Assistant
Hypothesis Tests...

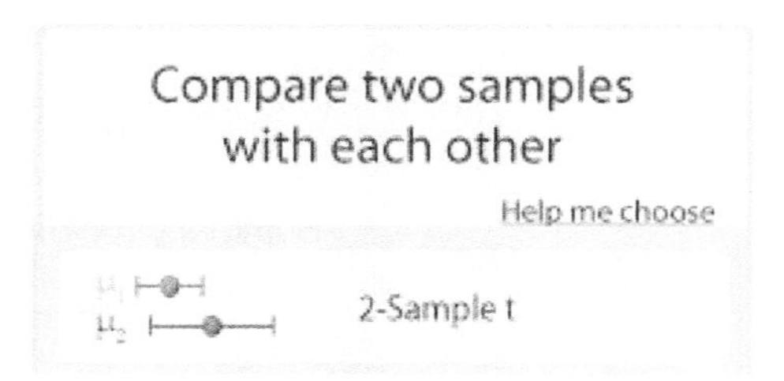

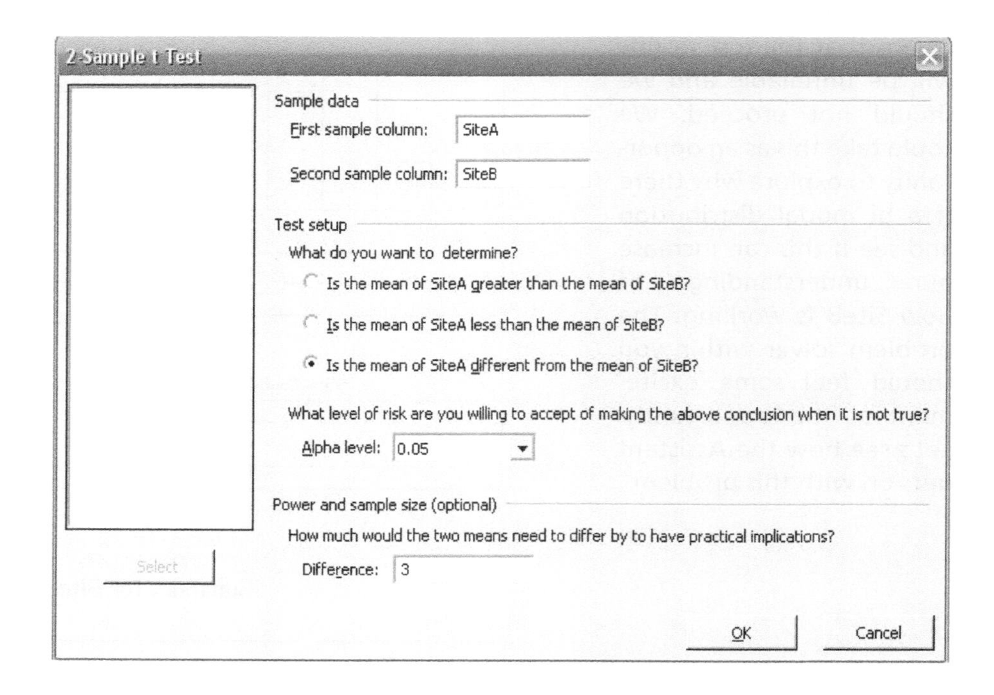

The Assistant has carried out the full test and reported that we cannot reject the null hypothesis due to a *P* value of 0.265.

As we did using the Classic Method it would be up to the user to recognise that the data did not meet the test criteria by looking at the histograms. On this occasion the Assistant does not warn us regarding the betrayal of our data.

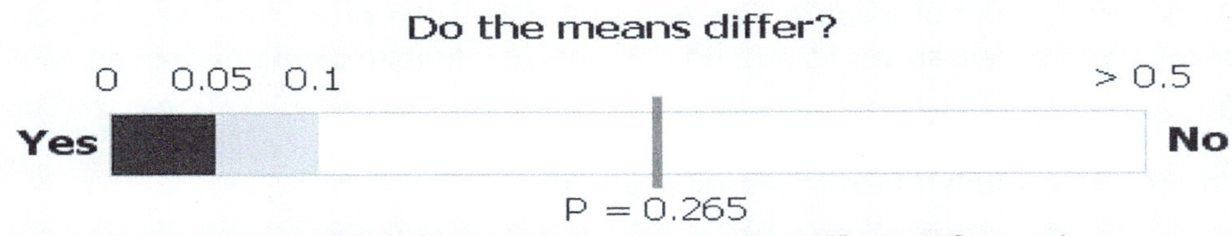

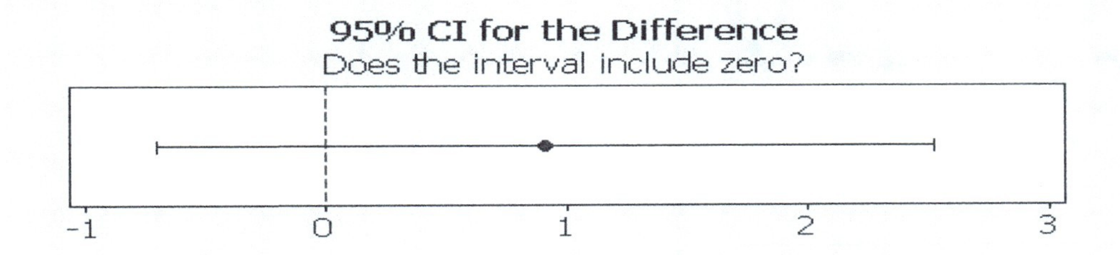

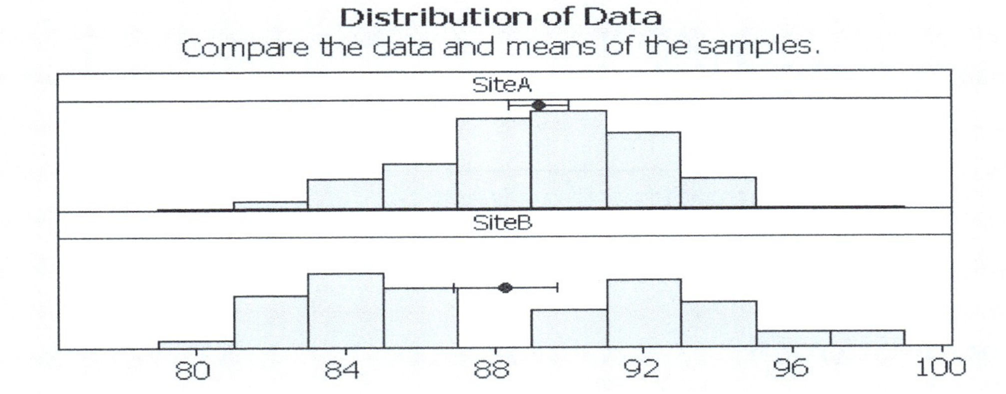

The Assistant does provide further investigative information in the form of the SPC charts. On this occasion the chart is not highlighting any unusual data points but having been presented with a bi modal distribution it would be the next logical step in the investigation to use SPC charts.

Even the report card does not show any warnings.

This example has not been used to underline weaknesses of the Assistant. It has been used to remind the problem solver that diligence is required when assessing the test data. With the Classic Method there are more checks as you work through the procedure. The Assistant makes the procedure very easy and this might make it harder for the problem solver to remain diligent. Remember, problems missed could actually be opportunities.

To finish off the chapter there are a couple of exercises for you to do.

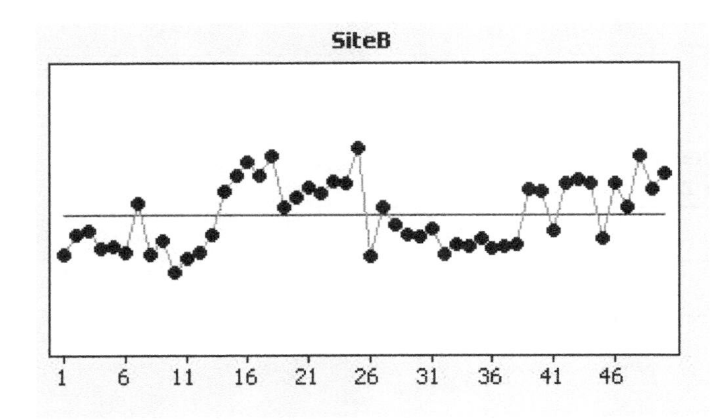

2-Sample t Test for the Mean of SiteA and SiteB
Report Card

Check	Status	Description
Unusual Data	✓	There are no unusual data points. Unusual data can have a strong influence on the results.
Normality	✓	Because both sample sizes are at least 15, normality is not an issue. The test is accurate with nonnormal data when the sample sizes are large enough.
Sample Size	✓	Although the test results are not significant, the power is adequate. Based on your sample sizes, standard deviations, and alpha, you have a 95.6% chance of detecting a difference of 3 between the means. Because the power is adequate, you can conclude that it is unlikely that there is a difference of 3 or larger.
Equal Variance	i	The 2-sample t used by Minitab's Assistant does not assume or require that the two samples have equal variances. Research shows that the test performs well with unequal variances, even when the sample sizes are not equal.

Exercise 3. 2 Sample *t* Test.

Hawk, the process engineer, works on a plant where they manufacture plasterboard. Hawk is involved with improving dryer efficiency. He believes that by adjusting inlet pressure he can improve the dryer efficiency. He intends to run the inlet pressure at two levels and then compare the dryer efficiency at those two levels. The levels are the normal setting of 10 mbar and the proposed new setting of 8 mbar.

If he wants to be able to detect a difference of 15 kWh/te and if the pooled StDev is 17, what power can be achieved with 30 samples in each group?

Calculate the power and then establish if the efficiency has changed between the two levels of inlet pressure used. Use both the Classic Method and the Assistant.

Worksheet: Efficiency.

A power of about 92% can be achieved.

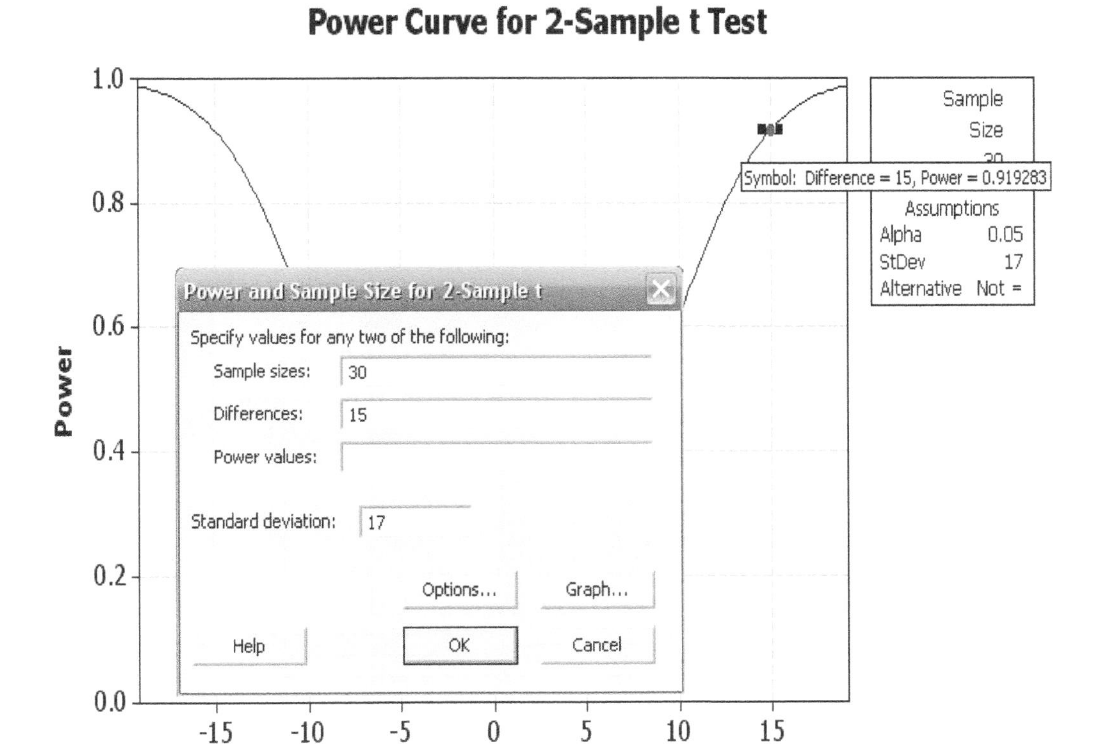

The data is reasonably symmetrical and unimodal and the *P* values indicate normality. The StDevs are 11 and 21 and the 95% CIs for StDevs do not overlap, indicating unequal variances. The 2 Variances Test confirms unequal variances (results shown below), which means that the tick box for 'Assume Equal Variances' in the 2 Sample t Test input menu box must be left blank.

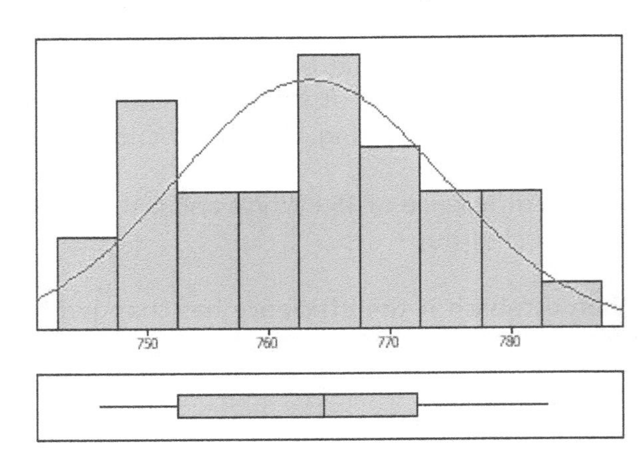

Anderson-Darling Normality Test	
A-Squared	0.37
P-Value	0.399
Mean	763.43
StDev	11.00
Variance	121.08
Skewness	-0.02430
Kurtosis	-1.11806
N	30
Minimum	746.00
1st Quartile	752.50
Median	764.50
3rd Quartile	772.25
Maximum	783.00
95% Confidence Interval for Mean	
759.32	767.54
95% Confidence Interval for Median	
757.23	771.09
95% Confidence Interval for StDev	
8.76	14.79

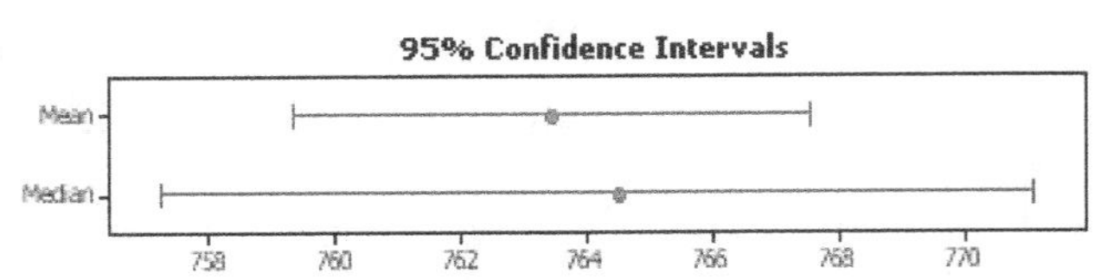

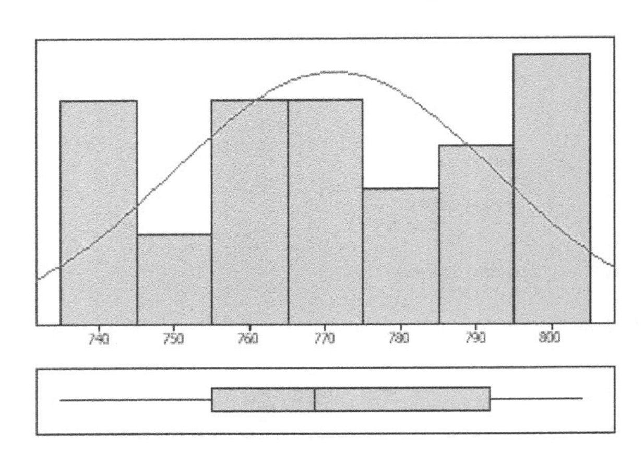

Anderson-Darling Normality Test	
A-Squared	0.50
P-Value	0.195
Mean	771.17
StDev	21.28
Variance	452.90
Skewness	-0.08182
Kurtosis	-1.21864
N	30
Minimum	735.00
1st Quartile	755.00
Median	768.50
3rd Quartile	791.75
Maximum	804.00
95% Confidence Interval for Mean	
763.22	779.11
95% Confidence Interval for Median	
761.23	787.17
95% Confidence Interval for StDev	
16.95	28.61

95% Confidence Intervals

Mean

Median

760 765 770 775 780 785 790

```
                                               Test
Method                          DF1  DF2  Statistic  P-Value
F Test (normal)                  29   29       0.27    0.001
Levene's Test (any continuous)    1   58      14.37    0.000
```

As the higher mean value indicates a poorer efficiency, it appears the change has made the dryer more inefficient. However we need to check the *P* value in the session window.

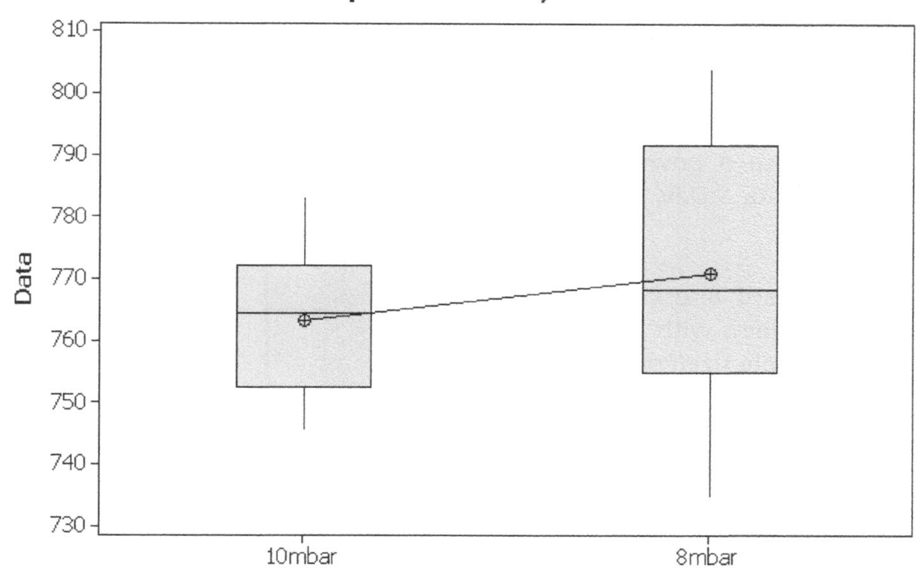

The *P* value is 0.084 and the 95% CI for the difference contains the zero value. Therefore we cannot say that the populations are different. The dryer did not change its efficiency when the inlet pressure was changed from 10 to 8 mbar.

Next we will look at the solution obtained using the Assistant.

Two-Sample T-Test and CI: 10mbar, 8mbar

```
Two-sample T for 10mbar vs 8mbar

        N   Mean  StDev  SE Mean
10mbar  30  763.4   11.0      2.0
8mbar   30  771.2   21.3      3.9


Difference = mu (10mbar) - mu (8mbar)
Estimate for difference:  -7.73
95% CI for difference:  (-16.55, 1.09)
T-Test of difference = 0 (vs not =): T-Value = -1.77  P-Value = 0.084  DF = 43
```

Within the Summary Report the Assistant also delivers a *P* value of 0.084 and we see that zero is within the 95% CI for the difference between the groups.

Within the Diagnostic Report we see that we would have a power of 91.8% using the sample data StDev. Also, there are no outliers.

The Report Card also confirms that there were no problems with Normality, Unusual Data and Sample size/Power.

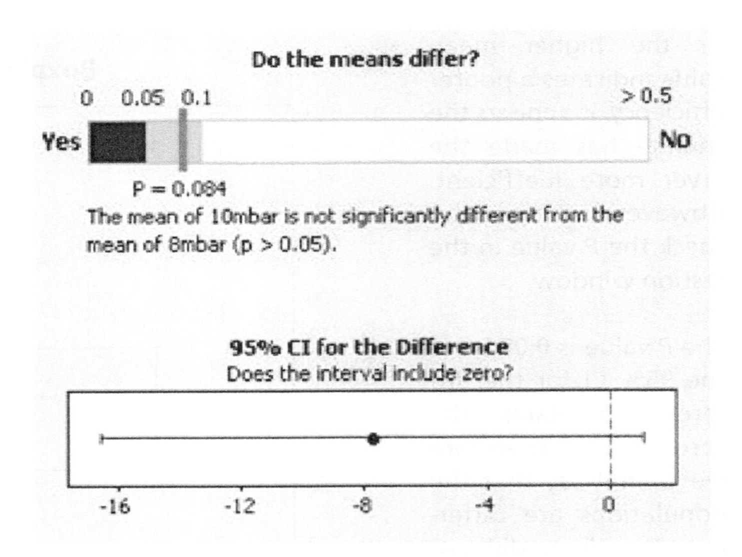

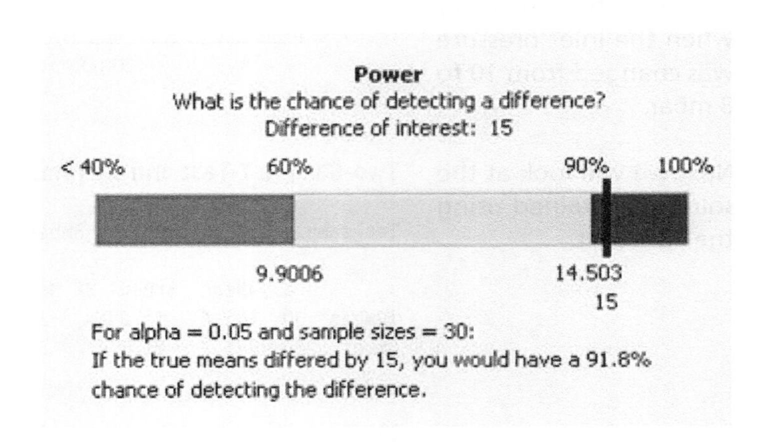

Exercise 4. 2 Sample *t* Test.

Beefy buys chips (fries) from two different chip shops. The chips at the two different shops cost the same but Beefy believes he gets a bigger portion at Chappo's chips. Over the next two weeks Beefy buys and eats a lot of chips so he can collect data for a 2 sample *t* test.

Beefy knows that he can achieve a power about 98% with 25 samples in each group, a pooled StDev of 13 g if the difference he wants to be able to detect is 15 g.

Establish which chip shop gives the biggest portion of chips.

Worksheet: Chips.

Summary for OK Haddock

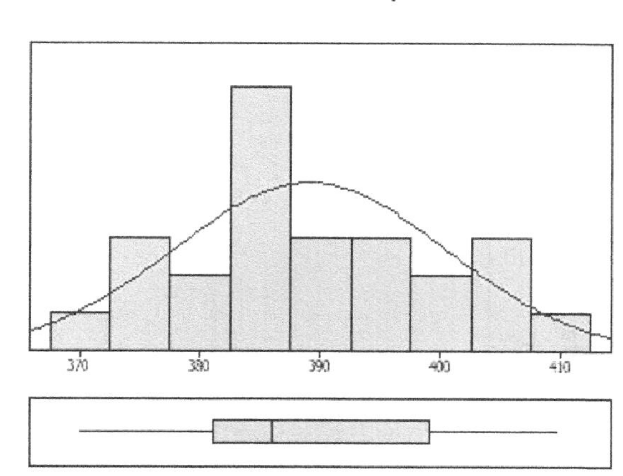

Anderson-Darling Normality Test	
A-Squared	0.34
P-Value	0.476
Mean	389.11
StDev	11.21
Variance	125.60
Skewness	0.152948
Kurtosis	-0.939179
N	25
Minimum	370.00
1st Quartile	381.00
Median	386.00
3rd Quartile	399.10
Maximum	409.70
95% Confidence Interval for Mean	
384.48	393.73
95% Confidence Interval for Median	
384.16	396.58
95% Confidence Interval for StDev	
8.75	15.59

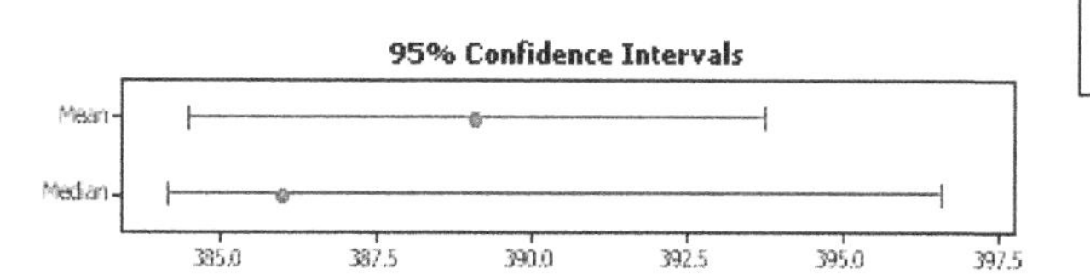

Summary for Chappos

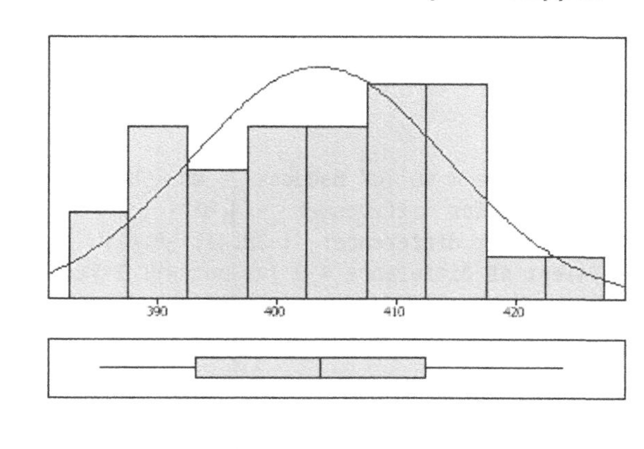

Anderson-Darling Normality Test	
A-Squared	0.33
P-Value	0.492
Mean	403.52
StDev	10.72
Variance	115.02
Skewness	-0.097893
Kurtosis	-0.980998
N	29
Minimum	385.20
1st Quartile	393.20
Median	403.60
3rd Quartile	412.45
Maximum	424.00
95% Confidence Interval for Mean	
399.44	407.60
95% Confidence Interval for Median	
398.59	410.23
95% Confidence Interval for StDev	
8.51	14.50

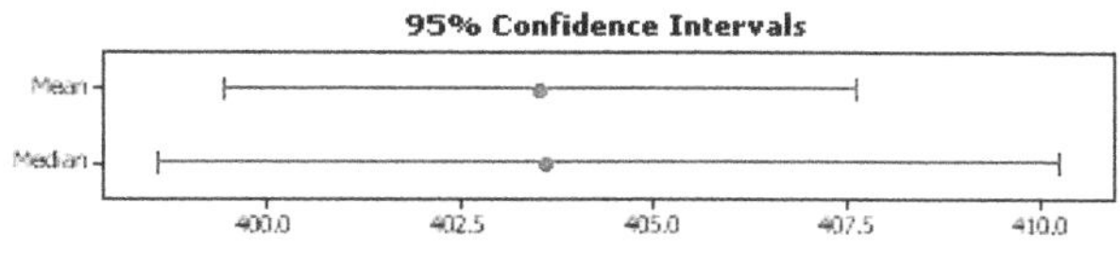

From the graphical summaries it can be seen that the data is reasonably symmetrical and uni modal and the *P* values indicate normality. The StDevs are 11.2 and 10.7 and the 95% CIs for StDevs overlap indicating equal variances. The 2 Variances Test confirms equal variances (results shown below), which means that the tick box for 'Assume Equal Variances' in the 2 Sample t Test input menu box must be ticked.

Method	DF1	DF2	Test Statistic	P-Value
F Test (normal)	24	28	1.09	0.817
Levene's Test (any continuous)	1	52	0.01	0.904

The boxplot indicates that Chappo's Chips provide a bigger portion than the OK Haddock.

The *P* value is 0.00 and the 95% CI for the difference does not contain zero. Therefore, we can conclude that the populations are different and that Chappo's Chips gives a bigger portion.

We will look at the solution obtained from using the Assistant.

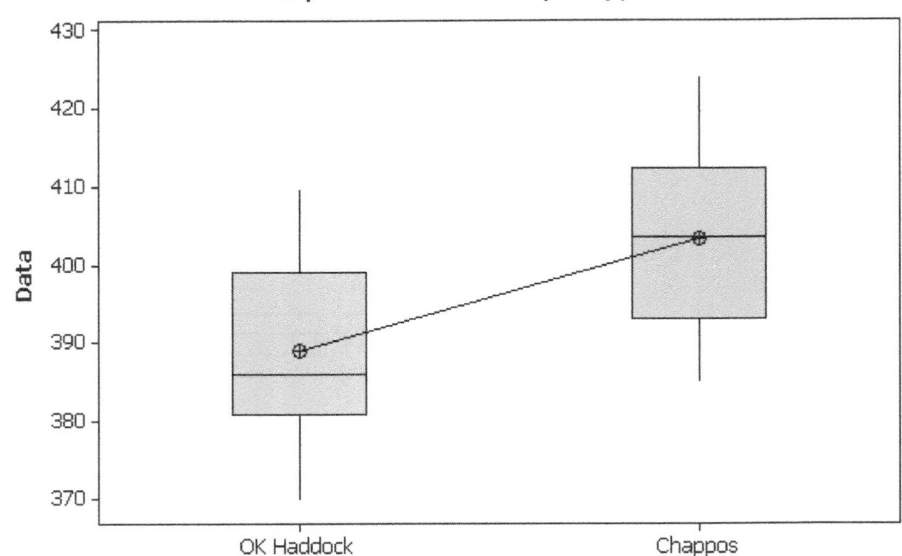


```
Two-sample T for OK Haddock vs Chappos

             N   Mean  StDev  SE Mean
OK Haddock  25  389.1   11.2      2.2
Chappos     29  403.5   10.7      2.0


Difference = mu (OK Haddock) - mu (Chappos)
Estimate for difference:  -14.41
95% CI for difference:  (-20.41, -8.41)
T-Test of difference = 0 (vs not =): T-Value = -4.82  P-Value = 0.000  DF = 52
Both use Pooled StDev = 10.9500
```

Within the Summary Report the Assistant also delivers a *P* value of 0.00 and we see that zero is not within the 95% CI for the difference between the groups.

Within the Diagnostic Report we see that we would have a power of 99.8% using the sample data StDev. Also, there are no outliers.

The Report Card also confirms that there were no problems with Normality, Unusual Data and Sample size/Power.

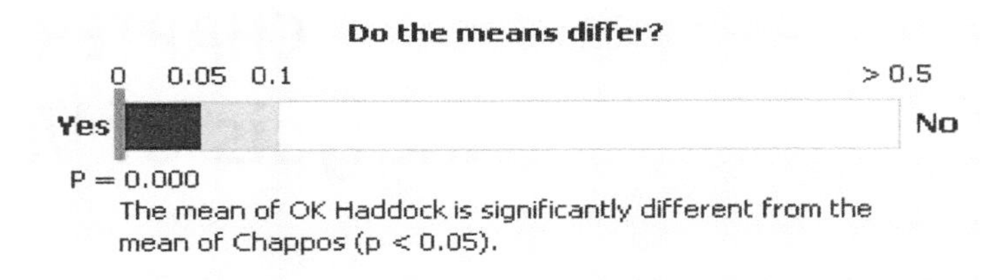

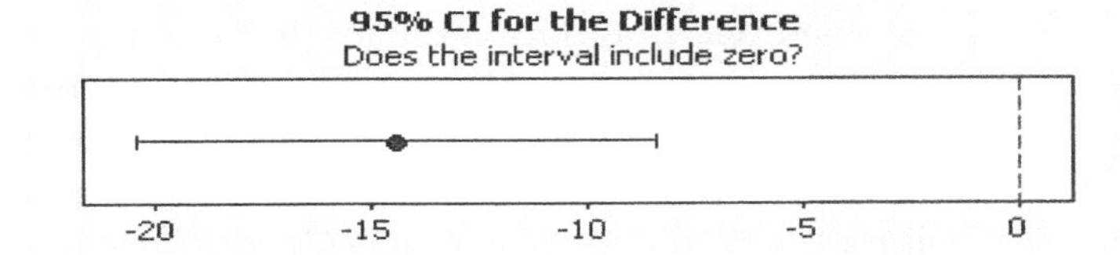

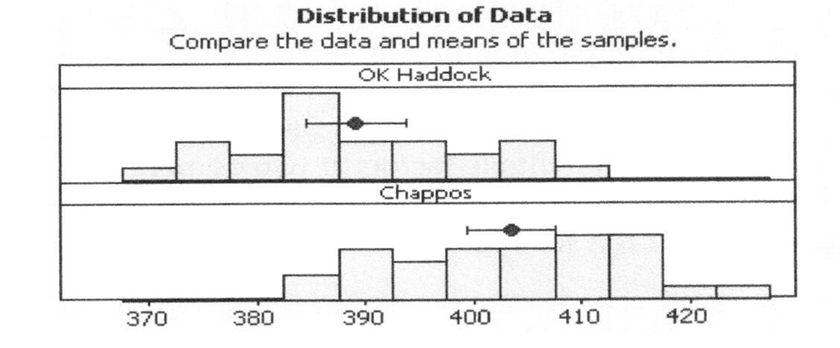

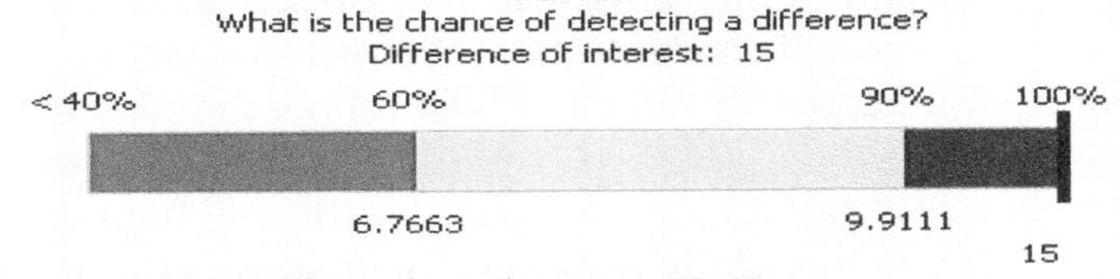

CHAPTER 5
Analysis of Variance

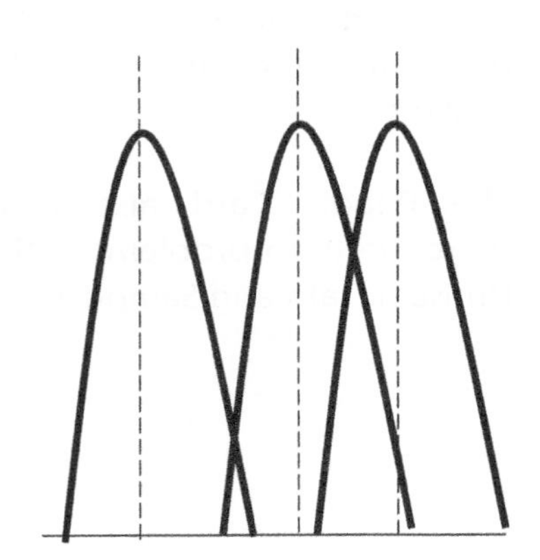

In the last chapter we used the 2 Sample t Test to analyse if there was a difference between two population means. In this chapter we are going to be using predominately the ANOVA procedure to test if there is a difference between two or more population means. ANOVA stands for ANalysis Of VAriance. We will start by looking at how ANOVA works. Then we will look at the One Way ANOVA procedure and look at how that is handled in the Classic menus and then within the Assistant. Amongst the examples we will also look at how to check for differences in Variances between multiple groups.

We will then move onto the ANOVA General Linear Model which allows the comparison of multiple factors at two or more levels.

All of the data sets for this chapter are in the spreadsheet 05 ANOVA.xls.

5.1 How ANOVA Works

In order to understand how ANOVA works let's pretend that we are investigating how yield responds to varying levels of concentration. We feel that yield is a function of concentration. We have looked at four levels of concentration and taken three samples at each level.

We will split this data into two parts as this will help us understand if there is going to be a difference in the populations of the response, Yield, as we alter the factor, Concentration.

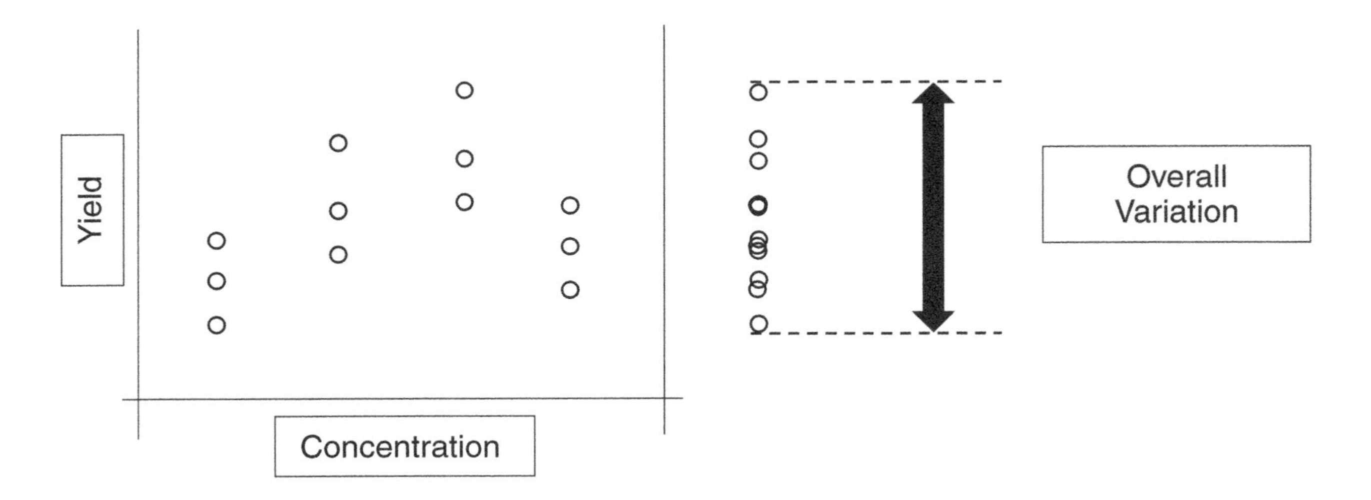

Problem Solving and Data Analysis using Minitab: A clear and easy guide to Six Sigma methodology, First Edition. Rehman M. Khan.

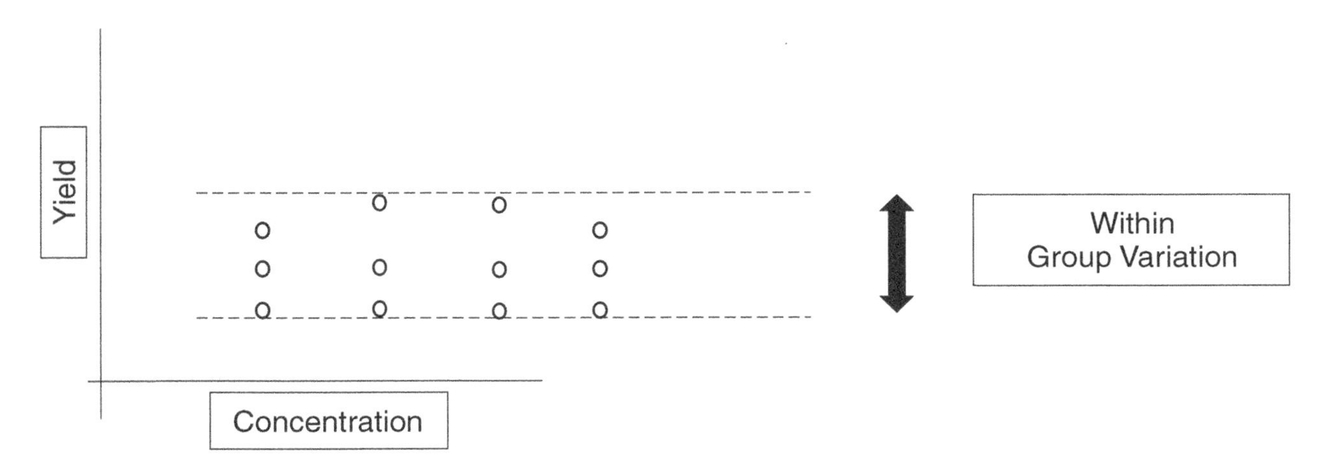

First, we will quantify the amount of variation at each level of concentration. This is usually called the Within Group Variation.

We then take the average value of each group. The variation of the averages is called the Between Group Variation

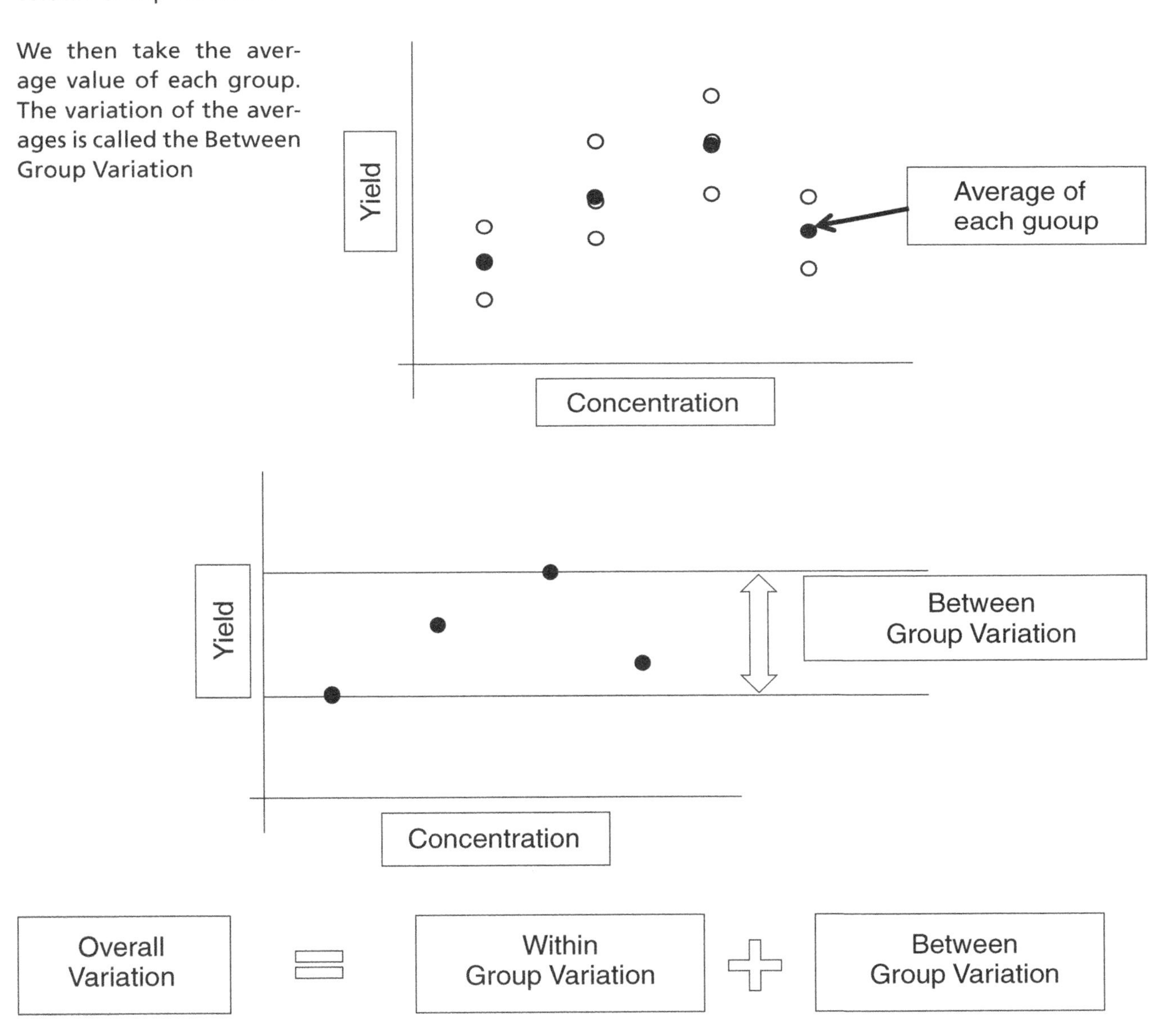

We say that Overall Variation is the sum of the Within Group and the Between Group Variation. ANOVA looks at the ratio of the Between/Within variation.

Between Group Variation
Within Group Variation

If the ratio is small then the Within Group Variation is larger and a difference between the group means does not exist.

If the ratio is large then the Between Group Variation is larger and a difference between the group means exists.

This ratio is called the F Statistic and within ANOVA the P value tells us if the ratio is significant.

Let's see how this works within the ANOVA procedure.

5.2 One Way ANOVA (Classic)

Procedure	• One Way ANOVA (Classic)
What's it used for?	• It is used directly to assess the significance of one factor or variable at two or more levels.
Assumptions and Limitations	• The underlying assumption for the procedure is that the data within the groups will come from a normally distributed population. The procedure is robust to slight departures from normality when the design is balanced (same number of samples at each level). • ANOVA is not robust to unequal variances when the design is balanced. • As a guide a balanced design with a large number of samples at each level (>20) will protect the validity of the results. • At the end of the procedure the residuals are analysed. They will show if the normality assumption and equal variance assumption were suitably met. Patterns in the residuals will point to problems with the analysis and any conclusions that were made. • The residuals must show Independence, a reasonably normally distribution and equal variance.

Example 1. One Way ANOVA (Classic).

Parky's crayon factory has seven machines making crayons. The ultimate aim of the exercise is to see if any of the machines are producing crayons with differing diameters and then to compare the mean diameters to a tolerance. We also want to check if the machines are delivering products with the same variance.

Initially, we will calculate the power of the test. The max difference of mean diameter that we want to be able to detect between any two machines is 0.25 mm, the estimated StDev is 0.3 mm and the intention is to take 40 samples.

After calculating the power, open Worksheet: Crayon and use this data to check the underlying assumptions required for the One Way ANOVA test. Then use One Way ANOVA to assess whether any of the machines are producing crayons with a different diameter.

The final part of the exercise is to check each of the machines against the tolerance for diameter. The required crayon diameter is actually 6.0 mm with a tolerance of ±0.25 mm. Any machine outside of this tolerance value should be targeted for improvement activities.

1. Click Stat <<P&SS << One Way ANOVA
2. Complete the menu box as shown. There are a couple of points to note here. The sample size listed should be for the smallest group as this will give us a safety margin. The StDev should be estimated from knowledge of the process, typically, the average value expected within one of the groups. The number of levels is seven due to the number of machines.
3. Then click OK.

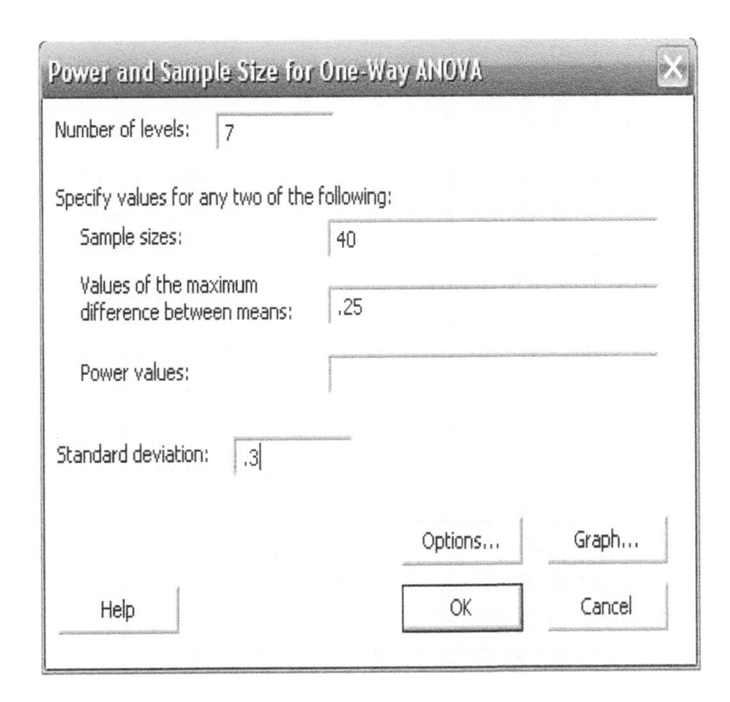

It can be seen that with a sample size of 40, a StDev of 0.3 and being able to detect a difference of about 0.25 between groups the power that we would achieve is 79.83%.

This is acceptable so we can continue with the test procedure.

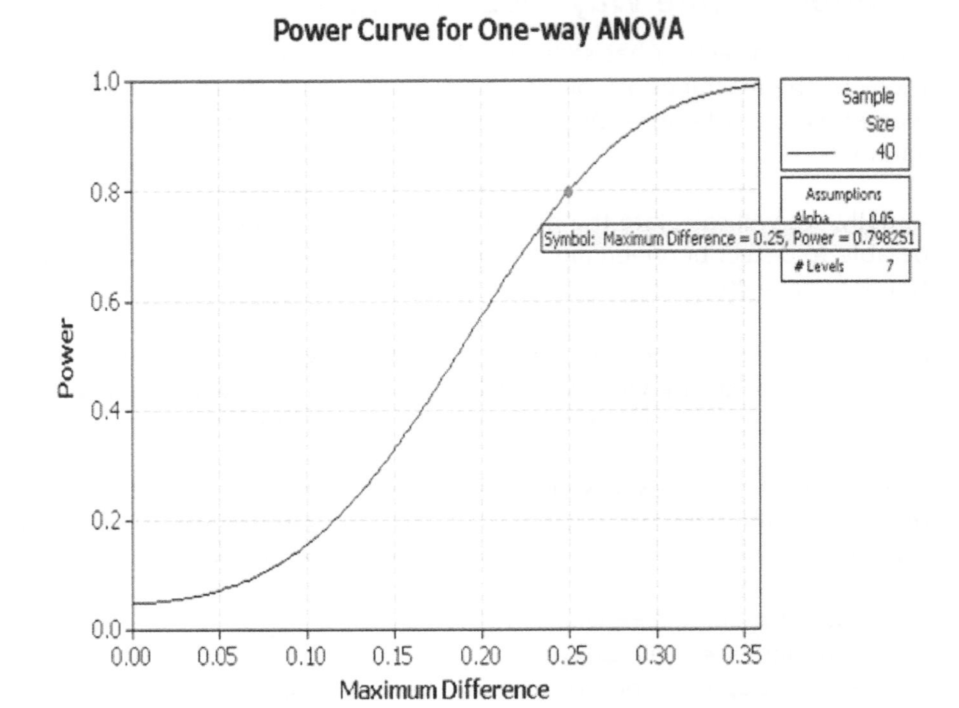

As we said in the last chapter it is always a good idea to see the data before you start any kind of analysis. We are going to use an Individual Value Plot in order to see the levels of the data and their distribution.

↓	C1	C2	C3	C4	C5	C6	C7
	Machine1	Machine2	Machine3	Machine4	Machine5	Machine6	Machine7
1	5.51	5.96	5.38	6.08	6.81	6.30	5.70
2	5.80	6.13	5.15	6.46	6.93	6.87	6.18
3	5.51	6.09	5.98	6.20	6.55	6.21	5.72
4	5.86	6.08	5.01	6.57	7.10	6.57	5.91
5	6.02	5.79	5.89	6.27	6.36	6.77	5.73

4. Import the data from Worksheet: Crayon into Minitab.
5. Click Graph <<Individual Value Plot. Select Multiple Ys, Simple.
6. Then Click OK.
7. Select Machine1 to Machine7 as the Graph variables. Then click OK.

Multiple Y's
Simple
Y1 Y2

Graph variables:
Machine1-Machine7

From the plot we can see how the sample data is distributed. It would appear that Machine3 is producing crayons of the smallest diameter and Machines 5 and 6 are producing crayons with the largest diameter.

We now need to check the underlying assumptions of normality and equal variance for the test.

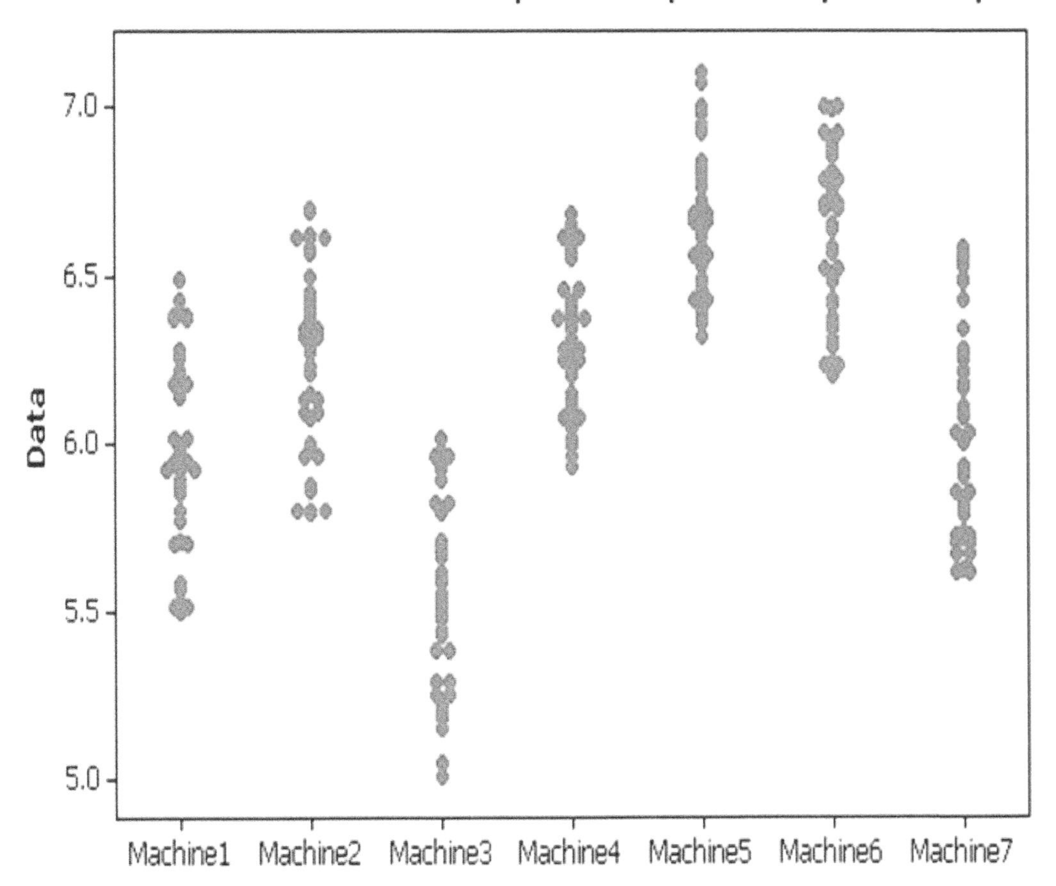

8. Produce a Graphical Summary for each of the Machines.

The graphical summary for Machine1 is shown and we can see that the distribution is deemed to be normal as the AD *P* value is >0.05. The StDev is 0.29. The table below summarises P values and StDevs.

From the table we see that Machine7 is not normally distributed. However, as the remaining groups are normal we will continue with the analysis. The StDevs vary from 0.21 to 0.31. We will need to test whether these groups have significantly differing variances.

Summary for Machine1

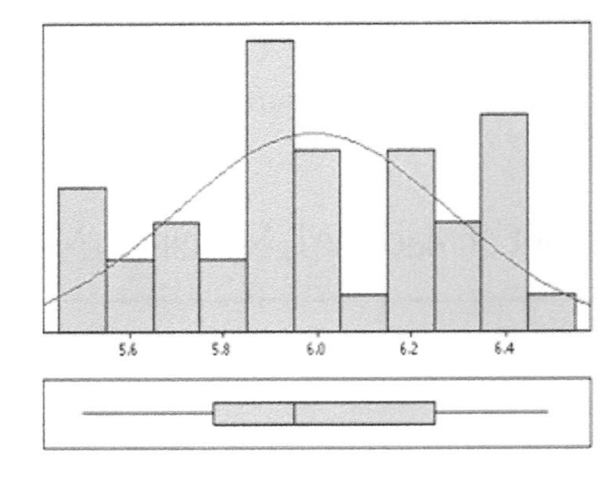

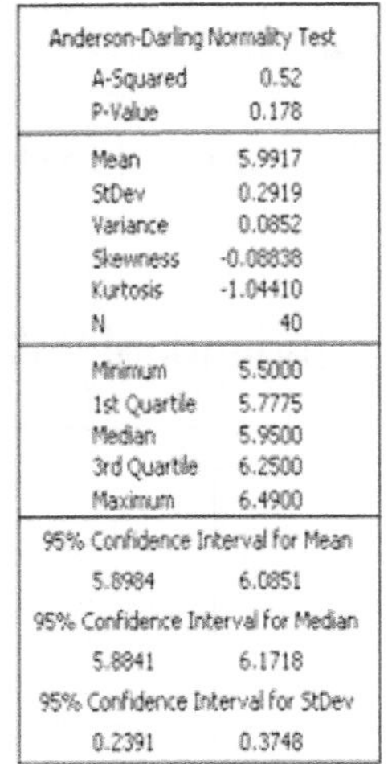

95% Confidence Intervals

Mean

Median

5.90 5.95 6.00 6.05 6.10 6.15 6.20

Machine	P-value	StDev
1	0.178	0.29
2	0.225	0.27
3	0.594	0.28
4	0.392	0.21
5	0.318	0.22
6	0.070	0.25
7	0.039	0.31

The worksheet that was copied into Minitab was in the unstacked format. This means that the labels or subscripts were at the top of each column of data. Minitab will also handle data in the stacked format. This is where the labels or subscripts are in one column and the data is in the adjacent column. When using the Classic Method to test for Equal Variances amongst groups the data needs to be in the stacked format so we must convert it.

9. Click Data <<Stack <<Columns.

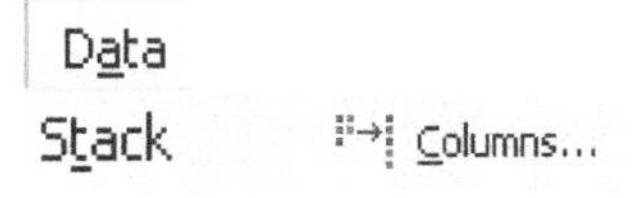

10. Complete the menu box as shown and click OK. We had the option of putting the stacked data into a new worksheet but we will put it into C9 and C10 of the existing worksheet.

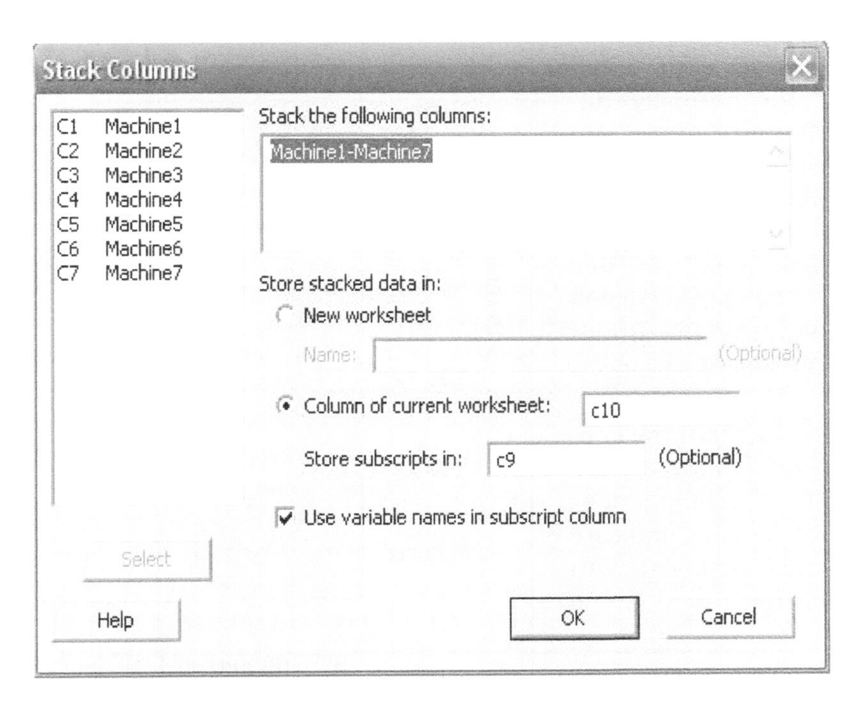

Notice how Minitab stacks all the data. C9 contains the labels C10 contains all the data.

11. Enter new labels for columns C9 and C10 as shown.

C9-T	C10
Subscripts	Diameter
Machine1	5.51
Machine1	5.80
Machine1	5.51
Machine1	5.86
Machine1	6.02

12. Click Stat <<ANOVA <<Test for Equal Variances
13. Complete the menu box as shown and click OK.

Stat
ANOVA Test for Equal Variances...

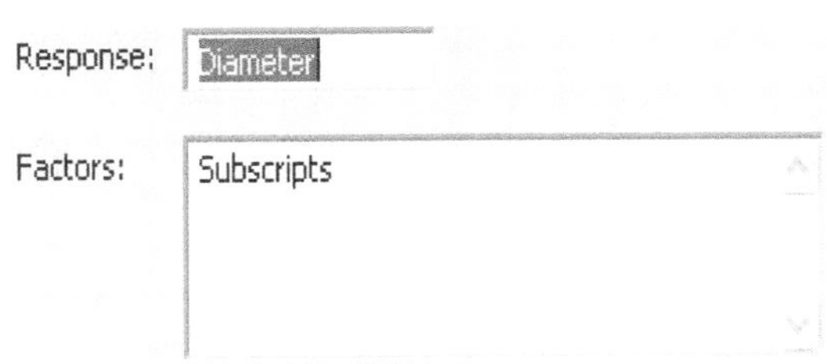

Minitab carries out a standard hypothesis test and also gives a graphical display.

Bartlett's Test result is for normally distributed data and Levene's for non normal data. As our data was mostly normally distributed we will use Bartlett's *P* value to say that the groups have equal variances.

The fact that the confidence intervals are all overlapping does not always indicate equal variances with this test.

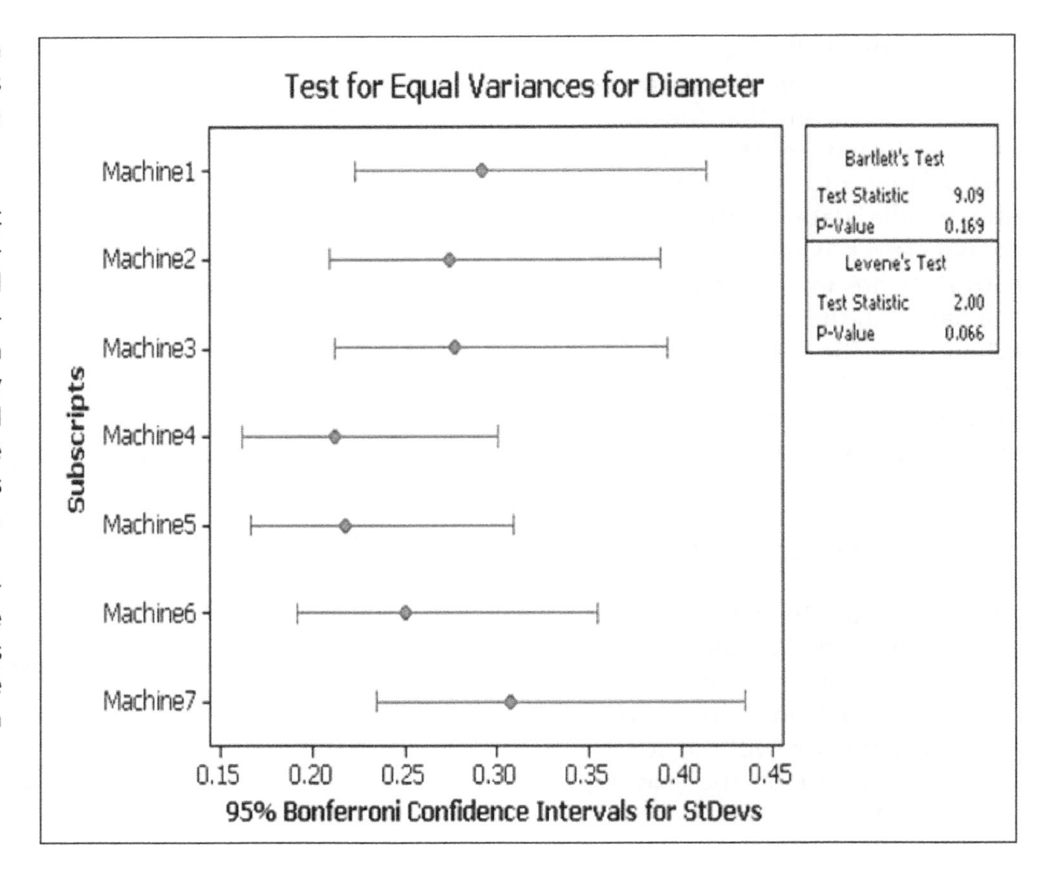

We will now conduct the One Way ANOVA test. There are two options for using One Way ANOVA, one for stacked data and one for unstacked. In this example we will use the stacked version

14. Click Stat <<ANOVA << One Way.

15. Complete the menu box as shown.

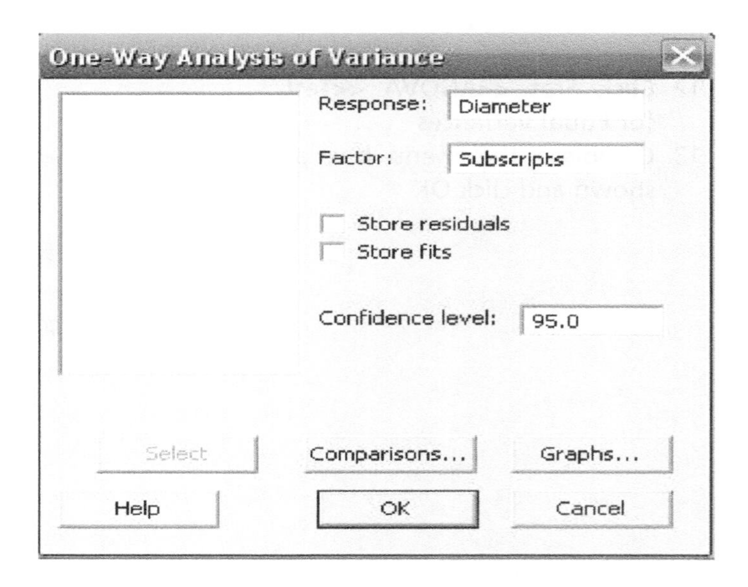

16. Then click Graphs and select the Four in one residual plots. If the Unstacked version of the test had been selected then we would not get the option of selecting the Residuals versus order plot and we would have a three in one plot.
17. Click OK and OK again.
18. Go to the session window to check the results.

```
Residual Plots
( ) Individual plots
    [ ] Histogram of residuals
    [ ] Normal plot of residuals
    [ ] Residuals versus fits
    [ ] Residuals versus order
(•) Four in one
```

Subscripts is the Between Group variation. The *F* value is our Between to Within ratio. The *P* value tells us if the *F* value is significant. In this case it is significant and we can reject the null hypothesis. This means that one or more of the groups are different.

Error is the Within Group variation.

The R-Sq(adj) tells us that 64.5% of the variation in diameter is due to changes in machine.

One-way ANOVA: Diameter versus Subscripts

```
Source       DF       SS      MS      F      P
Subscripts    6  35.6116  5.9353  85.50  0.000
Error       273  18.9517  0.0694
Total       279  54.5632

S = 0.2635   R-Sq = 65.27%   R-Sq(adj) = 64.50%
```

We can find out which groups are different by looking at the interval plot within the session window.

The interval plot shows the 95% CI for the mean crayon diameter. If any of the intervals do not overlap then we can definitely say that those groups are different, the converse is not always true. We can see that Machine3 is producing crayons with the lowest diameter. Machine5 and Machine6 are producing crayons with the highest diameter and because their CIs significantly overlap we cannot say that the mean diameter of these machines is different. It can also be seen that Machine1 and Machine7 could be paired. Also Machine2 and Machine4 can be paired. Occasionally, the CIs can be overlapping only very slightly but the P value indicates a difference. In these instances we go with the *P* value.

The original question asked us which of the machines were making crayons within a particular tolerance. To obtain this information we need to produce an Interval plot of our own.

```
                                   Individual 95% CIs For Mean Based on
                                   Pooled StDev
Level      N    Mean   StDev   ----+---------+---------+---------+-----
Machine1  40  5.9917  0.2919              (-*--)
Machine2  40  6.2538  0.2743                        (--*-)
Machine3  40  5.5548  0.2771   (--*-)
Machine4  40  6.3103  0.2118                           (-*--)
Machine5  40  6.6755  0.2175                                   (--*-)
Machine6  40  6.6012  0.2501                                (--*-)
Machine7  40  6.0313  0.3066                (-*--)
                               ----+---------+---------+---------+-----
                                 5.60      5.95      6.30      6.65

Pooled StDev = 0.2635
```

There are two methods of navigating to the Interval plot.

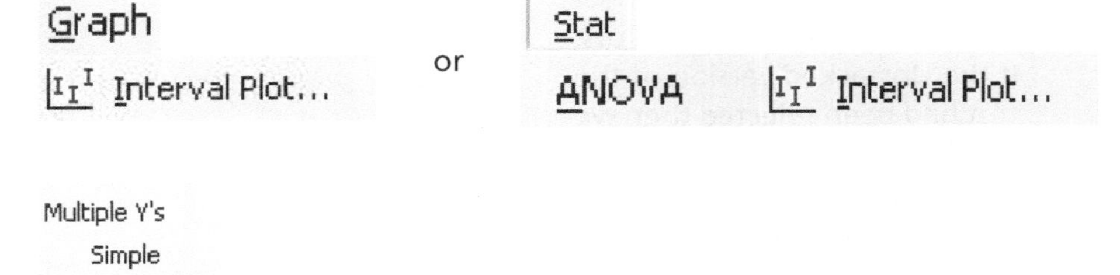

19. Click Stat <<ANOVA <<Interval Plot.

Or:

20. Click Graph <<Interval Plot.
21. Select Multiple Ys <<Simple. Click OK.
22. Select all the machines as the Graph variables.
23. Click OK.

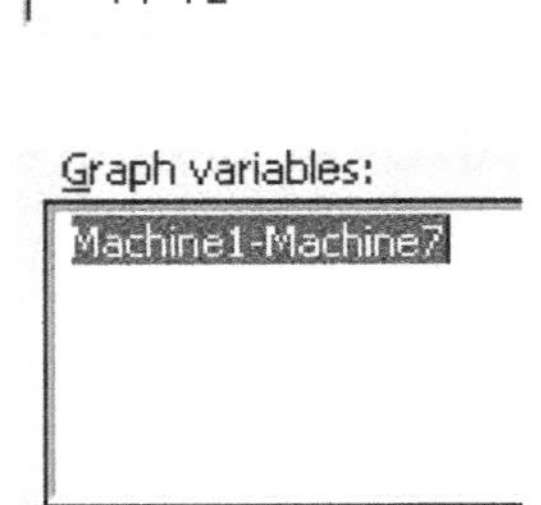

Graph variables:
Machine1-Machine7

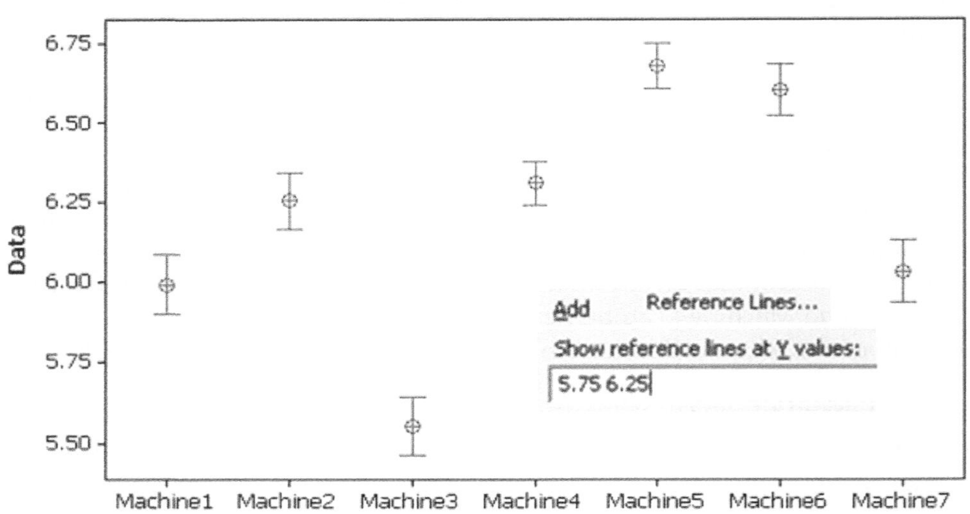

We get the same interval plot as previously produced within the session window, however, on this occasion the response data is on the y axis.

24. Right click onto the graph area.
25. Click on Add <<Reference Lines.
26. Enter the required tolerance values for the crayons.

As the CIs for machines 3, 5 and 6 are outside of the reference lines we know that their mean crayon diameter is not between 5.75 and 6.25 mm. Machines 3, 5 and 6 should be targeted for improvement activities.

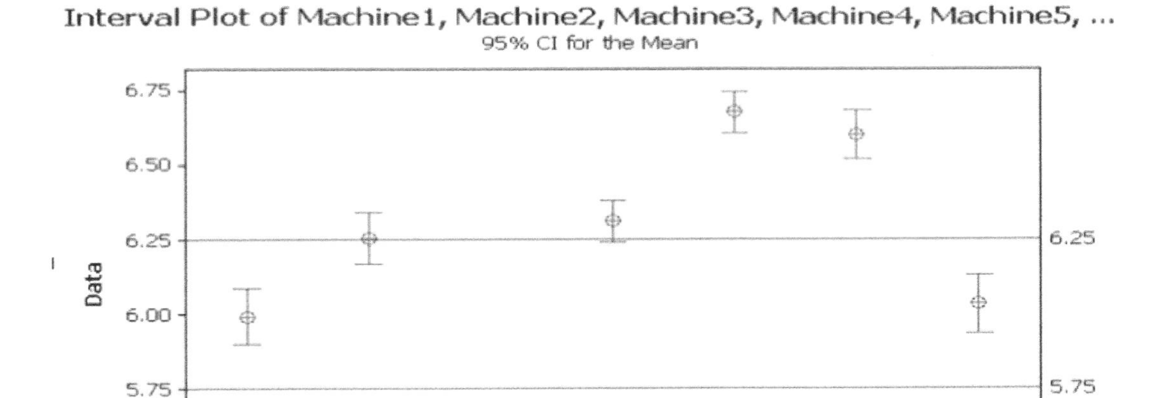

We will now check the residual plot and establish whether it validates our test procedure.

As shown earlier, as a part of the ANOVA calculation the mean of each group is calculated. The residual is the distance from a data point to the mean value. The residual for each point is calculated.

If our original data was not suitable for the One Way ANOVA test in terms of normality and equal variance this would show up as unusual patterns within the residuals. We need to check for patterns within the residuals to decide if the test is valid. If you look back we have requested the four in one residual plot from the Graphs menu when we conducted the One Way ANOVA test.

27. From the Graphs window find the four in one residual plot.

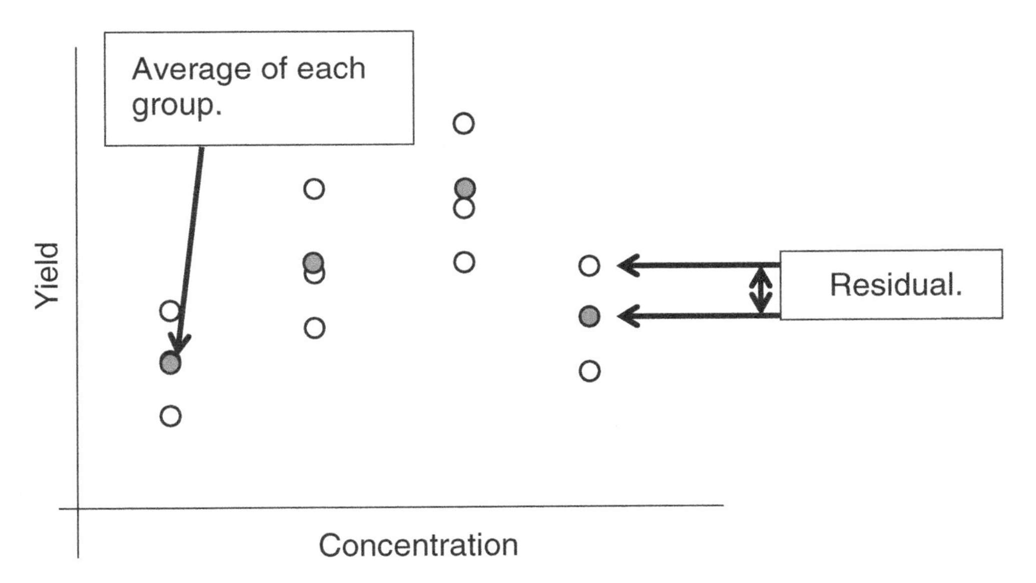

Within the test Assumptions and Limitations we said 'At the end of the procedure the residuals are analysed. They will show if the normality assumption and equal variance assumption were suitably met. Patterns in the residuals will point to problems with the analysis and any conclusions that were made.

The residuals must show Independence, a reasonably normally distribution and equal variance.'

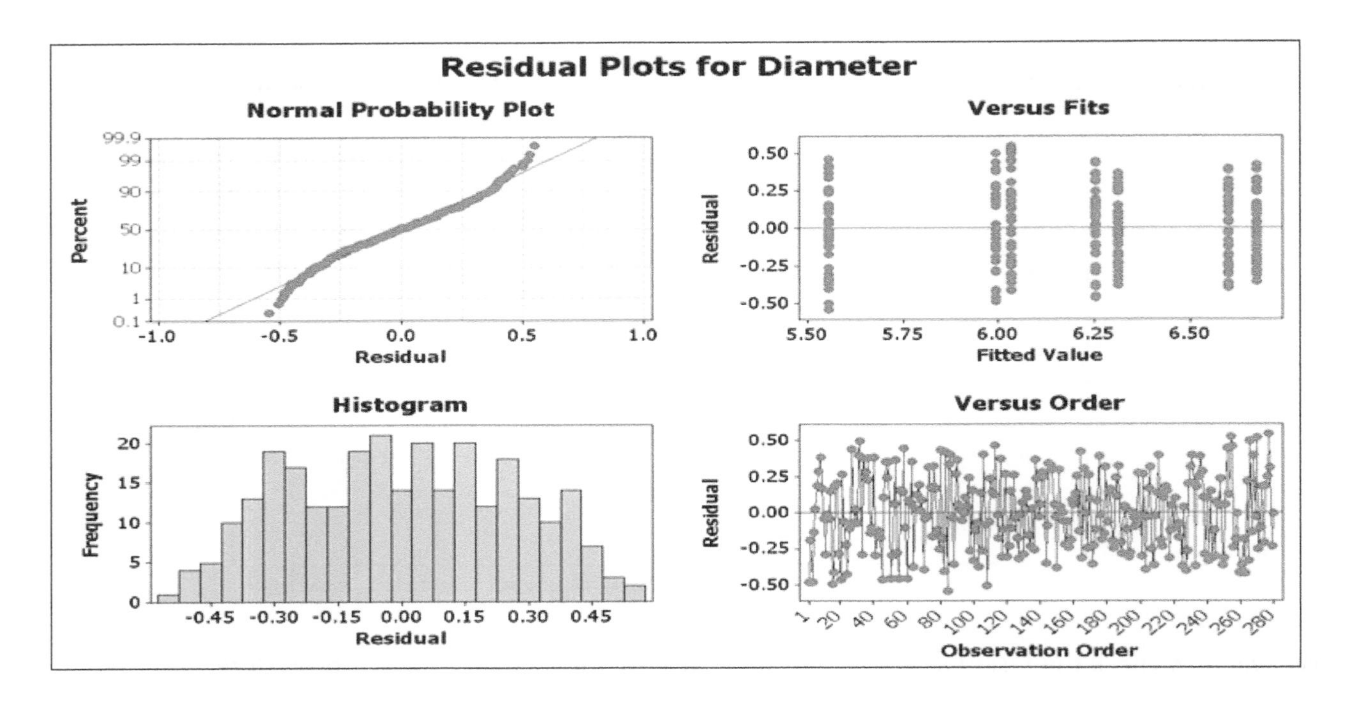

We will now examine the residual plots in greater detail.

We are only looking for departures from normality from the residuals within the Normal Probability plot. In this case the residuals would pass the 'fat pencil' test so there are no issues.

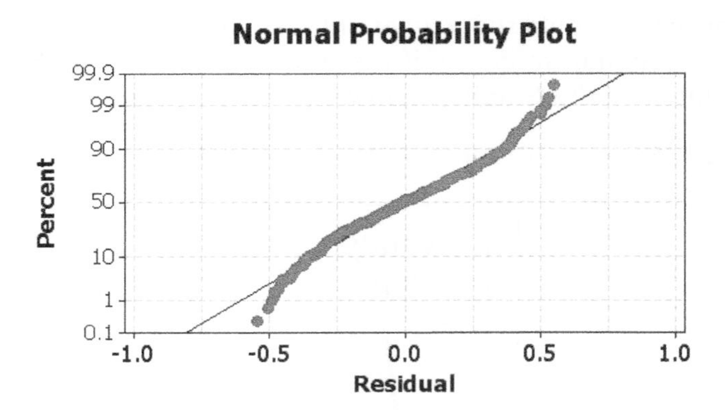

From the histogram we are looking to see if there is extreme skewness. Our distribution for the residuals looks quite good so there are no issues.

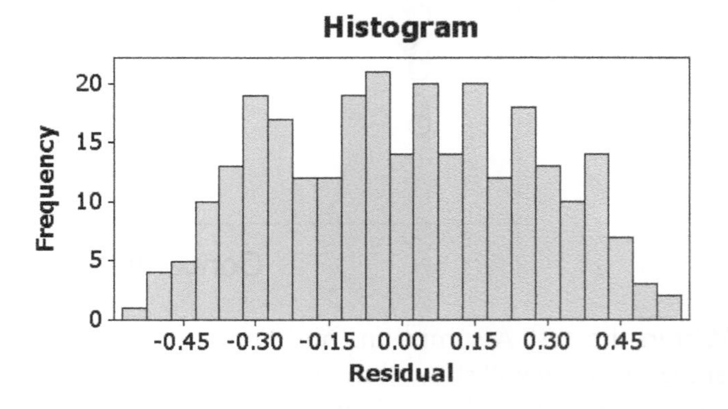

Within the Versus Fits plot we are looking to see if the residuals are equally spaced around the zero line. The residuals for our study are equally distant from the centre line.

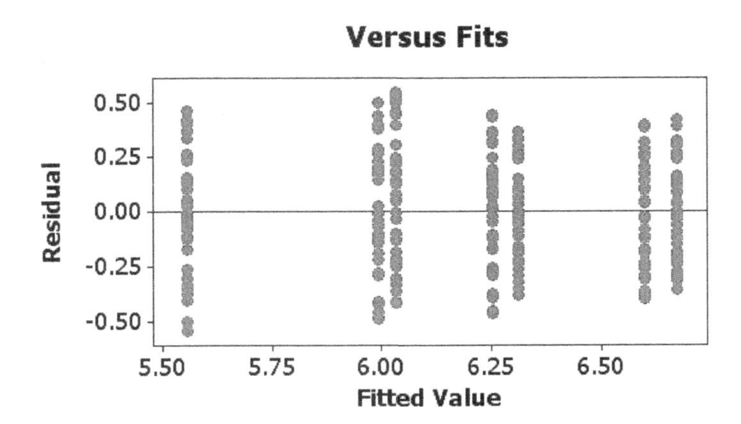

Within the Versus Order we are looking for unusual patterns or lack of randomness within the residuals. No patterns can be seen within the residuals from our study.

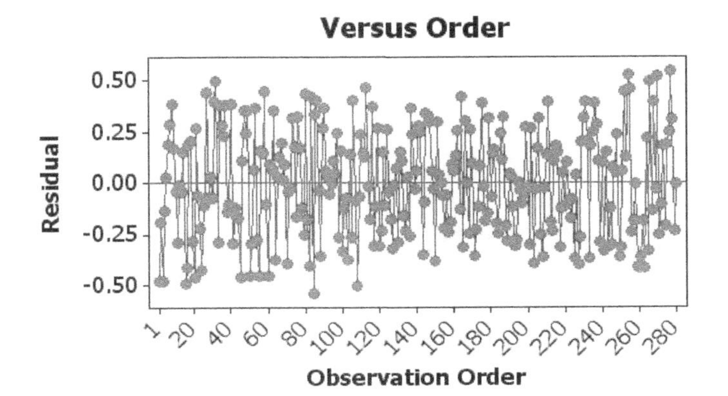

Honesty and discipline are required when checking the residuals. Don't ignore issues within the residuals which indicate that the results are flawed. Asking yourself why issues are occurring and then resolving the issues will probably lead to a more robust solution.

Now that we are satisfied that the residuals do not indicate any problems within our study and our objectives have been met we will repeat this problem using the Assistant.

5.3 One Way ANOVA with the Assistant

Procedure	• One Way ANOVA (Assistant)
What's it used for?	• It is used directly to asses the significance of one factor or variable at two or more levels.
Assumptions and Limitations	• The One Way ANOVA test within the Assistant uses a different method of calculation to the Classic Method. • It is robust to unequal variances but has a slightly lower power than the Classic Method when variances are equal. • After carrying out the test, if the null hypothesis was not rejected the Assistant will state whether the power was adequate. If the power was not sufficient the Assistant will recommend a sample size. The information provided by the Assistant will be more useful if you provide a value for the difference under investigation. • The test is robust to the normality assumption when the sample size is greater than 15 samples per group. • Residuals are not relied upon to validate the test results.

1. Click on Assistant <<Hypothesis Tests.
2. Click on One Way ANOVA within the 'Compare more than two samples' group.
3. The Assistant caters for stacked and unstacked data, this is set within the first drop down menu. I have opted to use the unstacked data. We will not change the significance level and we enter 0.25 as the difference between means that we are interested in. Then Click OK.

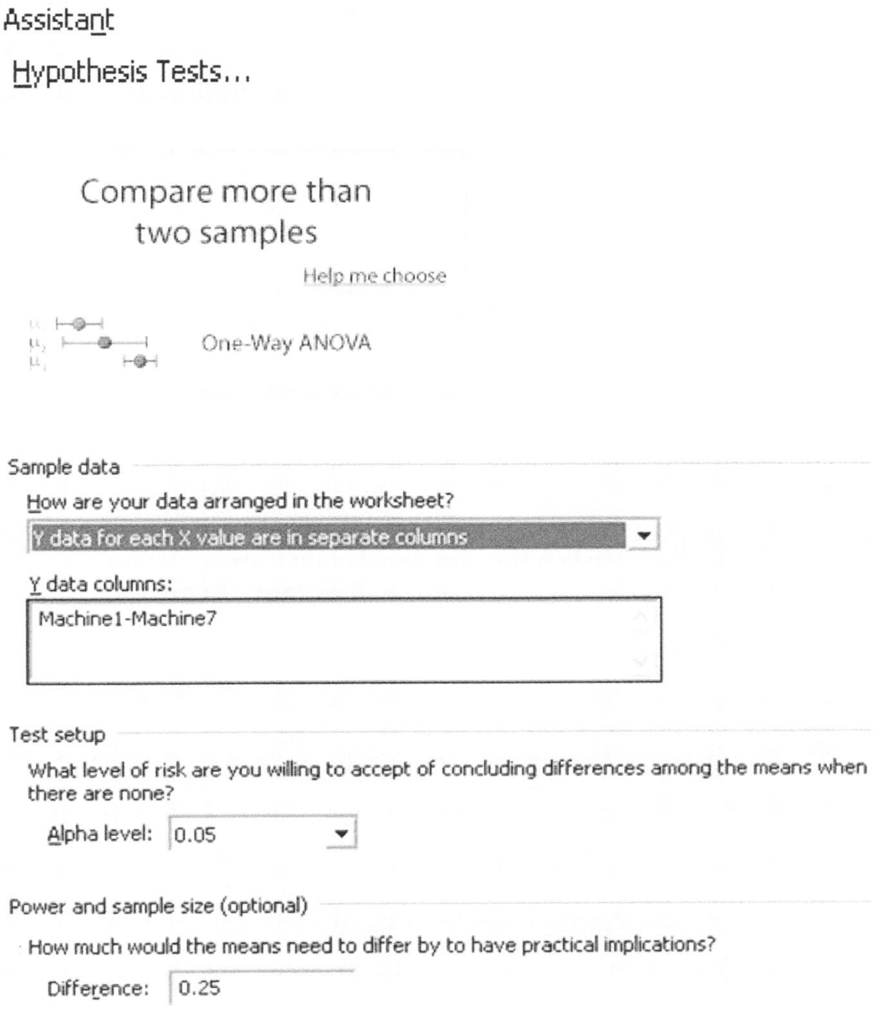

Assistant

Hypothesis Tests...

Compare more than two samples

Help me choose

One-Way ANOVA

Sample data

How are your data arranged in the worksheet?

Y data for each X value are in separate columns

Y data columns:

Machine1-Machine7

Test setup

What level of risk are you willing to accept of concluding differences among the means when there are none?

Alpha level: 0.05

Power and sample size (optional)

How much would the means need to differ by to have practical implications?

Difference: 0.25

A number of graphical report pages are produced. Go to the Summary Report as shown in the figure below.

The *P* value indicator in the top left indicates that there are differences between at least two of the groups.

The Means Comparison Chart shows the mean confidence intervals of the groups. Again, intervals that do not overlap indicate a difference between groups. However, on occasion intervals that do slightly overlap can also be different.

On the top right the Assistant provides a comparison table which makes it very easy to see which groups actually differ.

On the bottom right, within the comments section, you can read the conclusion of the test procedure. This can be edited.

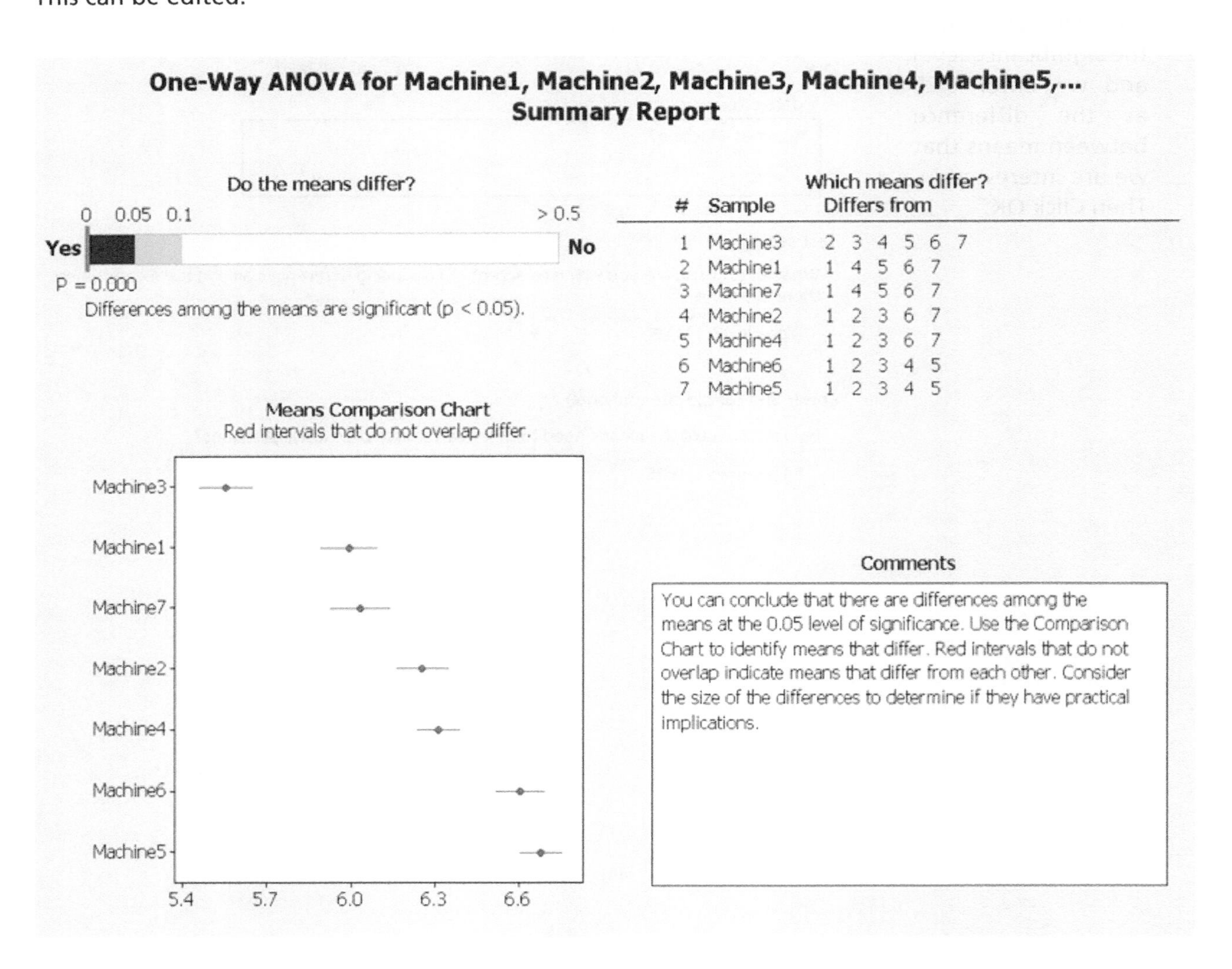

The next report page that we are going to look at is the Power Report. The power indicator on the top left gives a confidence interval for the power value. We have 78.44–98.3% chance of detecting a difference of 0.25 mm. This compares to a power of 79.83% obtained when using the Classic Method.

The bottom of the screen shows a table which breaks down the individual data based on each group.

One-Way ANOVA for Machine1, Machine2, Machine3, Machine4, Machine5,...
Power Report

Power
What is the chance of detecting a difference?
Difference of interest: 0.25

< 40% 60% 90% 100%

Based on your samples and alpha level (0.05), the chance of detecting a difference of 0.25 ranges from 78.44% to 98.30%.

What sample sizes are required to detect a difference of 0.25?

All Samples	Power
29	61.1 - 91.3
35	71.4 - 96.3
42	80.9 - 98.8
53	90.5 - 99.8

Power is a function of the sample sizes and the standard deviations. If the power is not satisfactory, consider increasing the sample sizes.

Statistics

Sample	Sample Size	Mean	Standard Deviation	Individual 95% CI for Mean
Machine1	40	5.9917	0.29188	(5.8984, 6.0851)
Machine2	40	6.2538	0.27425	(6.1660, 6.3415)
Machine3	40	5.5548	0.27714	(5.4661, 5.6434)
Machine4	40	6.3103	0.21175	(6.2425, 6.3780)
Machine5	40	6.6755	0.21755	(6.6059, 6.7451)
Machine6	40	6.6012	0.25007	(6.5213, 6.6812)
Machine7	40	6.0313	0.30664	(5.9332, 6.1293)

On the left of the Diagnostic Report we are presented with a histogram for each of the groups so that we can see spread, shape and location of each group.

On the right hand side we are given SPC charts. Unusual data points are marked as red dots. These could influence the results of the test so we are given the opportunity to locate the data points and then investigate them.

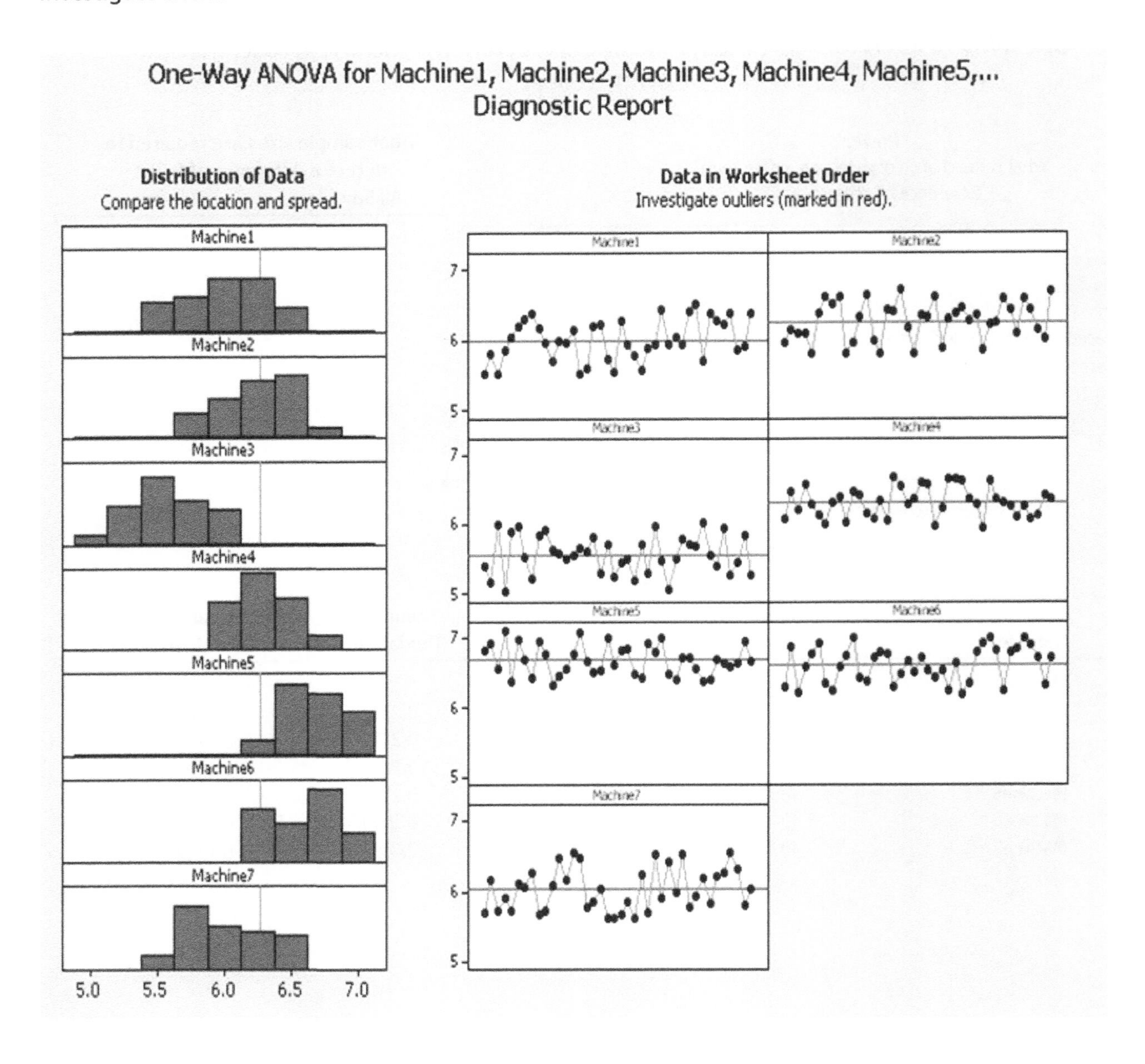

Finally, we have the Report Card. We are told that we had no unusual data points.

We are told that the test managed to detect a difference even though the power was low.

Finally, we are told that normality was not an issue as the sample size within each group was above 15.

One-Way ANOVA for Machine1, Machine2, Machine3, Machine4, Machine5,...
Report Card

Check	Status	Description
Unusual Data	✓	There are no unusual data points. Unusual data can have a strong influence on the results.
Sample Size	✓	The sample is sufficient to detect differences among the means. Because you entered a difference of interest, the Power Report provides a sample size evaluation for this difference. You do not need to be concerned that the power is low because the test detected a difference.
Normality	✓	Because all your sample sizes are at least 15, normality is not an issue. The test is accurate with nonnormal data when the sample sizes are large enough.

We are going to move onto another example which will help to reinforce our learning.

Example 2. One Way ANOVA.

A tyre manufacturer wants to investigate how the sulfur content of rubber affects the shear strength of the rubber. They vary the sulfur concentration at four different levels and check the shear strength of a number of samples at each level.

Initially calculate the power of the test. The max difference of mean strength that we want to be able to detect between any two levels is 2.5 kN, the estimated StDev is 3.2 kN and we intend to take 40 samples at each level.

After calculating the power, open Worksheet: Tyre and use this data to check the underlying assumptions required for the One Way ANOVA test. Then use One Way ANOVA within the Classic Method and the Assistant to assess whether changing the concentration has a statistically significant effect upon the shear strength of the rubber.

1. Click Stat <<P&SS << One Way ANOVA.
2. Complete the menu box as shown. The sample size listed should be for the smallest group as this will give us a safety margin. The StDev should be estimated from knowledge of the process, typically, the average value expected within one of the groups.
3. Then click OK.

Stat
Power and Sample Size ▸ One-Way ANOVA...

Power and Sample Size for One-Way ANOVA

Number of levels: 4

Specify values for any two of the following:

Sample sizes: 40

Values of the maximum difference between means: 2.5

Power values:

Standard deviation: 3.2

Options... Graph...

Help OK Cancel

It can be seen that with a sample size of 40, a StDev of 3.2 and being able to detect a difference of about 2.5 between groups the power that we would achieve is 83.7%.

This is acceptable so we can continue with the test procedure.

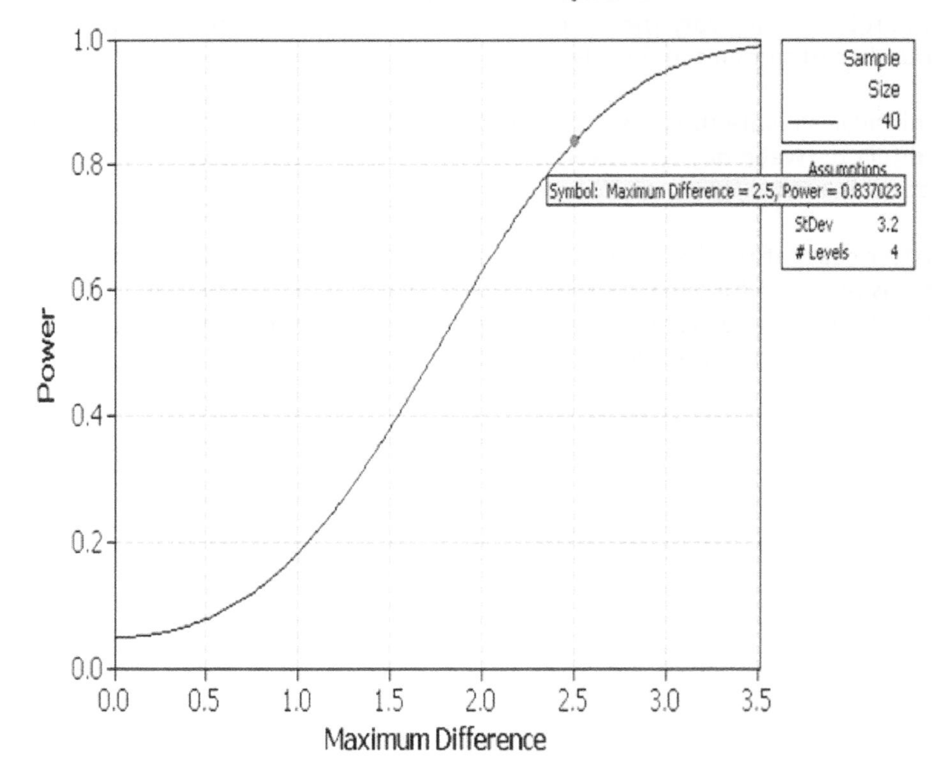

We want to see the data before we start the analysis. Last time we used an Individual Value Plot but this time we are going to use a boxplot for a change.

↓	C1	C2	C3	C4
	10ppm	20ppm	30ppm	40ppm
1	79.49	72.79	76.44	81.58
2	84.35	77.42	79.63	86.59
3	83.20	81.45	82.57	84.61
4	77.09	80.53	77.58	81.33
5	78.07	78.76	85.74	81.66

4. Import the data from Worksheet: Tyre into Minitab.
5. Click Graph <<Boxplot. Select Multiple Ys, Simple.
6. Click OK.
7. Select 10–40 ppm as the Graph variables. Then click OK.

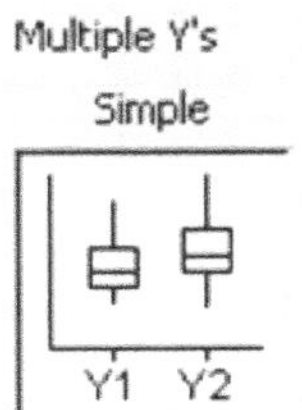

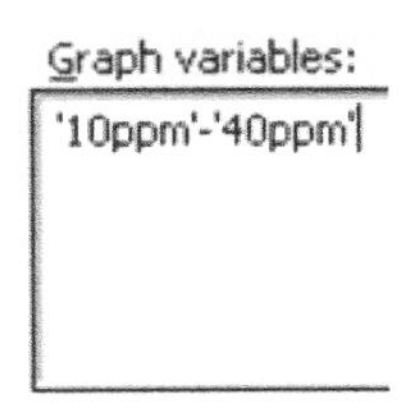

From the plot we can see how the sample data is distributed. It would appear that there is a positive correlation between concentration and shear strength. However, we don't know if there is actually a significant change within the populations.

Before we carry out the One-Way ANOVA we need to check the underlying assumptions of normality and equal variance for the test.

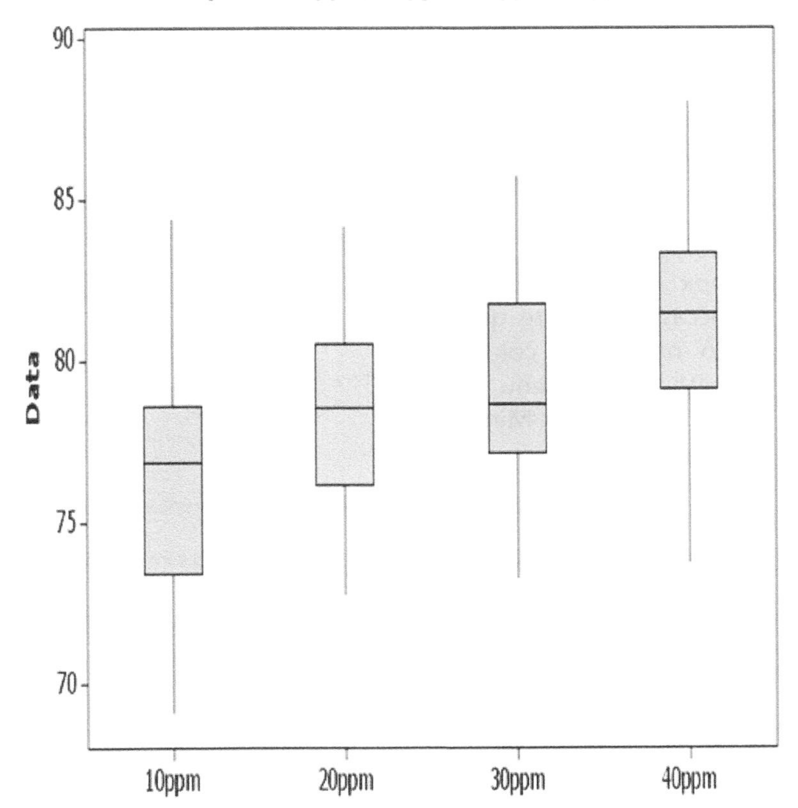

8. Produce a Graphical Summary for each of the concentrations.

The table below summarises the *P* values for the Anderson–Darling Normality Test and StDevs.

The data clearly meets the normality requirement. The StDevs vary from 2.93 to 3.94, which appears to be a significant difference. However, we need to check using the equal variances test.

Summary for 10ppm

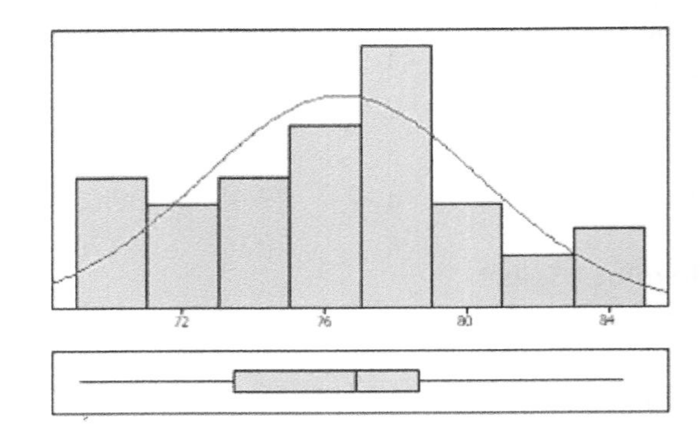

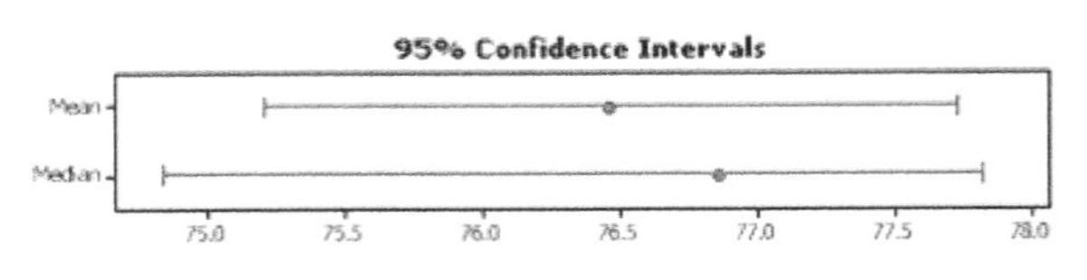

Anderson-Darling Normality Test	
A-Squared	0.25
P-Value	0.740
Mean	76.462
StDev	3.944
Variance	15.554
Skewness	0.067816
Kurtosis	-0.501591
N	40
Minimum	69.128
1st Quartile	73.426
Median	76.855
3rd Quartile	78.625
Maximum	84.351
95% Confidence Interval for Mean	
75.201	77.723
95% Confidence Interval for Median	
74.844	77.815
95% Confidence Interval for StDev	
3.231	5.064

Concentration	*P*-value	StDev
10 ppm	0.74	3.94
20 ppm	0.91	2.94
30 ppm	0.63	2.93
40 ppm	0.95	3.3

As our original data is in the unstacked format we need to stack the data to use the Equal Variances test.

9. Click Data <<Stack <<Columns.
10. Complete the menu box as shown. On this occasion we are going to name the columns within the menu box and we will let Minitab locate them on the worksheet.
11. Click OK.
12. Check the location of the stacked data it should be in columns C5 and C6.

Data
Stack   Columns...

Stack the following columns:
'10ppm'-'40ppm'

Store stacked data in:
New worksheet
Name: (Optional)
Column of current worksheet: Strength
Store subscripts in: Concentration (Optional)
Use variable names in subscript column

C5	C6-T
Strength	Concentration
79.49	10ppm
84.35	10ppm
83.20	10ppm
77.09	10ppm
78.07	10ppm

13. Click Stat <<ANOVA <<Test for Equal Variances.
14. Enter Strength as the Response and Concentration as the Factors.
15. Click OK.

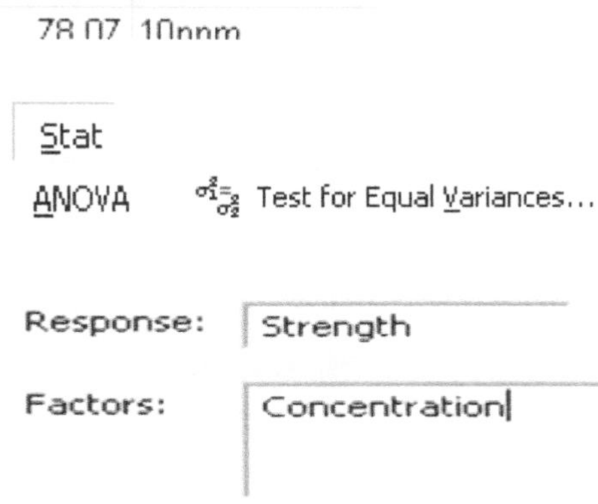

From the output graph we see that the *P* value for Barlett's Test (which is for normal data) is 0.193, indicating that we cannot reject the null hypothesis and we cannot say that the groups have unequal variances.

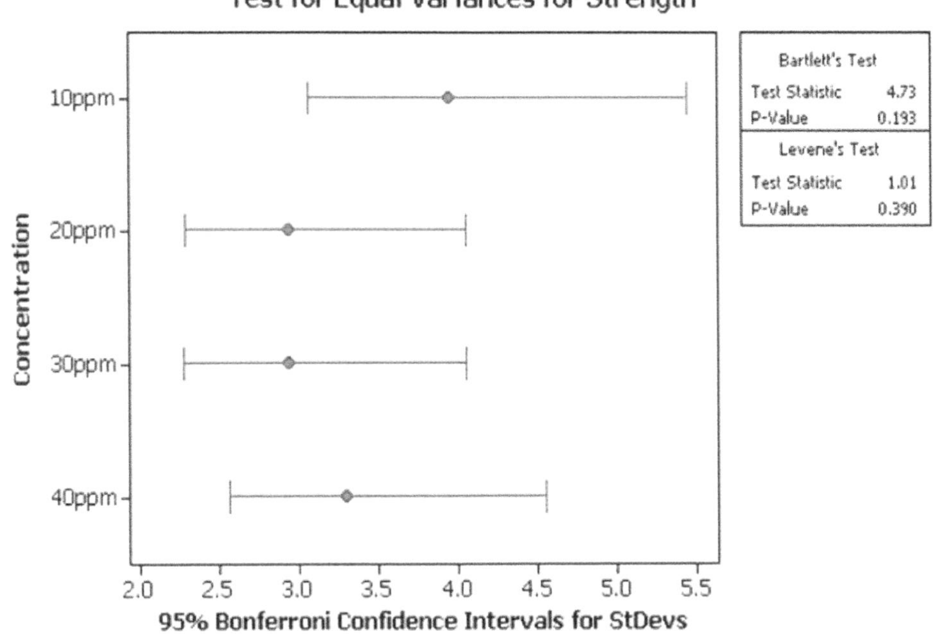

Last time we used the stacked version of One Way ANOVA. This time we will use the unstacked version just for a change.

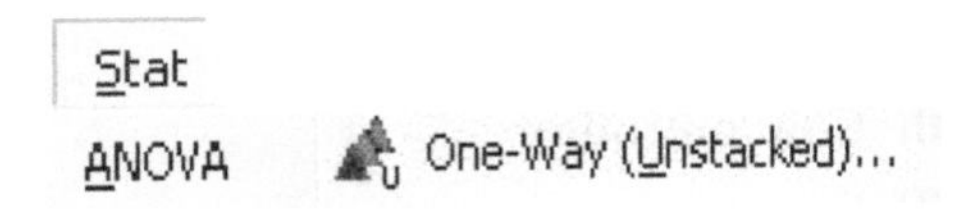

16. Click Stat <<ANOVA<< One Way (Unstacked).
17. Complete the menu box as shown.
18. Then click Graphs and select the Three in one residual plots.
19. Click OK and OK again.
20. Go to the session window to check the results.

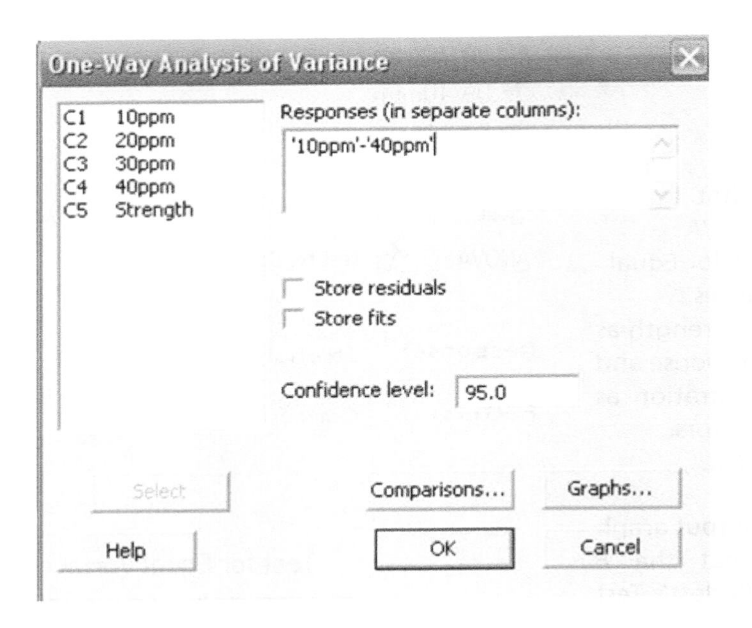

The *P* value for the test is low so the null must go, meaning that we can reject the null hypothesis. A difference has been detected between at least two of the groups.

Source	DF	SS	MS	F	P
Factor	3	447.5	149.2	13.67	0.000
Error	156	1702.2	10.9		
Total	159	2149.8			

S = 3.303 R-Sq = 20.82% R-Sq(adj) = 19.30%

The interval plot tells us that the 10 ppm level is definitely different to all the other levels. The 20 and 30 ppm levels significantly overlap so they must be the same. There is a bit of doubt about the 30 and 40 ppm levels as they could be slightly overlapping. We will repeat the One Way ANOVA in order to demonstrate a slightly different comparison method.

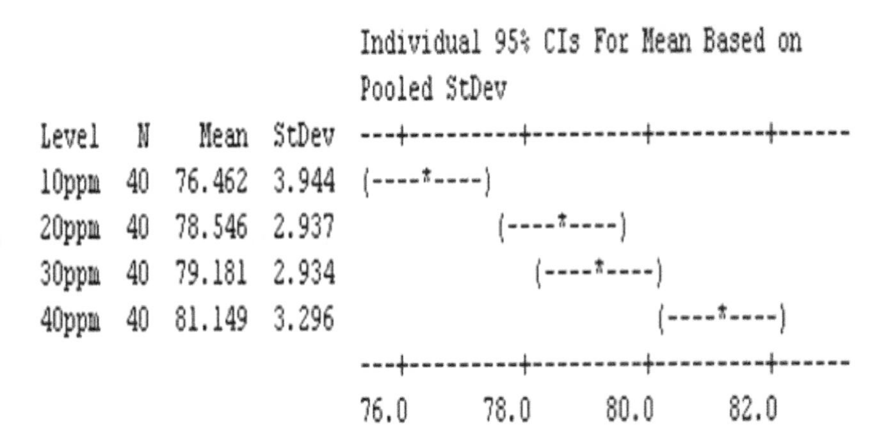

```
                            Individual 95% CIs For Mean Based on
                            Pooled StDev
Level   N    Mean  StDev  ---+---------+---------+---------+------
10ppm  40  76.462  3.944  (----*----)
20ppm  40  78.546  2.937            (----*----)
30ppm  40  79.181  2.934               (----*----)
40ppm  40  81.149  3.296                         (----*----)
                          ---+---------+---------+---------+------
                           76.0      78.0      80.0      82.0
```

Pooled StDev = 3.303

21. Click on the Edit Last icon.

22. Click on Comparisons.

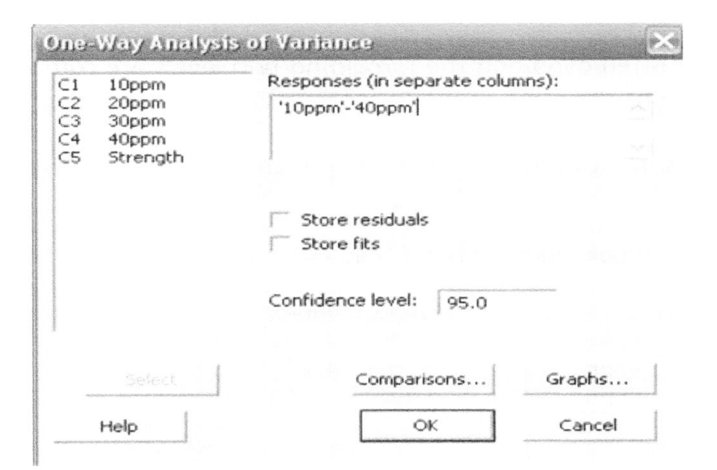

23. Select Tukey's.
24. Click OK and OK again.
25. Go to the session window to interpret the test results.

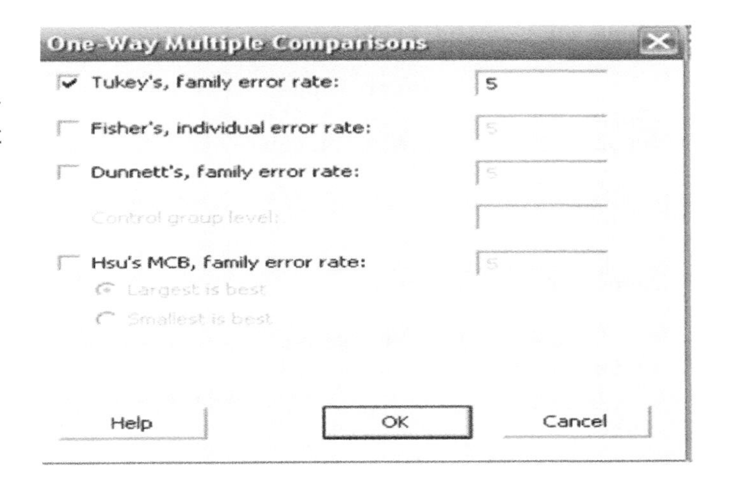

The Grouping Information table gives us an efficient summary of the results, it is an M16 enhancement.

As the 40 ppm level is the only member of Group A it is different to all other levels. Similarly the 10 ppm level is the only member of Group C. The 20 and 30 ppm levels, as stated earlier, are the same and are both in group B.

```
Grouping Information Using Tukey Method

        N    Mean  Grouping
40ppm  40  81.149  A
30ppm  40  79.181    B
20ppm  40  78.546    B
10ppm  40  76.462      C

Means that do not share a letter are significantly different.
```

Tukey's method of comparison is a step wise method. One of the levels is subtracted from the group and the remainders are compared to the subtracted level. Initially, the 10 ppm level is removed the remaining levels are compared to the 10 ppm level. If their CI does not include zero then they are different to 10 ppm.

After that 20 ppm is removed and the remaining two levels are compared to 20 ppm. 20 ppm and 30 ppm are deemed to be the same as the CI contains zero, whereas 20 ppm and 40 ppm are different.

In the last comparison it can be seen that 30 ppm and 40 ppm are different.

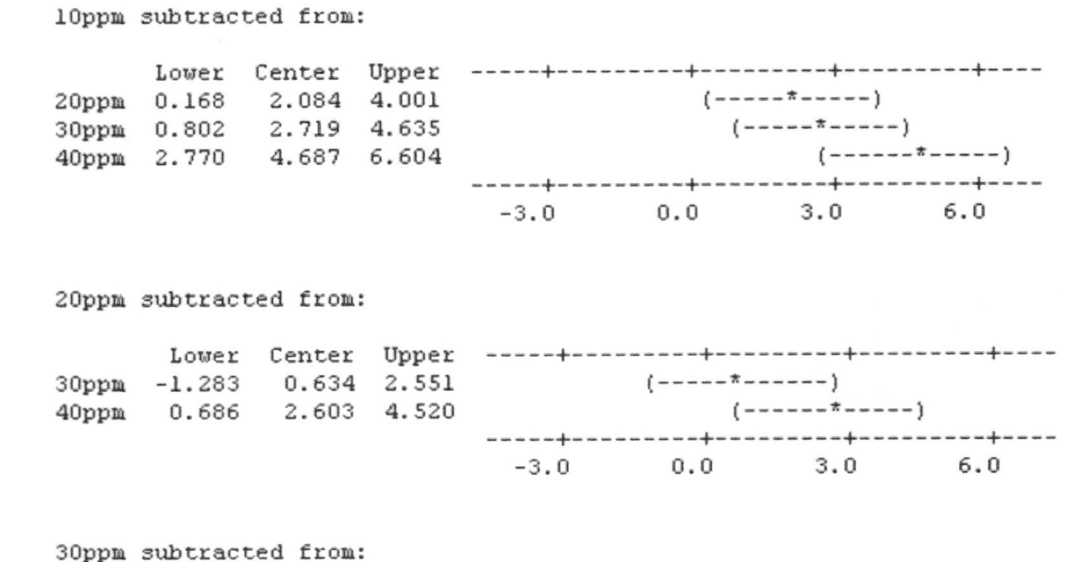

```
10ppm subtracted from:

       Lower  Center  Upper  -----+---------+---------+---------+----
20ppm  0.168   2.084  4.001                  (-----*-----)
30ppm  0.802   2.719  4.635                     (-----*-----)
40ppm  2.770   4.687  6.604                           (------*-----)
                             -----+---------+---------+---------+----
                                -3.0       0.0       3.0       6.0

20ppm subtracted from:

        Lower  Center  Upper  -----+---------+---------+---------+----
30ppm  -1.283   0.634  2.551             (-----*------)
40ppm   0.686   2.603  4.520                   (------*-----)
                              -----+---------+---------+---------+----
                                 -3.0       0.0       3.0       6.0
```

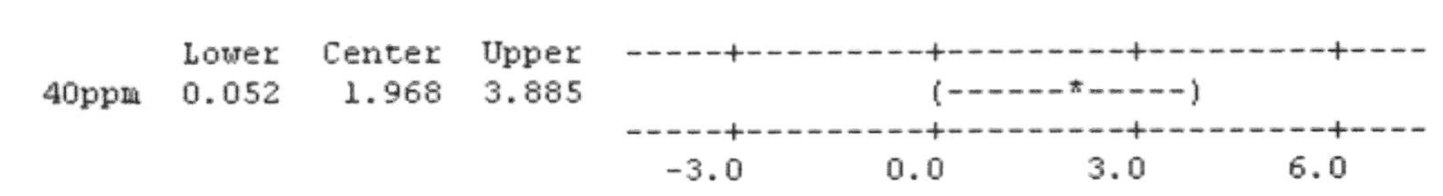

```
30ppm subtracted from:

       Lower  Center  Upper  -----+---------+---------+---------+----
40ppm  0.052   1.968  3.885                (------*-----)
                             -----+---------+---------+---------+----
                                -3.0       0.0       3.0       6.0
```

Before checking the residuals, let's have a quick look at the other comparison methods we could have used.

Tukey's, family error rate:

Fisher's, individual error rate:

Dunnett's, family error rate:

Hsu's MCB, family error rate:

Family and Individual error rate relate to Type I Errors when multiple comparisons are involved, which is incorrectly rejecting the null hypothesis. A family significance level of 5% for 20 elements is equivalent to 5/20 or 0.25% individual error rate. As family error rates take less alpha risk their confidence intervals are wider. The error rate that you select should be appropriate for the risk that you are willing to take.

The remaining comparison methods will be demonstrated. This is done only to show how they work not to help us with the example.

Tukey's and Fisher's comparison methods are both pairwise methods. The difference is that Tukey's uses a family error rate and Fisher's uses an individual error rate. As Tukey's method uses a lower alpha risk the confidence intervals are wider. For example, when comparing the 10 ppm level to the 20 ppm level using Tukey's method the CI was 0.168 to 4.001 whereas using Fisher's method we have 0.625 to 3.543. However, for this example the findings are identical using either of these two methods, that is the Grouping Information tables are the same using either method.

```
Grouping Information Using Fisher Method

        N    Mean  Grouping
40ppm  40  81.149  A
30ppm  40  79.181    B
20ppm  40  78.546    B
10ppm  40  76.462      C

Means that do not share a letter are significantly different.


Fisher 95% Individual Confidence Intervals
All Pairwise Comparisons

Simultaneous confidence level = 79.82%


10ppm subtracted from:

       Lower  Center  Upper  ----+---------+---------+---------+-----
20ppm  0.625   2.084  3.543                 (----*----)
30ppm  1.260   2.719  4.178                   (----*----)
40ppm  3.228   4.687  6.146                         (----*---)
                             ----+---------+---------+---------+-----
                               -3.0       0.0       3.0       6.0
```

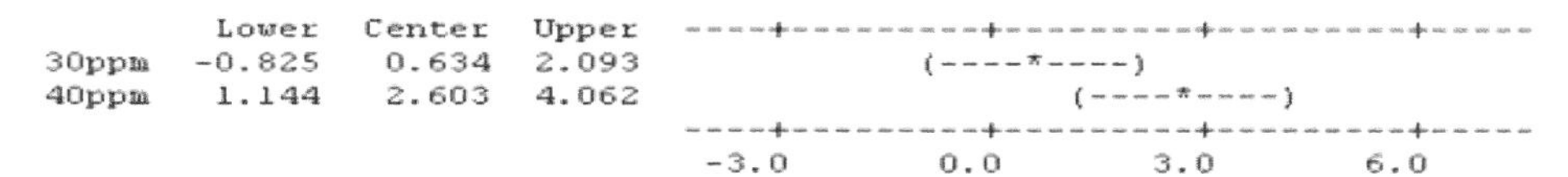

```
20ppm subtracted from:

        Lower  Center  Upper  ----+---------+---------+---------+-----
30ppm  -0.825   0.634  2.093                (----*----)
40ppm   1.144   2.603  4.062                      (----*----)
                              ----+---------+---------+---------+-----
                                -3.0       0.0       3.0       6.0


30ppm subtracted from:

       Lower  Center  Upper  ----+---------+---------+---------+-----
40ppm  0.509   1.968  3.427                 (----*---)
                             ----+---------+---------+---------+-----
                               -3.0       0.0       3.0       6.0
```

When using Dunnett's comparison method we are first required to select a control group. For the figure shown the 20 ppm level has been selected as the control group and this means that all other levels will be compared to 20 ppm. The Group Information table only tells us which groups are the same as the control and which are not.

Even the Interval plot only allows us to compare to the 20 ppm level. We cannot see that the 30 and 40 ppm groups are different.

☑ Dunnett's, family error rate: 5

Control group level: c2

```
Grouping Information Using Dunnett Method

Level            N    Mean  Grouping
20ppm (control)  40  78.546  A
40ppm            40  81.149
30ppm            40  79.181  A
10ppm            40  76.462

Means not labeled with letter A are significantly different from control level mean.


Dunnett's comparisons with a control

Family error rate = 0.05
Individual error rate = 0.0189

Critical value = 2.37

Control = 20ppm

Intervals for treatment mean minus control mean
```

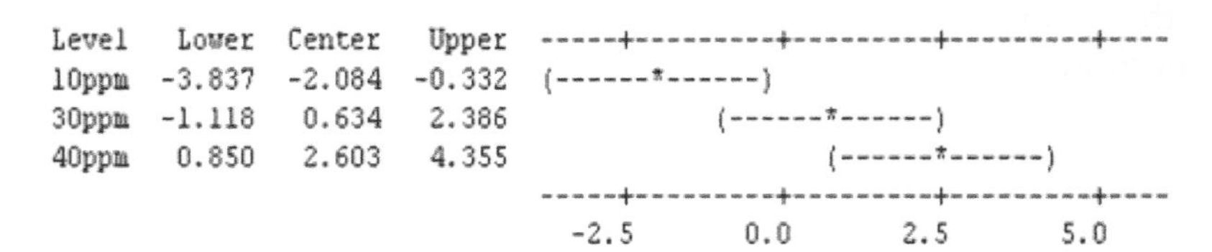

```
Level   Lower  Center   Upper  -----+---------+---------+---------+----
10ppm  -3.837  -2.084  -0.332  (------*------)
30ppm  -1.118   0.634   2.386            (------*------)
40ppm   0.850   2.603   4.355                   (------*------)
                               -----+---------+---------+---------+----
                                  -2.5       0.0       2.5       5.0
```

When using Hsu's MCB comparison method we are first required to select the best group. There is a radio button for stating whether this has the Largest or Smallest of the values.

For the example 'Largest is best' has been selected. This means that the 40 ppm is selected as the control group and all other groups are compared to it. For this case, we can confirm that all

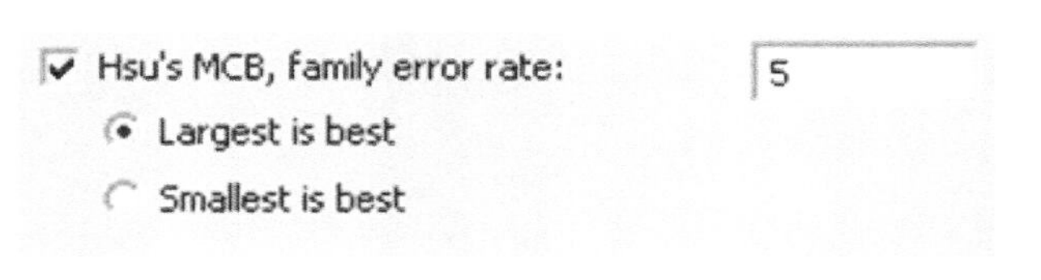


```
Hsu's MCB (Multiple Comparisons with the Best)

Family error rate = 0.05
Critical value = 2.08

Intervals for level mean minus largest of other level means
```

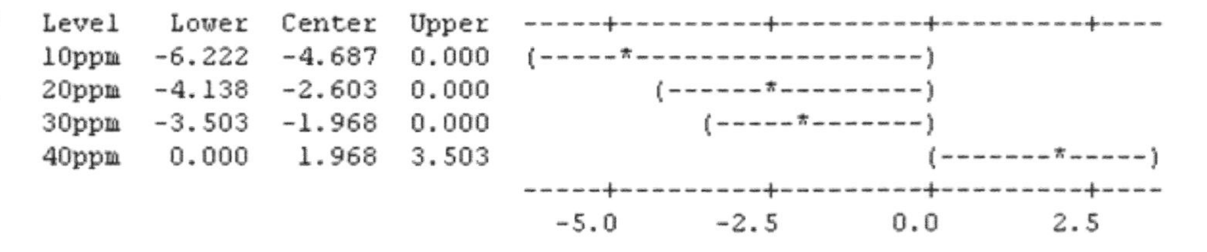

```
Level   Lower  Center  Upper  -----+---------+---------+---------+----
10ppm  -6.222  -4.687  0.000  (-----*------------------)
20ppm  -4.138  -2.603  0.000          (------*---------)
30ppm  -3.503  -1.968  0.000             (-----*-------)
40ppm   0.000   1.968  3.503                          (-------*-----)
                              -----+---------+---------+---------+----
                                 -5.0      -2.5       0.0       2.5
```

the other groups are different to the control group because none of the other intervals extend into positive values.

We can now complete the example using the Classic Method by reviewing the residuals and checking whether our test assumptions were met.

The residuals follow a fairly straight line so they can confidently be said to follow a normal distribution.

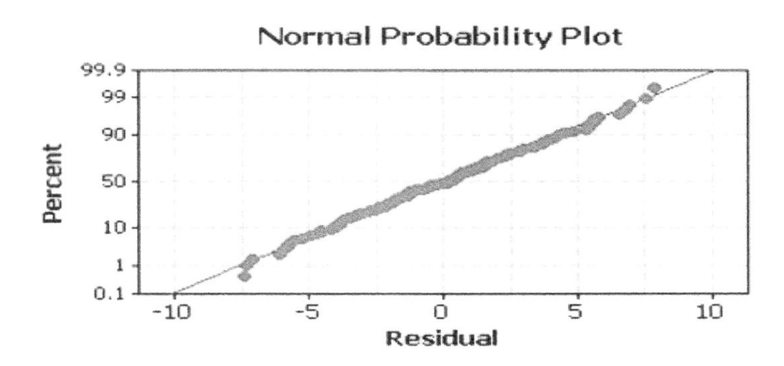

The histogram does not appear to show any skewness.

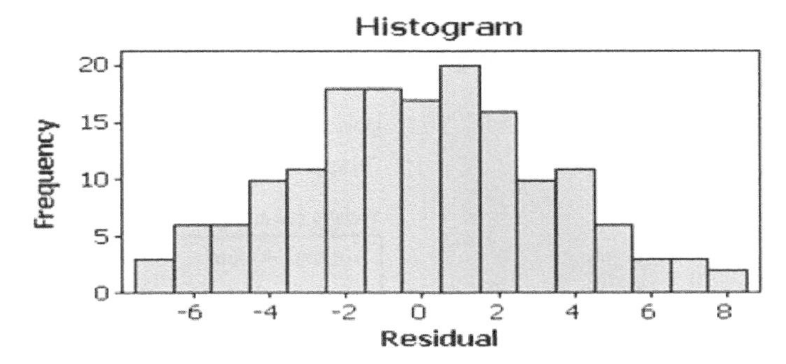

The Versus Fits plots shows that the residuals are equally distributed about the zero line. The residuals do not show any departures from normality that would lead us to question the validity of the test.

Let's now look at using the Assistant to solve this problem.

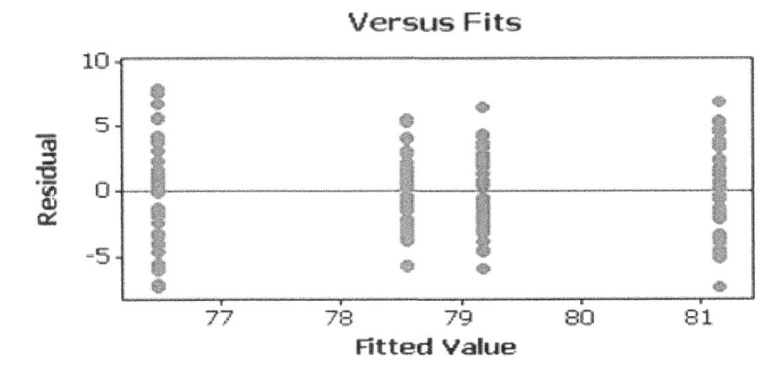

Example 2 with the Assistant

1. Click on Assistant <<Hypothesis Tests.
2. Click on One Way ANOVA within the 'Compare more than two samples' group.
3. Complete the menu box as shown and then Click OK. Note that we have again used the unstacked data and entered 2.5 as the difference that we are interested in detecting.

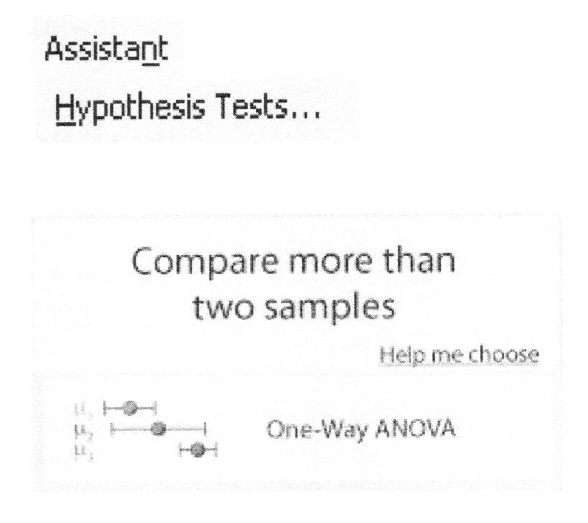

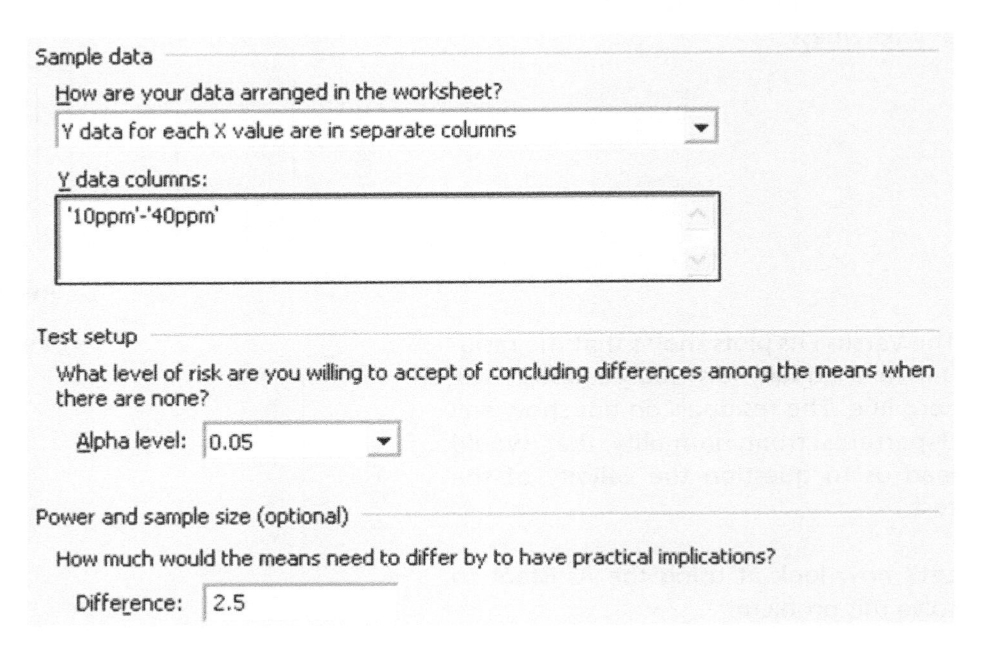

The Summary Report is shown below. The P value indicator in the top left indicates that there are differences between at least two of the groups.

The Means Comparison Chart and the Group Comparison Table both show the same results for the groups as previously calculated using the Classic Method.

On the bottom right, within the comments section, you can read the conclusion of the test procedure.

One-Way ANOVA for 10ppm, 20ppm, 30ppm, 40ppm
Summary Report

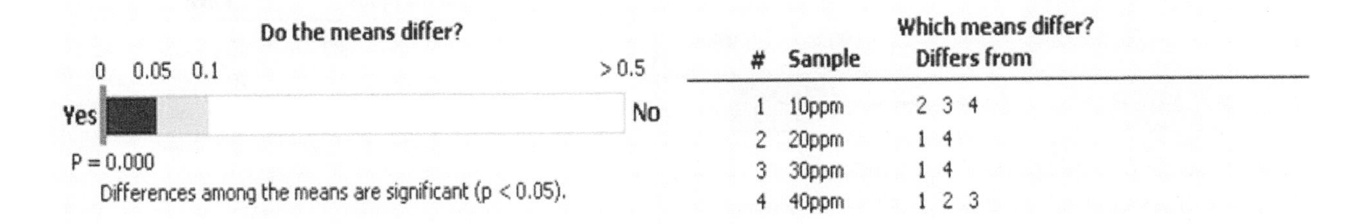

Means Comparison Chart
Red intervals that do not overlap differ.

10ppm
20ppm
30ppm
40ppm
75.0 77.5 80.0 82.5

Comments

You can conclude that there are differences among the means at the 0.05 level of significance. Use the Comparison Chart to identify means that differ. Red intervals that do not overlap indicate means that differ from each other. Consider the size of the differences to determine if they have practical implications.

On the Power Report shown below, the power indicator on the top left gives a confidence interval for the power value. We have 71.39–89.26% chance of detecting a difference of 2.5. This compares to a power of 83.7% obtained when using the Classic Method.

The bottom of the screen shows a table which breaks down the individual data based on each group.

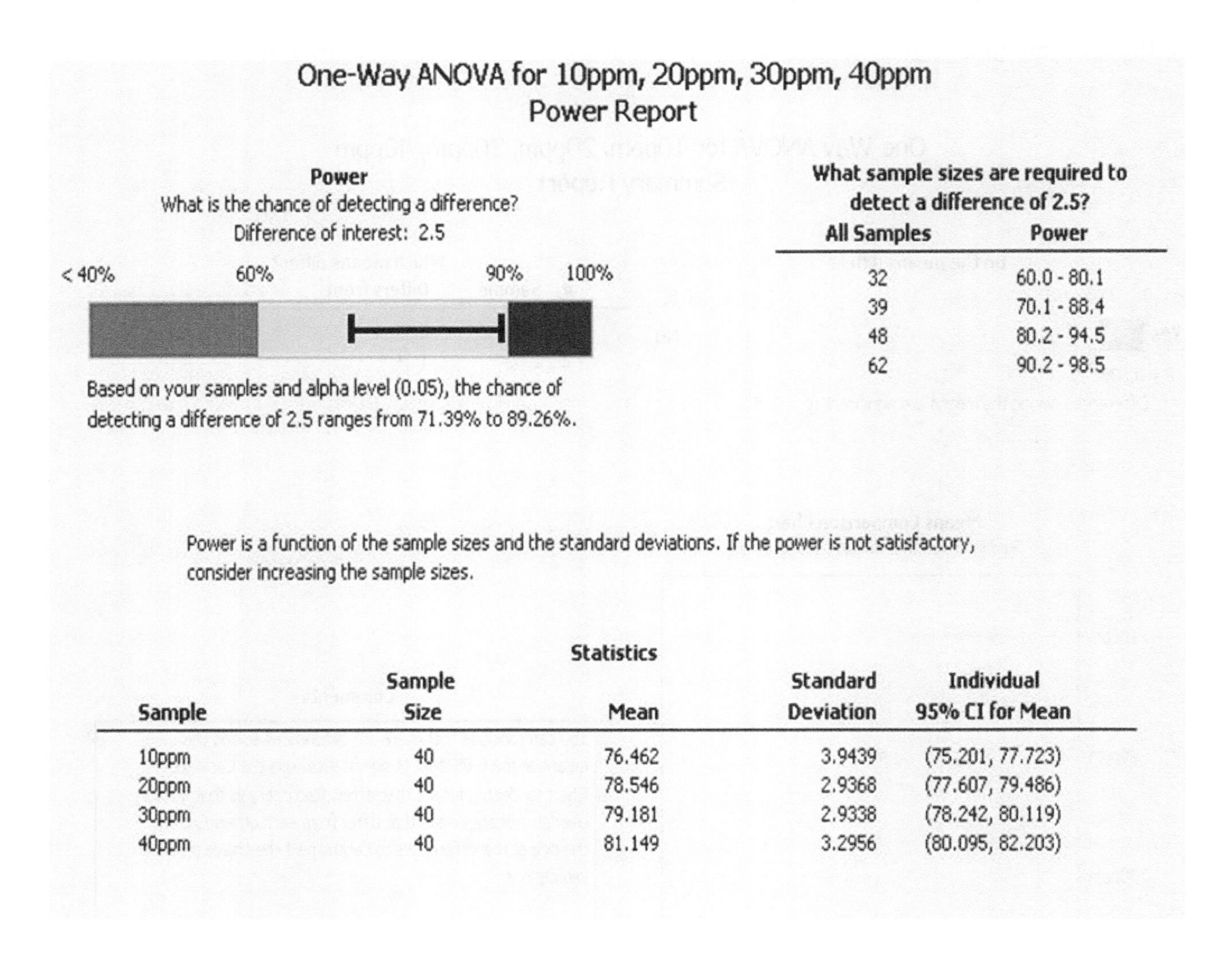

One-Way ANOVA for 10ppm, 20ppm, 30ppm, 40ppm
Power Report

Power
What is the chance of detecting a difference?
Difference of interest: 2.5

< 40% 60% 90% 100%

Based on your samples and alpha level (0.05), the chance of detecting a difference of 2.5 ranges from 71.39% to 89.26%.

What sample sizes are required to detect a difference of 2.5?

All Samples	Power
32	60.0 - 80.1
39	70.1 - 88.4
48	80.2 - 94.5
62	90.2 - 98.5

Power is a function of the sample sizes and the standard deviations. If the power is not satisfactory, consider increasing the sample sizes.

Statistics

Sample	Sample Size	Mean	Standard Deviation	Individual 95% CI for Mean
10ppm	40	76.462	3.9439	(75.201, 77.723)
20ppm	40	78.546	2.9368	(77.607, 79.486)
30ppm	40	79.181	2.9338	(78.242, 80.119)
40ppm	40	81.149	3.2956	(80.095, 82.203)

On the left of the Diagnostic Report we are presented with a histogram for each of the groups so that we can see spread, shape and location of each group.

The SPC charts on the right hand side do not show any outliers.

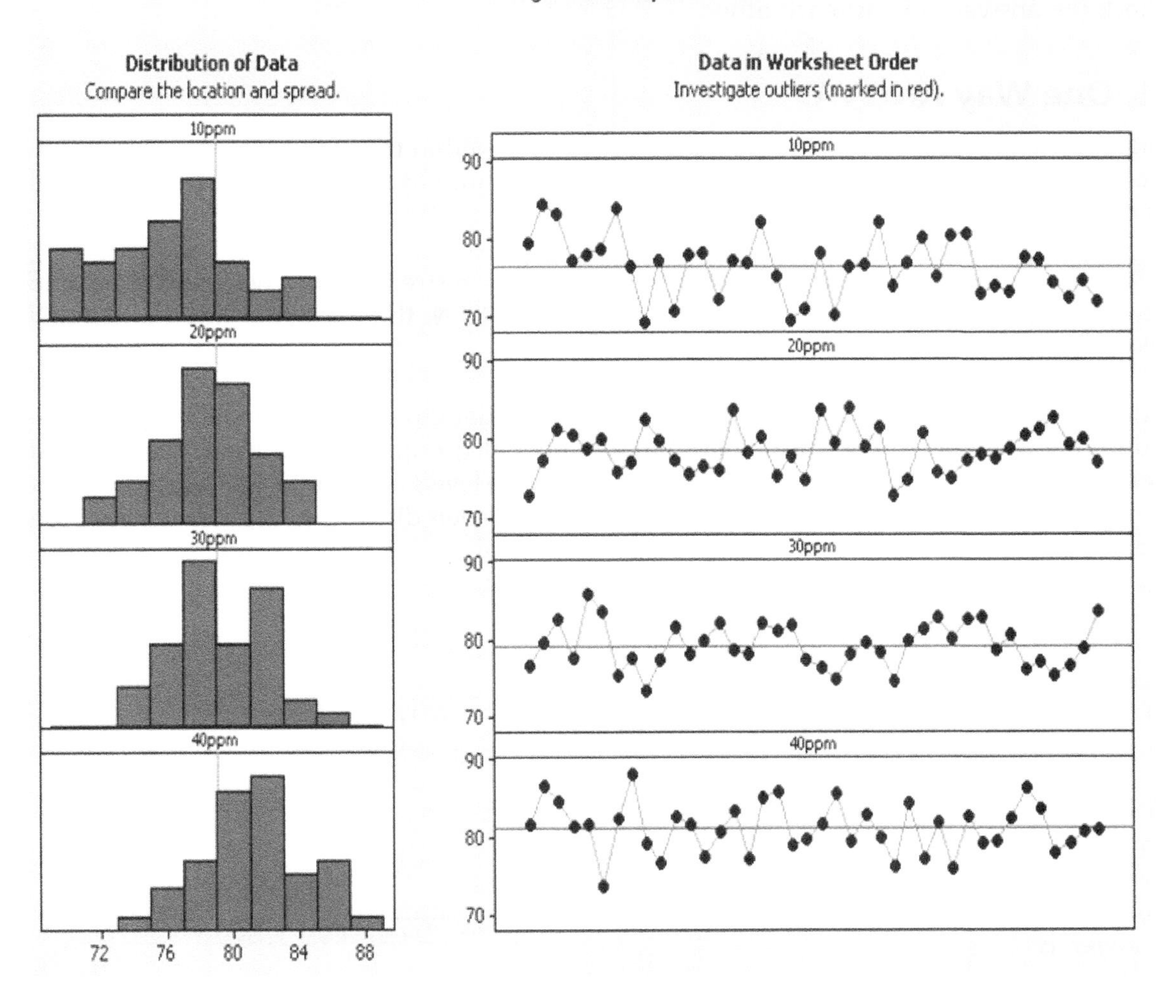

Finally, we have the Report Card. We are told that we had no unusual data points.

We are told that the test managed to detect a difference even though the power was low.

Finally, we are told that Normality is not an issue as the sample size within each group was above 15.

One-Way ANOVA for 10ppm, 20ppm, 30ppm, 40ppm
Report Card

Check	Status	Description
Unusual Data	✓	There are no unusual data points. Unusual data can have a strong influence on the results.
Sample Size	✓	The sample is sufficient to detect differences among the means. Because you entered a difference of interest, the Power Report provides a sample size evaluation for this difference. You do not need to be concerned that the power is low because the test detected a difference.
Normality	✓	Because all your sample sizes are at least 15, normality is not an issue. The test is accurate with nonnormal data when the sample sizes are large enough.

There is a short exercise for you to try. Initially use the Classic Method and then use the Assistant. Check the assumptions for the Classic Method prior to conducting the test. When both test methods have been completed check the answers against each other.

Exercise 1. One Way ANOVA.

At the plasterboard plant, Beefy has been varying the starch level within the plasterboard and monitoring board strength as his response. Beefy wants to know if the strength of the plasterboard has changed whilst he has been carrying out his starch alterations.

Initially calculate the power of the test. The max difference of mean strength that we want to be able to detect between any two levels is 8 N, the estimated StDev is 17.5 N, there are four levels and Beefy intends to take 81 samples.

After calculating the power, open Worksheet: Starch and use this data to check the underlying assumptions required for the One Way ANOVA test. Check the spread of all groups together using an Individual Value Plot. Then use One Way ANOVA to assess whether any of the levels of Starch used have a different mean strength. Use Tukey's comparison method to check how the levels differ. Don't forget to check the residuals to see if the test results are valid.

Use the Assistant to repeat the test if using M16.

It can be seen that with a sample size of 81, a StDev of 17.5 and being able to detect a difference of about 8 between groups the power that we would achieve is 67.6%. Even though the power is low Beefy wants to go ahead with the analysis.

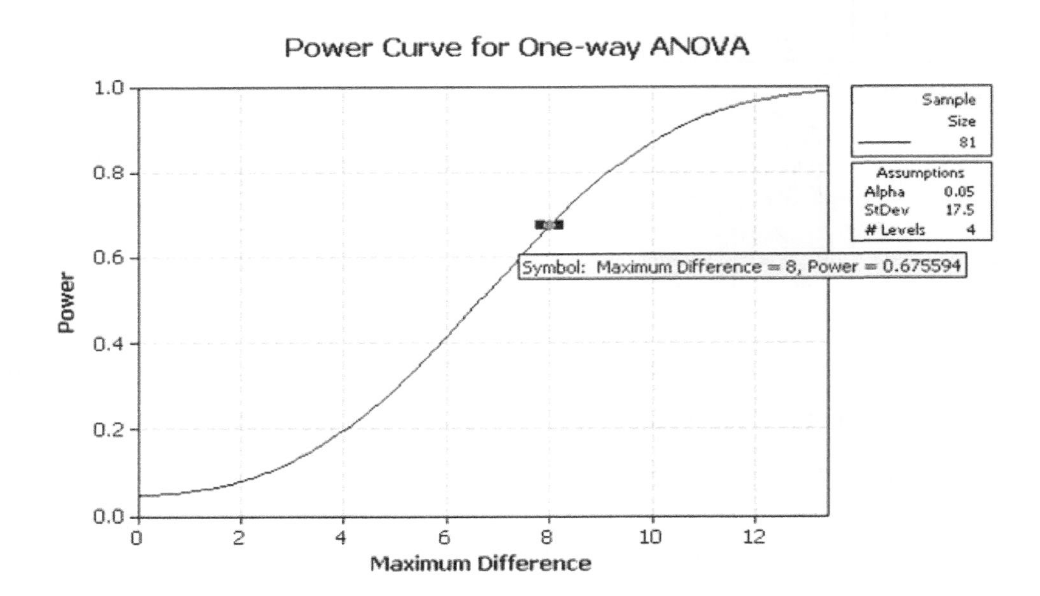

From the Individual Value Plot we can see how the sample data is distributed. It looks like starch at the 40 ppm level gives the highest strength but it is not clear which group has the lowest strength. It could be 50 or 20 ppm.

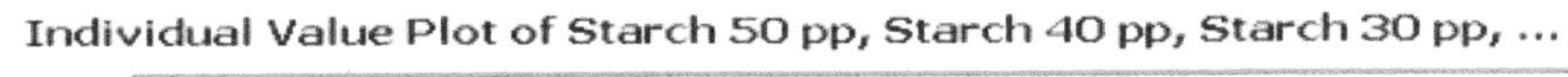

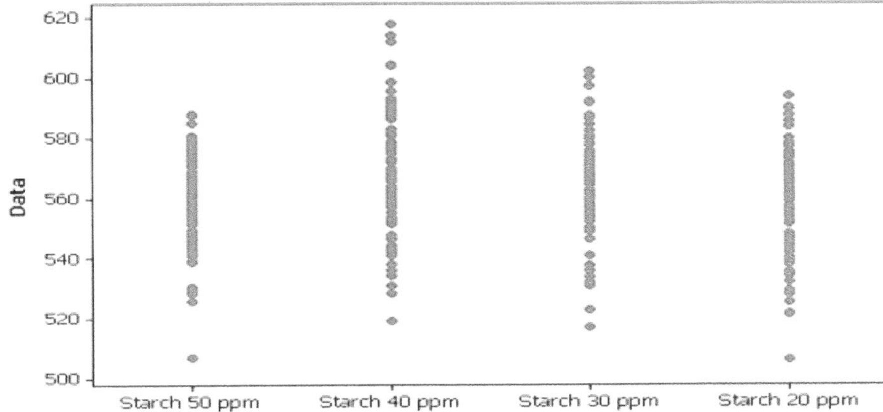

We produce all the graphical summaries. The table below shows the *P* values for the Anderson–Darling Test and the StDevs.

The data clearly meets the normality requirement. The StDev varies from 15.56 to 20.24.

Summary for Starch 20 ppm

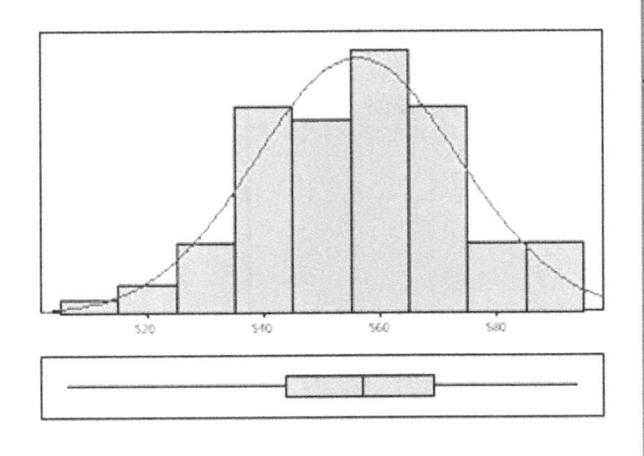

Concentration	*P*-value	StDev
20 ppm	0.959	17.52
30 ppm	0.317	17.50
40 ppm	0.918	20.24
50 ppm	0.711	15.56

The levels have equal variances as the P value for Barlett's Test is 0.133.

The normality and the equal variance assumptions have both been met.

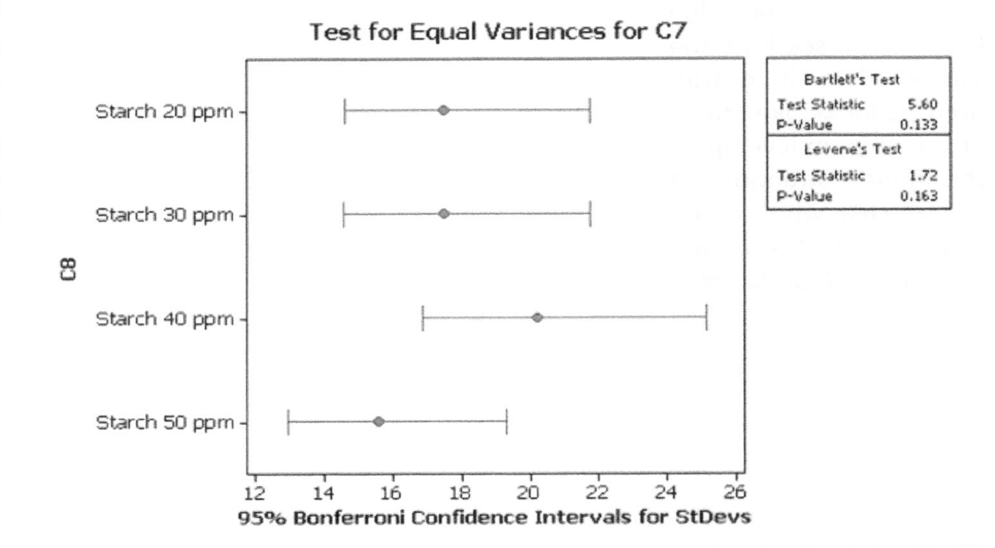

The P value for the One Way ANOVA test is zero. This means that a difference has been detected between at least two of the groups. The interval plot shows that the 20 and 40 ppm groups and the 50 and 40 ppm groups are definitely different.

Source	DF	SS	MS	F	P
Factor	3	6120	2040	6.45	0.000
Error	320	101209	316		
Total	323	107328			

S = 17.78   R-Sq = 5.70%   R-Sq(adj) = 4.82%

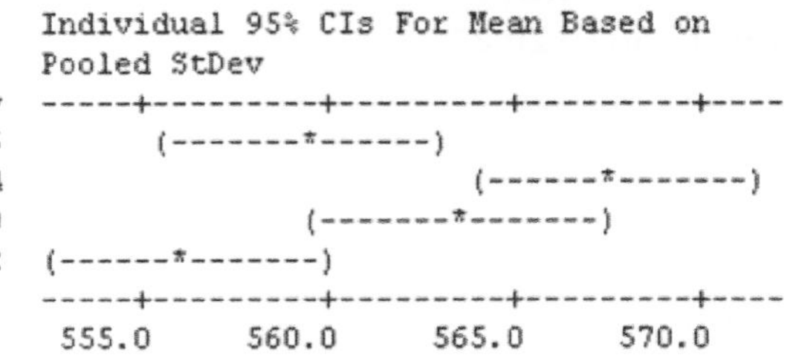

```
                                   Individual 95% CIs For Mean Based on
                                   Pooled StDev
Level           N    Mean  StDev  -----+---------+---------+---------+----
Starch 50 ppm  81  559.31  15.56        (-------*------)
Starch 40 ppm  81  567.67  20.24                         (------*-------)
Starch 30 ppm  81  563.63  17.50                (-------*-------)
Starch 20 ppm  81  556.18  17.52  (------*-------)
                                  -----+---------+---------+---------+----
                                     555.0     560.0     565.0     570.0

Pooled StDev = 17.78
```

In addition to what we have already established from the Interval Plot from Tukey's method we see that the 20 and 30 ppm groups are also different. This can be seen from the Grouping Information table.

```
Grouping Information Using Tukey Method

                 N    Mean  Grouping
Starch 40 ppm   81  567.67  A
Starch 30 ppm   81  563.63  A B
Starch 50 ppm   81  559.31    B C
Starch 20 ppm   81  556.18      C

Means that do not share a letter are significantly different.
```

The last section of the pairwise comparison also shows that the 30 and 20 ppm groups are different as the confidence interval does not include zero.

Starch 50 ppm subtracted from:

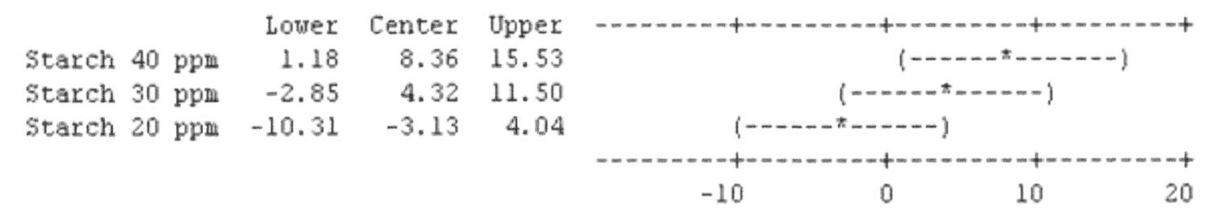

```
                 Lower  Center  Upper  ---------+---------+---------+---------+
Starch 40 ppm     1.18    8.36  15.53                       (------*-------)
Starch 30 ppm    -2.85    4.32  11.50                   (------*------)
Starch 20 ppm   -10.31   -3.13   4.04            (------*------)
                                       ---------+---------+---------+---------+
                                              -10         0        10        20
```

Starch 40 ppm subtracted from:

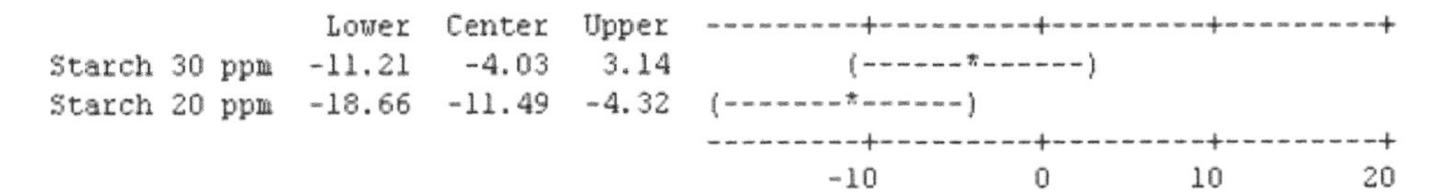

```
                 Lower  Center  Upper  ---------+---------+---------+---------+
Starch 30 ppm   -11.21   -4.03   3.14          (------*------)
Starch 20 ppm   -18.66  -11.49  -4.32  (-------*------)
                                       ---------+---------+---------+---------+
                                              -10         0        10        20
```

Starch 30 ppm subtracted from:

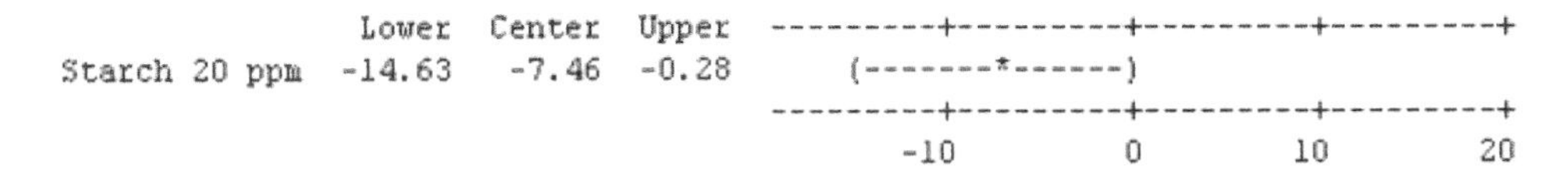

```
                 Lower  Center  Upper  ---------+---------+---------+---------+
Starch 20 ppm   -14.63   -7.46  -0.28      (-------*------)
                                       ---------+---------+---------+---------+
                                              -10         0        10        20
```

The residuals follow a fairly straight line so they can confidently be said to follow a normal distribution.

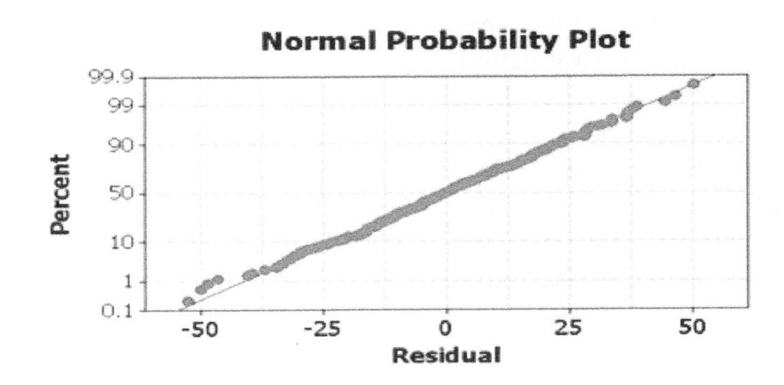

The histogram does not appear to show any skewness.

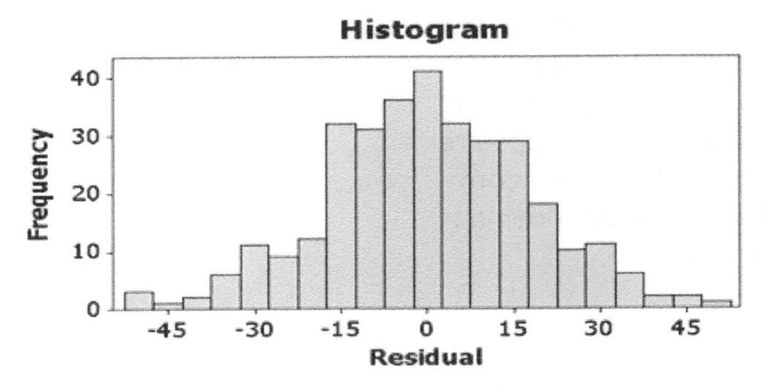

The Versus Fits plots shows that the residuals are equally distributed about the zero line. The residuals do not show any departures from normality that would lead us to question the validity for the test.

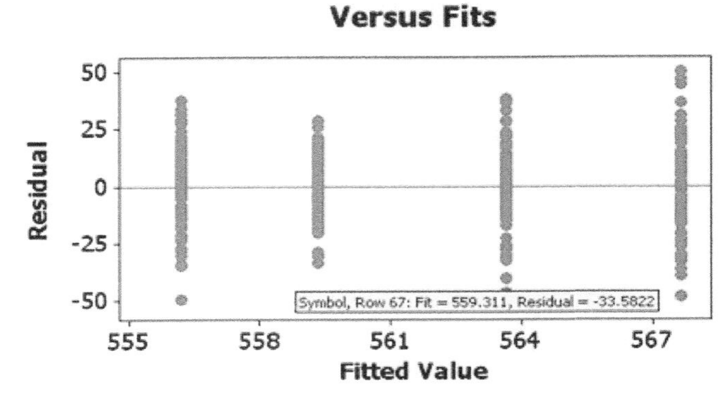

Let's now look at using the Assistant to solve this problem.

The *P* value indicator in the top left indicates that there are differences in at least two of the groups.

The Means Comparison Chart and the Group Comparison Table both show the same results for the groups as previously calculated using the Classic Method. Although, be careful as Minitab has reordered our groups.

On the bottom right, within the comments section, you can read the conclusion of the test procedure.

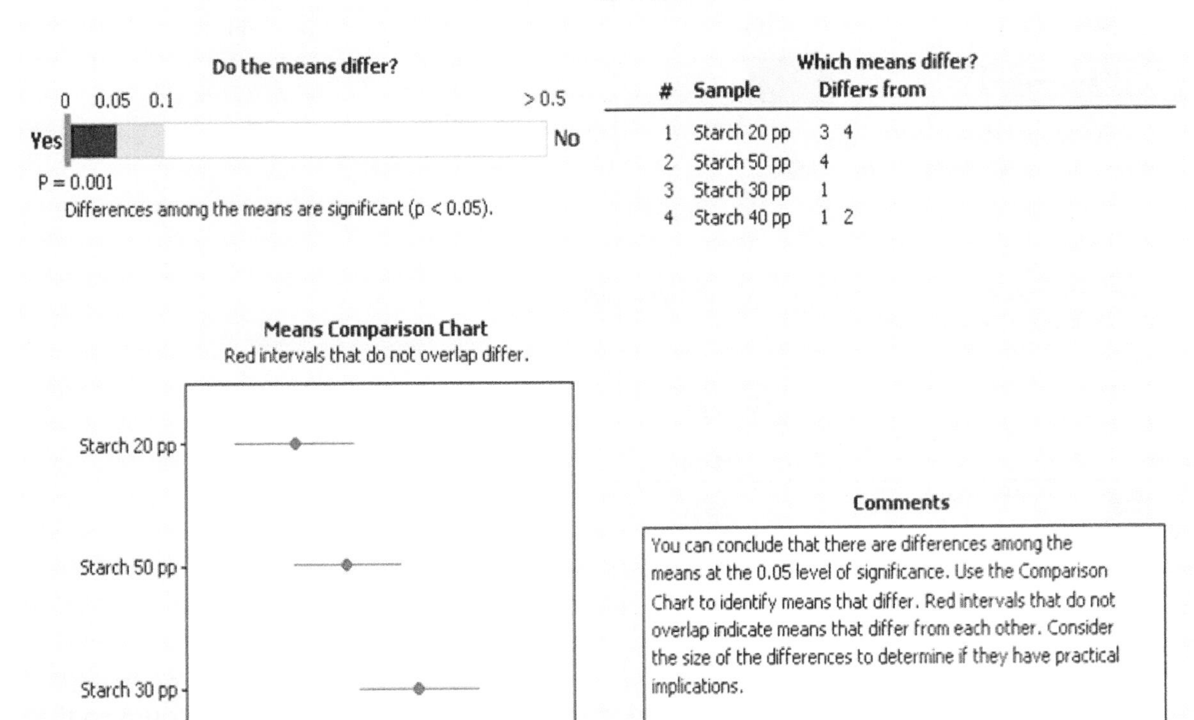

The power indicator on the top left gives a confidence interval for the power value. We have 59.62–72.54% chance of detecting a difference of eight.

The bottom of the screen shows a table which breaks down the individual data based on each group.

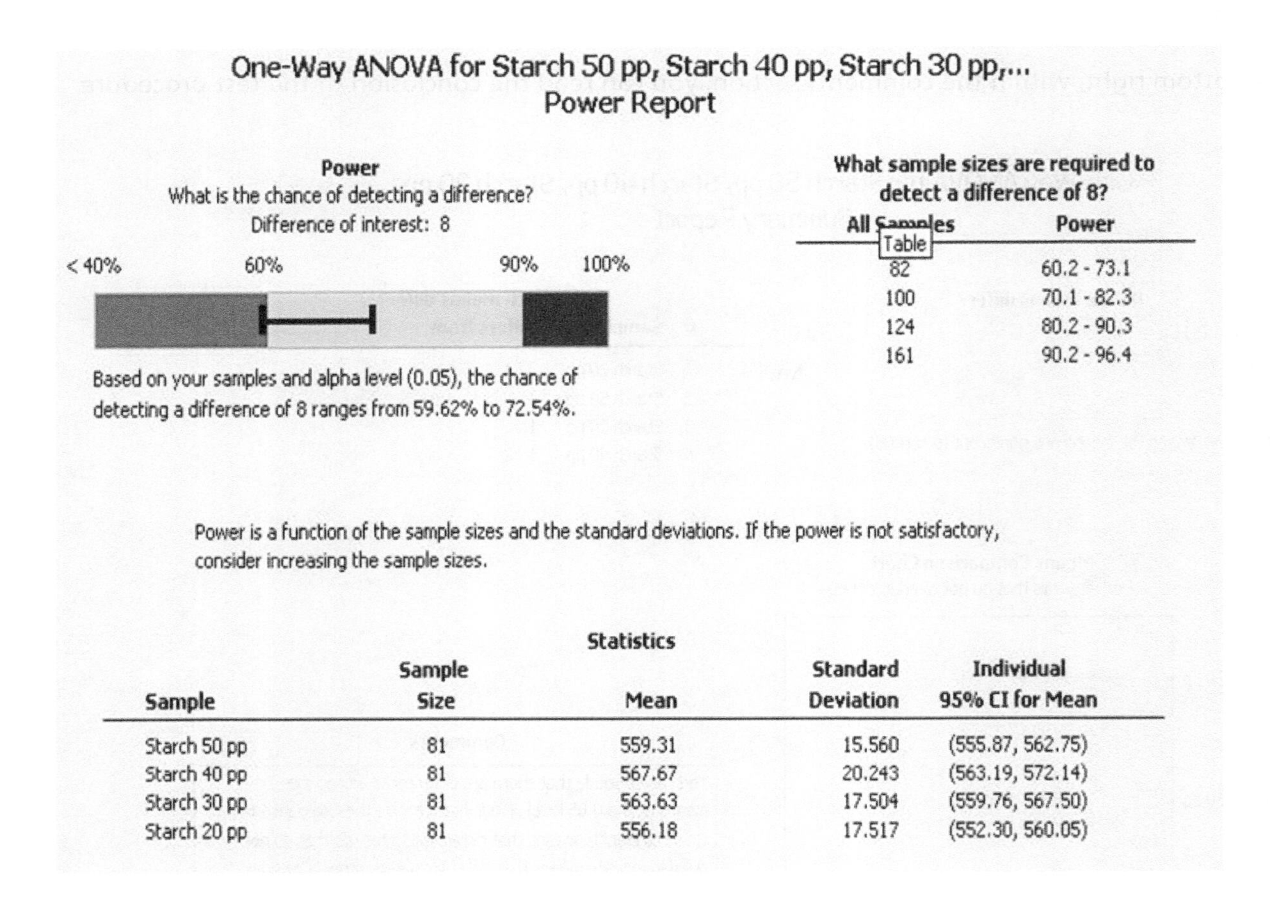
One-Way ANOVA for Starch 50 pp, Starch 40 pp, Starch 30 pp,...
Power Report

Power
What is the chance of detecting a difference?
Difference of interest: 8

< 40% 60% 90% 100%

Based on your samples and alpha level (0.05), the chance of detecting a difference of 8 ranges from 59.62% to 72.54%.

What sample sizes are required to detect a difference of 8?

All Samples	Power
82	60.2 - 73.1
100	70.1 - 82.3
124	80.2 - 90.3
161	90.2 - 96.4

Power is a function of the sample sizes and the standard deviations. If the power is not satisfactory, consider increasing the sample sizes.

Statistics

Sample	Sample Size	Mean	Standard Deviation	Individual 95% CI for Mean
Starch 50 pp	81	559.31	15.560	(555.87, 562.75)
Starch 40 pp	81	567.67	20.243	(563.19, 572.14)
Starch 30 pp	81	563.63	17.504	(559.76, 567.50)
Starch 20 pp	81	556.18	17.517	(552.30, 560.05)

On the left of the Diagnostic Report we are presented with a histogram for each of the groups so that we can see spread, shape and location of each group.

The SPC charts on the right hand side show a couple of outliers. However, two outliers, which are not extreme, in all this data is nothing to worry about.

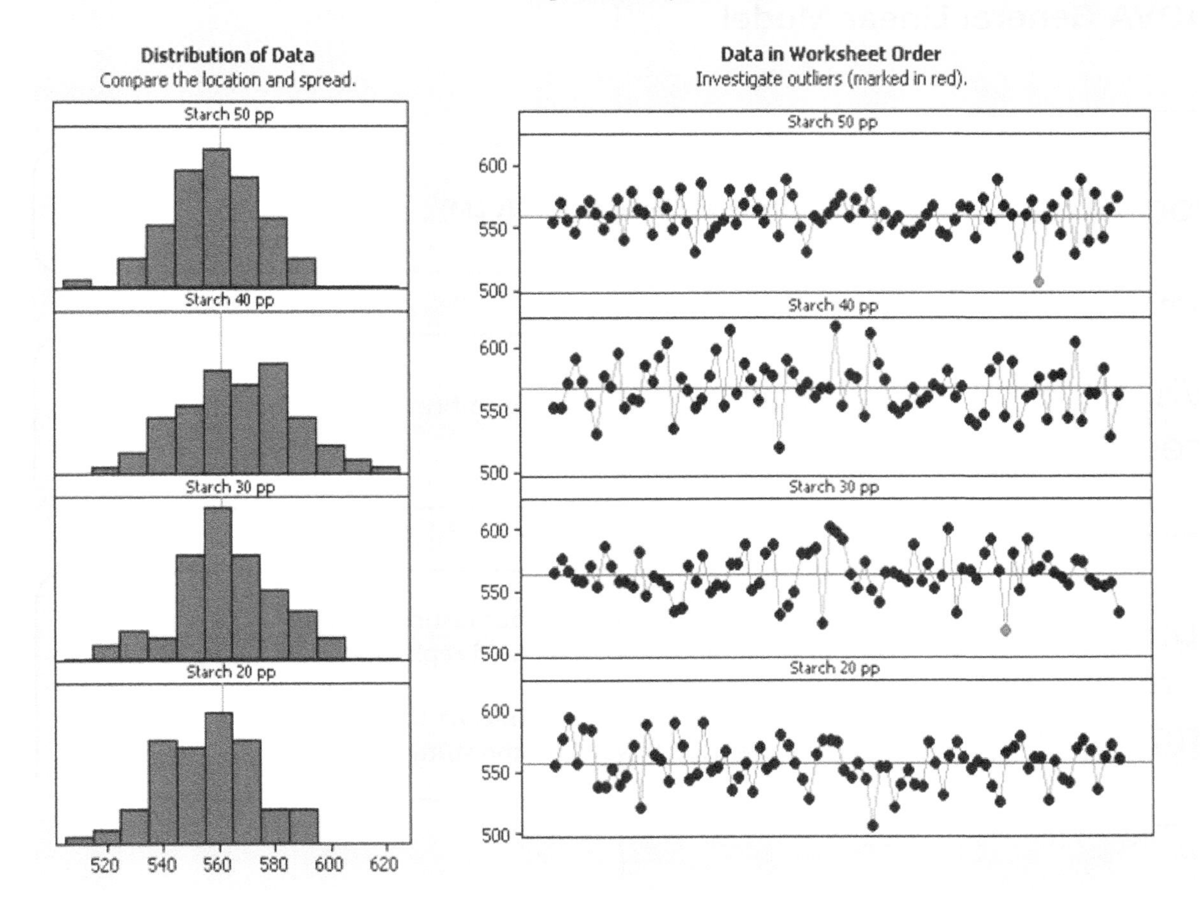

Finally, we have the Report Card and we are informed about unusual data points.

We are told that the test managed to detect a difference even though the power was low. On this occasion, Beefy's gamble of using a low power paid off.

Finally, we are told that normality is not an issue as the sample size within each group was above 15.

One-Way ANOVA for Starch 50 pp, Starch 40 pp, Starch 30 pp,...
Report Card

Check	Status	Description
Unusual Data	⚠	Some of the data points are unusual compared to the others in the same sample. Because unusual data can have a strong influence on the results, you should try to identify the cause of their unusual nature. These points are marked in red on the Diagnostic Report. You can hover over a point or use Minitab's brushing feature to identify the worksheet row. Correct any data entry or measurement errors. Consider removing data that are associated with special causes and repeating the analysis.
Sample Size	✓	The sample is sufficient to detect differences among the means. Because you entered a difference of interest, the Power Report provides a sample size evaluation for this difference. You do not need to be concerned that the power is low because the test detected a difference.
Normality	✓	Because all your sample sizes are at least 15, normality is not an issue. The test is accurate with nonnormal data when the sample sizes are large enough.

We are going on to look at the ANOVA General Linear Model for the remainder of this chapter.

5.4 ANOVA General Linear Model

Procedure	• ANOVA General Linear Model (GLM)
What's it used for?	• It is used directly to assess the significance of multiple factors at two or more levels.
Assumptions and Limitations	• There must always be continuous response data. • For our purposes the ANOVA GLM replaces the Two Way ANOVA and the Balanced ANOVA. • The ANOVA GLM is only available via the Classic Method. • Residuals are used to validate the study.

The Terminology of ANOVA GLM

The GLM can handle both balanced and unbalanced designs. A balanced design has the same number of runs for each combination of factors.

The GLM can handle multiple comparisons.

Predictors can be fixed factors or random factors. Fixed factors are discrete variables (e.g. low, medium, high). The GLM will check whether the means of the response data is different at these levels. Random factors are also discrete variables but they are selected from a larger population.

Continuous predictors can be entered as covariates. Covariate predictors can either be items that are controllable or items that are not controllable. The addition of a covariate can change the results of a study, as seen within the Rotary Kiln example.

Example 3. ANOVA GLM.

Aman wants to buy a new car stereo. He decides to carry out some research into prices of stereo systems. He varies three factors that he is interested in and then records the price of the systems. The factors are number of speakers, number of CDs the CD player can hold and whether or not the stereo has an MP3 input. Study the data in Worksheet: Car and decide which of the factors are significant to Cost. Do any of the factors interact with each other? Are the levels of each of the factors significantly different from each other? Finally, check the residuals from the study to establish if there are any problems with the validity of the study.

The discrete levels for these factors are as follows:

Speakers: 4, 8, 12

CD: Single, Double, Multi

MP3 Input: Yes, No

Previously, it was said that it always a good idea to view the data prior to conducting the test procedure and we used the Graphical Summary to do this. For the ANOVA GLM we use the Main Effects & Interactions plots to look at our data before we conduct the GLM procedure.

C1	C2-T	C3-T	C4
Speakers	CD	MP3 Input	Cost
4	Single	Yes	123.075
4	Single	No	96.075
4	Double	Yes	103.425
4	Double	No	87.925

1. Transfer the data from Worksheet: Car into Minitab.
2. Click Stat <<ANOVA <<Main Effects Plot.
3. Enter Cost as the response and Speaker, CD and MP3 Input as the Factors. Click on OK.

Stat

ANOVA Main Effects Plot...

Responses:
Cost

Factors:
Speakers CD 'MP3 Input'

The Main Effects Plot isolates each of the factors and shows how the response variable changes due to only changes in the factor being studied. The steeper the line the greater the effect of the change. However, we do not know if the factors are significant at this point.

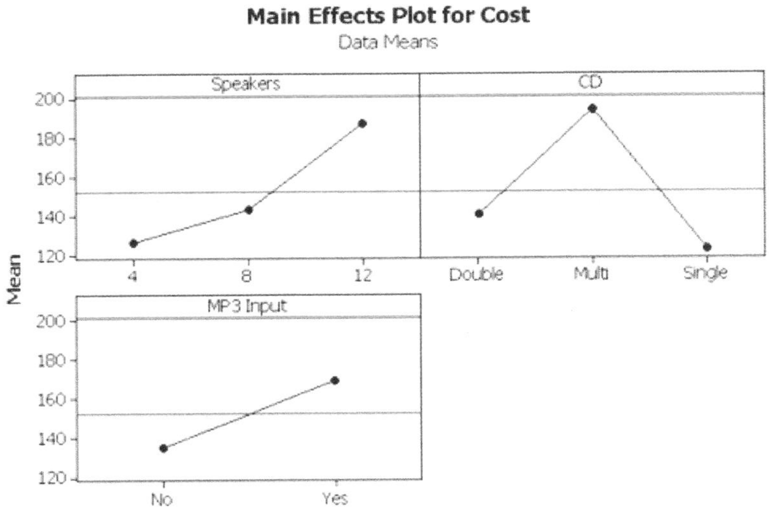

Main Effects Plot for Cost
Data Means

Speakers: 4, 8, 12 — CD: Double, Multi, Single — MP3 Input: No, Yes — Mean: 120, 140, 160, 180, 200

4. Click Stat <<ANOVA <<Interactions Plot.

Stat

ANOVA Interactions Plot...

5. Enter Cost as the response and Speaker, CD and MP3 Input as the Factors. Click on OK (the full interaction plot just reverses the axis of the graphs and displays them again).

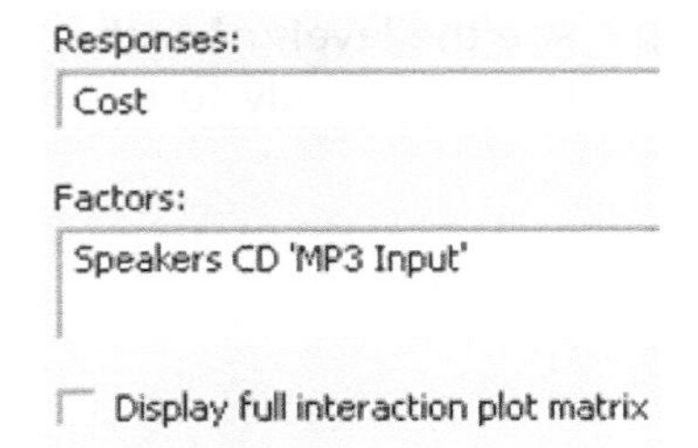

An interaction between the factors produces a change different to the level of change we were expecting. No interaction manifests itself as parallel lines. There does not appear to be an interaction between Speakers and CD because the lines are almost parallel. The same can be said for Speakers and MP3 Input. However, there does appear to be an interaction between CD and MP3 Input and we will find out if this is significant when we carry out the analysis.

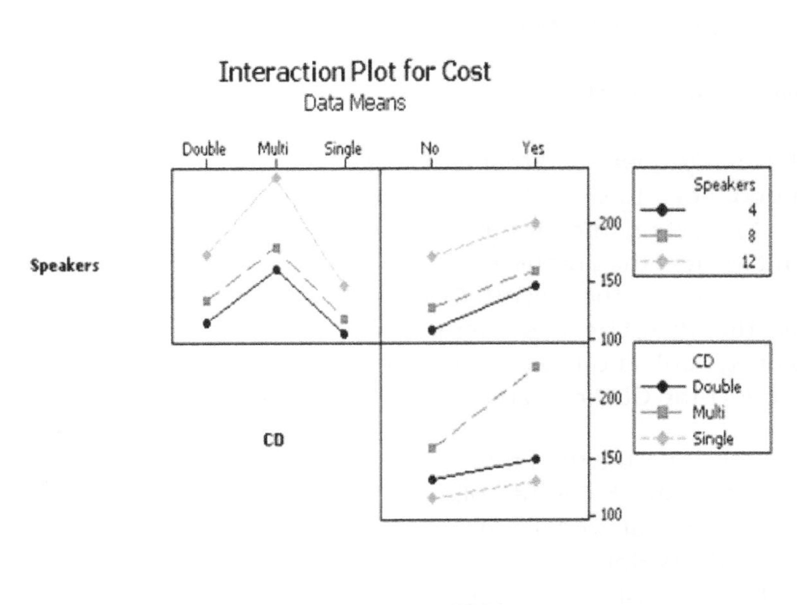

Now that we have had a look at the data we will carry out the ANOVA GLM procedure.

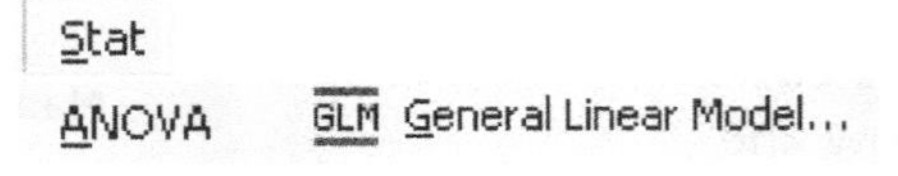

6. Click Stat <<ANOVA <<General Linear Model
7. Complete the menu box as shown. Note there are two ways of adding the factors. The vertical bar indicates the main effect as well as all the interactions and using it saves a lot of typing. If typing out in full we use a * to indicate an interaction term.
8. Click OK.

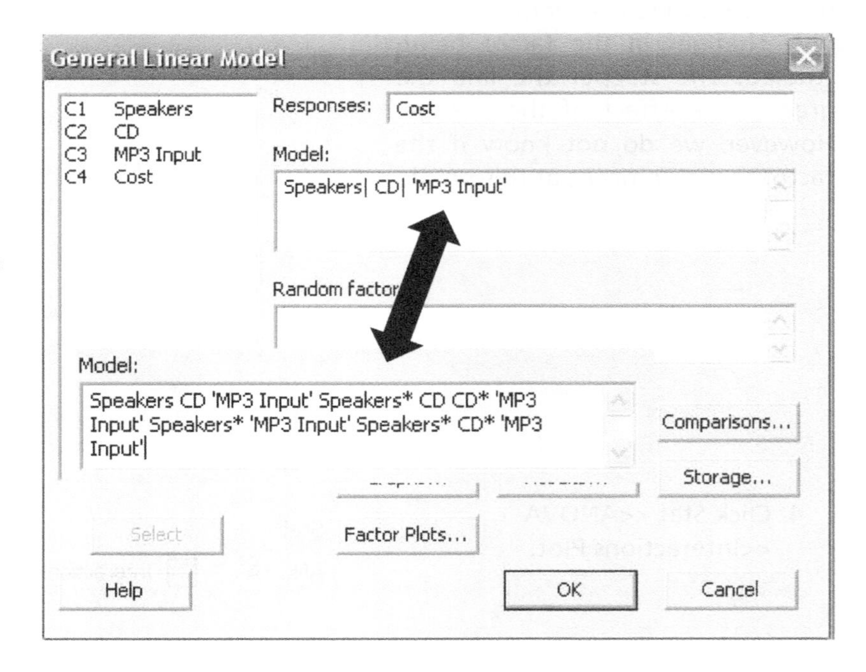

Within the Session Window we see the factors and interactions listed down the left hand side. Reading along the table we see the *P* value for each factor. The factors with a *P* value below our significance level of 0.05 are significant. The R-Sq(adj) value means that 86.88% of the changes within our response can be explained by changes within our factors. R-Sq(adj) is a value that shows the fit and allows equal comparison of models with different numbers of terms. We will now reduce our model which means remove the terms that are not significant.

Analysis of Variance for Cost, using Adjusted SS for Tests

Source	DF	Seq SS	Adj SS	Adj MS	F	P
Speakers	2	23107.8	23107.8	11553.9	38.33	0.000
CD	2	33106.9	33106.9	16553.5	54.91	0.000
MP3 Input	1	10448.8	10448.8	10448.8	34.66	0.000
Speakers*CD	4	1650.7	1650.7	412.7	1.37	0.284
CD*MP3 Input	2	6009.9	6009.9	3005.0	9.97	0.001
Speakers*MP3 Input	2	176.2	176.2	88.1	0.29	0.750
Speakers*CD*MP3 Input	4	480.9	480.9	120.2	0.40	0.807
Error	18	5426.0	5426.0	301.4		
Total	35	80407.2				

S = 17.3621 R-Sq = 93.25% R-Sq(adj) = 86.88%

As we will only reduce the model once we will produce our residual plot and also the Grouping Information Table with the reduced model.

9. Click the Edit last Dialog icon.
10. Complete the menu box as shown, on this occasion it is not worth using the vertical bar shortcut.
11. Click on Comparisons.

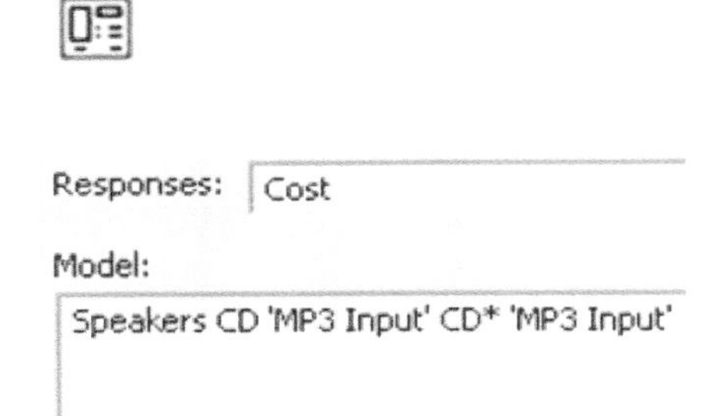

We are going to use pairwise comparisons to see how the levels within our factors differ. We are going to look at the Grouping Information Tables to check the comparison. M15 users will have to use Confidence Intervals for the comparisons.

12. Select Speakers, CD and MP3 Input as the factors for which we are going to display tables. Click OK and then click on graphs.

General Linear Model - Comparisons

C1 Speakers
C2 CD
C3 MP3 Input
C4 Cost

Pairwise comparisons
Comparisons with a control
Terms: Speakers CD 'MP3 Input'
Control levels:
Method: Tukey, Dunnett, Bonferroni, Sidak
Alternative: Less than, Not equal, Greater than
Grouping information
Confidence interval
Test
Confidence level: 95.0
Select
Help OK Cancel

13. Click Four in One for Residual plots.

14. Click OK and OK again.

Residual Plots
Individual plots
Histogram of residuals
Normal plot of residuals
Residuals versus fits
Residuals versus order
Four in one

From the session window we can confirm that the remaining factors and single interaction are significant. Notice that the R-Sq(adj) has increased as we now have a better model.

Analysis of Variance for Cost, using Adjusted SS for Tests

Source	DF	Seq SS	Adj SS	Adj MS	F	P
Speakers	2	23108	23108	11554	41.83	0.000
CD	2	33107	33107	16553	59.93	0.000
MP3 Input	1	10449	10449	10449	37.83	0.000
CD*MP3 Input	2	6010	6010	3005	10.88	0.000
Error	28	7734	7734	276		
Total	35	80407				

S = 16.6195   R-Sq = 90.38%   R-Sq(adj) = 87.98%

Each of our factors is given its own Grouping Information Table so we can see if the levels within the factors are actually different from each other. The Main Effects plot for each factor has been placed next to its relevant table. All the levels within the CD and MP3 Input factor are different. For the speaker group, the factors that cannot be said to be different have been grouped.

Now that we have seen which factors and levels are different we will have a look at the residuals.

Speakers	N	Mean	Grouping
12	12	187.0	A
8	12	143.2	B
4	12	127.0	B

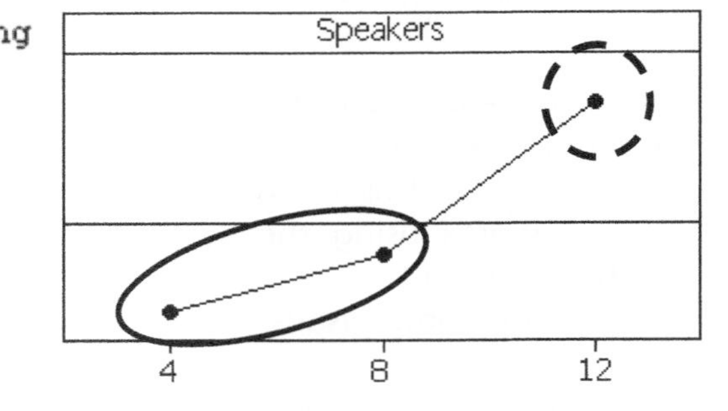

CD	N	Mean	Grouping
Multi	12	194.0	A
Double	12	140.6	B
Single	12	122.6	C

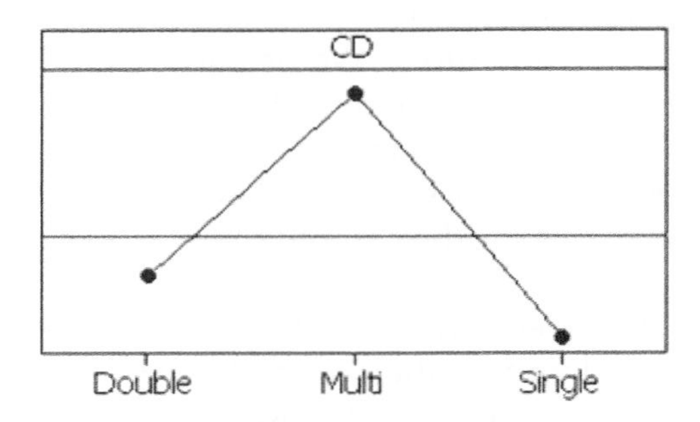

MP3 Input	N	Mean	Grouping
Yes	18	169.4	A
No	18	135.4	B

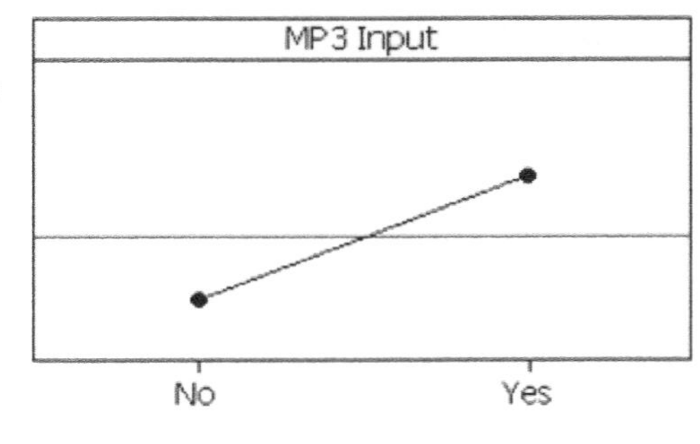

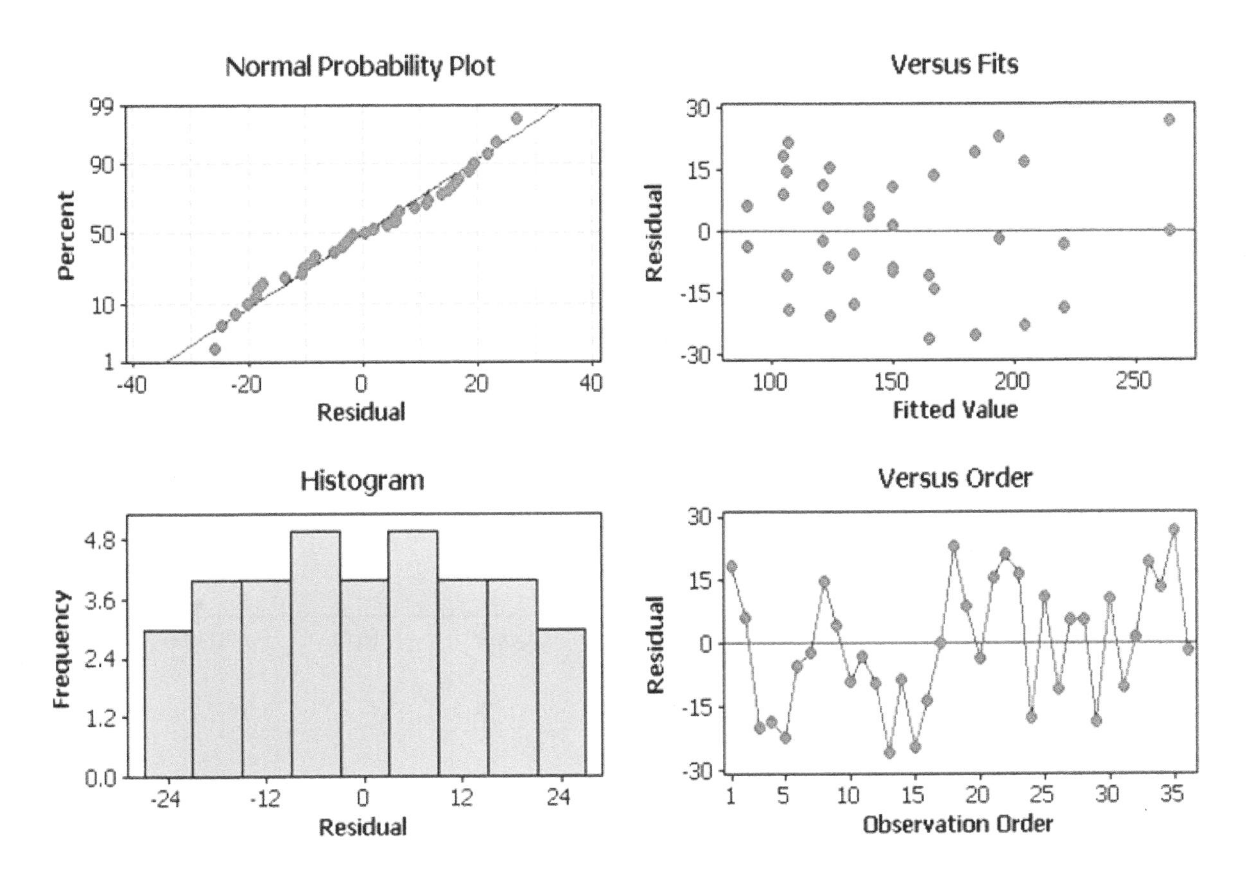

We check the validity of the model by looking for patterns in the 4 in 1 Residual Plot. Starting from the top and going from left to right:

- The Normal Probability Plot shows that the Residuals are normally distributed.
- The Versus Fits Plot shows that the Residuals are equally distributed about the centre line.
- The Histogram Plot shows that the Residuals do not show extreme skewness.
- The Versus Order Plot shows that the Residuals do not display any patterns or ordered behaviour.

In summary, there are no issues with the residuals within this study.

Aman can conclude that Speakers, CD and MP3 Input are all significant factors when it comes to stereo system cost. The cost of a system with both Multi CD and MP3 Input goes up considerably, this comes from the interaction term.

There is not a significant cost difference if you buy a system with four or eight speakers whereas there is a difference if you buy a system with 12 speakers.

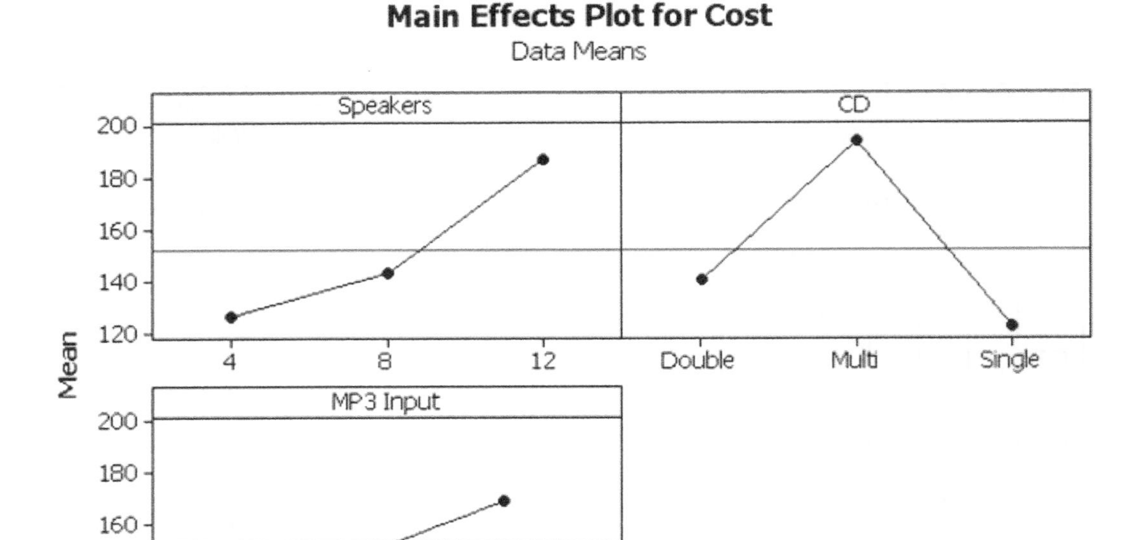

Example 4. ANOVA GLM.

Within the nuclear industry UF6 gas is converted to UO2 powder inside a rotary kiln. The UF6 is reacted with steam at high temperature and pressure. The UO2 produced is a powder but other non uranic impurities are present. The quality of the UO2 is measured by checking for residual fluorides within the product. Using the factors of steam pressure, kiln temperature and revolutions per minute (rpm) assess which factors are significant in the production of fluorides. Are there any interactions present?

Worksheet: Kiln.

We will look at the data before we carry out the GLM procedure by producing Main Effects and Interactions plots. We will not be using the RoomTemp data immediately.

C1-T	C2	C3	C4	C5
SteamPress	RPM	KilnTemp	Fluorides	RoomTemp
Low	50	200	183.07	14
Low	50	200	202.40	21
Low	50	200	190.89	18

1. Transfer the data from Worksheet: Kiln into Minitab.

2. Click Stat <<ANOVA <<Main Effects Plot.

3. Enter Fluorides as the response and SteamPress, RPM and KilnTemp as the Factors. Click on OK.

Responses:
Fluorides

Factors:
SteamPress RPM KilnTemp

The Main Effects Plot shows that all the factors appear to be having an effect upon the response variable. KilnTemp appears to be having the greatest effect and then SteamPress.

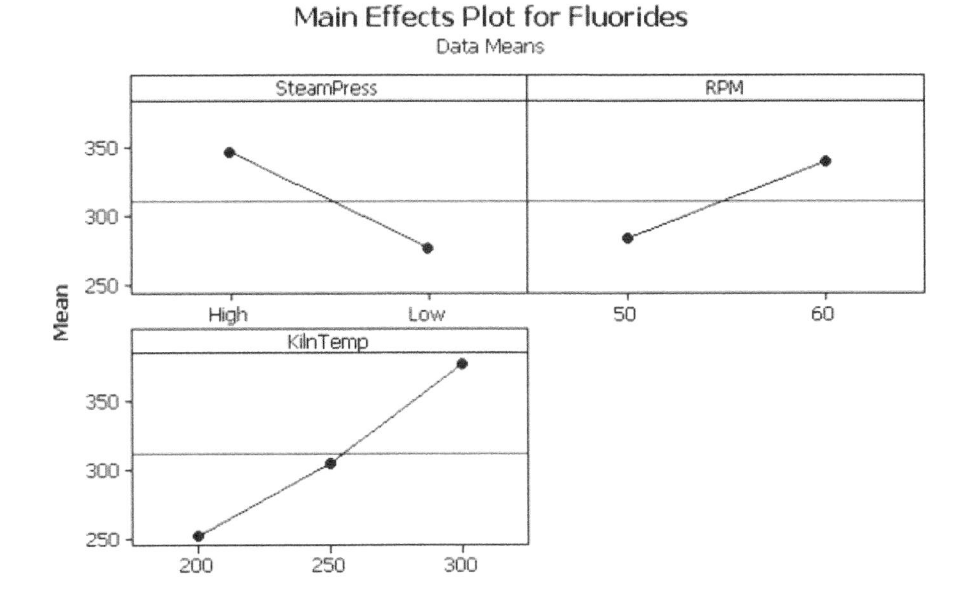

4. Click Stat <<ANOVA <<Interactions Plot.
5. Enter Fluorides as the response and SteamPress, RPM and KilnTemp as the Factors. Click on OK.

Stat
ANOVA
Interactions Plot...

Responses:
Fluorides

Factors:
SteamPress RPM KilnTemp

On the Interaction Plot lines that are not parallel indicate an interaction. There only appears to be a slight interaction present for the interactions involving KilnTemp.

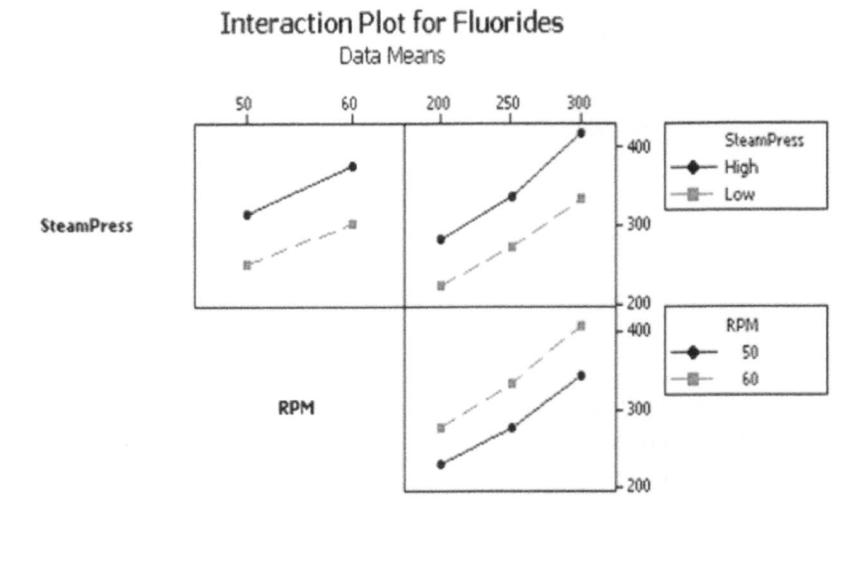

Now that we have had a look at the data we will carry out the ANOVA GLM procedure.

Stat

ANOVA GLM General Linear Model...

6. Click Stat <<ANOVA <<General Linear Model
7. Complete the menu box as shown.
8. Click OK. Then go to the Session window to look at the results table.

Responses: Fluorides

Model:

SteamPress| RPM| KilnTemp

All of the factors and all of the two way interactions are significant. The three way interaction is not significant.

We will now reduce the model and remove the nonsignificant three way interaction.

Source	DF	Seq SS	Adj SS	Adj MS	F	P
SteamPress	1	88191	88191	88191	692.24	0.000
RPM	1	57038	57038	57038	447.70	0.000
KilnTemp	2	187179	187179	93590	734.61	0.000
SteamPress*RPM	1	629	629	629	4.94	0.030
SteamPress*KilnTemp	2	2229	2229	1115	8.75	0.000
RPM*KilnTemp	2	868	868	434	3.40	0.040
SteamPress*RPM*KilnTemp	2	118	118	59	0.46	0.632
Error	60	7644	7644	127		
Total	71	343897				

S = 11.2872 R-Sq = 97.78% R-Sq(adj) = 97.37%

9. Click the Edit Last Dialog icon.

10. Complete the menu box as shown. Note, that we can use the minus sign to remove the three way interaction.

Responses: Fluorides

Model:

SteamPress RPM KilnTemp SteamPress* RPM SteamPress* KilnTemp RPM* KilnTemp

or

Responses: Fluorides

Model:

SteamPress| RPM| KilnTemp- SteamPress* RPM* KilnTemp

11. Click on Graphs and select the Four in one residual plots.

12. Click OK and OK again. Then go to the session window.

Residual Plots

Individual plots

Histogram of residuals

Normal plot of residuals

Residuals versus fits

Residuals versus order

Four in one

Make a note of the significant factors and interactions because in the next exercise we are going to look at the effects of adding in a continuous factor as a covariate.

For now we still need to check the residuals.

Source	DF	Seq SS	Adj SS	Adj MS	F	P
SteamPress	1	88191	88191	88191	704.46	0.000
RPM	1	57038	57038	57038	455.61	0.000
KilnTemp	2	187179	187179	93590	747.58	0.000
SteamPress*RPM	1	629	629	629	5.03	0.029
SteamPress*KilnTemp	2	2229	2229	1115	8.90	0.000
RPM*KilnTemp	2	868	868	434	3.46	0.037
Error	62	7762	7762	125		
Total	71	343897				

S = 11.1888   R-Sq = 97.74%   R-Sq(adj) = 97.42%

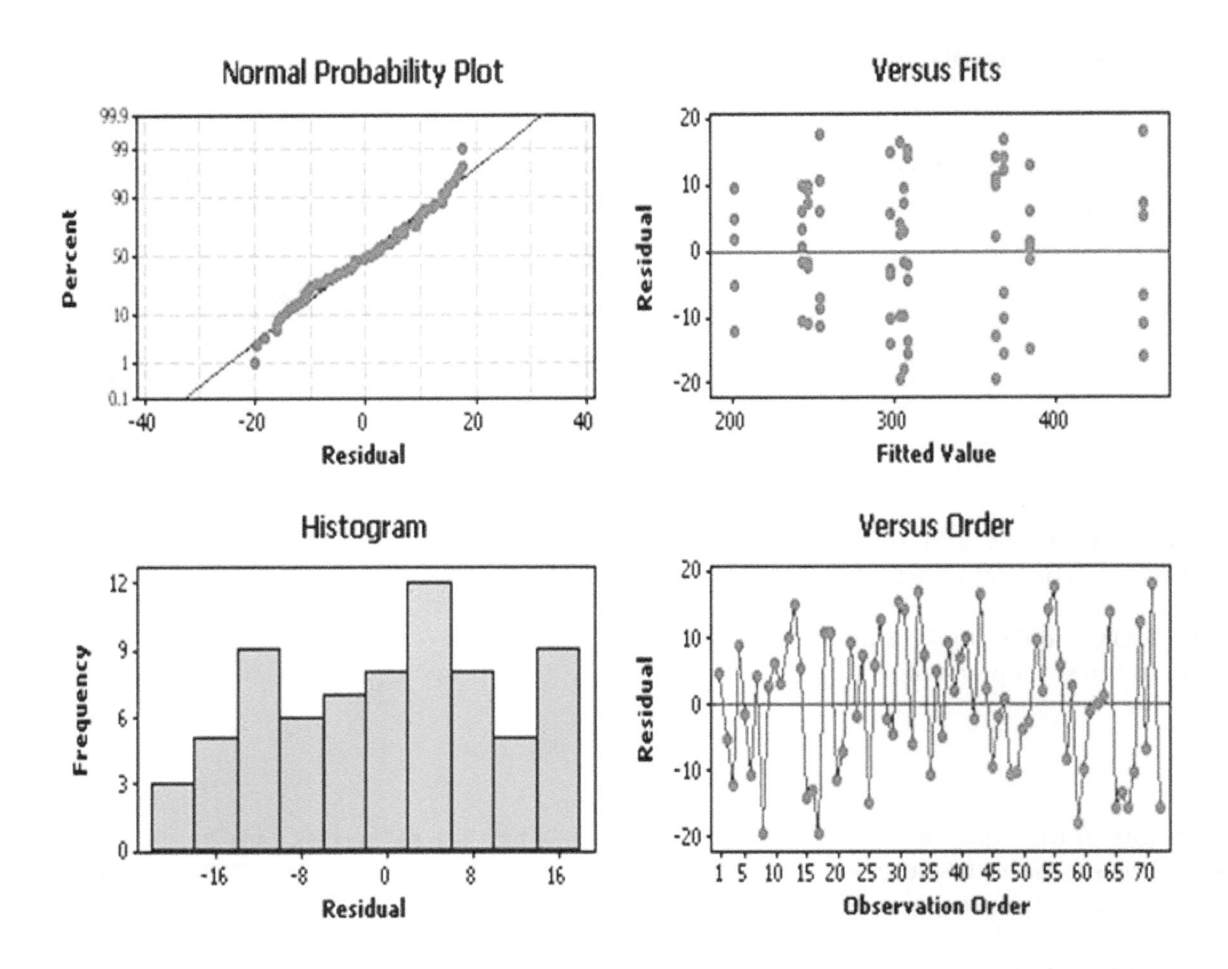

As in the earlier example we check the validity of the model by looking for patterns in the Four in one Residual Plot.

Starting from the top and going from left to right:

- The Normal Probability Plot shows that the Residuals are normally distributed.
- The Versus Fits Plot shows that the Residuals are equally distributed about the centre line.
- The Histogram Plot shows that the Residuals do not show extreme skewness.
- The Versus Order Plot shows that the Residuals are not displaying any patterns or ordered behaviour.

In summary, there are no issues with the residuals within this study.

Example 4a. ANOVA GLM.

We have assessed factors, with fixed levels, that affect residual fluorides within UO2 powder. Let's say for the purposes of illustrating a point that after carrying out this work it becomes apparent that the room temperature may also be related to the presence of residual fluorides. We will repeat the exercise with the original data but on this occasion we are going to use the continuous data within the RoomTemp column as a covariate. Will using RoomTemp as a covariate change the results of our study?

Worksheet: Kiln.

Continuous factors are added to the study as covariates. However, Covariates must have a linear relationship with the response. We are going to be covering Correlation and Regression in C9.

1. In order to confirm the linear relationship click Stat <<Regression <<Fitted Line Plot.
2. Enter Fluorides as the Response and RoomTemp as the Predictor.
3. Click OK.

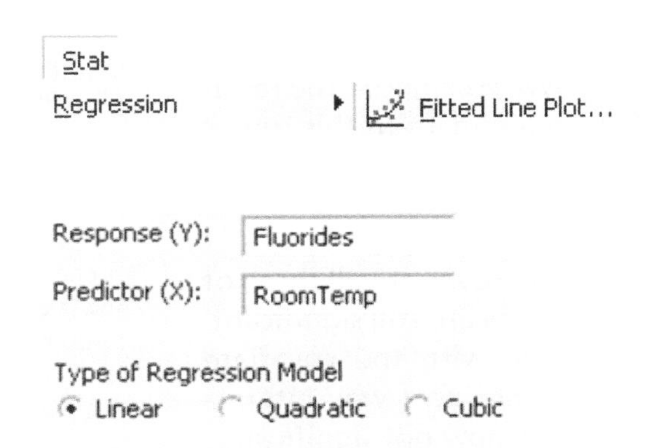

The graph shows a roughly linear relationship and the R-sq(adj) value indicates a strong linear fit between the variables. This confirms that we can use RoomTemp as a covariate within this study.

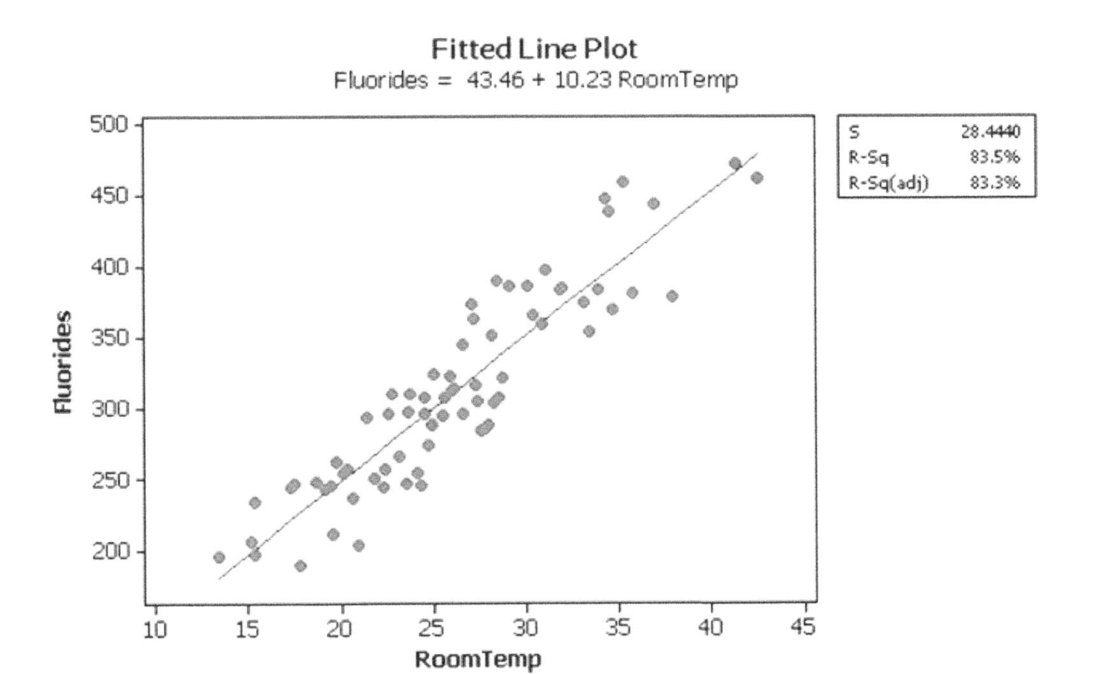

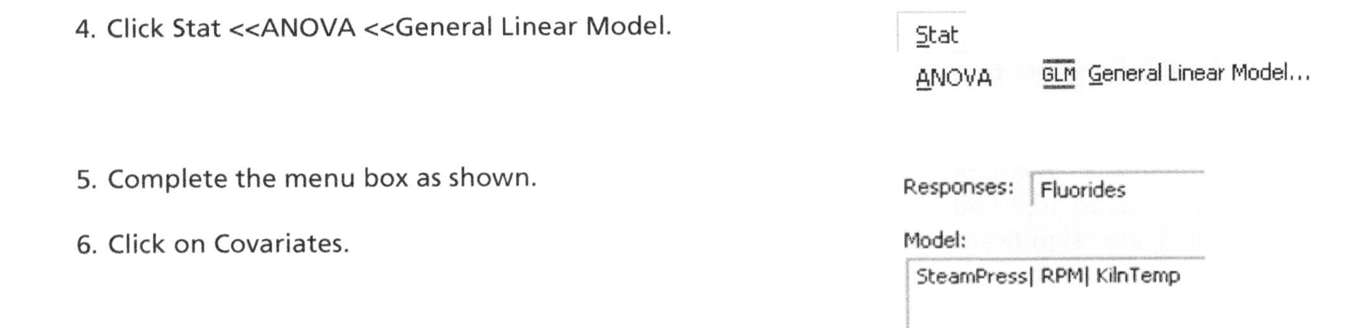

4. Click Stat <<ANOVA <<General Linear Model.

5. Complete the menu box as shown.

6. Click on Covariates.

7. Select RoomTemp as the Covariate.
8. Click OK and OK again. (We will ignore the residual plot from this run as we still need to reduce the model.)

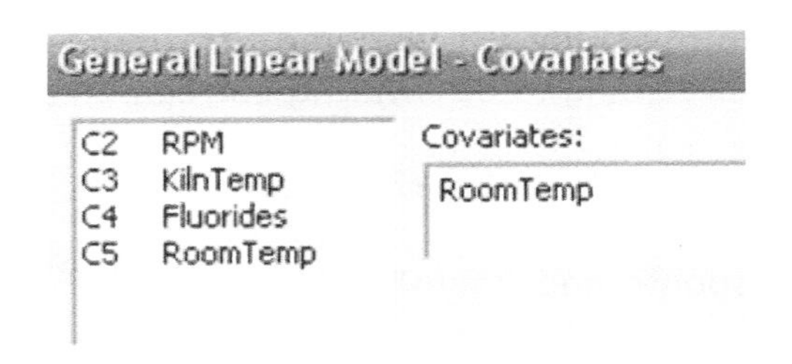

With the covariate all three of our factors are still significant. However, with the covariate two of our two way interactions are now not significant.

We will remove nonsignificant items from the model.

Source	DF	Seq SS	Adj SS	Adj MS	F	P
RoomTemp	1	287262	923	923	8.11	0.006
SteamPress	1	4829	33322	33322	292.54	0.000
RPM	1	3400	25048	25048	219.89	0.000
KilnTemp	2	38717	40604	20302	178.23	0.000
SteamPress*RPM	1	331	380	380	3.34	0.073
SteamPress*KilnTemp	2	1920	1950	975	8.56	0.001
RPM*KilnTemp	2	597	597	298	2.62	0.081
SteamPress*RPM*KilnTemp	2	121	121	60	0.53	0.591
Error	59	6721	6721	114		
Total	71	343897				

S = 10.6728   R-Sq = 98.05%   R-Sq(adj) = 97.65%

9. Click the Edit last Dialog icon.
10. Complete the menu box as shown.
11. Click on OK and OK again. Go to the Session window to check the results.

Responses: Fluorides

Model:

SteamPress RPM KilnTemp SteamPress* KilnTemp

Above is the table from the study using the RoomTemp as a covariate and below is the original study without the covariate.

Source	DF	Seq SS	Adj SS	Adj MS	F	P
RoomTemp	1	287262	1464	1464	12.02	0.001
SteamPress	1	4829	32433	32433	266.31	0.000
RPM	1	3400	24190	24190	198.63	0.000
KilnTemp	2	38717	39786	19893	163.34	0.000
SteamPress*KilnTemp	2	1894	1894	947	7.78	0.001
Error	64	7794	7794	122		
Total	71	343897				

S = 11.0357   R-Sq = 97.73%   R-Sq(adj) = 97.49%

It is always important to identify all the input factors that are affecting the response. This exercise has highlighted how identifying a covariate will alter the factors that you might think are significant and change the results of a study.

We need to finish this example off by checking the residuals.

Source	DF	Seq SS	Adj SS	Adj MS	F	P
SteamPress	1	88191	88191	88191	704.46	0.000
RPM	1	57038	57038	57038	455.61	0.000
KilnTemp	2	187179	187179	93590	747.58	0.000
SteamPress*RPM	1	629	629	629	5.03	0.029
SteamPress*KilnTemp	2	2229	2229	1115	8.90	0.000
RPM*KilnTemp	2	868	868	434	3.46	0.037
Error	62	7762	7762	125		
Total	71	343897				

S = 11.1888   R-Sq = 97.74%   R-Sq(adj) = 97.42%

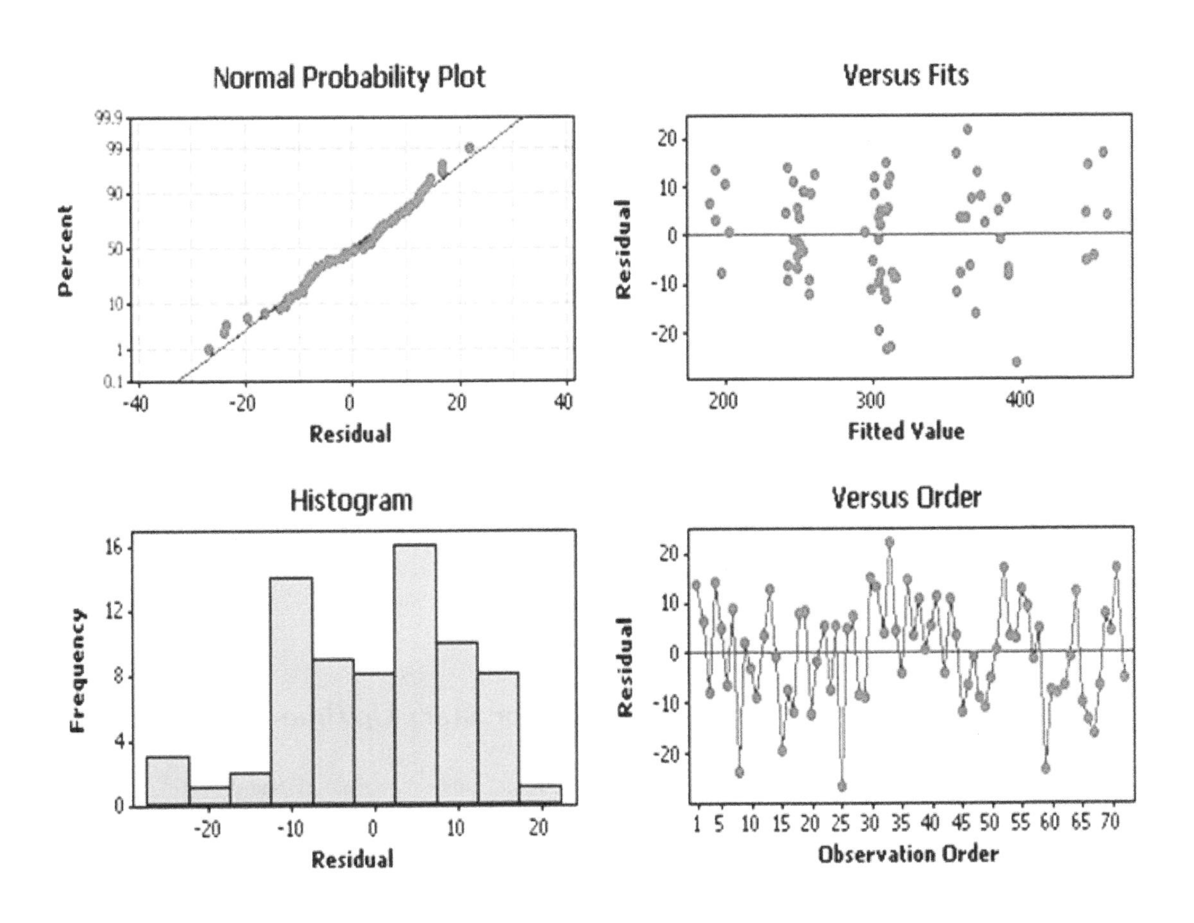

There are no issues with the residuals within this study.

Exercise 2. ANOVA GLM.

Bear wants to talk about buying a new digital camera. He wants to discuss start up time and the factors that affect it. He collects start up time data for different cameras and records the following factors: Make, Number of Sensors and Cost.

Produce a Main Effects plot and Interaction plot. Use the ANOVA GLM to establish which factors and interactions are significant. Explore if the levels within the factors are actually different.

Worksheet: Camera.

The Main Effects Plot shows that the number of sensors and cost of the camera appear to have quite a linear relationship with camera start up time. It would appear that Pony cameras take longer to start than the other makes.

There appears to be interactions within each pane but it is not yet possible to tell if they are significant.

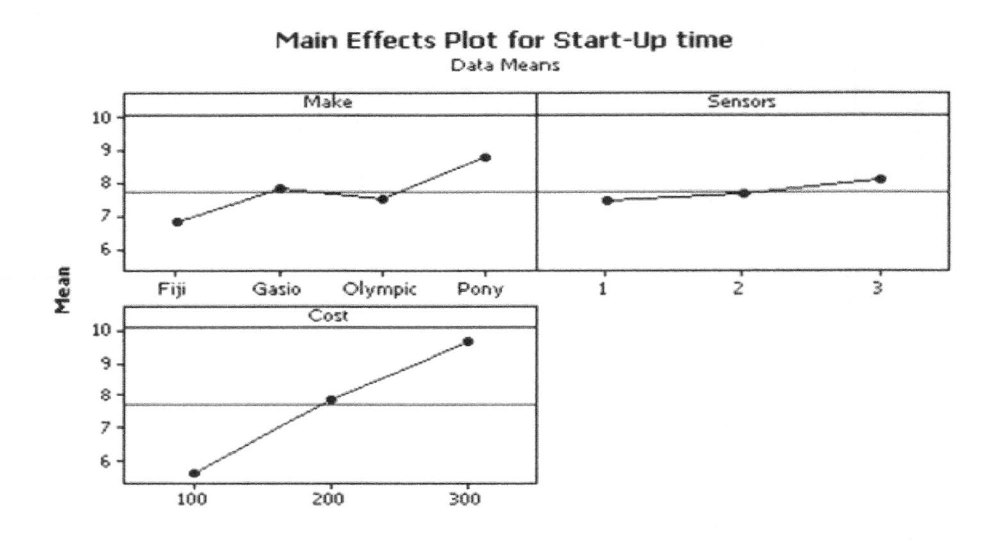

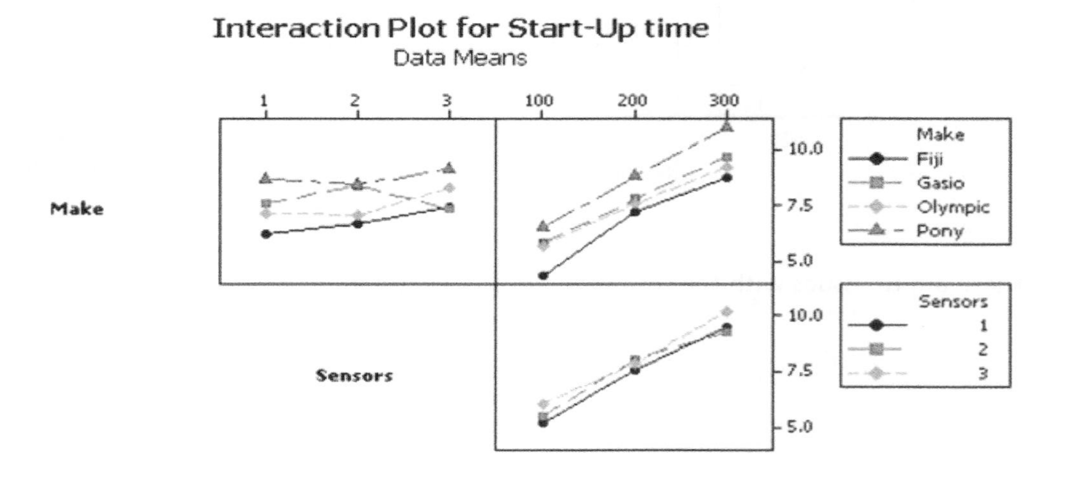

The top results table is produced on the initial run and the table below is the reduced model showing only significant items.

Source	DF	Seq SS	Adj SS	Adj MS	F	P
Make	3	36.060	36.060	12.020	8.69	0.000
Sensors	2	4.960	4.960	2.480	1.79	0.181
Cost	2	201.869	201.869	100.934	73.00	0.000
Make*Sensors	6	9.898	9.898	1.650	1.19	0.332
Make*Cost	6	2.903	2.903	0.484	0.35	0.905
Sensors*Cost	4	2.765	2.765	0.691	0.50	0.736
Make*Sensors*Cost	12	11.341	11.341	0.945	0.68	0.756
Error	36	49.773	49.773	1.383		
Total	71	319.569				

S = 1.17583   R-Sq = 84.43%   R-Sq(adj) = 69.28%

The bottom table shows that only the factors of Make and Cost were significant. None of the interactions were significant.

Source	DF	Seq SS	Adj SS	Adj MS	F	P
Make	3	36.060	36.060	12.020	9.72	0.000
Cost	2	201.869	201.869	100.934	81.60	0.000
Error	66	81.640	81.640	1.237		
Total	71	319.569				

S = 1.11219   R-Sq = 74.45%   R-Sq(adj) = 72.52%

The Grouping Information Table for Make confirms that Pony cameras are different to the other makes in terms of start up speed.

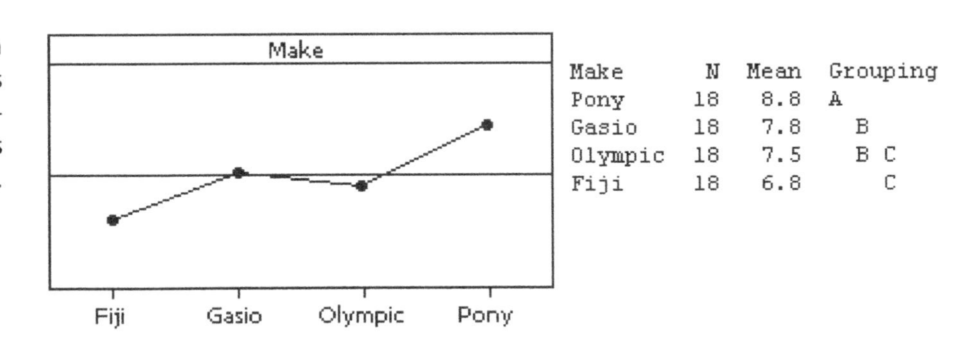

Make	N	Mean	Grouping
Pony	18	8.8	A
Gasio	18	7.8	B
Olympic	18	7.5	B C
Fiji	18	6.8	C

For Cost we see that each of the levels is different.

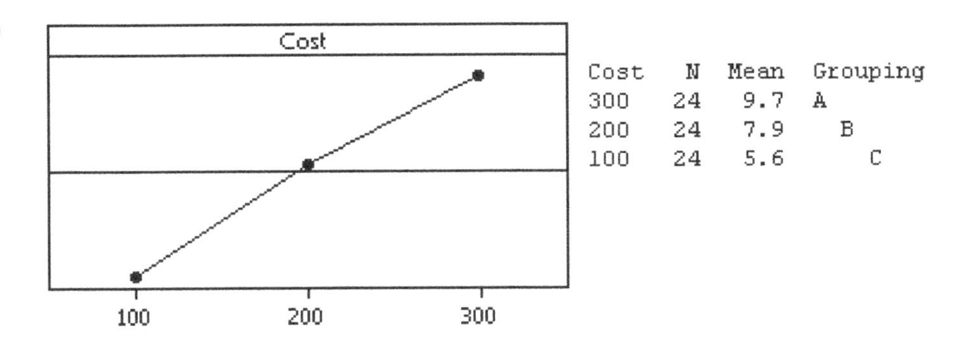

Cost	N	Mean	Grouping
300	24	9.7	A
200	24	7.9	B
100	24	5.6	C

The four in one residual plot validates the study. We don't see anything within any of the plots that indicates a problem with the GLM procedure.

We have successfully explored the factors that affect camera start up speed and given Bear plenty to talk about.

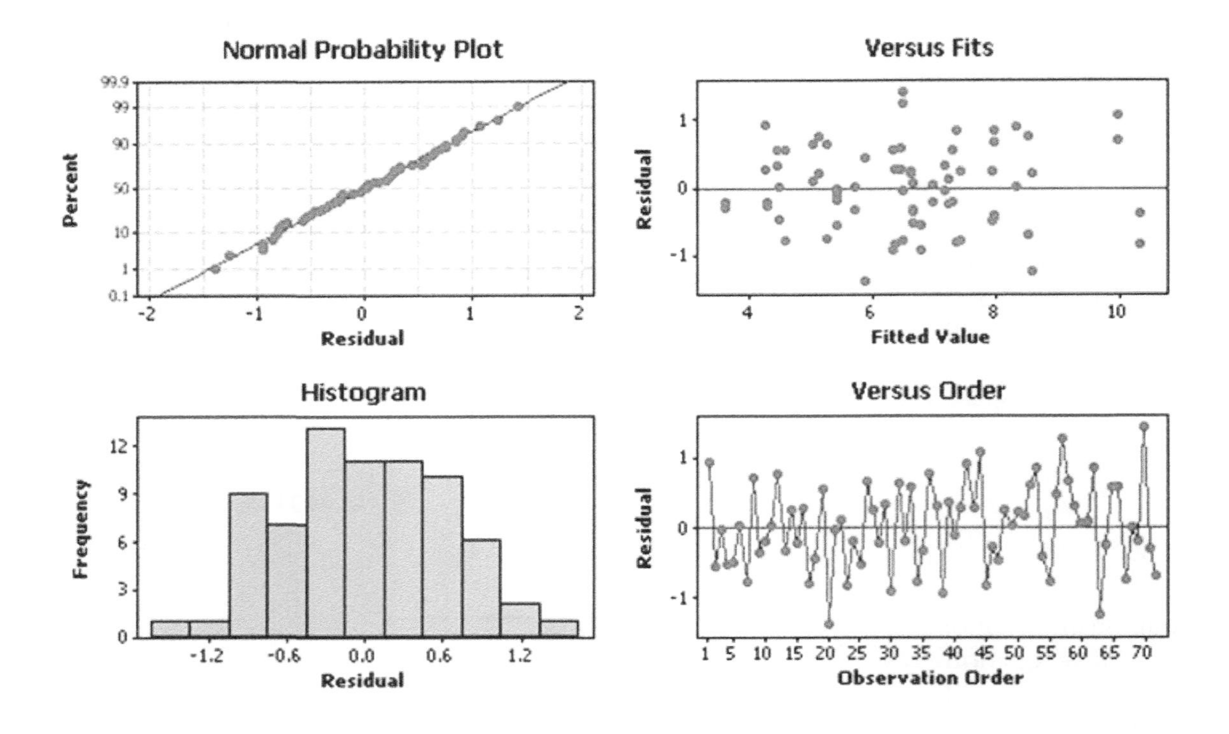

CHAPTER 6
Measurement System Analysis

6.1 The Importance of Measurement Systems

Whatever our process, our measurement systems are the equivalent of our senses. They measure the process and bring us data.

At the start of every project you should check and verify the measurement systems related to your key output. If you do this it will create a solid foundation for the project which you can then build upon. If you don't do this you will always be on shaky ground.

Here are a couple of predictions for you.

1. At the start of every project your measurement systems will be inadequate, most of the time.
2. Even if the measurement system is deemed to be acceptable at some point in the project you will go back and question the measurement system again.

Even though you probably won't heed my warning, until it has happened to you a couple of times, do not rush into a project without first having checked the measurement system as thoroughly as possible. Rushing in is usually a false economy.

6.2 How Measurement Systems Affect Data

The variation within the measurement system is usually incorporated within process variation.

Consider the following two processes;

- One has a small part to part variance with a higher measurement system variance.
- The other has a high part to part variance but a low measurement system variance.

If we look at the output measurements of the processes they both appear to be identical. Without Measurement System Analysis (MSA) we won't know whether we should be trying to reduce the measurement system variation or the process variation.

Problem Solving and Data Analysis using Minitab: A clear and easy guide to Six Sigma methodology, First Edition. Rehman M. Khan.

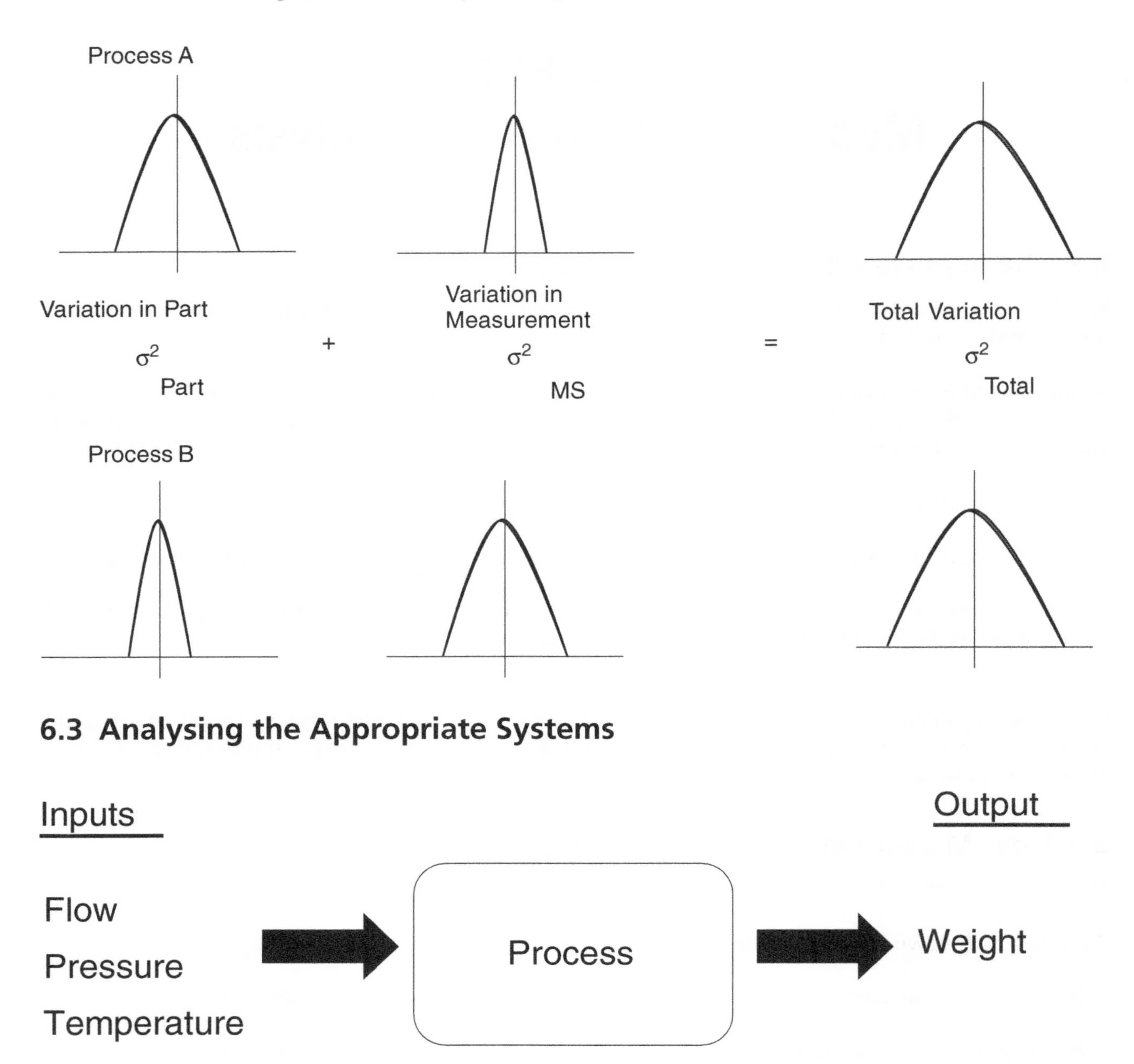

6.3 Analysing the Appropriate Systems

Earlier it was said that, at the start of the project, you should confirm that your measurement systems are adequate. At the start of the project it should be easy enough to find the systems that are measuring the output that you are interested in. However, at that point in the project you may not know which inputs affect the output. I would suggest that you only confirm the input measurement systems later in the project when you have established which inputs are significant to the output.

Let's have a deeper look at the types of measurement system error and learn some of the terminology.

6.4 Types of Measurement Systems Error

1. Precise and accurate

2. Precise but not accurate

3. Not Precise but accurate

4. Not Precise and not accurate

Traditionally the terms accuracy and precision are explained by showing shots fired at the centre of a target. Precision relates to the spread of the shots and accuracy relates to how well the mean of the shots aligns to the centre of the target. The first four shots are accurate and precise because the spread is low and the mean is aligned to the centre. The second four shots have a tight spread so they are described as precise but the mean value is not aligned to the centre so the shots are not accurate.

Let's have a look how precision and accuracy relate to measurement systems.

Precision errors can be further broken down into the categories of repeatability and reproducibility.

Repeatability errors relate to the measurement variation generated by the same person repeatedly measuring the same item.

Reproducibility errors relate to the measurement variation generated by different people measuring the same part.

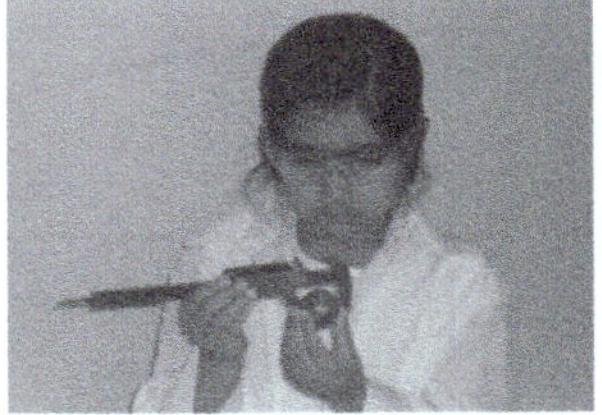
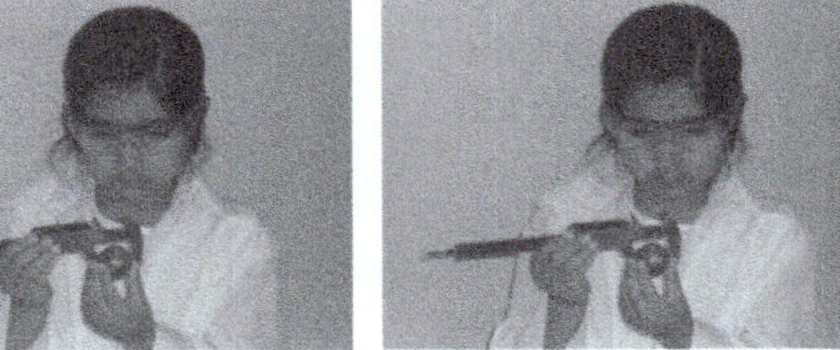
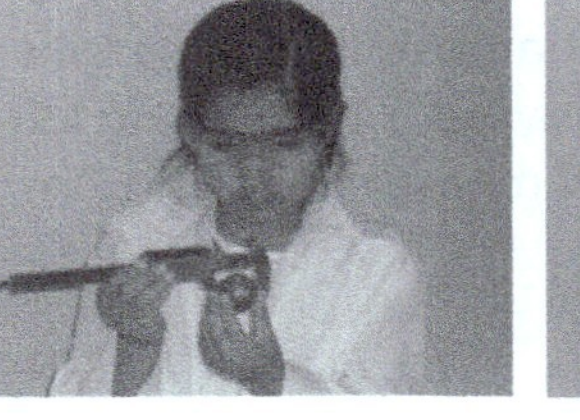
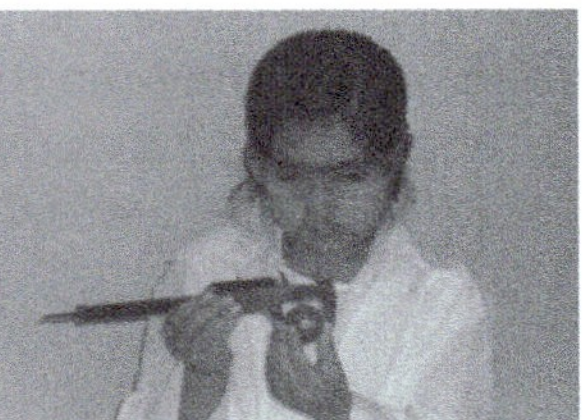

Repeatability (same person)

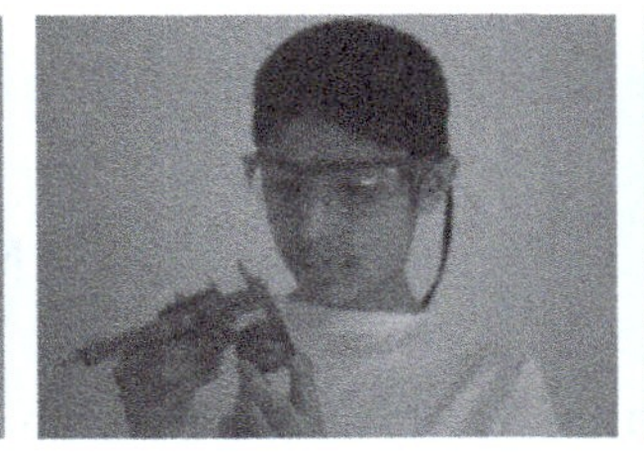
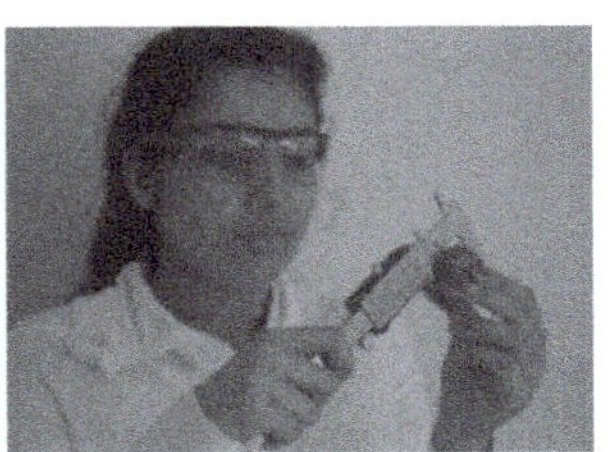

Reproducibility (different people)

If you are learning the terms a good way of remembering the terms is that 'it takes more than one to reproduce'.

Accuracy errors can be further broken down into the categories of linearity, bias and stability.

A linearity error displays an offset between the measured value and the real value. However, it does not have a constant offset or bias.

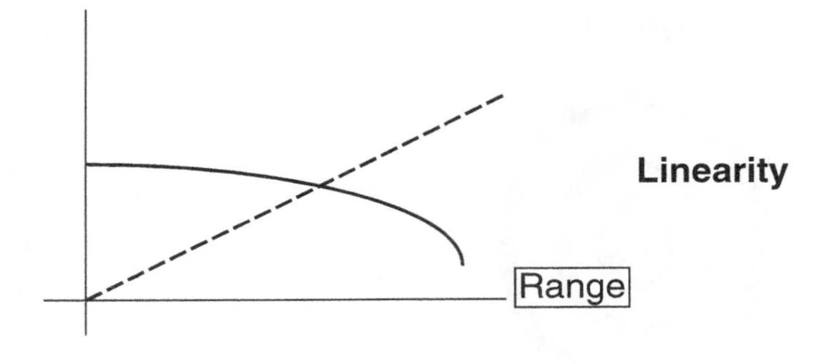

A bias error has a constant offset between the measured value and the real value over the range of the measurement. It could be argued that bias is a special type of linearity error.

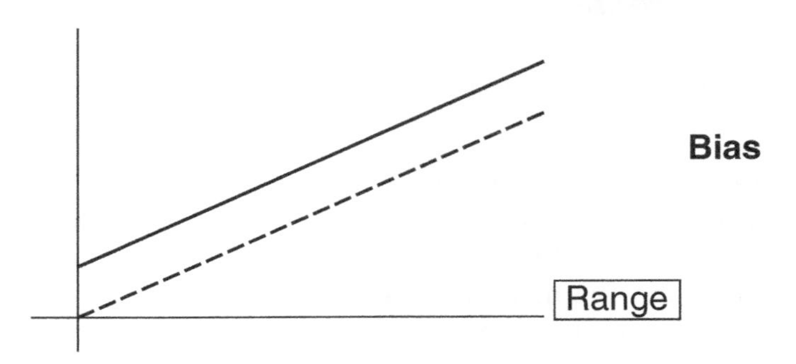

Stability errors manifest themselves as an offset; however, they appear over time rather than range. The time that stability errors take to appear will obviously vary depending on the process and type of measurement system.

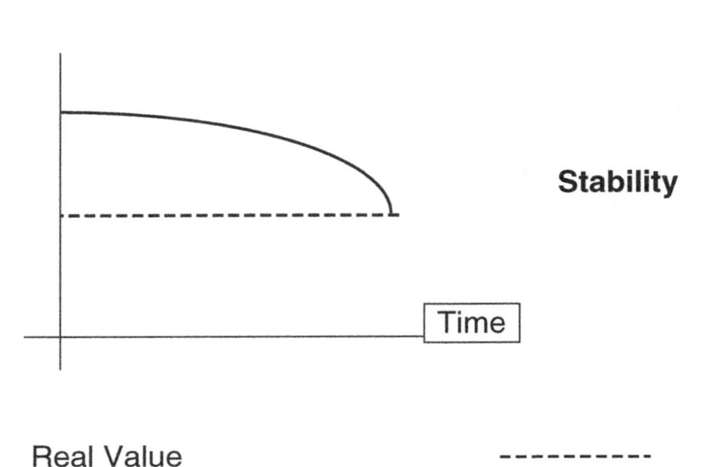

Real Value ----------
Measured Value ———————

Resolution or Discrimination is also a type of measurement error. On occasions it is overlooked in statistical texts as it is something that should have been considered when setting up or purchasing the measurement system.

Imagine that you have to time an event. You could employ the following techniques:

1. Use a clock.
2. Use an analogue stopwatch with a second hand.
3. Use a digital stopwatch which gives times to one 100th of a second.

Each successive method gives a better resolution.

There is a rule of thumb for deciding what the resolution for your measurement system should be. It brings in another term that we must learn: Tolerance.

To explain tolerance, let's pretend that we are making ring doughnuts and there is a specification for the size of the hole. The specification for the minimum diameter of the hole is 20 mm and the maximum diameter is 30 mm. We say that the Tolerance or Tolerance Width is 30 – 20 = 10 mm.

The rule of thumb is that the resolution of the measurement system should be a tenth of the tolerance. In this case 1 mm. The measurement system should be able to measure clearly between every 1 mm change. We say it should have a resolution of 1 mm.

6.5 Measurement Systems Toolbox

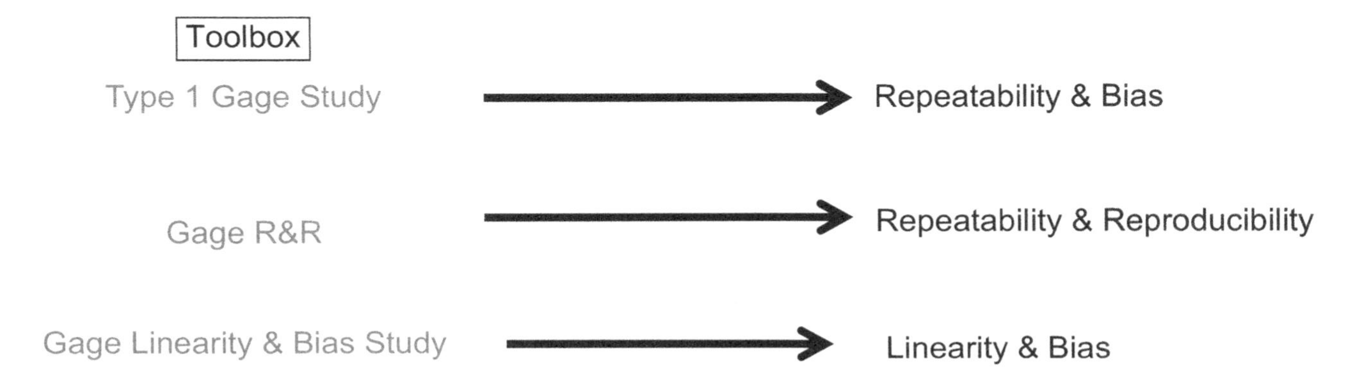

Underneath toolbox you can see the techniques that are available in Minitab and that we will be going through within this chapter. You might notice that Stability is not present. This is because, if all other sources of error had been eliminated and stability was thought to be an issue, it would be investigated by using observational techniques.

People familiar with U.K. English will have noticed that Minitab uses the U.S. spelling of gage. Even though this course is written in U.K. English we will spell 'Gauge' as 'Gage'.

In the next section we will begin to work through MSA examples and exercises. All of the example data is within separate worksheets in the file 06 MSA.xls.

6.6 Type 1 Gage Study

Procedure	• Type 1 Gage Study
What's it used for?	• A reference part is measured repeatedly to make an initial assessment of Repeatability and Bias Error of a measurement system.
Assumptions and Limitations	• The term reference part means that parameter being measured within the part has been confirmed using highly precise and accurate methods. • The test runs should be conducted by a single person as no distinction is made for the Reproducibility term.

Example 1. Type 1 Gage Study.

Bear decides to buy a new digital vernier gauge. He wants to confirm how well his new vernier measures objects so he takes two metal blocks and measures their lengths on his CPU controlled laser profiling equipment. The first object has a length of 50 mm and Bear gives himself a tolerance width of 0.5 mm for measuring this block in the future. Bear measures the length of this object 50 times and records the results in Worksheet: Type 1GS. The results are in the Block 1 column.

Use the results and carry out a Type 1 Gage Study to establish if the measurement system is repeatable and if a bias is present.

1. Copy the data from column A of theExcel file and copy it into Minitab.

Block 1
49.963
50.004
49.958
50.003

2. Click on Stat <<Quality Tools <<Gage Study <<Type 1 Gage Study.

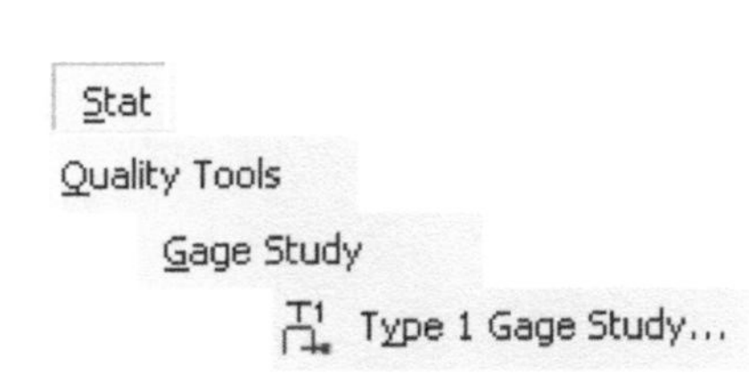

3. Within the menu box select Block 1 as the measurement data. Bear has accurately measured the part and knows the reference length is 50. The tolerance width that he is using is 0.5. Therefore, acceptable parts in the future will be within 50 ± 0.25 mm.

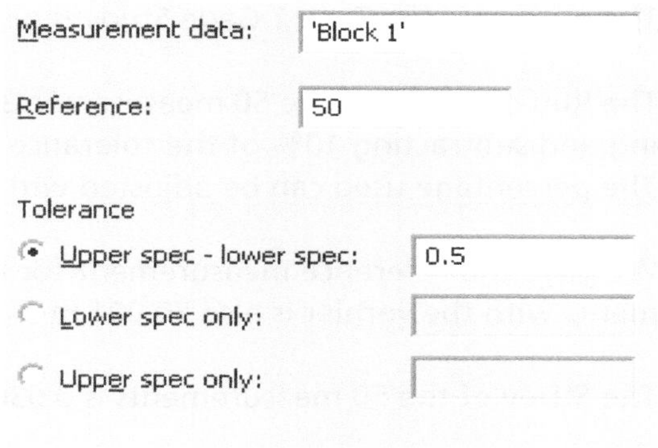
Measurement data: 'Block 1'
Reference: 50
Tolerance
Upper spec - lower spec: 0.5
Lower spec only:
Upper spec only:

4. Click on Gage Info. It is always a good idea to enter relevant information about the study. It will probably prove useful at some point in the future. Click OK and OK again.

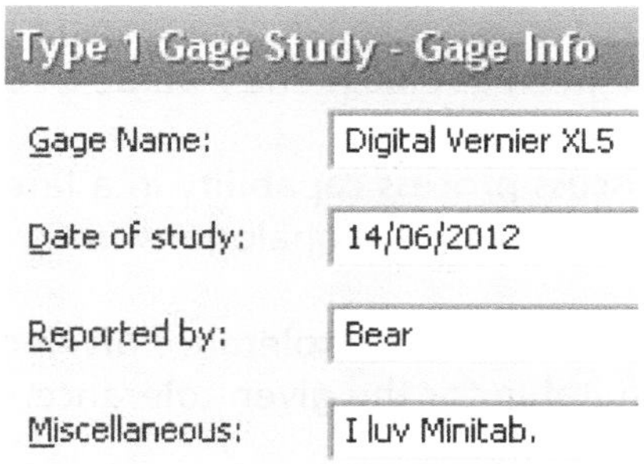
Type 1 Gage Study - Gage Info
Gage Name: Digital Vernier XL5
Date of study: 14/06/2012
Reported by: Bear
Miscellaneous: I luv Minitab.

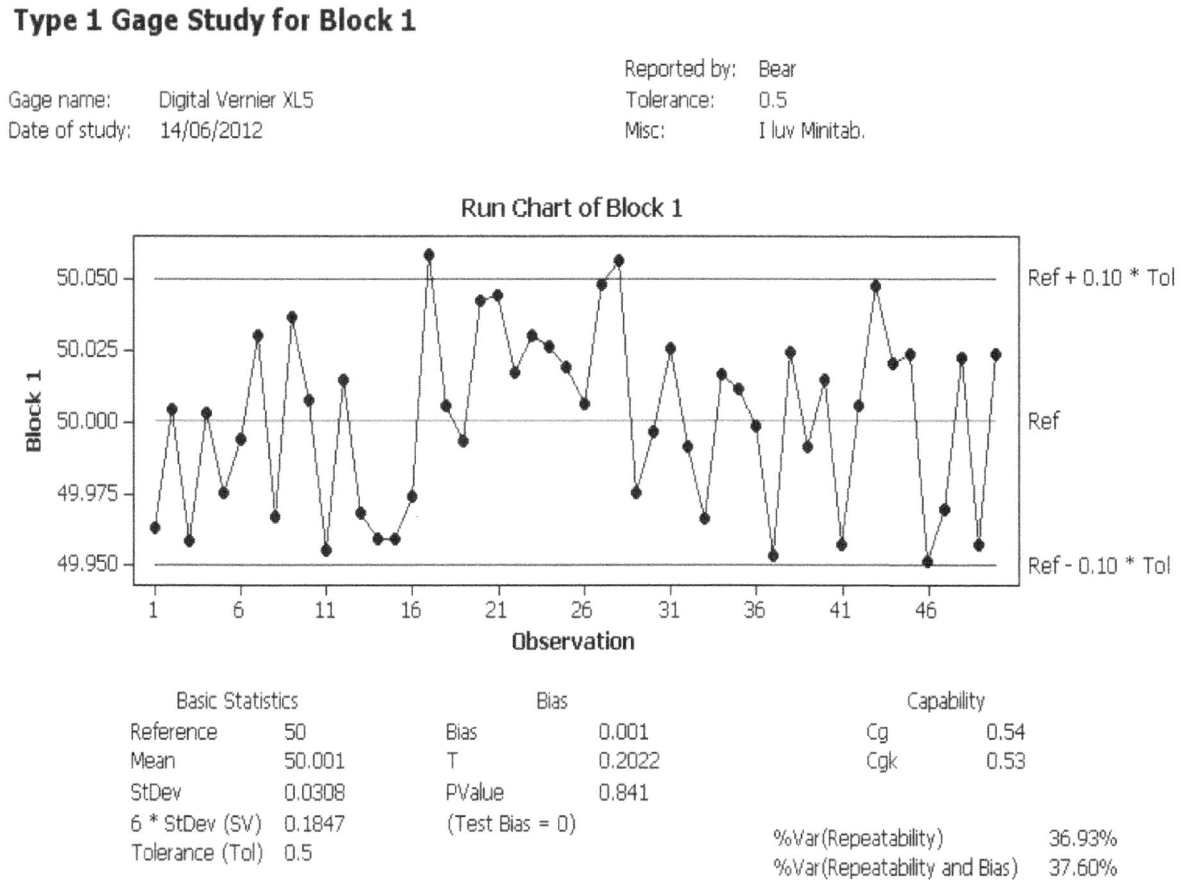
Type 1 Gage Study for Block 1
Gage name: Digital Vernier XL5
Date of study: 14/06/2012
Reported by: Bear
Tolerance: 0.5
Misc: I luv Minitab.
Run Chart of Block 1
Block 1
50.050
50.025
50.000
49.975
49.950
Ref + 0.10 * Tol
Ref
Ref - 0.10 * Tol
1 6 11 16 21 26 31 36 41 46
Observation
Basic Statistics
Reference 50
Mean 50.001
StDev 0.0308
6 * StDev (SV) 0.1847
Tolerance (Tol) 0.5
Bias
Bias 0.001
T 0.2022
PValue 0.841
(Test Bias = 0)
Capability
Cg 0.54
Cgk 0.53
%Var(Repeatability) 36.93%
%Var(Repeatability and Bias) 37.60%

The output of the Type 1 Gage Study is given wholly within the graph shown.

The Run chart shows the 50 measurements that were recorded. The control limits are calculated by adding and subtracting 10% of the tolerance to the reference value (not the mean of the measurements). The percentage used can be adjusted within the options menu of the test.

We gave the reference measurement for the part as 50. It turns out that the mean of the 50 measurements with the vernier is also 50.001 mm.

The StDev of the 50 measurements is 0.0308.

The tolerance width, which was stated by Bear, for this part being measured was 0.5mm.

The Bias is given as 0.001. The *P* value of 0.841 confirms that the bias is not significant.

We will discuss process capability in a later chapter, where the terms Cp and Cpk will be discussed and defined. Cp and Cpk are analogous to Cg and Cgk.

Cg relates to 20% of the tolerance divided by 6*StDev (0.1847). Its value would be >1 if the gage precision was suitable for the given tolerance.

Cgk relates to the distance to the closest control limit divided by 3*StDev. Its value would be >1 if the gage precision and accuracy was suitable for the given tolerance.

The %Var values tells us how much of the tolerance is wasted by Repeatability and then by Repeatability and Bias. This value should be less than 10%, ideally.

Exercise 1. Type 1 Gage Study.

The second object has a length of 64.5 mm and Bear gives himself a tolerance width of 0.8 mm for measuring this block in the future. Bear measures the length of this object 75 times and records the results in Worksheet: Type 1GS. The results are in the Block 2 column.

Use the results and carry out a Type 1 Gage study to establish if the measurement system is repeatable and if a bias is present. Is the measurement system suitable?

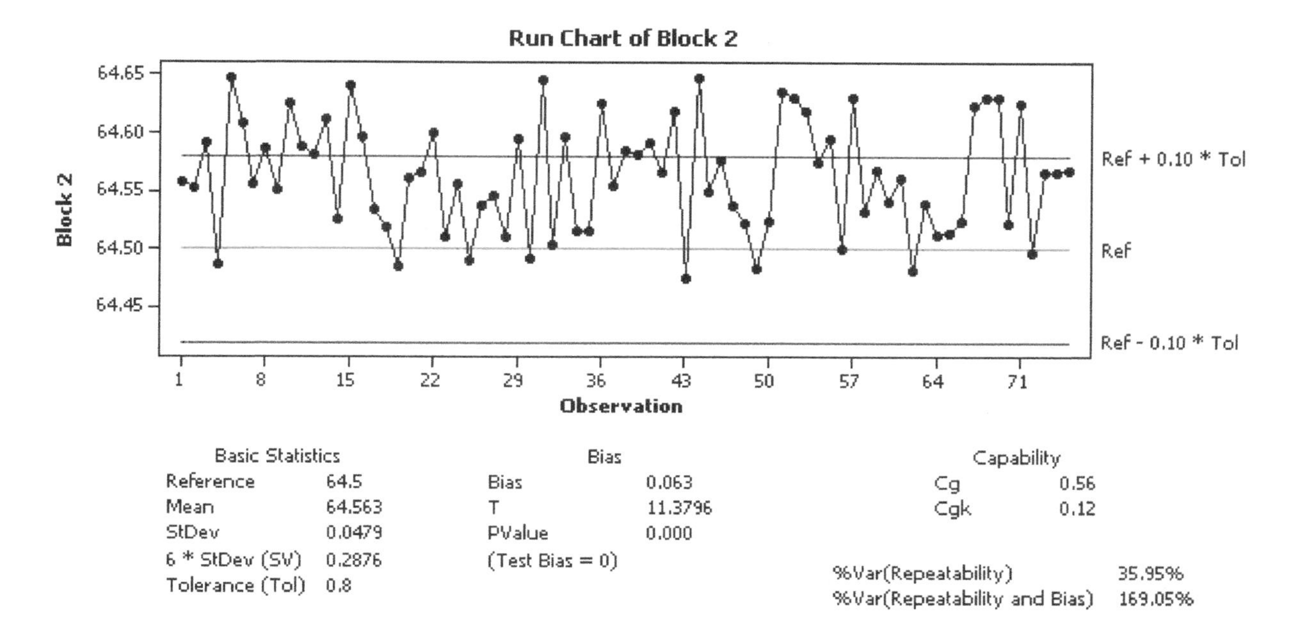

We see that there is a significant Bias within this study of 0.063 mm. The Repeatability takes up 35.95% of the tolerance. What really makes this measurement system a failure is the Bias which contributes an additional 133% to the %Variation.

We can conclude that, although Bear's new vernier was poor at measuring Block 1, it is absolutely dire at measuring Block 2.

Next, we are going to move on to look at Gage R&R studies.

6.7 Gage Repeatability and Reproducibility Studies

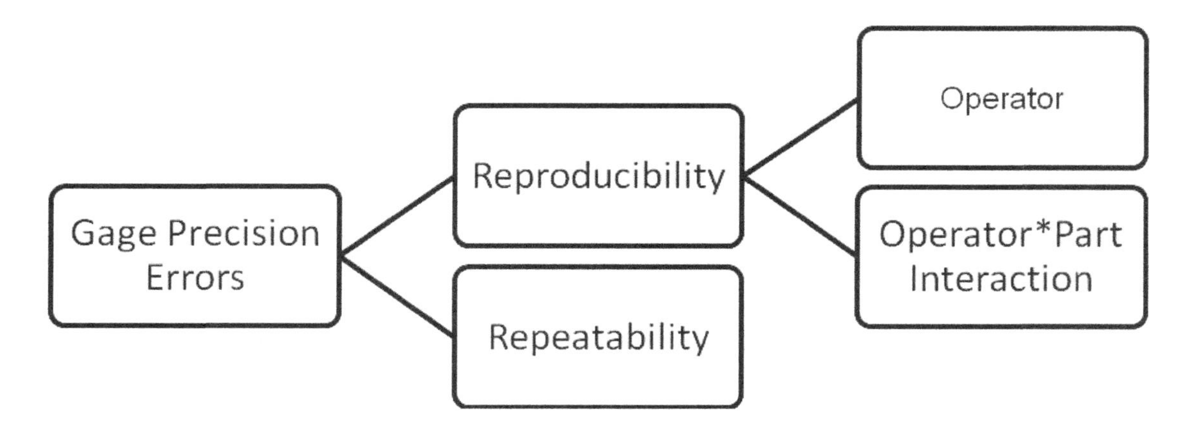

As we are now going to concentrate on Gage Repeatability and Reproducibility (Gage R&R) studies we need to understand a bit more about the two types of study. Initially we will look at the Gage R&R Crossed Study. The diagram above shows how the total gage error is broken down into its components for a crossed study. The other type of study is the Nested Study. It has almost the same breakdown of components as a Crossed Study but we don't get the Operator*Part Interaction term. You will get to see why later in the section.

In order to understand the difference between the Crossed and the Nested Study take a look at the figure on the right. The diagram shown is for illustrative purposes and the number of parts and operators shown are not as they would be in a real study. However, what it does show is that each operator measures each part. That is why this is a Crossed Study.

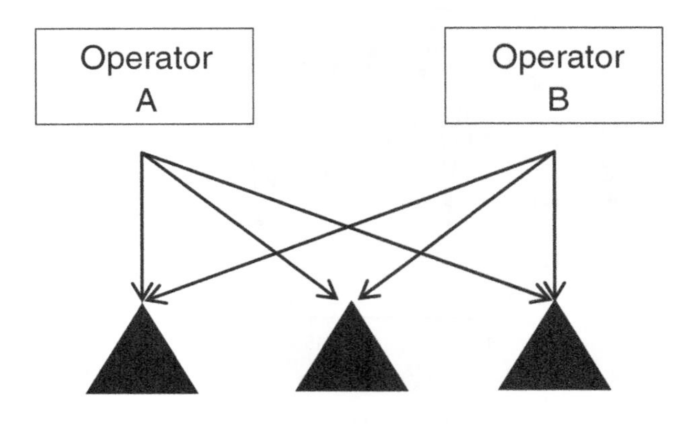

On occasion the testing is destructive and the part being measured is altered or destroyed by the measuring process, as in breaking strength tests, for example. For these types of tests we still need to be able to estimate repeatability.

We do this by making batches of parts that are identical. If they are identical then all the measurement variation can be attributed to repeatability. In order to use the Crossed Study for destructive testing the batch of parts has to be big enough so that different operators can measure from the same batch and essentially repeat measurements can be made.

On occasion it is not possible to produce a batch where there are enough samples or pieces for every operator to be able to measure the sample. On these occasions we use a Nested Study.

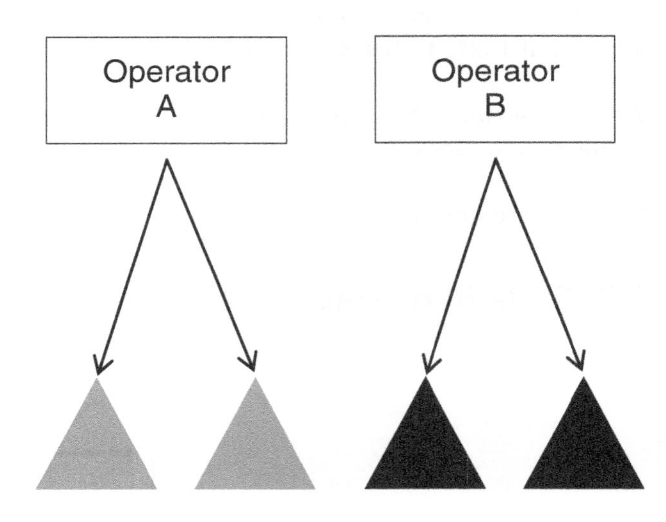

The diagram shows that the parts are nested within the operators. Operator A will measure the two parts from the first batch and Operator B two parts from the second batch.

If batches, with the same properties, cannot be produced a Gage R&R study cannot be carried out.

The parts that you select for the study can have a massive impact on the results of the study. The parts that you select should be parts that are deemed to be good parts with respect to your process specification/tolerance. If you use reject parts you may exaggerate the part to part variation and that will make your study metric better than it should be.

On occasion it may seem like a good idea to make yourself a tester in order to increase the number of operators you have. If you are not a regular tester/operator you may inadvertently increase both types of measurement error.

If you can, keep the part identifications hidden from the operators/testers and randomise the studies. This is because operators will use their judgement as well as the measurement system.

Apart from the Assistant there are a couple of upgrades under the bonnet of M16 when using the Classic method. The first is the option of calculating the confidence intervals for the study metrics. Give this a go when you are carrying out the exercises. You will find the button for using confidence intervals in the Gage R&R input menu. However, a word of warning you will find the results frightening!

The other addition is the Gage R&R Study (Expanded). This procedure allows you to add additional factors to the analysis and check if they impact on the measurement system.

The first example that we are going to look at is how to create a simple worksheet which can be used to collect data for a Gage R&R Crossed Study.

6.8 Create Gage R&R Study Worksheet

Example 2. Gage R&R Study Worksheet.

My three kids are not very good at baking cakes. Before we address the problem we need to ensure the measurement system is satisfactory.

Create a randomised Study Worksheet so we can collect data for a Gage R&R study; there are four parts.

No Data File

1. Click Stat <<Quality Tools <<Gage Study <<Create Gage R&R Study Worksheet.

2. Complete the page as shown. Note we must have 'Replicates' >1 for the Repeatability term and more than one operator for our Reproducibility term.
3. Then click Options.

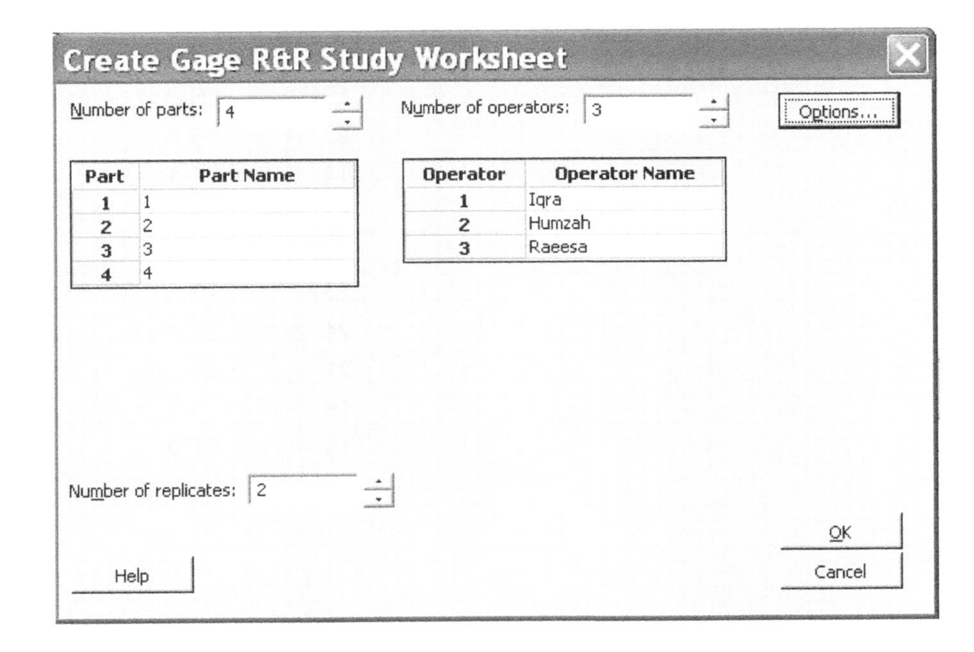

4. Select 'Randomize runs within operators'.
5. Then Click OK and OK.

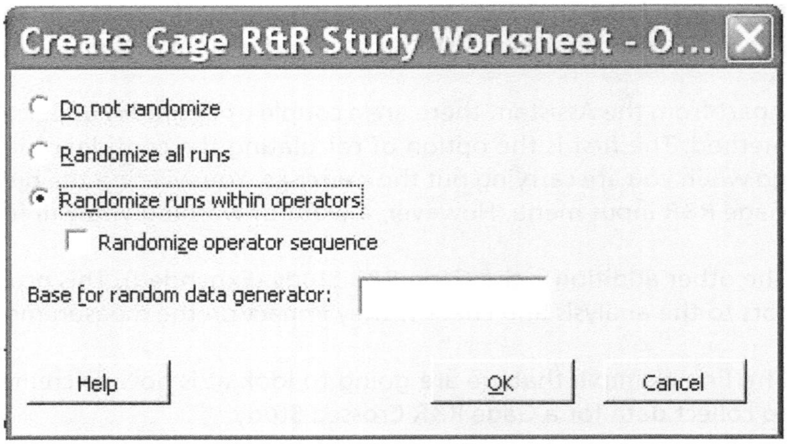

Normally, you would randomise the runs as far as possible but this depends on the time and effort required to set up the study.

Create Gage R&R Study Worksheet - O...

- Do not randomize
- Randomize all runs
- Randomize runs within operators
  - Randomize operator sequence

Base for random data generator:

Help | OK | Cancel

Minitab creates the study worksheet. If we were using this worksheet we would carry out the study as per the run order.

C1	C2	C3-T	C4-T
StdOrder	RunOrder	Parts	Operators
10	1	4	Iqra
4	2	2	Iqra
1	3	1	Iqra
7	4	3	Iqra
5	5	2	Humzah
11	6	4	Humzah
2	7	1	Humzah
8	8	3	Humzah
3	9	1	Raeesa
6	10	2	Raeesa
9	11	3	Raeesa
12	12	4	Raeesa
19	13	3	Iqra
13	14	1	Iqra
16	15	2	Iqra
22	16	4	Iqra
14	17	1	Humzah
23	18	4	Humzah
17	19	2	Humzah
20	20	3	Humzah
24	21	4	Raeesa
21	22	3	Raeesa
15	23	1	Raeesa
18	24	2	Raeesa

6.9 Gage R&R (Crossed)

Procedure	• Gage R&R (Crossed)
What's it used for?	• It is used to compare the reproducibility and repeatability elements of the measurement system with the overall study variation and, if required, tolerance.
Assumptions and Limitations	• Parts selection for the study can be critical. The parts should not be deliberately selected to be at the extremes of the process. The parts should be selected so that they represent normal operations. • Human behavior can affect the results. Therefore, it is best if the runs can be randomised and the parts kept blind. • The crossed study is used to assess systems where all operators can measure all parts.

6.10 Gage R&R Crossed Studies

We have just shown you how to create a study worksheet but that example did not use the recommended number of parts. The Automotive Industry Action Group recommend using 10 parts, two or three operators and two or three replicates when carrying out a Gage R&R Crossed Study. However, 10 parts may not give a precise representation of the actual part to part variation. Therefore, it is strongly recommended within the Assistant that a historical standard deviation of the measurement is entered.

We will now start looking at examples of the Gage R&R Crossed Study. The study examples will be covered using both the Classic Method and the Assistant. The first example will show a simple study and will introduce the basic analysis. Once we have understood the basic analysis we will move onto a more in depth example, where we can look in more detail at the metrics that we can choose to evaluate our measurement system. You can expect some exercises along the way as well.

If you are using the Classic method there are five possible metrics you could use to assess the measurement system. These are %Contribution, %Study Variation, %Tolerance, %Process and Number of Distinct Categories (NoDC). NoDC should not be used as a metric only as an indicator. %Tolerance will only be given within the results if a tolerance value is entered. %Process will only be given within the results if a historical StDev is entered.

If you are using the Assistant there are only two possible metrics: %Process and %Tolerance. The part to part variation used to calculate the %Process will be calculated from the parts within the study or a historical StDev of the measurements. Minitab strongly recommends you enter a historical StDev. %Tolerance will only be given within the results if a tolerance value is entered.

For both the Assistant and the Classic Methods:

If we wanted to improve parts to a specification we would use %Tolerance as our metric.

If we wanted to reduce part to part variation we would use %Study Variation or %Process as the metric.

6.11 Gage R&R (Crossed) Study

Example 3. Gage R&R Study.

My three kids, Iqra, Humzah and Raeesa are not very good at baking cakes. I give them 10 packets of flour and ask them to weigh the packets as a part of a Gage R&R Study. Each of my kids measures each packet three times, so there are 10 parts, three operators, three repeats of each measurement making 90 runs in total. My kids have short attention spans so I did not need to randomise the study. The cake baking process at my house does not have a specification so I choose not to use a tolerance for this analysis.

Using the results table in Worksheet: Cakes carry out a Gage R&R (Crossed) study.

Use the study results to learn if our measurement system is suitable.

1. Copy the data from the Excel file into Minitab. There were 10 bags of flour and each is designated a number from 1 to 10, this can be seen in the part column. The measurement column is used to record the value of the measurement operation, weighing. Person is simply the name for the person carrying out the measurement. I always find it easier to use column labels that Minitab will be expecting within its menus, it's one less thing to think about.

Part	Measurement	Person
1	49.998	Iqra
2	50.264	Iqra
3	49.940	Iqra
4	50.078	Iqra
5	49.853	Iqra

2. As stated in earlier chapters it is always a good idea to get a feel for your data before conducting the test procedure. For Gage R&R studies we are going to use a Gage Run Chart to do this. Click Stat <<Quality Tools <<Gage Study <<Gage Run Chart.

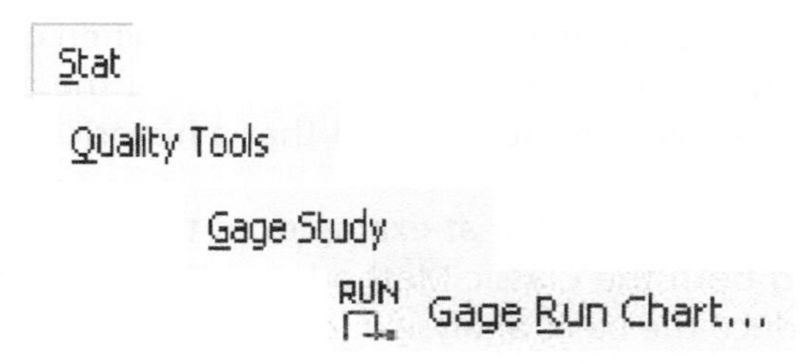

3. Complete the menu box as shown.
4. Click on Gage Info. When carrying out a lot of studies it's worth recording some extra information for the future.

Part numbers: Part
Operators: Person
Measurement data: Measurement
Trial numbers: (optional)
Historical mean: (optional)
Gage Info...
Options...

5. Then click OK and OK again.

Gage name:	Kitchen Weighscale
Date of study:	19th March 20120
Reported by:	Kids
Gage Tolerance:	No tolerance
Miscellaneous:	Dad torturing us again.

What we get is a graph showing all the results of the measurements. The measurements are initially split by part and then by operator. It can be seen that the variability within the operators is low, indicating a good repeatability. The measurement variability between operators is also low indicting good reproducibility. The range of the part to part variation appears to be given by parts two and nine. Both types of operator measurement variability appears much lower than the part to part variation.

To carry out the actual test procedure we are first going to use the Classic Method and then are going to use the Assistant.

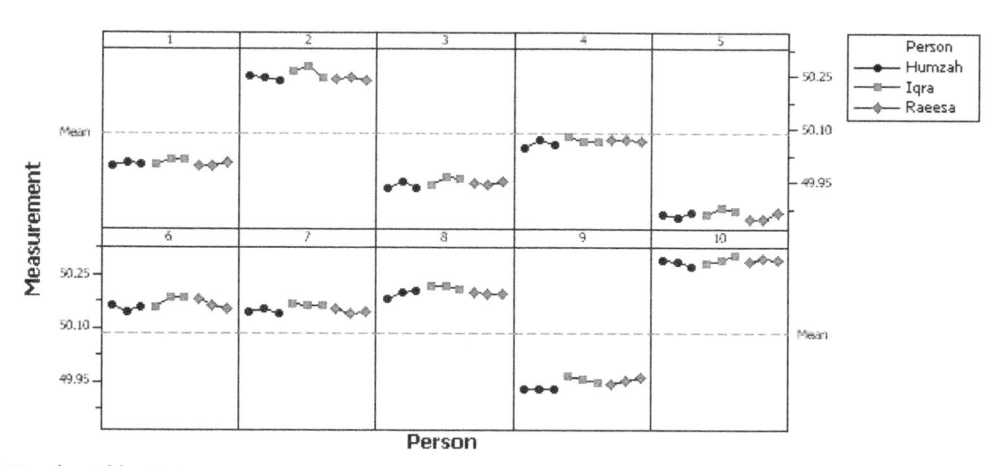

6. Now that we have got a feel for the measurement data we are going to start the test procedure. Click Stat <<Quality Tools <<Gage Study <<Gage R&R Study (Crossed).

7. Complete the menu as shown. I have already gone into Gage Info and entered some details about this study. We will always use the ANOVA method for this procedure.
8. Click OK to execute the procedure.

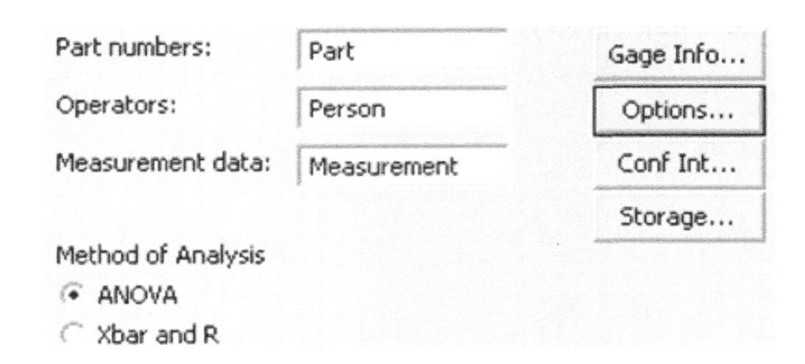

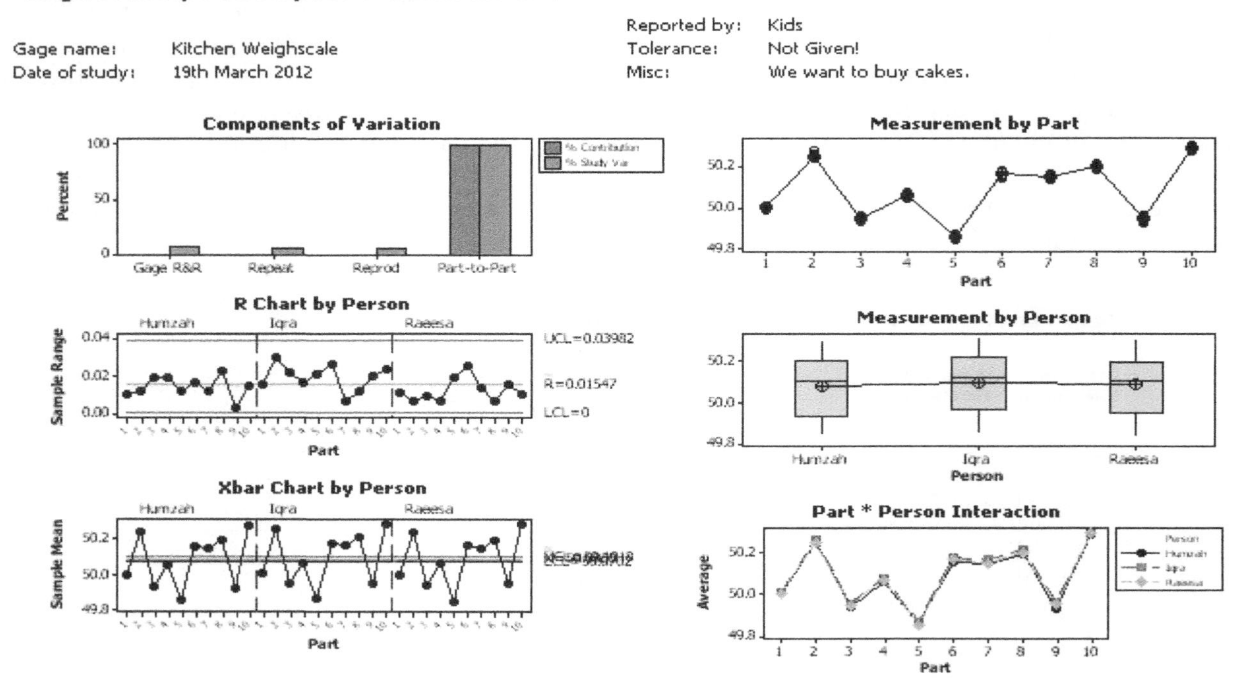

Minitab produces this set of six graphs and also information within the session window. We shall look at the results in detail and look at how they helps us to understand our measurement system.

Four sets of bar graphs are within the Components of Variation bar chart. The first set is the Gage R&R bar graph. This is then broken into its separate components of Reproducibility and Repeatability. The Part to Part variation is shown as the fourth set of bars.

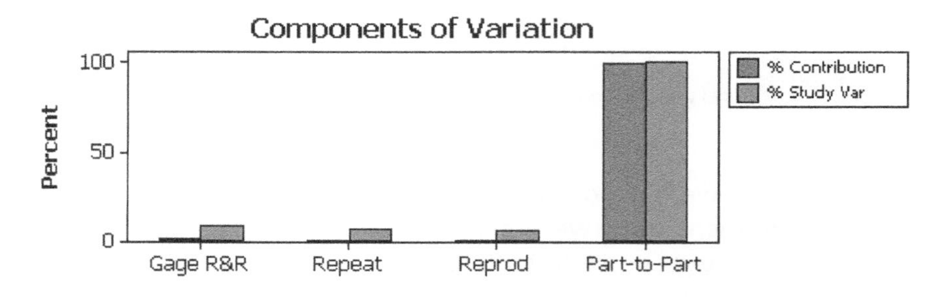

Each Bar graph is broken down into two components %Contribution and %Study Variation. Normally, we would choose %Study Variation to decide if the measurement system was acceptable by comparing it to the values shown in the table below.

If we wanted to improve parts to a specification we would use %Tolerance as our metric. In order to display %Tolerance within the graph we would have needed to enter a tolerance figure within the options menu. As I did not provide a tolerance figure for my kids this is not applicable within this example.

	%Contribution	%Study Variation
Acceptable	<1%	<10%
Marginal	≤9%	≤ 30%
Unacceptable	>9%	>30%

Therefore, we will use %Study Variation as our metric in deciding whether our measurement system is acceptable.

In our study we can see that the Gage R&R error is probably acceptable using the %Study Variation metric. Hovering over the bars with the mouse pointer will actually show the value of each of the bars.

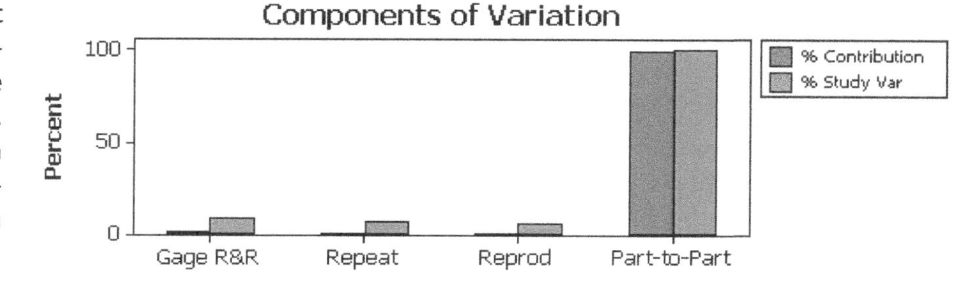

The repeatability and the reproducibility errors appear to be about the same.

The gage error is much smaller than the part to part variation. This is good news as we would want the variation between the parts to be much greater than our Gage R&R error.

	%Contribution	%Study Variation
Acceptable	<1%	<10%
Marginal	≤9%	≤ 30%
Unacceptable	>9%	>30%

The R chart shows the range of measurements for each part in turn by each person. This chart highlights repeatability issues.

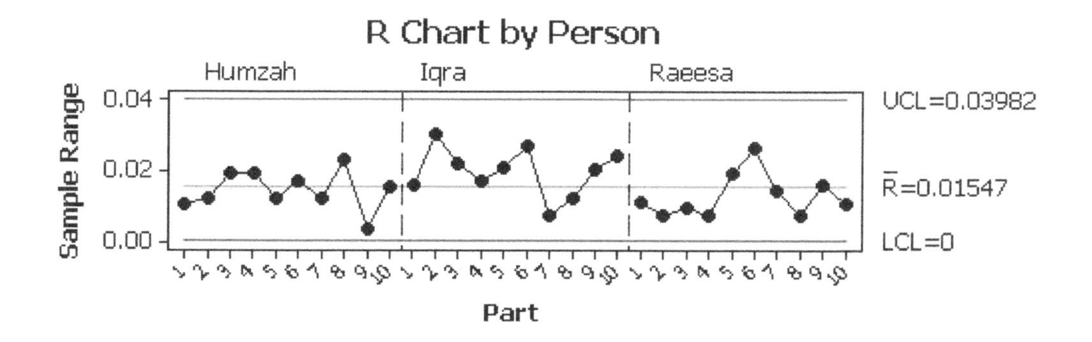

All points should be within the control lines and the pattern of the data should be the same.

A point outside the control lines indicates that a person is having trouble measuring a particular part. If this occurs, the problems with the measurement system should be studied and then corrected. The study should then be repeated.

Control lines are calculated using the average range of each part. Essentially, this chart and the next are an Xbar-R chart of the measurement data. You will learn more about this type of chart within the SPC module.

The XBar chart shows the average measurements for each part in turn for each person.

The Control lines on this chart represent the repeatability variation of the study.

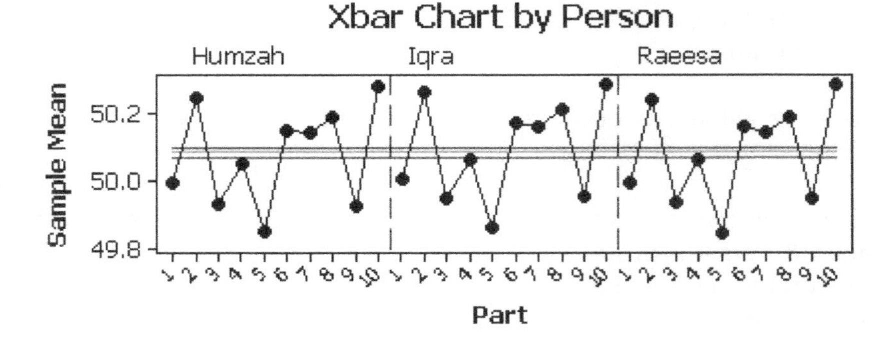

The data points should be outside the control lines and the pattern of the data should be the same. Having the points outside the control lines indicates that the part to part variation is greater than the measurement system repeatability. Minitab suggests that at least 50% of the data points should be outside the control lines.

The Measure by Part chart shows all the measured values for each part and the average value for each part.

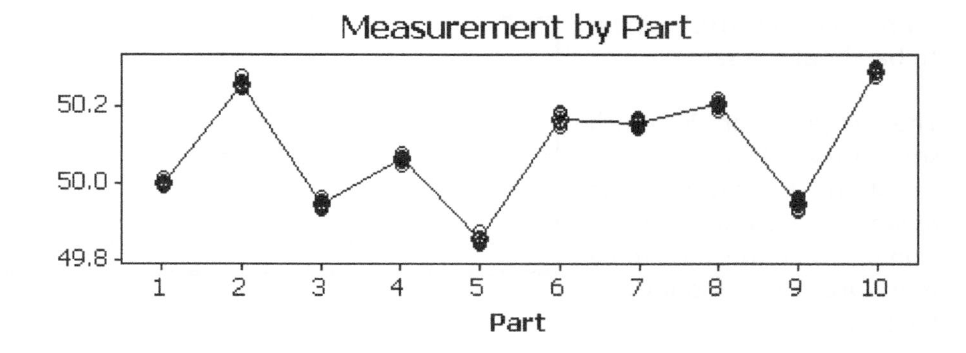

From this chart it would be easy to see if the operators were having trouble measuring any particular part. The range of values around that part would be greater than any other. This problem does not seem to be apparent in our study.

The Measure by Operator chart shows the spread of measured values for each person and the average value for each person.

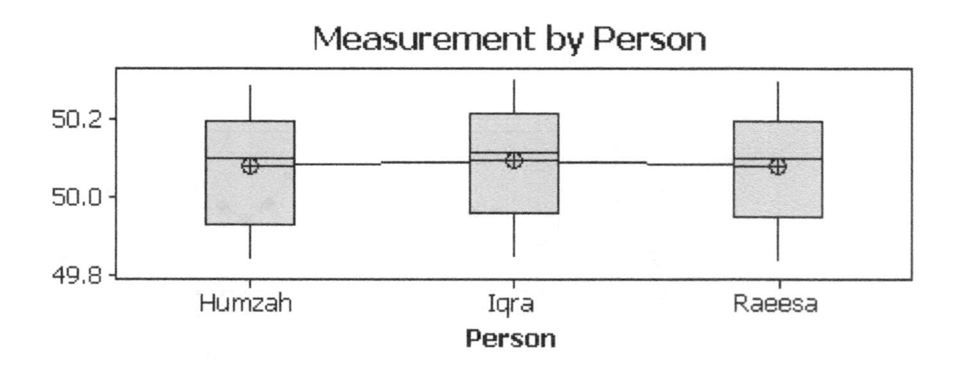

From this graph it can be seen if any of my kids are

getting different results to any of the others. That is, you would be able to tell if one of them was measuring consistently higher or lower than the others or if their spread of measurements was greater.

In our study it can just be seen that Iqra is measuring a slightly higher average value than Humzah or Raeesa.

For the Operator* Part interaction chart we want all the lines to be the same and overlaid on top of each other.

If one line is consistently higher or lower than the rest this indicates that the person is consistently over measuring or under measuring.

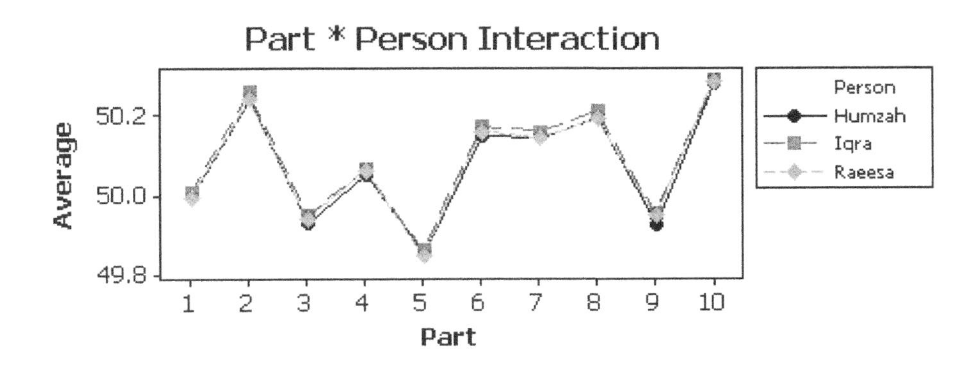

If the lines cross this may indicate an operator part interaction i.e. the person's ability to measure is dependent upon the part. The ANOVA table within the Session Window would confirm the strength of the interaction.

Since my kids did such a splendid job of measuring the lines are overlaid and we don't seem to be seeing any Operator*Part interaction.

We are now going to look at the output within the Session Window. This is where we get the numbers to be able to judge our measurement system.

Minitab uses the ANOVA procedure to calculate the components of variance and then uses those components to calculate the percentage variation of the different components of the measurement system.

In the first table the *P* value for the interaction term is calculated at 0.395. As this is above 0.25, Minitab removes the interaction term from the ANOVA calculation and carries out the test again. In effect, Minitab is saying that there is no Operator*Part interaction.

As the *P* value for person is zero it means that the difference of measurements is statistically significant.

Two Way ANOVA Table With Interaction

Source	DF	SS	MS	F	P
Part	9	1.73911	0.193235	2311.42	0.000
Person	2	0.00422	0.002109	25.23	0.000
Part * Person	18	0.00150	0.000084	1.08	0.395
Repeatability	60	0.00465	0.000078		
Total	89	1.74949			

Alpha to remove interaction term = 0.25

Two Way ANOVA Table Without Interaction

Source	DF	SS	MS	F	P
Part	9	1.73911	0.193235	2447.02	0.000
Person	2	0.00422	0.002109	26.71	0.000
Repeatability	78	0.00616	0.000079		
Total	89	1.74949			

As %Contribution is based on Variance the Part to Part variation and Total Gage R&R adds up to 100%. When the tables are based on StDevs the total will no longer add up to 100%.

Total Gage R&R is made up of Repeatability and Reproducibility. Reproducibility is made up of Operator (Person) and the Operator*Part interaction. However, as the P value for the Operator*Part interaction was above 0.25 it was removed from the calculations.

Gage R&R

```
                                   %Contribution
Source               VarComp     (of VarComp)
Total Gage R&R     0.0001466             0.68
  Repeatability    0.0000790             0.37
  Reproducibility  0.0000677             0.31
    Person         0.0000677             0.31
Part-To-Part       0.0214618            99.32
Total Variation    0.0216084           100.00


                               Study Var  %Study Var
Source             StdDev (SD)  (6 * SD)       (%SV)
Total Gage R&R        0.012110  0.072659        8.24
  Repeatability       0.008886  0.053318        6.05
  Reproducibility     0.008227  0.049361        5.60
    Person            0.008227  0.049361        5.60
Part-To-Part          0.146498  0.878990       99.66
Total Variation       0.146998  0.881988      100.00


Number of Distinct Categories = 17
```

In terms of %Study Variation we see that the Total Gage R&R is 8.24% which means that our measurement system is acceptable. The Repeatability component is 6.05% and the Reproducibility component is slightly lower at 5.6%.

	%Contribution	%Study Variation
Acceptable	<1%	<10%
Marginal	≤9%	≤ 30%
Unacceptable	>9%	>30%

Number of Distinct Categories gives an overall grade to the measurement system but lacks detail. This value should be 6 or above. Sometimes this value is taken to indicate the number of bins that your measurement system can resolve a part into.

So if Number of Distinct Categories was two, the measurement system could only grade the item as a pass or fail.

Only use this as an indicator not a metric for assessing the measuring system.

Now we will have a look at using the Assistant to handle this analysis.

1. Click on Assistant <<Measurement Systems Analysis.

Assistant

Measurement Systems Analysis (MSA)...

2. The selection flowchart appears. We will only be looking at the Measurement side which deals with continuous data, which is on the left. Click on the button for Gage R&R Study Crossed.

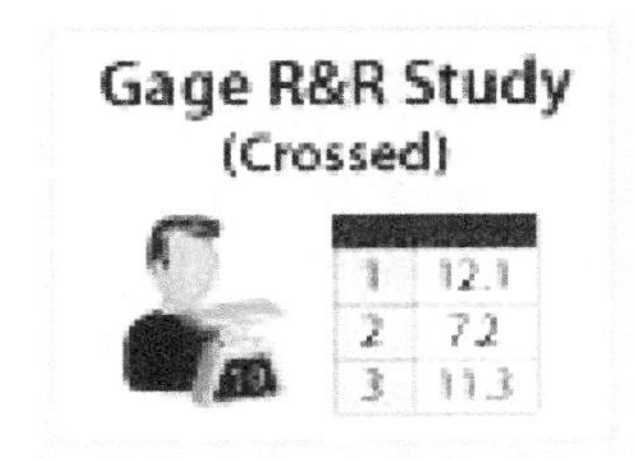

3. Complete the menu as shown. Note, that the Assistant strongly recommends that we use a historical StDev. Also, we will not be entering a tolerance as I did not specify one for my kids. We will rely on Percentage of Process as our metric for improving the measurement system, in this example.
4. Click OK.

Measurement data

Operators: Person (optional)

Parts: Part

Measurements: Measurement

Process variation

How would you like to estimate the process variation?

Use historical standard deviation: (strongly recommended)

Estimate from parts in the study

Tolerance (optional)

Would you like to evaluate the capability of the measurement system to accurately accept or reject parts?

Enter at least one specification limit

Lower spec:

Upper spec:

Enter tolerance width

Upper spec - lower spec:

Starting in the top left, the Summary Report tells us that 8.2% of the observed variation is coming from the measurement system. Since this is below the 10% level the measurement system is deemed to be acceptable.

Below that is a bar chart showing the Variation Breakdown. We can see that Repeatability and Reproducibility are very similar in terms of percentage of process variation.

The comments section gives us a more precise breakdown of the Repeatability and Reproducibility components. Remember, the components won't add to 100% as they are based on StDevs.

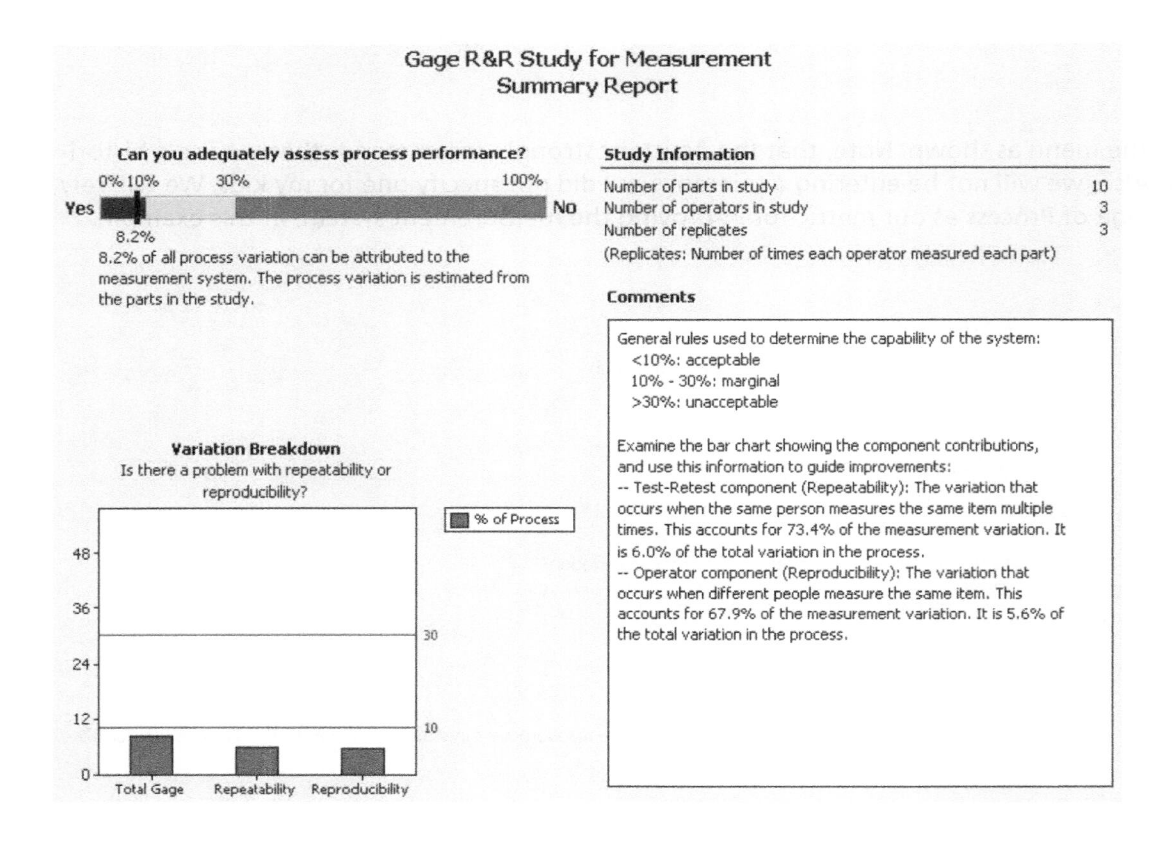

Within the Variation Report we are given the relevant information presented within the same format as the Classic method. We see that the XBar Chart, R Chart, Operator*Part Interaction Chart and Measurement by Person Chart all appear again. On the bottom right we see the breakdown of the components of variation. %Process is similar to %Study Variation which we saw within the Classic Method. Note, the '(data)' below the %Process heading indicates that the Assistant calculated the Process Variation from the measurement data, that is 10 parts.

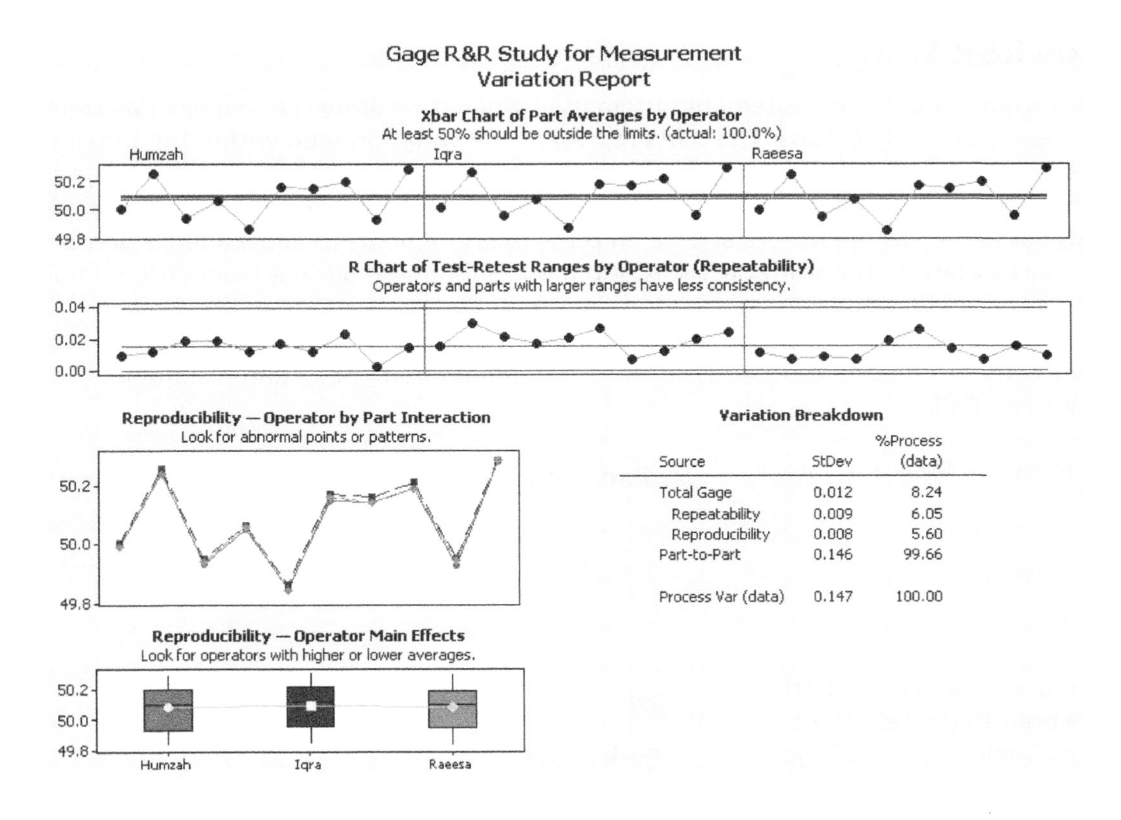

Source	StDev	%Process (data)
Total Gage	0.012	8.24
Repeatability	0.009	6.05
Reproducibility	0.008	5.60
Part-to-Part	0.146	99.66
Process Var (data)	0.147	100.00

On the Report Card we are given three pieces of key information about this type of study.

The first is regarding our choice to ask the Assistant to calculate the part to part variation from the study data, 10 parts, and not provide a historical figure based on many more parts. The general advice would be to enter a historical StDev wherever possible, otherwise the data for 10 parts, three operators and two or three repeats is the minimum that you should have.

Let's have a look at the next example where we enter a tolerance and calculate the part to part variation from historical data.

Gage R&R Study for Measurement Report Card

Check	Status	Description
Amount of Data	i	To determine if a measurement system is capable of assessing process performance, you need good estimates of the process variation and the measurement variation. -- Process variation: Comprised of part-to-part and measurement variation. It can be estimated from a large sample of historical data, or from the parts in the study. You chose to estimate from the parts. Although the number of parts (10) satisfies the typical requirement of 10, the estimate may not be precise. If the selected parts do not represent typical process variability, consider entering a historical estimate or using more parts. -- Measurement variation: Estimated from the parts, it is broken down into Reproducibility and Repeatability. The number of parts (10) and operators (3) meets the typical requirement of 10 parts and 3 operators. This is usually adequate for estimating Repeatability, but the estimate of Reproducibility is less precise. If the %Process for Reproducibility estimate is large, you may want to examine the differences between operators and determine if these differences are likely to extend to other operators.
Xbar Chart	i	The control limits are based on Repeatability. Ideally, the variation from repeated measurements is much less than the variation between parts. Guidelines suggest that approximately 50% or more should fall outside the limits. In this study, 100.0% are outside.
R Chart	i	Each point is the range of the measurements for a part. In this study, no points are above the upper control limit, indicating all parts were measured with similar consistency.

Example 4. Gage R&R Study.

Bads has set up a study examining the measurement system used when measuring the caliper of plasterboard. He uses three operators and 10 parts. The parts represent the variation seen within the process and not the extremes of the process. The tolerance for the caliper is 0.6 mm.This MSA is part of a project which is intended to reduce process waste when the product goes out of the caliper tolerance required by the customer. Bads has also collected historical data on these measurements to allow a more precise calculation of part to part variation. The study data was originally randomised but has been ordered for our convenience. Incidentally, Minitab does not need the data to be ordered.

Using the results table carry out a Gage R&R Study to learn if the measurement system is suitable. Use the data within Worksheet: Caliper

1. Copy the data from the Excel file into Minitab.

Part	Measurement	Operator	Historical Data
1	13.607	Aamir	13.449
2	13.613	Aamir	13.420
3	13.518	Aamir	13.413

2. We are going to use a Gage Run Chart to view our data prior to the test procedure. Click Stat <<Quality Tools <<Gage Study <<Gage Run Chart.

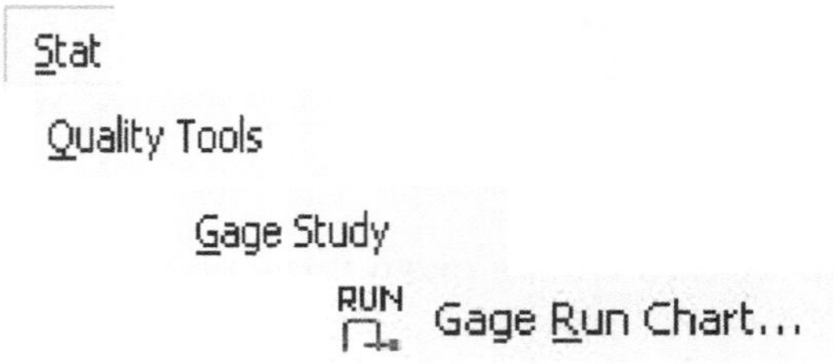

Stat

Quality Tools

Gage Study

RUN Gage Run Chart...

3. Select the following columns into the appropriate section of the Gage Run Chart menu.
4. Record some appropriate information into the Gage Info menu and then click OK and OK again.

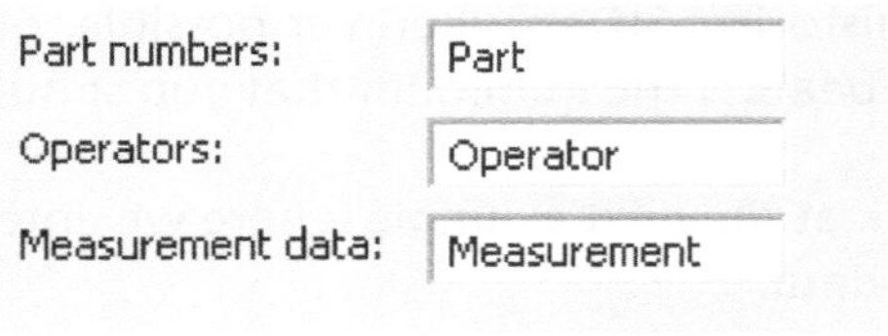

Part numbers: Part

Operators: Operator

Measurement data: Measurement

For this example it can be seen that there are both repeatability and reproducibility issues. It is hard to say which is greater. It is also difficult to say whether any particular operator is consistently having the same type of problem.

To carry out the actual test procedure we are first going to use the Classic Method and then are going to use the Assistant.

Gage Run Chart of Measurement by Part, Operator

		Reported by:	Bads
Gage name:	Calliper NCSEA 1701	Tolerance:	0.6mm
Date of study:	12th August 2012	Misc:	Calliper Reject Project

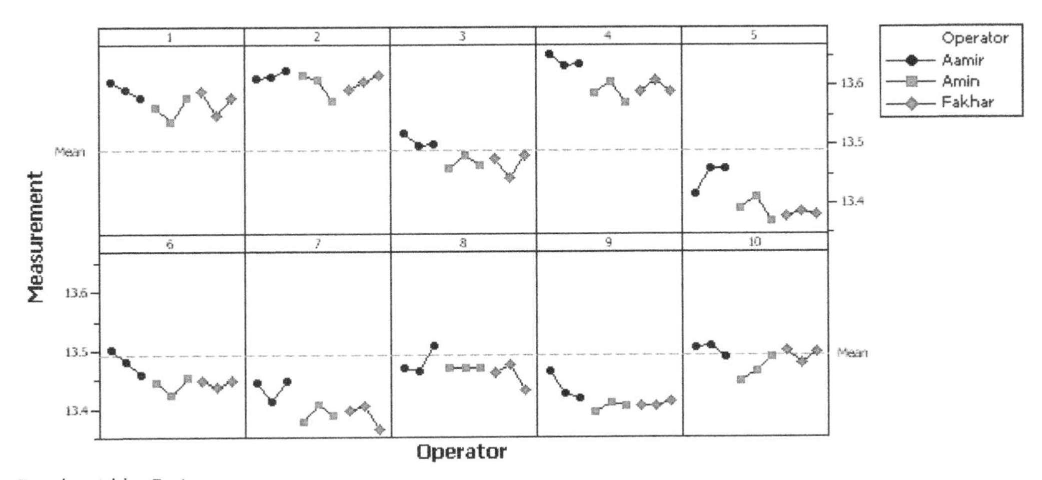

Panel variable: Part

There are 500 values of measurement data in the column labeled Historical Data. We need to calculate the StDev of this data.

5. Click Stat <<Basic Statistics <<Display Descriptive Statistics.
6. Select the Historical Data column.
7. Click on Statistics.
8. Clear the default selection and then select Standard Deviation.
9. Then click OK and OK again.
10. The result is given within the session window. This is the value we will use for historical standard deviation.

Stat

Basic Statistics

Display Descriptive Statistics...

Variables:

'Historical Data'

Mean

SE of mean

Standard deviation

Variance

Coefficient of variation

```
Variable         StDev
Historical  Data  0.0762
```

11. In order to conduct the test procedure click Stat <<Quality Tools <<Gage Study <<Gage R&R Study (Crossed).
12. Complete the menu as shown. I have already gone into Gage Info and entered some details about this study.
13. Click on Options.

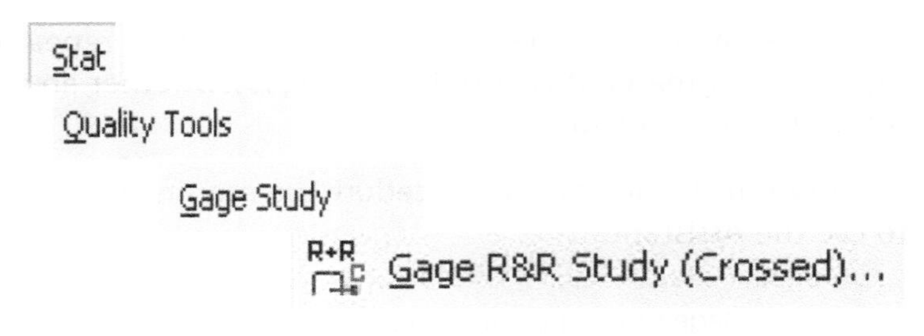

Part numbers: Part
Operators: Person
Measurement data: Measurement
Gage Info...
Options...
Conf Int...
Storage...
Method of Analysis
(•) ANOVA
( ) Xbar and R

14. Enter the tolerance of the process and the Historical Standard Deviation. Note, that from within this menu we can also change the number of StDevs that are used within the study, normally we would leave this at 6. We also have the option of changing when we will remove the Operator*Part interaction from the ANOVA table.
15. Click OK and OK again to execute the procedure.

Study variation: 6 (number of standard deviations)
Process tolerance
( ) Enter at least one specification limit
Lower spec:
Upper spec:
(•) Upper spec - Lower spec: 0.6
Historical standard deviation: 0.0762
Alpha to remove interaction term: 0.25

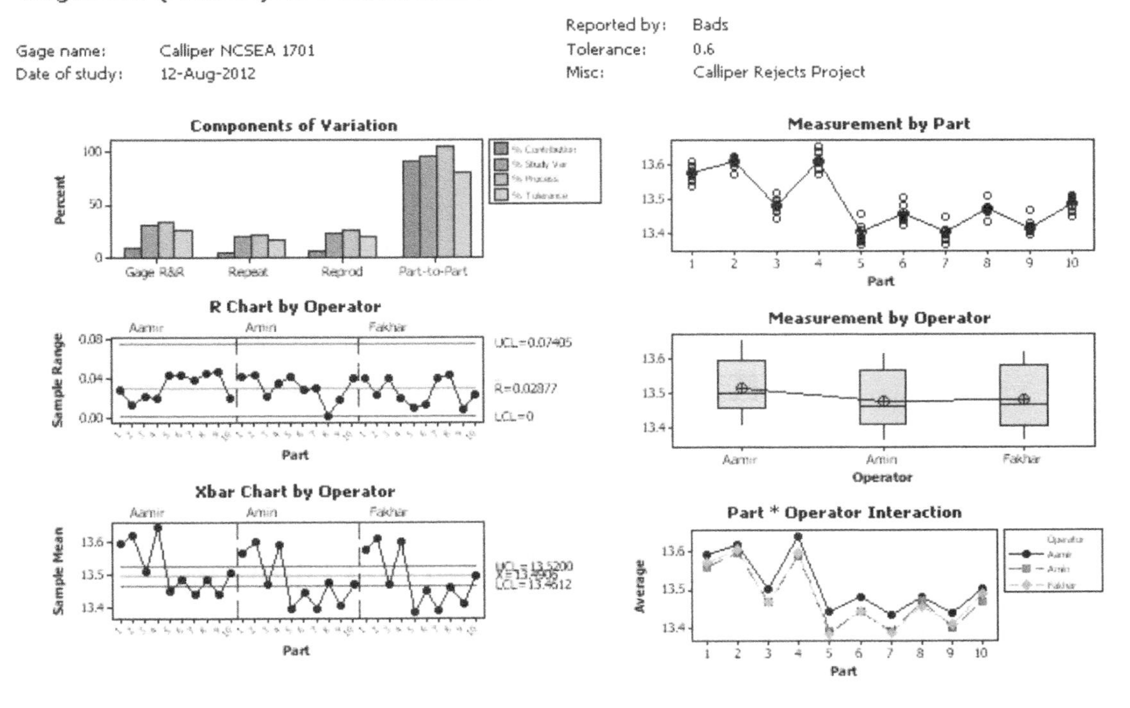

Minitab produces this set of six graphs and also information within the session window. We shall look at the results in detail and look at how it helps us to understand our measurement system.

On this occasion it can be seen that the components of variation bar graph has expanded to include new metrics. As we entered a tolerance we get a bar for %Tolerance so we can see the components of variation as a percentage of the tolerance. We also entered a historical StDev and from this we get the components of variation as a percentage of the process variation.

On this occasion we want to improve parts to a specification, therefore, we would use %Tolerance as our metric.

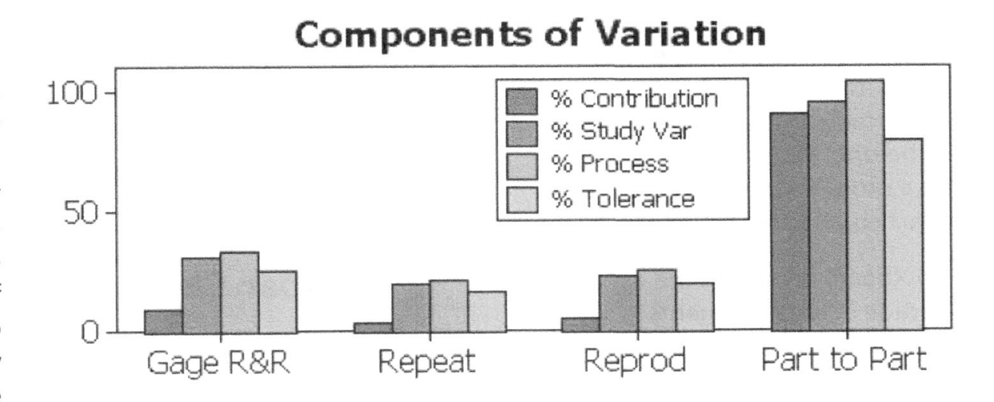

	%Contribution	%Study Variation %Process %Tolerance
Acceptable	<1%	<10%
Marginal	≤9%	≤ 30%
Unacceptable	>9%	>30%

We see that %Study Variation, %Process and %Tolerance all use the same scale to grade the measurement system.

The value of the %Tolerance within the total Gage R&R chart can be seen by clicking onto or hovering over the relevant bar within the graph.

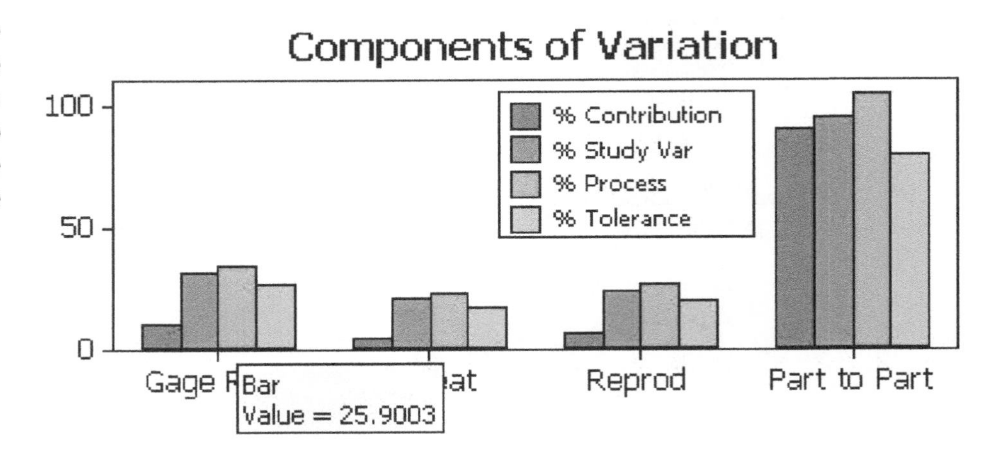

The R chart shows the range of measurements for each part in turn by each operator.

All points should be within the control lines and the pattern of the data should be the same.

All of the points are within the control lines so we can say that none of the operators are having a problem measuring a particular part.

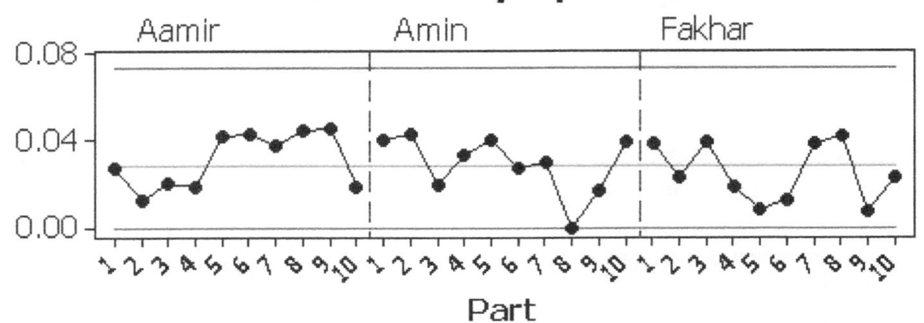

The Xbar chart shows the average measurements for each part in turn for each operator.

It can be seen that the Xbar values are inside and outside of the control lines, which is what we want. Looking more closely we can see a trend in that Aamir appears to be measuring a higher average value than the other two operators. This looks like the beginnings of a reproducibility issue.

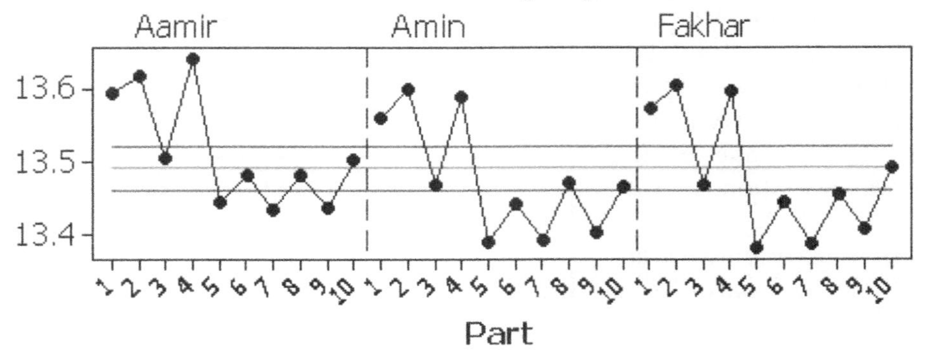

The Measure by Part chart shows all the measured values for each part and the average value for each part.

We can see that Part 2 has the least spread of measurements. To find the part with the greatest spread of measurements you can hover over the measurements at the extremes for each part and take the difference as the range for that part.

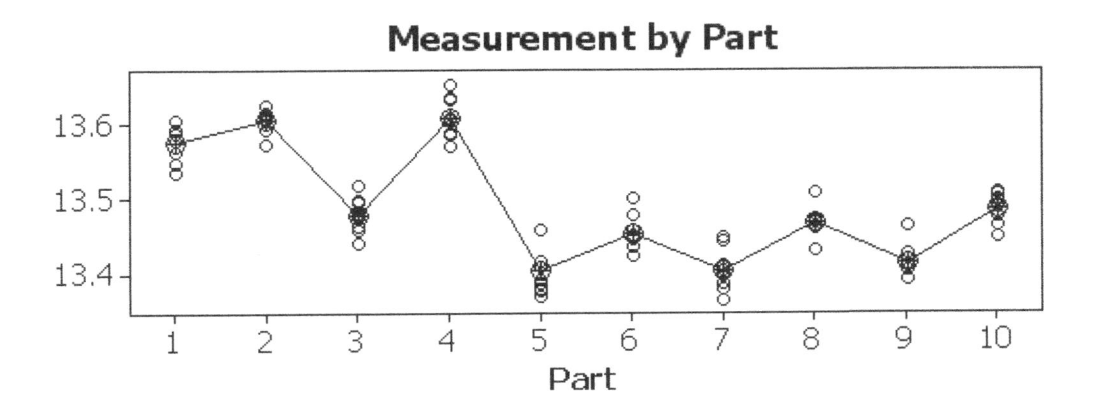

The Measure by Operator chart shows the spread of measured values for each operator and the average value for each operator.

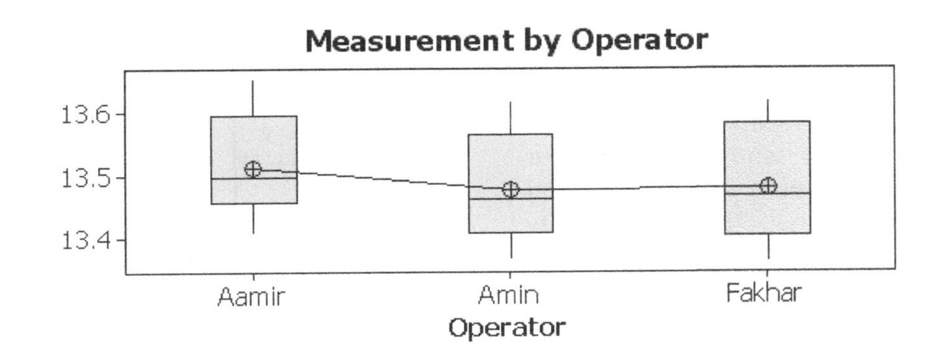

We can again see that Aamir is on average reporting a higher value than the other two operators. We cannot say that Aamir has a positive bias as that would imply we know the real values for the parts. It is plausible that Aamir is reading the correct value and Amin and Fakhar have a negative bias but we just don't know. And that is why we carry out Bias and Linearity Studies as discussed later.

For the Part * Operator interaction chart we want all the lines to be the same and overlaid on top of each other.

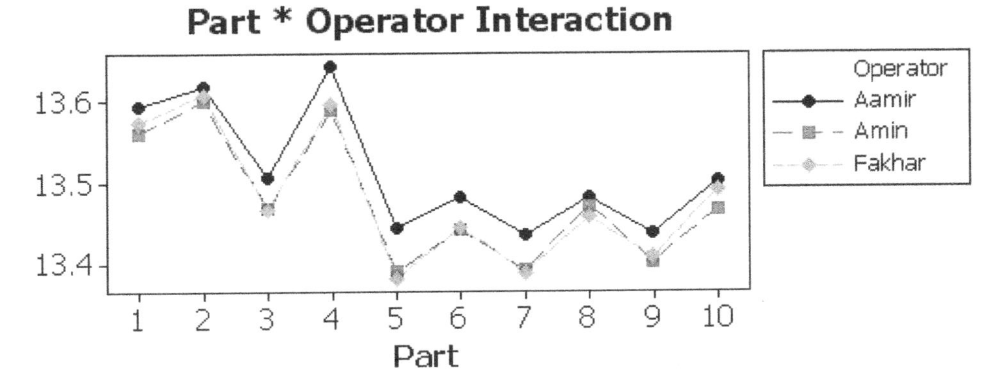

Again we see that Aamir has a higher average reading.

We don't see many lines that are not parallel. This indicates that the Part * Operator interaction may be low.

In the Session Window Minitab uses the ANOVA procedure to calculate the components of variance and then uses those components to calculate the percentage variation in the measurement system.

Two-Way ANOVA Table With Interaction

Source	DF	SS	MS	F	P
Part	9	0.513866	0.0570962	191.909	0.000
Operator	2	0.023952	0.0119760	40.253	0.000
Part * Operator	18	0.005355	0.0002975	1.078	0.395
Repeatability	60	0.016562	0.0002760		
Total	89	0.559735			

Alpha to remove interaction term = 0.25

In the first table the P value for the interaction term is calculated at 0.395. As this is above 0.25, Minitab removes the interaction term from the ANOVA calculation and carries out the test again.

Two-Way ANOVA Table Without Interaction

Source	DF	SS	MS	F	P
Part	9	0.513866	0.0570962	203.196	0.000
Operator	2	0.023952	0.0119760	42.621	0.000
Repeatability	78	0.021917	0.0002810		
Total	89	0.559735			

We are concerned with improving parts to a tolerance within this exercise and so we are using %Tolerance as the metric for this MSA. The value of %Tolerance is 25.9% and therefore our measurement system is graded as Marginal. The Repeatability component is slightly less than the Reproducibility component. All of the Reproducibility is made up from the differences between operators and none of it is made up from the Part * Operator interaction.

Gage R&R

Source	VarComp	%Contribution (of VarComp)
Total Gage R&R	0.0006708	9.61
Repeatability	0.0002810	4.02
Reproducibility	0.0003898	5.58
Operator	0.0003898	5.58
Part-To-Part	0.0063128	90.39
Total Variation	0.0069836	100.00

Process tolerance = 0.6
Historical standard deviation = 0.0762

If our objective had been to improve part to part variation we would have used %Process as our metric. If we had done that we would have said that the measurement system was graded as bad.

Source	StdDev (SD)	Study Var (6 * SD)	%Study Var (%SV)	%Tolerance (SV/Toler)	%Process (SV/Proc)
Total Gage R&R	0.0259003	0.155402	30.99	25.90	33.99
Repeatability	0.0167628	0.100577	20.06	16.76	22.00
Reproducibility	0.0197442	0.118465	23.63	19.74	25.91
Operator	0.0197442	0.118465	23.63	19.74	25.91
Part-To-Part	0.0794532	0.476719	95.08	79.45	104.27
Total Variation	0.0835681	0.501409	100.00	83.57	109.67

Number of Distinct Categories = 4

If this was your measurement system would you try and improve it or would you proceed with the project? Remember, at best the measurement system has been graded as marginal. Again, it comes down to the risk that you are willing to take and the potential consequence.

Let's have a look at how to handle this study within the Assistant.

1. Click on Assistant <<Measurement Systems Analysis.

Assistant

Measurement Systems Analysis (MSA)...

2. Click on the button for Gage R&R Study Crossed.

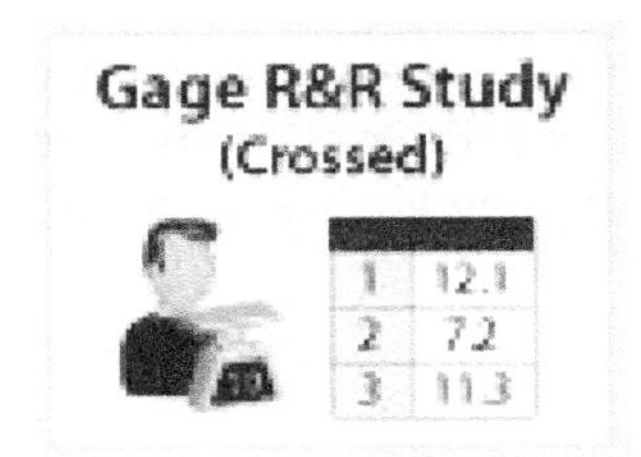

3. Complete the menu as shown. We enter our historical standard deviation that we previously calculated at 0.0762. We also enter the tolerance of 0.6 mm.
4. Then click OK.

Measurement data

Operators: Operator (optional)

Parts: Part

Measurements: Measurement

Process variation

How would you like to estimate the process variation?

(•) Use historical standard deviation: 0.0762 (strongly recommended)

( ) Estimate from parts in the study

Tolerance (optional)

Would you like to evaluate the capability of the measurement system to accurately accept or reject parts?

( ) Enter at least one specification limit

Lower spec:

Upper spec:

(•) Enter tolerance width

Upper spec - lower spec: 0.6

Starting in the top left, the Summary Report tells us that 34% of the process variation is coming from the measurement system.

As we entered a tolerance there is also a scale which asks whether we can tell good parts from bad. We are told that the measurement system variation is 25.9% of the tolerance. This is within the marginal band.

We see the breakdown of the Total Gage variation in the Variation Breakdown Chart. The comments section gives us a more precise breakdown of the Repeatability and Reproducibility components. Remember, the components won't add to 100% as they are based on StDevs.

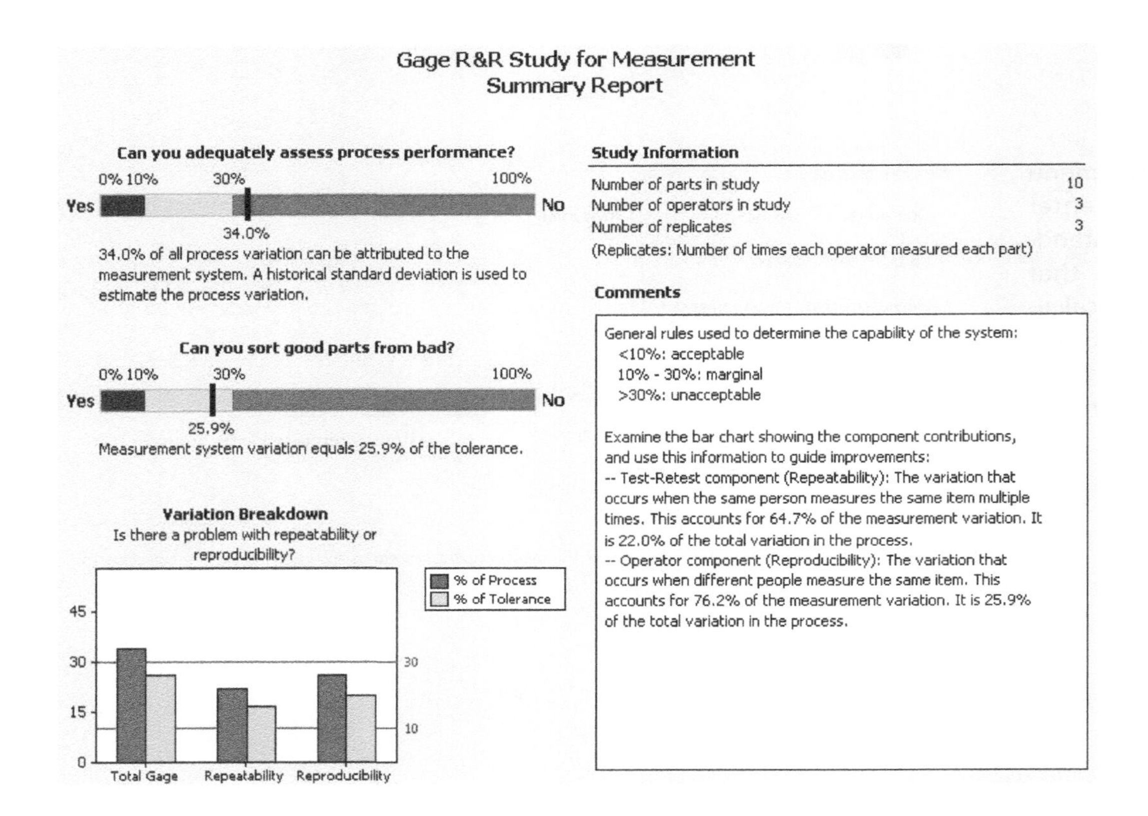

Gage R&R Study for Measurement
Summary Report
Can you adequately assess process performance?
0% 10%
30%
100%
Yes
No
34.0%
34.0% of all process variation can be attributed to the measurement system. A historical standard deviation is used to estimate the process variation.
Can you sort good parts from bad?
0% 10%
30%
100%
Yes
No
25.9%
Measurement system variation equals 25.9% of the tolerance.
Variation Breakdown
Is there a problem with repeatability or reproducibility?
% of Process
% of Tolerance
45
30
15
0
30
10
Total Gage
Repeatability
Reproducibility
Study Information
Number of parts in study 10
Number of operators in study 3
Number of replicates 3
(Replicates: Number of times each operator measured each part)
Comments
General rules used to determine the capability of the system:
<10%: acceptable
10% - 30%: marginal
>30%: unacceptable
Examine the bar chart showing the component contributions, and use this information to guide improvements:
-- Test-Retest component (Repeatability): The variation that occurs when the same person measures the same item multiple times. This accounts for 64.7% of the measurement variation. It is 22.0% of the total variation in the process.
-- Operator component (Reproducibility): The variation that occurs when different people measure the same item. This accounts for 76.2% of the measurement variation. It is 25.9% of the total variation in the process.

Within the Variation Report we are given the relevant information presented within the same format as the Classic Method. We see that the Xbar Chart, R Chart, Operator by Part Interaction Chart and measurement by person (Operator Main Effects) chart all appear again.

In the bottom right we see the Variation Breakdown table. Notice that there are separate columns for %Process based on the historical StDev we provided and that measured from the 10 parts used within the study.

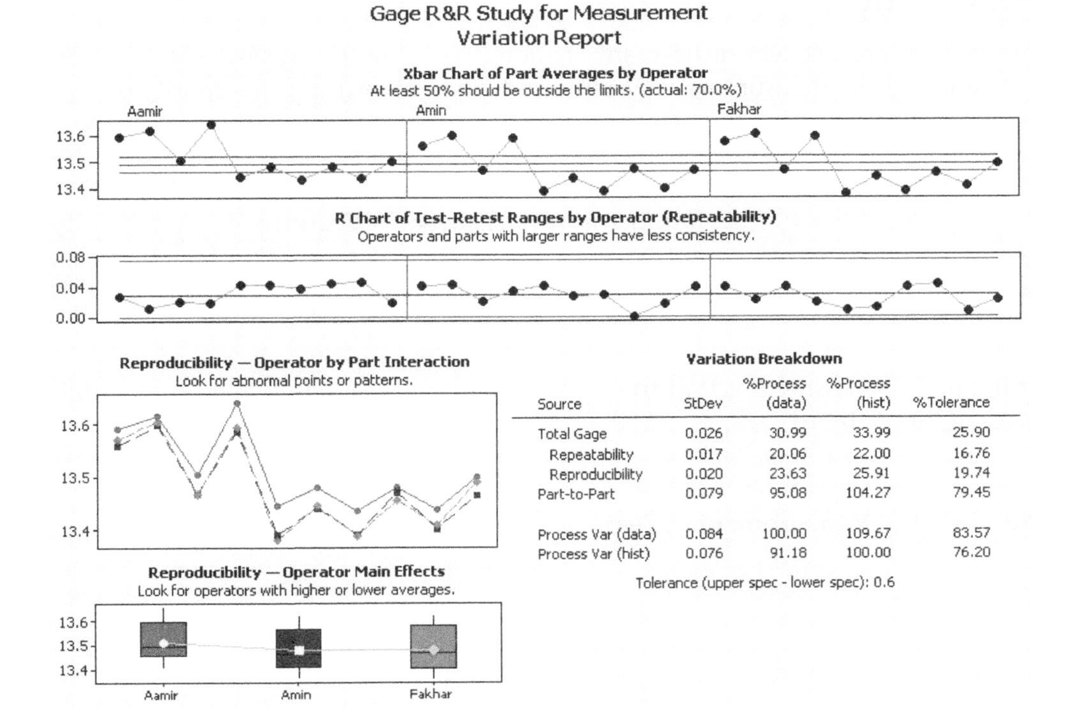

Source	StDev	%Process (data)	%Process (hist)	%Tolerance
Total Gage	0.026	30.99	33.99	25.90
Repeatability	0.017	20.06	22.00	16.76
Reproducibility	0.020	23.63	25.91	19.74
Part-to-Part	0.079	95.08	104.27	79.45
Process Var (data)	0.084	100.00	109.67	83.57
Process Var (hist)	0.076	91.18	100.00	76.20

On the Report Card we are given three pieces of key information about this type of study.

We are given information that our estimate of Reproducibility may not be that precise. Our value for Reproducibility was 25.95% under %Process (hist) which was not particularly large.

It's exercise time again.

Gage R&R Study for Measurement
Report Card

Check	Status	Description
Amount of Data	i	To determine if a measurement system is capable of assessing process performance, you need good estimates of the process variation and the measurement variation. -- Process variation: Comprised of part-to-part and measurement variation. It can be estimated from a large sample of historical data, or from the parts in the study. You entered a historical standard deviation so both estimates are available. You can compare them to see how well they agree. Although the number of parts in this study (10) satisfies the typical requirement of 10, the historical value should provide a more precise estimate of the process variation. -- Measurement variation: Estimated from the parts, it is broken down into Reproducibility and Repeatability. The number of parts (10) and operators (3) meets the typical requirement of 10 parts and 3 operators. This is usually adequate for estimating Repeatability, but the estimate of Reproducibility is less precise. If the %Process for Reproducibility estimate is large, you may want to examine the differences between operators and determine if these differences are likely to extend to other operators.
Xbar Chart	i	The control limits are based on Repeatability. Ideally, the variation from repeated measurements is much less than the variation between parts. Guidelines suggest that approximately 50% or more should fall outside the limits. In this study, 70.0% are outside.
R Chart	i	Each point is the range of the measurements for a part. In this study, no points are above the upper control limit, indicating all parts were measured with similar consistency.

Exercise 2. Gage R&R Study.

Omer is running an improvement project where he wants to improve a metric against specification levels. He sets up a study examining the measurement system used when measuring the density of fuel pellets. He uses three operators and 10 parts. The parts represent the variation seen within the process. The tolerance for the products is (USL-LSL) 0.6 kg/m^3. The historical StDev is given as 0.0753.

Check the results using a Gage Run Chart. Then using the results given in Worksheet: Fuel Density carry out a Gage R&R to establish if the measurement system is deemed to be acceptable for tolerance. Marginal is not good enough within the nuclear industry. Use the Classic Method and the Assistant, if available.

From the Gage Run Chart the initial assessment is that the measurement system is displaying repeatability and reproducibility issues, a lot of them!

Gage Run Chart of Measurement by Part, Operator

Gage name: Densometer 451
Date of study: 14 June 2012

Reported by: Uzzy
Tolerance: 0.6
Misc: Pellets feel warm!

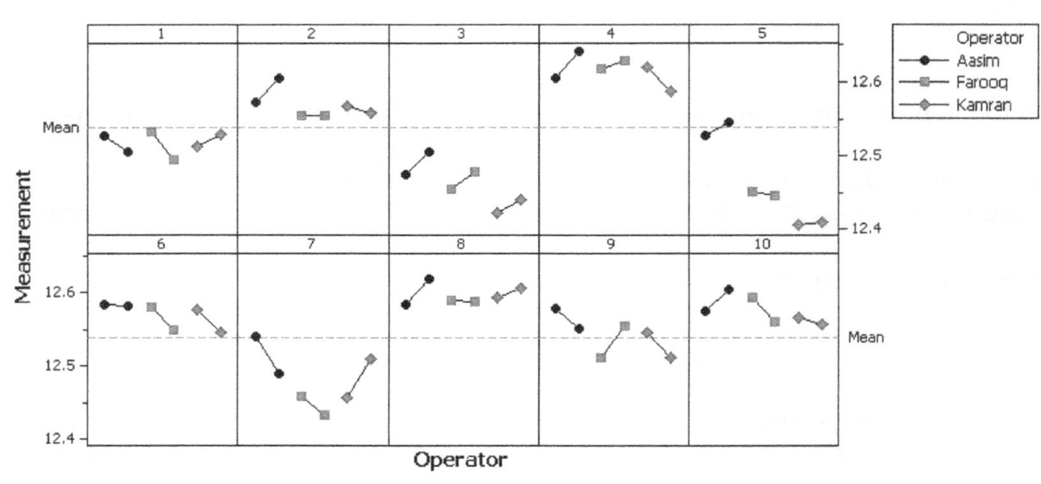

Panel variable: Part

The Components of Variation graph tells us that the %Tolerance is classed as bad if we hover over the %Tolerance bar relating to Gage R&R.

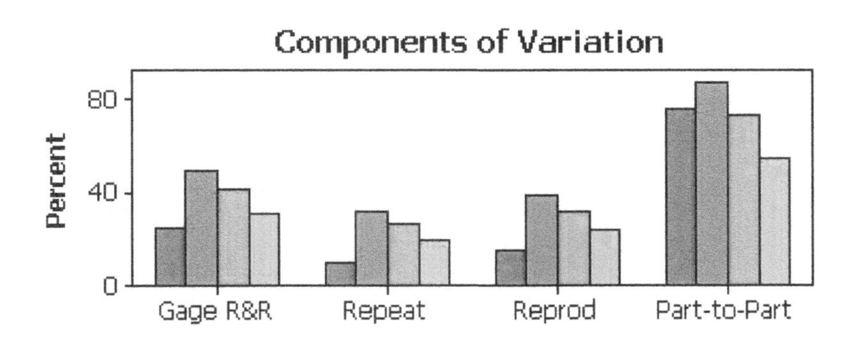

Although the Gage Run chart showed us issues with both repeatability and reproducibility the R chart shows that the repeatability variation is within control.

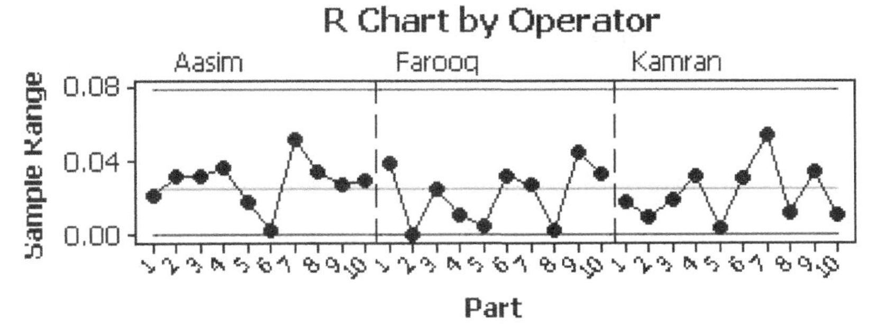

The Xbar chart shows points both inside and outside the control lines. We appear to be on the 50% limit.

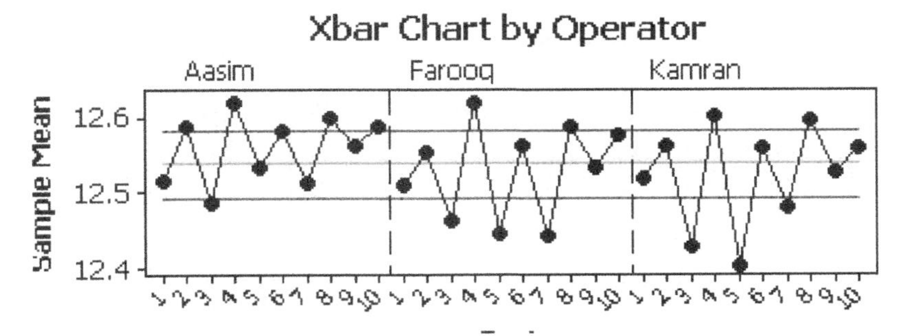

The Measure by Part chart shows that Part 5 had the greatest spread of measurements and Part 1 had the least spread.

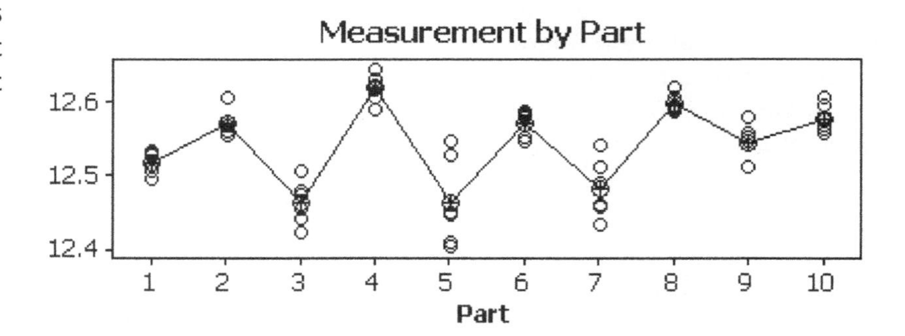

The Measurement by Operator chart shows that on average Aasim has the highest measurements.

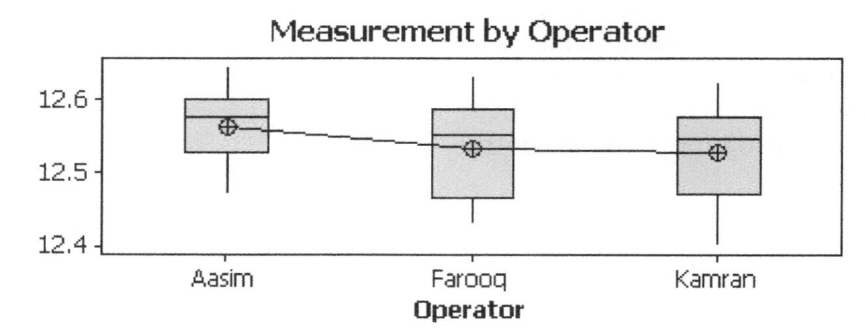

Interactions in terms of lines crossing can be seen within the Part * Operator interaction chart. However, we don't know if they are significant.

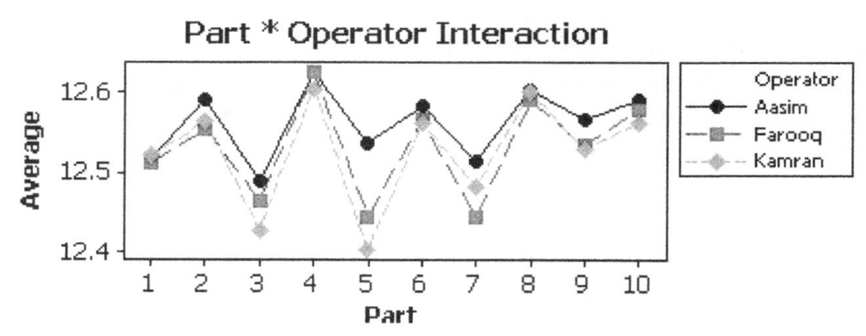

For the first time we see the Part * Operator interaction is significant and has been carried through into the Gage R&R table.

The %Tolerance for the Total Gage R&R is 31.11%. This means that 31.11% of the tolerance is taken up by measurement system variation. This means that this measurement system is graded as unacceptable.

Had we used %Process or %Study Variation as the study metric the results would have been even worse.

Gage R&R

Source	VarComp	%Contribution (of VarComp)
Total Gage R&R	0.0009677	24.48
Repeatability	0.0003864	9.77
Reproducibility	0.0005813	14.70
Operator	0.0003082	7.80
Operator*Part	0.0002731	6.91
Part-To-Part	0.0029856	75.52
Total Variation	0.0039533	100.00

Process tolerance = 0.6
Historical standard deviation = 0.0753

Source	StdDev (SD)	Study Var (6 * SD)	%Study Var (%SV)	%Tolerance (SV/Toler)	%Process (SV/Proc)
Total Gage R&R	0.0311080	0.186648	49.48	31.11	41.31
Repeatability	0.0196571	0.117942	31.26	19.66	26.10
Reproducibility	0.0241103	0.144662	38.35	24.11	32.02
Operator	0.0175550	0.105330	27.92	17.55	23.31
Operator*Part	0.0165267	0.099160	26.28	16.53	21.95
Part-To-Part	0.0546408	0.327845	86.90	54.64	72.56
Total Variation	0.0628755	0.377253	100.00	62.88	83.50

Number of Distinct Categories = 2

The Assistant tells us that this measurement system is unacceptable for process improvement or for improving the parts to the tolerance.

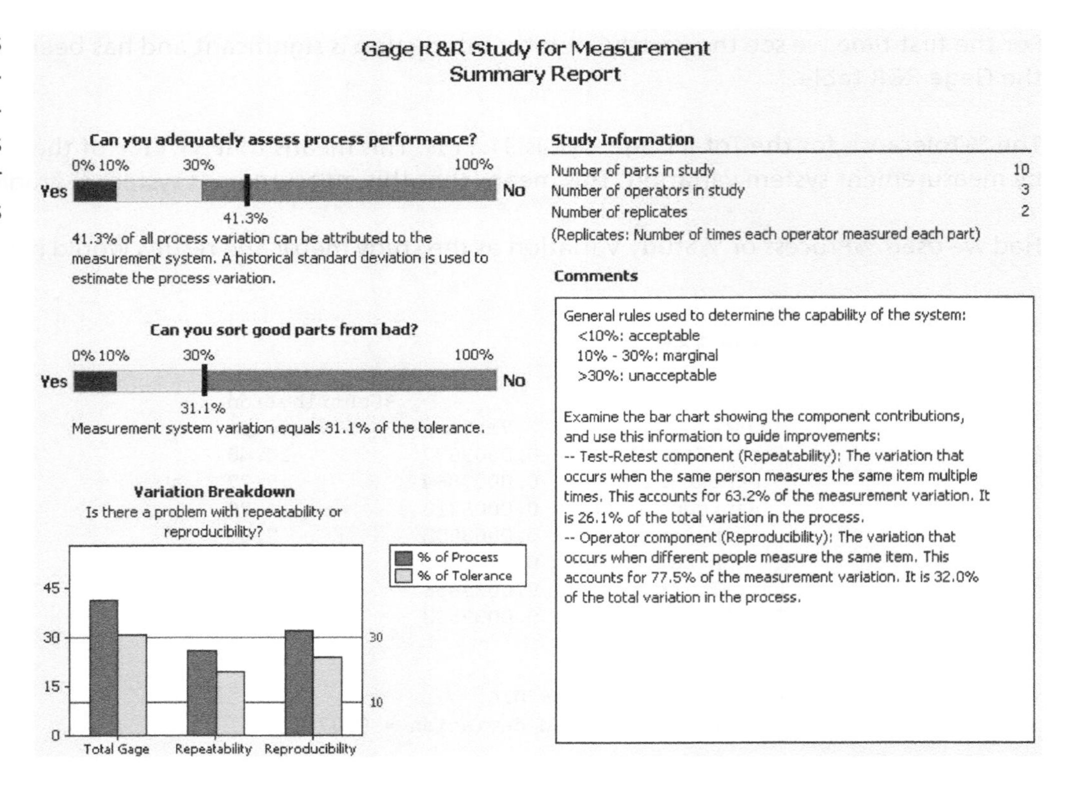

Again the Variation Report reflects the information presented within the Classic Method. The one difference being that it is possible to see the difference between %Process(data) and %Process(hist).

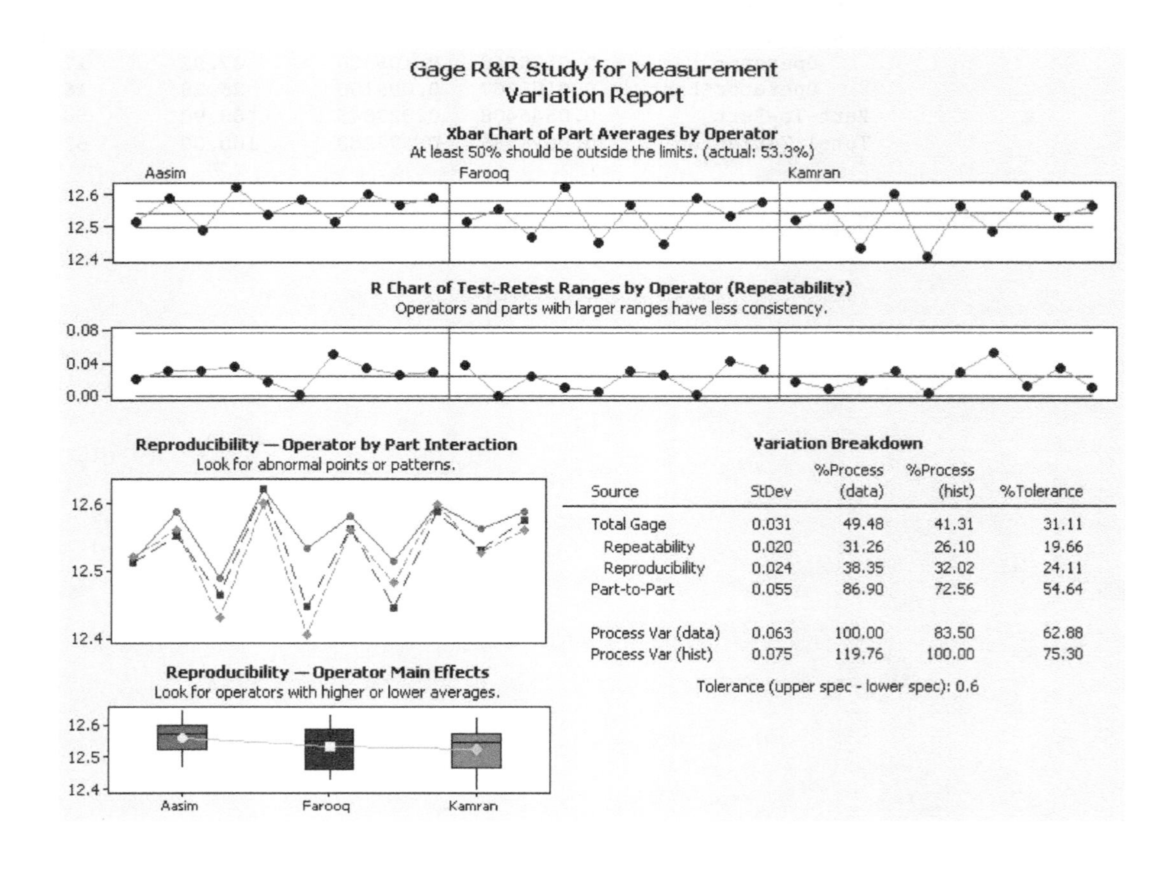

Source	StDev	%Process (data)	%Process (hist)	%Tolerance
Total Gage	0.031	49.48	41.31	31.11
Repeatability	0.020	31.26	26.10	19.66
Reproducibility	0.024	38.35	32.02	24.11
Part-to-Part	0.055	86.90	72.56	54.64
Process Var (data)	0.063	100.00	83.50	62.88
Process Var (hist)	0.075	119.76	100.00	75.30

Tolerance (upper spec - lower spec): 0.6

6.12 Gage R&R (Nested)

Procedure	• Gage R&R (Nested)
What's it used for?	• It is used to compare the reproducibility and repeatability elements of the measurement system with the overall study variation and, if required, tolerance.
Assumptions and Limitations	• The Nested study is used to assess measurement systems associated with destructive testing where large batches of identical parts cannot be produced. However, small batches must be available to allow a single operator to make repeated measurements from the same batch. • Parts are randomly assigned to operators. • It is better to have a large number of parts within the study as this reduces the chance that a particular operator will receive a part at the extreme of the measurement range. • There is not an option with the Assistant for Nested Studies.

Example 5. Gage R&R Study.

Mahwish has set up a study examining the measurement system used when measuring the fire resistance of plasterboard. The measurement given is time taken for the plasterboard to fail when exposed to a set flame. She uses three operators and six parts, each part is split into three pieces. The same operator tests the pieces from the same part. The parts represent typical parts seen within the process.

Using the results table carry out a Gage R&R Nested study to learn if the measurement system is suitable. The data is in Worksheet: Fire Test.

1. Copy the data into Minitab.

Part	Operator	Measurement
A	Maryam	110.8
A	Maryam	111.0
A	Maryam	110.9
B	Maryam	109.6

2. Click Stat <<Quality Tools <<Gage Study <<Gage R&R Study (Nested).

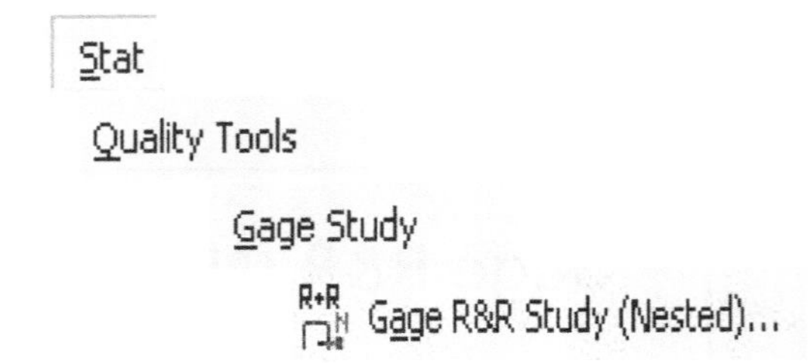

3. Complete the menu box as shown. I have entered some relevant information about the test into Gage Info.
4. Then click OK.

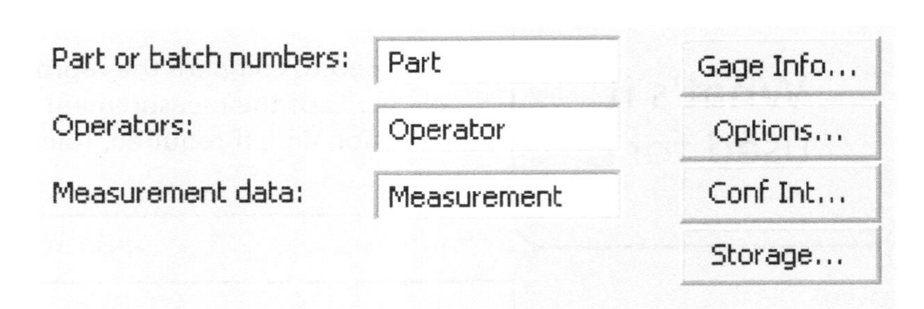

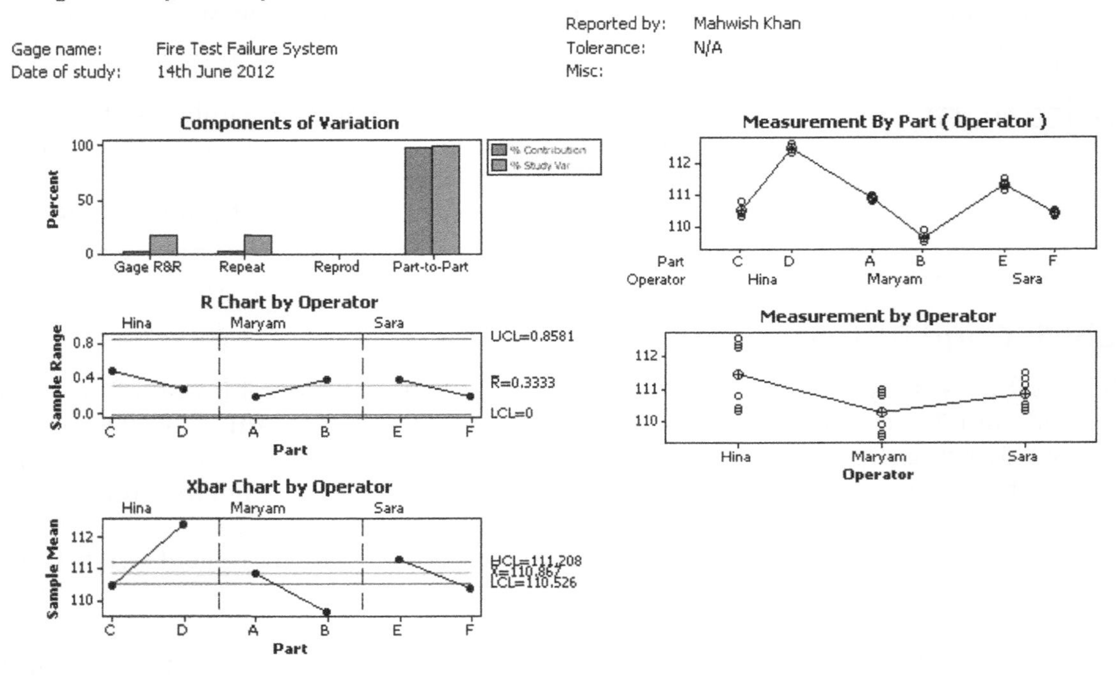

Minitab produces this set of five graphs and also information within the session window. Note: the Operator * Part graph is not available as all operators do not measure all parts.

The Components of Variation Chart shows us that the Gage variation is made up completely of Repeatability. However, do not assume that Nested studies will not show Reproducibility errors.

We can also see that the Gage Error is smaller than the Part to Part error.

As we did not enter a tolerance we will not see any tolerance metrics.

Components of Variation

% Contribution
% Study Var
Gage R&R Repeat Reprod Part to Part

	%Contribution	%Study Variation
Acceptable	<1%	<10%
Marginal	≤9%	≤ 30%
Unacceptable	>9%	>30%

The R chart shows that the measurement data for each part is under control.

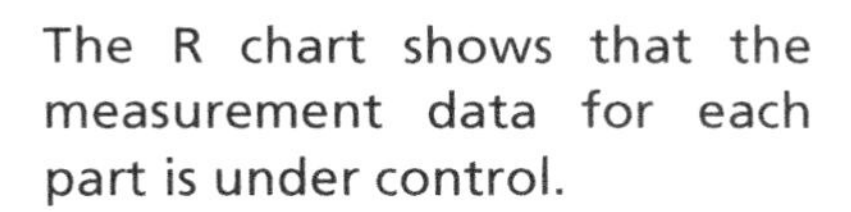

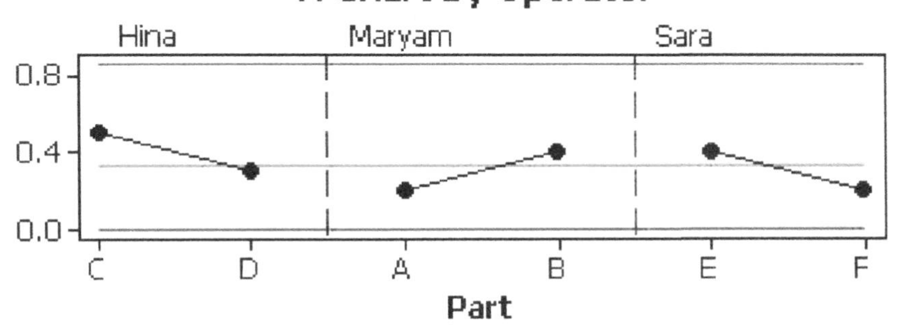

As the parts were distributed randomly we would expect the ranges for the operators to be almost the same. A higher range from one operator would indicate a Repeatability issue.

It can be seen that a number of the points are outside of the control lines. This indicates that the part to part variation is greater than the repeatability variation.

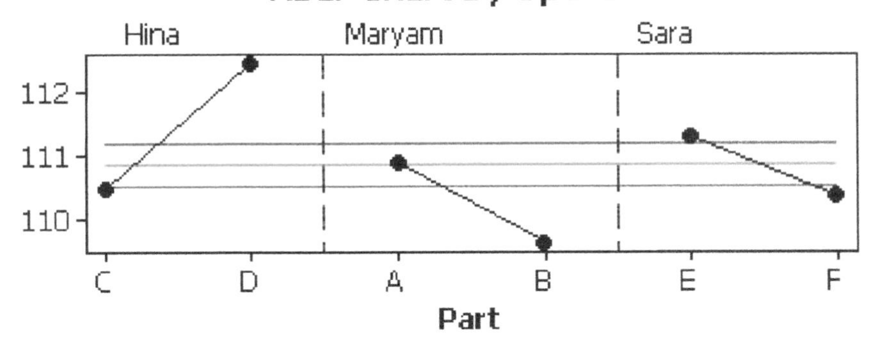

This chart shows the individual plots for each part by operator. Minitab compares the between batch variation to the within batch.

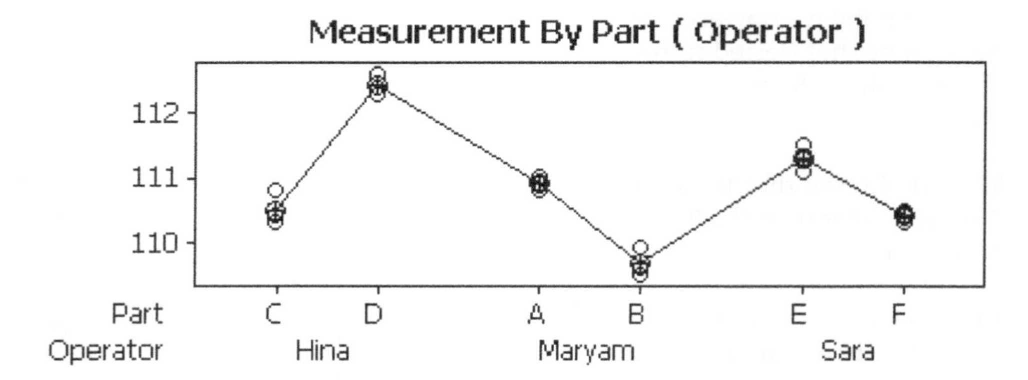

If one operator is measuring values higher than the others it could be due to the parts that they received. We therefore need to assign parts randomly to check if the differences are due to operators. The more parts we have the better.

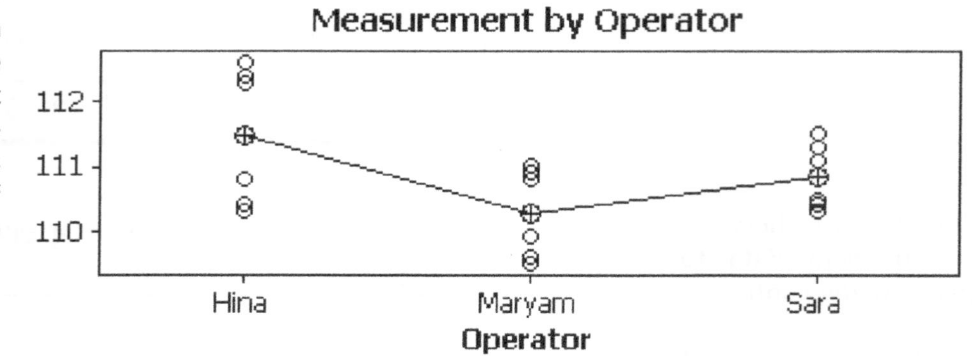

In our graph Maryam appears to be measuring a lower value than the other operators. This could be just due to parts A and B.

The %Study Variation is 17.81% for the total GR&R. All of this is from Repeatability Errors.

Number of Distinct Categories is seven.

We can say that our measurement system is Marginal.

Next, there is an exercise for you to complete.

Gage R&R

Source	VarComp	%Contribution (of VarComp)
Total Gage R&R	0.03278	3.17
Repeatability	0.03278	3.17
Reproducibility	0.00000	0.00
Part-To-Part	1.00056	96.83
Total Variation	1.03333	100.00

Source	StdDev (SD)	Study Var (6 * SD)	%Study Var (%SV)
Total Gage R&R	0.18105	1.08628	17.81
Repeatability	0.18105	1.08628	17.81
Reproducibility	0.00000	0.00000	0.00
Part-To-Part	1.00028	6.00167	98.40
Total Variation	1.01653	6.09918	100.00

Number of Distinct Categories = 7

Exercise 3. Gage R&R (Nested) Study.

Han has set up a study examining the measurement system used when measuring the impact resistance of steel. He uses three operators and nine parts, and each part is split into three pieces. The same operator tests the pieces from the same part. The parts represent the variation seen within the process. The tolerance for this measurement system is 3 kN and the aim of this study is to improve parts to the tolerance.

Using the results table carry out a Gage R&R Nested Study to learn if the measurement system is suitable. The data is in Worksheet: Impact Test.

Gage R&R (Nested) for Measurement

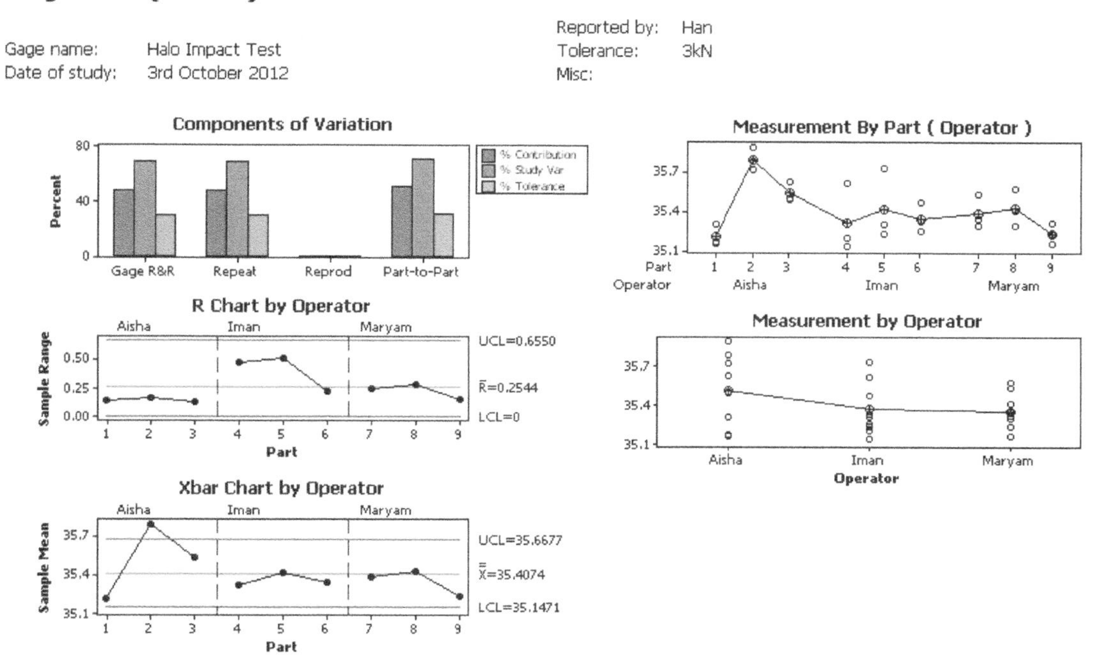

Again the Gage variation is made up completely of Repeatability and the same warning goes out again: do not assume that nested studies will not show Reproducibility errors.

We see that the part to part variation and the total gage variation have similar values. This is a bad sign.

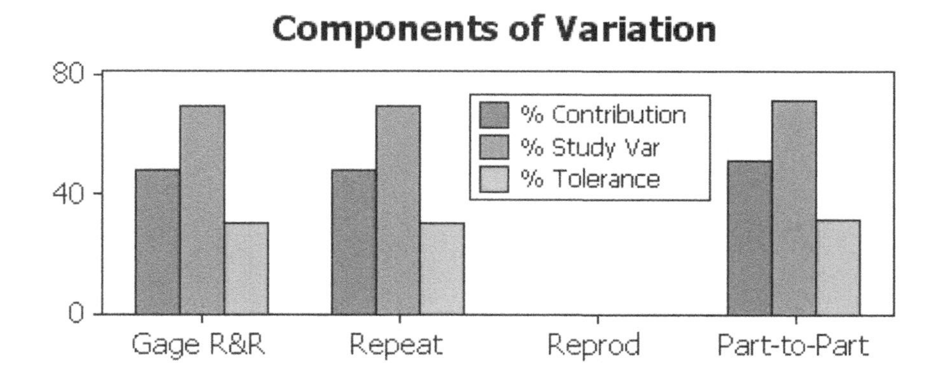

The %Study Variation looks very high but this study is focused around the %Tolerance value.

	%Contribution	%Study Variation %Process %Tolerance
Acceptable	<1%	<10%
Marginal	≤9%	≤ 30%
Unacceptable	>9%	>30%

The R chart shows that the measurement data for each part is under control.

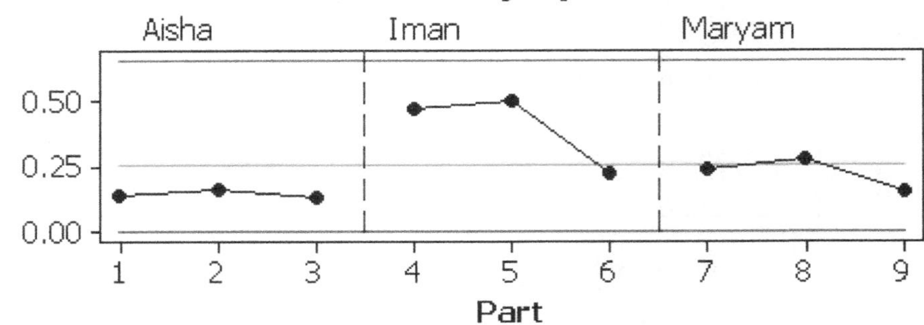

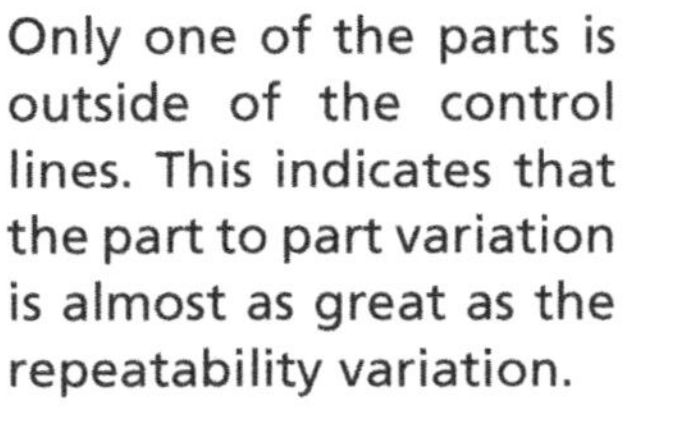
Only one of the parts is outside of the control lines. This indicates that the part to part variation is almost as great as the repeatability variation.

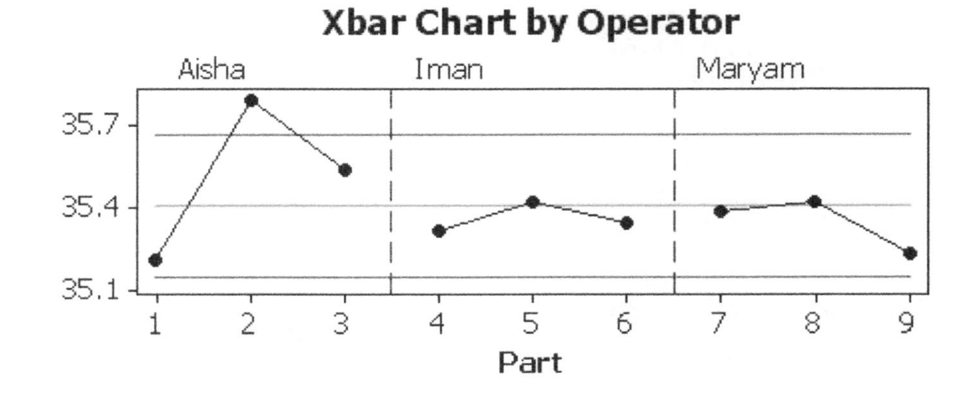

We see that the spread of measurements around Parts 4, 5 and 6 is quite large compared to the other parts. Iman may be having trouble measuring consistently.

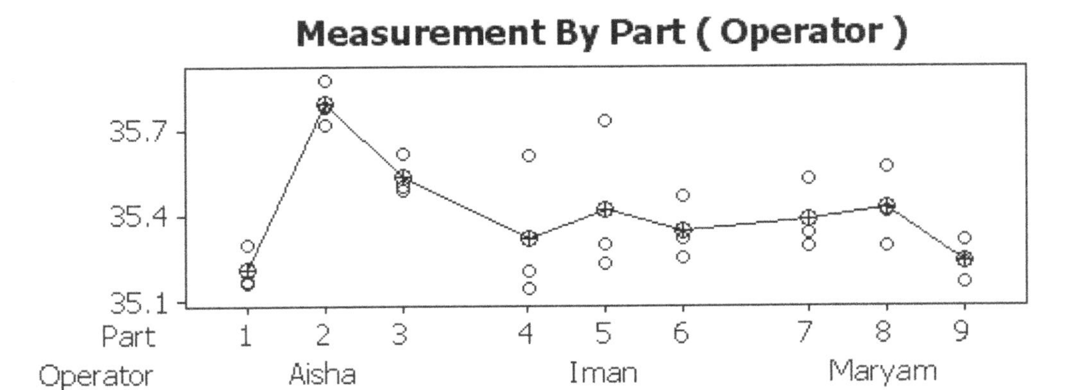

The spread shown by the operators here may also be due to the parts.

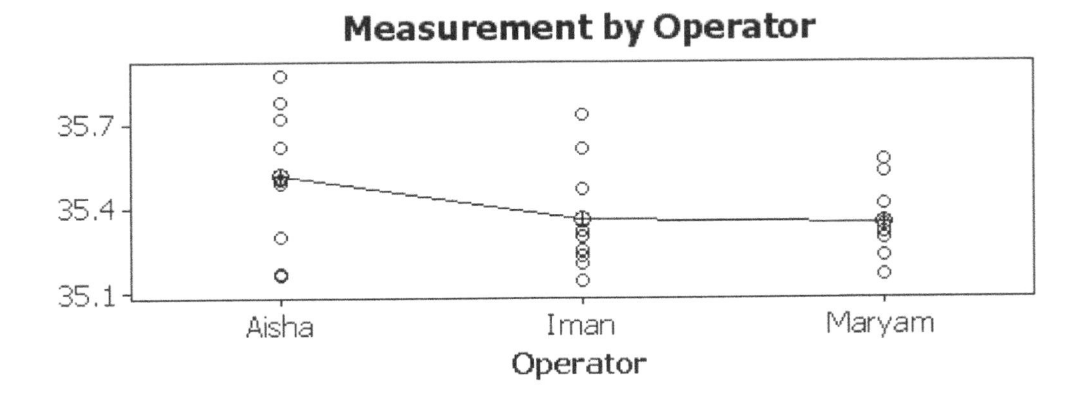

In terms of %Tolerance this measurement system is almost marginal. The %Tolerance figure is only close to being Marginal as the tolerance is quite wide. This is something that Han should consider.

Number of Distinct Categories is one.

We can say that the measurement system is unacceptable in terms of %Tolerance and needs to be improved.

Gage R&R

Source	VarComp	%Contribution (of VarComp)
Total Gage R&R	0.0234037	48.33
Repeatability	0.0234037	48.33
Reproducibility	0.0000000	0.00
Part-To-Part	0.0250198	51.67
Total Variation	0.0484235	100.00

Process tolerance = 3

Source	StdDev (SD)	Study Var (6 * SD)	%Study Var (%SV)	%Tolerance (SV/Toler)
Total Gage R&R	0.152983	0.91790	69.52	30.60
Repeatability	0.152983	0.91790	69.52	30.60
Reproducibility	0.000000	0.00000	0.00	0.00
Part-To-Part	0.158176	0.94906	71.88	31.64
Total Variation	0.220053	1.32032	100.00	44.01

Number of Distinct Categories = 1

6.13 Gage Bias and Linearity Study

Procedure	• Gage Bias and Linearity Study
What's it used for?	• It is used to assess the Bias and Linearity of a Gage over its operating range.
Assumptions and Limitations	• At least five parts that cover the operating range of the gage must be used. There must be at least 10 readings from each part. • The parts must be reference parts with known readings for each part. • It is best if the runs can be randomised and the parts kept blind. • There isn't an option within the Assistant for Gage Bias and Linearity Studies.

Example 6. Gage Bias and Linearity Study.

Uzzy wants to carry out a Gage Bias and Linearity study on the measurement system used to measure the caliper of plasterboard. The reference parts representing each of the standard sizes of board are used. The reference parts are 7.5, 9.5, 12.5, 15.0 and 19.0 mm. Each one is measured 12 times in a randomised study. It is known that six times the historical StDev is 0.38.

Using the results table carry out a Gage Bias and Linearity Study to learn if the measurement system is suitable.

Worksheet: Caliper B&L

1. Copy the data from the Excel file and paste it into Minitab.

Part	Reference	Measurement
1	7.5	7.886678482
1	7.5	7.831444021
1	7.5	7.84686318
1	7.5	7.916339125
1	7.5	7.813096127
1	7.5	7.925552069
1	7.5	7.889115978
1	7.5	7.871259865
1	7.5	7.944487935
1	7.5	7.855861467
1	7.5	7.923716655
1	7.5	7.905462991
2	9.5	9.967571597
2	9.5	9.883420579
2	9.5	9.928312155
[illegible]	[illegible]	[illegible]

2. Click Stat <<Quality Tools <<Gage Study <<Gage Linearity and Bias Study.

Stat

Quality Tools

Gage Study

LIN Gage Linearity and Bias Study...

3. Complete the menu box as shown.
4. Then click OK.

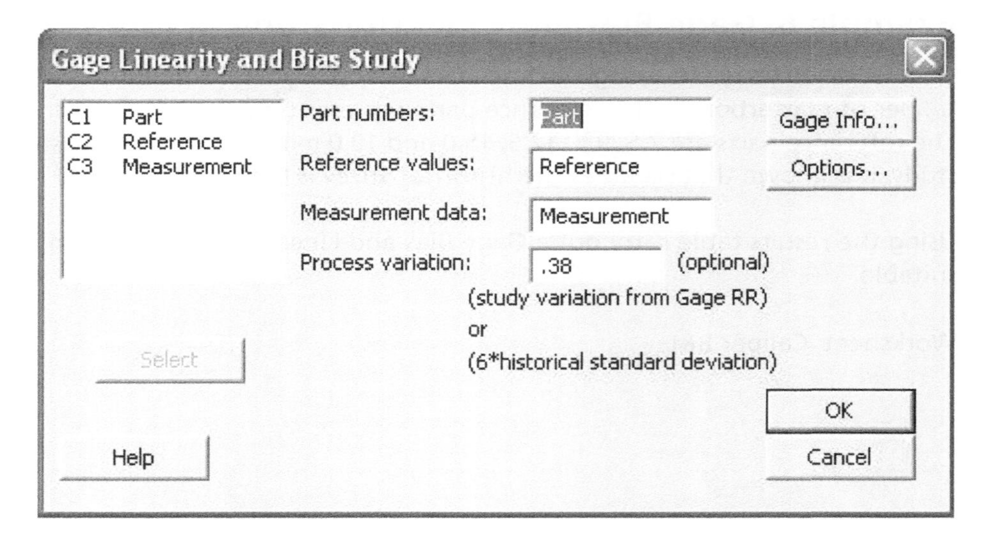

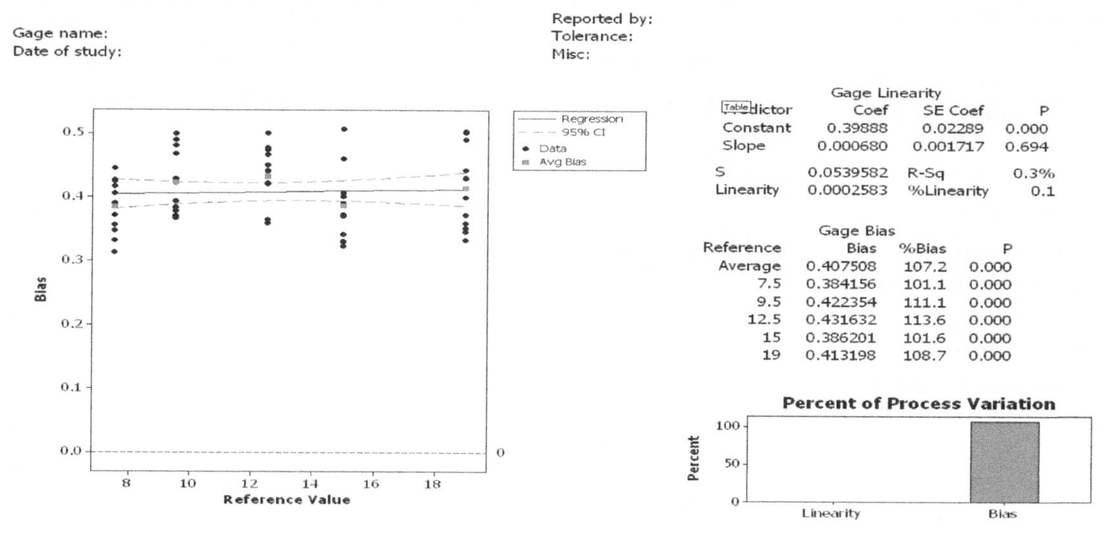

Gage Linearity

Predictor	Coef	SE Coef	P
Constant	0.39888	0.02289	0.000
Slope	0.000680	0.001717	0.694

S	0.0539582	R-Sq	0.3%
Linearity	0.0002583	%Linearity	0.1

Gage Bias

Reference	Bias	%Bias	P
Average	0.407508	107.2	0.000
7.5	0.384156	101.1	0.000
9.5	0.422354	111.1	0.000
12.5	0.431632	113.6	0.000
15	0.386201	101.6	0.000
19	0.413198	108.7	0.000

Minitab produces this output window with no information within the session window. We shall look at the results in detail and see how it helps us to understand our measurement system.

The scatterplot shows the bias (measured value – reference value) for each part. The red dot shows the average for each part.

Using Linear regression a line is fitted through the bias data points on the scatter plot.

The regression line has a confidence band around it. If the zero line is outside of the confidence band then we have bias in our measurement system when there are no linearity issues.

As all points in this study are above the zero line we have a positive bias.

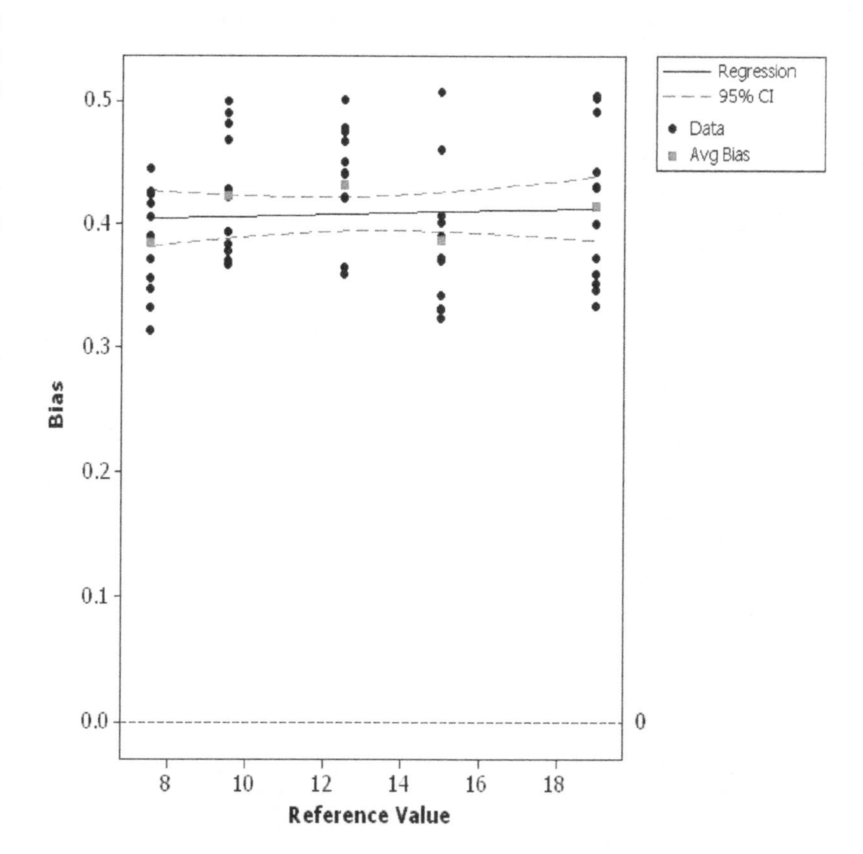

Minitab carries out two hypothesis tests on the regression line, one to check whether the constant (intercept) of the line is zero and the other to check whether the slope of the line is zero.

We find that we can reject that the intercept of the line is zero; that is it does not go through the origin.

More importantly we cannot reject that the slope is zero. This tells us that the study does not have a linearity issue.

The Linearity as a percentage of process variation is 0.1%.

Gage Linearity

Predictor	Coef	SE Coef	P
Constant	0.39888	0.02289	0.000
Slope	0.000680	0.001717	0.694

S	0.0539582	R-Sq	0.3%
Linearity	0.0002583	%Linearity	0.1

The gage bias table shows the average bias for each part and the values as a percentage of process variation. The P values shows whether the bias was significant for that value.

To check for overall Bias we would check the P value for the average of the points.

Gage Bias

Reference	Bias	%Bias	P
Average	0.407508	107.2	0.000
7.5	0.384156	101.1	0.000
9.5	0.422354	111.1	0.000
12.5	0.431632	113.6	0.000
15	0.386201	101.6	0.000
19	0.413198	108.7	0.000

Finally, we get a graphical display of linearity and bias as a percentage of process variation.

In summary, the real issue for this measurement system is bias.

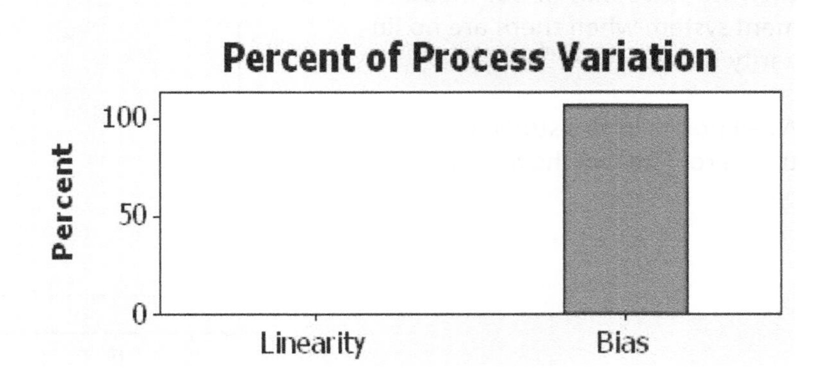

Exercise 4. Gage Bias and Linearity Study.

Parky wants to carry out a Gage Bias and Linearity study on a weigh scale. Reference weights are used which cover the range of the weigh scale. These are 100, 200, 300, 400, and 500 kg. Each one is measured 10 times in a randomised study. Process Variation is 1.5 kg.

Using the results table carry out a Gage Bias and Linearity Study to learn if the measurement system is suitable.

Worksheet: Weights B&L

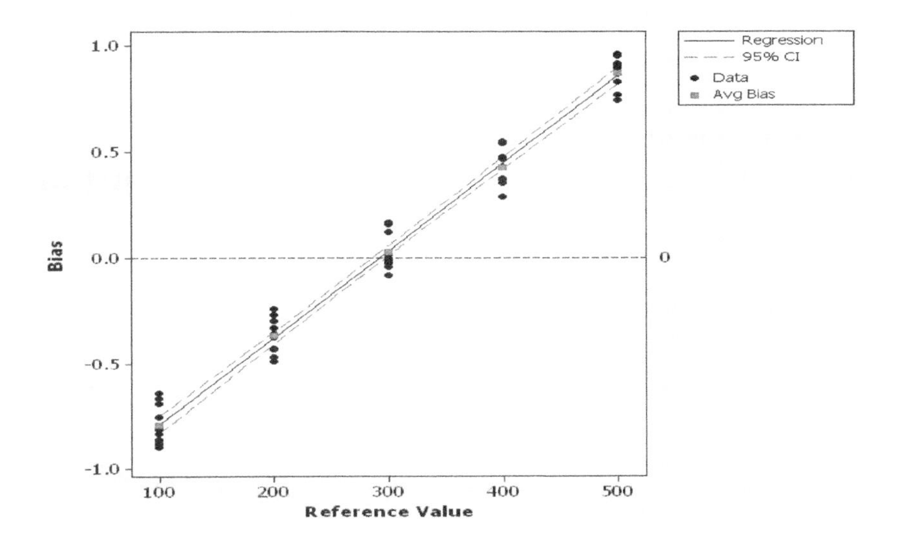

The slope of the regression line is significant so we have a linearity issue. This is also seen in the scatter plot.

Table

Gage Linearity

Predictor	Coef	SE Coef	P
Constant	-1.20233	0.02766	0.000
Slope	0.00411762	0.00008339	0.000

S	0.0833935	R-Sq	98.1%
Linearity	0.0061764	%Linearity	0.4

For the Average of the reference values we find that the *P* value is significant. This means that if the linearity issue were resolved there would still be a bias of 0.03 on average.

Gage Bias

Reference	Bias	%Bias	P
Average	0.032954	2.2	0.006
100	−0.791816	52.8	0.000
200	−0.368434	24.6	0.000
300	0.028791	1.9	0.272
400	0.426900	28.5	0.000
500	0.869328	58.0	0.000

As a percentage of process variation Bias is the bigger issue.

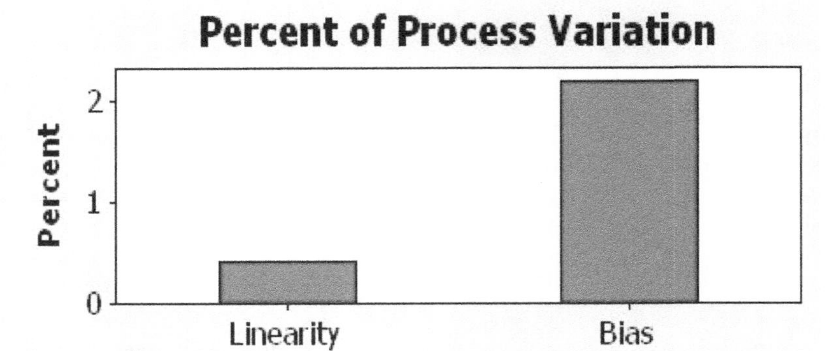

CHAPTER 7
Statistical Process Control

7.1 The Origins of Statistical Process Control

Dr Walter A. Shewart is credited as being the father of Statistical Process Control (SPC). It was in the 1920s when Shewart wrote to his boss and proposed his ideas. At that time he worked for the Western Electric Company at the Hawthorne Works. (Incidentally, this is the same Hawthorne Electrical Works where the famous Hawthorne Effect which relates to industrial psychology was recognised.)

Later William Demming applied the Statistical Control techniques to the production of munitions and essentials during WWII. Demming also worked in postwar Japan applying the same techniques that helped shape Japan as an industrial giant.

It was not until much later that control charts were used in nonmanufacturing environments, beginning with Computer Software.

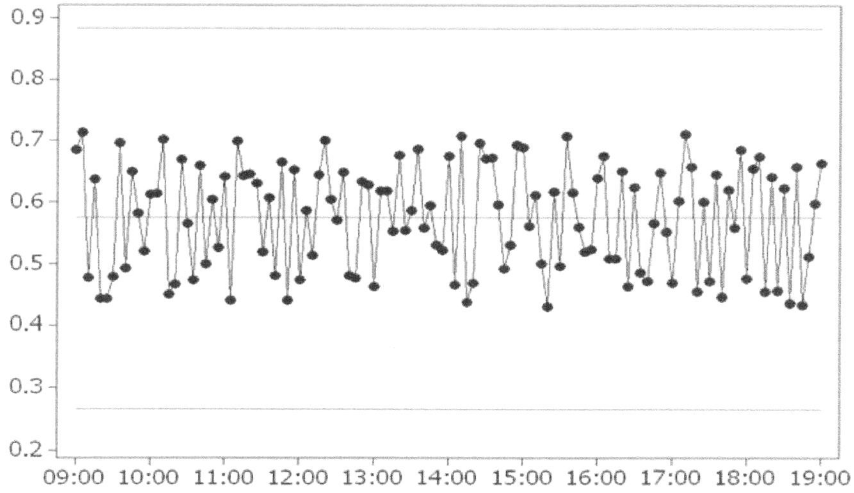

Dr Shewart recognised the importance of reducing variation in manufacturing processes. He also concluded that continual process adjustments by operators would in all likelihood increase variation and result in more defects.

He looked at problems in terms of common cause and special cause variation. He concluded that every process displays common cause variation and set limits to when the variation was caused by new or additional factors which he called special cause variation.

Dr Shewart used ±3 StDevs as the control limits to separate common cause variation from special cause variation. This figure shows a typical control chart indicating a process that is in control.

Problem Solving and Data Analysis using Minitab: A clear and easy guide to Six Sigma methodology, First Edition. Rehman M. Khan.

7.2 Common Cause and Special Cause Variation

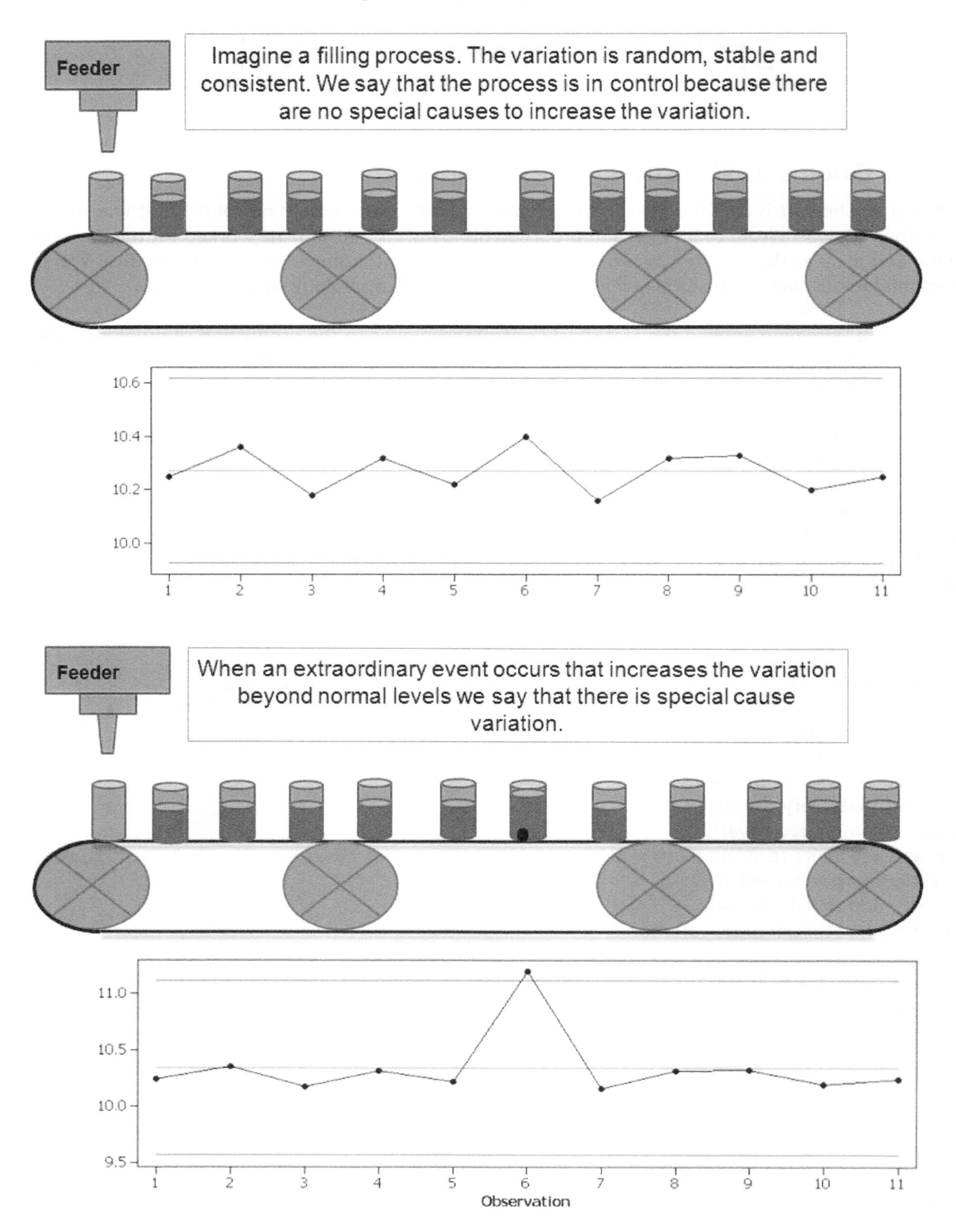

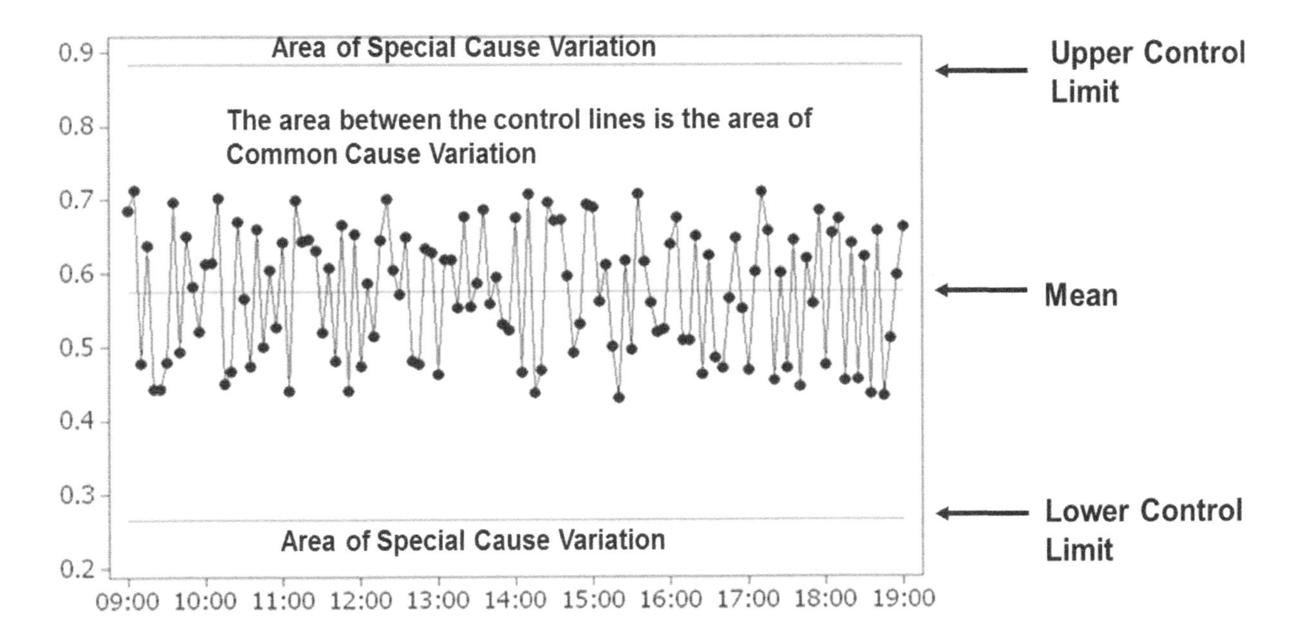

The control chart provides a simple way of detecting special cause variation. Usually the Control Limits that are calculated will be within the tolerance limits of the process. The aim is to react when the control limits are exceeded by investigating and rectifying the special cause. If we do this before we go out of the tolerance limits we avoid defects and waste.

7.3 Detection Rules for Special Causes

When using the Classic Method Minitab provides eight separate tests for detecting special cause variation. Test 1, which is the original used by Dr Shewart, is always on by default. However, you can change the settings to your individual needs as can be seen from the menu. Minitab also allows you to set the value of K which will alter the settings at which the individual Test works. You can change the default options within the Tools <<Options menu.

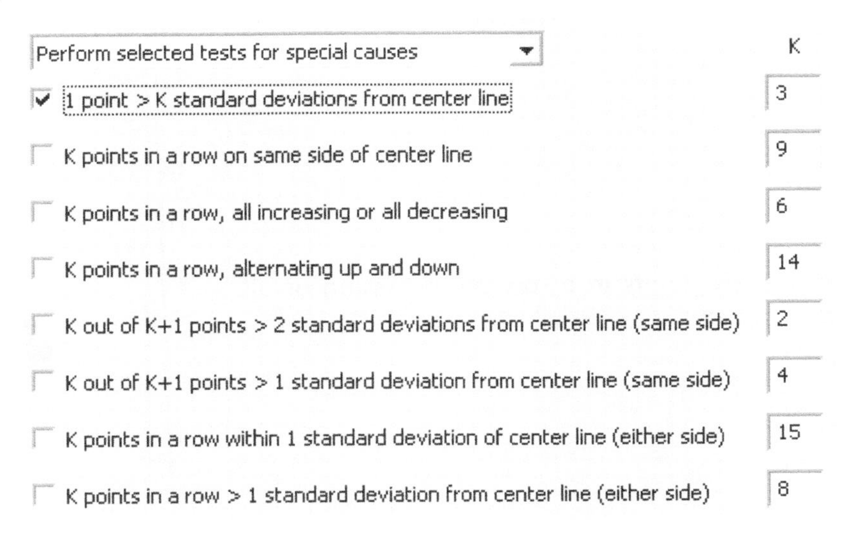

When using the Assistant the tests used to detect special causes are set by default. In fact, the Assistant only uses Tests 1, 2 and a modified version of Test 7 for the charts it produces. You can read about the tests used within the Assistant in the Minitab White Paper for Control Charts.

In order to show the special causes that Minitab can detect we are going to use the following control chart. It has a mean of 100 and StDev of 10. Note, where there is a sequence of test points involved in a test that has failed Minitab only highlights the last point in the sequence.

Test 1: one point more than three StDevs from the centre line (outside the control limits).

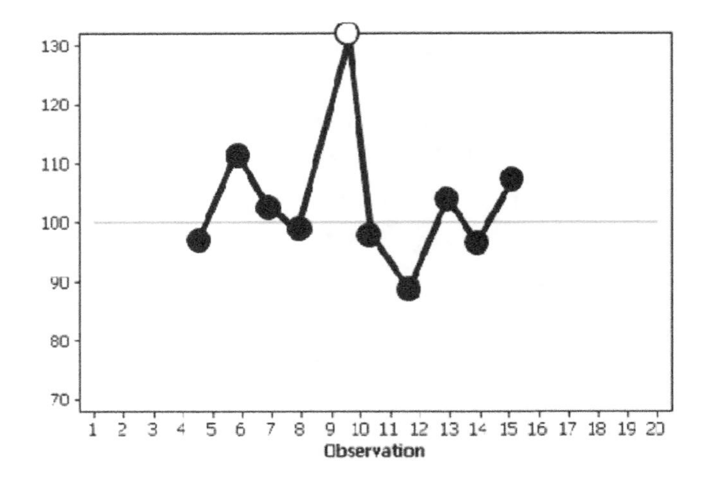

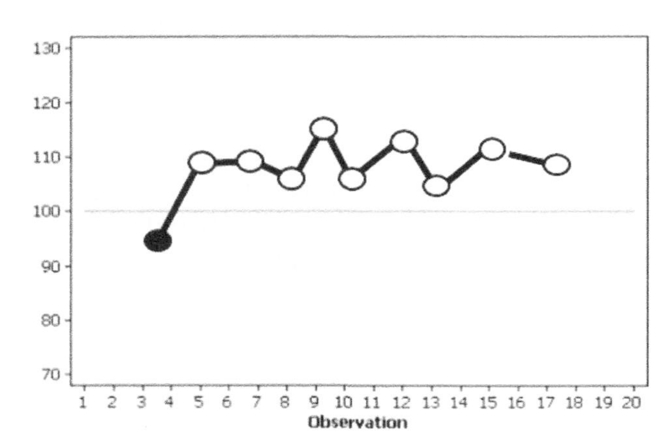

Test 2: nine points in a row, all on the same side of the centre line.

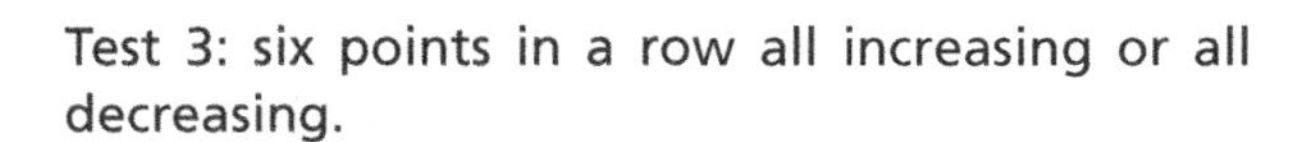

Test 3: six points in a row all increasing or all decreasing.

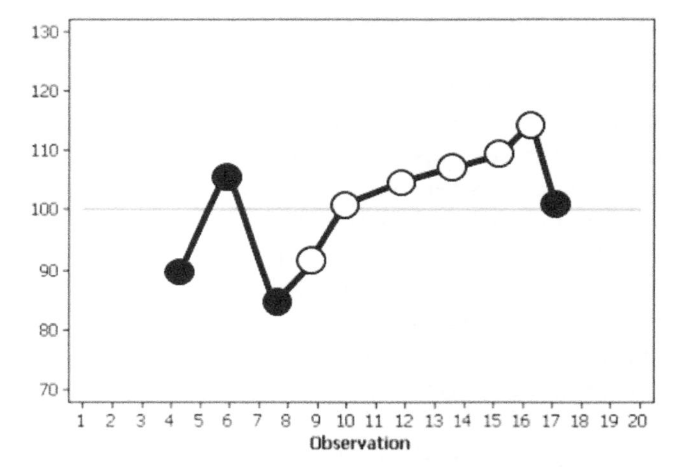

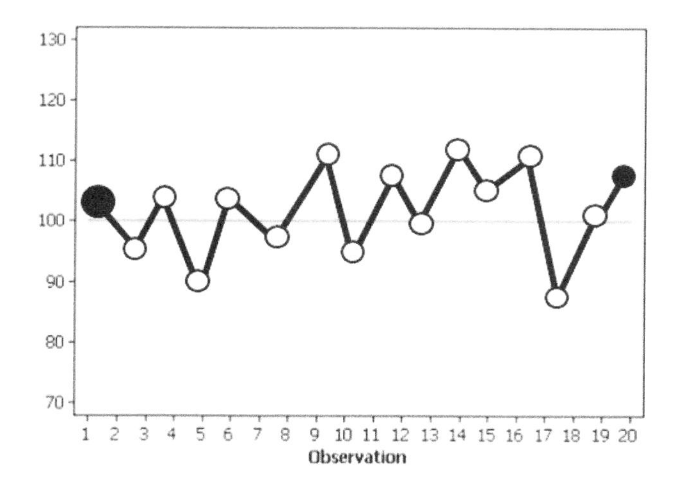

Test 4: 14 points in a row alternating up and down.

Test 5: two of three points more than two StDevs from the centre line on the same side.

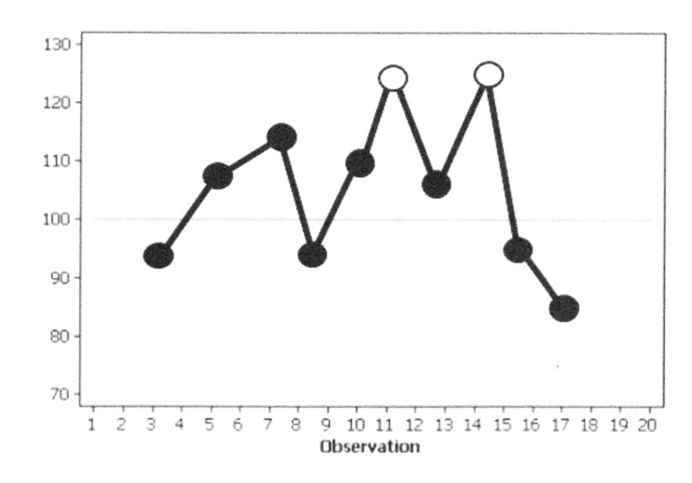

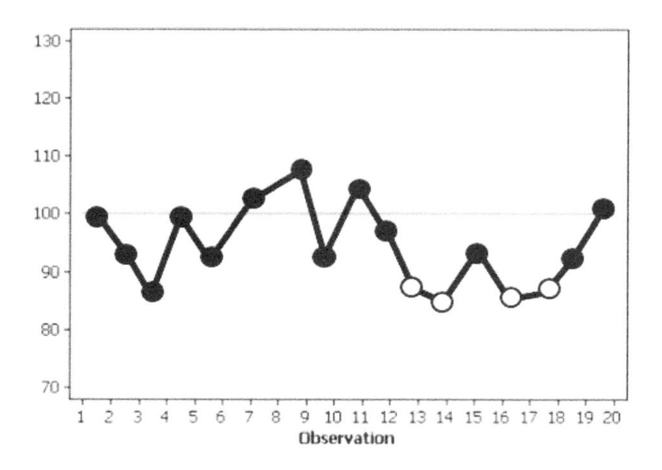

Test 6: four of five points more than one StDev from the centre line on the same side.

Test 7: 15 points in a row within one StDev of the centre line (either side)

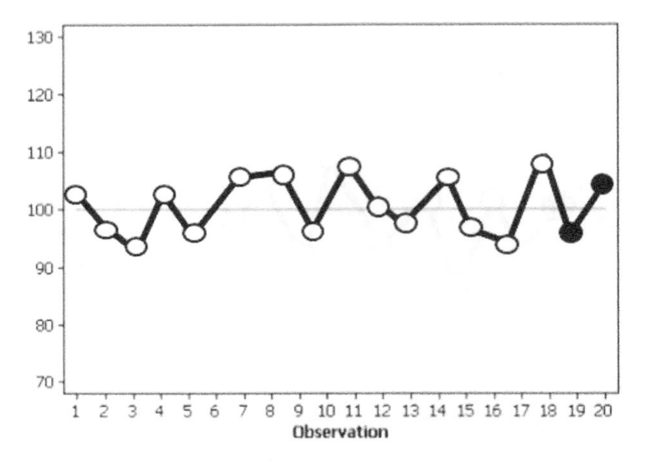

Test 8: eight points in a row more than one StDev from the centre line (either side).

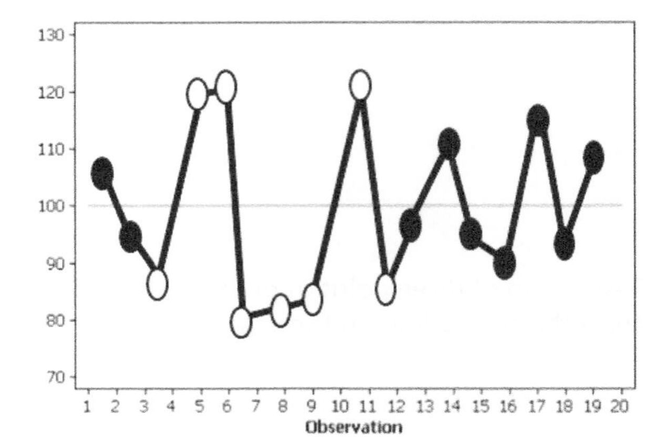

7.4 False Alarms

You may have already realised that there could be a slight problem. Even when you have a normally distributed process by definition only 99.73% of the points will be within three StDevs of the mean. Therefore, if you produce a control chart of a normally distributed process 0.27% of the points will be outside the control limits, we call these false alarms. This is because these points are not the result of a special cause but due to the intrinsic variation within the process that has resulted in the point being outside the control limits. 0.27% is the equivalent of one point in 370. If the process has extended tails or we add Tests 2–8 there will be additional false alarms. And remember the expectation is that all special cause variation will be investigated.

Therefore, we must set a balance between the detection rules and generating alarms so that we don't miss any important special causes and we don't generate too many false alarms.

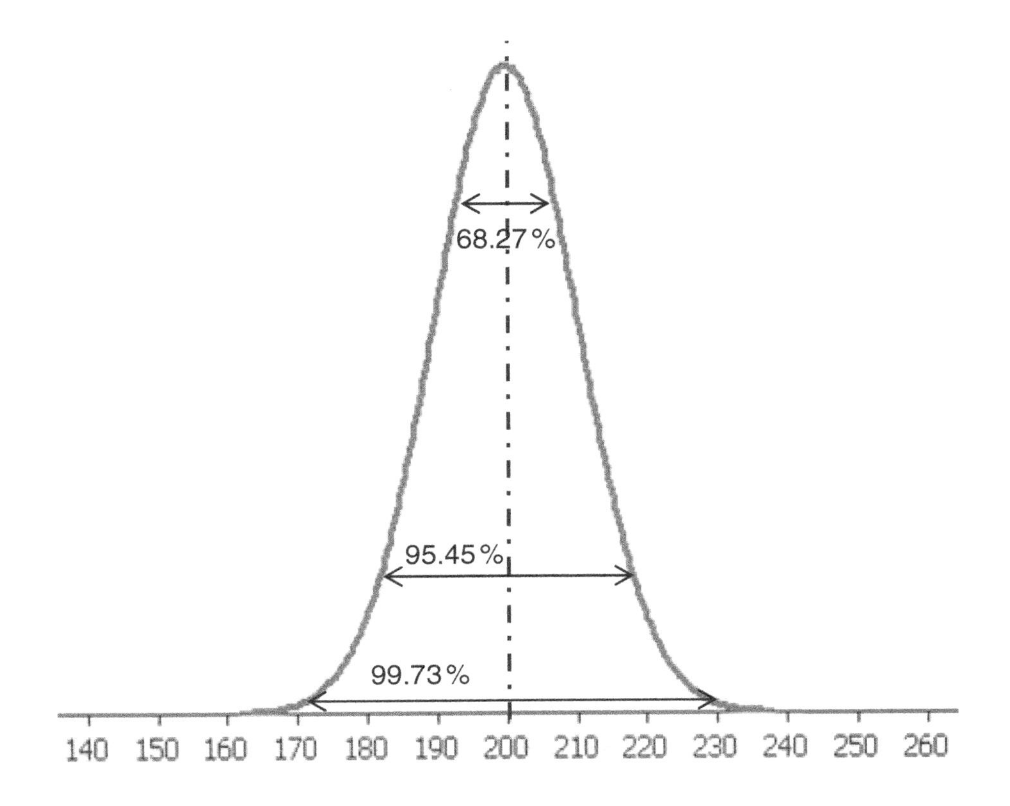

7.5 When Should We Use SPC Charts?

There is a massive number of potential uses for control charts. Most of the uses can be broken down into four main categories.

Monitoring This is the obvious one. We want to be able to monitor our process and identify special cause variation. When we see a control chart we can be confident and feel secure that our process is being monitored and should things go out of control a reaction plan will come into play. This will alert the right people so that they can investigate and rectify the situation. It should be an awful lot more than a display of data.

Process Improvement Typically when we start improvement work our system is out of control. Initially we should try to achieve stability before we try to understand our system and then finally improve. The control chart can be an invaluable tool when it comes to bringing stability to the system and it helps us to understand and eliminate special cause variation. As stated in the Minitab White papers for tests related to Hypothesis Testing, including ANOVA, outliers or data related to special cause variation can have a strong influence on the results of the tests. Therefore, if we are working on an improvement project we should aim to make the process stable, in control, before we use hypothesis testing and other characterisation procedures.

Characterisation As we are trying to achieve stability and start to identify special cause variation we can trend multiple variables to identify the root cause of the special cause variation. This type of characterisation work is shown in Exercise 1.

Defect Reduction Usually our control limits will be within our tolerance limits for the process. Therefore, if we are using control charts to monitor the process we can react and rectify the process before the specification limits are breached. This will reduce defects.

7.6 Subgrouping

Before we start to discuss the types of chart we will be using we need to understand subgrouping. Imagine a process where we are filling cans at a rate of 5000 per hour and require a consistent fill level. In order to monitor the process we decide to take six consecutive cans every hour and check their fill level. This gives us a subgroup size of six. From our process knowledge we know that the variation between these six cans represents the short term variation of the process and the variation within the subgroup is probably the best we can achieve.

Due to the deliberate manner in which we are taking six consecutive samples from the line we can assume that when special cause variation takes place it will occur in the time period between sampling. It is due to the design of our sampling/subgrouping that we can make the link between Special Cause Variation and Between Group Variation.

Remember in the ANOVA chapter we said that 'Overall Variation is the sum of the Within Group and the Between Group Variation'. Due to our foresight we have set up our sampling regime so that the variation within the subgroup will be representative of common cause variation. The variation between subgroups will capture the special cause variation.

7.7 The Appropriate Chart

We will be looking at two common types of Control Chart which use subgroups. They are the Xbar-R chart and the Xbar-S chart. Both of these charts are split into two sections. The Xbar chart is always at the top and this displays the average of the subgroup. The Xbar chart is usually accompanied by either the R chart or the S chart, where R denotes the range of the subgroup and S denotes the StDev of the subgroup. It can be seen that on both charts the Xbar is displaying the average of the subgroup and the R or S charts are attempting to display the within subgroup variation.

Charts which use subgroups give greater protection from false alarms generated by non normal distributions. As the subgroup size increases the greater the level of protection.

When it has not been possible to sample and analyse subgroups we use the I-MR chart. This is made up of an individual value chart and a Moving Range chart. The I-MR chart would probably be used where the sampling or analysis was particularly complex or onerous.

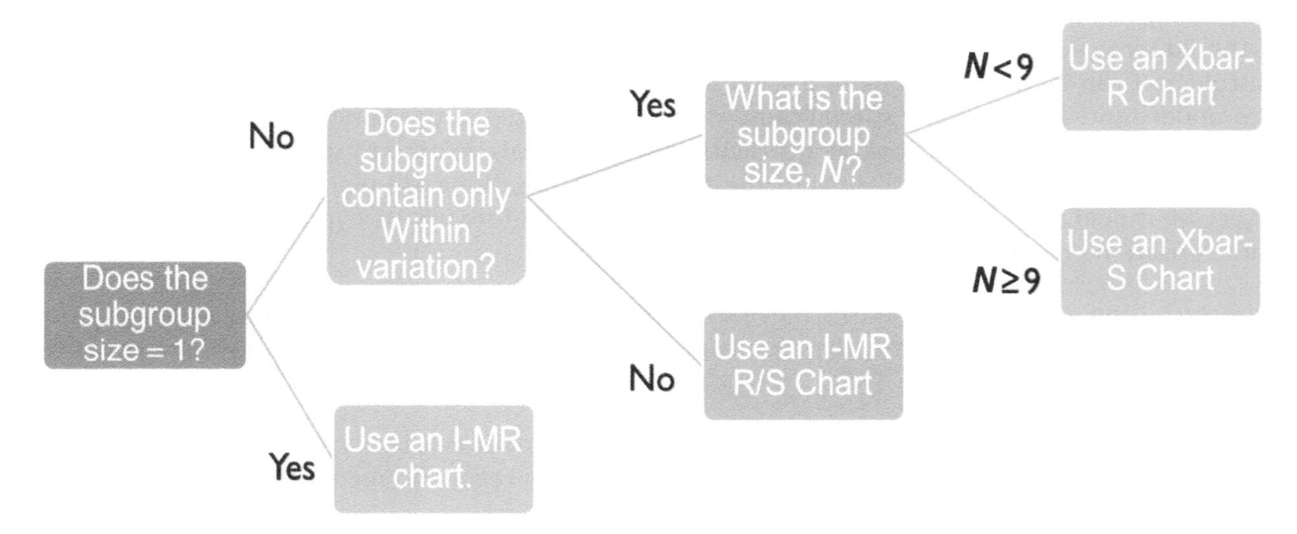

Having a subgroup size of one indicates individual data points without any subgrouping. If $N = 1$, then the choice is simple use an I-MR chart.

On occasion there will be more than the within variation trapped inside the subgroup. On these occasions we use an I-MR R/S chart.

If we have established that we have subgrouping and that it only contains Within variation the recommendation has always been that if you have a subgroup of less than nine you would use the Xbar-R chart and for anything above that use the Xbar-S chart. Note that the R chart only uses the points at the extremes to represent within group variation whereas the S chart uses all the points within the subgroup.

7.8 The I-MR Chart

Procedure

- I-MR Chart (Classic Method)

What's it used for?

- It is used to detect changes in time ordered data where individual values are being examined.

Assumptions and Limitations

- I-MR charts are usually used where it is not possible to sample and analyse subgroups.
- The false alarm rate can be four to five times higher when the distribution is not normal.
- Using optional additional tests is likely to increase the false alarm rate.
- Control limits are calculated using the average moving range.
- Help <<Methods and Formulas shows how the control limits are calculated.

Procedure	• I-MR Chart (Assistant)
What's it used for?	• It is used to detect changes in time ordered data where individual values are being examined.
Assumptions and Limitations	• The recommendation from the Minitab White Paper on Variable control charts is to have at least 100 data points. • The Assistant will only conduct Tests 1 and 2 for the I Chart and only Test 1 for the MR Chart. • If the data is generating a lot of false alarms and certain conditions are met, the Assistant will offer to transform the data for you; see example 2.

Example 1. I-MR Chart.

Willy Winker is developing a new Chocolate within his R&D laboratory. He takes a sample from the line every 5 min and tests it for crunchiness. Testing for crunchiness is a difficult and time consuming process therefore subgroups are not analysed.

Produce an I-MR Chart of Crunchiness and explore any process changes.

All data sets for this module are in file 08 SPC.xls. This example uses Worksheet: Crunchiness.

1. Copy all the data from Worksheet: Crunchiness into a new Minitab project worksheet.

Date	Time	Crunchiness	Nougat	Caramel	Sugar	Water	Gelling Agent
3/10/2010	9:00:00 AM	0.946	0.3730	0.685714	34.44	36.84	0.680
3/10/2010	9:05:00 AM	0.736	0.4355	0.712857	33.12	37.08	0.623
3/10/2010	9:10:00 AM	0.697	0.4665	0.477857	36.92	36.96	0.989
3/10/2010	9:15:00 AM	0.721	0.4390	0.637143	20.00	38.20	0.780
3/10/2010	9:20:00 AM	0.650	0.4845	0.442857	32.64	36.48	0.799
3/10/2010	9:25:00 AM	0.891	0.4580	0.442143	26.84	39.20	0.723

2. Click on the Information icon so we can check the amount of data and whether the formatting is correct.

3. We see that there is no missing data and the column formatting is correct. There are 121 rows of data, which are for 121 samples.

Name	Id	Count	Missing	Type
Date	C1	121	0	D
Time	C2	121	0	D
Crunchiness	C3	121	0	N
Nougat	C4	121	0	N
Caramel	C5	121	0	N
Sugar	C6	121	0	N
Water	C7	121	0	N
Gelling Agent	C8	121	0	N

4. Click Stat <<Control Charts <<Variables for Individuals <<I-MR.

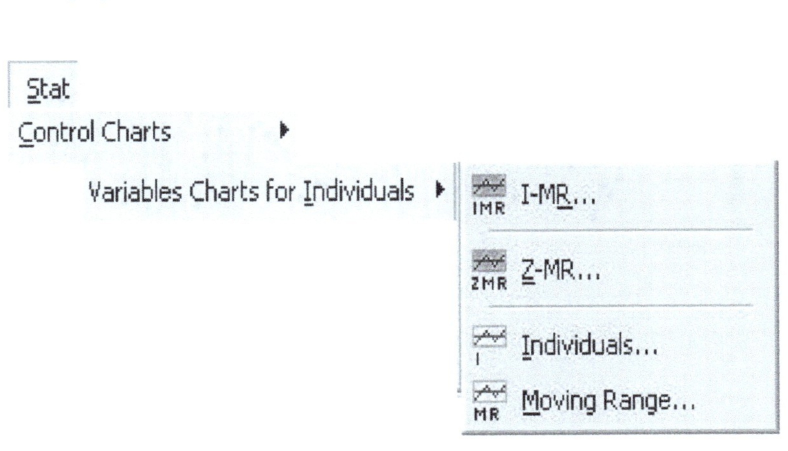

The I-MR Chart produces the Individual and Moving Range charts on the same sheet. As can be seen within the full menu there are options available which will allow the production of only one of these charts if required. This flexibility is only available using the Classic Method.

5. Select Crunchiness as the variable to be plotted.
6. Then Click on Scale. The Scale option allows us to use our own labels for the x axis. The default setting is just an indexed number for each point.

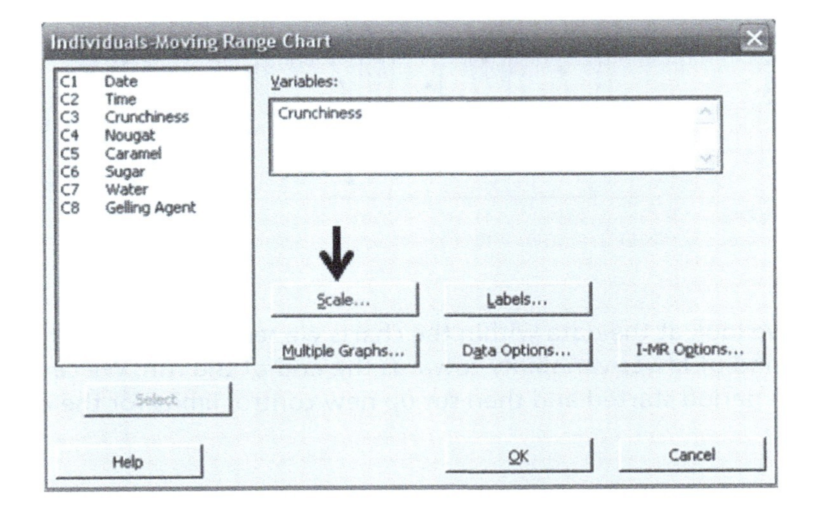

7. For the X Scale click on the Stamp radio button.

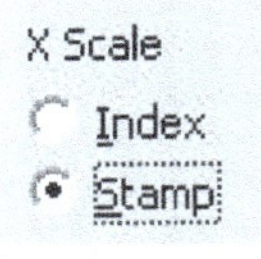

8. Select Time first then Date. We can add three layers of labels. Minitab is telling us to put the innermost first, in our case we want Time to be immediately next to the x axis.
9. Click OK and OK again.

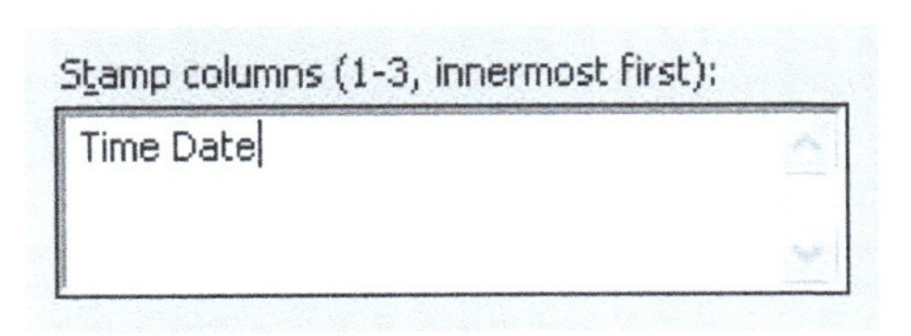

Our I-MR chart is split into two sections. The top shows the individual values, there should be 121 of these. The bottom is the Moving Range chart. This shows the difference in range of sequential points. As it takes the difference of sequential points there are 120 points.

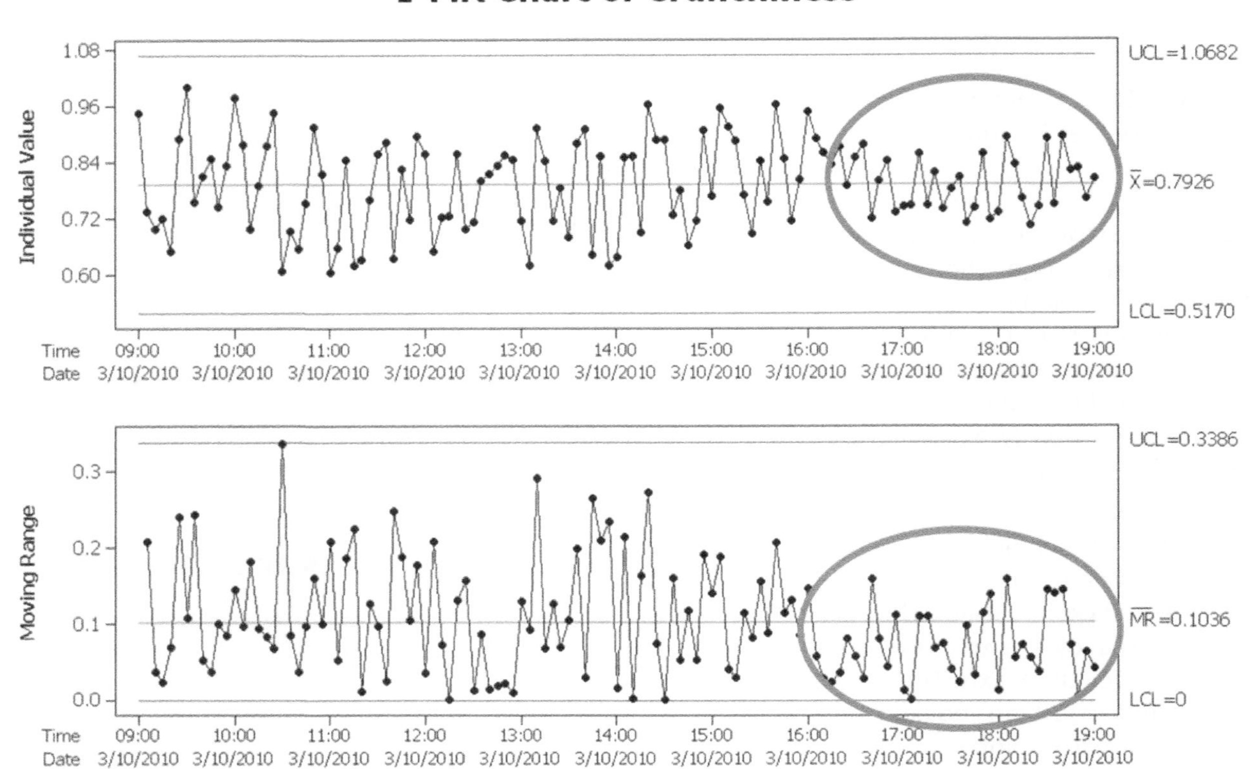

If we look at the data within the charts we see that none of the points are out of control. We also see a period of lower variability towards the end of the run. We can hover over points to get the time when this period started and then set up new control limits for the end of the run.

10. In order to set up new control limits for the period of lower variability, which was at the end of the run, click on the Edit Last Dialog Box icon.

11. Click on I-MR Options.

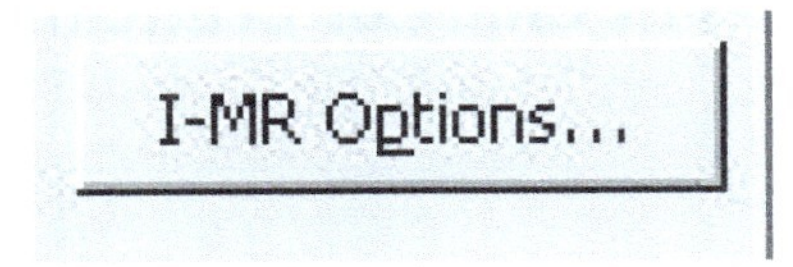

12. Click on the Stages tab.

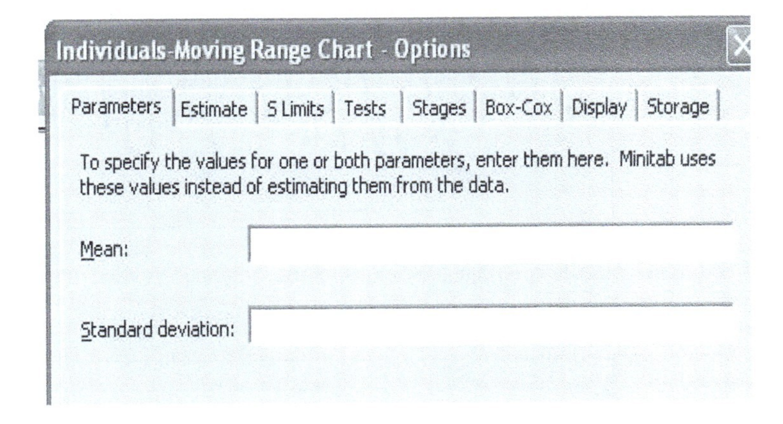

13. Use Time to define the Stage.

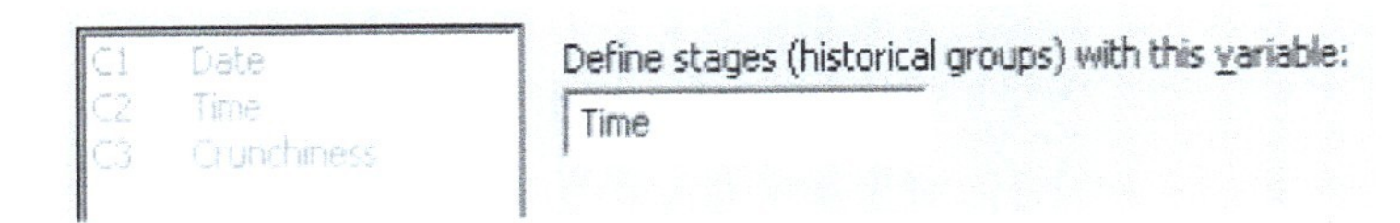

14. Use 16:45 as the time to start a new stage. 16:45 is the time that the period of lower variability is estimated to have started.
15. Click OK and OK again.

We now have two sets of control lines and two mean lines on each chart as Minitab has calculated new control values for each period. However, the values for Xbar, MRbar, UCL and LCL only relate to the latter part of the chart. We can ask Minitab to display these parameters for each stage.

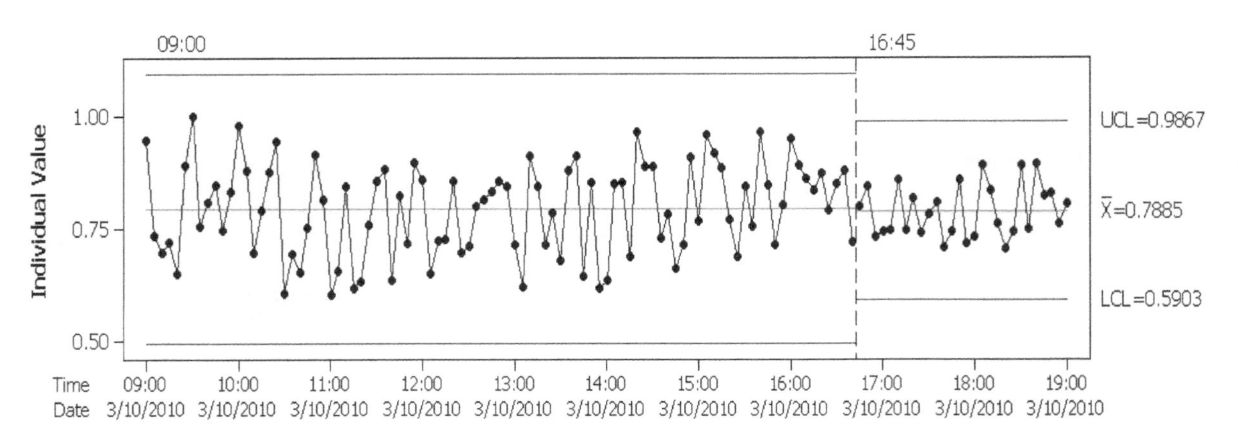

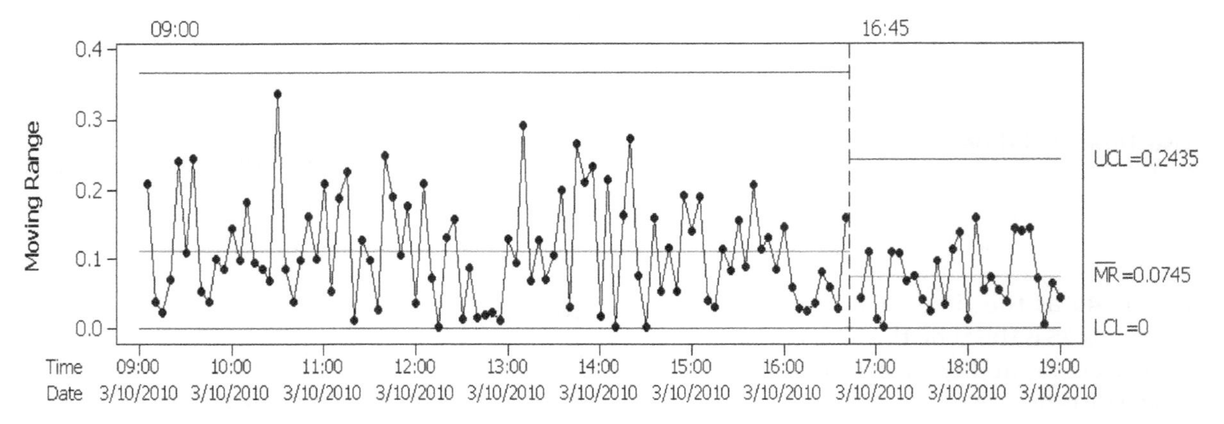

16. Click on the edit last dialog box icon.

17. Click on I-MR Options.

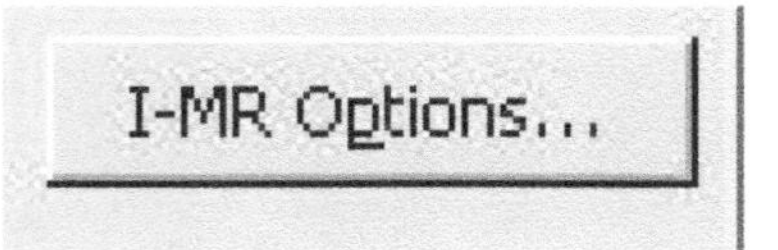

18. Click on the Display tab.

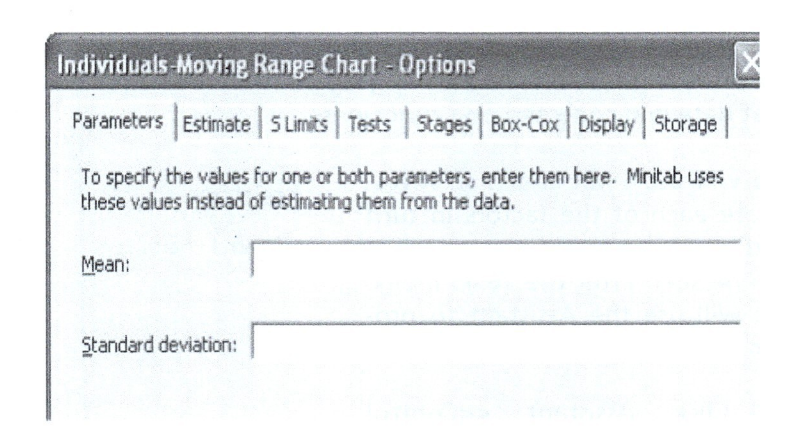

19. In order to display all the stage data click into the box for Display control limit/center line labels for all stages.
20. Click OK and OK again.

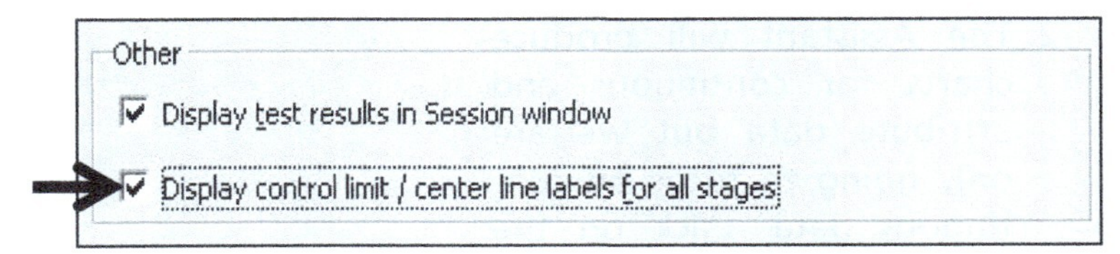

We now have control limits and centre line values for both sections of the chart. If it does look untidy or you cannot read the values as they are being obscured you can always drag the values to a more suitable location.

We have identified a trend within the common cause region and we will use additional I-MR charts to investigate additional factors that might be involved with the period of lower variability for Crunchiness.

I-MR Chart of Crunchiness by Time

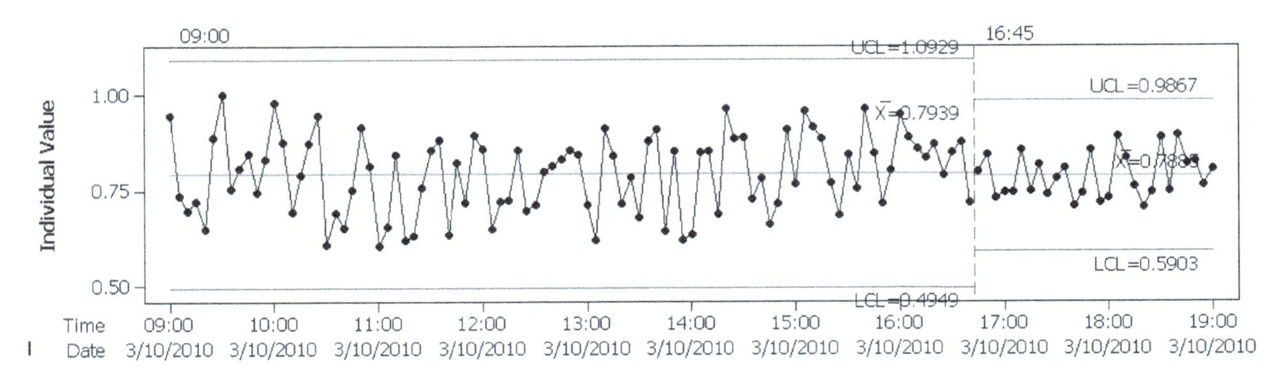

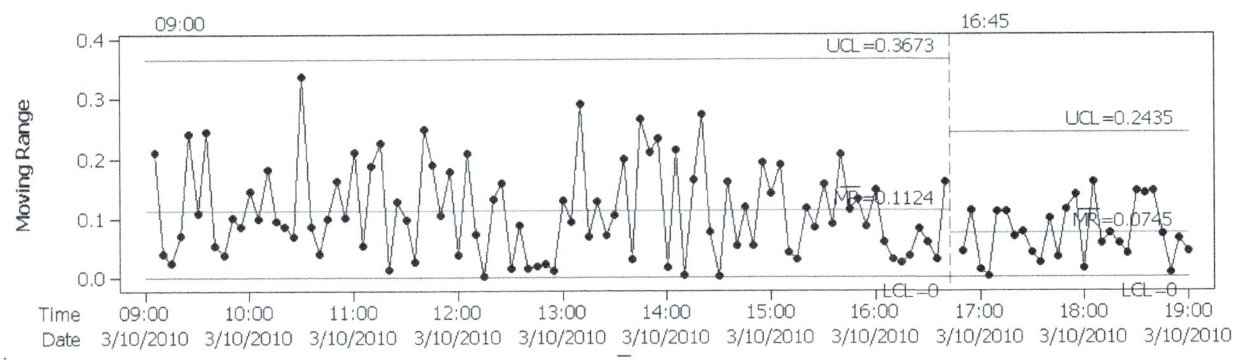

Next to the Crunchiness column there are also columns of data for the following factors: Nougat, Caramel, Sugar, Water and Gelling Agent. Data was also collected on these parameters at the same time that data was collected on Crunchiness.

We will use I-MR charts to investigate each of the factors in turn and check to see if any changes can be seen after the 16:45 index. We will use the Assistant to produce an I-MR chart for Nougat.

Assistant

Control Charts...

1. Click Assistant <<Control Charts.
2. The Assistant will produce charts for continuous and attribute data but we are only going to focus on continuous data. Click on the Icon for the I-MR chart.

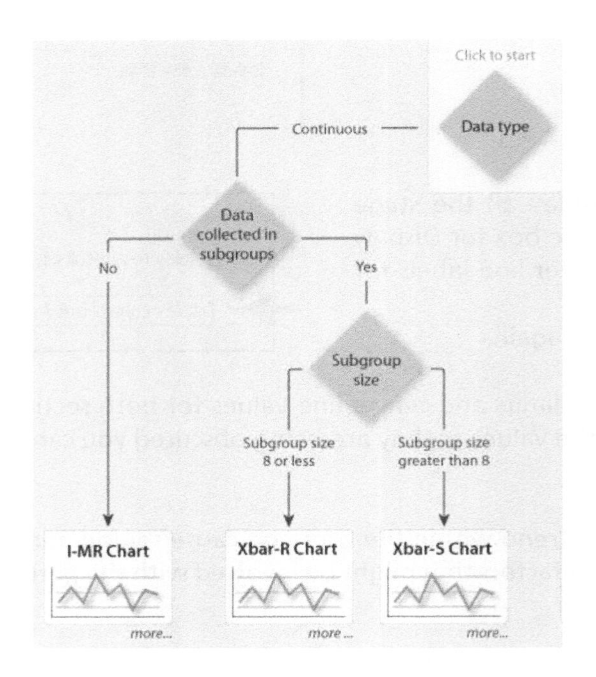

3. Select Nougat as the Data column.
4. Notice that the Assistant gives us the option of specifying control limits from known values. We might want to use these if we were producing charts for an already established process with established control limits. As we don't want this we will ask Minitab to calculate control limits from the Data.
5. Prior to producing the chart the Assistant scans the data and alerts us if any points are out of control or unstable and gives us the option of omitting them from the data. We will leave these points in. The Assistant actually carries out a number of checks these

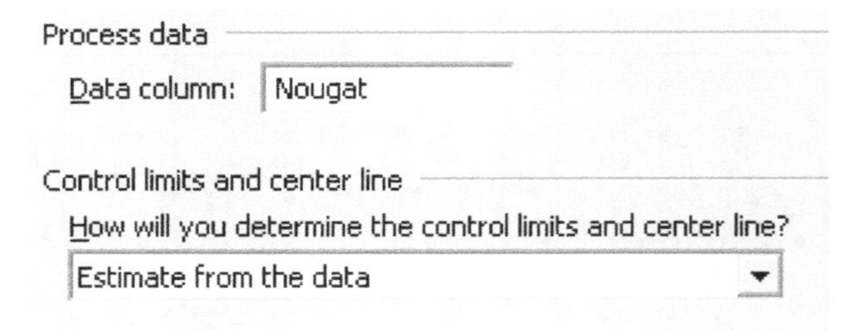

Estimate from the data
Use known values

Minitab has determined that some data points are out of control. Because control limits should be calculated from a stable process, you should identify which points have special causes and omit them from the calculations.

If you omit a point, it is excluded from the calculations for both charts.

Omit	Point	Chart	Reason
☐	89 - 100	I	Shift in mean
☐	102 - 121	I	Shift in mean

include checks for normality, Auto correlation and amount of data. The results of which will be shown in the Report Card
6. Click OK to produce the chart.

On the top left of the Summary Report we see an indicator bar that should now be familiar to users of the Assistant. It is asking whether the process mean is stable and the answer is no because 13.2% of the points are out of control.

On the right hand side we see the I and MR charts. We can see that there has been a shift in the Nougat level near the time we were interested in. Disappointingly, we cannot set the index within the Assistant.

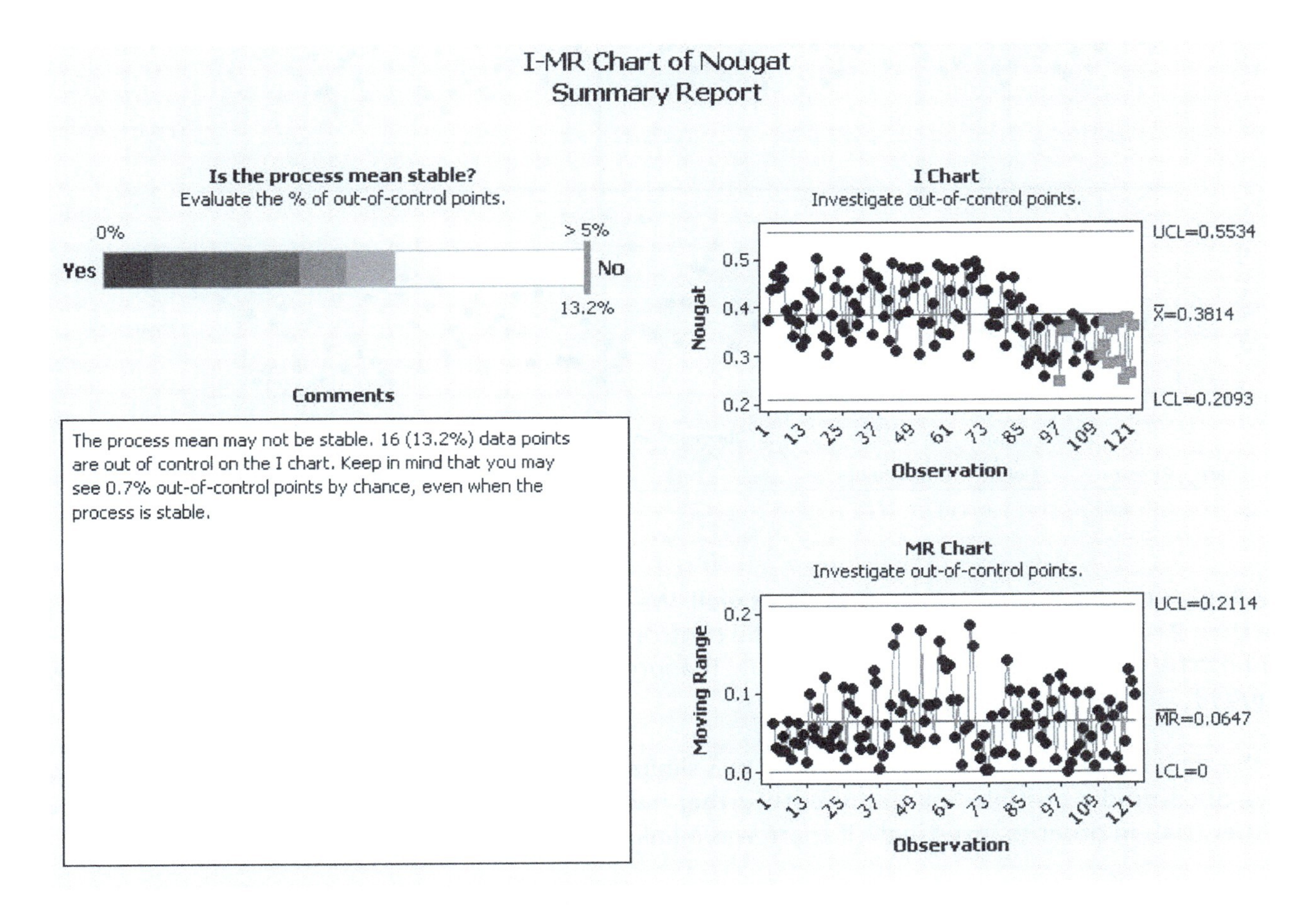

The Stability Report enlarges the I-MR charts allowing us to look for patterns more easily. At the bottom of the page we see the points that were deemed to be unstable are listed.

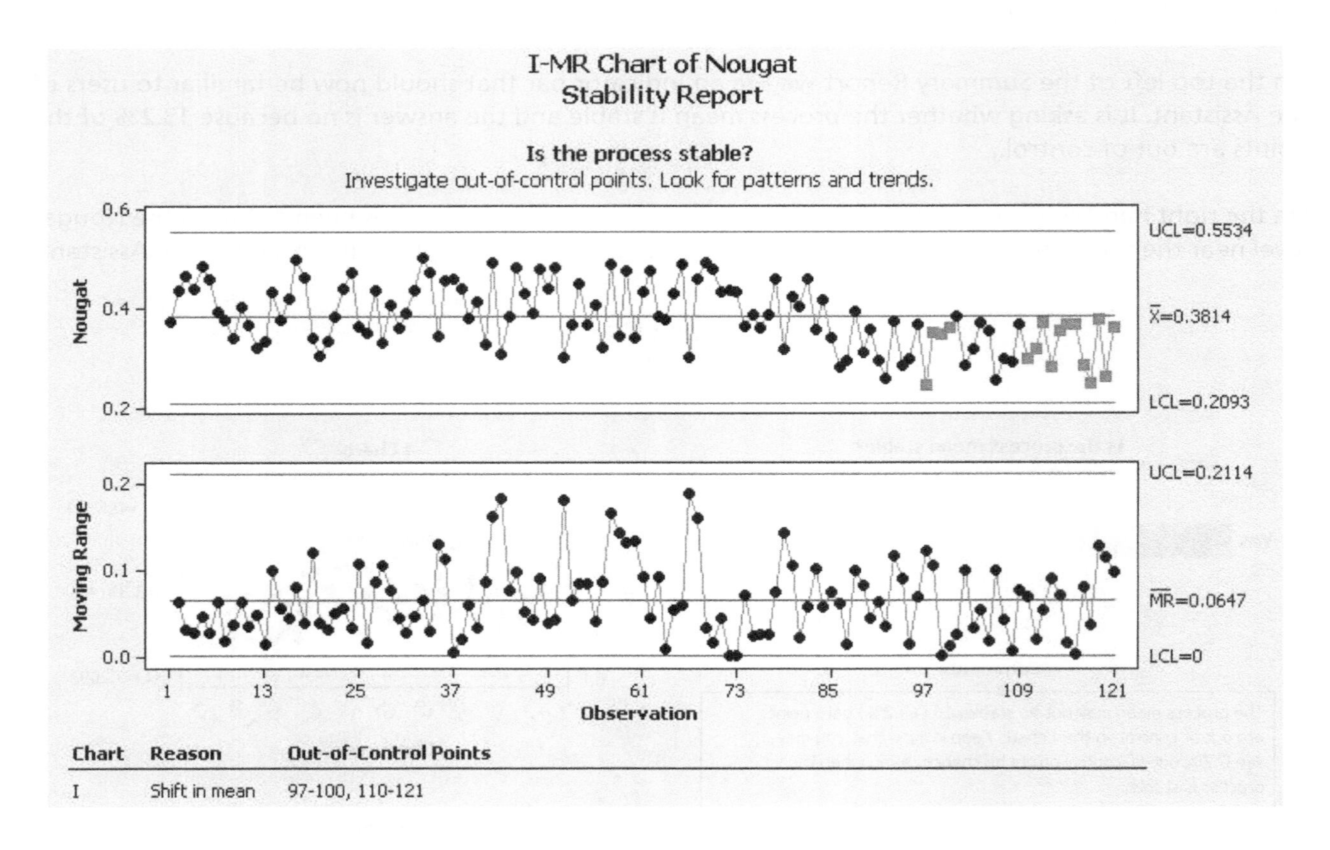

The Report Card follows a familiar format as well. We are told that normality was not an issue because less than 2% of the points were outside of the control limits. If this condition was not true the Assistant would carry out a normality check on the data. If appropriate it would offer to transform our data if we required.

As the time when the Nougat level reduced was similar to the reduction in variation in Crunchiness we have discovered a possible link that merits further investigation. In the real world we would carry out further trials in order to investigate if there was a link.

I-MR Chart of Nougat
Report Card

Check	Status	Description
Normality	✓	If the data are nonnormal, you may see an increased number of false alarms. Because less than 2% of the points are outside the control limits on the I chart, the normality test is not needed.
Stability	⚠	The process variation is stable. No points are out of control on the MR chart. However, the process mean may not be stable. 16 (13.2%) points are out of control on the I chart (you may see 0.7% out-of-control points by chance, even when the process is stable). You should investigate out-of-control points and omit those with special causes from the calculations.
Amount of Data	✓	You do not need to be concerned about the precision of your control limits because 100 or more data points are included in the calculations.
Correlated Data	✓	If the data are correlated, you may see an increased number of false alarms. Because less than 2% of the points are outside the control limits on the I chart, the correlation test is not needed.

Exercise 1. I-MR Chart.

We wanted to investigate the change that we noticed in Crunchiness and we have produced an I-MR chart for Nougat. The I-MR chart for Nougat showed us that the level of Nougat reduced during the period we were interested in. There are four remaining factors that need to be investigated; Caramel, Sugar, Water and Gelling Agent. Use the Classic Method to produce I-MR charts for Caramel and Sugar, use Time as the x axis scale. Use the Assistant to produce I-MR charts for Water and Gelling Agent. Look for trends within the charts that could help explain the change in Crunchiness.

Worksheet: Crunchiness.

The I-MR chart for Caramel is consistently in control. No unusual trends can be seen within the Individuals chart or the Moving Range chart.

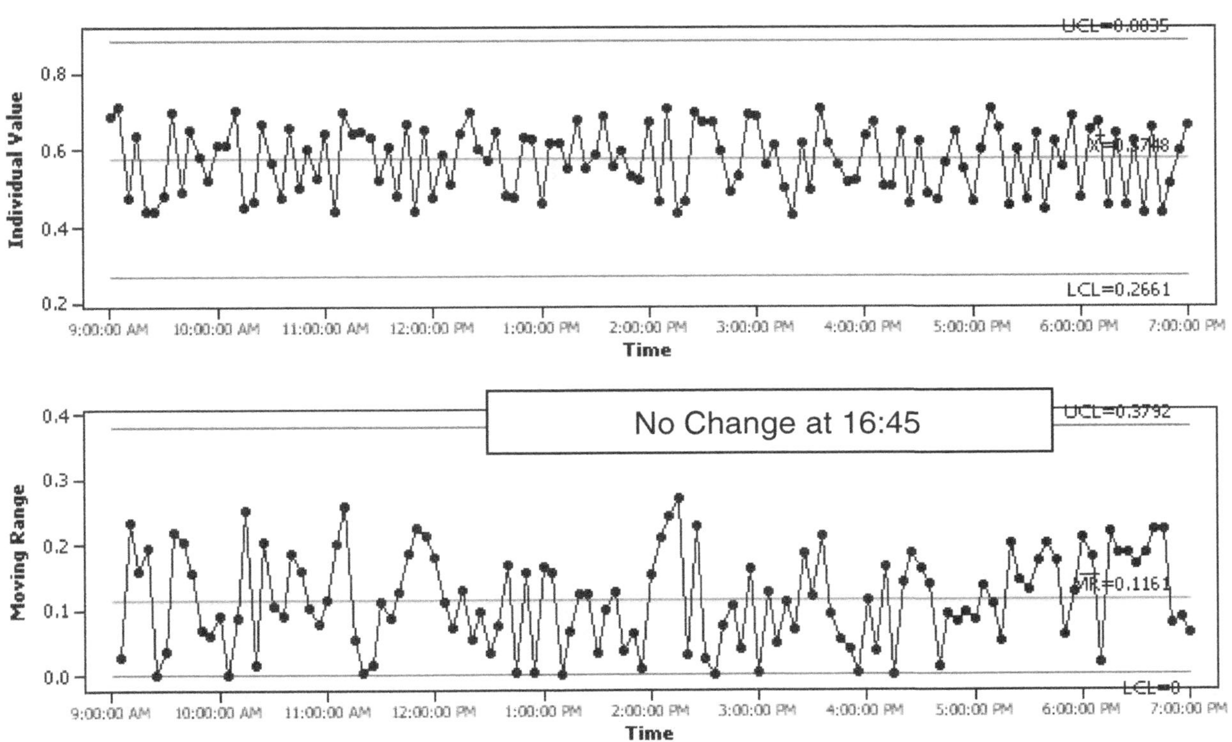

The I-MR chart for Sugar is consistently in control. No unusual trends can be seen within the Individuals chart or the Moving Range chart.

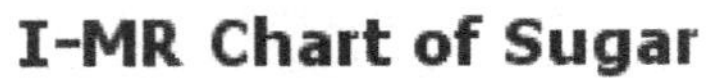

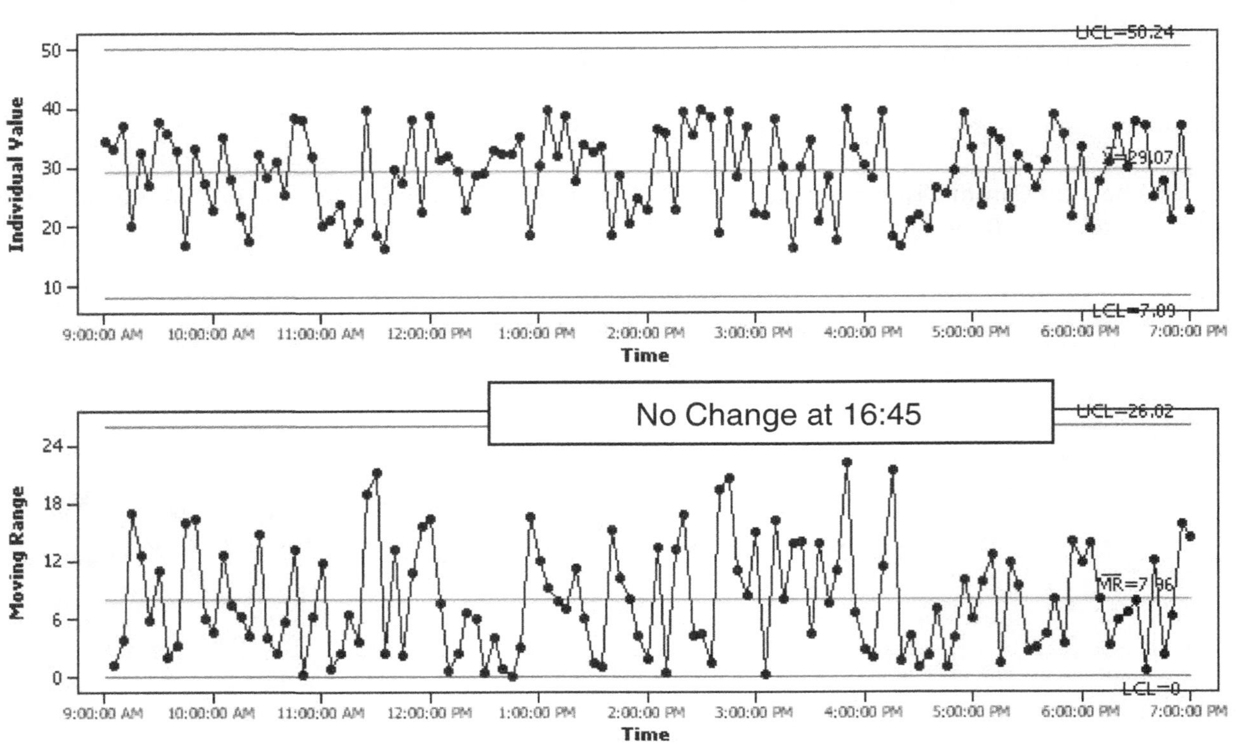

The Summary Report for Water states that the process mean is stable with no points going out of control. The I-MR chart for Water is consistently in control. No unusual trends can be seen within the Individuals chart or the Moving Range chart.

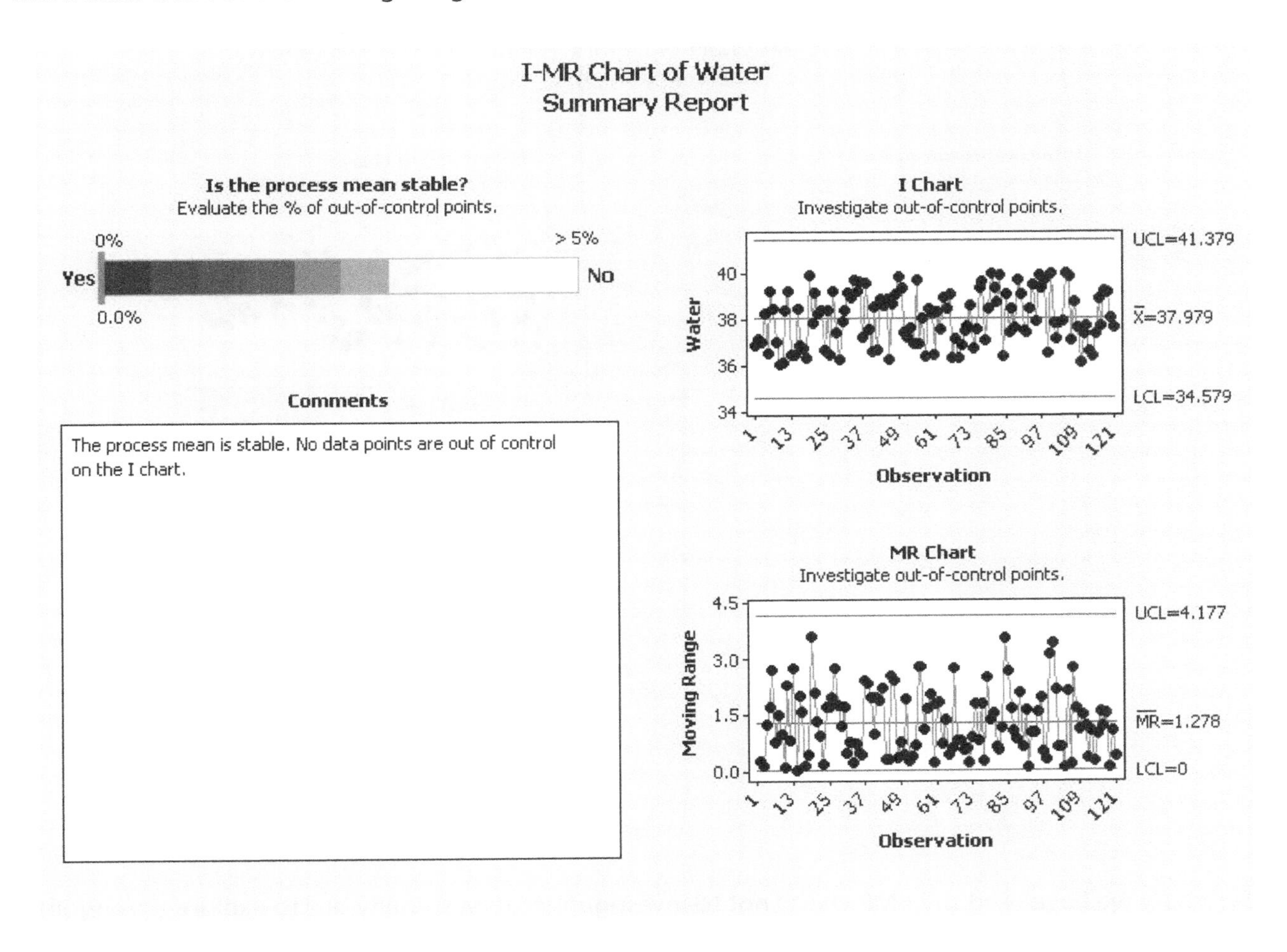

The Summary Report for Gelling Agent states that the process mean is stable with no points going out of control. The MR chart does show two points that are out of control. Within the I chart we see that there is reduced variability during our target time which we could only identify by looking at the charts.

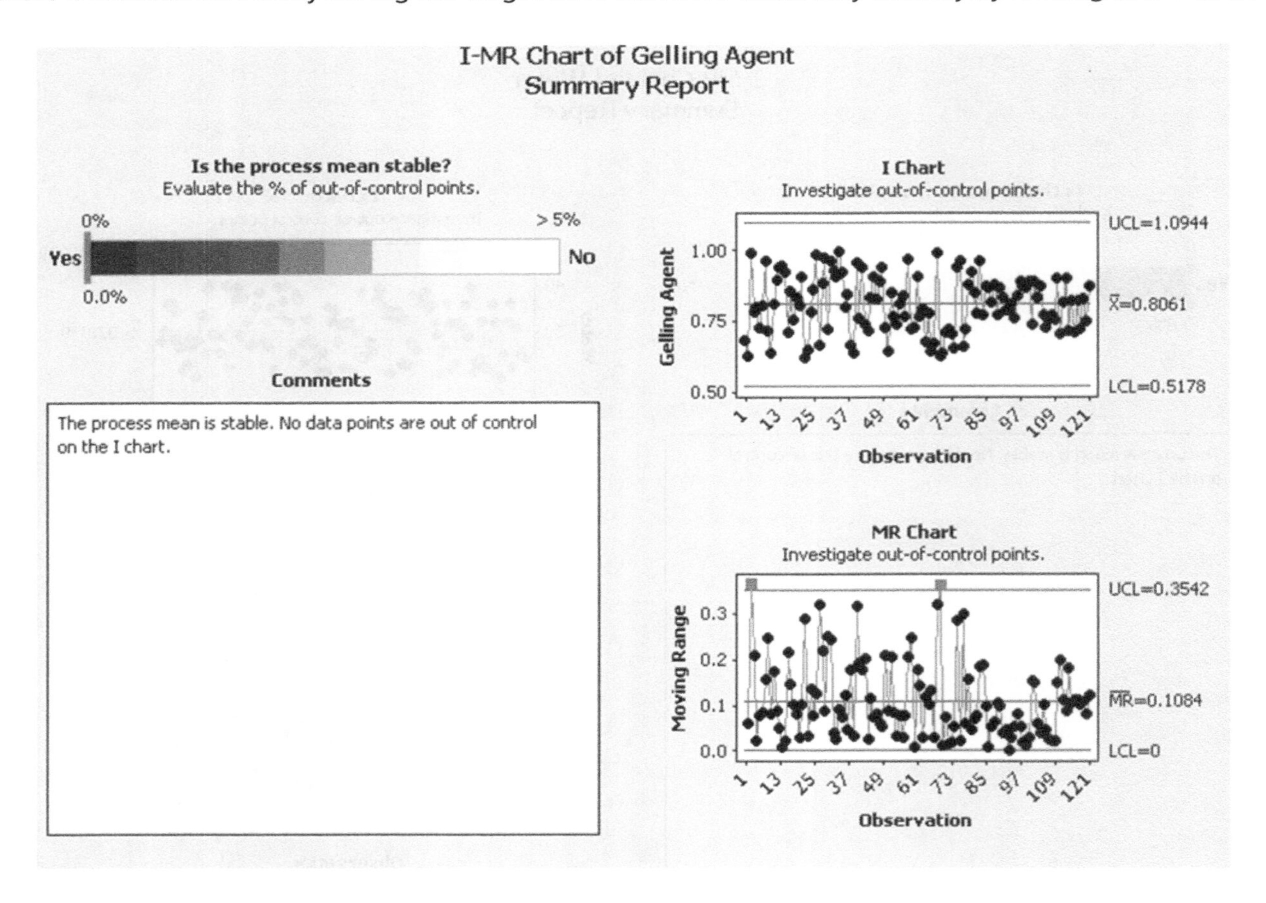

In Summary we have used the I-MR charts not to investigate process stability but to explore a trend that was of interest. We managed to link the factors of Crunchiness with Nougat and Gelling Agent. In the real world this would only be a start and we then might want to set up further experiments or further investigate the link within our factors by using hypothesis testing.

Example 2. I-MR Chart.

As an experiment Humzah decides to check how long it takes him to roll two sixes. He starts a timer and then continuously rolls the dice until he throws two sixes. He repeats this 120 times. Produce an I-MR chart of the data to assess whether the process is in control.

Worksheet: Dice.

1. Copy the single column of data from Worksheet: Dice into a new Minitab project worksheet.

Time
18.897
6.866
10.519
27.367
11.809

2. Click Stat <<Control Charts <<Variables for Individuals <<I-MR.

3. Select Time as the variable to be charted.
4. Click on OK to produce the chart.

Variables:

Time

5. Inspect the graph that has been produced.

A couple of things stand out about the graph. The first is that there are a number of points out of control on both sections of the graph.

We need to have a look at the type of distribution we are dealing with so we can make some better decisions. We will produce a histogram of the data.

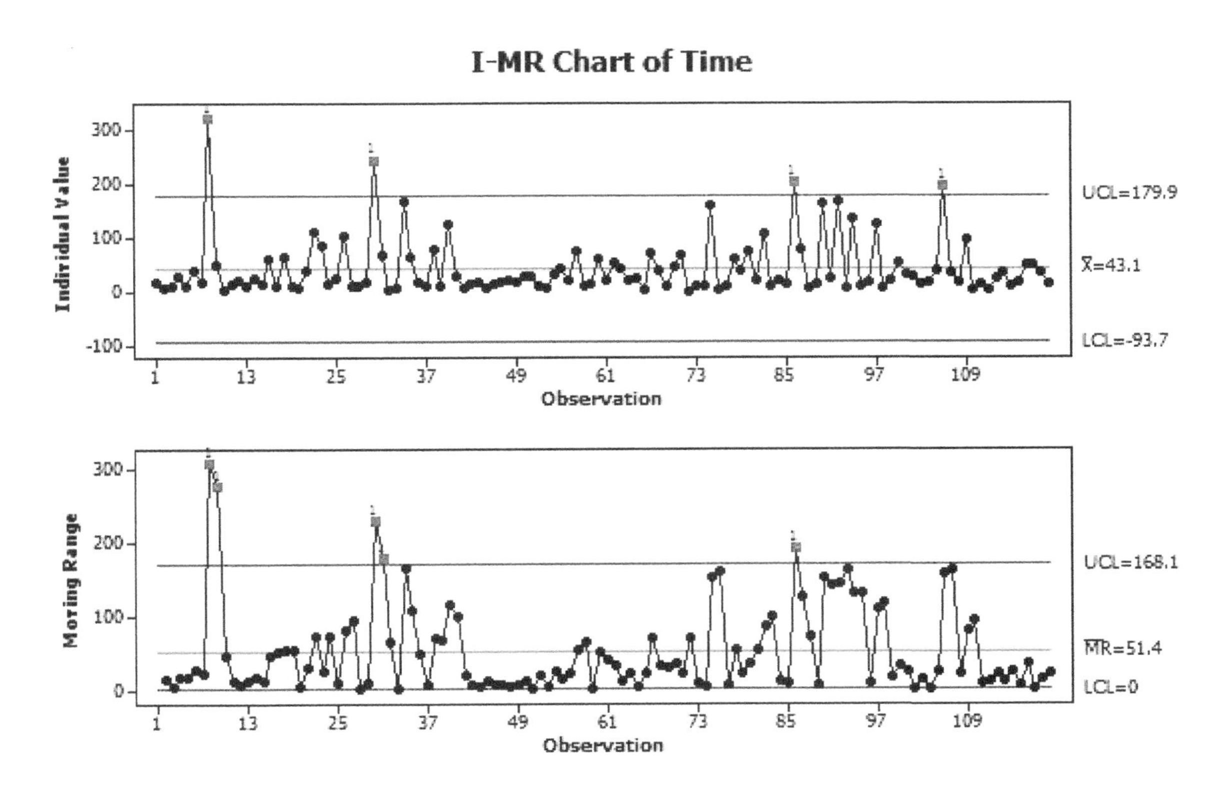

6. Select Graph <<Histogram.

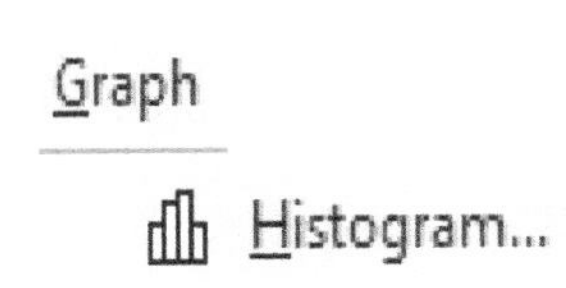

7. Select the option for producing a histogram With Fit.

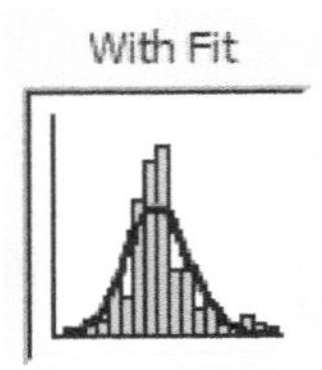

8. Select Time as the variable to be plotted. Click OK.

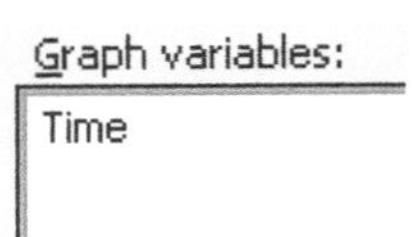

9. Inspect the histogram.

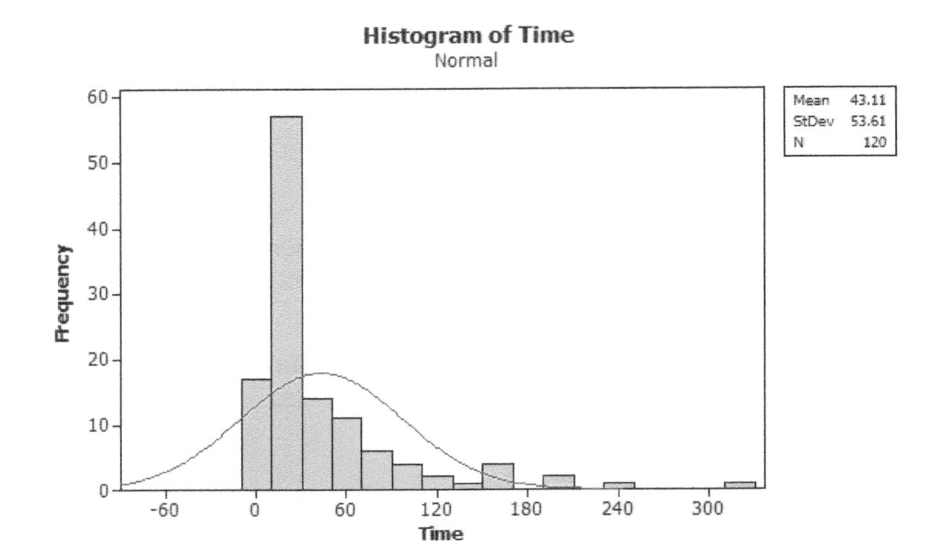

It can be seen that the histogram is right skewed. There are now two possibilities, one is that the distribution should be normal and the out of control points that have been identified will help us correct the process. The other possibility is that the process does not follow a normal distribution and the out of control points are false alarms.

Due to the fact that we are dealing with a very simple process with a fairly large sample size we can assume that the skewed distribution is representative of the process. After making this decision we can use a Box–Cox Transformation to transform this distribution into a normally distributed one. For this transformation to work the distribution must be positively (or right) skewed and all values must be greater than zero.

We should not take the decision to transform lightly. We should only transform the data if we are sure that out of control points are due to the natural distribution of the process and not due to special causes influencing the data. This means that we should not transform the data if we were expecting a normally distributed process.

The transformation process is two step. Initially we need to find the optimal value of the conversion factor, lambda.

10. Click Stat <<Control Charts <<Box–Cox Transformation. Ensure the selector is set to all values in one column. Select Time as the Variable and Subgroup size as 1.
11. Click on OK.

All observations for a chart are in one column:

Time

Subgroup sizes: 1 (enter a number or ID column)

12. The Box–Cox Plot is produced. We are given the 95% confidence interval for the optimal conversion factor. The value used in the chart is the Rounded Value. The rounded value is usually a value that is easy to use within the transformation, such as 0 which corresponds to the natural log or 0.5 which corresponds to the square root. We have been given a rounded value of 0.00 for this conversion.

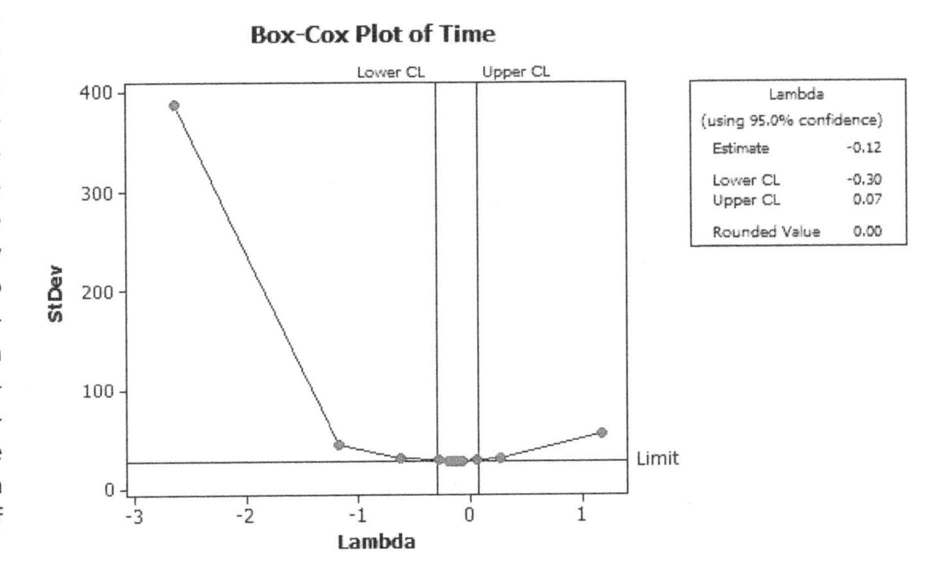

13. Click Stat <<Control Charts <<Variables for Individuals <<I-MR.

14. Click on I-MR Options
15. Click on the Box–Cox tab.

I-MR Options...

16. Activate the Use a Box–Cox Transformation tick box. Having analysed for the optimal value of Lambda we know that we want to use the default value of Lambda = 0.
17. Click on OK and OK again.

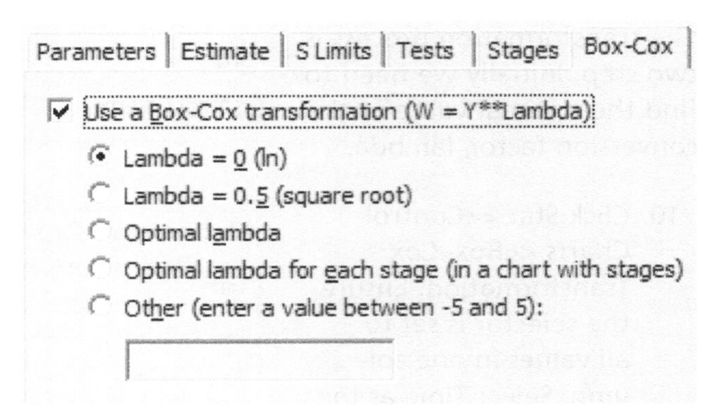

Minitab produces a control chart using the transformed data. We can see that both sections of this chart are in control. Notice that the control limits are also in equivalent transformed units.

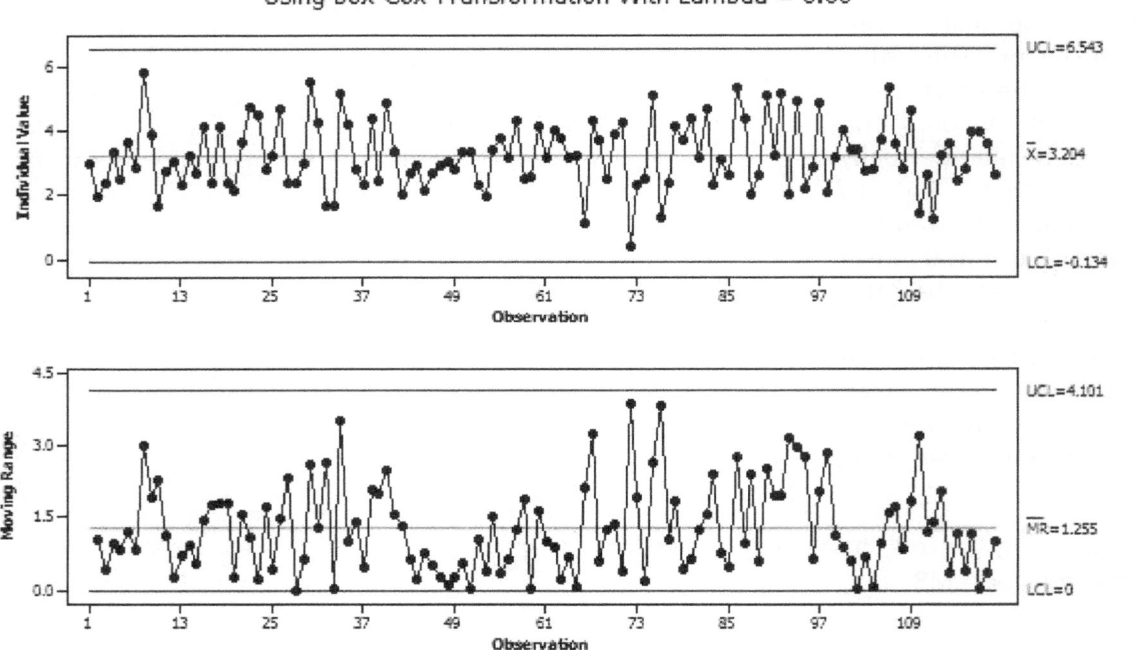

1. In order to produce the Control Charts for this data using the Assistant click Assistant <<Control Charts.

Assistant

Control Charts...

2. Click on the Icon for the I-MR chart.

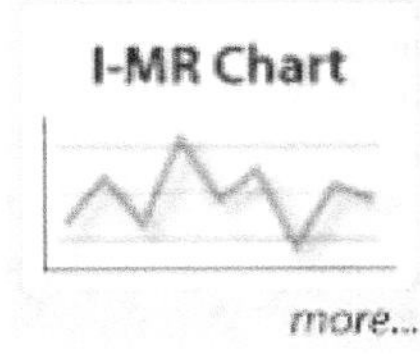

3. In the Menu box that opens select Time as the Data Column and set the selector box to estimate control limits and centre line from the data.

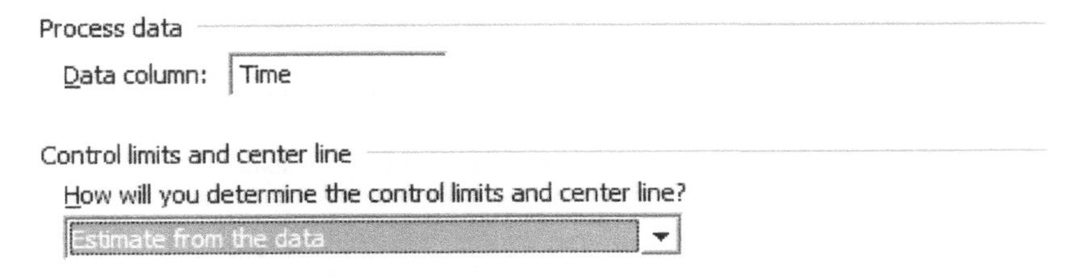

4. Minitab will ask if we wish to omit any of the out of control points from the data but we want to leave them all in. Click OK.

Minitab has determined that some data points are out of control. Because control limits should be calculated from a stable process, you should identify which points have special causes and omit them from the calculations.

If you omit a point, it is excluded from the calculations for both charts.

Omit	Point	Chart	Reason
☐	8	Both	Outside control limits or shift in mean
☐	9	MR	Above upper control limit
☐	30	Both	Outside control limits or shift in mean
☐	31	MR	Above upper control limit
☐	41 - 56	I	Shift in mean
☐	86	Both	Outside control limits or shift in mean

Due to the number of out of control points the Assistant carries out a normality check on our data. We are warned that our data is non normal and due to this we may see an increased number of false alarms. We are asked if we want to transform our data with the guidance that we should not transform the data if we were expecting our process to produce normal data.

5. Click on Yes to transform the data.

Your data fails the normality test. You may see an increased number of false alarms.

Consider a transformation as it will correct the nonnormal condition.

-- If your process naturally produces nonnormal data, click Yes.

-- If your process naturally produces normal data, click No.

Do you want to transform your data?

Yes No

The Stability Report shows us the same charts that we saw previously when we used the Classic Method.

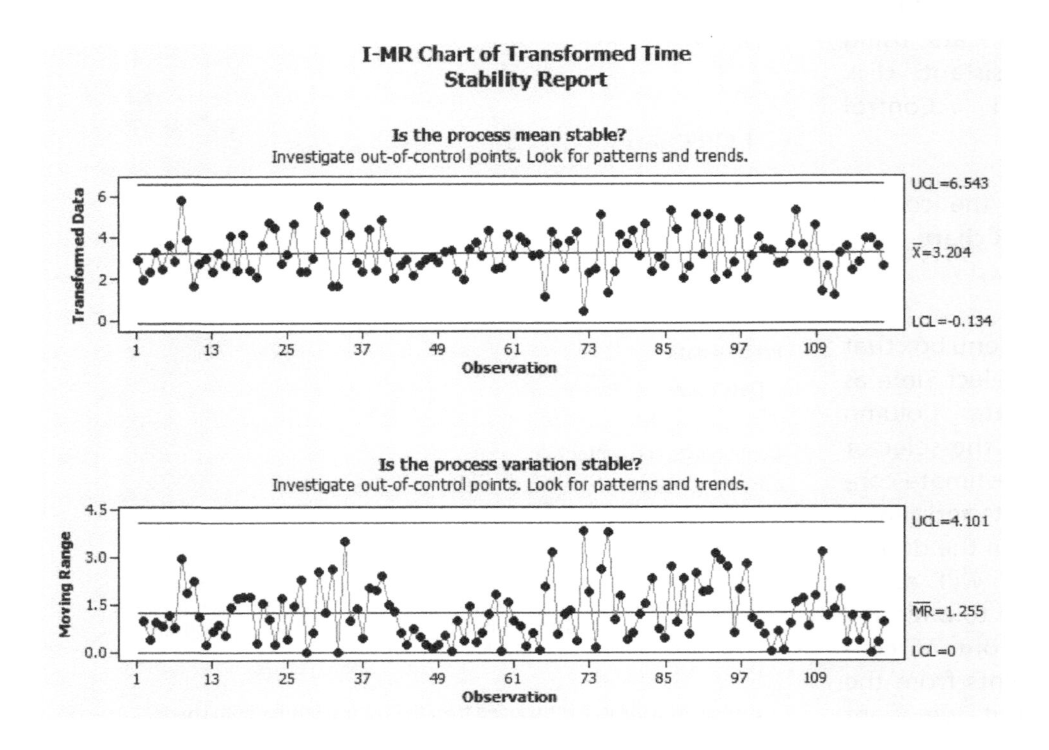

The Report Card gives us the all clear in terms of the checks that it carries out.

I-MR Chart of Transformed Time
Report Card

Check	Status	Description
Normality	✓	The transformed data passed the normality test.
Stability	✓	The process mean and variation are stable. No points are out of control on either chart of the transformed data.
Amount of Data	✓	You do not need to be concerned about the precision of your control limits because 100 or more data points are included in the calculations.
Correlated Data	✓	If the data are correlated, you may see an increased number of false alarms. Because less than 2% of the points are outside the control limits on the I chart, the correlation test is not needed.

Exercise 2. I-MR Chart.

Iqra measures the concentration of micro-organisms on industrial air conditioning filters. From historical data she knows that the distribution is naturally skewed. Help Iqra produce an I-MR chart of her results. Initially produce a histogram of this data and visually check if the distribution is positively skewed and all values are greater than zero. Then use the Classic Method and the Assistant to produce an I-MR chart of the transformed data.

Worksheet: Air.

From the histogram it can be seen that the distribution is positively skewed. A check of the data shows that all values are greater than zero.

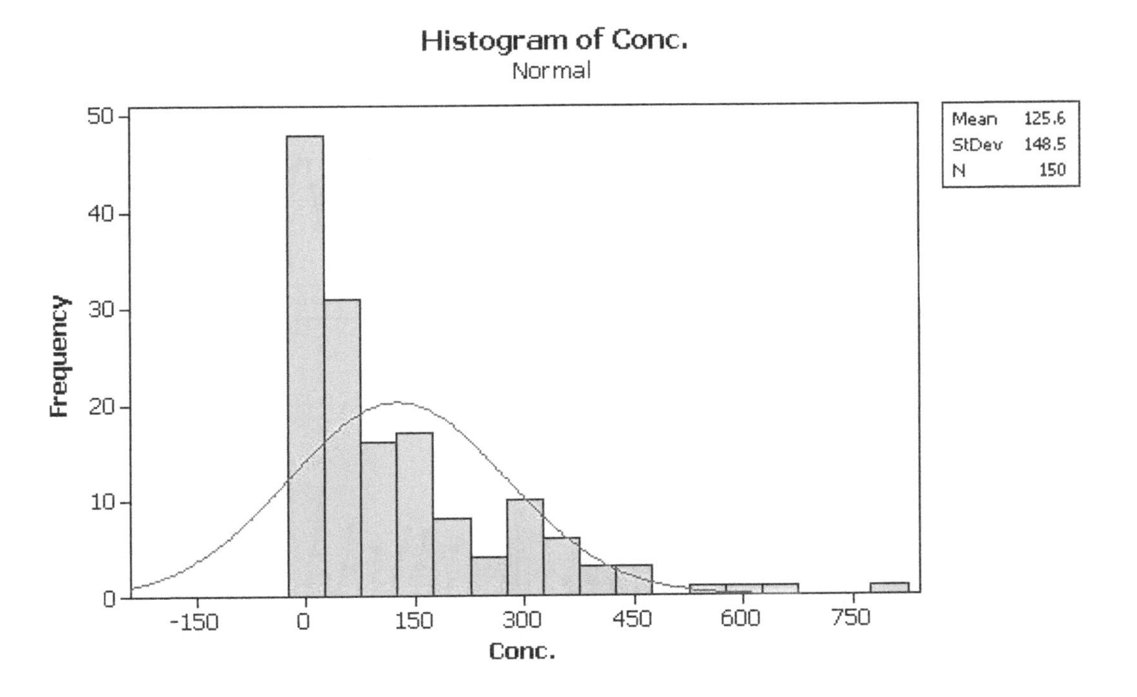

The Box–Cox Plot is produced and gives us a Rounded Value for Lambda of 0.26.

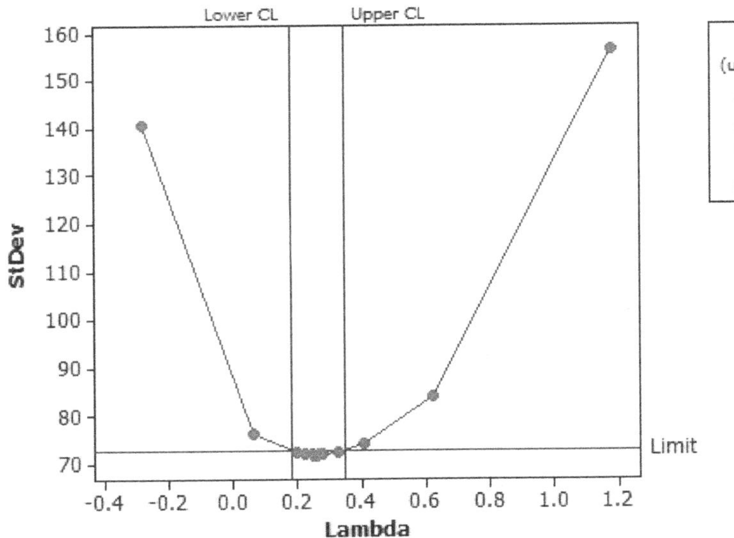

When you go to the Box–Cox tab and engage the selector for the transformation also select the radio button for the Optimal lambda.

☑ Use a Box-Cox transformation (W = Y**Lambda)

- ○ Lambda = 0 (ln)
- ○ Lambda = 0.5 (square root)
- ◉ Optimal lambda
- ○ Optimal lambda for each stage (in a chart with stages)
- ○ Other (enter a value between -5 and 5):

With the transformed data we can see that both the I chart and the MR chart are in control.

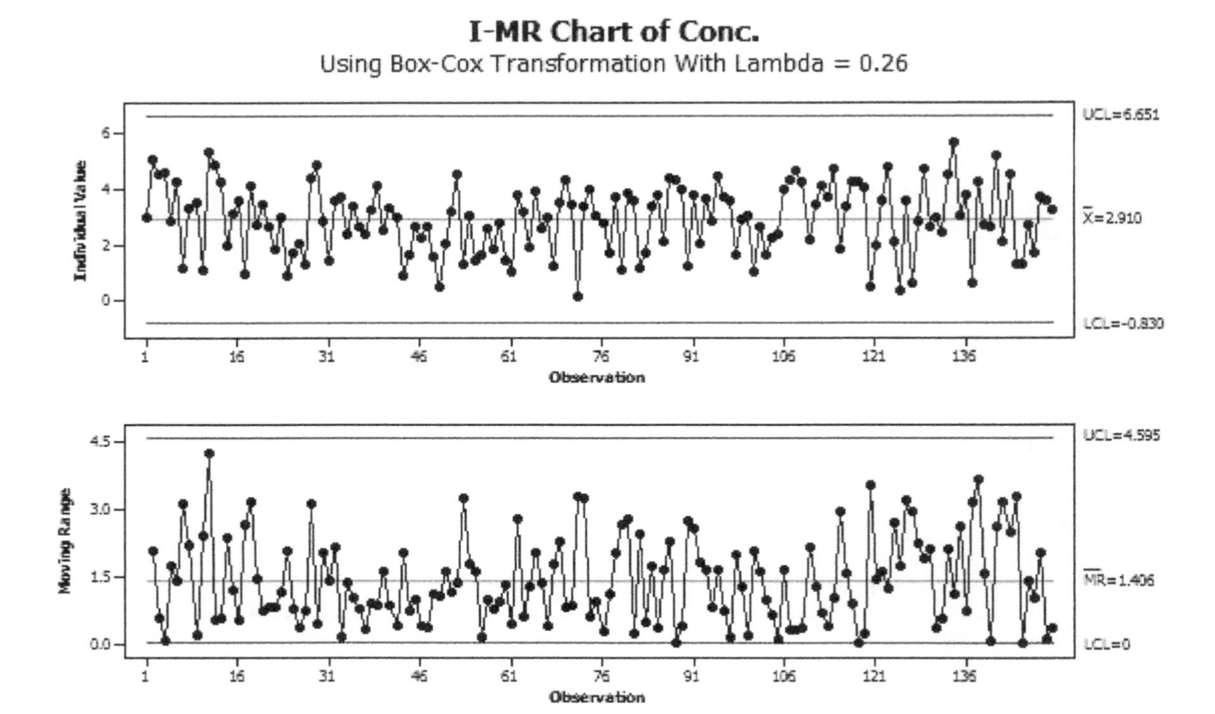

When we use the Assistant to complete the same procedure the Stability Report should show us the same charts that we saw previously when we used the classic method.

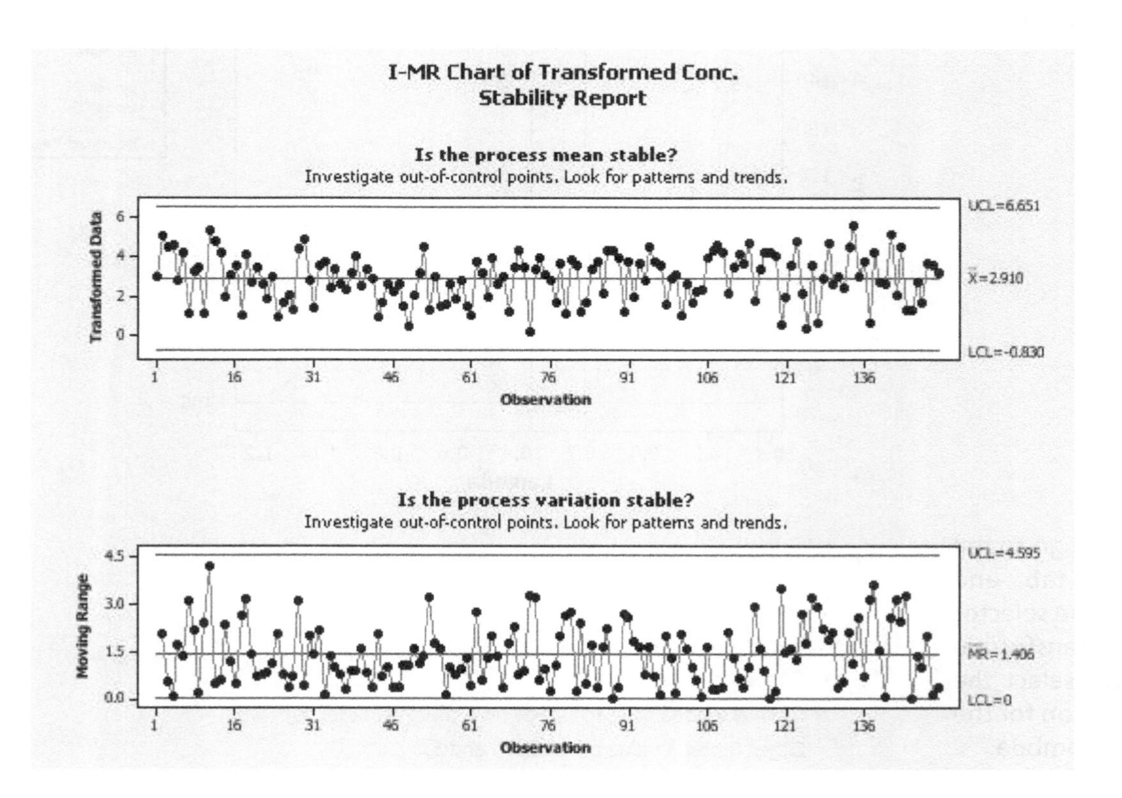

The Report Card should not give us any warnings.

Check	Status	Description
Normality	✓	The transformed data passed the normality test.
Stability	✓	The process mean and variation are stable. No points are out of control on either chart of the transformed data.
Amount of Data	✓	You do not need to be concerned about the precision of your control limits because 100 or more data points are included in the calculations.
Correlated Data	✓	If the data are correlated, you may see an increased number of false alarms. Because less than 2% of the points are outside the control limits on the I chart, the correlation test is not needed.

7.9 The Xbar-R Chart

Procedure	• Xbar-R Chart
What's it used for?	• It is used to detect changes in time ordered data where subgroups are being examined. Subgroup sizes should be between two and eight, inclusive.
Assumptions and Limitations	• Used when there are subgroups and not individual data points. • The subgroup (within group) variability must be in control before examining the Xbar values (between groups). This is because the subgroup variability is used to calculate the control limits for the Xbar control chart. • Help <<Methods and Formulas shows how the control limits are calculated.

Example 3. Xbar-R Chart.

On the chocolate bunny line Willy Winker is taking five samples at the start of every hour and weighing them. Initially four sets of samples were taken in the startup phase but these are not deemed to be a part of the production run.

Produce an Xbar-R chart of bunny weight and explore any process changes.

Worksheet: Bunny.

1. Copy all columns of data from Worksheet: Bunny into a new Minitab project worksheet. The hourly times that the samples were taken are in the Time column and within the same row are the five weights that were recorded. Therefore, we will consistently have a subgroup size of five within this chart.

Time	Weight1	Weight2	Weight3	Weight4	Weight5
AA	3.7	5.6	5.1	4.3	5.9
BB	9.5	5.4	5.6	5.7	0.4
CC	5.5	4.5	1.2	9.6	5.5
DD	6.2	7.2	8.6	5.6	7.6
00:00	40.00	20.80	20.80	37.80	32.60
01:00	7.00	22.00	45.00	33.60	30.60

2. Click Stat <<Control Charts <<Variables for Subgroups <<Xbar-R.

Stat
Control Charts ▸
Variables Charts for Subgroups ▸ Xbar-R...

3. The Xbar-R chart menu opens. Normally Minitab takes data in columns so we must tell it that the subgroup data is across the rows. Select 'Subgroup Observations in Rows'

Observations for a subgroup are in one row of columns:
All observations for a chart are in one column:
Observations for a subgroup are in one row of columns:

4. Select columns Weight1 to Weight5 as the data to be charted.

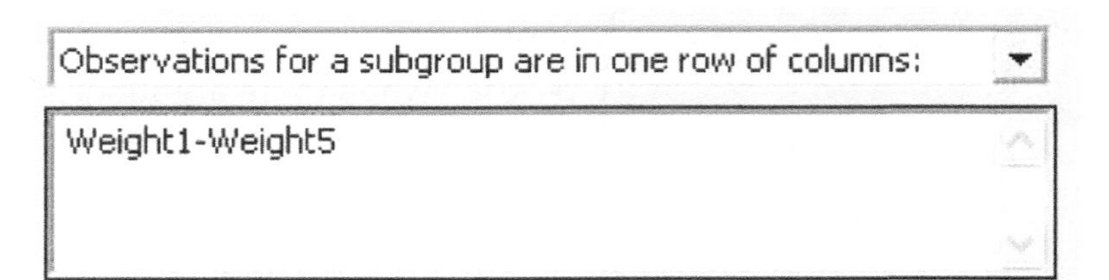

5. Then Click on Scale. For the X Scale click on the Stamp radio button. Select Time.
6. Click OK and OK again.

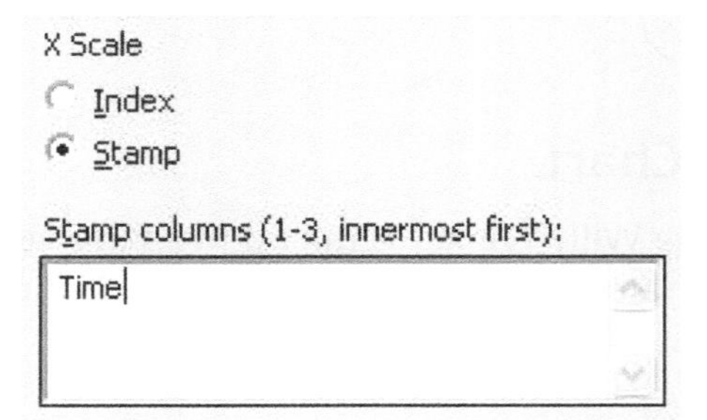

We have produced an initial Xbar-R chart but really we should have removed the startup data before we started. We actually did this to demonstrate the use of the Data Options menu which can be used within any of our chart types when using the Classic Method. We are going to use the Data Options Menu to remove the first four subgroups.

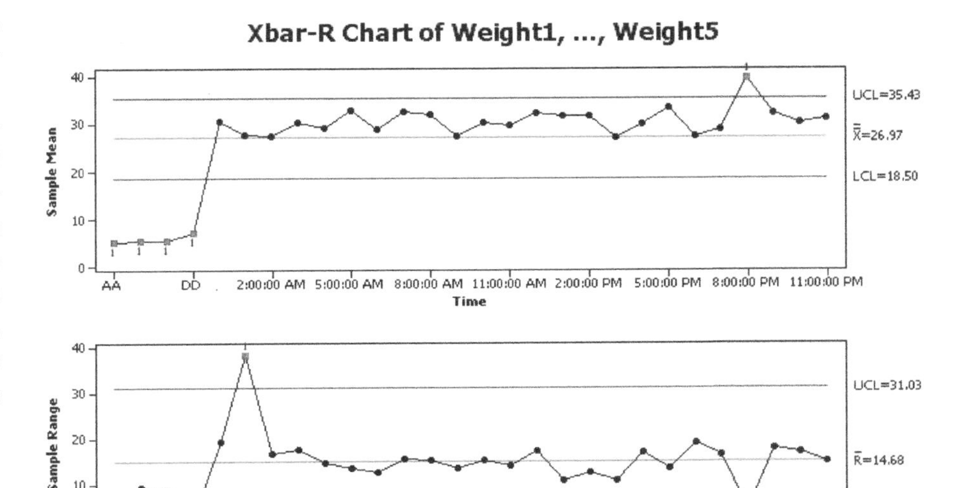

7. Click on the edit last dialog box icon.

8. Then click on the Data Options button.

Data Options...

The menu works by either specifying which rows to include or exclude, depending on which is easier for the user. There are also a number of options on specifying the rows. It can be done by row number, brushed rows or we can set up logical conditions. When specifying items Minitab uses the same system all the time. 1:4 means 1 to 4 inclusive. 2:6 10 would mean 2 to 6 inclusive plus 10.

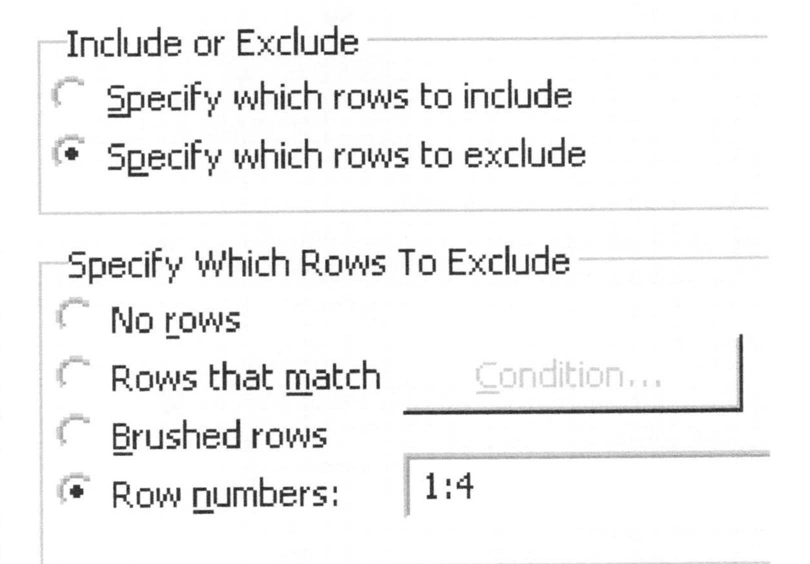

9. Select the radio button to specify which rows to exclude.
10. Select the radio to specify selected row numbers. Enter 1:4.
11. Click OK and OK again.

There is a single point on the R chart that is out of control. We must address this first before we consider the Xbar chart. This is because the within subgroup variation is used to calculate the Xbar control limits on the Xbar-R and Xbar-S charts.

We will now have a look at the data within the 01:00 subgroup.

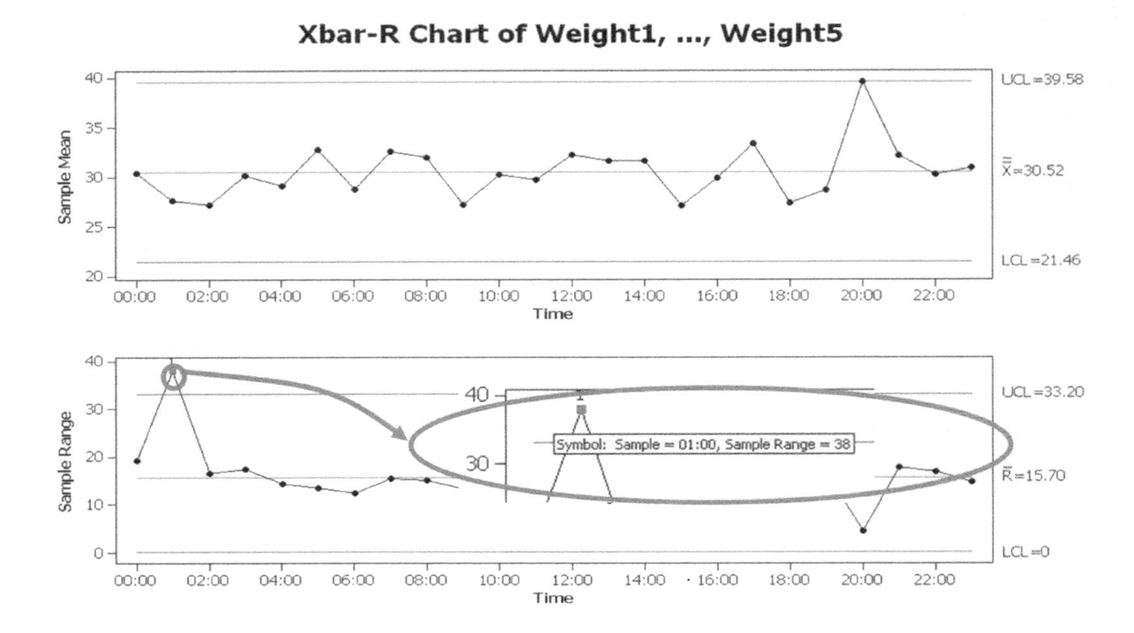

12. Click on the Show Worksheets icon.

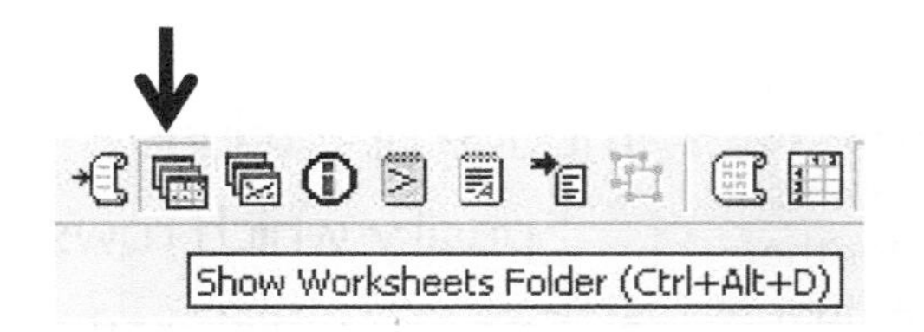

We see that there is an unusual value for Weight1 at 01:00 of 7.0. After a thorough and exhausting investigation we found that the value was a typing error; it should be 27.

13. Correct the value from 7 to 27 by editing the value on the worksheet.

14. We are going to update the graph using a shortcut rather than go through the menus again. Click on the Show Graphs icon.

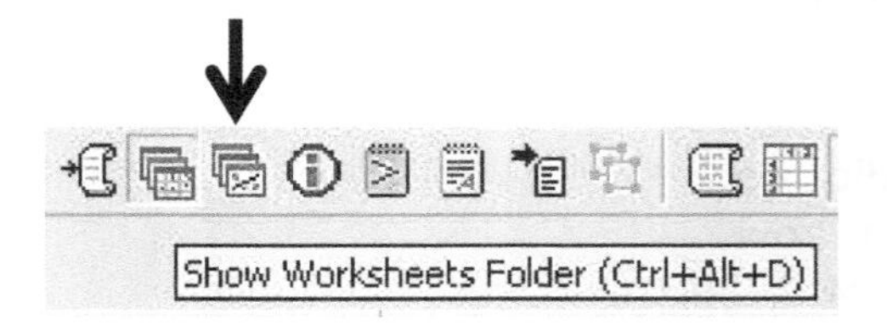

15. Right click on the graph and select Update Graph Automatically.

Having selected this option the graph will now automatically reflect the data within the worksheet. There is also the option to Update Graph Now which is a single update.

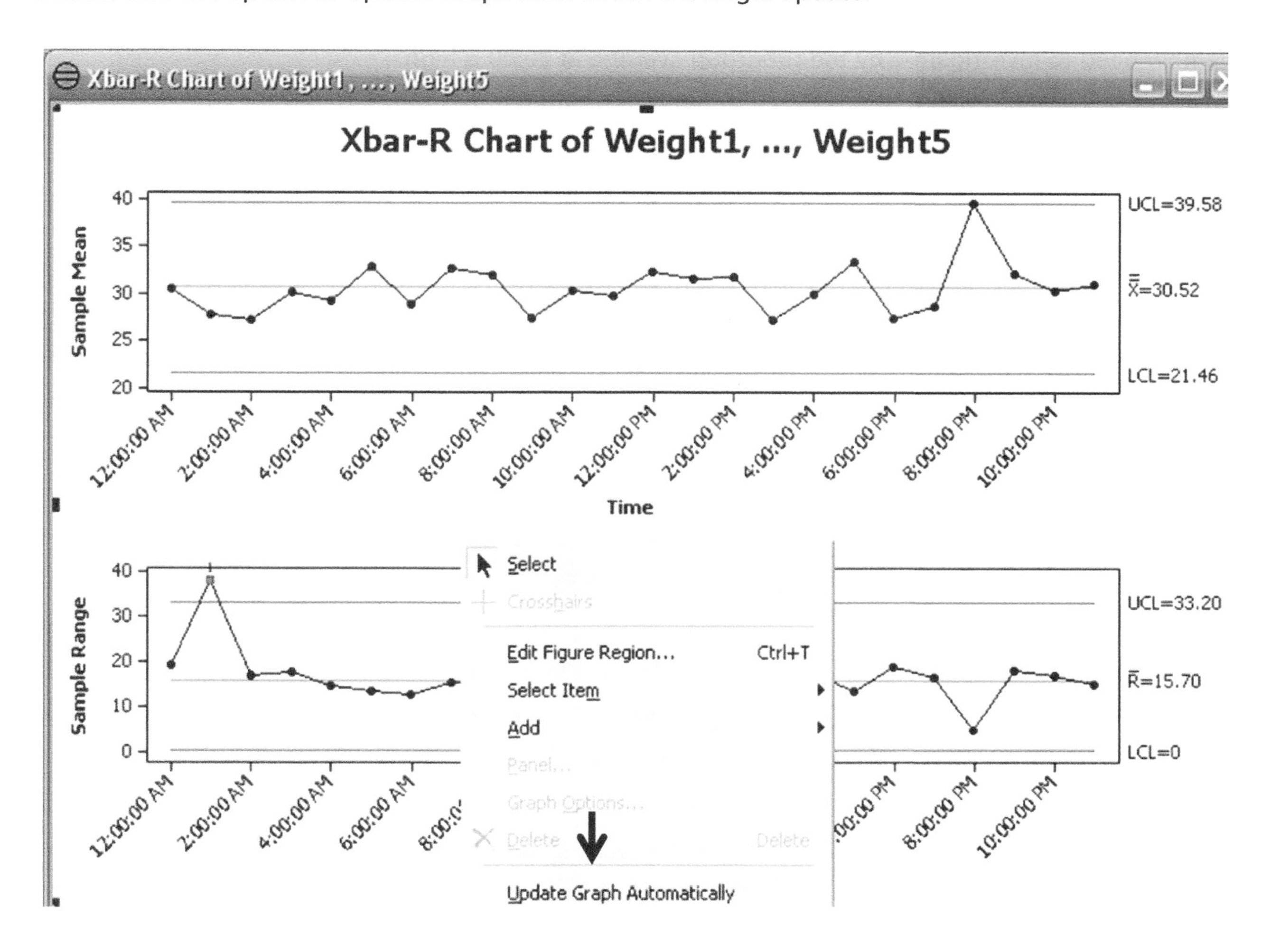

We see that the R chart is now in control. However, a different point on the Xbar chart is now out of control. The reason for this is that the high subgroup range generated by the data at 01:00 widened the control limits for the Xbar chart. Since that has been corrected the Xbar control limit width has become smaller.

We now need to investigate why the subgroup average at 20:00 is high.

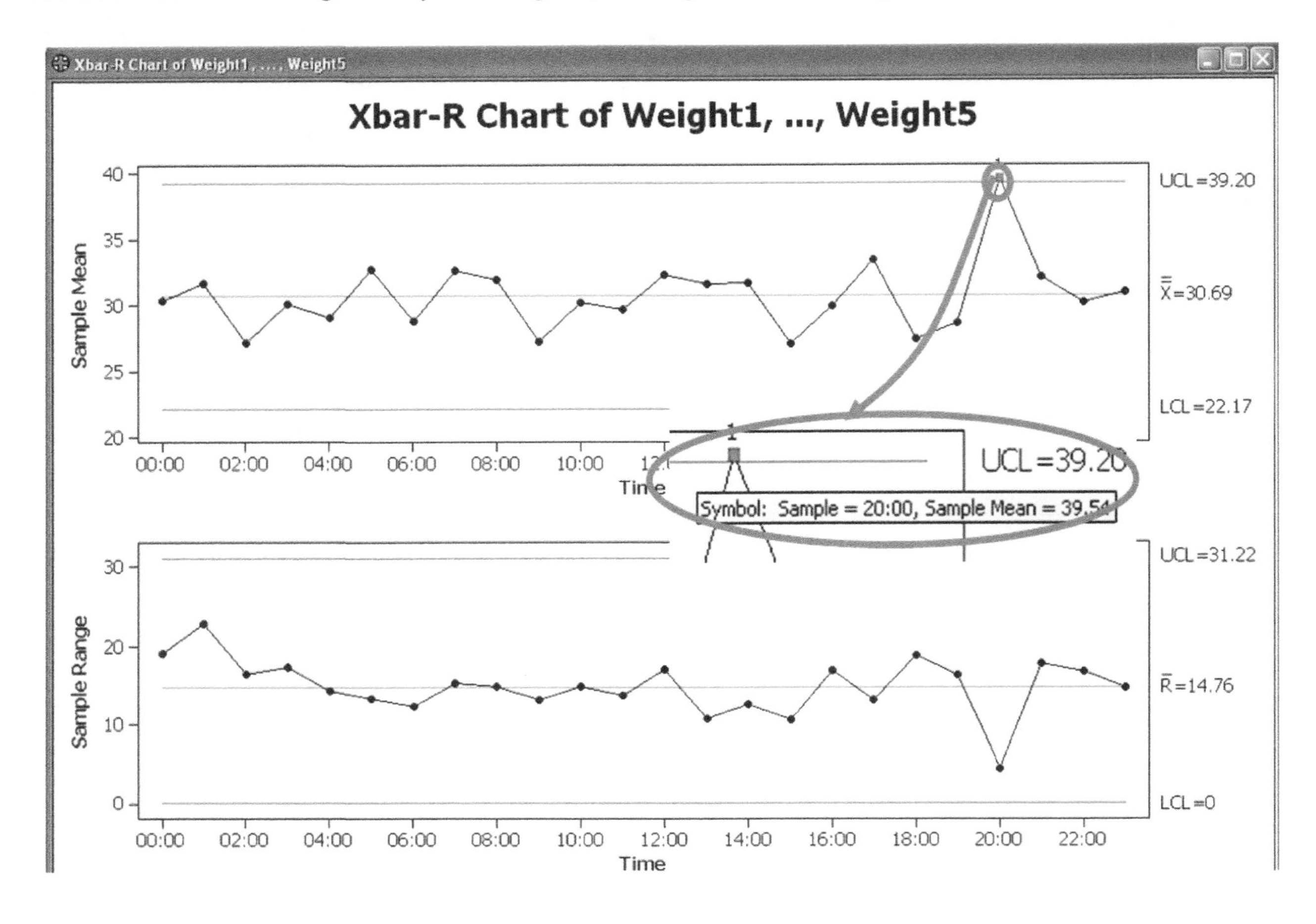

Looking at the 20:00 subgroup we see that all the values are at the top end of the range giving a high average value. Although the point is out of control it is deemed to be acceptable by Willy Winker.

8:00:00 PM	38.9	39.8	37.2	41.6	40.2

This example is intended to highlight that the correct procedure is always to address the control limits on the R chart prior to addressing issues on the Xbar graph. This is because the within subgroup variation is used to calculate the Xbar control limits on Xbar-R and Xbar-S charts. On I-MR charts the average moving range is used to calculate the control limits on the Individuals chart.

We will now chart the same data set using the Assistant. As we have changed the data set it would probably be a good idea to reload the data set into a new Minitab project.

1. Click Assistant <<Control Charts.

Assistant

Control Charts...

2. Click on the Icon for the Xbar-R chart.

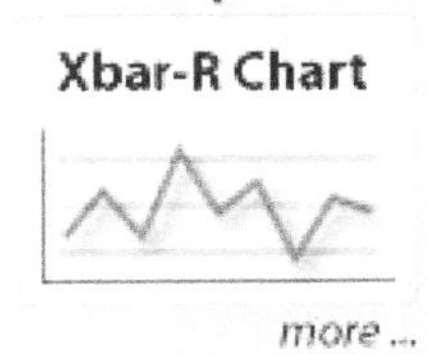

3. Again, we must tell Minitab how the data is arranged. Select Data are in Multiple Columns from the drop down box.

How are your data arranged in the worksheet?

Data are in one column for all subgroups

Data are in one column for all subgroups
Data are in multiple columns, one row for each subgroup

4. Select the Data Columns as Weight1 to Weight5 and use the next selector to tell Minitab that we want control limits to be Estimated from the Data.

Process data

How are your data arranged in the worksheet?

Data are in multiple columns, one row for each subgroup

Data columns:

Weight1-Weight5

Control limits and center line

How will you determine the control limits and center line?

Estimate from the data

As soon as we tell Minitab how we want the control limits to be calculated we are asked to omit the points which are out of control.

5. Click the selector to omit the first four points which were not part of the production run.
6. Click OK.

If you omit a subgroup, it is excluded from the calculations for both charts.

Omit	Subgroup	Chart	Reason
☑	1	XBar	Below lower control limit
☑	2	XBar	Below lower control limit
☑	3	XBar	Below lower control limit
☑	4	XBar	Below lower control limit
☐	6	R	Above upper control limit
☐	5 - 24	XBar	Shift in mean

On the Stability Report we see both the Xbar and R charts. It looks as though Minitab has not omitted the first four points. In actual fact it has omitted the first four points from the calculations of the control limits and the mean. In order to tidy up the chart we will delete the data points that we don't want from the actual worksheet.

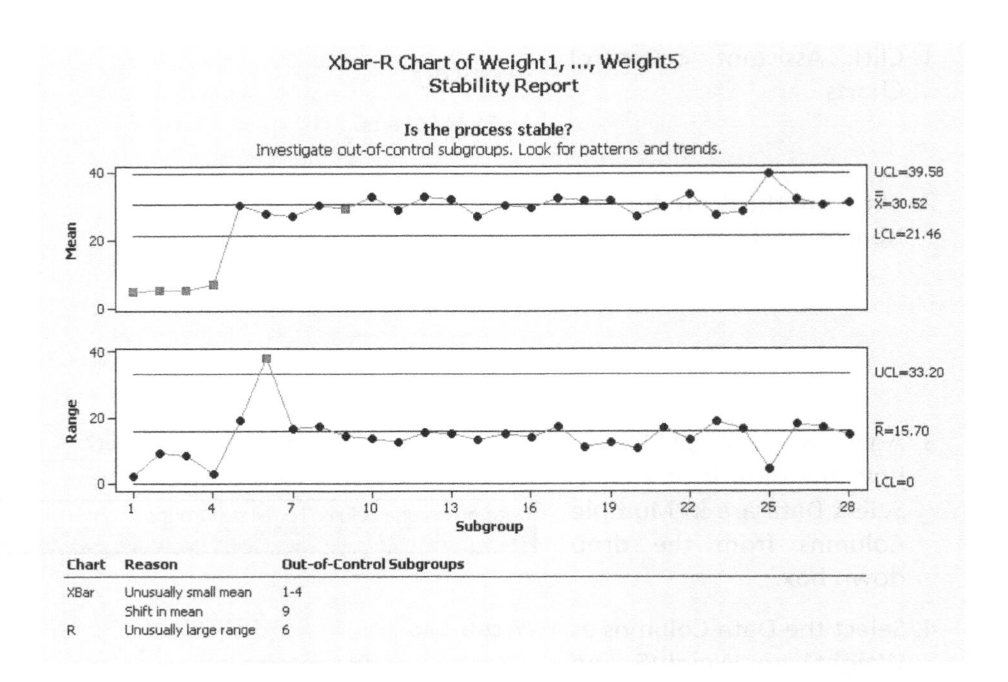

7. Select the first four rows of data by left clicking and dragging.

↓	C1-T	C2	C3	C4	C5	C6
	Time	Weight1	Weight2	Weight3	Weight4	Weight5
1	AA	3.7	5.6	5.1	4.3	5.9
2	BB	9.5	5.4	5.6	5.7	0.4
3	CC	5.5	4.5	1.2	9.6	5.5
4	DD	6.2	7.2	8.6	5.6	7.6

8. Click Edit <<Delete Cells.

Edit

Delete Cells

9. Correct the Typo, change 7 to 27, which is on the first column of the 01:00 subgroup.

1:00:00 AM	27.0	22.0	45.0	33.6	30.6

The Assistant does not allow the chart to be updated as per the Classic Method. When using the Assistant the chart must be redrawn from the Assistant menu.

10. Redraw the Xbar-R chart using the Assistant menu.

Once the chart is redrawn the Stability Report should look like the one shown here. This is exactly the same as the one we produced using the Classic Method.

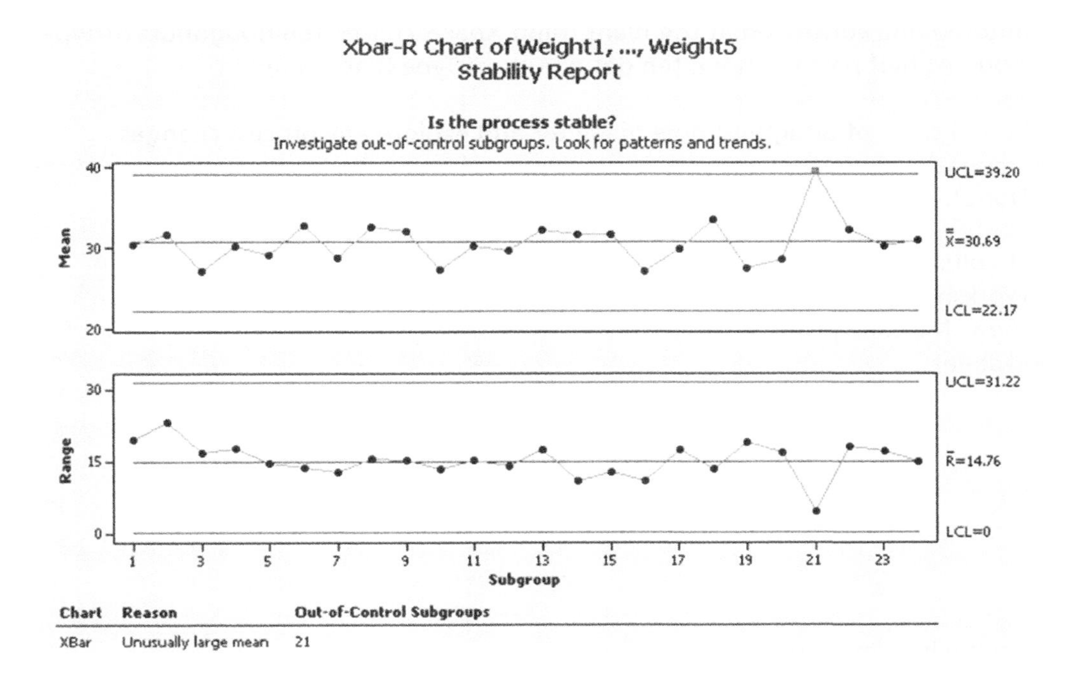

7.10 The Xbar-S Chart

Procedure

- Xbar-S Chart

What's it used for?

- It is used to detect changes in time ordered data where subgroups are being examined. The chart is for subgroup sizes of nine and greater.

Assumptions and Limitations

- Used when there are subgroups and not individual data points.
- The subgroup (within group) variability must be in control before examining the Xbar values (between groups). This is because the subgroup variability is used to calculate the control limits for the Xbar control chart.
- Help <<Methods and Formulas shows how the control limits are calculated.

Example 4. Xbar-S Chart.

Willy Winker 's secret to making a good doughnut is the internal hole diameter of the doughnut. This is carefully monitored and controlled at the plant using Xbar-S charts. Ten doughnuts of Type A are sampled on the hour. At half past the hour, ten doughnuts of Type B are sampled.

Produce an Xbar-S chart of doughnut hole diameter and explore any process changes.

Worksheet: Donut.

1. Copy all columns of data from Worksheet: Donut into a new Minitab project worksheet.

Shift	Type	Dia01	Dia02	Dia03	Dia04	Dia05	Dia06	Dia07	Dia08	Dia09	Dia010
0:00	A	31.5	26.75	26.1	25.9	27.05	22.55	31.4	28.9	32.35	24.85
0:30	B	28	29.65	25.6	27.35	27.05	29.3	29.05	29.95	27.7	27.2
1:00	A	28.3	24.2	31.25	30.5	28.75	22.55	25.05	30.35	29.3	26.35
1:30	B	25.05	26.9	29.45	29.75	27.25	28.65	29.55	27.75	25.95	28.5
2:00	A	31.9	27.9	26.55	32.05	30.6	30.5	29.05	25.05	28	29.1
2:30	B	25.9	29.7	25.35	29.3	26.7	26.75	26.35	26	29.9	25.65
3:00	A	32.2	31.15	25.9	29.35	27.4	27.75	24.95	29.85	23.4	31.5
3:30	B	25.35	28.05	29.7	25.55	28.75	29.3	27.1	28.8	28.5	25.4

2. Click onto the Information icon and check the data formatting.

3. We see that there are 48 rows of data. There are ten columns of data on internal doughnut hole diameter which will form the subgroups. The Shift column contains the times that the doughnuts were sampled.

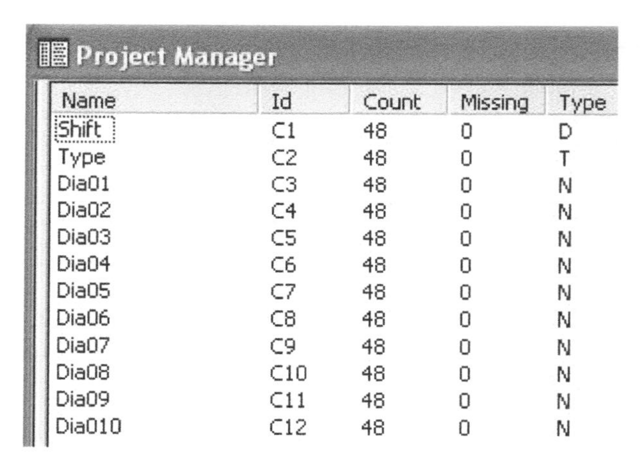

Project Manager

Name	Id	Count	Missing	Type
Shift	C1	48	0	D
Type	C2	48	0	T
Dia01	C3	48	0	N
Dia02	C4	48	0	N
Dia03	C5	48	0	N
Dia04	C6	48	0	N
Dia05	C7	48	0	N
Dia06	C8	48	0	N
Dia07	C9	48	0	N
Dia08	C10	48	0	N
Dia09	C11	48	0	N
Dia010	C12	48	0	N

4. Click Stat <<Control Charts <<Variables Charts for Subgroups <<Xbar-S.

5. Select 'Observations for a subgroup are in one row'.

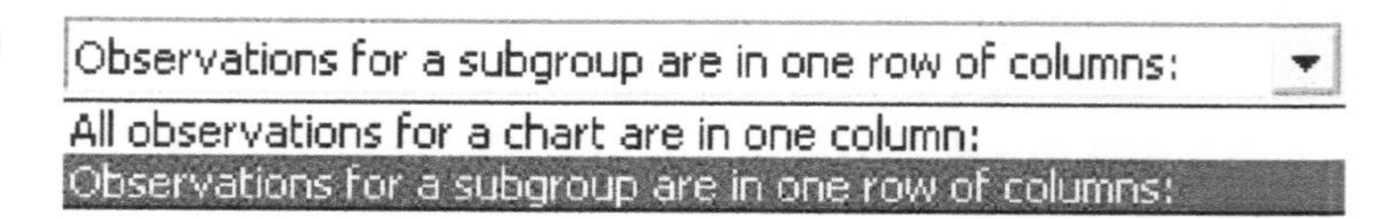

6. Select the data from Dia01 to Dia10 to form the subgroups to be plotted.
7. Then Click on Scale.

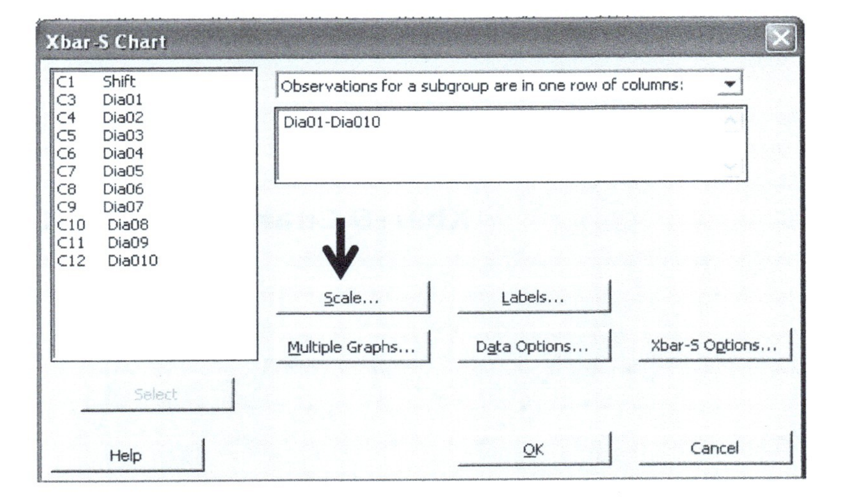

8. Click on the Stamp radio button and then select Shift as the Stamp. Click OK and OK again.

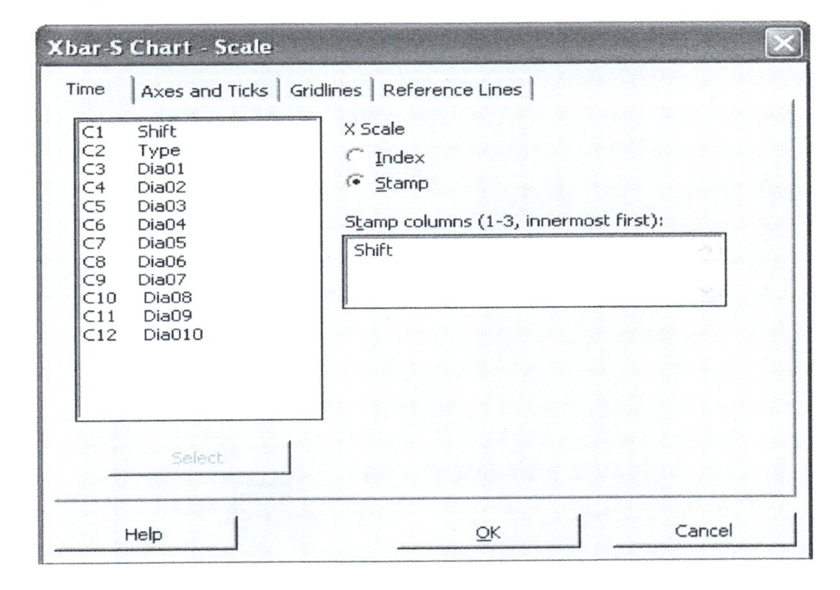

On the S chart we see one point that is out of control and we see an unusual pattern within the Sbar chart.

We will produce the chart again but on this occasion we will switch the additional tests on.

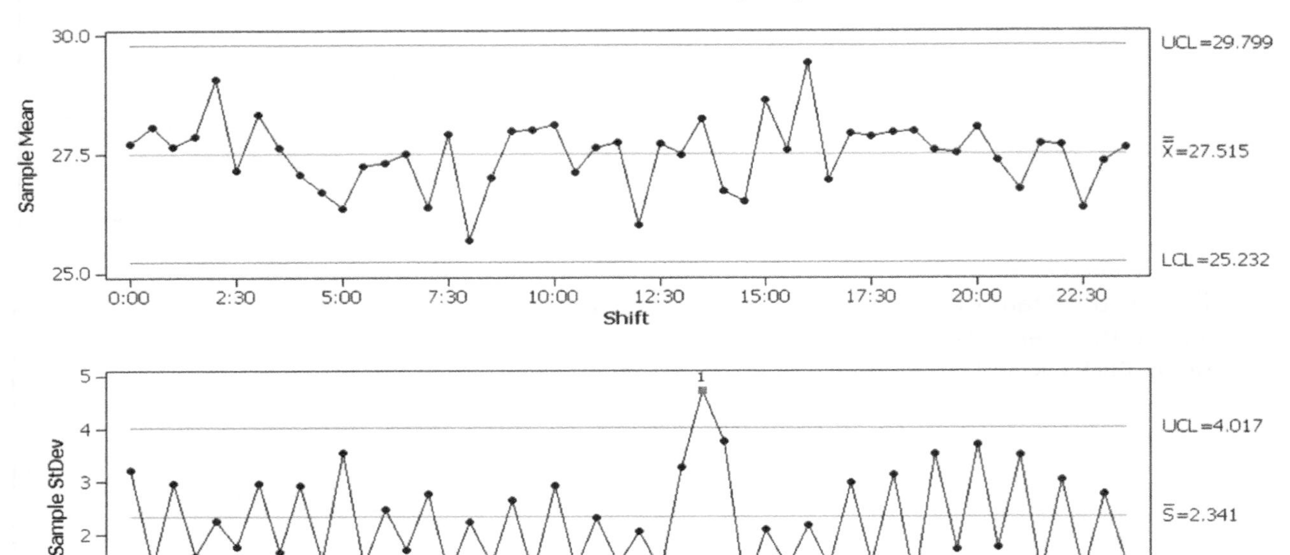

9. Click onto the edit last dialog box icon.

10. Click onto Xbar-S Options.

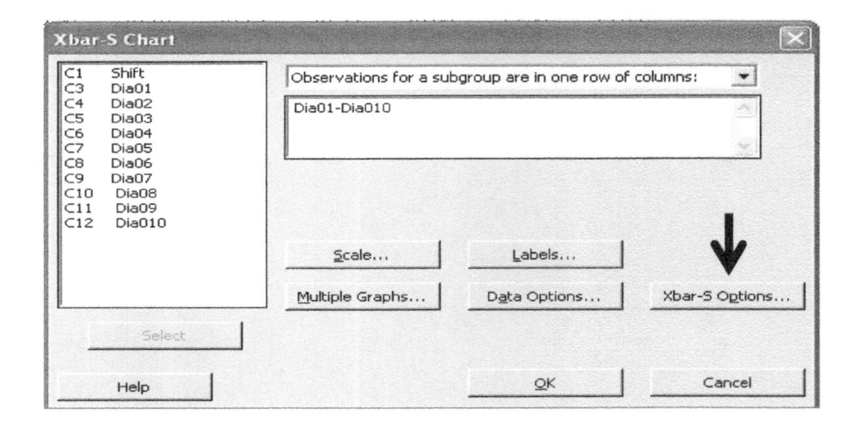

11. Click onto the Tests tab.
12. Click onto the selector tab and choose Perform all tests for special causes.
13. Click OK and OK again.

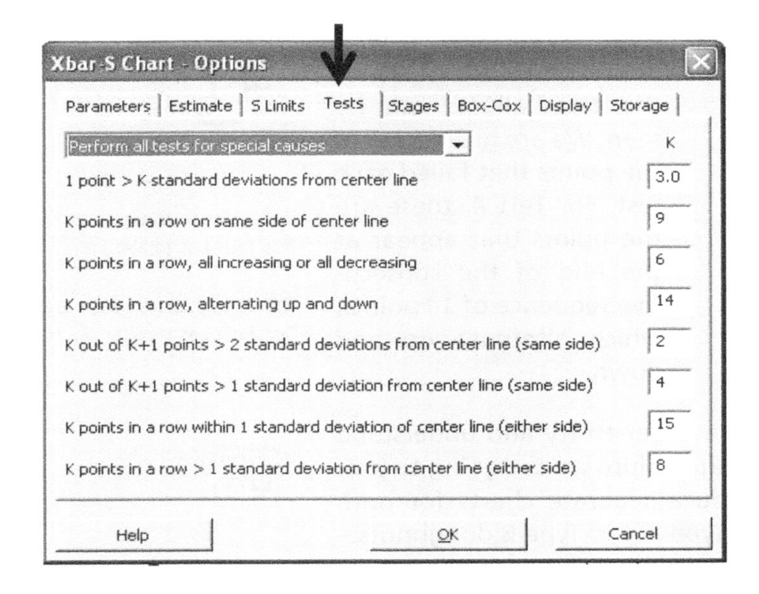

We see that the Sbar chart fails Test 4 in addition to Test 1. Test 4 is failed due to 14 points alternating up and down.

We need to investigate the reason for the additional failure of Test 4.

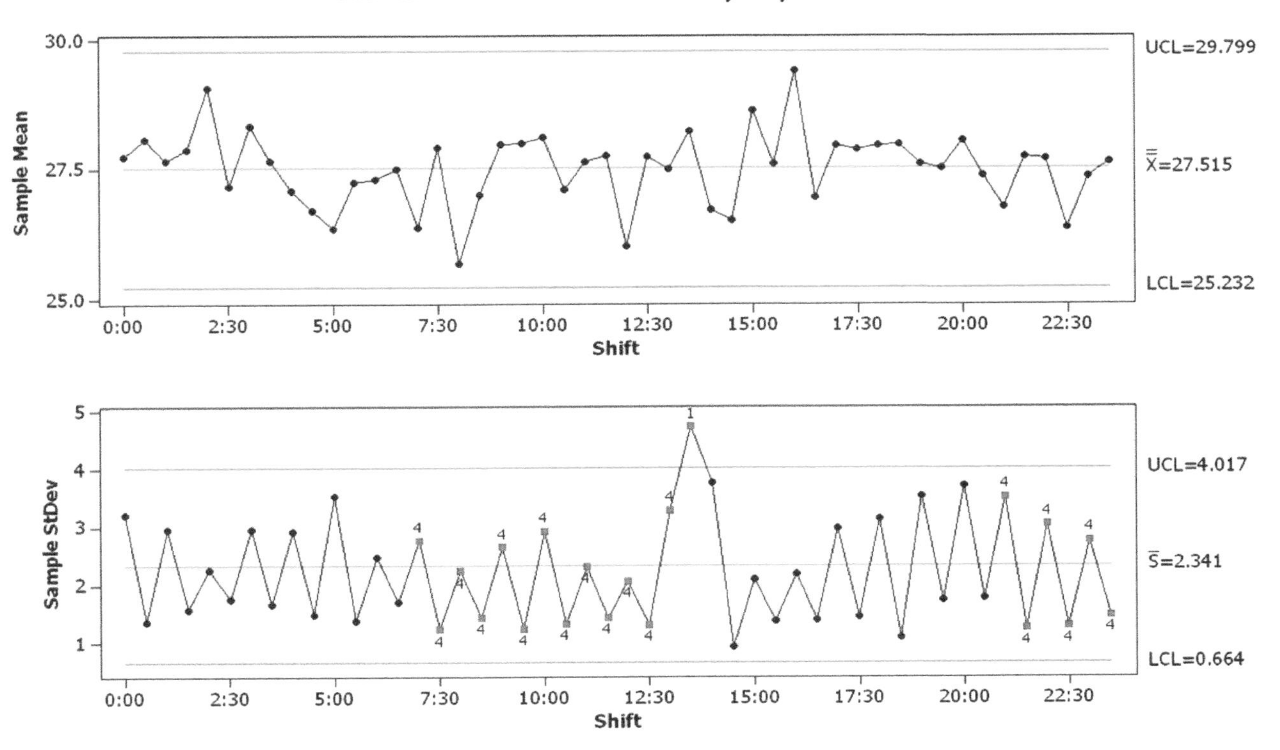

14. Click on the Session window icon. More detail of the test failures can be seen. We are given a list of the points that failed each test. For Test 4, these are the points that appear at the end of the consecutive sequence of 14 points, which alternate up and down.

```
TEST 1. One point more than 3.00 standard deviations from center line.
Test Failed at points:  28

TEST 4. 14 points in a row alternating up and down.
Test Failed at points:  15, 16, 17, 18, 19, 20, 21, 22, 23, 24, 25, 26, 27, 43, 44, 45, 46, 47, 48
```

In order to try and understand the failure we are going to produce separate charts for both Type A and Type B doughnuts.

15. Click on the Edit Last Dialog box icon.
16. Click on Data Options.

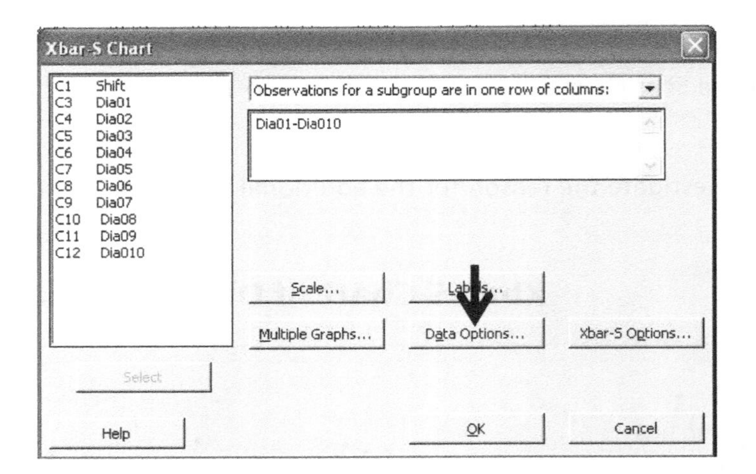

We are going to use a logical condition to select the type of donut that we are going to chart.

17. Under Specify Which Rows to Include, click on the radio button for Rows that match. Then click on Condition.

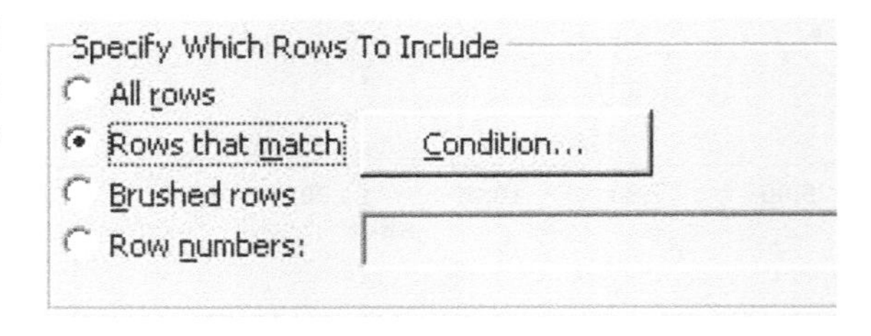

18. Within the Condition box enter Type = 'A' as shown. This means that only rows of type A doughnut will be selected for the chart. The conditional selector is quite powerful and allows the logical conditions to be extended by using And, Or and Not.
19. Click OK, OK and OK again.

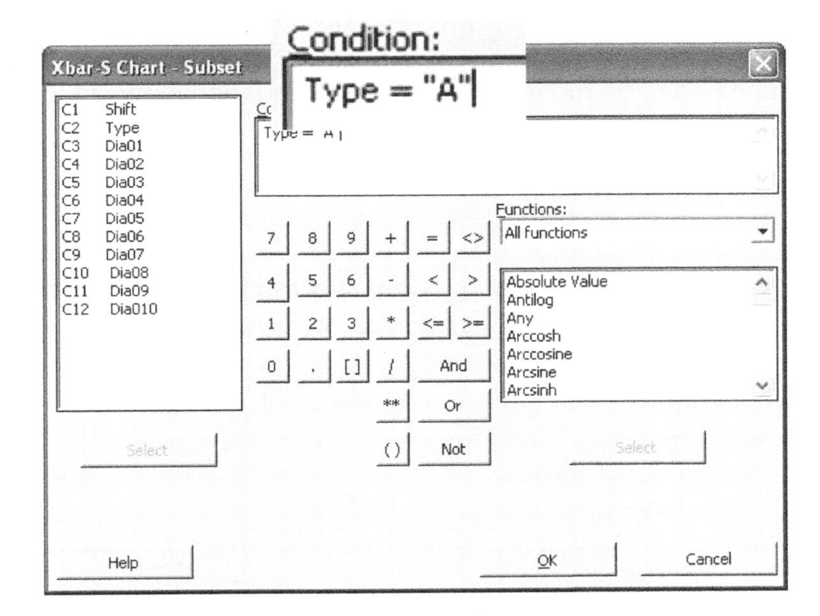

20. Having multiple charts with the same title can get confusing so double click onto the chart title and edit it to include the doughnut type.

Both the Xbar and S chart are in control for Type A.

21. As a mini exercise produce the Xbar-S for Type B.

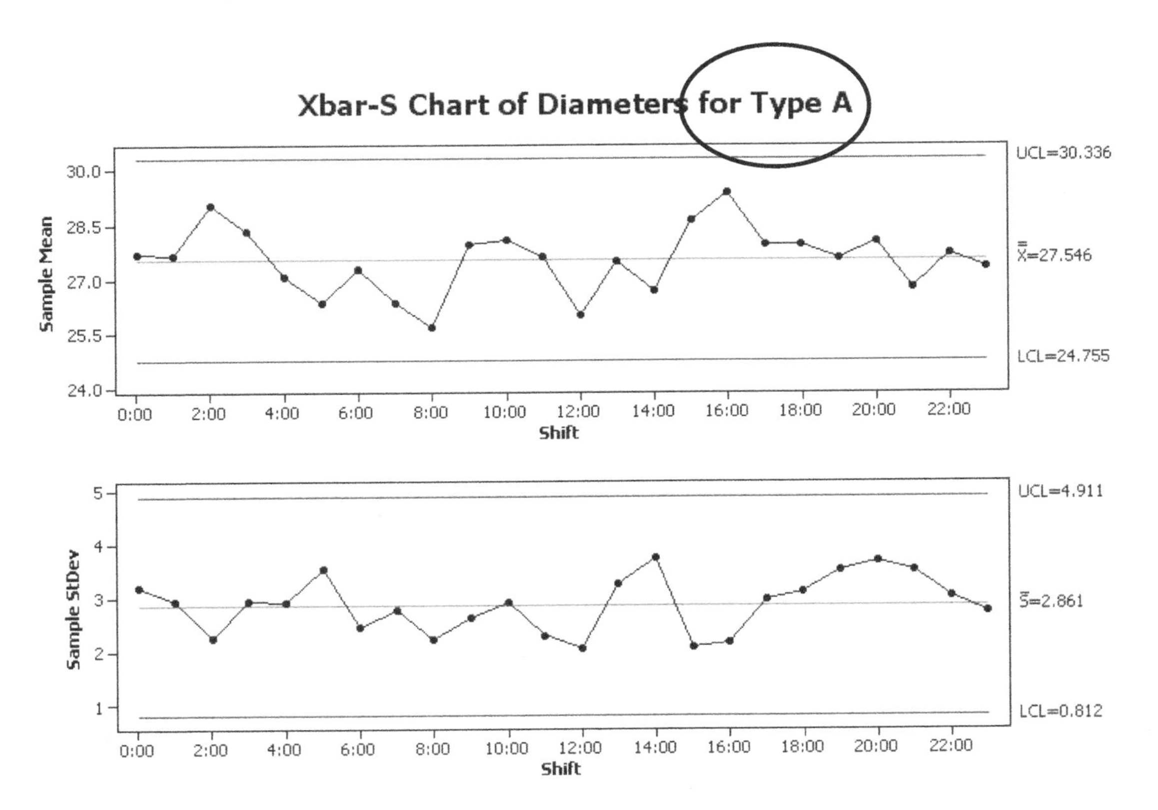

We see that most of our problems have been cured by separating Type A and Type B doughnuts. The problem was simply having two different items on the chart. The two types of doughnut trapped what appeared to be too much special cause variation between the subgroups.

For Type B doughnuts we still see one point that is out of control, this is the subgroup for 13:30. We must investigate this before making conclusions about the Xbar graph.

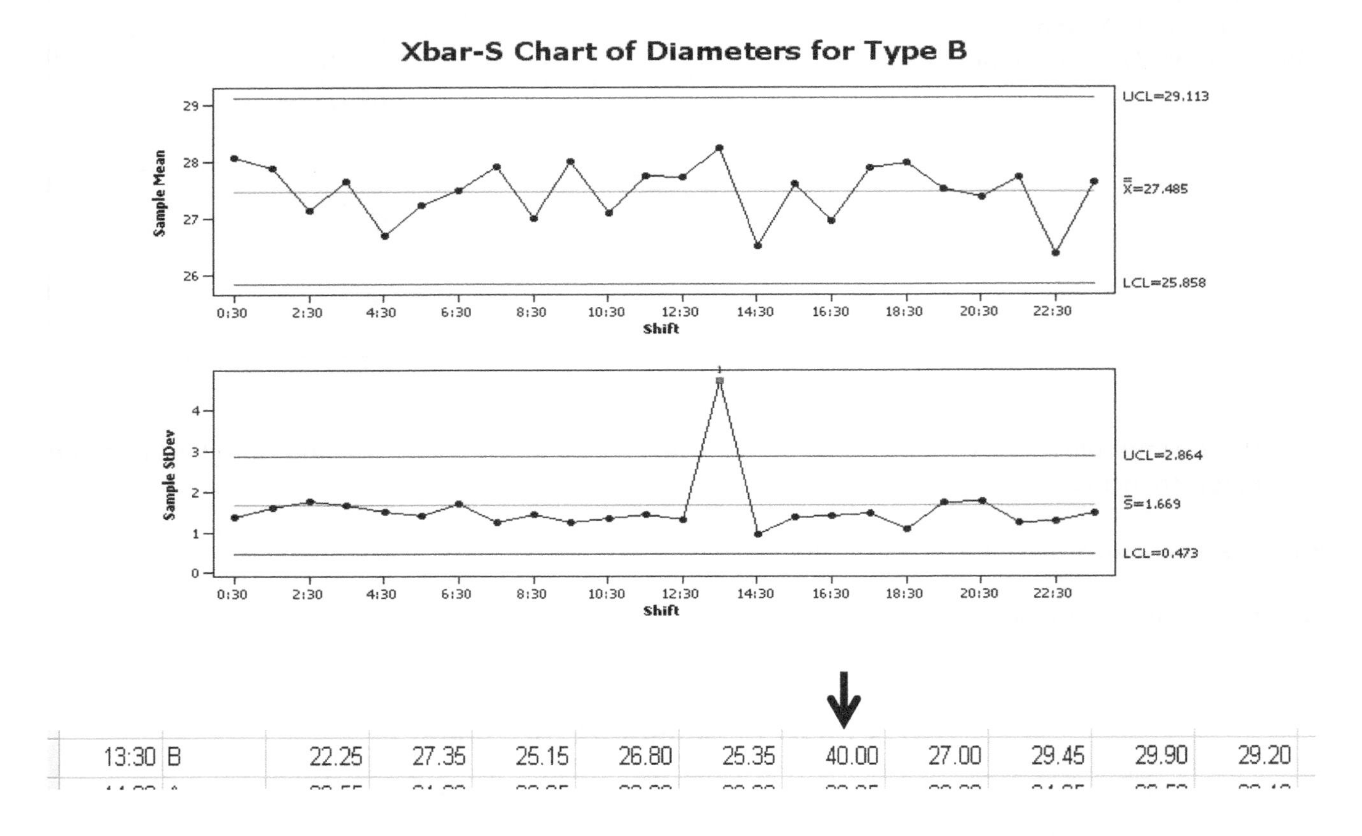

13:30	B	22.25	27.35	25.15	26.80	25.35	40.00	27.00	29.45	29.90	29.20

The S chart is out of control on a single point. However, since the Xbar chart control limits will be based on the within group variation, we must first address the issues on the S chart before we continue.

Looking at the data for time 13:30 Willy finds another typing error. Dia6 should be 30 not 40.

22. Change the value of Dia6 and update the chart for Type B doughnuts and check if there are any further points out of control.

28	13:30	B	22.25	27.35	25.15	26.80	25.35	30.00	27.00	29.45	29.90	29.20

With the correction the S chart is now under control and we can now conclude that the Xbar chart is also under control.

Within the example we saw the importance of correctly setting up subgroups. Although the doughnuts of Type A and Type B have a similar average hole diameter, the StDev within the subgroups is very different. Type A has an Sbar of 2.861 and Type B of 1.465.

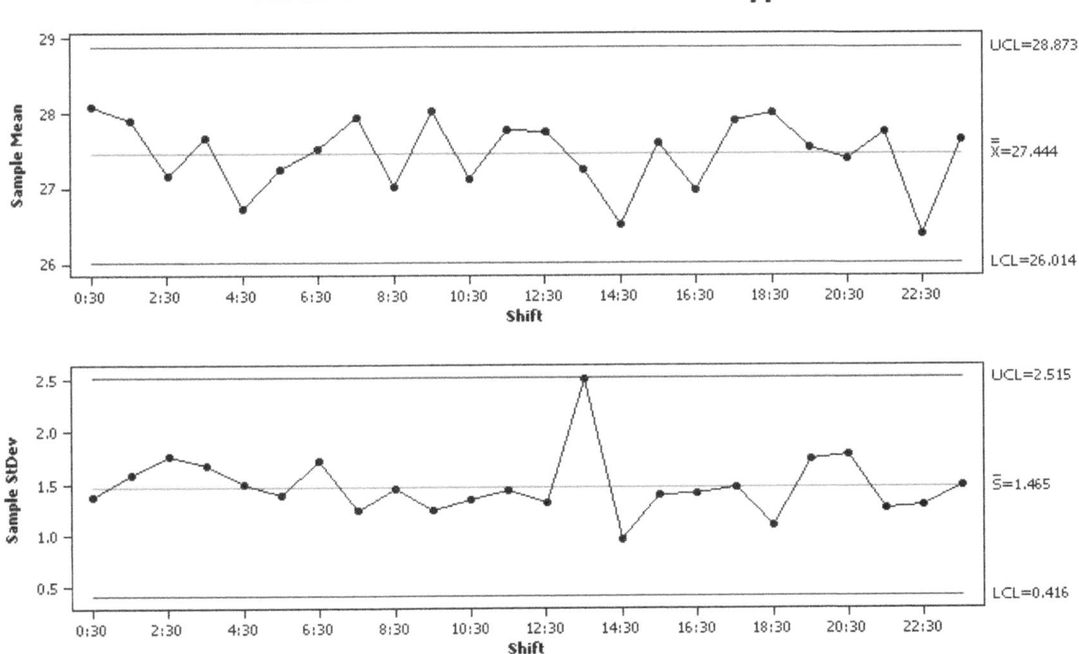

We will not repeat this example using the Assistant as the approach would be quite different and there would not be any new learning.

7.11 SPC Exercise

Exercise 3.

Willy Winker is developing a new chocolate coin and wants to understand the properties of the foil that the coins will be wrapped in. He understands that the pulse time of the electroplaters (P_Time) and concentration of the electrolyte (Conc) both affect the wear resistance of the gold plate on the foil wrapping.

In order to ensure both of these properties are in control and predictable before further development work is done Willy Winker wants confirmation using the appropriate control chart.

Produce control charts for both P_Time and Conc using the data provided within Worksheet: Coin. The data has already been put into subgroups. Use the Classic Method and the Assistant to produce the charts.

As P_Time has a subgroup size of 10 we should use an Xbar-S chart to check if the data is in control. We see that for P_Time both the Xbar and S charts are in control. Using the Classic Method or the Assistant produces the same charts.

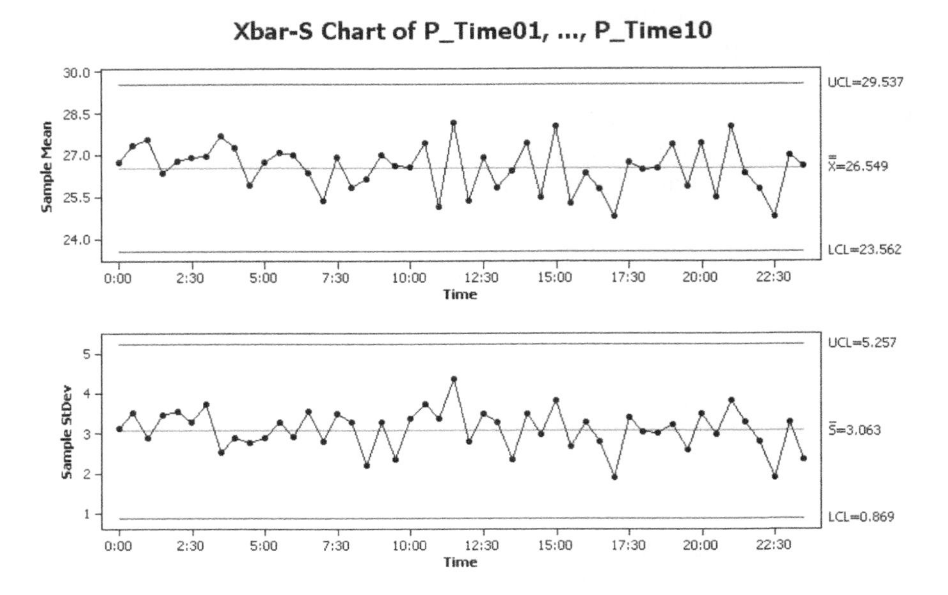

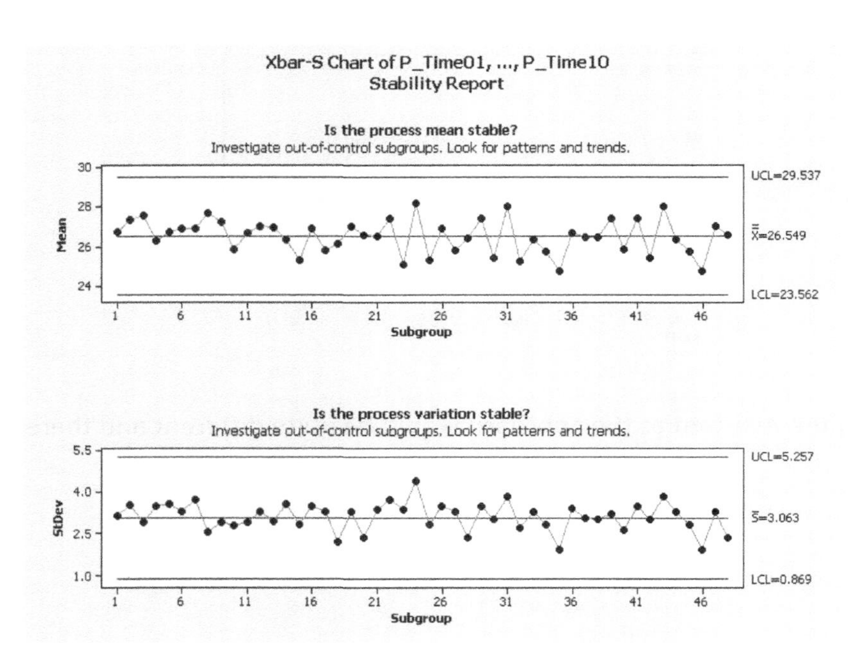

As Conc has a subgroup size of five we should use an Xbar-R chart to check if the data is in control. For Conc we see that both the Xbar and R charts are in control.

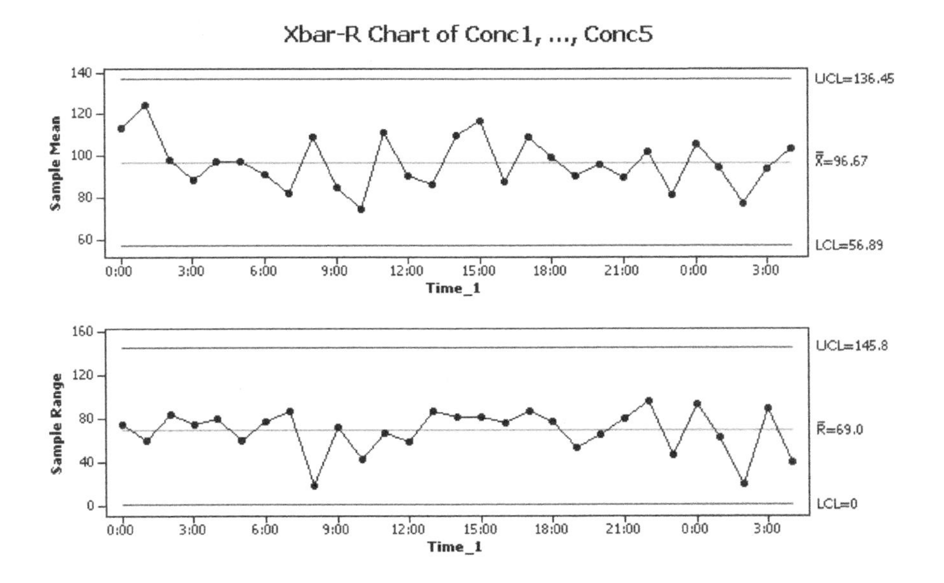

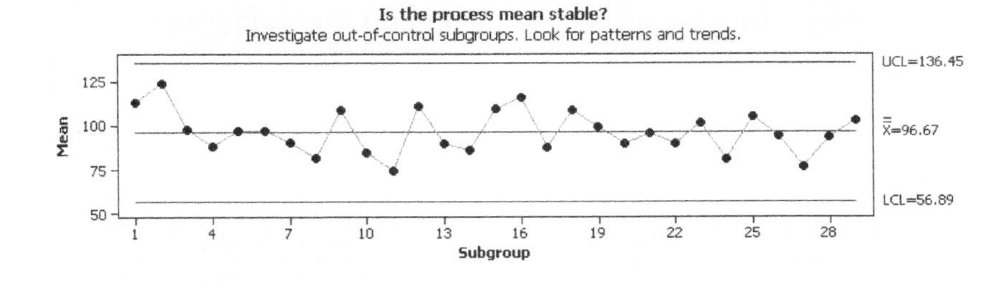

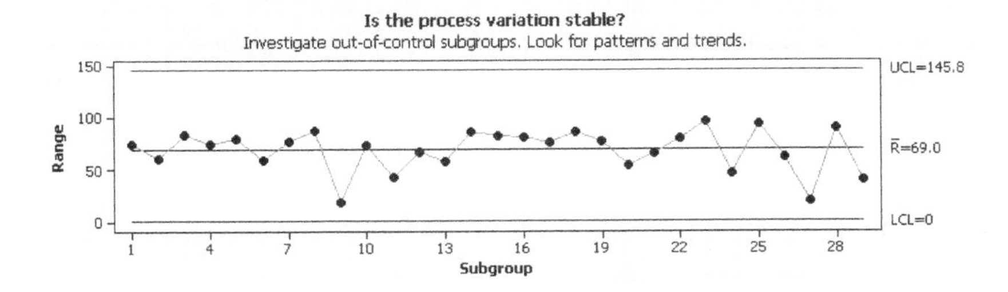

Knowing that there isn't any special cause variation that affects these products will help produce a stable foundation for further development work.

7.12 The I-MR-R/S Chart

Procedure	• I-MR-R/S Chart
What's it used for?	• It is used to detect changes in time ordered data where subgroups are being examined.
Assumptions and Limitations	• The I-MR-R/S chart is used when the data contains subgroups. However, the subgroups contain more variation than just the Within variation. • Using a Xbar-R or Xbar-S with this type of data could hide out of control points as the Xbar variation is based on the exaggerated within Subgroup variation. • The I-MR-R/S chart is not available in the Assistant.

Example 5. I-MR-R/S Chart.

In a bid to break Willy Winker's monopoly on cream cakes, Mahwish starts to mass produce her double jammed cream scone. She carefully monitors her process by checking one scone every hour. The diameter of the sample is checked in three planes, 120° apart and the results are recorded. Produce an Xbar-R chart of scone diameter and establish if the diameter is out of control. Repeat the exercise using an I-MR-R/S chart.

Worksheet: Scone.

1. Copy all columns of data from Worksheet: Scone into a new Minitab project worksheet.

Diameter1	Diameter2	Diameter3
26.187	24.906	26.049
26.157	27.210	26.019
25.894	24.628	26.038
[illegible]	[illegible]	[illegible]

2. Click Stat <<Control Charts <<Variables for Subgroups <<Xbar-R

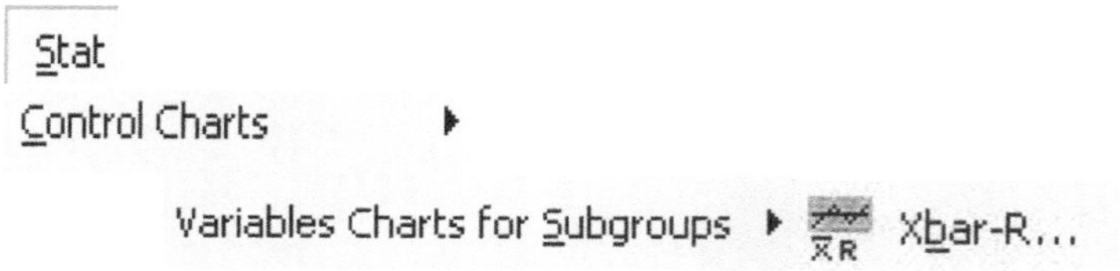

Stat
Control Charts ▸
Variables Charts for Subgroups ▸ Xbar-R...

3. Select 'Subgroup Observations in Rows' and then select the three columns of data as shown.
4. Click OK to produce the chart.

Observations for a subgroup are in one row of columns:

Diameter1-Diameter3

We see that the Xbar-R chart shows that the process is control in terms of the sample range and the subgroup mean.

However, on this occasion we know that the subgroup is capturing common cause variation and the variation between the three positions where diameter is measured. Therefore, the Within Group variation estimate is exaggerated. As we know that the limits on the Xbar chart are calculated using the Within Group Variation we know they will also be exaggerated. That is why we use the I-MR-R/S chart and it also highlights the need to be mindful when selecting subgroups.

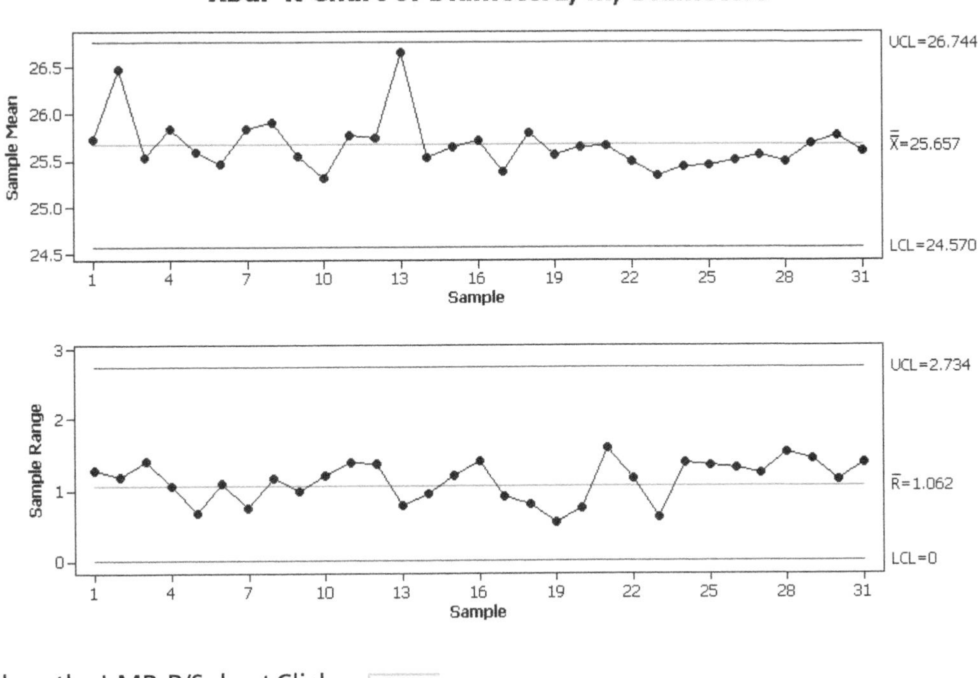

5. To produce the I-MR-R/S chart Click Stat <<Control Charts <<Variables for Subgroups <<I-MR-R/S.

Variables Charts for Subgroups

I-MR-R/S (Between/Within)...

6. Select 'Observations for a subgroup are in one row' and then select the three columns of data as shown. Notice that the menu and the sub-menus still carry the same format as the other charts we have produced using the Classic Method.
7. Click OK to produce the chart.

Observations for a subgroup are in one row of columns:

Diameter1-Diameter3

Scale... Labels...

Multiple Graphs... Data Options... I-MR-R/S Options...

The I-MR-R/S chart shows that the process is out of control. The chart is made up of three subgraphs. The top one shows the subgroup mean. The middle one shows the moving range of the subgroup mean. It is the short term variation of the moving mean that is used to calculate the control limits of the top graph.

The bottom chart gives the subgroup sample range. It is the same graph at the bottom of the Xbar R chart we previously produced. Minitab displays an R chart or an S chart depending on the chosen estimation method and the size of the subgroup.

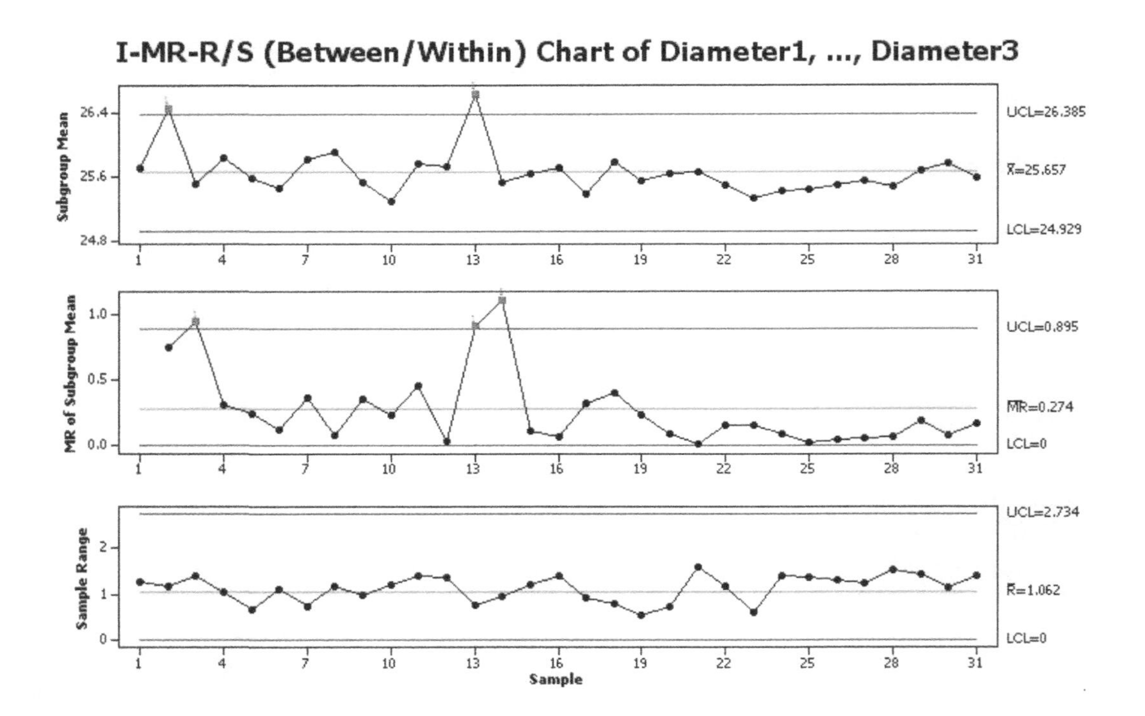

CHAPTER 8
Process Capability

8.1 The Basics of Process Capability

Process Capability studies are necessary to ensure that our process outputs meet our customer specifications.

We want to ensure the distribution produced by our process, Voice Of the Process (VOP) lies completely within the specification of the customer, Voice Of the Customer (VOC).

$$\text{Capability} = \frac{\text{Voice of the Customer}}{\text{Voice of the Process}} = \frac{\text{VOC}}{\text{VOP}}$$

Equation (8.1)

We use the following to measure this type of Process Capability:

$$C_p, C_{pk}, P_p, P_{pk}$$

Other measures include DPMO (%).

In order to introduce the basic concepts we will work through the first example and learn the terminology used within Process Capability.

Example 1. Dam Busters.

Sir Barnes Wallis developed the bouncing bomb in order to attack the Ruhr dams. There was a lot of development work done to ensure the best design was used. For this example let's pretend that we want to explore the process capability of the bomb design. We will consider the cylindrical bomb and the spherical bomb.

Explore the capability of both types of dam busting bomb when fired from a Lancaster bomber.

The distribution of bomb spread is given for the optimal firing distance, therefore, we need only consider the *x* dimension.

No data file.

Problem Solving and Data Analysis using Minitab: A clear and easy guide to Six Sigma methodology, First Edition. Rehman M. Khan.

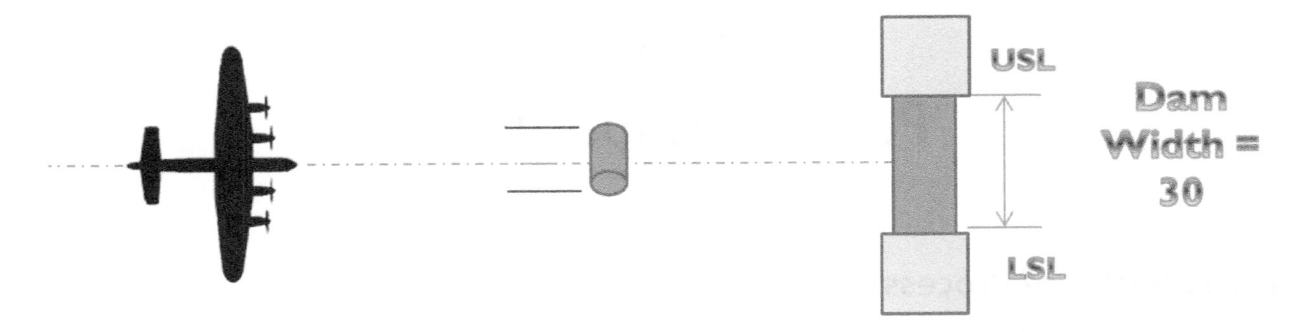

Let's start by looking at the cylindrical bomb. Let's say that the width of the dam is 30 m; we can say that Upper Specification Limit (USL) – Lower Specification Limit (LSL) is 30 m. This is the Voice Of the Customer (VOC).

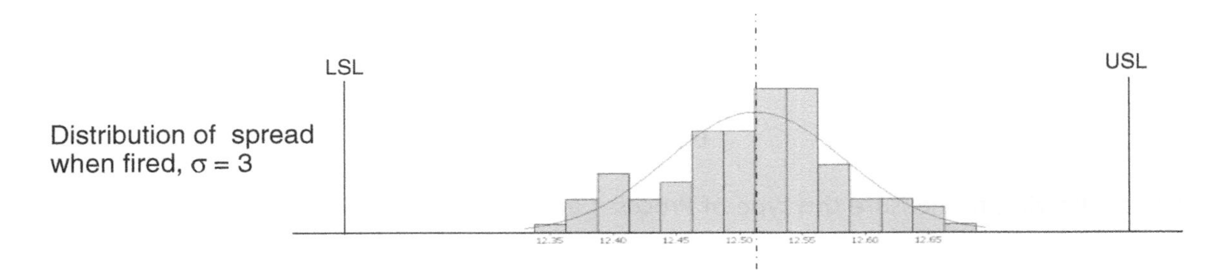

Following extensive development work it has been found that there is a spread of impact locations that the bomb will produce when it is fired from the optimal firing distance. This spread or distribution has been checked and it has been found that the StDev is 3 m. Let's say that the process we are looking at is normally distributed. This means that the Voice of the Process (VOP) is represented by 6σ. If this was not a normally distributed process we would not easily be able to predict the process spread. For this reason we are going to stick to normally distributed processes for the first half of this module.

$$C_p = \frac{VOC}{VOP} = \frac{USL - LSL}{6\sigma} = \frac{30}{6 \times 3} = 1.67$$

Equation (8.2)

Since we know the VOC and the VOP we can now calculate the capability. We get a value of 1.67.

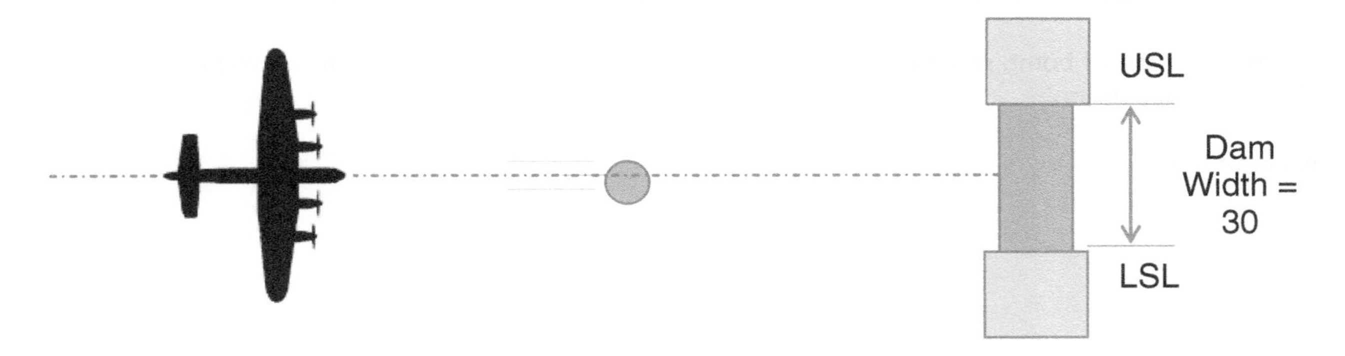

We will now calculate the capability of the spherical bomb. The VOC remains at 30 m because the dam width stays the same.

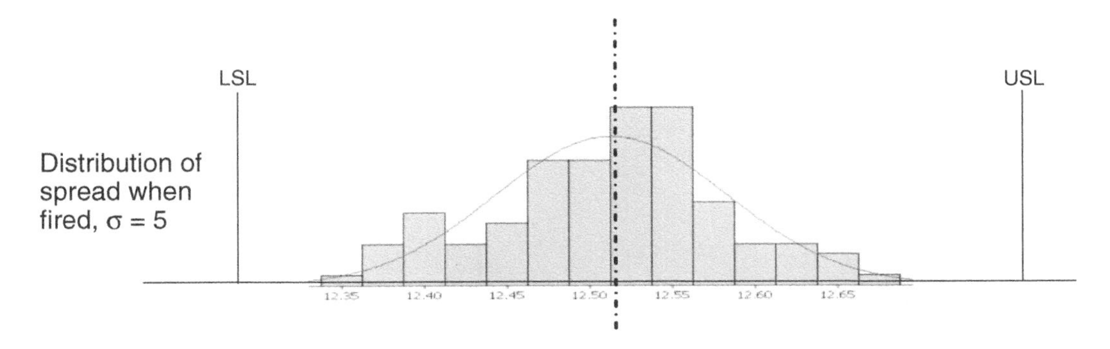

For the spherical bomb the distribution is checked and it is found that the StDev is 5 m. As the VOP is represented by 6σ it also has a value of 30 m.

$$C_p = \frac{VOC}{VOP} = \frac{USL - LSL}{6\sigma} = \frac{30}{6 \times 5} = 1.0$$

Equation (8.3)

We can calculate the capability and we get a value of 1.0. As VOC and VOP have the same value this means that the distribution just fits into the customer requirements. We are just going to break out of the example for a moment and give you some more detail on C_p levels and what they mean.

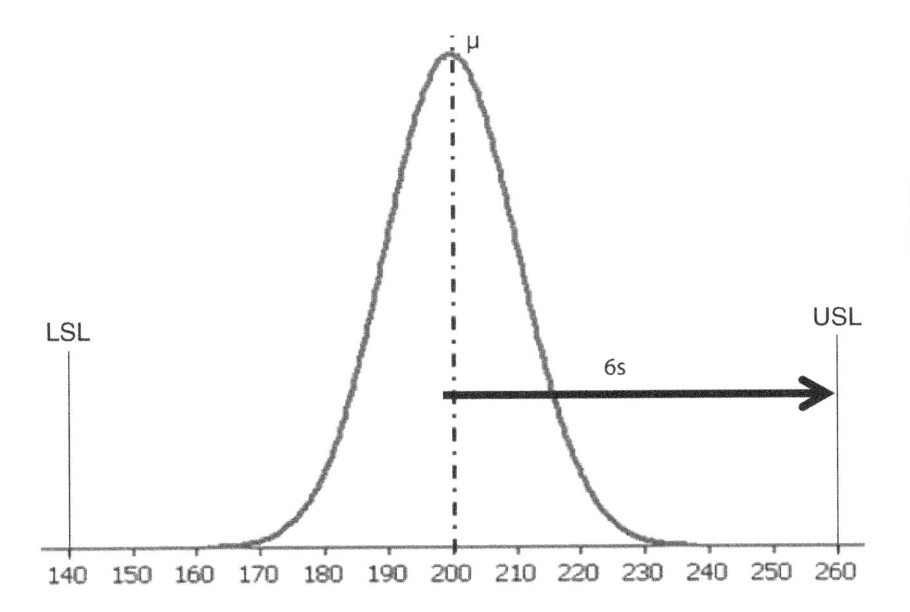

Sigma Level	C_p	DPMO
2	0.67	308 450
3	1.0	66 800
4	1.33	6 210
5	1.67	233
6	2.0	3.4

The diagram above shows a normal process distribution with a mean of 200 and StDev of 10. The Upper Specification Limit (USL) for this process is 260 and the LSL is 140. These both happen to be six StDevs from the mean. If we were lucky enough to own such a process we could declare our process to be 6σ capable or call it a 6σ process. By definition a 6σ process has a C_p of 2 and will produce 3.4 defects per million opportunities.

The table shows additional figures of varying sigma level and the corresponding changes in C_p and DPMO. Many industries use 1.33 as the target for C_{pk} or P_{pk} and this is a good starting target but the target should be process and industry specific.

We will get back into our example and consider what happens when the plane is not flying along the centre of the dam.

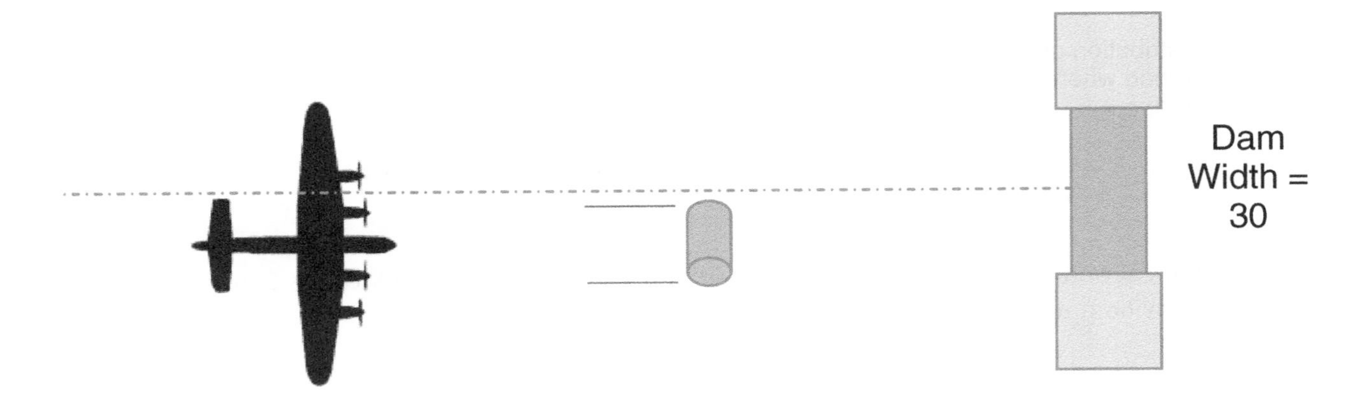

We are now going to go back to the cylindrical bomb and are going to consider what happens when the plane is no longer flying along the centre of the dam.

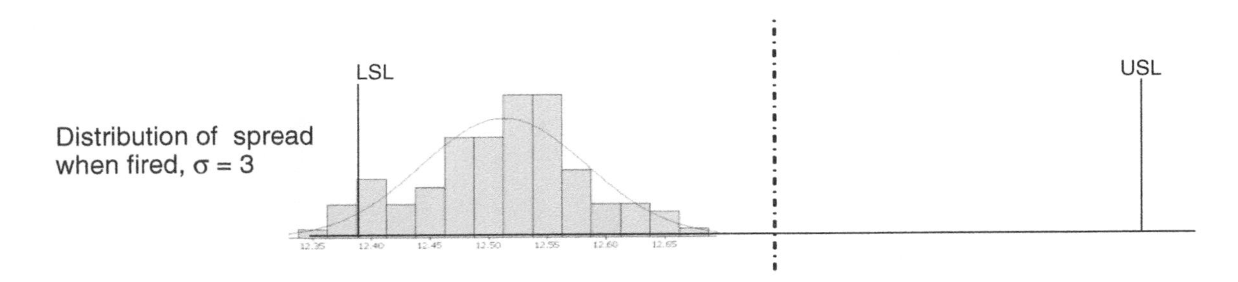

The width of the distribution is the same but as it is now off centre it cuts the lower specification limit.

$$C_p = \frac{VOC}{VOP} = \frac{USL - LSL}{6\sigma} = \frac{30}{6 \times 3} = 1.67$$

Equation (8.4)

However, the Capability Value is the same. This isn't good; we have a good capability value but as the process is not centred it will give results that are outside the specification limits. We need a new measurement for processes that are not centred.

$$C_{pu} = \frac{USL - \bar{x}}{3\sigma}$$

Equation (8.5)

$$C_{pl} = \frac{\bar{x} - LSL}{3\sigma}$$

Equation (8.6)

In order to get a more appropriate measure for processes that are not centred we split the calculation and measure the capability to each of the specification limits.

$$C_{pk} = \min(C_{pu}, C_{pl})$$

Equation (8.7)

C_{pk} then takes the value of whichever is lower out of C_{pu} is C_{pl}. As C_{pk} is a measure that looks at the nearest specification limit it is more appropriate when the distribution is not centered.

Let's have a look at some simple distributions and have a look at how C_p and C_{pk} vary for these distributions.

The diagram shows the case when the distribution just fits within the spec limits. Both C_p and C_{pk} are equal to 1.

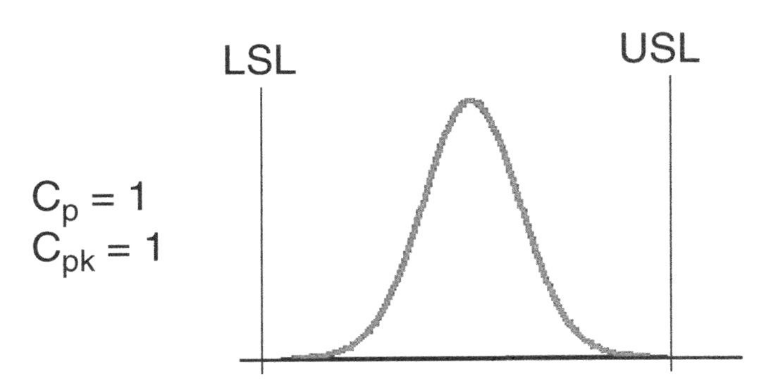

The diagram now shows the case when the mean of distribution is in line with the Upper Spec Limit. C_p is still equal to 1 because the VOP and VOC have not changed. However, C_{pk} is now equal to 0 because it reports the minimum value of either (USL – Xbar)/3σ or (Xbar – LSL)/3σ. In the case of the diagram the difference between the mean and the USL is zero. When C_{pk} is zero, yield is 50%.

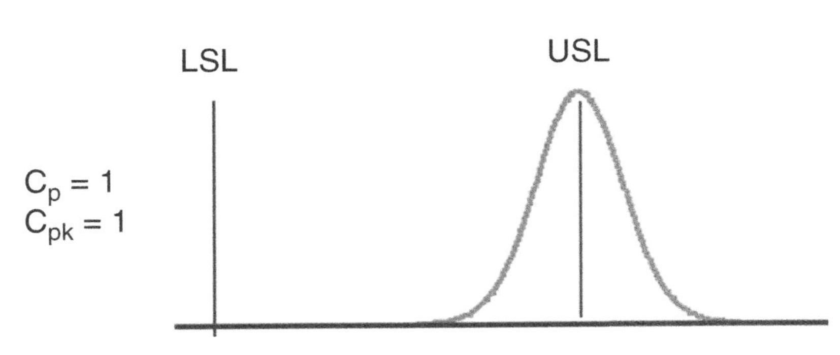

The diagram now shows that the distribution has gone just beyond the USL. C_{pk} is now −1 because the mean is 3σ away from the USL. Therefore, USL – Xbar = −3σ.

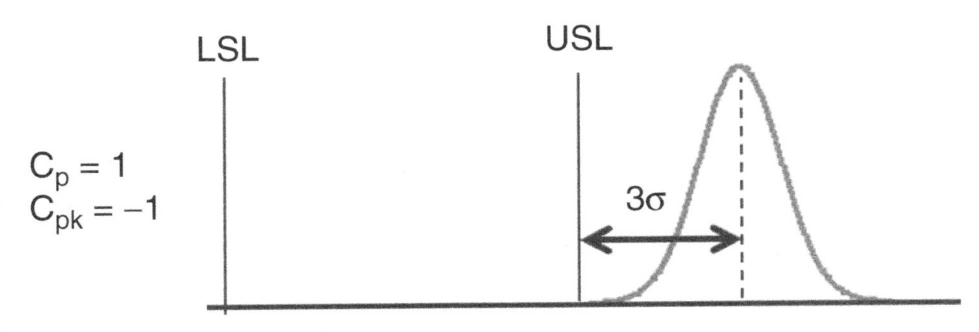

At this point everything that is being made is rejected.

8.2 Short Term and Overall Capability

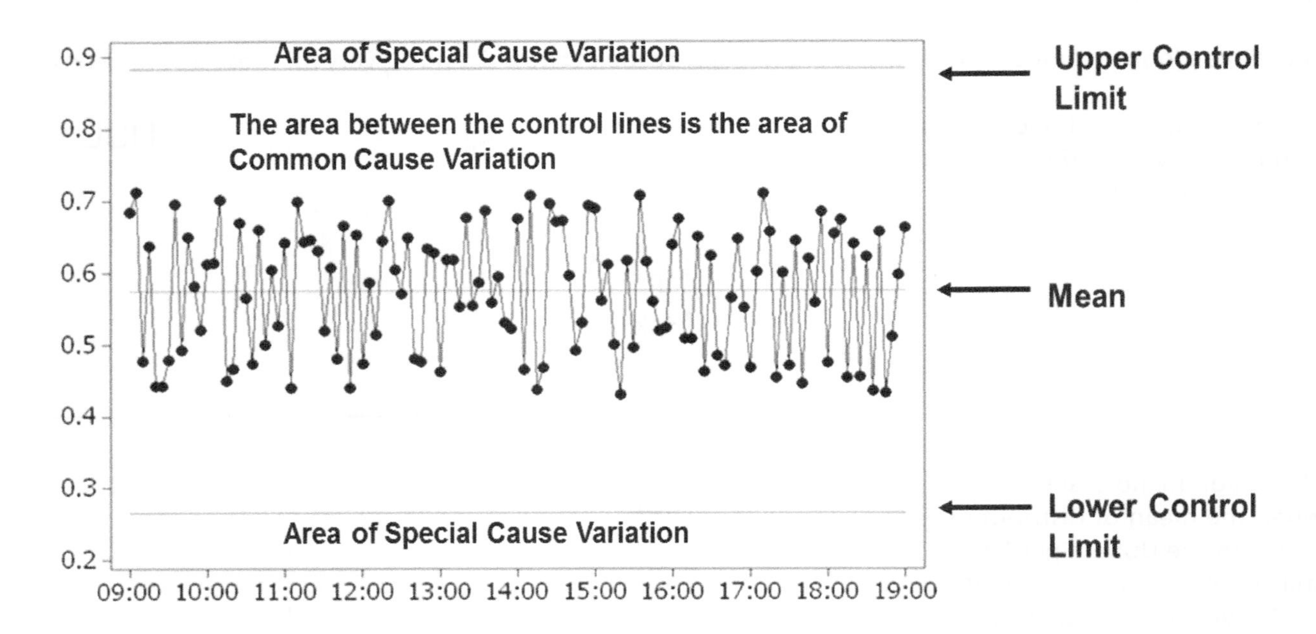

Process Capability has strong links with SPC, indeed you will see a lot of control charts used within this module. One of the main principles that is carried over is that of common cause and special cause variation. As stated earlier due to our clever sampling using subgroups we will see the short term variation captured within the subgroups and the special cause variation should be captured between the groups.

When measuring Process Capability C_p and C_{pk} are measures of short term capability and P_p and P_{pk} are measures of overall capability.

If a process is displaying only common cause variation with no special cause variation then $C_{pk} = P_{pk}$, approximately. If the process is displaying special cause variation then C_p and C_{pk} figures are only achievable if we manage to eliminate all the special cause variation.

We use Process Capability to benchmark our processes and sometimes make estimates of the yield we can expect. If there is special cause variation then the process is not stable and not predictable. In which case the capability figures only apply to the data set they were derived from. To make predictions there must be predictability within the process and that means no special cause variation.

8.3 Capability Analysis for Normal Data

Procedure	• Capability Sixpack and Analysis (Normal)
What's it used for?	• The Sixpack is used in order to help us understand whether the process data is normal and in control. • Capability Analysis (Normal) is used to calculate capability metrics for normal data.
Assumptions and Limitations	• The data should be time ordered. • Having the data in subgroups may give a better understanding of within and overall capability. • The Sixpack uses Control Charts and Probability plots to ensure the validity of the capability results. It is usually used prior to the Capability Analysis procedure to ensure the process is normal and in control. • In order to use the results for a given data set to predict future performance the process must be in control.

Example 2. Normal Capability.

Willy Winker's caramel filled chocolate hearts must have a size between 32 and 28 mm, inclusive, so that they can fit into the box correctly. A number of samples have been taken from the line and measured. The data has a subgroup size of three and is time ordered.

Use the capability Sixpack and then Normal Capability Analysis to assess the process capability. Use both the Classic Method and the Assistant.

All data sets are in spreadsheet 08 Process Capablity.xls

Worksheet: Hearts.

1. Open the Excel file and transfer the data into a new Minitab project worksheet.

C1	C2	C3
Heart Width1	Heart Width2	Heart Width3
29.03	29.58	29.42
29.85	29.51	29.22
29.90	30.40	30.64
30.13	30.72	29.71

Initially, we do not know if the data has a normal distribution. Luckily the Capability Sixpack contains a normal probability plot as well as control charts to tell us if the process is in control. Therefore, it makes a good starting point for the Capability Analysis when using the Classic Method.

2. Click Stat <<Quality Tools << Capability Sixpack << Normal.

Stat

Quality Tools ▸

Capability Sixpack Normal...

3. Our data is laid out with the subgroups within rows so we need to select the appropriate radio button and then select our data columns.
4. We enter the values of the lower and upper spec limits and then click OK to produce the Capability Sixpack.

Data are arranged as

Single column:

Subgroup size:

(use a constant or an ID column)

Subgroups across rows of:

'Heart Width1'-'Heart Width3'

Lower spec: 28

Upper spec: 32

Historical mean: (optional)

Historical standard deviation: (optional)

Transform...

Tests...

Estimate...

Options...

OK

Cancel

The Sixpack has been split into two sections to allow the inclusion of the explanation against the sections. We see an Xbar and R chart of our data. We can see that our data is in control. As the data is in control the overall variation should contain predominantly short term variation. This means that the values for the short term metrics and overall metrics should be similar.

Below the control charts we have a chart displaying the actual values of the points within the last 25 subgroups. We can inspect for outliers within the subgroups and whether they are equally spaced around the mean. We can also check for changes in distribution within the groups.

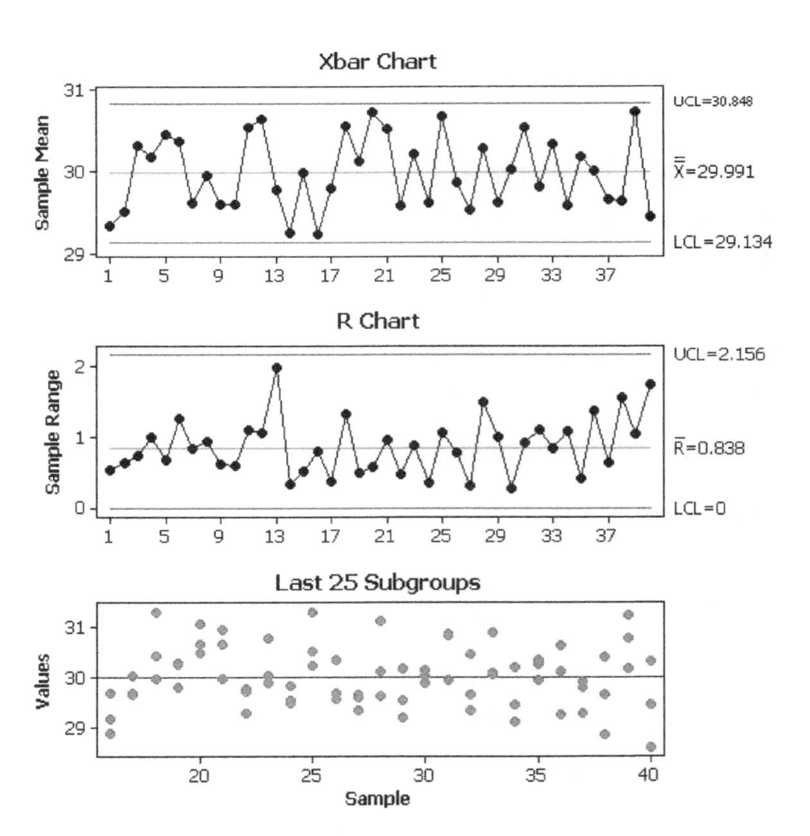

The Capability Histogram shows our distribution which is within the Spec limits. The distribution appears to be centred within the spec limits. These are both good signs. There are also two lines of fit, one is a black dashed line which represents the overall estimate and the red unbroken line represents the within estimate.

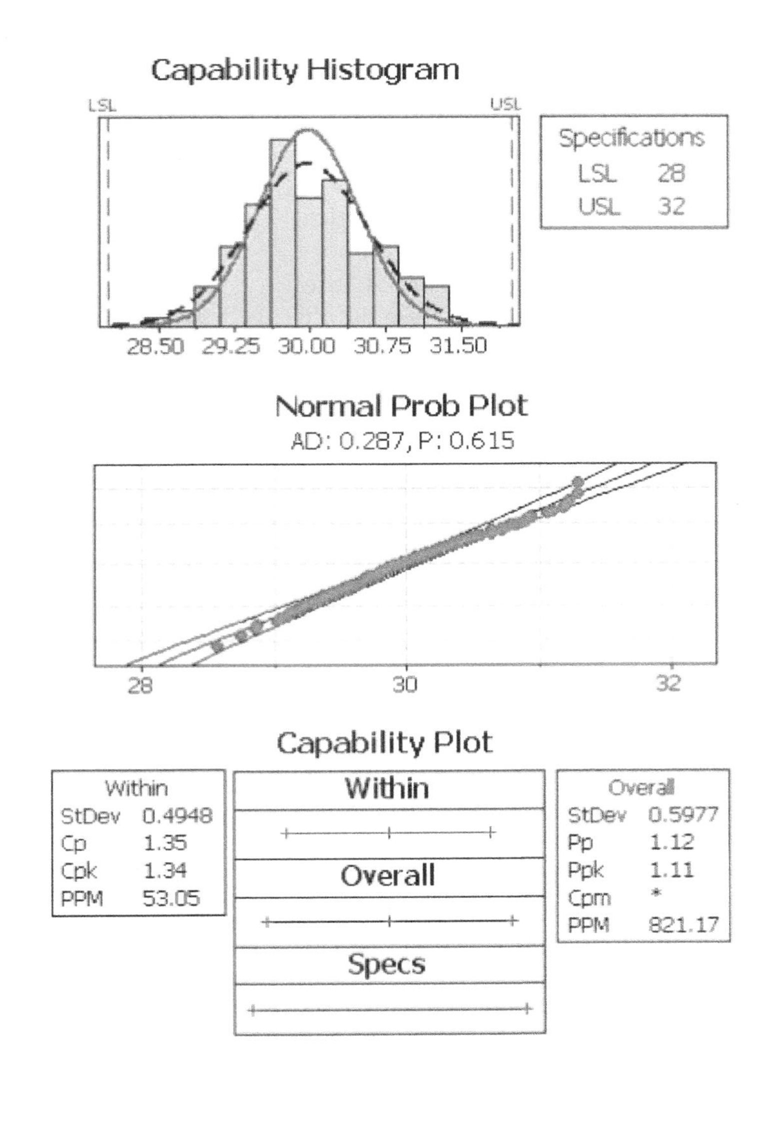

The probability plot shows that we are dealing with a normal distribution as the *P* value above the plot is 0.615. This validates our initial decision to use the Sixpack for normal distributions.

The summary data gives us our capability metrics. In the centre we see lines representing the width of the process and the specs which are representative of the widths seen on the Capability Histogram. We then have a box on the left containing the within metrics and one on the right containing the overall metrics. The within StDev is 0.4948 which is estimated from the within subgroup variation. The C_p and C_{pk} are calculated from this. The PPM is the number of defective parts per million we would get based on the within variation. The overall StDev is higher as it is calculated globally and therefore P_{pk} is lower than C_{pk}. As it stands, we would expect to have 821 defective parts per million but we don't know whether they would fail the upper or the lower specification limit.

Now that we know we have a normal distribution we can use the Capability Analysis for normal distributions to get more detailed information on the capability metrics.

5. Click Stat <<Quality Tools <<Capability Analysis << Normal.

Stat

Quality Tools ▶

Capability Analysis Normal...

6. Our data is laid out with the subgroups within rows so we need to select the appropriate radio button and then select our data columns.
7. Enter the LSL and USL. Notice that here we have the opportunity to set either of the specs as a Boundary, we didn't have that option with the Sixpack. Ticking the Lower spec Boundary tick box will tell Minitab that it is not physically possible to have parts below 28, therefore it won't calculate defective parts for the boundary. It won't calculate any capability metrics associated with that limit either.
8. Then click on options.

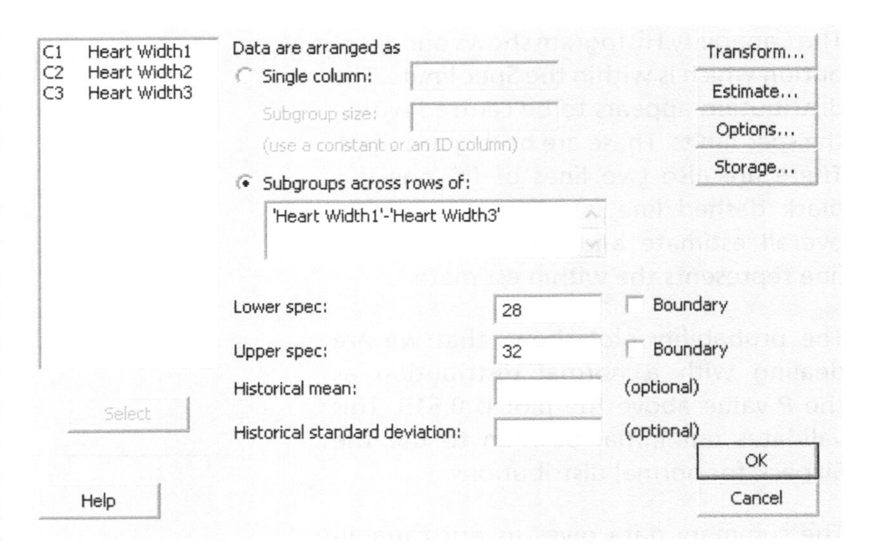

9. Under Display set the radio button onto Percents (this is a personal preference as I prefer percentage data over PPM but I suppose the choice is also dependent on how well your processes work). There is also an option here to include the confidence intervals for P_{pk} and C_{pk}. We can also enable the Cpm metric which is a new overall metric which measures the central location our distribution.
10. Then click OK and OK again.

Target (adds Cpm to table):
Use tolerance of K*sigma for capability statistics K = 6
Perform Analysis
Within subgroup analysis
Overall analysis
Display
Parts per million
Percents
Capability stats (Cp, Pp)
Benchmark Z's (sigma level)
Include confidence intervals

The detailed capability analysis is given here. The main difference between this and the Capability Sixpack is that information on Observed and Expected performance is given. We also see the breakdown of whether most of our losses are at the LSL or USL. It appears that the expected overall performance attributes the losses equally between the specs. It is interesting to note that the observed losses were 0% at either spec. We should also note that the expected performance is a valid prediction of future performance as our process is in control.

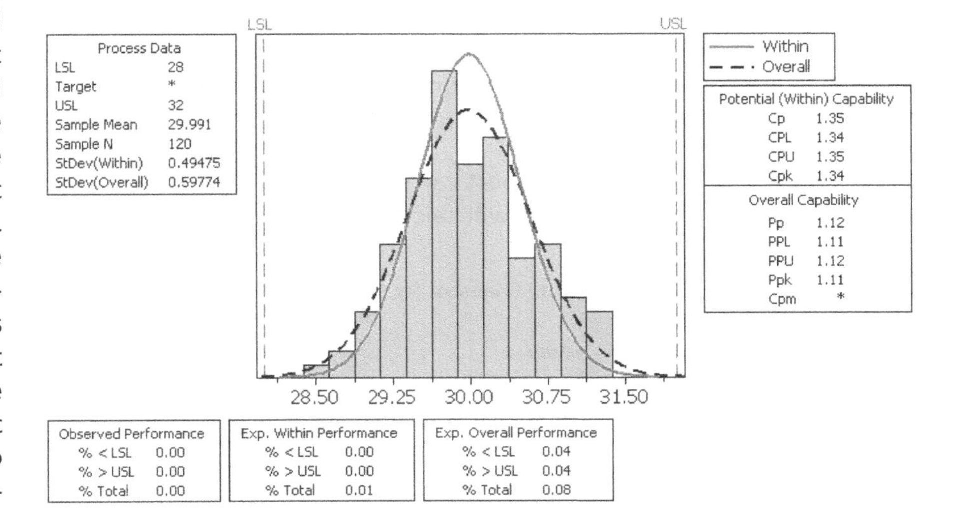

We will now use the Assistant to carry out the same capability study. The Assistant offers two types of analysis for continuous data. It offers Capability Analysis which gives results similar to those we have just seen. It also offers Capability Comparison which is used to compare two sets of capability data. This is obviously very useful when showing the improvements made within a project.

Assistant

Capability Analysis...

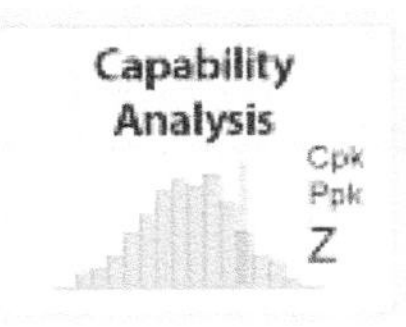

1. Click Assistant <<Capability Analysis.
2. Click on the Capability Analysis Icon.
3. We have a choice of two types of analysis. For now we will select the Snapshot.
4. Select Data in multiple columns then enter the columns where the data is stored. The snapshot does not calculate the within metrics so it will treat the three columns as one.
5. Enter the LSL and USL. Then click OK.

Type of analysis

Complete

Snapshot (only use with individual data that are not in time order)

Process data

How are your data arranged in the worksheet?

Data are in multiple columns, all data are considered one sample

Columns: 'Heart Width1'-'Heart Width3'

Specification limits (at least 1 required)

Lower spec: 28

Upper spec: 32

Mean test (optional)

Enter the target value for the process mean.

Target:

The Snapshot shows us a histogram of our data within the spec limits. Below that is a Normality Plot with the results of the AD normality test.

The Process Characterization gives us all the metrics that we have seen previously with the exception of Z.Bench. This is in simple terms defined as the capability experienced by the customer in units of σ. Where 6 is excellent and 0 is 100% defects.

The Snapshot also generates a report card which is not shown here. For a more detailed analysis we will use the Complete Analysis which will calculate Within Metrics, as shown next.

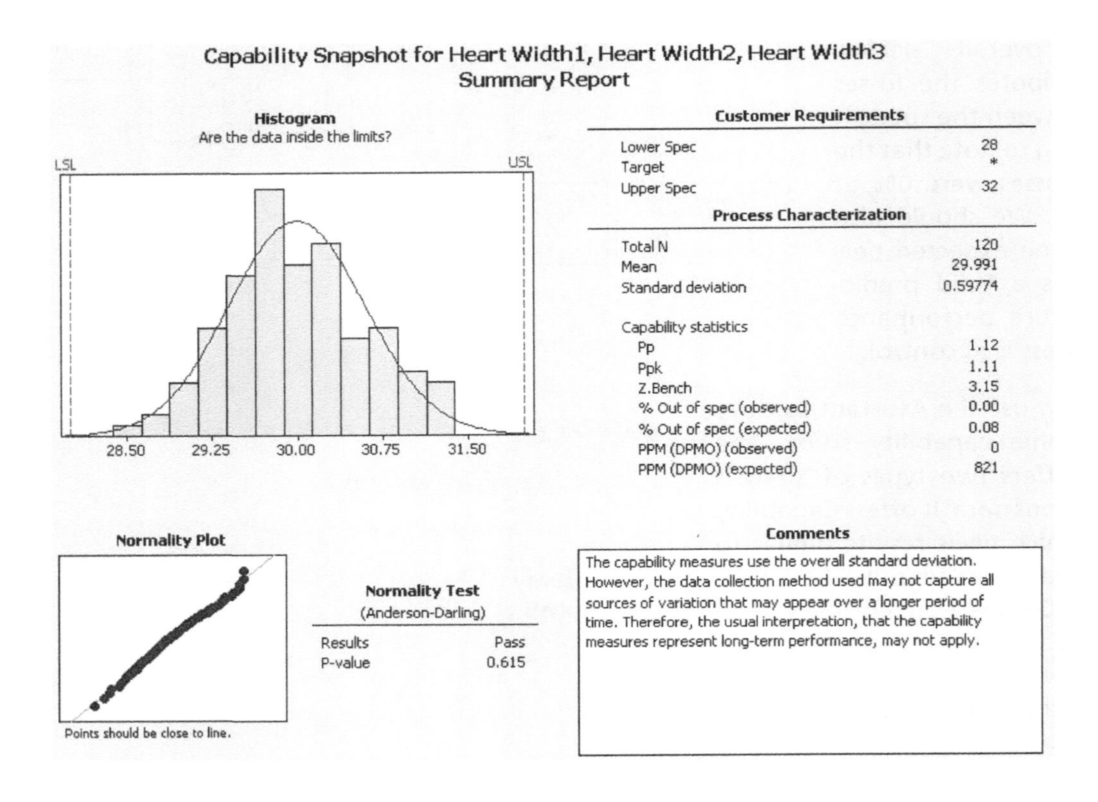

1. Click Assistant <<Capability Analysis.

Assistant

Capability Analysis...

2. Click on the Capability Analysis Icon.

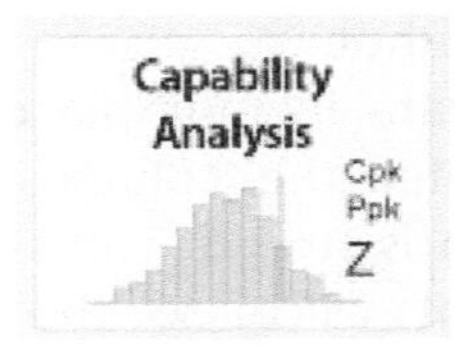

3. In order to complete the comparison we will use the Complete Analysis on this occasion.
4. Select Data in multiple columns then enter the columns where the data is stored.
5. Enter the LSL and USL. Then click OK.

Type of analysis

- (•) Complete
- ( ) Snapshot (only use with individual data that are not in time order)

Process data

How are your data arranged in the worksheet?

Data are in multiple columns, one row for each subgroup

Columns: 'Heart Width1'-'Heart Width3'

Specification limits (at least 1 required)

Lower spec: 28

Upper spec: 32

The Summary Report is very similar to the Snapshot Summary Report. The exception is the Z.Bench indicator. We see that our process is about average on this indicator. The Summary Report again gives metrics on the Overall Capability.

Capability Analysis for Heart Width1, ...
Summary Report

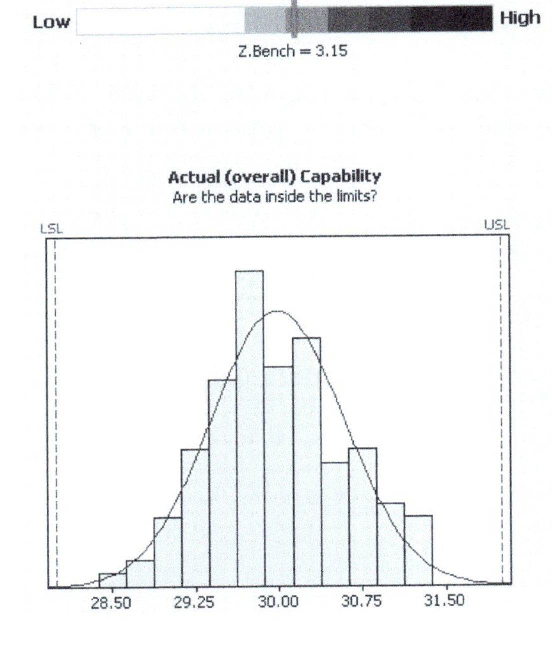

Customer Requirements

Upper Spec	32
Target	*
Lower Spec	28

Process Characterization

Mean	29.991
Standard deviation	0.59774
Actual (overall) capability	
Pp	1.12
Ppk	1.11
Z.Bench	3.15
% Out of spec	0.08
PPM (DPMO)	821

Comments

Conclusions
-- The defect rate is 0.08%, which estimates the percentage of parts from the process that are outside the spec limits.

Actual (overall) capability is what the customer experiences.

The Process Performance Report shows the histogram again with both the Overall and Within fits. Capability Metrics for both Within and Overall are given on the opposite side.

The Diagnostic Report which is not shown here shows us the SPC charts and Normality test results.

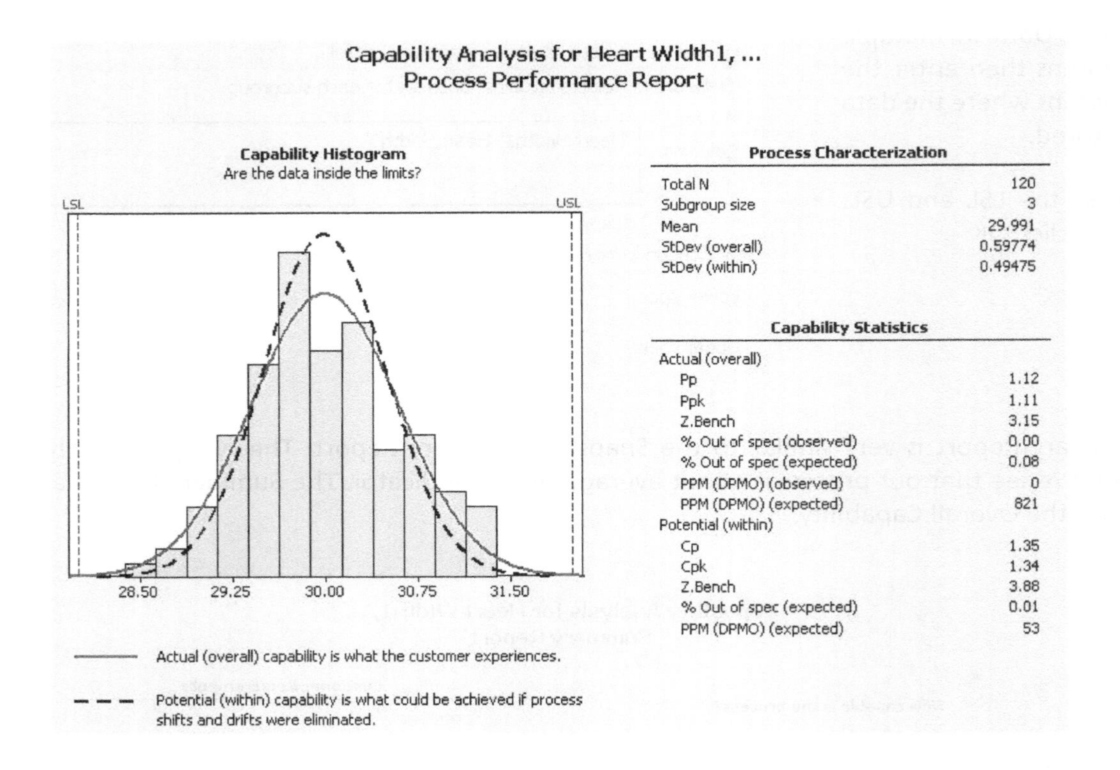

The Report Card gives information about the validity of the test procedure. As we have provided more than 100 observations in 40 subgroups Minitab is satisfied with the likely precision of our results.

Capability Analysis for Heart Width1, ...
Report Card

Check	Status	Description
Stability	✓	The process mean and variation are stable. No points are out of control.
Number of Subgroups	i	You have 40 subgroups. For a capability analysis, this is usually enough to capture the different sources of process variation when collected over a long enough period of time.
Normality	✓	Your data passed the normality test. As long as you have enough data, the capability estimates should be reasonably accurate.
Amount of Data	✓	The total number of observations is 100 or more. The capability estimates should be reasonably precise.

Exercise 1. Normal Capability.

Omer checks the weight of nuclear fuel rods. He collects data from two samples every hour for a capability study. LSL = 19.50 kg and USL = 20.50 kg.

Use the capability Sixpack to establish if the process is in control and normally distributed. Then conduct the full Capability Analysis using the Classic Method. Then carry out the same task using the Complete Analysis on the Assistant.

Worksheet: Fuel Rods.

As indicated in the question the subgroup size needs to be set to two. From the Sixpack we see that there are no issues with the process being out of control. The Capability Histogram looks worrying as the right edge of the distribution intersects with the USL.

We also see that the process is normally distributed with a *P* value of 0.562.

The Capability Plot also indicates that the process may not be centred.

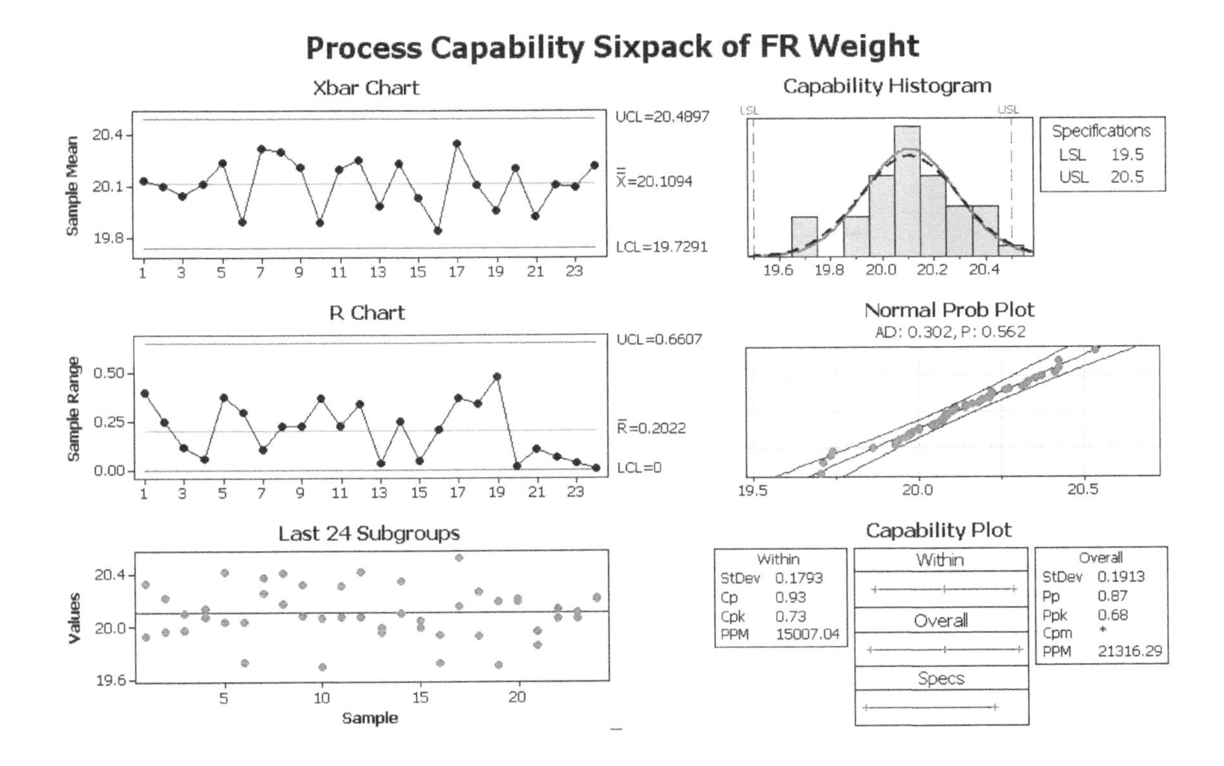

It can be seen that both P_p and C_p are less than one, indicating that the distribution will not fit within the specification limits even if it was centred. The Observed Performance is better than the Expected Performance. The Expected Overall Performance indicates that 2.06% of parts would be rejected due to the USL and 0.07% would be rejected due to the LSL.

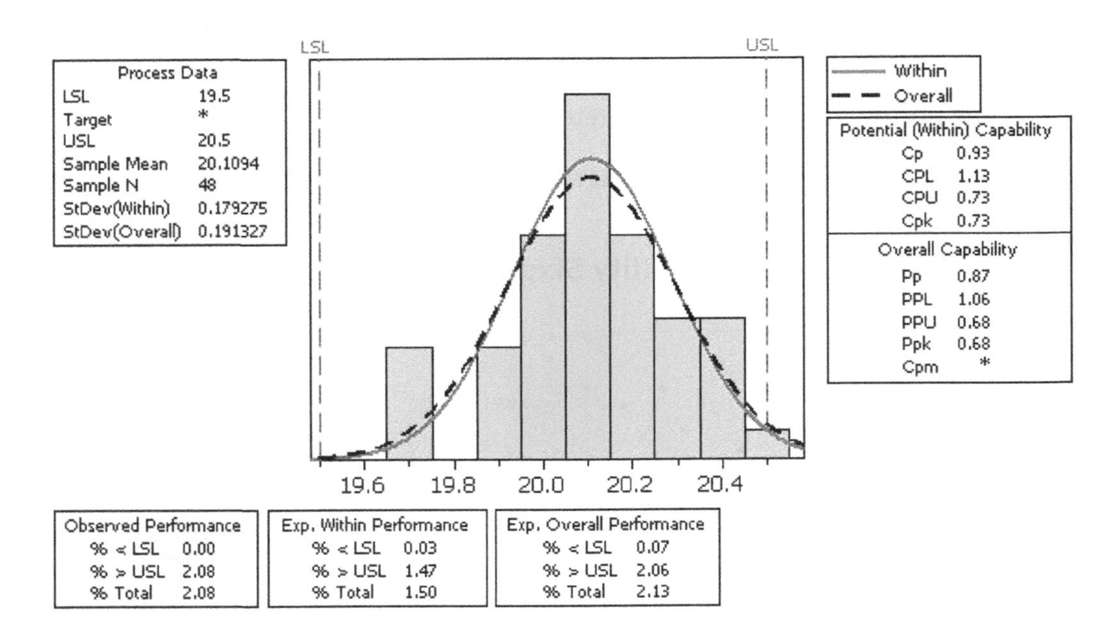

The Process Performance Report shows the histogram again with both the Overall and Within fits. Capability Metrics for both Within and Overall are given on the opposite side. As our C_{pk} is significantly less than the C_p this confirms that the distribution is not centered between the spec limits.

The Diagnostic Report, which is not shown here, shows us the SPC charts and Normality test results.

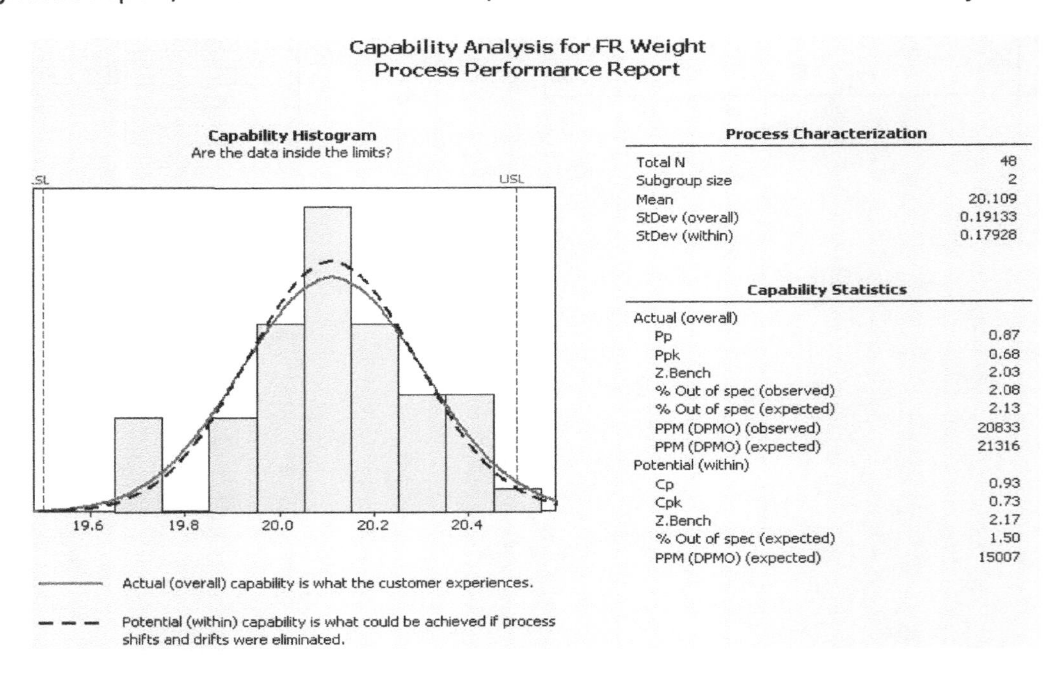

On this occasion the Report Card gives us a warning about the number of observations. We are also given information about the number of subgroups being slightly low. Stability and Normality are not an issue.

Capability Analysis for FR Weight
Report Card

Check	Status	Description
Stability	✓	The process mean and variation are stable. No points are out of control.
Number of Subgroups	i	You only have 24 subgroups. For a capability analysis, it is generally recommended that you collect at least 25 subgroups over a long enough period of time to capture the different sources of process variation.
Normality	✓	Your data passed the normality test. As long as you have enough data, the capability estimates should be reasonably accurate.
Amount of Data	!	The total number of observations is less than 100. You may not have enough data to obtain reasonably precise capability estimates. The precision of the estimates decreases as the number of observations becomes smaller.

8.4 Capability Analysis for Non Normal Data

As we saw in the SPC module if we are faced with non normal data we first have a critical decision to make. Using our skill, judgement and experience we must make a decision about the nature of our process. Is it a process that should produce normal data but it has some special causes acting on it or does this process align to a different type of distribution?

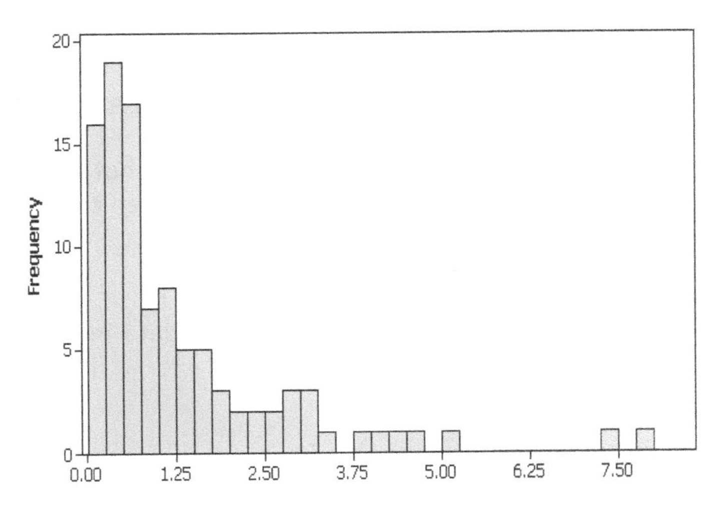

If we think that our data should be normal and it isn't close to being normal, as in the diagram shown, we should not be calculating capability metrics for it. We should be finding out what is doing this to the process.

To highlight the point, let's say you are using non normal data techniques to calculate capability metrics within an improvement project. Without making any major changes to the process you collect more data and then find that process requires a different transformation than the one you started with or the data turns normal. This means you probably got it wrong to start with. You have special cause variation influencing your process.

The remainder of this section deals with distributions that produce non normal data inherently.

The problem with the Normal Capability checks performed on non normal data is that to calculate the Voice of the Process the data is fitted with a normal curve. This means that the non normal distribution will be a fitted with a curve that does not fit.

In order to get past this we have a couple of options, we can transform the data and then try and fit the data with a normal curve. Minitab uses the Johnson and the Box–Cox transformations. The Johnson transformation is a complex function but it usually finds a suitable transformation. The Box–Cox transformation is much simpler but it does not always find a suitable transformation.

Or we have the option of applying a different type of distribution to our data. Both methods will be illustrated within the examples.

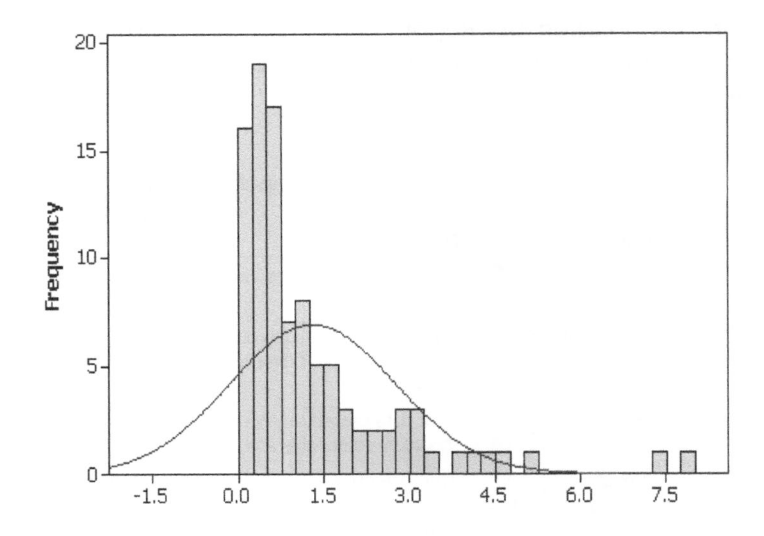

Example 3. Non Normal Capability.

Prawn is carrying out a project to improve the yield of fire rated plasterboard. He has collected data at the start of the project and wants to conduct an initial capability analysis. The thickness specification of fire rated plasterboard is very strict and is between 12.80 and 13.00 mm.

It is known from previous studies that this process does not produce normal data.

Initially, use the Assistant and check whether that will transform the data. Also use the Classic Method to find the appropriate transformation and then conduct a Non Normal Capability Analysis.

Worksheet: PB Thickness.

1. Open the Excel file and transfer the data into a new Minitab Project Worksheet.

PB Thick
12.863
12.895
12.852
12.870
12.954

2. For this example we are going to begin by using the Assistant. Click Assistant <<Capability Analysis.

Assistant

Capability Analysis...

3. Click on the Capability Analysis button.

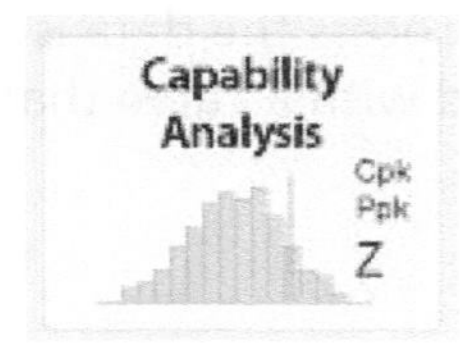

4. Select the Complete Analysis. Then complete the menu box as shown. Note, the data was not collected in subgroups, therefore the Subgroup size is listed as 1.
5. Enter the Spec limits and then click OK.

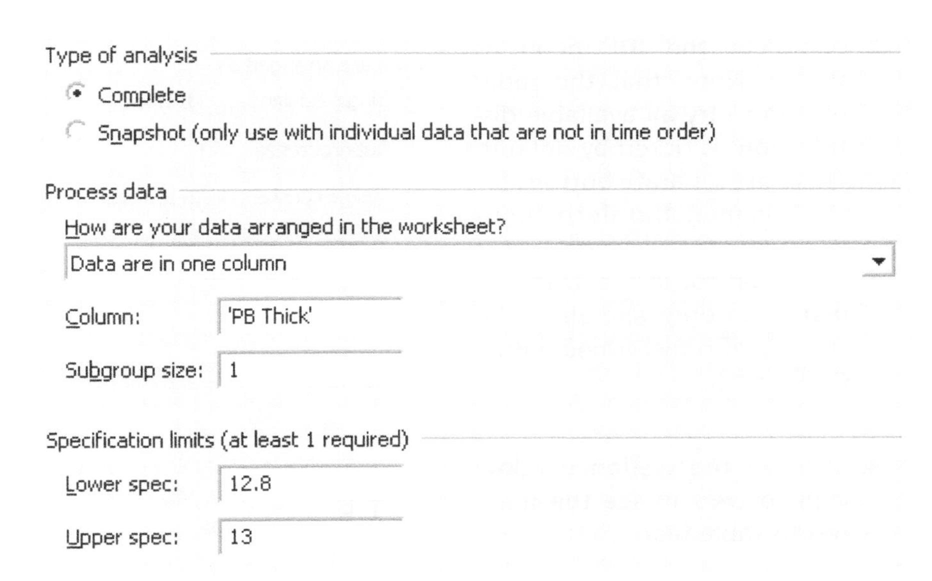

6. Notice that the Assistant has not transformed the data for this procedure. We know this as it did not display a confirmation box for the transformation. Also, look at the Normality Plot on the Diagnostic Report, the data is clearly not normal. The Assistant did not transform the data as the Box–Cox transformation would not have delivered a suitable transformation.

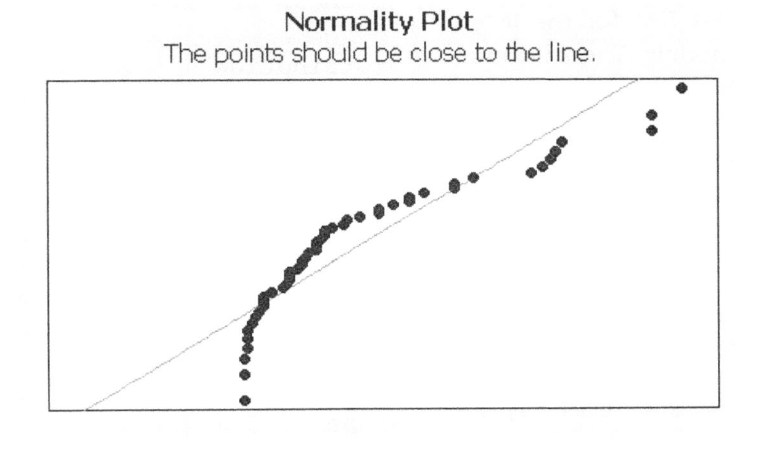

Your data failed the normality test. A transformation will not correct the problem. Get help to determine next steps because the capability estimates may be inaccurate.

The Report Card also generates the warning message displayed in the figure above. The warning message is not strictly true as it only refers to the Box–Cox transformation.

We will now use the Individual Distribution Identification (IDI) procedure using the Classic Method. This procedure tests the fit of a number of distribution models and tests the two common transformations.

1. Click Stat <<Quality Tools <<Individual Distribution Identification

2. Complete the IDI menu as shown. Note, that the radio button to try all available distributions is ticked by default. There are separate buttons for the common transformations in order to allow the user more control in the transformation settings and allow the storage of transformed data.
3. Click on OK.

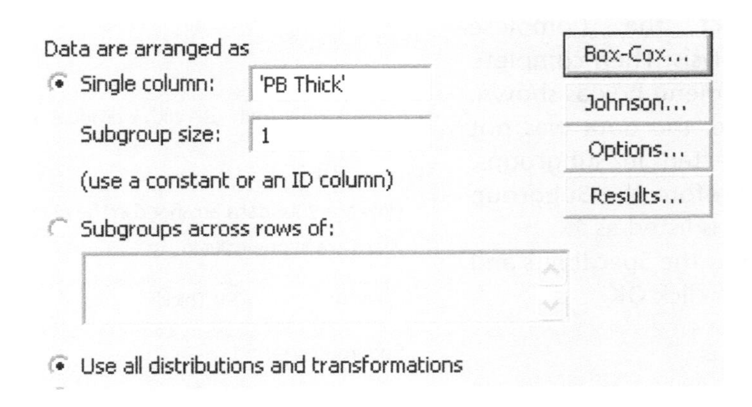

4. Click on the session window icon so we can see the main results table first.

The *P* values given here are for the Goodness of Fit Test for the listed distribution models. These include the two common transformations. We can see that the Johnson transformation gives the best *P* value, 0.671. Initially, we will use this transformation for the capability analysis. Although, there are other models which give an acceptable fit like the three parameter Weibull which has a *P* value of >0.500.

The Likelihood Ratio Test (LRT) is a hypothesis test used to differentiate between distributions which have a different number of parameters. For example, a *P* value of 0.0 indicates that the three parameter Weibull was better than the two parameter.

Goodness of Fit Test

Distribution	AD	P	LRT P
Normal	2.795	<0.005	
Box-Cox Transformation	2.731	<0.005	
Lognormal	2.785	<0.005	
3-Parameter Lognormal	0.546	*	0.000
Exponential	22.835	<0.003	
2-Parameter Exponential	0.479	>0.250	0.000
Weibull	4.045	<0.010	
3-Parameter Weibull	0.326	>0.500	0.000
Smallest Extreme Value	4.058	<0.010	
Largest Extreme Value	1.528	<0.010	
Gamma	2.811	<0.005	
3-Parameter Gamma	0.312	*	0.000
Logistic	2.259	<0.005	
Loglogistic	2.251	<0.005	
3-Parameter Loglogistic	0.764	*	0.000
Johnson Transformation	0.268	0.671	

5. Click on the Graphs icon to see the graphical output of the IDI procedure.

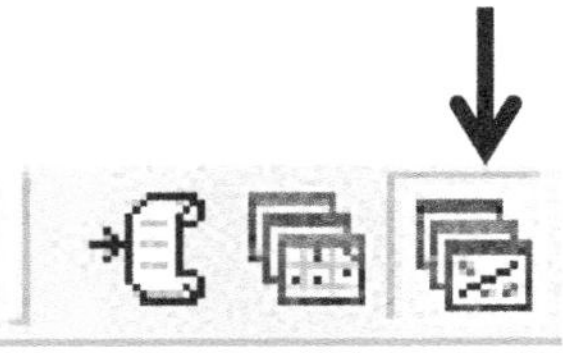

There are four pages of graphical output. Only one of the pages is shown here.

It can be seen that the data has an almost normal fit after the Johnson transformation.

Now that we have identified that we wish to use the Johnson transformation we will carry out an initial assessment using the Capability Sixpack.

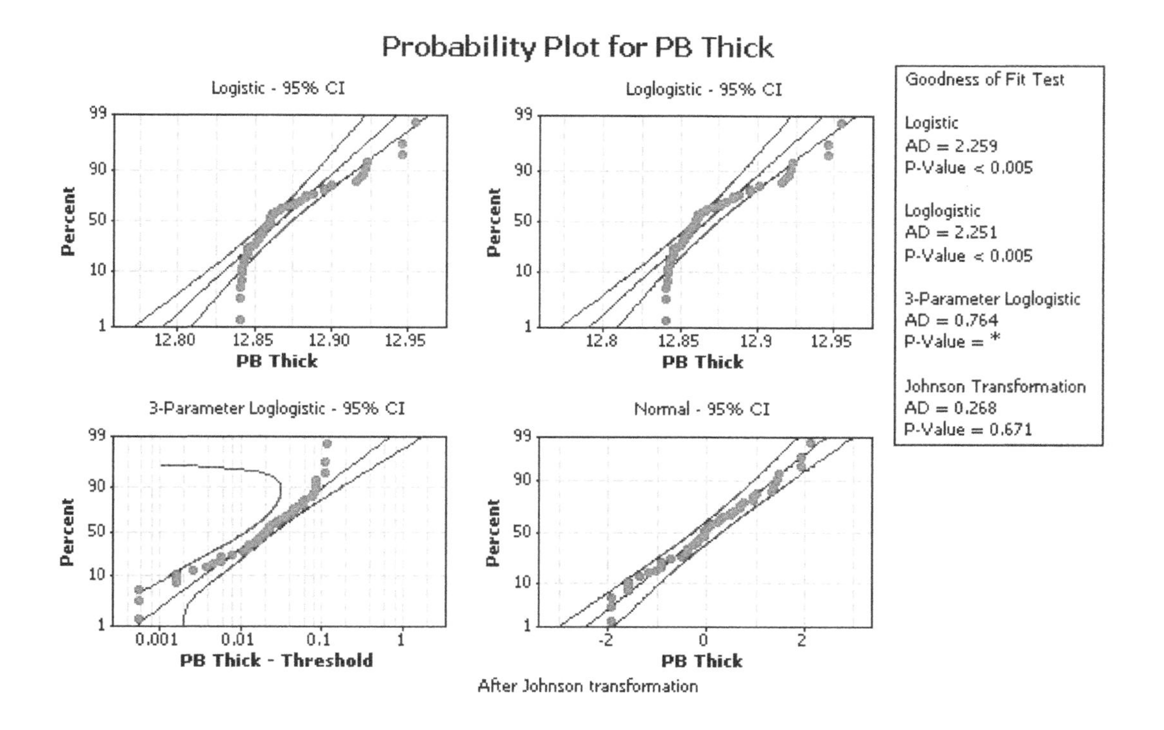

If you are using M16 you will find the Transformations within the Normal menus. M15 users will find the Transformations with the non normal distribution models in the Non normal menus.

6. Click Stat<<Quality Tools<<Capability Sixpack << Normal (for M16).
7. Complete the menu box as shown. Subgroup size is 1, as there was no indication of any subgrouping. Enter the specification limits.

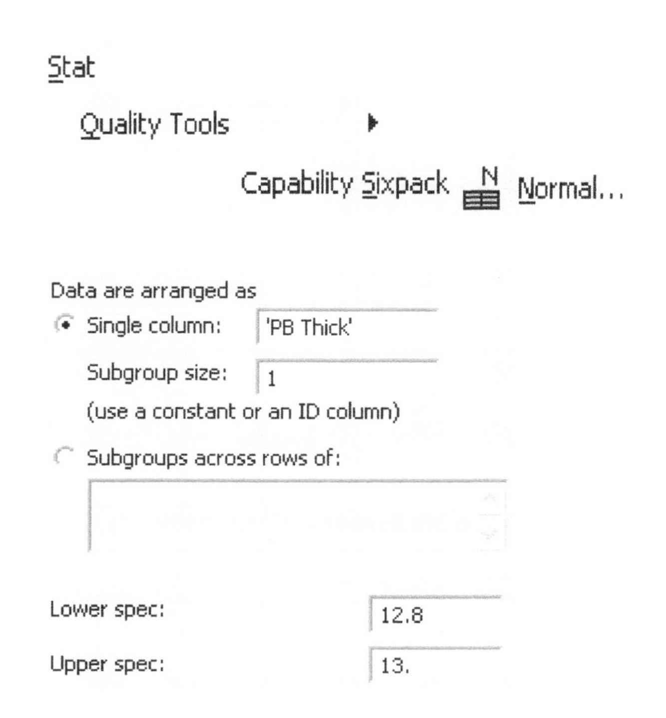

8. Click on the Transformations button.

Transform...

9. Select the radio button for Johnson transformation. Then click OK and OK again.

No transformation
Box-Cox power transformation (W = Y**Lambda)
Use optimal lambda
Lambda = 0 (ln)
Lambda = 0.5 (square root)
Other (enter a value between -5 and 5):
Johnson transformation (for overall analysis only)
P-Value to select best fit: 0.10

It is impressive to see the complexity of the Johnson transformation but you would not want to work out the transformation values with a calculator.

When we get to the Capability Histogram we begin to see that there is a problem, there are no specification limits. Looking into the Session Window we see that the problem was that the specification limits were outside of the transformation function. We will have to go plan C and try the three parameter Weibull.

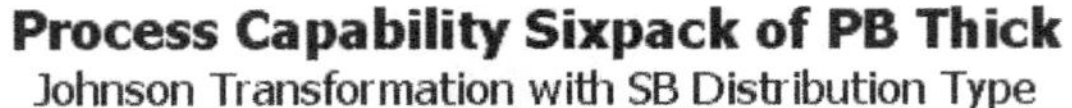

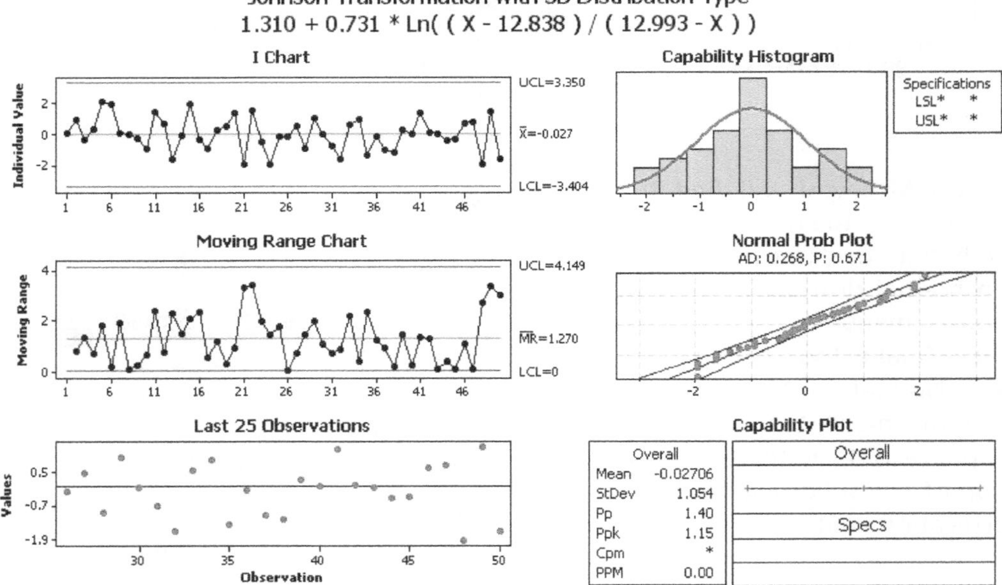

Process Capability Sixpack of PB Thick

```
* NOTE * Specification limit or target for PB Thick outside range of transformation function.
```

M15 and M16 both put the Non normal distributions in the Non normal menu.

10. Click Stat <<Quality Tools << Capability Sixpack << Non normal.

11. Complete the menu box as shown. Ensure the selector for Fit distribution is changed to 3-parameter Weibull. Then click OK.

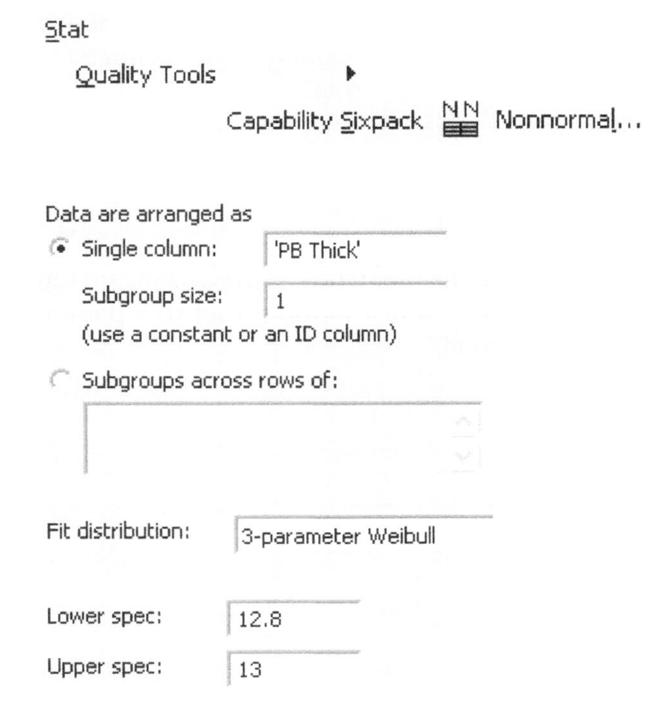

Having fitted the three parameter Weibull distribution we see that the process is in control.

The capability histogram displays our specification limits with our data and the fitted distribution.

As our P_{pk} is significantly less than the P_p this means that the distribution is not centered between the spec limits.

The PPM defect of 9594 indicates that 0.96% of the product would breach the USL.

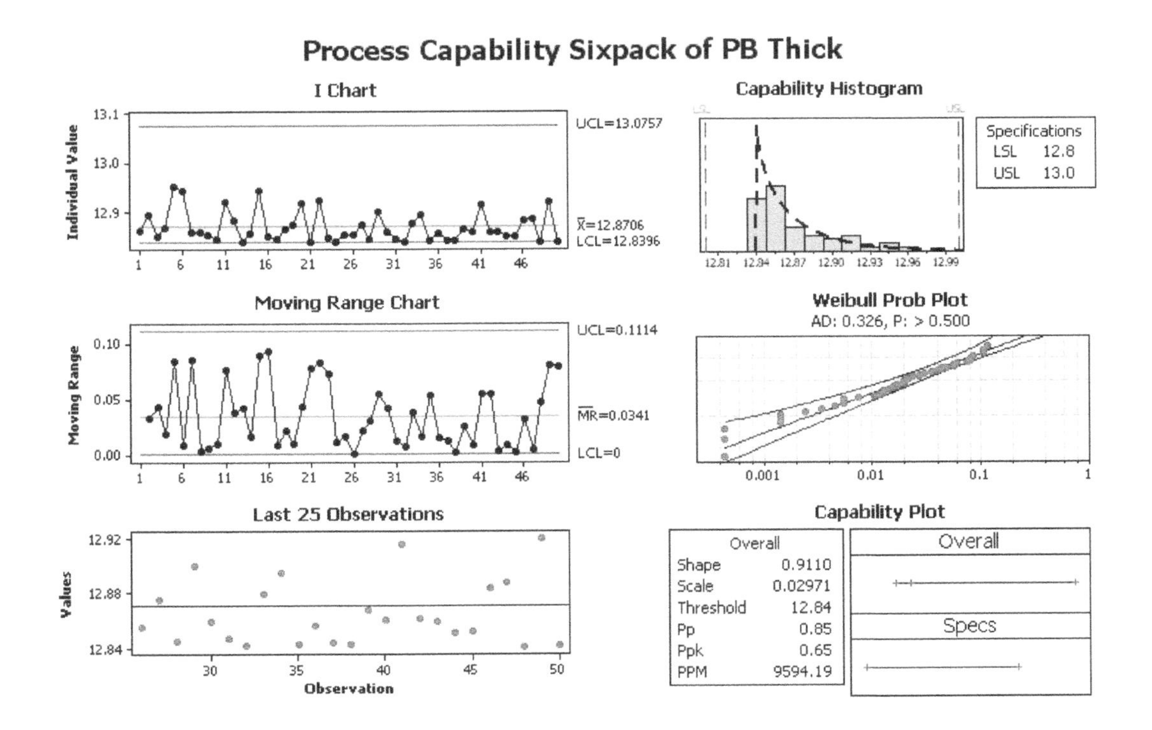

12. In order to carry out a detailed Capability Analysis click Stat <<Quality Tools <<Capability Analysis <<Non Normal.

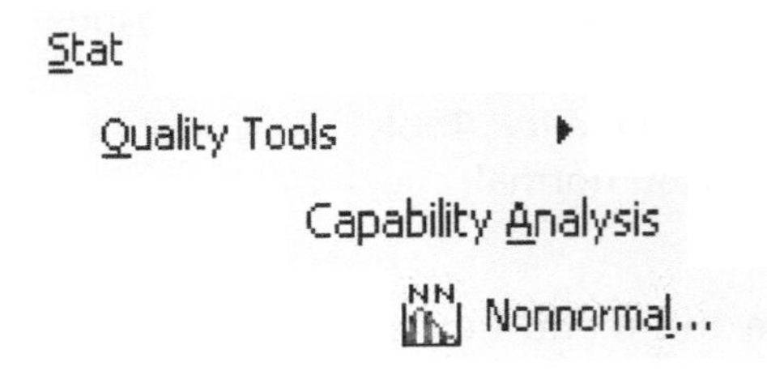

13. Complete the menu box as shown. Ensure the selector for Fit distribution is set to 3-parameter Weibull.

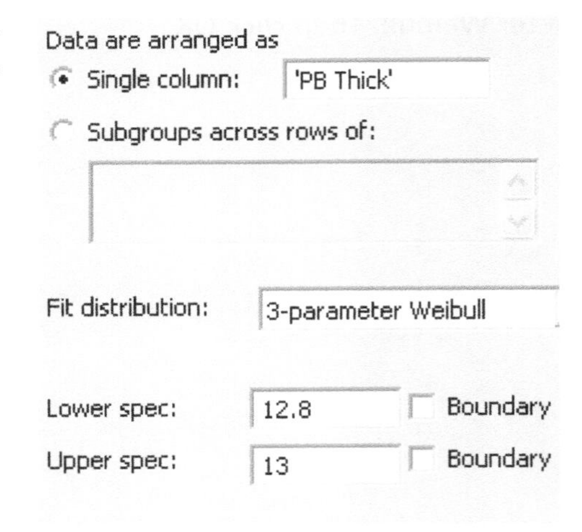

14. In order to change the data display into percentages rather than parts per million click on Options and then select the Percents radio button. Then click OK and OK again.

Display
Capability stats (Pp)
Parts per million
Benchmark Z's (sigma level)
Percents

The overall capability is 0.65. The observed performance, from the given data set does not produce any defects. However, from the expected overall performance we would expect 0.96% of product to be above the USL.

Within capability metrics are not given for non normal data distributions.

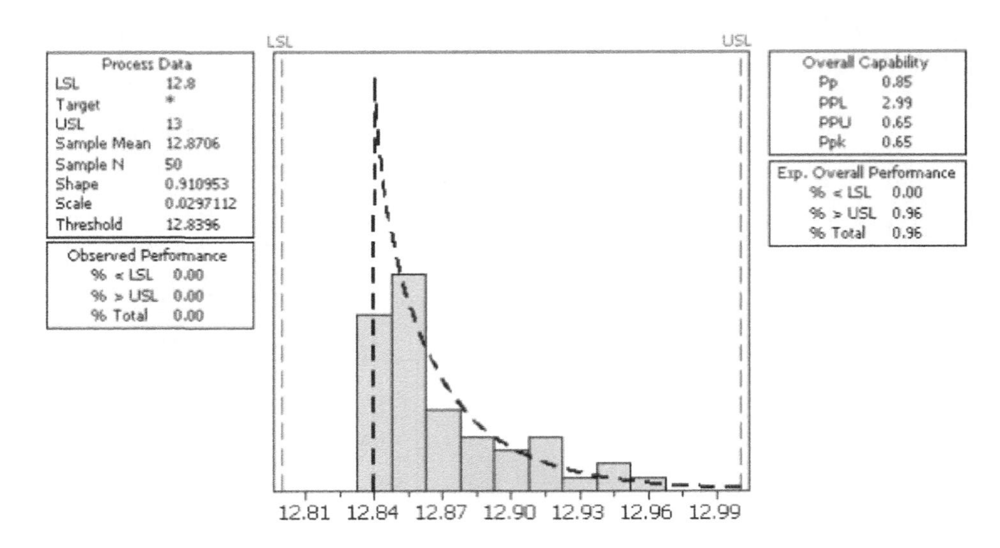

Exercise 2. Non Normal Capability

The Quality Department want to introduce a specification for banana curvature. Bananas with an outside curvature of 20–50° will be subject to an internal quality defect.

Assess the historical data and establish how many bananas will be deemed defective due to the new specification.

Initially, use the Assistant and establish if the Box–Cox transformation within the Assistant suitably transforms the data. Then use the Classic Method and use the IDI procedure to identify the best three possible fits/transformations. Use the top three to conduct the capability study (Sixpack only). If the Box–Cox is one of the top three you do not need to repeat the work already done by the Assistant.

Worksheet: Banana.

We know the data has been transformed as we pressed the Yes button at the prompt which asked us if we wanted to transform the data.

From the Summary Report we see that 0.44% of our data would be out of spec due to the top spec limit. If the distribution had been centred it would have fitted within the spec limits.

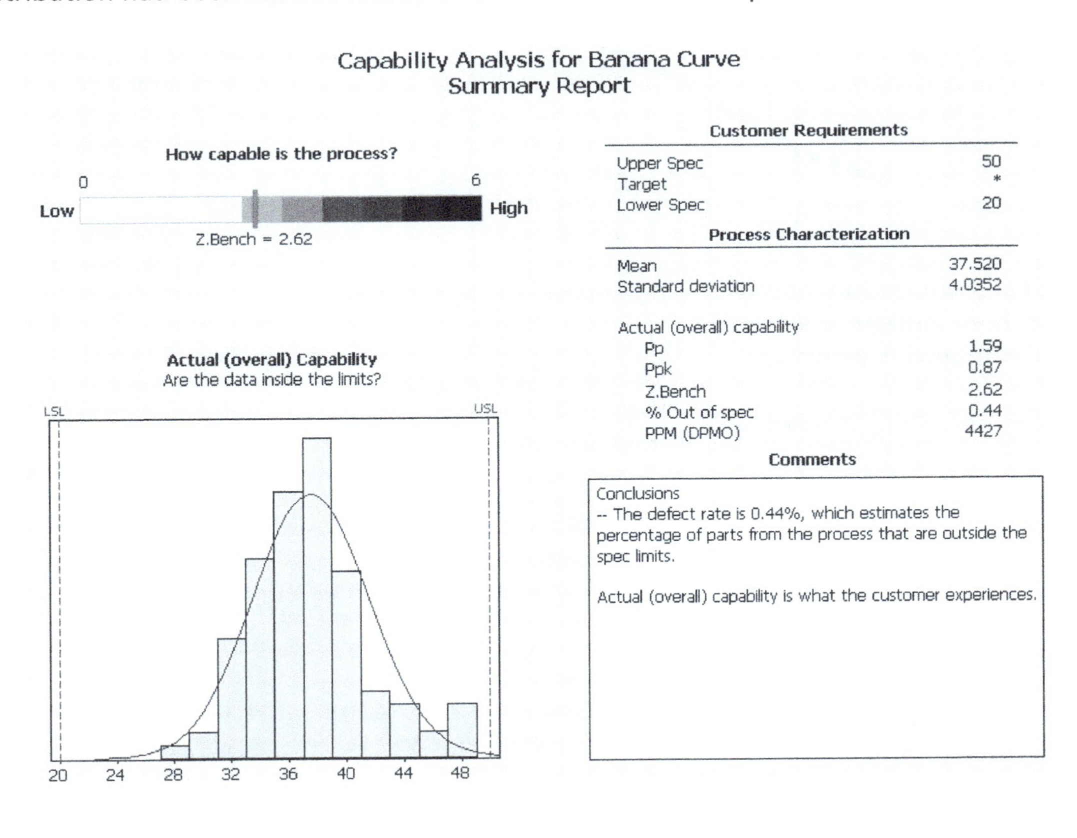

The Diagnostic Report shows that the data is in control. Therefore, we can use this data to make predictions about future performance.

We also see the *P* values of the original and transformed data. The Normality Plot header tells us that a lamda of –0.5 was used in the transformation.

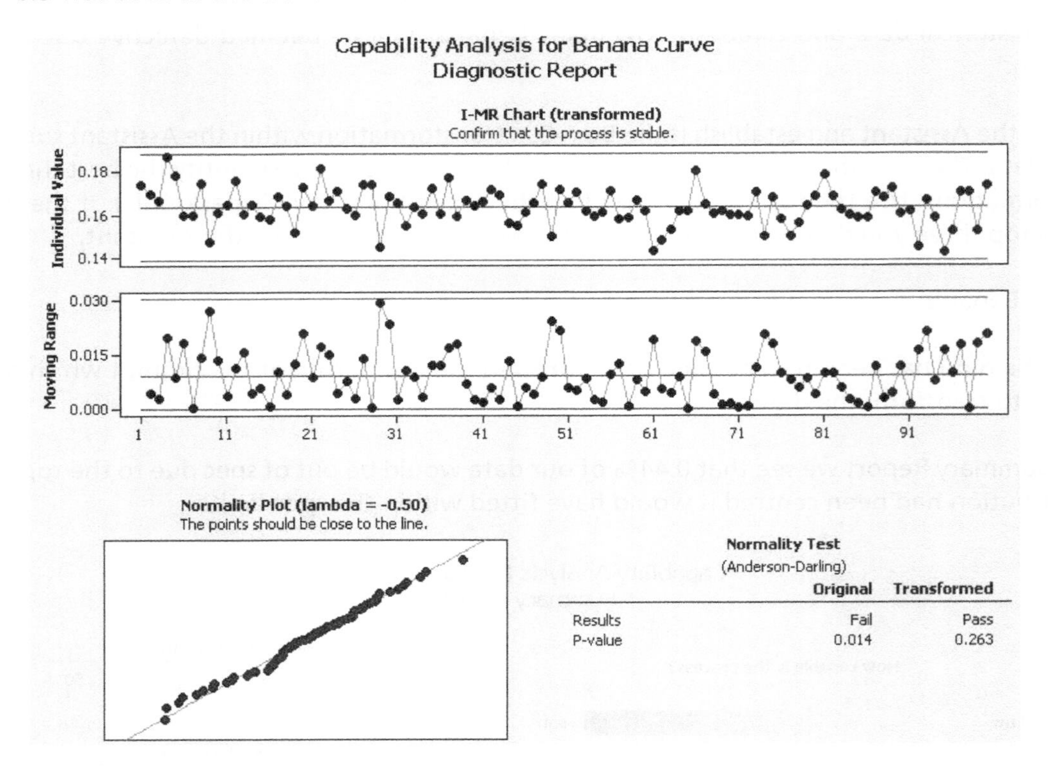

The Individual Distribution Identification procedure indicates that the best transformations/fits would be:

Johnson transformation, *P* value = 0.402

Box–Cox transformation, *P* value = 0.263

Loglogistic distribution fit, *P* value > 0.25

From the top three we need to check the Johnson transformation and the Loglogistic distribution fit using the Classic Method.

Goodness of Fit Test

Distribution	AD	P	LRT P
Normal	0.974	0.014	
Box-Cox Transformation	0.455	0.263	
Lognormal	0.561	0.143	
3-Parameter Lognormal	0.430	*	0.269
Exponential	37.088	<0.003	
2-Parameter Exponential	14.601	<0.010	0.000
Weibull	2.798	<0.010	
3-Parameter Weibull	0.745	0.042	0.000
Smallest Extreme Value	4.051	<0.010	
Largest Extreme Value	0.627	0.099	
Gamma	0.665	0.086	
3-Parameter Gamma	0.453	*	0.134
Logistic	0.604	0.078	
Loglogistic	0.421	>0.250	
3-Parameter Loglogistic	0.368	*	0.321
Johnson Transformation	0.378	0.402	

If you were using M16, did you remember to access the Johnson transformation via the Normal menu?

The Sixpack using the Johnson transformation is shown. Again, the data is in control. Therefore, we can use this data to make predictions about future performance.

Using the Johnson transformation we predict that 0.81% of Bananas will be defective as they breach the upper spec limit.

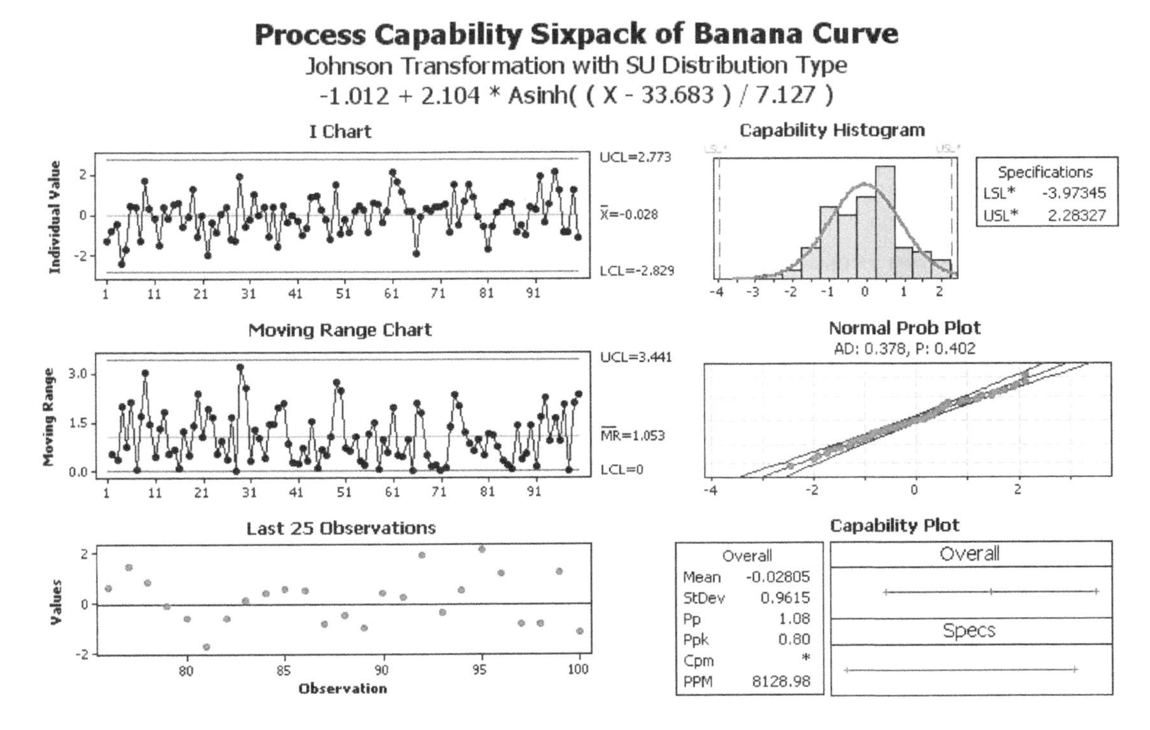

The Sixpack using the Loglogistic fit is shown. This time we see that one data point is out of control on the MR chart. In order to be able to make future predictions we would need to investigate the cause of this variation.

Using the Loglogistic fit the procedure predicts that 0.67% of bananas will be defective as they will breach the upper spec limit.

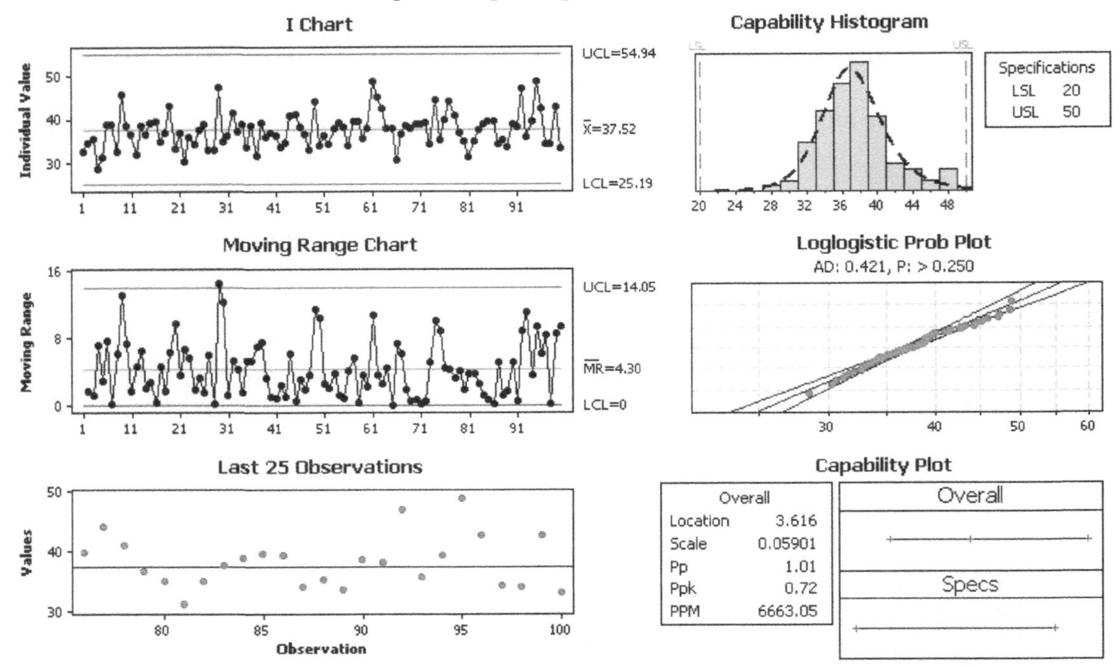

8.5 Capability Comparison using the Assistant

Procedure	• Capability Comparison (Assistant)
What's it used for?	• It is used to compare capability metrics for two sets of data, ideally before and after a process change.
Assumptions and Limitations	• The procedure will calculate if there has been a statistically significant shift in mean or StDev. • The data should be time ordered and normally distributed. • In order to use the results for a given data set to predict future performance the process must be in control.

Example 4. Capability Comparison.

Zaf has almost completed his first 6σ project. He has been reducing the order lifetime within a busy industrial components supplier. His target has been to reduce the order lifetime to between 65 and 75 days. This target has been in place for a long period of time but never achieved.

Data on order lifetimes is in two columns one labelled Before and the other After. Due to the nature of the process subgrouping was not attempted.

Use the Capability Comparison to establish if Zaf has been successful in his project. How have the capability metrics changed?

Worksheet: Order Time.

1. Open the Excel file and transfer the data into a new Minitab Project Worksheet.

Before	After
71.32	70.70
72.61	70.47
71.44	69.26
71.70	68.20

2. Click Assistant <<Capability Analysis.

Assistant

Capability Analysis...

3. Click on the Capability Comparison Icon.

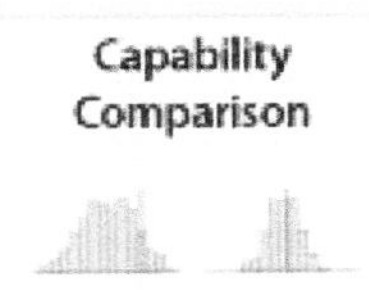

We enter the Baseline and the Improved data as before. Minitab gives us the option of having subgroups in rows again. We need to make sure that the selector is set to Data are in one Column for both sets of data.

4. Enter the Column Data
5. Enter the LSL and USL as 65 and 75, respectively. Note that we only have the option of entering one set of specification limits and they will be used for both sets of data.
6. Click on OK.

Baseline process data

How are your data arranged in the worksheet?

Data are in one column

Column: Before

Subgroup size: 1

Improved process data

How are your data arranged in the worksheet?

Data are in one column

Column: After

Subgroup size: 1

Process requirements

Lower spec: 65 (at least 1 spec limit is required)

Upper spec: 75

The Summary Report statistically confirms that Zaf has not reduced the process StDev but has changed the mean. Zaf has reduced %Out of Spec from 2.26 to 0%.

The histograms show how the distribution has shifted so that it is more centred and not crossing the USL.

On the right hand side we can see the overall capability metrics and how they have changed.

P_p has reduced from 1.64 to 1.58. But P_{pk} has gone from 0.67 to 1.52.

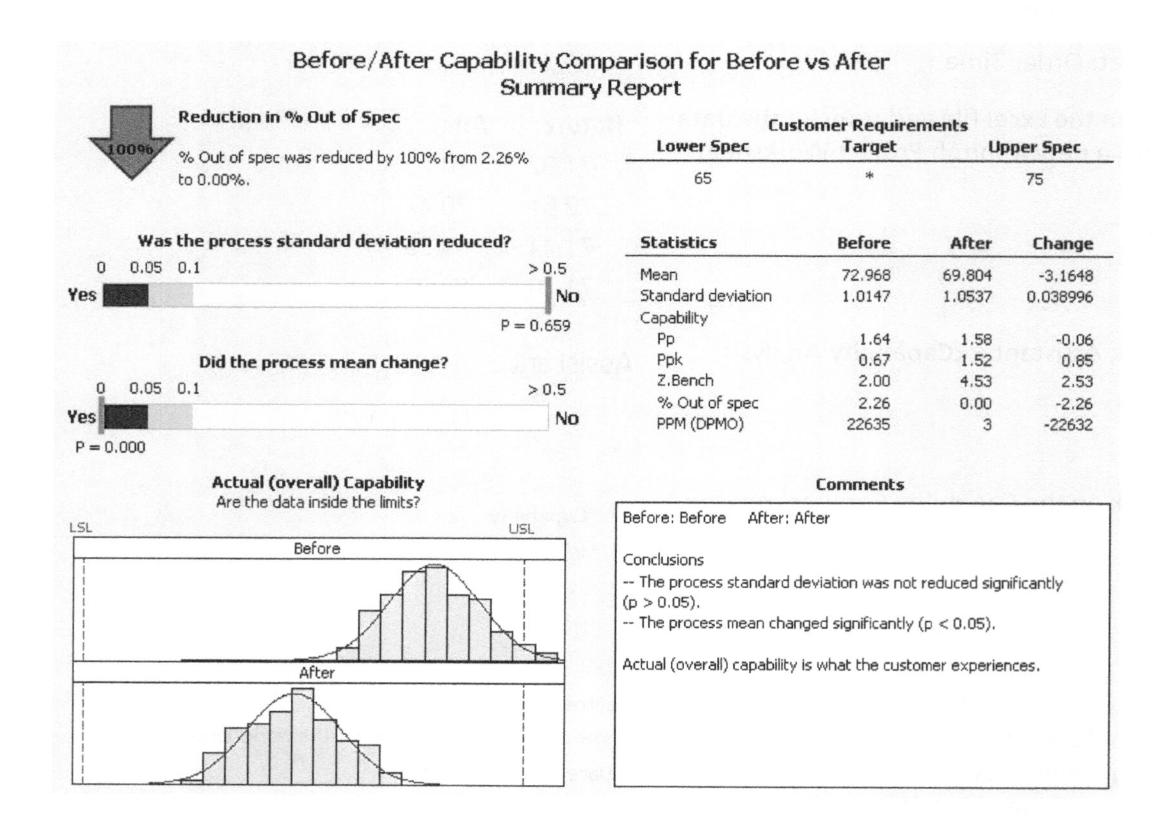

The Diagnostic Report looks at the normality and stability requirements of the data. We see that both the Before and the After data is in control. This allows us to use these metrics to predict future performance.

Notice, that both SPC charts are using the same y scale.

We also see that both sets of data passes the normality test.

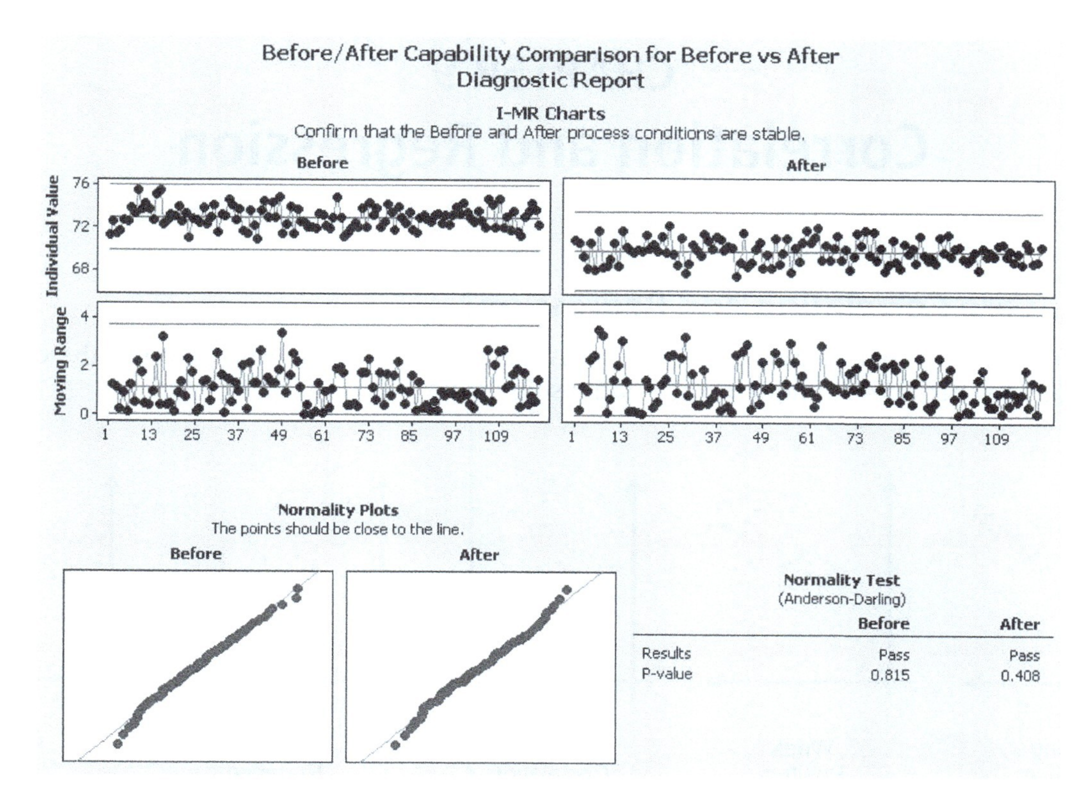

	Before	After
Results	Pass	Pass
P-value	0.815	0.408

The Process Performance Report gives us more detail on the capability metrics. We get Overall and Within metrics and we also get observed and expected metrics.

Finally, we get the report card, which is not shown here. That gives us the all clear on stability, normality, sample size and number of subgroups.

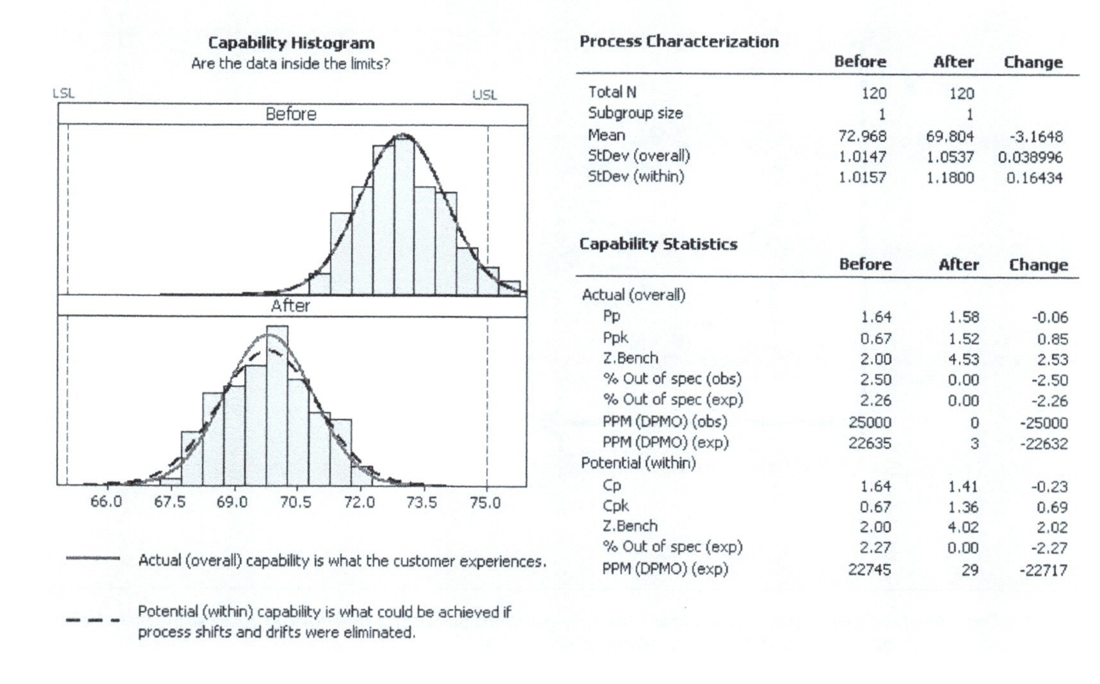

Process Characterization

	Before	After	Change
Total N	120	120	
Subgroup size	1	1	
Mean	72.968	69.804	-3.1648
StDev (overall)	1.0147	1.0537	0.038996
StDev (within)	1.0157	1.1800	0.16434

Capability Statistics

	Before	After	Change
Actual (overall)			
Pp	1.64	1.58	-0.06
Ppk	0.67	1.52	0.85
Z.Bench	2.00	4.53	2.53
% Out of spec (obs)	2.50	0.00	-2.50
% Out of spec (exp)	2.26	0.00	-2.26
PPM (DPMO) (obs)	25000	0	-25000
PPM (DPMO) (exp)	22635	3	-22632
Potential (within)			
Cp	1.64	1.41	-0.23
Cpk	0.67	1.36	0.69
Z.Bench	2.00	4.02	2.02
% Out of spec (exp)	2.27	0.00	-2.27
PPM (DPMO) (exp)	22745	29	-22717

CHAPTER 9
Correlation and Regression

9.1 What are Correlation and Regression?

We use correlation to check whether two variables have a linear relationship. We use the correlation coefficient (r) to check the strength of the relationship.

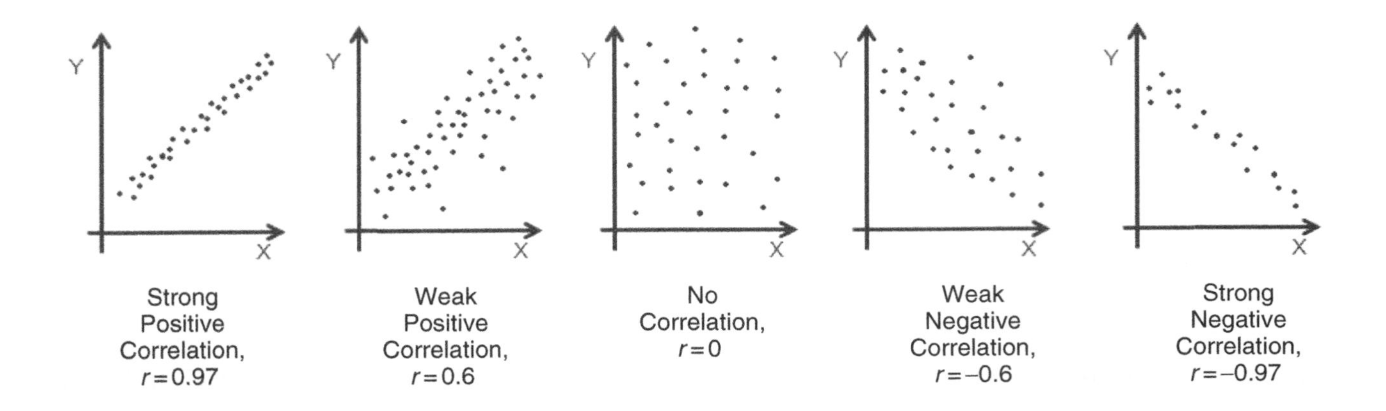

The correlation coefficient, r, varies from −1 to 1. Zero implies that there is no linear correlation. However, there can be nonlinear patterns when the correlation coefficient is zero. Therefore, it is advisable to graphically check the data before carrying out the test procedure.

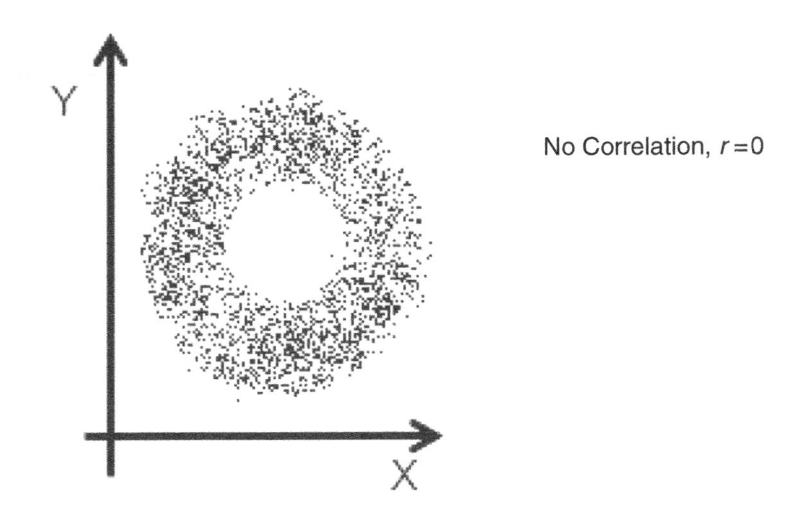

No Correlation, $r = 0$

Problem Solving and Data Analysis using Minitab: A clear and easy guide to Six Sigma methodology, First Edition. Rehman M. Khan.

Regression is more about building a mathematical model which describes the relationship between one or more predictors and a single response variable.

The graph shows the data points collected as tyre pressure is varied and then fuel efficiency is measured. A line of best fit has been drawn through the points using a cubic relationship to form the model. Even though we have used a cubic relationship there is only one factor/predictor and one dependent response.

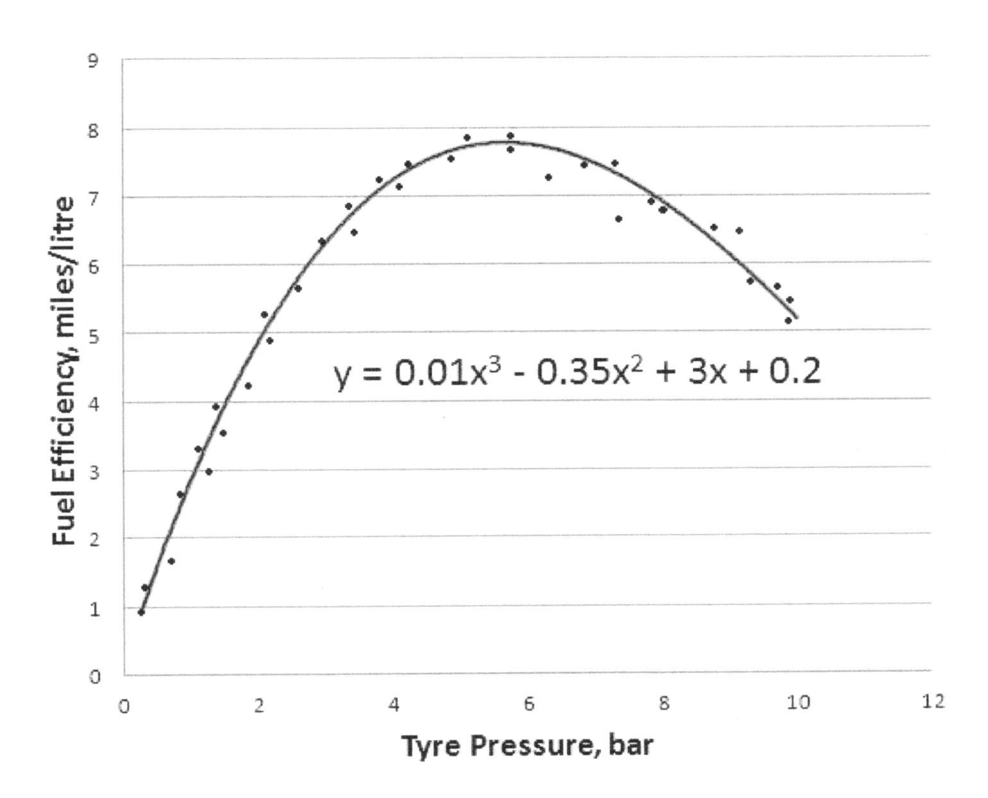

Once you have a regression equation it can be used to predict the response when the predictor is known.

Regression goes beyond correlation by adding this prediction capability.

Unfortunately, we can't expect our equations to perfectly model the real world so there will always be some error.

All data sets for this chapter are in the Excel file 09 Correlation and Regression.xls.

9.2 Correlation

Test	• Correlation
What's it used for?	• It is used to test whether two sets of variables have a linear relationship.
Assumptions and Limitations	• The data should be checked using a scatter plot prior to conducting the test in order to establish if a non linear relationship is present. • It is up to the tester to ensure there is a real world relationship between the factor and the response. Correlation does not mean causation. • Data for the study needs to be carefully collected so that it is representative of the population. • Testing for Correlation is only available via the Classic Menus.

Example 1. Correlation.

Spud has been collecting data on the sugar content of cake versus the taste rating (0–100) of a number of samples of cake.

Analyse the data Spud has collected and check whether there is a linear correlation between the variables. What is the strength of the correlation?

Worksheet: Cake.

1. The first thing we need to do before we conduct the procedure is to ask ourselves if we believe there is a real world link between our predictor and our response. It is our responsibility to ensure 'damn lies and statistics' are not linked. In this case the link between the predictor and the response does sound plausible.
2. Open the Excel file and transfer the data into Minitab.
3. We need to view the graphical relationship between the parameters in case a nonlinear relationship is present. Click Graph <<Scatterplot.

	A	B
1	Sugar Content	Taste Rating
2	8	34.40
3	20	84.00
4	9	36.90
5	14	67.20

Graph

Scatterplot...

4. Select Simple and then click OK.

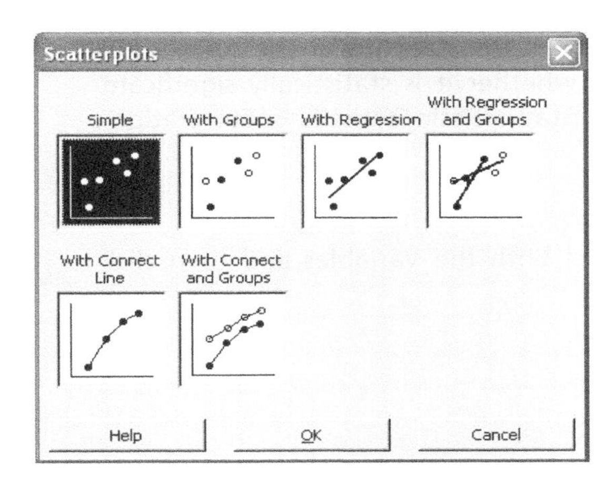

5. Since Taste Rating is the response we will use that as the Y variable. Enter Sugar Content as the X variable and click OK.

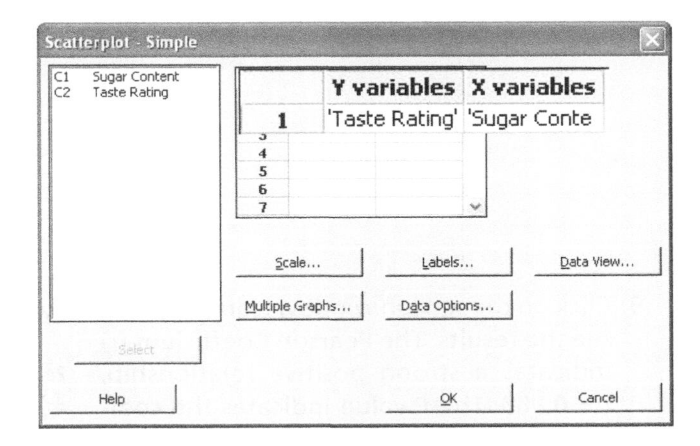

6. From the scatterplot there appears to be a positive correlation but we don't know if it is statistically significant. Note: we conduct a graphical check because checking the correlation coefficient alone would not detect if there was a strong nonlinear relationship.

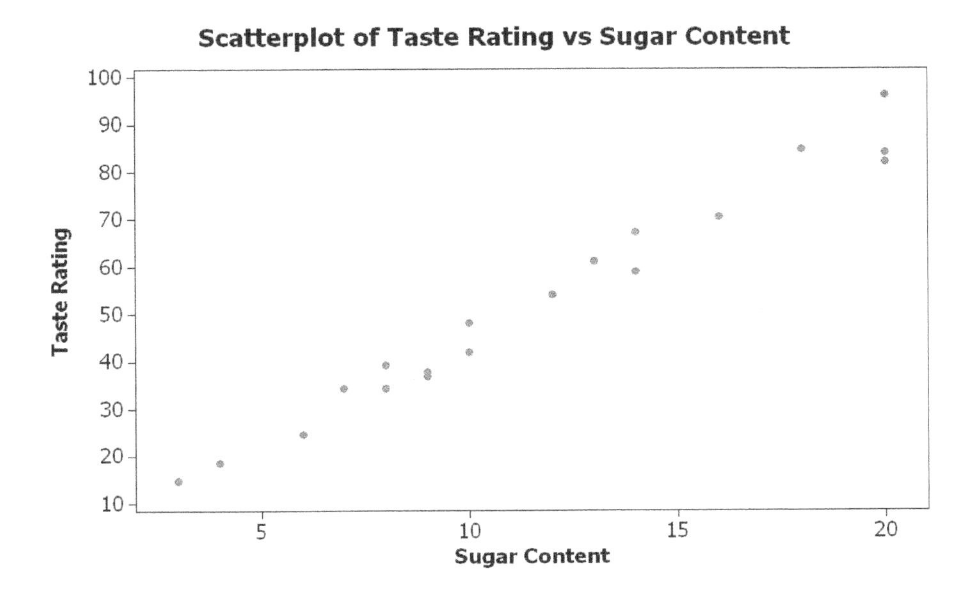

7. To check the strength of the correlation and whether it is statistically significant click Stat <<Basic Statistics <<Correlation.

8. Select both the Variables and then click OK.

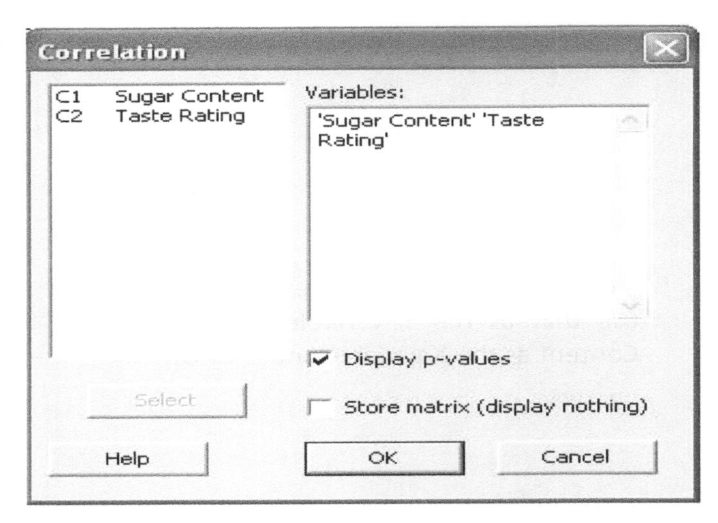

9. Click on the session window icon so we can see the results. The Pearson Coefficient (r) indicates a strong positive relationship, $r = 0.988$. The P value indicates the coefficient is significant.

Correlations: Sugar Content, Taste Rating

```
Pearson correlation of Sugar Content and Taste Rating = 0.988
P-Value = 0.000
```

Exercise 1. Correlation.

Aman is a keen gamer, however, his kids keep beating him on his favourite games. Aman is worried that reaction time declines with age. He collects data on age versus the completion time of a new video game to test the theory.

Analyse the data and check whether there is a linear correlation between the variables. What is the strength of the correlation? Should Aman be worried?

Worksheet: Reactions.

The scatter plot does appear to show a linear relationship. It would appear that as age increases the time taken to finish the game is longer.

The results in the session window are shown below the graph. The Pearson coefficient indicates a strong positive relationship, r = 0.946. The P value suggest the coefficient is significant.

Aman is so worried by the results that he bans his children from playing further video games.

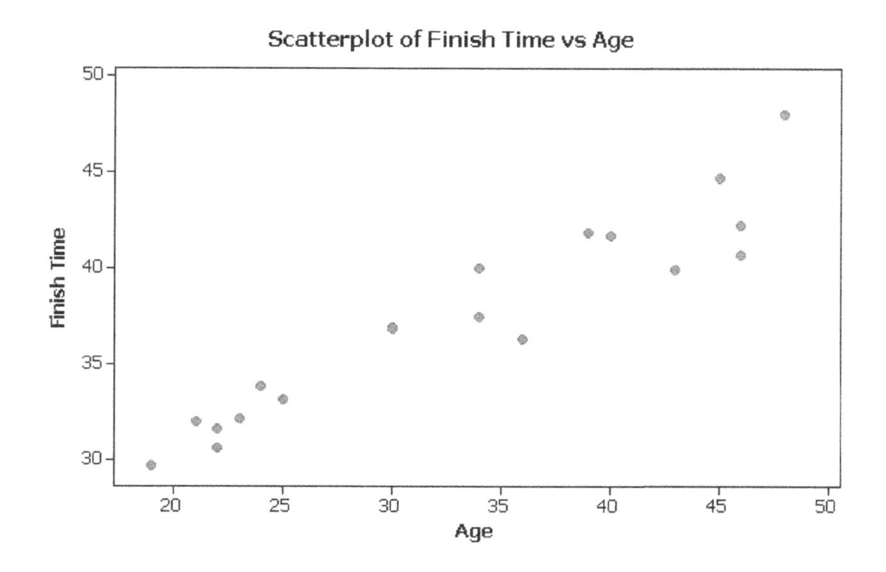

Correlations: Finish Time, Age

```
Pearson correlation of Finish Time and Age = 0.946
P-Value = 0.000
```

9.3 Multiple Correlations

Example 2. Multiple Correlations.

Eight presses are used to manufacture steel nuts. Samples are taken every hour and the across flats dimension is measured. If there is a strong correlation between any of the presses it may give an opportunity to reduce sampling.

Analyse the data and check whether there is a linear correlation between any of the presses. What is the strength of the correlation?

Worksheet: Nut.

1. Open the Excel file and transfer the data into Minitab.

	A	B	C	D	E	F	G	H
1	Press1	Press2	Press3	Press4	Press5	Press6	Press7	Press8
2	25.059	25.048	25.004	25.122	25.05	25.148	25.046	25.043
3	25.163	25.078	25.023	25.093	25.182	25.084	25.02	25.052
4	25.084	25.115	25.104	25.198	25.082	25.018	25.191	25.124
5	25.13	25.052	25.049	25.014	25.112	25.087	25.075	25.073
6	25.099	25.111	25.081	25.014	25.111	25.194	25.111	25.105

2. As with single correlations we need to view the graphical relationship between the parameters in case of a nonlinear relationship. Click Graph <<Matrix Plot.

Graph

Matrix Plot...

3. Select Simple and then click OK.

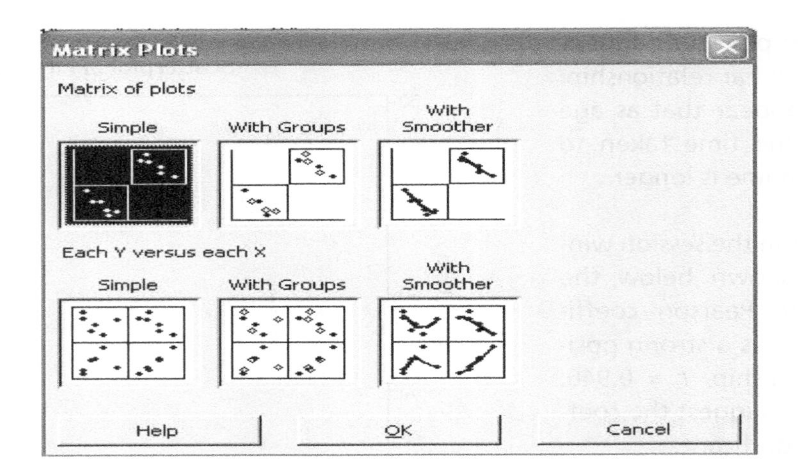

4. Select all the data columns as the graph variables. Then click on Matrix Options.

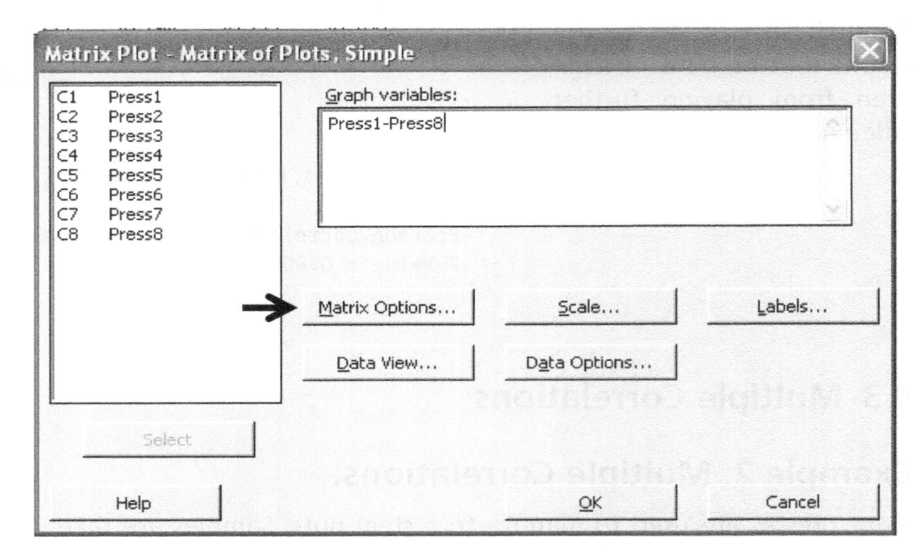

5. Under Matrix Display select the radio button for Lower Left. (This is a personal preference which I think makes the graph easier on the eye.)
6. Click OK and OK again.

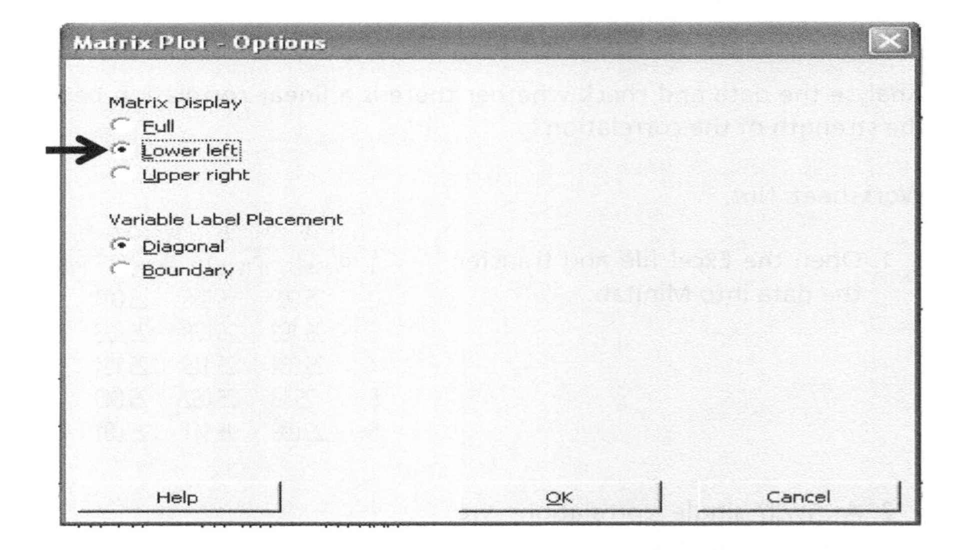

All possible combinations of scatterplot are presented here.

We can visually check for nonlinear relationships.

We can see that Press1 & Press5 and Press2 & Press8 appear to be correlated. As before we need to check if the correlation is significant.

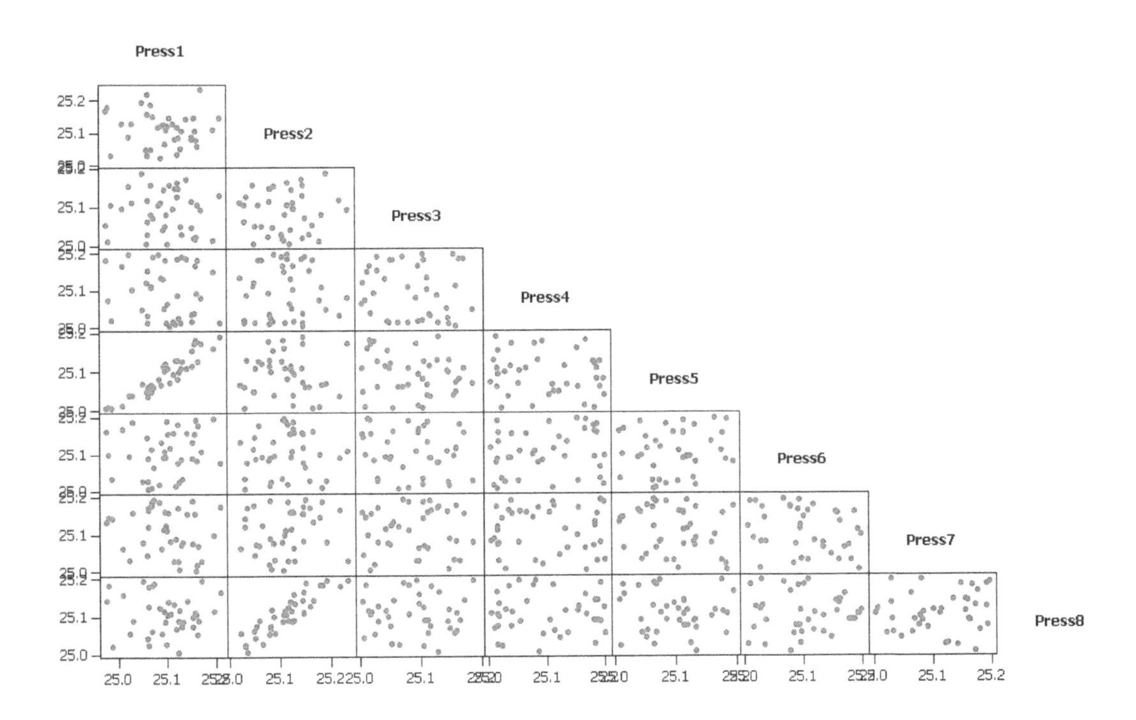

1. To check the strength of the correlation and whether it is statistically significant click Stat <<Basic Statistics <<Correlation.

2. Select all the Variables and then click OK.

3. Select the Session Window icon to see the results.

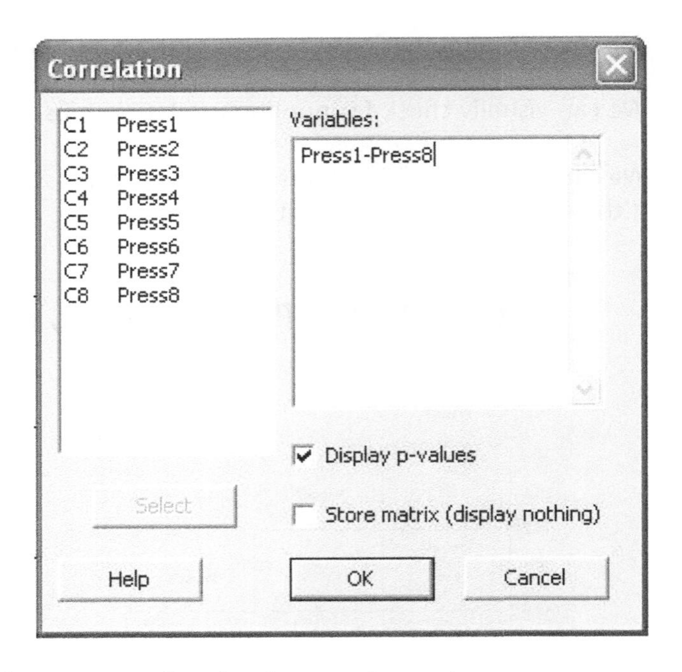

All combinations of correlation coefficient with their corresponding *P* value are shown here.

We can confirm that Press1 & Press5 and Press2 & Press8 are correlated and that correlation is statistically significant.

Knowing which presses are correlated may allow us to reduce sampling.

Correlations: Press1, Press2, Press3, Press4, Press5, Press6, Press7, Press8

```
        Press1  Press2  Press3  Press4  Press5  Press6  Press7
Press2  -0.074
         0.655

Press3  -0.049   0.026
         0.766   0.877

Press4  -0.135   0.014  -0.122
         0.413   0.931   0.459

Press5   0.933  -0.043  -0.190  -0.145
         0.000   0.794   0.247   0.379

Press6   0.165   0.026  -0.163   0.202   0.126
         0.315   0.876   0.322   0.218   0.445

Press7  -0.210   0.168   0.029  -0.032  -0.217  -0.251
         0.200   0.308   0.862   0.845   0.185   0.123

Press8  -0.041   0.895   0.011   0.028  -0.061   0.033   0.143
         0.805   0.000   0.945   0.866   0.711   0.843   0.384


Cell Contents: Pearson correlation
               P-Value
```

Exercise 2. Multiple Correlations.

Ten machines are used to manufacture 50 μF capacitors. Samples are taken every hour and the capacitance is measured. If there is a strong correlation between any of the machines it may give an opportunity to reduce sampling.

Analyse the data and check whether there is a linear correlation between any of the machines. What is the strength of the correlation?

Worksheet: Capacitor

We can see that all combinations of machines appear to be correlated. As before we need to check if the correlation is significant.

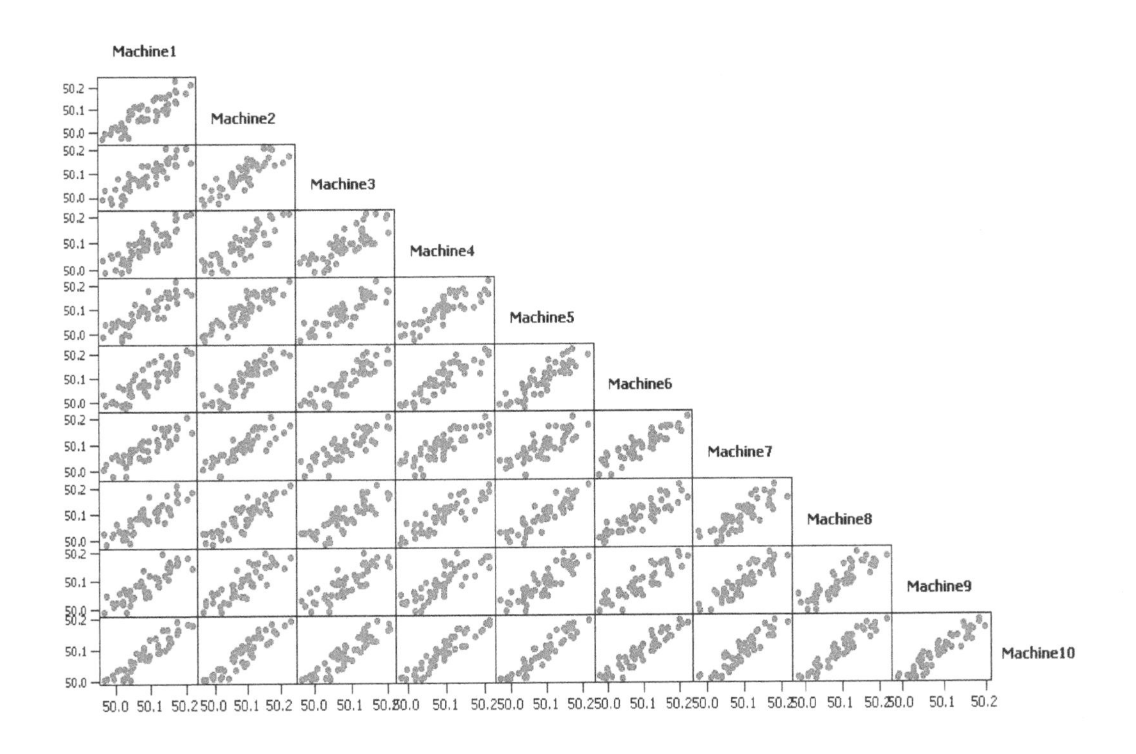

The *P* value tells us that all the machines are correlated.

This may be good in terms of reducing sampling but in process terms it would be very beneficial to understand why the machines are behaving in this way.

Correlations: Machine1, Machine2, Machine3, Machine4, Machine5, Machine6, Machine7, Machine8, Machine9, Machine10

	Machine1	Machine2	Machine3	Machine4	Machine5	Machine6	Machine7	Machine8	Machine9
Machine2	0.853								
	0.000								
Machine3	0.853	0.851							
	0.000	0.000							
Machine4	0.842	0.838	0.791						
	0.000	0.000	0.000						
Machine5	0.849	0.873	0.856	0.830					
	0.000	0.000	0.000	0.000					
Machine6	0.846	0.870	0.871	0.879	0.869				
	0.000	0.000	0.000	0.000	0.000				
Machine7	0.846	0.840	0.837	0.814	0.794	0.880			
	0.000	0.000	0.000	0.000	0.000	0.000			
Machine8	0.862	0.866	0.818	0.865	0.883	0.863	0.809		
	0.000	0.000	0.000	0.000	0.000	0.000	0.000		
Machine9	0.873	0.832	0.838	0.825	0.864	0.872	0.885	0.868	
	0.000	0.000	0.000	0.000	0.000	0.000	0.000	0.000	
Machine10	0.933	0.917	0.900	0.918	0.933	0.942	0.913	0.938	0.936
	0.000	0.000	0.000	0.000	0.000	0.000	0.000	0.000	0.000

Cell Contents: Pearson correlation
P-Value

9.4 Introduction to Regression

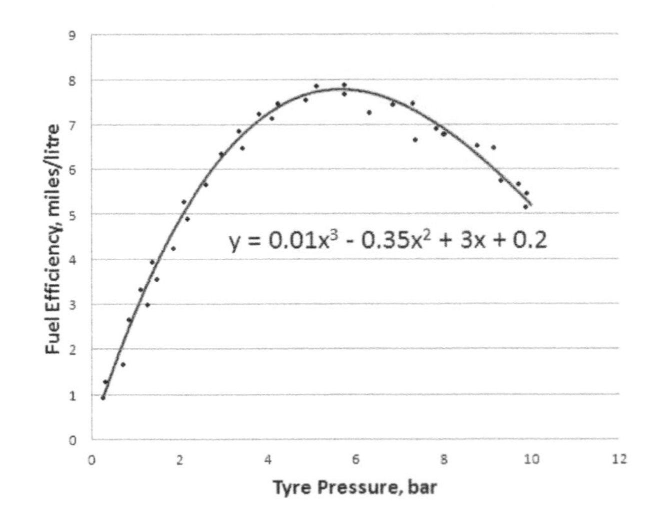

As stated earlier, Regression is more about building a mathematical model which describes the relationship between one or more predictors and a single response variable.

This course will split the topic of Regression into Single Predictor Regression and Multiple Predictor Regression. You saw the example given earlier when we took Tyre Pressure as a single factor and showed a cubic equation of how the response, Fuel Efficiency, would vary with Tyre Pressure. This was single predictor regression. If we wanted to reduce the error in the model we could add additional predictors such as vehicle weight, engine size and drag factor. That would be multiple predictor regression.

Later we will also take a quick look at another type of regression which is Nonlinear Regression. This is M16 only.

For Single Predictor Regression Minitab will work with the user to try and fit an equation. You have already seen the cubic equation, the other types are linear and quadratic. We measure how well these equations model our data and we call that the 'fit'.

On occasion, there isn't much difference in the achieved fit between the different types of equation. Sometimes, the simpler model is preferred as it is easier for subsequent work.

When using the Classic Menus we will use the Fitted Line Plot procedure. In M16, the Assistant offers us Single Predictor Regression but not Multiple Predictor Regression.

Another key difference is that when using the Classic Method we validate the model using the Residuals. The Assistant validates the model using the methods described in the Minitab White Paper for Regression. That is by checking the sample size and checking for unusual data points. Advice on getting a valid model can be obtained when starting the Assistant procedure. After the test procedure the summary of the validity checks are given in the Report Card, as always.

The White Paper recommends a sample size of ≥40 in order to obtain a precise estimate of the strength of the relationship and states that for this sample size the normality of the residuals is not an issue. There is no reason for us not to use these guidelines to help ourselves when using the Classic Method.

9.5 Single Predictor Regression

Test	• Single Predictor Regression
What's it used for?	• It is used to model an independent predictor variable against a dependent response variable.
Assumptions and Limitations	• Assumptions should not be made about data that is extrapolated outside the study range. • Data for the study needs to be carefully collected so that it is representative of the population. • The residuals are used to validate the model when using the Classic Method. The Assistant informs us of the reliability of our model within the Report Card.

Example 3. Fitted Line Plot.

Researchers believe that the diameter of an apple determines its sugar concentration.

Analyse the data to establish if any kind of relationship exists. What is the strength of the relationship?

For an apple of 90 mm diameter could the mean sugar content be 32 mg/g?

For an apple of 120 mm diameter could the sugar content be 32 mg/g?

For an apple of 50 mm diameter could the sugar content be 24 mg/g?

Worksheet: Apple.

1. Open the Excel file and transfer the data into a new Minitab project worksheet.

	A	B
1	Diameter	Sugar Content
2	105.1	30.50
3	134.4	24.22
4	71.9	31.56
5	73	31.68
6	117.4	29.58

2. Click on Stat <<Regression <<Fitted Line plot.

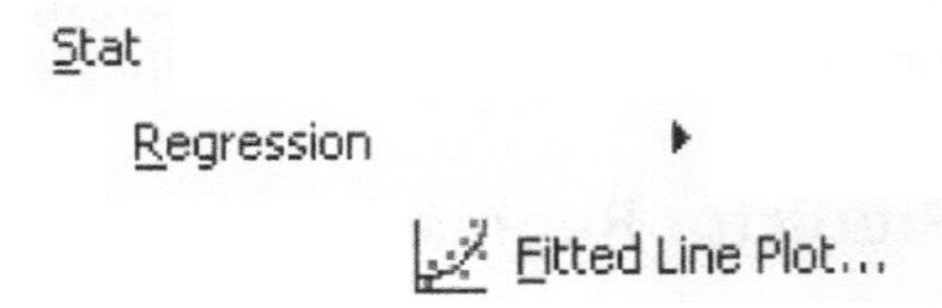

3. Select Sugar Content as the Response and Diameter as the Predictor. We have a one in three chance of guessing the correct model. If we want to stack the odds in our favour we should select the Cubic model first. Click OK.

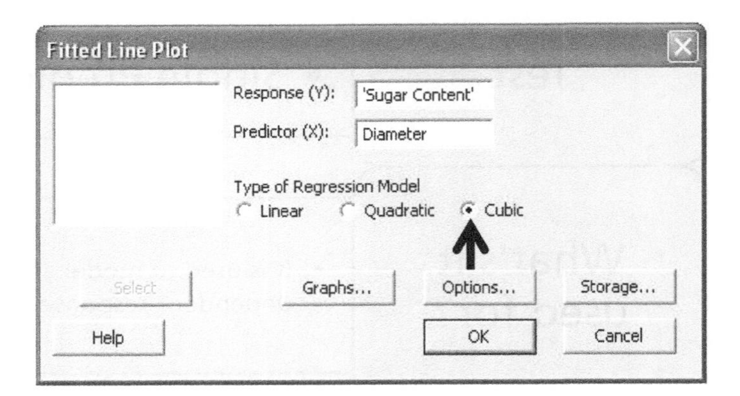

4. The Cubic model gives an R-Sq(adj) of 81.9%. This means that 81.9% of changes in the response can be explained by changes in the predictor.

Fitted Line Plot
Sugar Content = - 16.78 + 1.276 Diameter
- 0.01006 Diameter**2 + 0.000021 Diameter**3

S	1.01219
R-Sq	83.3%
R-Sq(adj)	81.9%

Sugar Content
Diameter

5. Go to the Session Window and check the results of the Sequential Analysis of Variance. We see that the Cubic term is not actually significant. As the next highest term, the quadratic, is significant we should select that model instead.

Sequential Analysis of Variance

Source	DF	SS	F	P
Linear	1	56.408	13.05	0.001
Quadratic	1	125.510	119.88	0.000
Cubic	1	1.855	1.81	0.187

6. Click on the Edit Last button.

7. Select the radio button for the Quadratic model.

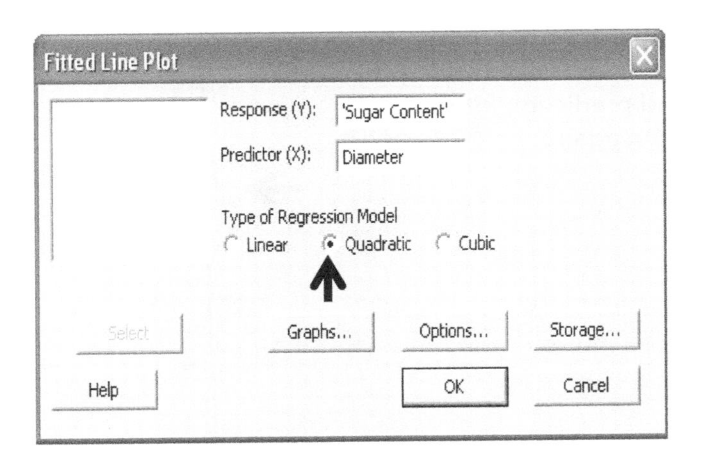

8. Click on Graphs and select the radio button for Four in one Residual Plots.
9. Click on OK and OK again.

Residual Plots
Individual plots
Histogram of residuals
Normal plot of residuals
Residuals versus fits
Residuals versus order
Four in one

10. Find the graph with the quadratic model on it. Note that the quadratic model has a slightly lower R-Sq(adj) than the cubic model of 81.5%.

The R-Sq(adj) of 81.5% indicates that there is a strong relationship between Diameter and Sugar Content.

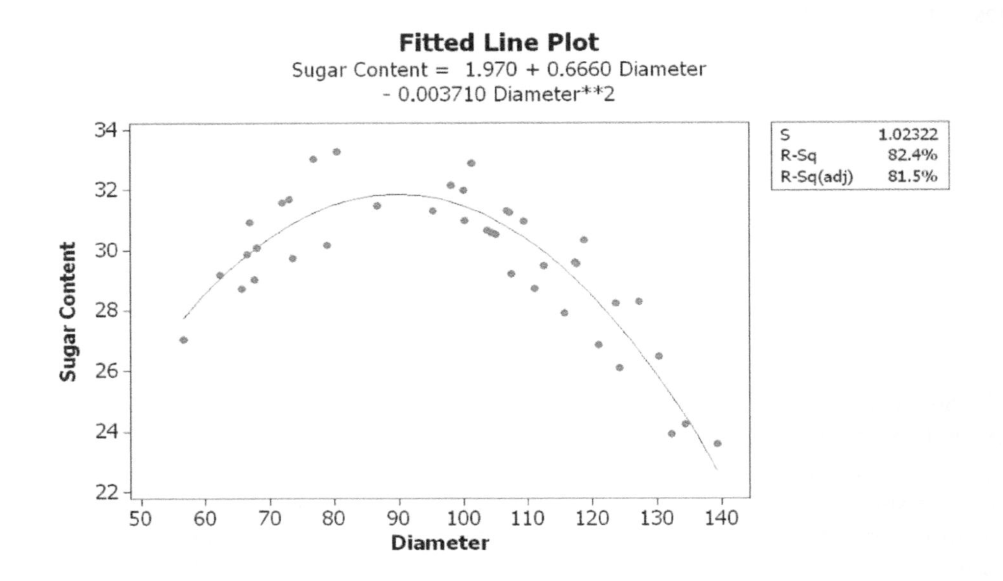

We will now move onto the second part of the question.

11. Click on the Edit Last button.

12. Then click on the Options button.
13. Select the tick boxes for displaying the confidence and prediction intervals.
14. Click OK and OK again.

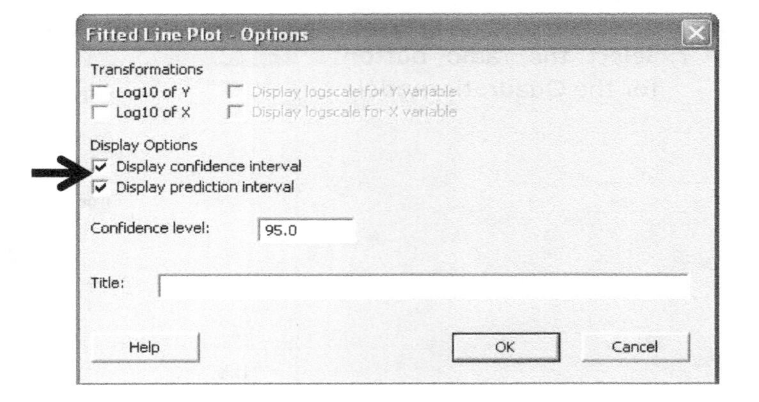

15. The red dashed line is the 95% confidence interval for the mean. The prediction interval gives a confidence boundary for individual points. We can be 95% confident that an individual point will fall within the outer green dashed line. We can be 95% confident that if apple diameter was kept constant the mean would fall within the inner red dashed line.

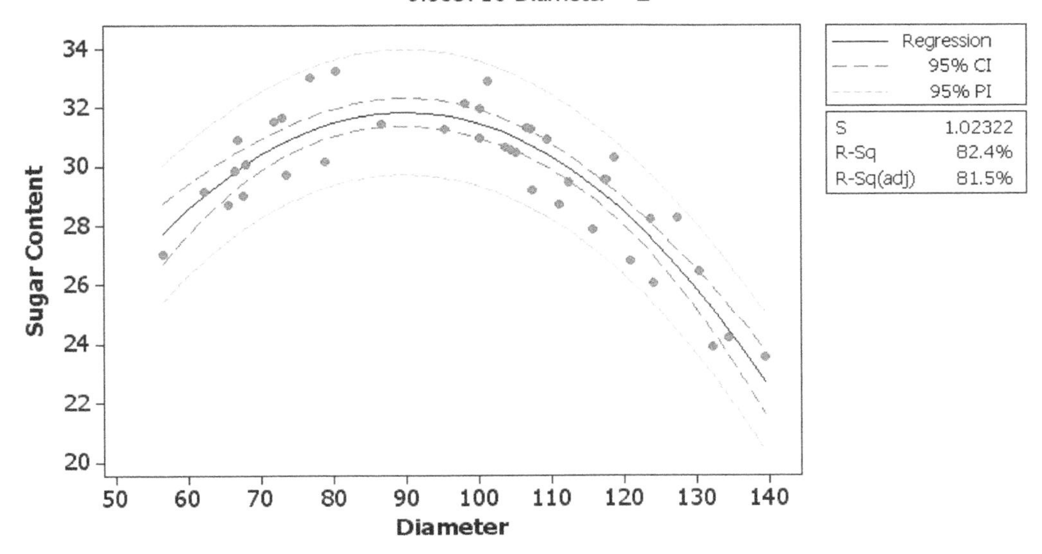

16. In order to help us answer the questions about predicting means and individual points we can add reference lines to the graph. Right click on the graph panel and select Add <<Reference Lines

Add
Reference Lines...

17. Enter the values as shown and then click OK.

Show reference lines at Y values:
32

Show reference lines at X values:
90 120

Reference lines appear on the quadratic plot and we can use them to help us answer the second part of the question.

We can say with 95% confidence that a diameter of 90 can give a mean sugar content of 32 as the intersection of the relevant reference lines lies within the confidence interval for the mean.

We can also say that a diameter of 120 is not likely give a single sugar content of 32. The intersection of the relevant reference lines does not lie within the prediction interval.

An apple of 50 mm diameter falls outside the range of our data and we would be extrapolating. Therefore, we must decline to answer the third part of the problem.

Finally we must check the residuals to establish the validity of the model.

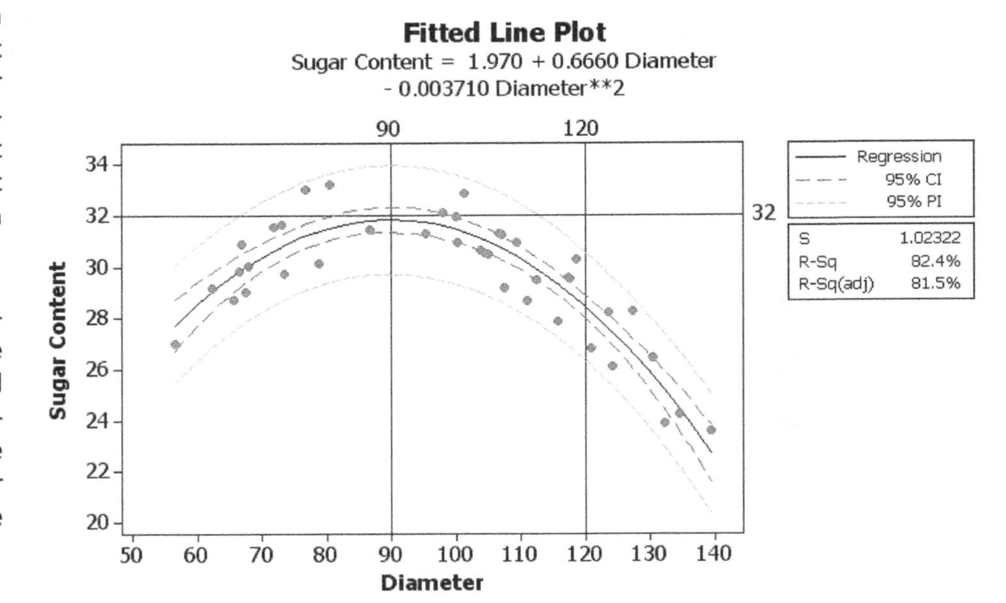

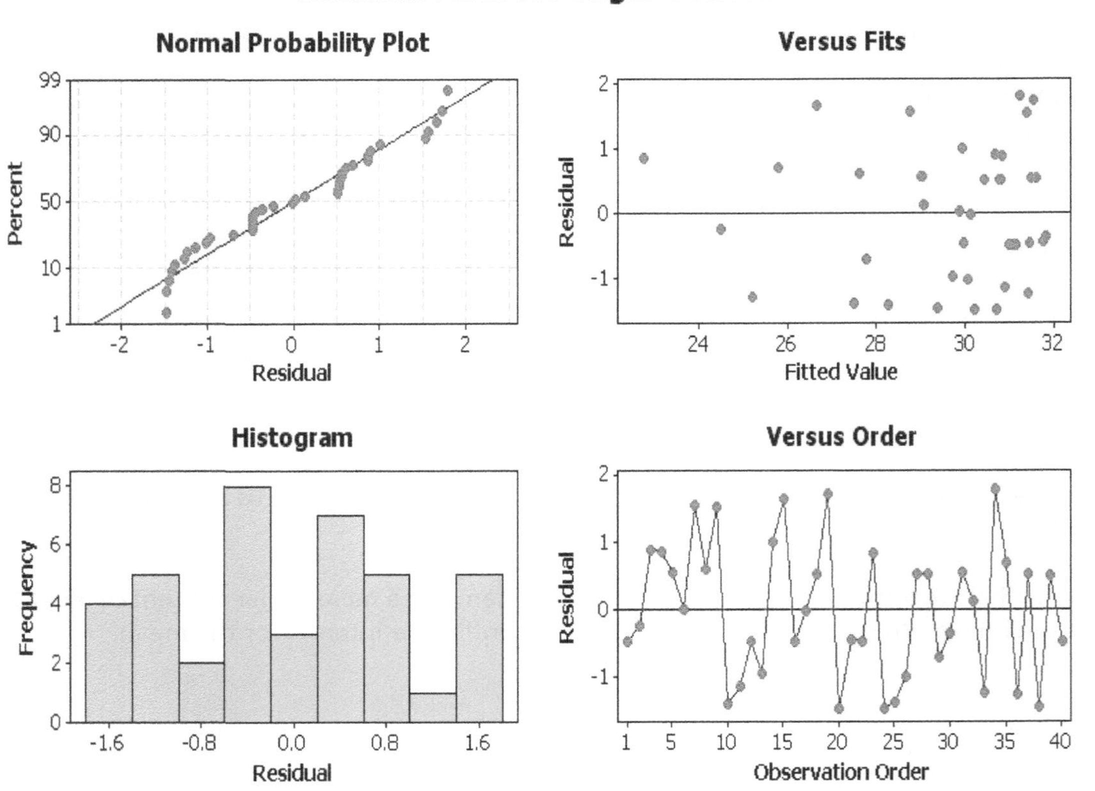

Starting from the top and going from left to right:

- The Normal Probability Plot shows that the Residuals are normally distributed.
- The Versus Fits Plot shows that the Residuals are equally distributed about the centre line.
- The Histogram Plot shows that the Residuals do not show extreme skewness.
- The Versus Order Plot shows that the Residuals are not displaying any patterns or ordered behaviour. This is only relevant if the results were recorded in the order that the experiment was carried out.

In summary, there are no issues with the residuals within this study.

Example 4. Regression with the Assistant.

Researchers believe that the diameter of an apple determines its sugar concentration. Assume the data has been collected and recorded in time order.

Use the Assistant to analyse the data in order to establish if any kind of relationship exists. What is the strength of the relationship?

(The Assistant has it strengths but it does not have as many extra options available as the Classic Method. At present the Assistant does not provide Confidence and Prediction Intervals as part of the analysis so we will have to omit this section from the examples when using the Assistant.)

Worksheet: Apple.

1. Click on Assistant << Regression.

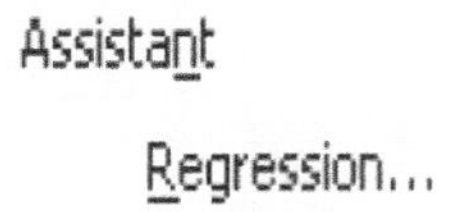

Assistant

Regression...

The main screen for regression with the Assistant is shown. It contains advice on collecting the data, evaluating the results and using the model. Clicking on any of the crosses will expand the selection and gives us the advice we want. The hidden information gives us the rules for ensuring the model is correctly validated and the rules to ensure we do not commit any of the common mistakes within Regression.

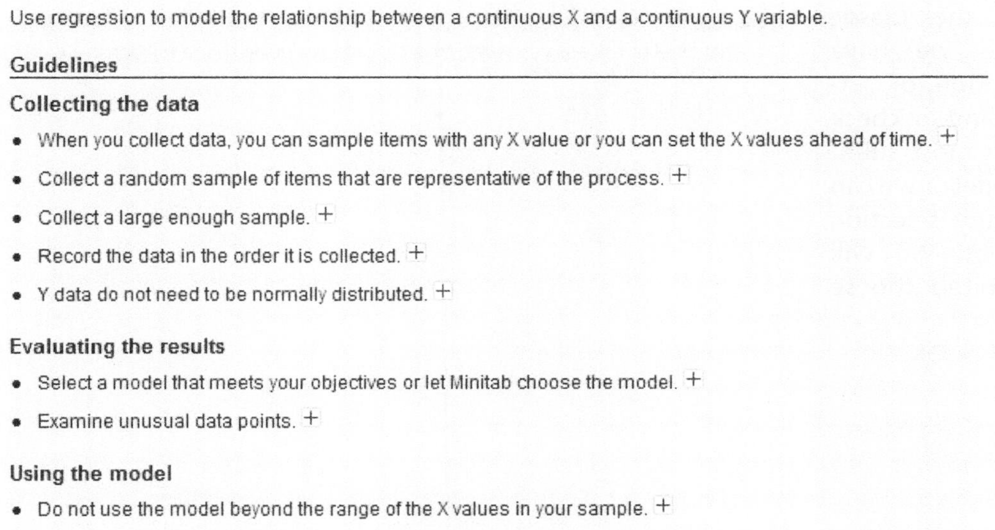

Use regression to model the relationship between a continuous X and a continuous Y variable.

Guidelines

Collecting the data

- When you collect data, you can sample items with any X value or you can set the X values ahead of time. +
- Collect a random sample of items that are representative of the process. +
- Collect a large enough sample. +
- Record the data in the order it is collected. +
- Y data do not need to be normally distributed. +

Evaluating the results

- Select a model that meets your objectives or let Minitab choose the model. +
- Examine unusual data points. +

Using the model

- Do not use the model beyond the range of the X values in your sample. +

An example is shown here in which we have expanded the very bottom selection.

Using the model

- Do not use the model beyond the range of the X values in your sample.

The regression model should only be used to predict Y for values of X that fall within the range of the X values in your sample data. Any predictions based on X values that are significantly outside of the values in your sample may be extremely inaccurate.

2. Click on the button in the top left corner labelled 'Click to perform analysis'.
3. Complete the menu box as shown. As we are saying that the Sugar Content is dependent upon the Diameter of the apple, the Diameter is the predictor and the Sugar Content is the response. Traditionally, the response is entered on the y axis.
4. Select the tick box for 'Data are recorded in time order'. If we don't tick the box we will not get Diagnostic Report 2.
5. Unlike the Classic Method we have the option of letting Minitab check for the most suitable model or we can make the selection ourselves. We will let Minitab choose for us.
6. Then click OK.

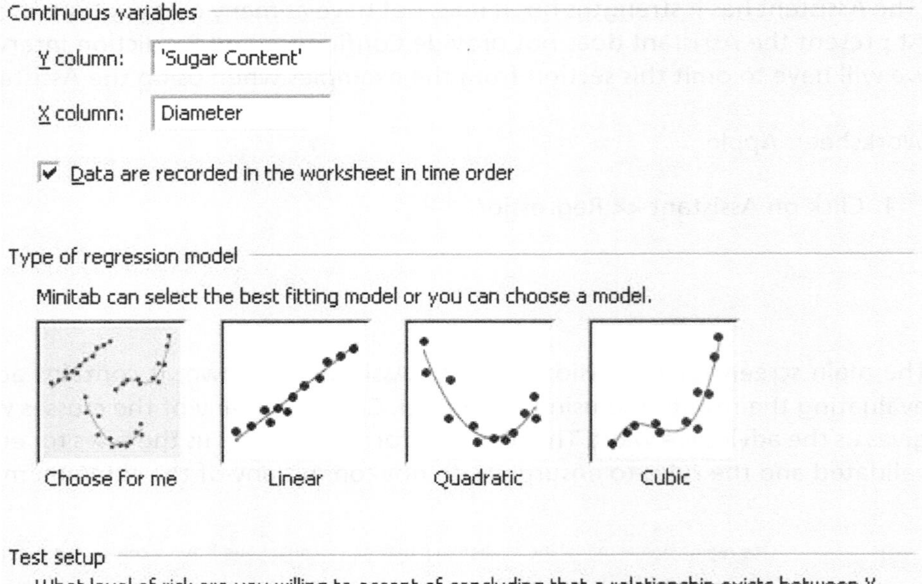

On the top left hand side of the Summary Report we are told that the *P* value is zero and that there is a relationship between X and Y. Below that we are told that the R-Sq (adj) is 81.5% meaning that 81.5% of the variation in Sugar Content can be explained by changes in Diameter.

The Assistant uses the same method as we did to select the best model. As it is the same method it has also selected the quadratic model as shown in the plot on the top right.

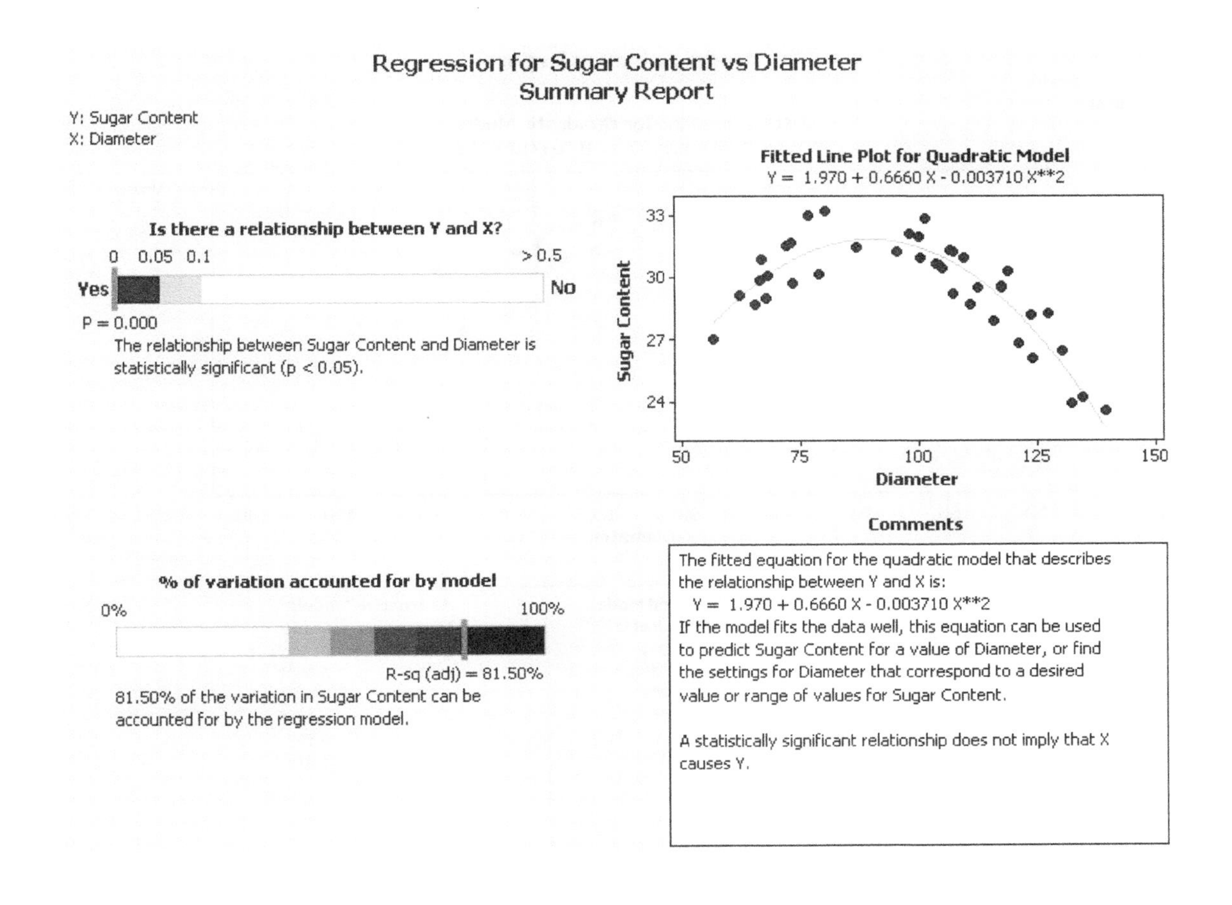

The Model Selection Report shows us the plot of the data with the selected fit. It has two points that are deemed to be unusual.

At the bottom of the page we have the model selection statistics shown in a table. The cubic model was rejected because the cubed term was not significant, with a *P* value of 0.187. The quadratic term in the model was the highest order significant term so it was selected to be the basis of the model.

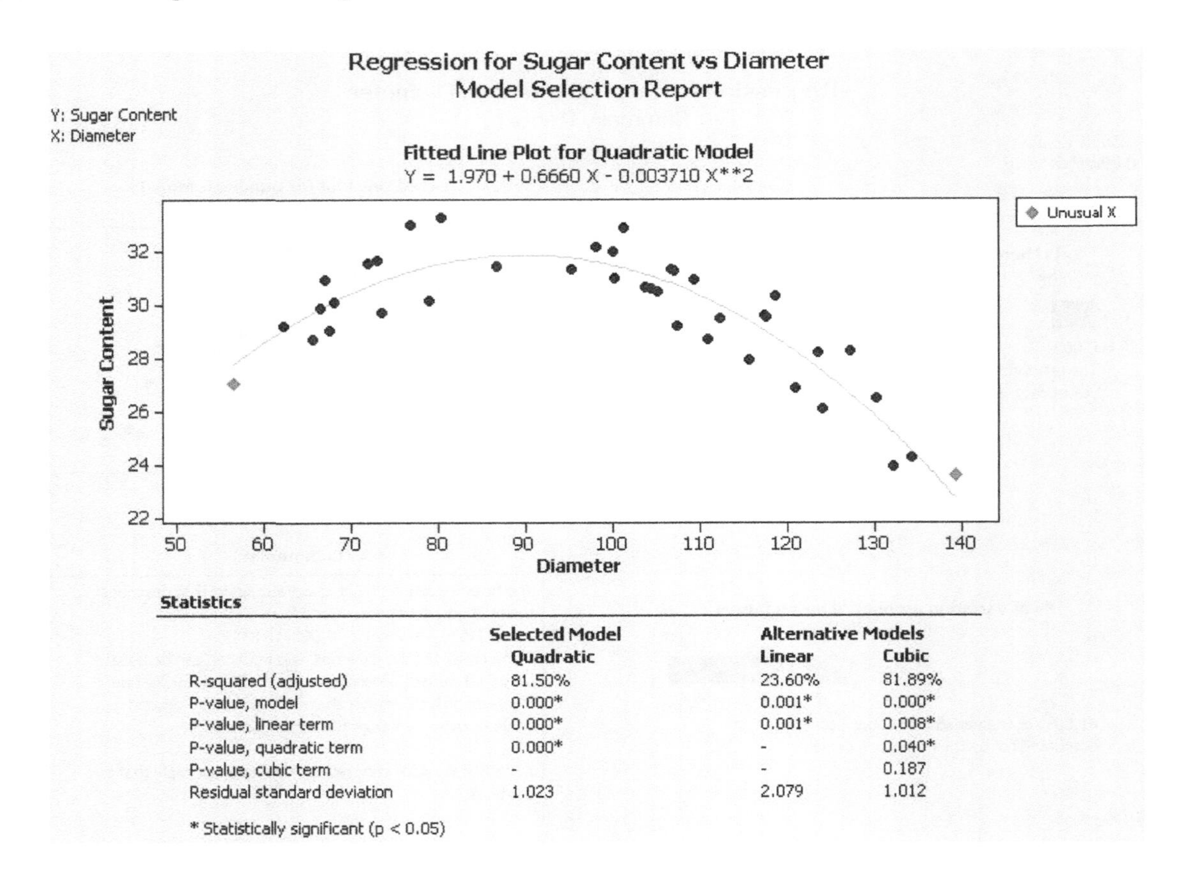

Statistics

	Selected Model	Alternative Models	
	Quadratic	Linear	Cubic
R-squared (adjusted)	81.50%	23.60%	81.89%
P-value, model	0.000*	0.001*	0.000*
P-value, linear term	0.000*	0.001*	0.008*
P-value, quadratic term	0.000*	-	0.040*
P-value, cubic term	-	-	0.187
Residual standard deviation	1.023	2.079	1.012

* Statistically significant (p < 0.05)

Diagnostic Report 1 displays the plot of Residual versus Fitted Values. When this plot was shown in the four in one residual plot we checked to make sure the Residuals were evenly distributed about the zero line. Additionally, the Assistant gives us examples of patterns that would indicate problems with our model.

Regression for Sugar Content vs Diameter
Diagnostic Report 1

Residuals vs Fitted Values

Look for large residuals (marked in red) and patterns.

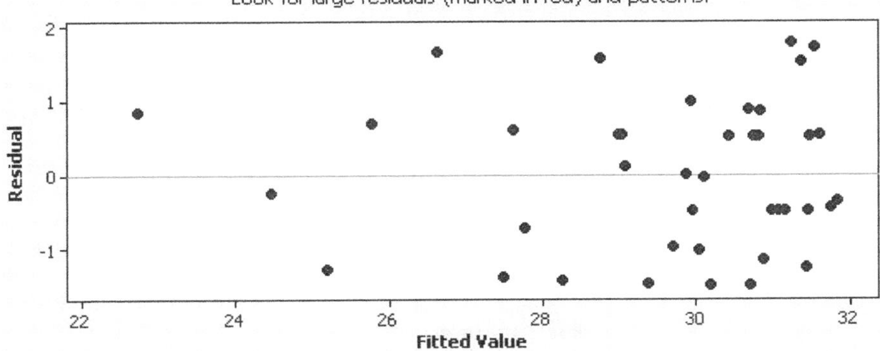

Examples of patterns that may indicate problems with the fit of the model:

Unequal variation

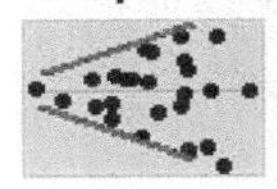

Uneven variability, such as when the spread of points increases as the fitted values increase. If the unequal variation is severe, get help to address the problem.

Strong curvature

Curve in the data that is not well explained by the regression model. If you are already using the best fitting model, get help to address the problem.

Clusters

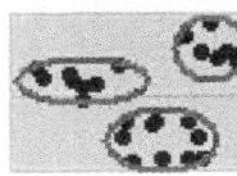

Groups of points that suggest there may be important X variables that were not included in the regression model. Get help to address the problem.

Large residuals

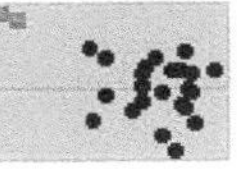

Points that are not well fit by the model. Try to understand why the points are unusual. Correct measurement or data entry errors and consider removing data that have special causes.

Diagnostic Report 2 displays the plot of Residual versus Observation Order. This plot is only displayed when we indicate that the data was in time order in the data entry menu box.

When this plot was shown in the four in one residual plot we checked to make sure the Residuals did not show any patterned behaviour.

Additionally, the Assistant gives us examples of patterns that would indicate problems with our model.

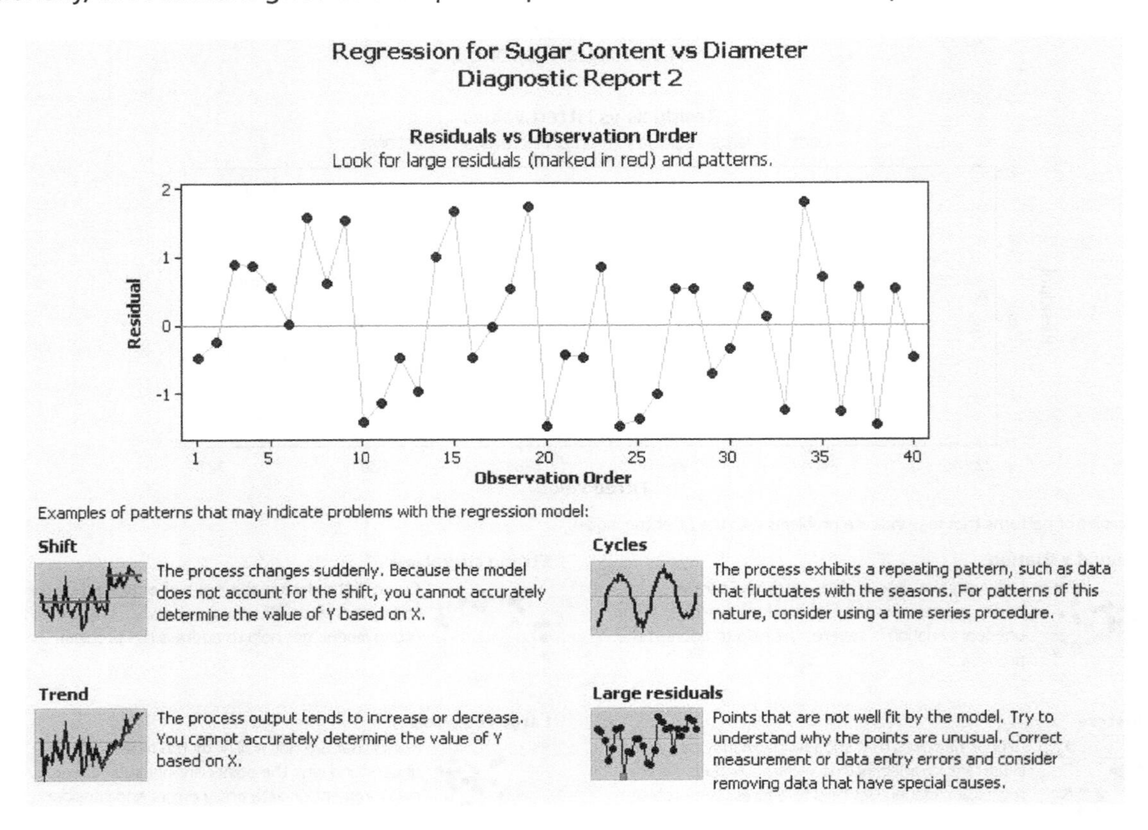

The Report Card validates our test procedure. We are told we had enough data to obtain a precise estimate of the strength of the relationship and for the normality of residuals not to be an issue.

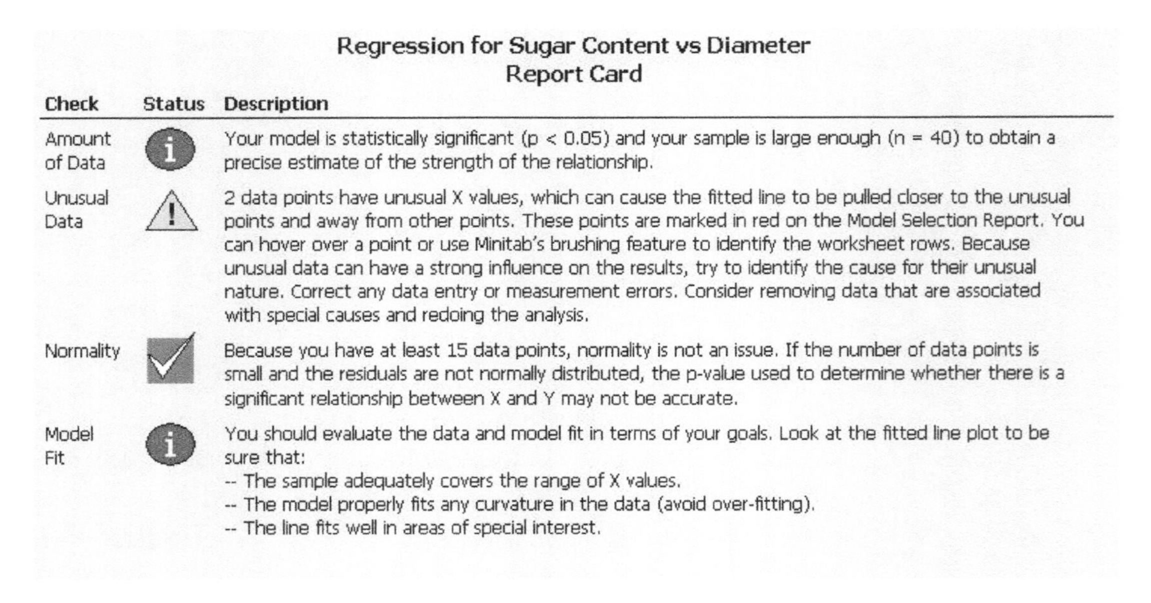

Regression for Sugar Content vs Diameter
Report Card

Check	Status	Description
Amount of Data	i	Your model is statistically significant (p < 0.05) and your sample is large enough (n = 40) to obtain a precise estimate of the strength of the relationship.
Unusual Data	!	2 data points have unusual X values, which can cause the fitted line to be pulled closer to the unusual points and away from other points. These points are marked in red on the Model Selection Report. You can hover over a point or use Minitab's brushing feature to identify the worksheet rows. Because unusual data can have a strong influence on the results, try to identify the cause for their unusual nature. Correct any data entry or measurement errors. Consider removing data that are associated with special causes and redoing the analysis.
Normality	✓	Because you have at least 15 data points, normality is not an issue. If the number of data points is small and the residuals are not normally distributed, the p-value used to determine whether there is a significant relationship between X and Y may not be accurate.
Model Fit	i	You should evaluate the data and model fit in terms of your goals. Look at the fitted line plot to be sure that: -- The sample adequately covers the range of X values. -- The model properly fits any curvature in the data (avoid over-fitting). -- The line fits well in areas of special interest.

We are given a warning that we had two unusual data points and that we should try and find the reason why these data points were unusual.

Exercise 3. Fitted Line Plot.

Scientists have discovered three new materials. A pulling force is applied to each material and then the length of the material is measured, both the force and the extension are recorded. This process is repeated with a continually increasing force until the material breaks.

Use the Classic Method and the Assistant to establish if any kind of relationship exists between the force applied and the extension of the material. What is the strength of the relationship?

Using the Classic Method establish if:

- A force of 1500 N on material A could give an increase in length of 150 mm?
- A force of 800 N on material C could give an increase in length of 250 mm?

Worksheet: Materials.

Material A

Initially, the cubic model was selected but the results in the session window showed that the cubic term was not significant. The quadratic model was then selected and this gave an R-Sq(adj) of 98.9%. We see that a force of 1500 N cannot give an increase in length of 150 mm as this point lies outside the prediction interval. The residual plots do not show any extreme patterns or behaviour.

Sequential Analysis of Variance

Source	DF	SS	F	P
Linear	1	438206	823.07	0.000
Quadratic	1	15552	122.97	0.000
Cubic	1	3	0.02	0.883

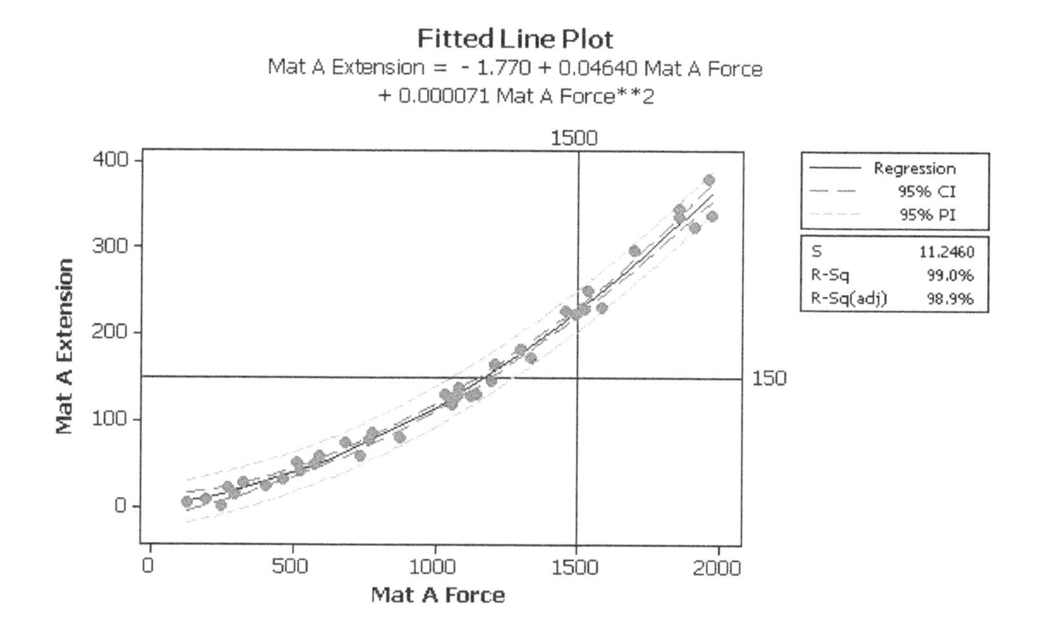

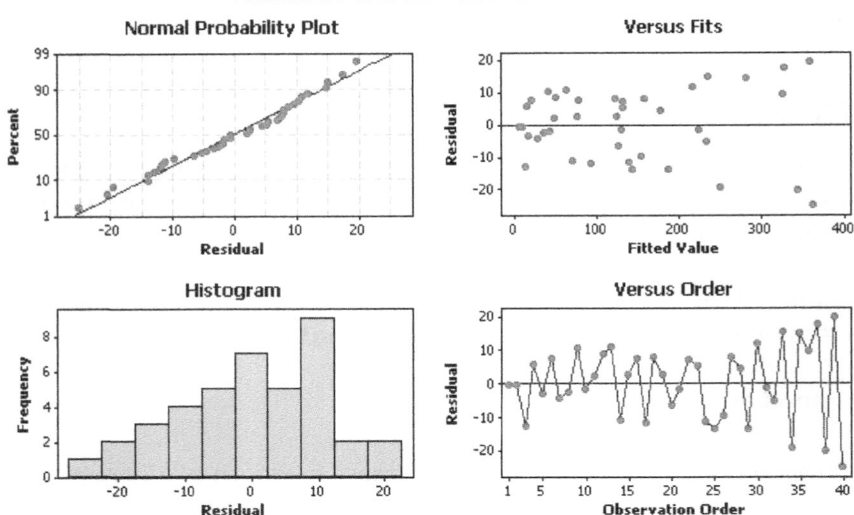

Material B

The results in the session window show that the linear model should be selected. With the linear model selected an R-Sq(adj) of 99.9% is achieved. The residual plots do not show any extreme patterns or behaviour.

Sequential Analysis of Variance

Source	DF	SS	F	P
Linear	1	3172237	38773.61	0.000
Quadratic	1	83	1.01	0.321
Cubic	1	49	0.59	0.448

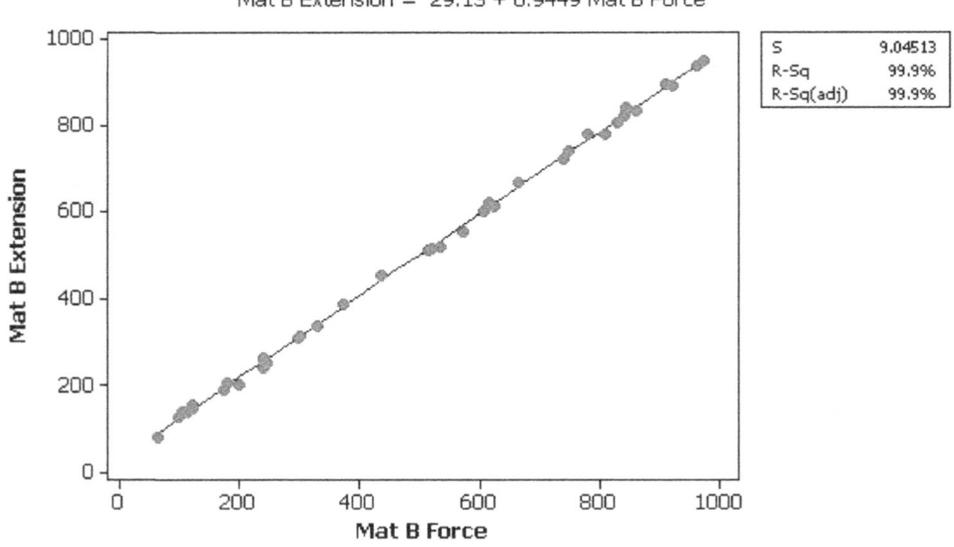

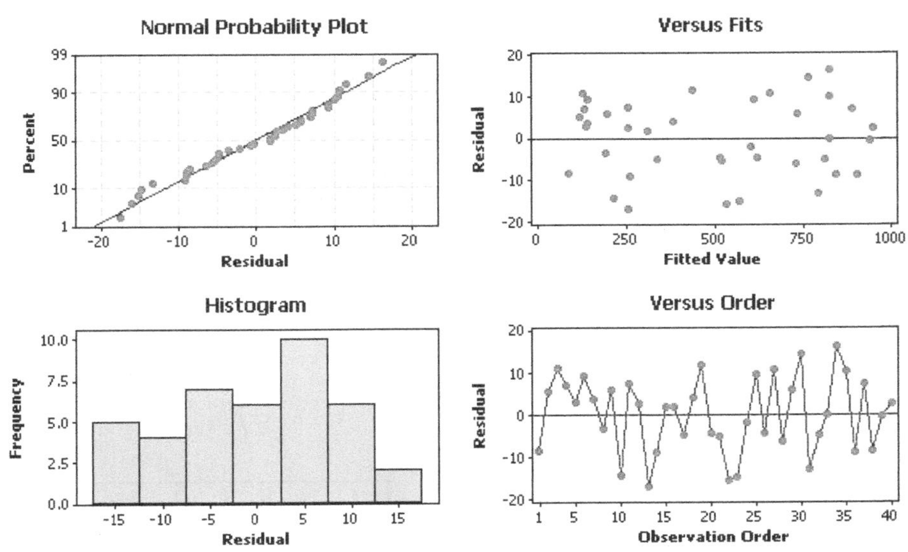

Material C

On this occasion the cubic model was selected and it turned out to be the correct choice. The cubic model gives an R-Sq(adj) of 99.5%. We see that a force of 800 N can give an increase in length of 250 mm as this point lies clearly within the prediction interval.

The residual plots do not show any extreme patterns or behaviour.

```
Sequential Analysis of Variance

Source      DF      SS       F      P
Linear       1  920121  939.38  0.000
Quadratic    1   10619   14.77  0.000
Cubic        1   22487  196.75  0.000
```

Fitted Line Plot

Mat C Extension = 130.1 - 0.9545 Mat C Force
+ 0.002193 Mat C Force**2 - 0.000001 Mat C Force**3

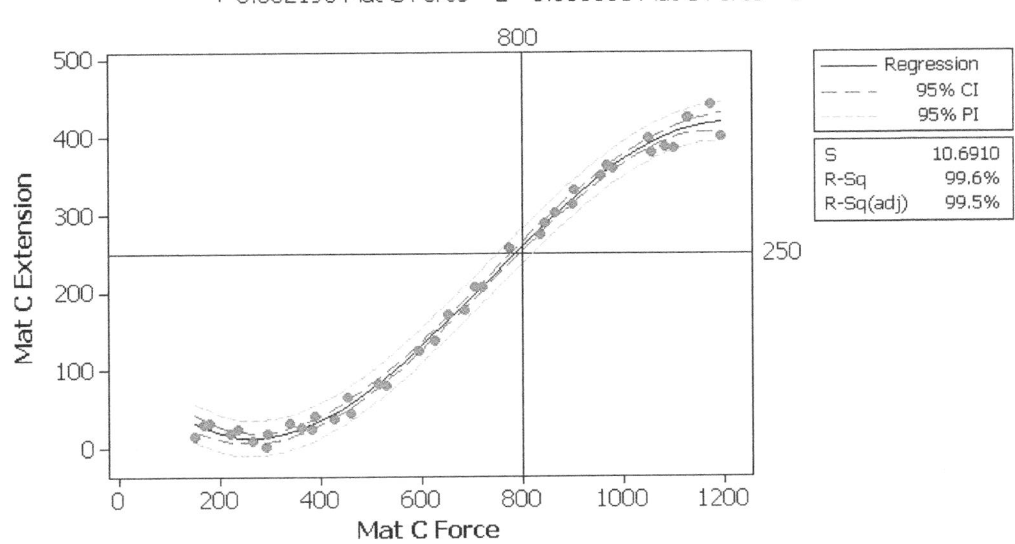

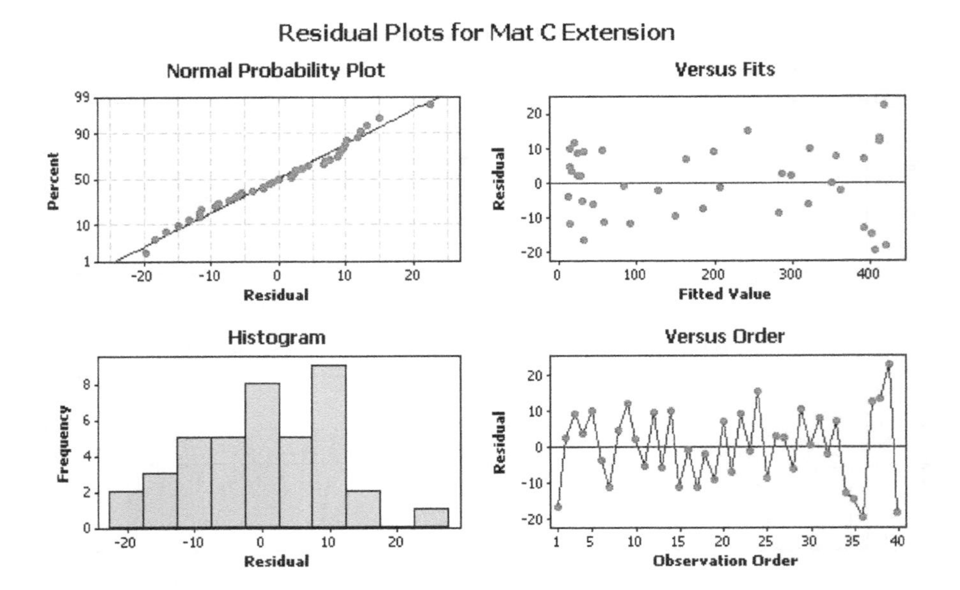

Material A

On the Summary Report we see that the Assistant also selected the quadratic model. We see that the relationship is significant and the R-Sq(adj) is the same as that calculated in the Classic Method. The Assistant has also fitted the same equation as the Classic Method.

The Model Selection Report, not shown, had one residual which was deemed to be unusual.

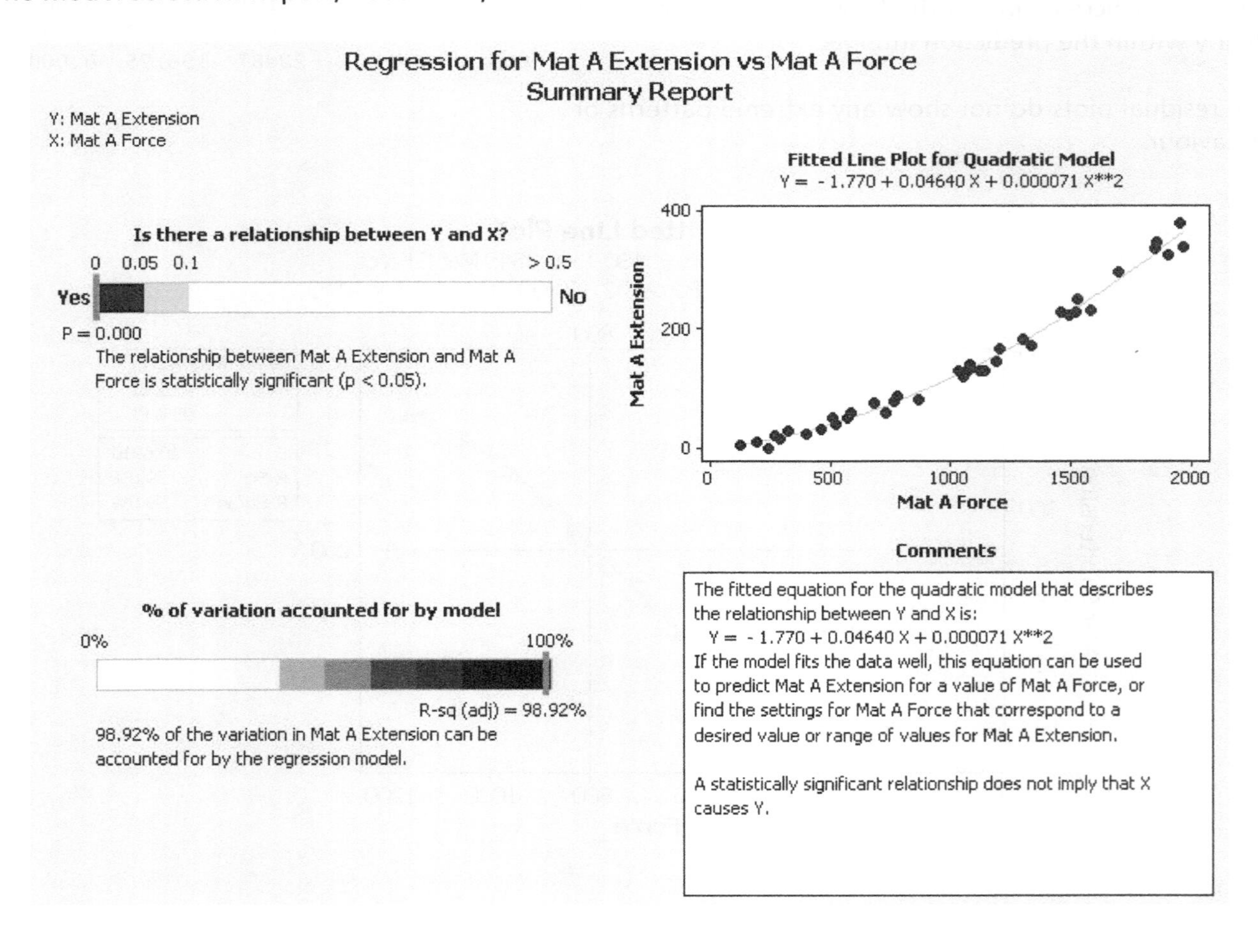

Material B

On the Summary Report we see that the Assistant also selected the linear model. We see that the relationship is significant and the R-Sq(adj) is the same as that calculated in the Classic Method. The Assistant has also fitted the same equation as the Classic Method.

As the model is linear the Assistant displays the value of the correlation coefficient. The result is $r = 1$.

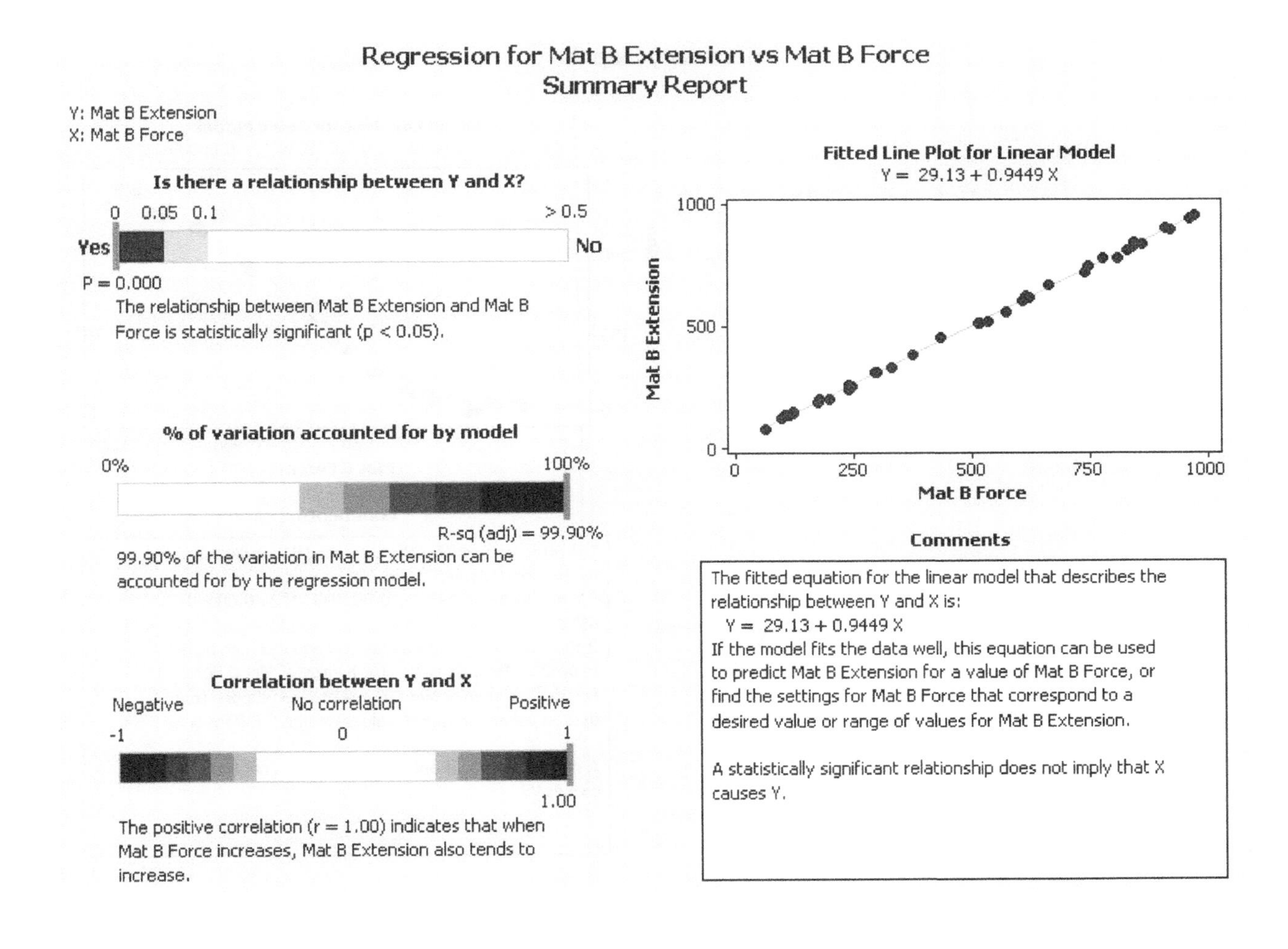

Material C

The Assistant has selected the cubic model. The key results and the fitted equation are again the same as the Classic Method.

The Model Selection Report shows that two residuals were deemed to be unusual.

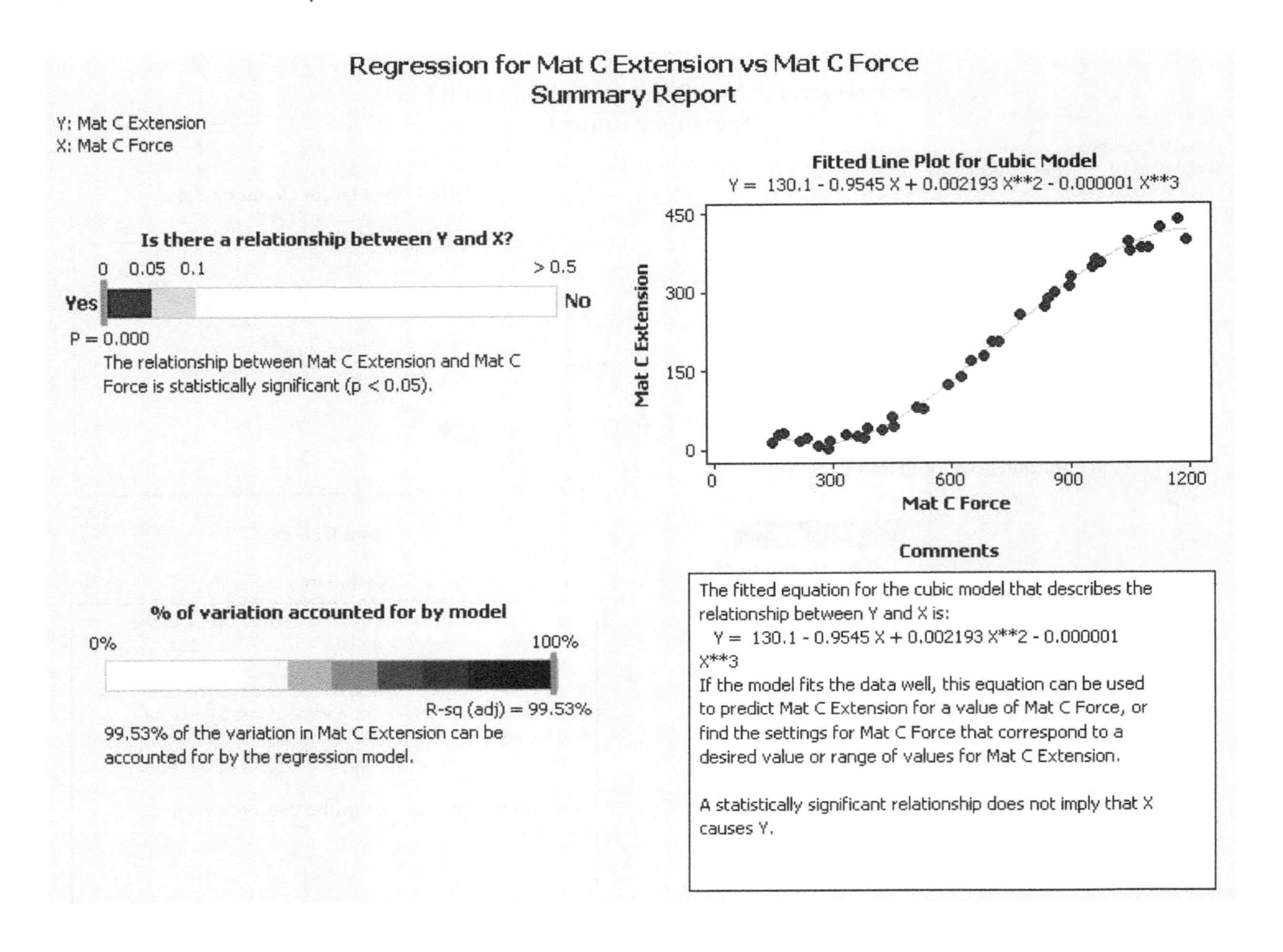

9.6 Introduction to Multiple Predictor Regression

We have used the Fitted Line Plot and the Assistant to form models for single factor regression. If we wanted to we could also have used the full regression models. However, in this course we have saved the full regression procedures for their intended purpose which is Multiple Predictor Regression.

Life would be too easy if there was only one version of this procedure so we must rejoice that there are two versions. M15 users must use the standard regression procedure found within Stat <<Regression << Regression. M16 users have both versions available to them but the version found under Stat <<Regression << General Regression is the updated version. It can handle categorical predictors and can directly model polynomial predictors and predictors with interaction. It is suggested that M16 users use this version. This text will also use the General Regression procedure but we will on one occasion show the M15 and M16 methodologies separately.

9.7 Multiple Predictor Regression

Test	• Regression
What's it used for?	• It is used to model multiple predictor variables against a single dependent response variable
Assumptions and Limitations	• Assumptions should not be made about data that is extrapolated outside the study range. • Data for the study needs to be carefully collected so that it is representative of the population. • If the predictors are dependent upon each other multi-co-linearity may exist and it may be challenging to reduce the model. • The Regression procedure will not automatically search for quadratic or cubic relationships.

Example 5.

The Admiralty have asked Captain Pilch to study factors which could affect the range of a standard naval cannon used in galleons.

Analyse the data to establish which of the predictors are significant to the response? Then form the regression equation.

The Admiralty have supplied nine sets of preset values of the predictors in columns K to P of the Excel file. They want these values to be used to produce nine estimates of the firing distance using the regression equation. Also, they want the 95% confidence limits (CL) and 95% prediction limits (PL) for the Firing Distance in each of the nine cases.

Worksheet: Cannon.

	A	B	C	D	E	F	G	H	I	J	K	L	M	N	O	P
1	Ball_Wt	Charge_W	Charge_Gr	Wadding_	Annulus_R	Barrel_Len		Distance Fired			Pre_Ball_Wt	Pre_Charge_W	Pre_Charge_G	Pre_Wadding_	Pre_Annulus_F	Pre_Barrel_Len
2	5	1177	7.7	9	2.1	112		274.2			6	1100	16	10	2.1	110
3	9.2	1132	19.1	8	2.1	116		329.3			6	1120	17	10	2.1	110
4	10.5	1174	19.2	6	2.6	117		326.8			6	1140	18	10	2.1	110

1. Open the Excel file and transfer all of the data into Minitab. Columns K to P contain nine rows of data that we will use to produce estimates of the firing distance after we have formed the regression equation.

2. We need to view the graphical relationship between the predictors to check whether they are independent. Click Graph <<Matrix Plot.

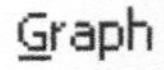

3. Select Simple and then click OK.

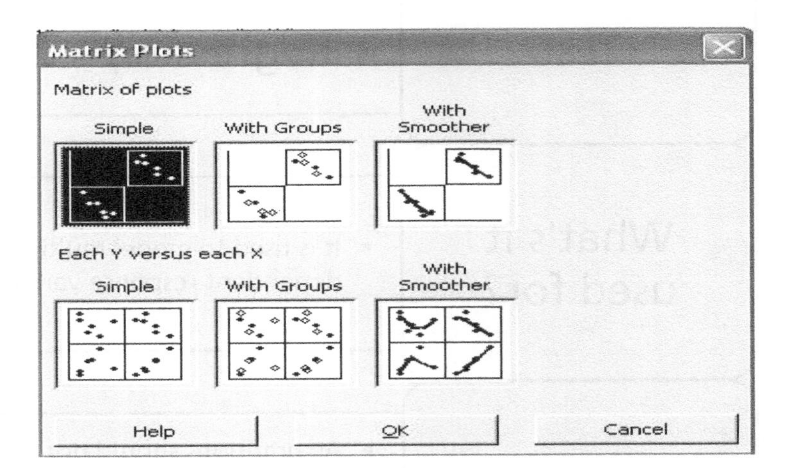

4. Select all the data columns as the Graph variables.

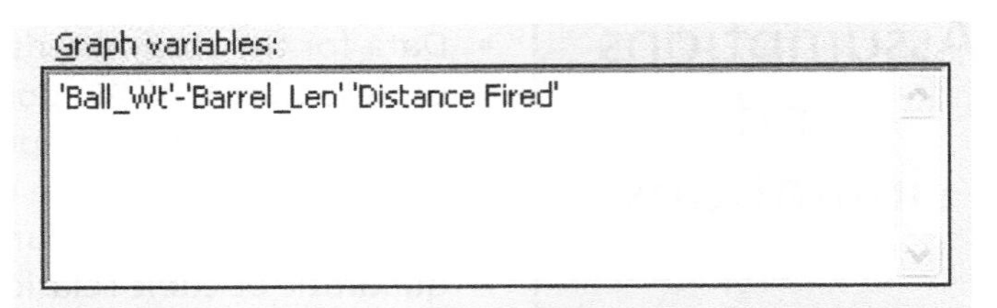

5. Click on Matrix Options.

6. Under Matrix Display select the radio button for Lower left (personal preference).
7. Click OK and OK again.

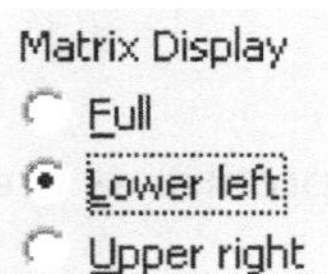

Distance Fired is our response variable.

It is okay if we see patterns between the response and the predictors. However, all the predictors should be independent from each other.

Graphically, it can be seen that the predictors are independent but we will check the correlation coefficients as well.

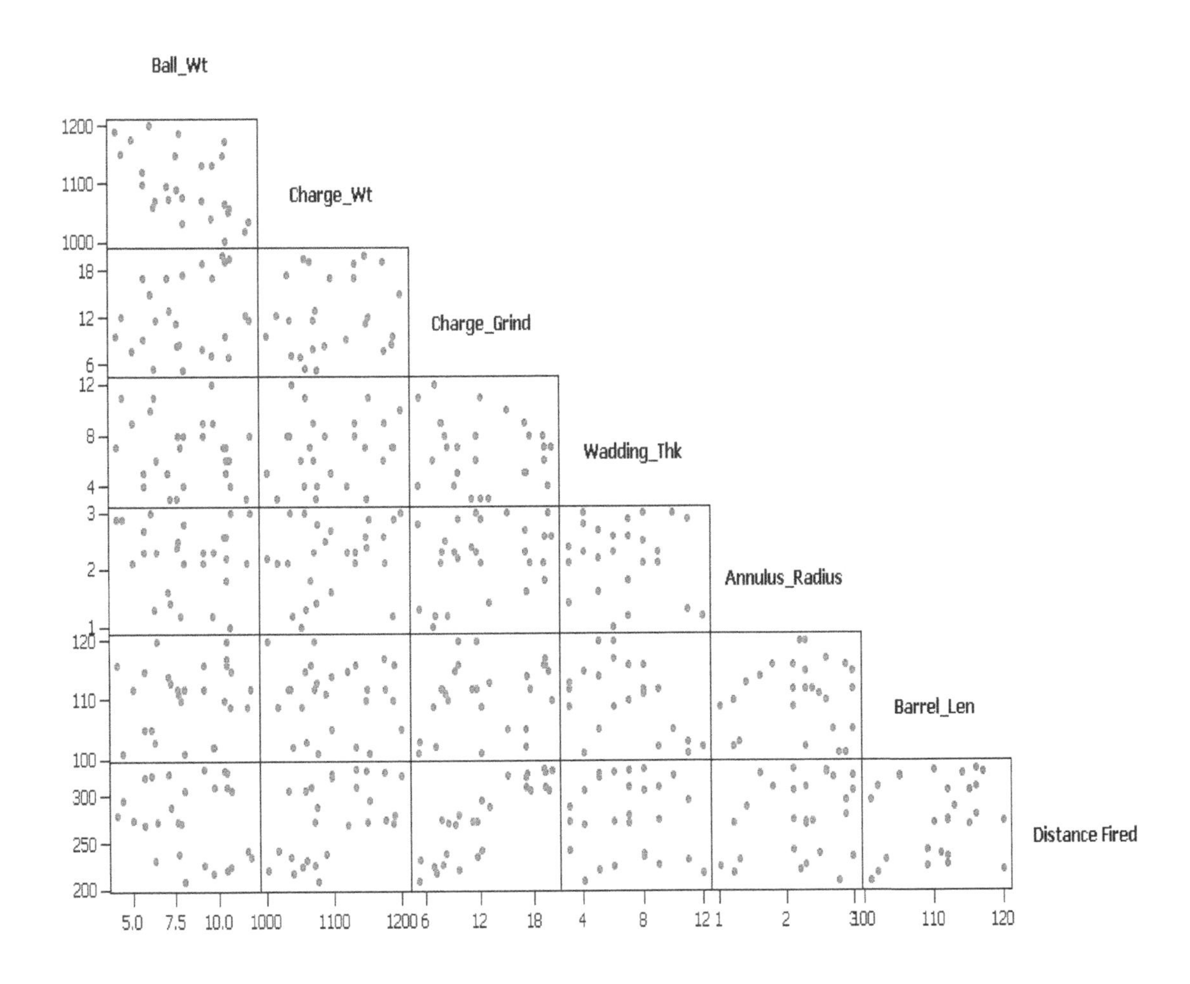

8. To check the strength of the correlation and whether it is statistically significant click Stat <<Basic Statistics <<Correlation.

9. Select all the Variables and then click OK.
10. Select the session window icon to see the results.

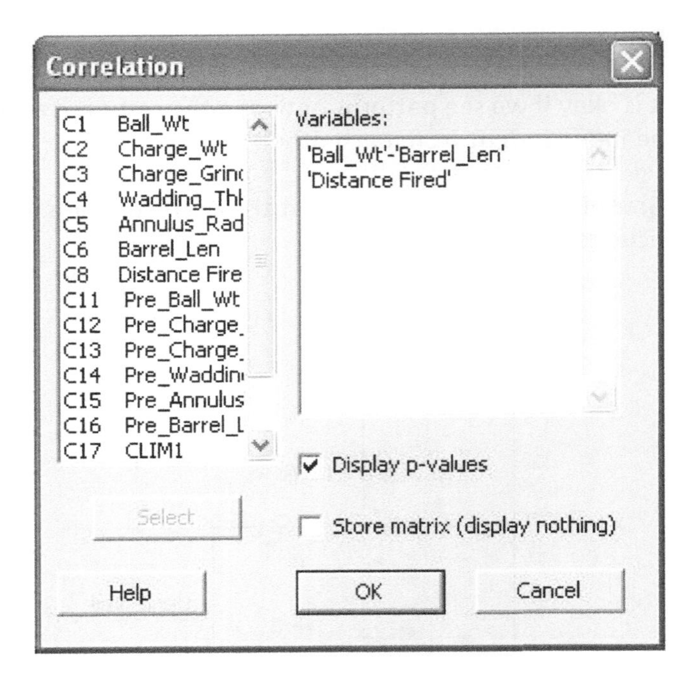

We see that Charge_Wt and Ball_Wt have a significant correlation. Also, Barrel_Len and Wadding_Thk have a significant correlation. This means that there may be instability within our model. When it comes to reducing our model we should only remove one predictor at a time.

We are now ready to carry out the regression analysis.

Correlations: Ball_Wt, Charge_Wt, Charge_Grind, Wadding_Thk, Annulus_Radius, Barrel_Len, Distance Fired

	Ball_Wt	Charge_Wt	Charge_Grind	Wadding_Thk	Annulus_Radius	Barrel_Len
Charge_Wt	-0.521					
	0.005					
Charge_Grind	0.267	0.144				
	0.178	0.474				
Wadding_Thk	-0.174	0.175	-0.156			
	0.386	0.382	0.436			
Annulus_Radius	-0.116	0.267	0.296	-0.076		
	0.564	0.178	0.135	0.707		
Barrel_Len	0.154	-0.072	0.266	-0.440	0.036	
	0.444	0.722	0.179	0.021	0.859	
Distance Fired	-0.168	0.521	0.878	-0.049	0.258	0.183
	0.402	0.005	0.000	0.806	0.193	0.361

Cell Contents: Pearson correlation
P-Value

11. Click on Stat <<Regression <<General Regression. M15 users will have to use Stat <<Regression <<Regression. The results of this example should be virtually the same for both test procedures.
12. Select Distance Fired as the Response and C1 to C6 as the predictors.
13. Click OK.
14. As yet there are no graphical outputs from this procedure. Click on the session window icon as all the results are in the session window. The regression equation is shown below.

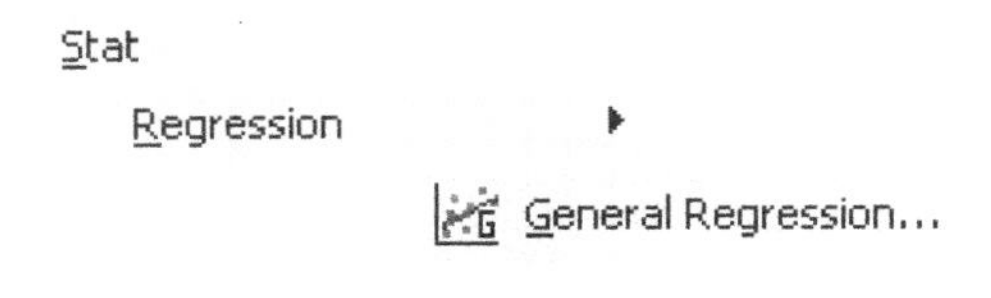

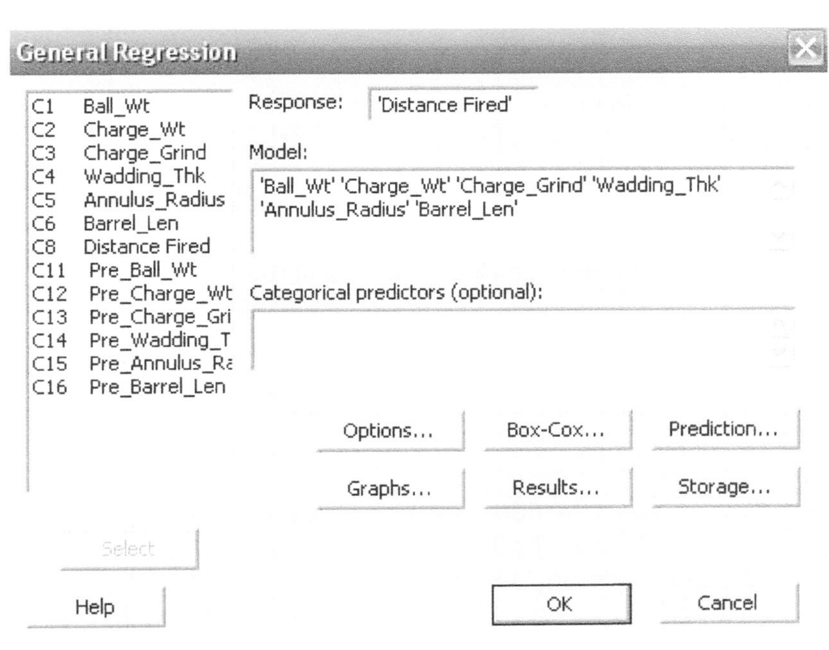


```
Distance Fired  =  46.534 - 5.42601 Ball_Wt + 0.182509 Charge_Wt + 8.00111 Charge_Grind - 0.176265 Wadding_Thk - 8.79371
                   Annulus_Radius - 0.0566777 Barrel_Len
```

15. The regression table shows all the predictors and whether they are significant or not in terms of the *P* value. We see that Wadding_Thk and Barrel_Len are not significant. Previously, we found these predictors to be correlated so we should be careful and remove them one at a time. We will go through the regression exercise again and remove Barrel_Len as a predictor. Also, note the R-Sq(adj) value, this is a good fit for the process.

Coefficients

Term	Coef	SE Coef	T	P
Constant	46.5340	21.0939	2.2060	0.039
Ball_Wt	-5.4260	0.3412	-15.9034	0.000
Charge_Wt	0.1825	0.0134	13.6148	0.000
Charge_Grind	8.0011	0.1477	54.1725	0.000
Wadding_Thk	-0.1763	0.2673	-0.6595	0.517
Annulus_Radius	-8.7937	1.0889	-8.0758	0.000
Barrel_Len	-0.0567	0.1223	-0.4633	0.648

Summary of Model

```
S = 3.07570        R-Sq = 99.55%         R-Sq(adj) = 99.42%
PRESS = 354.400    R-Sq(pred) = 99.17%
```

16. Click on the Edit Last button.

17. Change the selection for the predictors so that Barrel_Len is not included and Click OK .

Response: 'Distance Fired'

Model:

'Ball_Wt' 'Charge_Wt' 'Charge_Grind' 'Wadding_Thk' 'Annulus_Radius'

18. If multi colinearity had existed Wadding_Thk would have become significant but this has not happened. We can proceed and remove Wadding_Thk as well. Click on the Edit Last icon.

Predictor	Coef	SE Coef	T	P
Constant	39.98	15.35	2.60	0.017
Ball_Wt	-5.4279	0.3347	-16.22	0.000
Charge_Wt	0.18257	0.01315	13.88	0.000
Charge_Grind	7.9868	0.1417	56.36	0.000
Wadding_Thk	-0.1256	0.2392	-0.52	0.605
Annulus_Radius	-8.765	1.067	-8.22	0.000

19. Change the selection for the predictors so that both Barrel_Len and Wadding_Thk are not included.
20. Click OK.

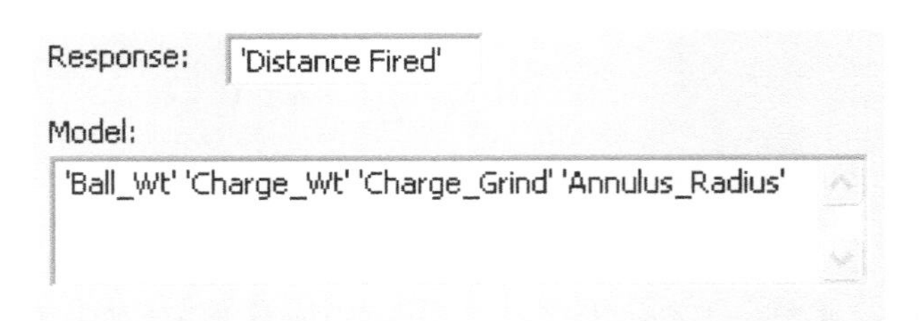

21. In the session window look at the regression table, it now only includes significant predictors and we have a very good fit of 99.5%.

Coefficients

Term	Coef	SE Coef	T	P
Constant	40.0673	15.0931	2.6547	0.014
Ball_Wt	-5.4220	0.3290	-16.4812	0.000
Charge_Wt	0.1815	0.0128	14.2166	0.000
Charge_Grind	7.9966	0.1381	57.8857	0.000
Annulus_Radius	-8.7163	1.0449	-8.3414	0.000

Summary of Model

```
S = 2.96754        R-Sq = 99.54%         R-Sq(adj) = 99.46%
PRESS = 295.563  R-Sq(pred) = 99.30%
```

Regression Equation

```
Distance Fired  =  40.0673 - 5.42203 Ball_Wt + 0.18145 Charge_Wt + 7.99658 Charge_Grind - 8.71625 Annulus_Radius
```

The final regression equation with only significant predictors is shown above. As the R-Sq(adj) is high we can be confident that our regression equation will give us an accurate model to predict distance fired.

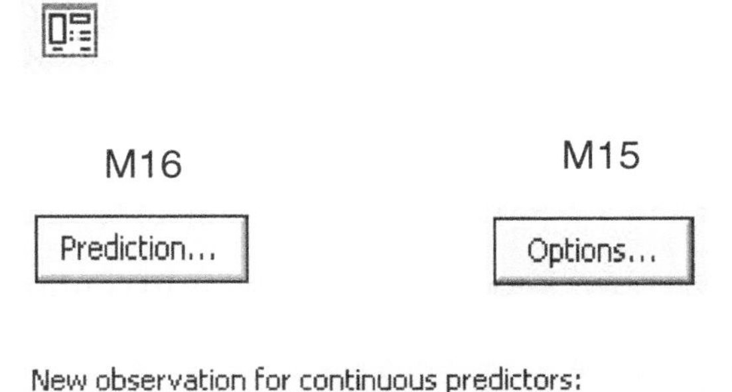

22. We can now ask Minitab to use the regression equation and our preselected values for the predictors to calculate what firing distance they will give us including the confidence and prediction limits. Click on the Edit Last icon and then the prediction button. M15 users need to click on the Options button.

New observation for continuous predictors:

harge_Wt' 'Pre_Charge_Grind' 'Pre_Annulus_Radius'

New observation for categorical predictors:

23. In the menu box for New Observations for continuous predictors enter the pre-defined values for our significant predictors. Under storage select the tick boxes for Confidence limits (CL) and Prediction limits (PL).

Storage
- Fits
- SEs of fits
- Confidence limits
- Prediction limits

24. Then click OK.

25. Again we must check the validity of the model by looking at the residuals. To do this click on Graphs.

Graphs...

26. Then select Four in one under Residual Plots. Click OK and OK again.

Residual Plots
- Individual plots
  - Histogram of residuals
  - Normal plot of residuals
  - Residuals versus fits
  - Residuals versus order
- Four in one

27. Since we had asked for the results of the CLs and PLs to be stored they appear in the current worksheet. They also appear in the session window. To view them click on the Session Window icon.

C17	C18	C19	C20
CLIM1	CLIM2	PLIM1	PLIM2
314.283	319.260	310.133	323.410
325.894	330.900	321.753	335.041
337.398	342.647	333.332	346.713
309.349	313.350	304.878	317.821
320.897	325.053	316.479	329.471
332.323	336.879	328.038	341.163
304.273	307.582	299.555	312.300
315.735	319.371	311.136	323.970
327.076	331.281	322.675	335.682

There were nine rows of pre-defined predictor values and we were asked to find the estimated Distance Fired and the CIs for the mean and PIs for additional points. Each row in the session window corresponds to each row of data, in order.

For our first set of data the Distance Fired would be 316.772, the mean firing distance would be between 314.3 and 319.3. The Prediction Interval for additional points is 310.1 and 323.4.

We will now check the residual plots to confirm the validity of the model.

New Obs	Fit	SE Fit	95% CI	95% PI
1	316.772	1.200	(314.283, 319.260)	(310.133, 323.410)
2	328.397	1.207	(325.894, 330.900)	(321.753, 335.041)
3	340.023	1.266	(337.398, 342.647)	(333.332, 346.713)
4	311.350	0.964	(309.349, 313.350)	(304.878, 317.821)
5	322.975	1.002	(320.897, 325.053)	(316.479, 329.471)
6	334.601	1.098	(332.323, 336.879)	(328.038, 341.163)
7	305.927	0.798	(304.273, 307.582)	(299.555, 312.300)
8	317.553	0.876	(315.735, 319.371)	(311.136, 323.970)
9	329.179	1.014	(327.076, 331.281)	(322.675, 335.682)

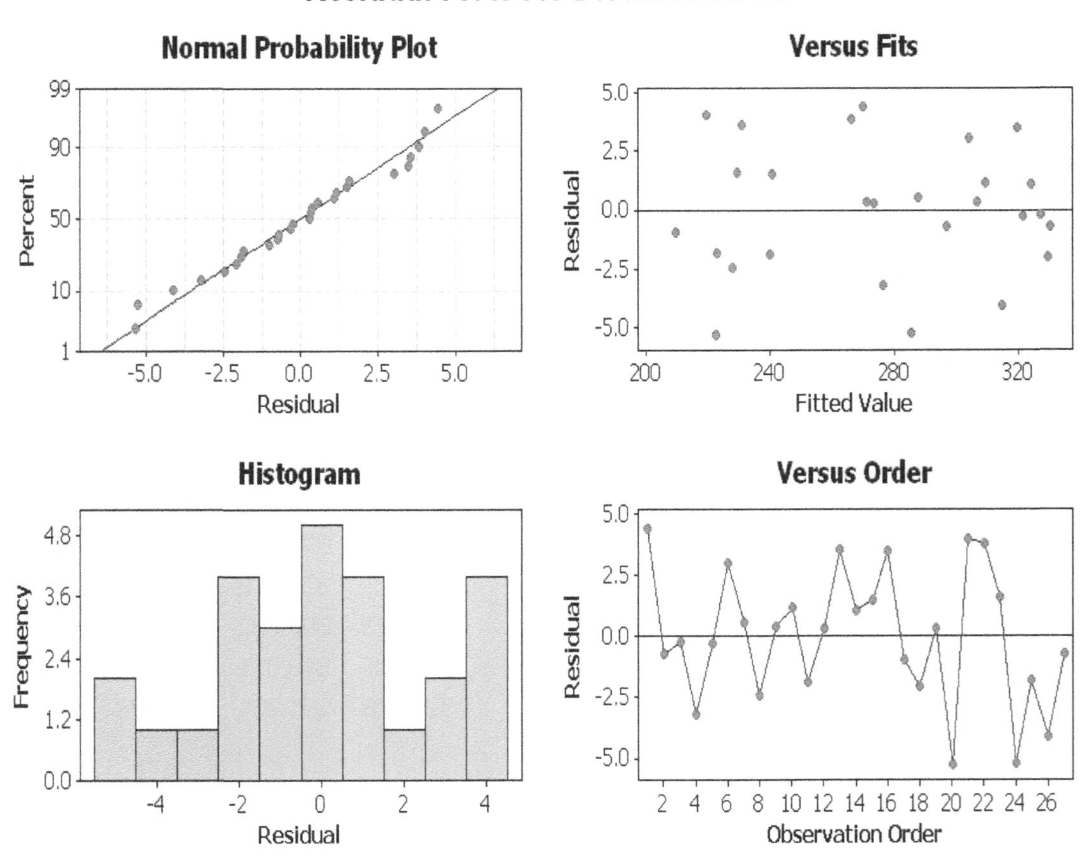

Starting from the top and going from left to right:

- The Normal Probability Plot shows that the Residuals are normally distributed.
- The Versus Fits Plot shows that the Residuals are equally distributed about the centre line.
- The Histogram Plot shows that the Residuals do not show extreme skewness.
- The Versus Order Plot shows that the Residuals are not displaying any patterns or ordered behaviour.

In summary, there are no issues with the residuals within this study.

Example 6.

Prawn wants to buy a car and is very interested in the power output of the engine. He knows that the power delivered by a car engine is affected by a number of factors and sets about collecting data on four key predictors that he feels will be significant.

Analyse the data Prawn has collected in order to establish which of the predictors are significant to the response?

Form the regression equation.

Worksheet: Engine Power.

1. Open the Excel file and transfer the data into Minitab.

	A	B	C	D	E	F
1	Comp ratio	EngTemp	MPG	Air Density		Power
2	16.2	64.5	29	0.7143		1346.7
3	30	70	26	0.8571		4588.5
4	8.8	87.7	36	1.4286		573.4
5	10.4	78.8	51	0.4286		430.1
6	14.4	44.5	33	0.8571		1155.6
7	27	54.1	32	0.2857		3574.8
8	17	90.7	35	1.4286		1633.7

2. Click Graph <<Matrix Plot.

Graph

Matrix Plot...

3. Select Simple and then click OK.

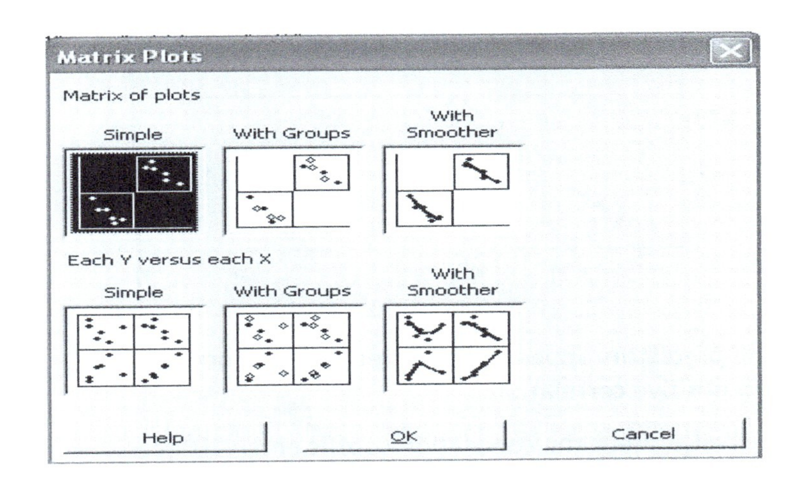

4. Select all the data columns as the graph variables.

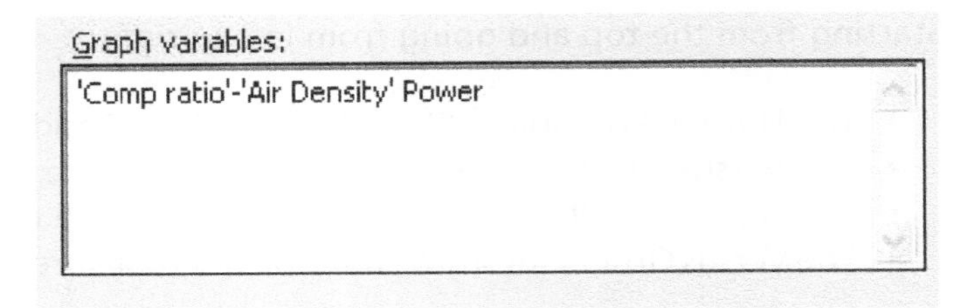

5. Click on Matrix options.

6. Under Matrix Display select the radio button for Lower left (personal preference).
7. Click OK and OK again.

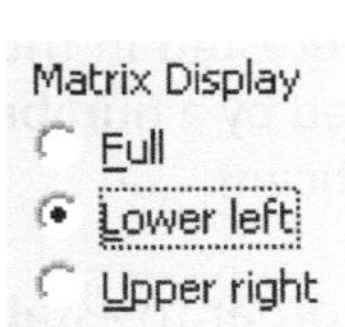

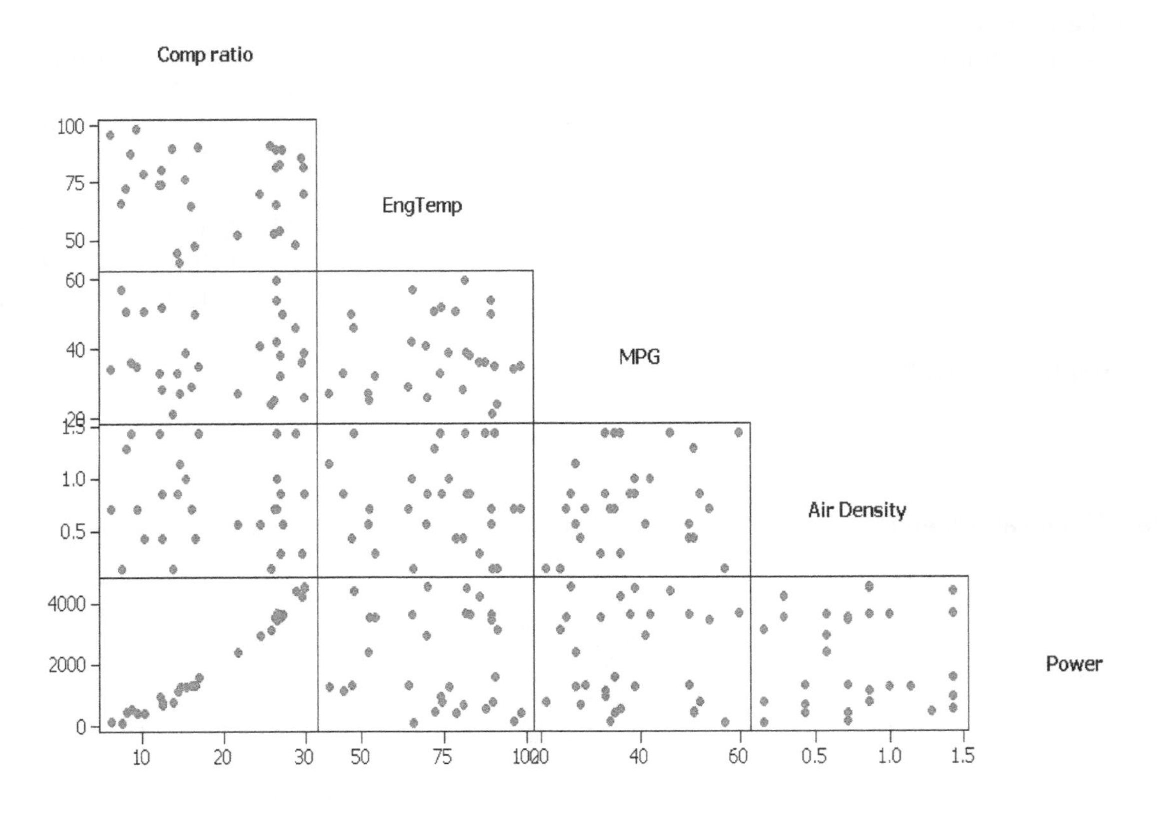

All the predictors appear to be independent from each other. Comp ratio and Power appear to have a strong positive correlation.

We will also check the correlation coefficients.

8. To check the strength of the correlation and whether it is statistically significant click Stat <<Basic Statistics <<Correlation.

Stat

Basic Statistics ▸

COR Correlation...

9. Select all the Variables and then click OK.
10. Select the session window icon to see the results.

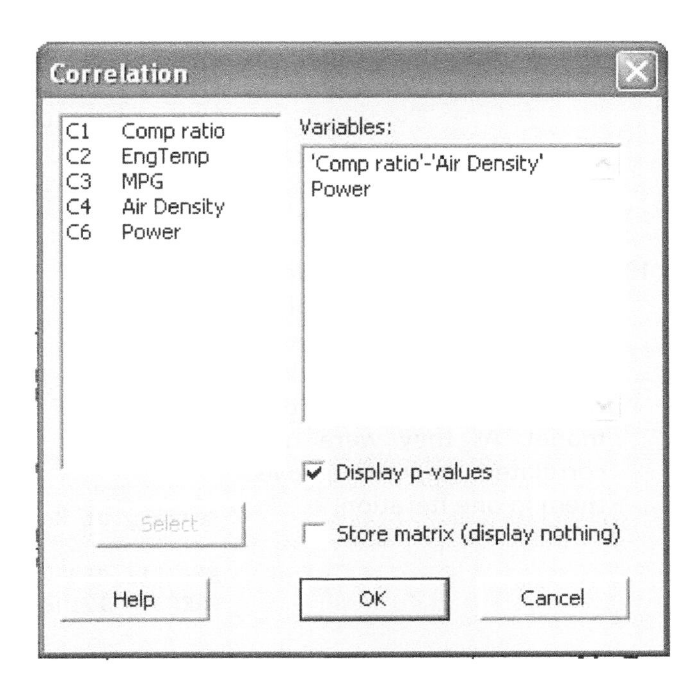

We see that Comp ratio and Power have a significant correlation.

None of the predictors are correlated so when it comes to reducing our model we can remove more than one predictor at a time.

We are now ready to carry out the regression analysis.

Correlations: Comp ratio, EngTemp, MPG, Air Density, Power

	Comp ratio	EngTemp	MPG	Air Density
EngTemp	-0.113			
	0.553			
MPG	-0.041	0.072		
	0.829	0.705		
Air Density	-0.055	-0.076	0.179	
	0.772	0.690	0.344	
Power	0.988	-0.094	-0.010	0.023
	0.000	0.620	0.959	0.904

Cell Contents: Pearson correlation
P-Value

11. Click on Stat <<Regression <<General Regression.

Stat
Regression ▸
General Regression...

12. Select Power as the response and all the other variables as predictors. Click OK.

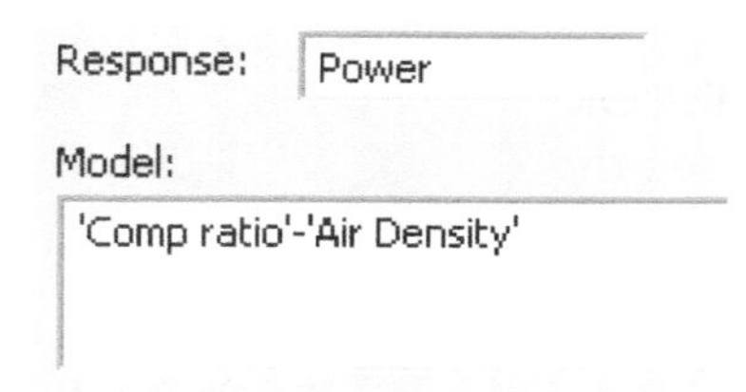

13. Click on the session window icon. The regression table shows that EngTemp and MGP are not significant. We can remove them from the model. As they were not correlated we can remove them in one iteration.

```
Term              Coef  SE Coef        T      P
Constant      -1953.18  258.540  -7.5546  0.000
Comp ratio      190.96    4.935  38.6956  0.000
EngTemp           2.10    2.380   0.8811  0.387
MPG               2.25    3.729   0.6033  0.552
Air Density     290.62   98.345   2.9551  0.007


Summary of Model

S = 211.754       R-Sq = 98.37%          R-Sq(adj) = 98.11%
PRESS = 1713645   R-Sq(pred) = 97.51%
```

14. Click on the Edit Last icon.

15. Change the selection of the predictors to Comp ratio and Air Density. Then click OK.

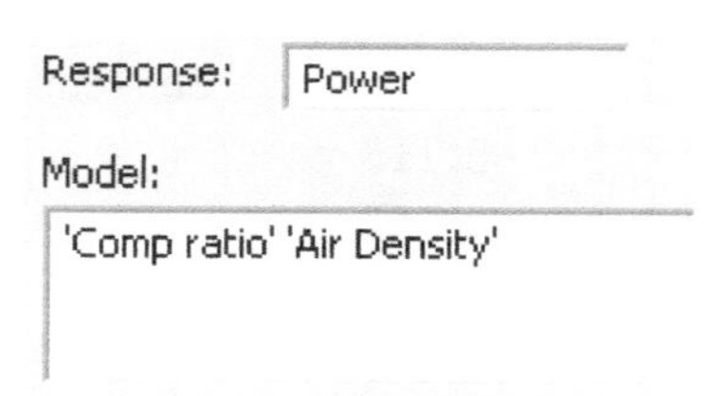

16. Click on the session window icon. The regression table only includes significant parameters. The R-Sq(adj) has improved from 98.1 to 98.2%.

```
Power  =  -1704.22 + 190.35 Comp ratio + 294.035 Air Density


Coefficients

Term              Coef  SE Coef         T      P
Constant      -1704.22  127.638  -13.3519  0.000
Comp ratio      190.35    4.830   39.4105  0.000
Air Density     294.03   94.969    3.0961  0.005
```

17. In order to check the residuals click on the Edit Last icon.

18. Then click on Graphs.

Graphs...

19. Under Residual Plots select Four in one.
20. Click OK and OK again.

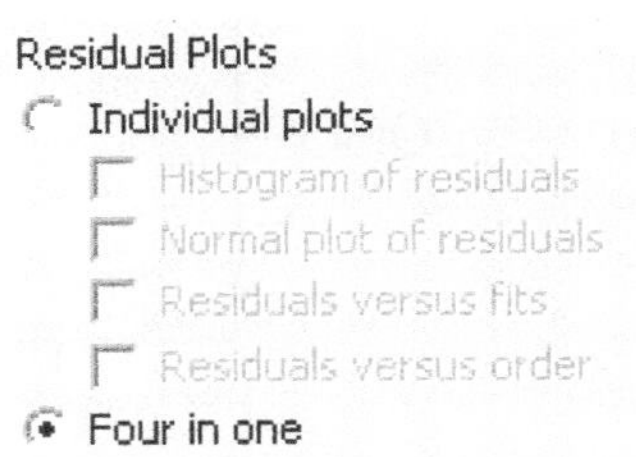

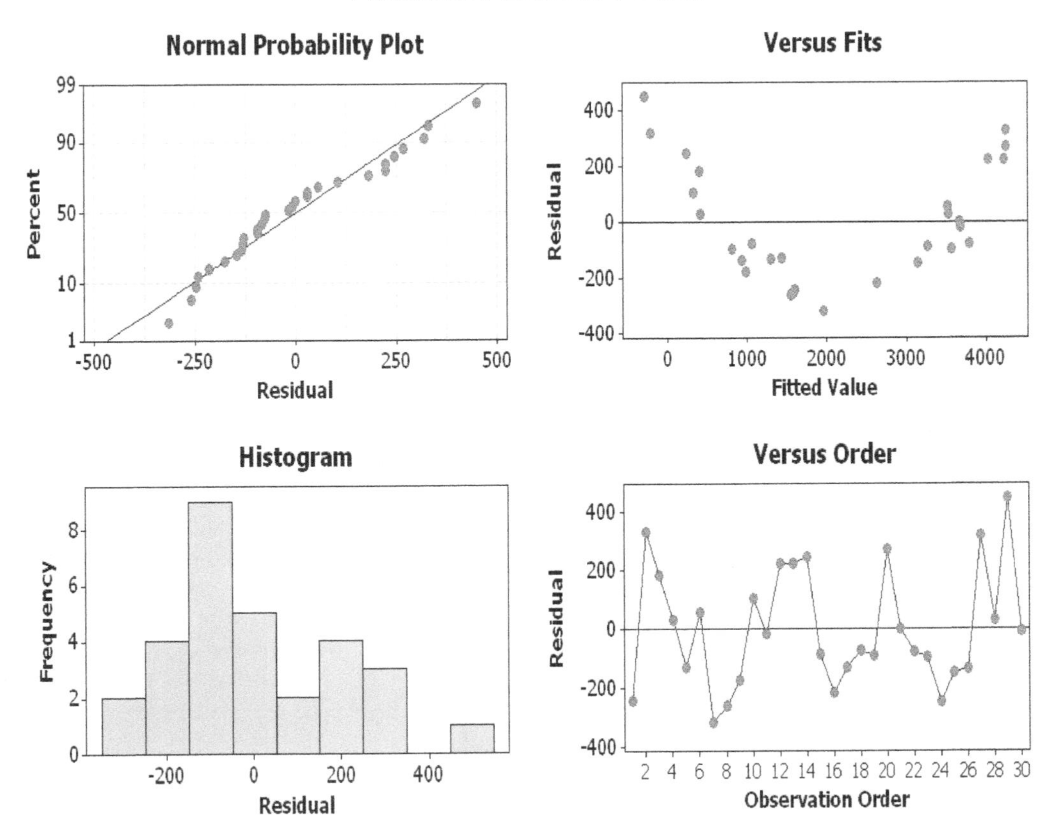

Starting from the top and going from left to right:

- The Normal Probability Plot shows that the Residuals are normally distributed.
- The Versus Fits Plot is showing patterned behaviour. We need to investigate why this is happening. Strong curvature indicates that the data is not well explained by the regression model.
- The Histogram Plot shows that the Residuals do not show extreme skewness.
- The Versus Order Plot shows that the Residuals are not displaying any patterns or ordered behaviour.

In summary, the Versus Fits plot is indicates that the model does not fit well. We need to go back and investigate why this is happening.

If we go back to the matrix plot we can see that the relationship is not quite linear. It may be quadratic. We can explore this in the regression model. The General Regression procedure in M16 has made this very easy. Initially, we will go through the method that M15 users need to use to investigate interactions and higher powers as predictors. The start of each step will indicate whether it is for M15 or M16 in order to highlight the version methodology.

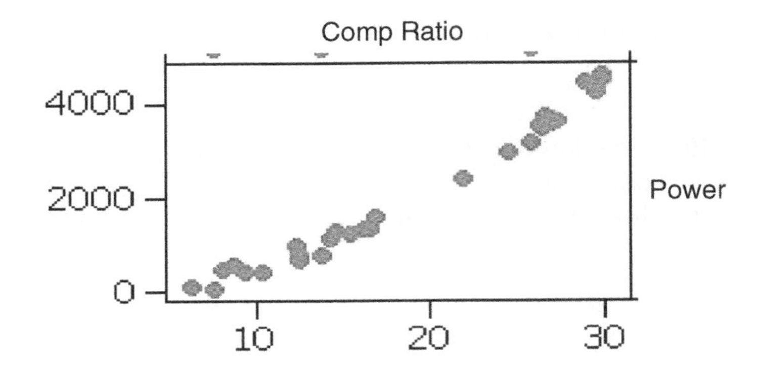

21. M15. Change the title of column C8 to Comp Ratio^2. '^' is to designate raising Comp ratio to the power of two.

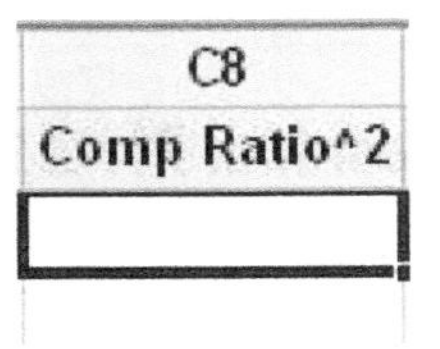

22. M15. Click on Calc <<Calculator.

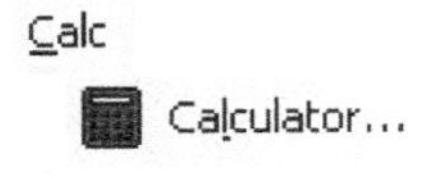

23. M15. In the Calculator menu store the results in column C8 and enter the expression Comp ratio*Comp ratio.
24. M15. Then click OK.

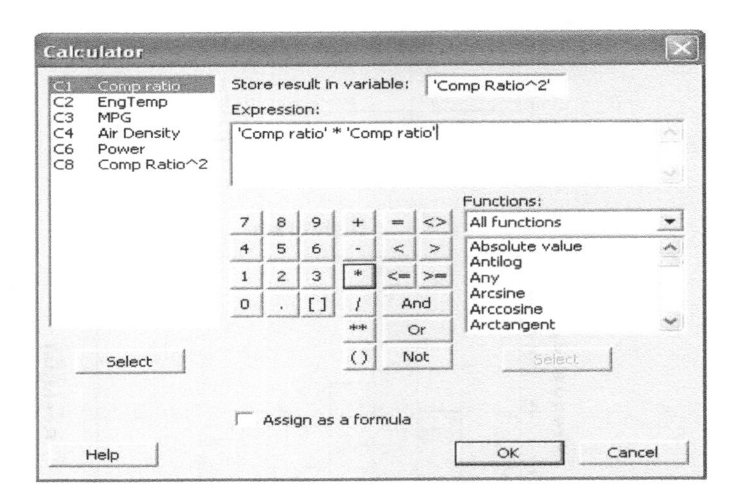

25. M15. We have created our Comp ratio squared term in C8.

C8
Comp Ratio^2
262.44
900.00
77.44

26. M15. Click on Stat <<Regression <<Regression.

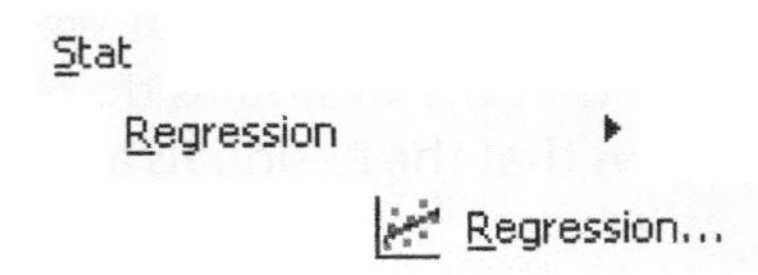

27. M15. Change the selection of the predictors to Comp Ratio, Comp Ratio^2 and Air Density. Then click OK.

Response: Power
Predictors: 'Comp ratio' 'Air Density' 'Comp Ratio^2'

For M15 we have to manually create a column if we want to enter a polynomial or an interaction term. With M16's General Regression menu we can enter the terms directly into the menu as shown next.

28. M16. Click on Stat <<Regression << General Regression.

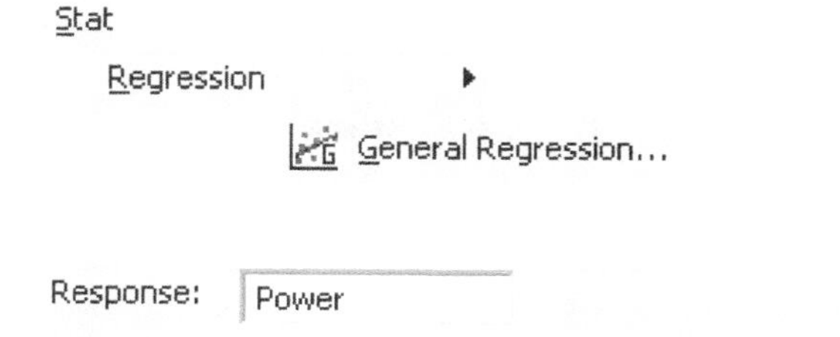

29. M16. Change the selection of the predictors to Comp ratio, Comp ratio*Comp ratio and Air Density. Then click OK.

Response: Power
Model:
'Comp ratio' 'Air Density' 'Comp ratio'* 'Comp ratio'

The results in the session window are the same for both methods. The regression table shows that the quadratic term is significant. Also note the improvement in R-Sq(Adj).

Let's check the residuals and see if they have improved.

```
Term                           Coef  SE Coef        T      P
Constant                   -332.414  57.2958  -5.8017  0.000
Comp ratio                   15.455   6.6641   2.3192  0.029
Air Density                 292.104  18.2838  15.9761  0.000
Comp ratio*Comp ratio         4.611   0.1740  26.5038  0.000

S = 40.1885      R-Sq = 99.94%         R-Sq(adj) = 99.93%
PRESS = 53575.3  R-Sq(pred) = 99.92%
```

```
Power  =  -332.414 + 15.4553 Comp ratio + 292.104 Air Density + 4.61091 Comp ratio*Comp ratio
```

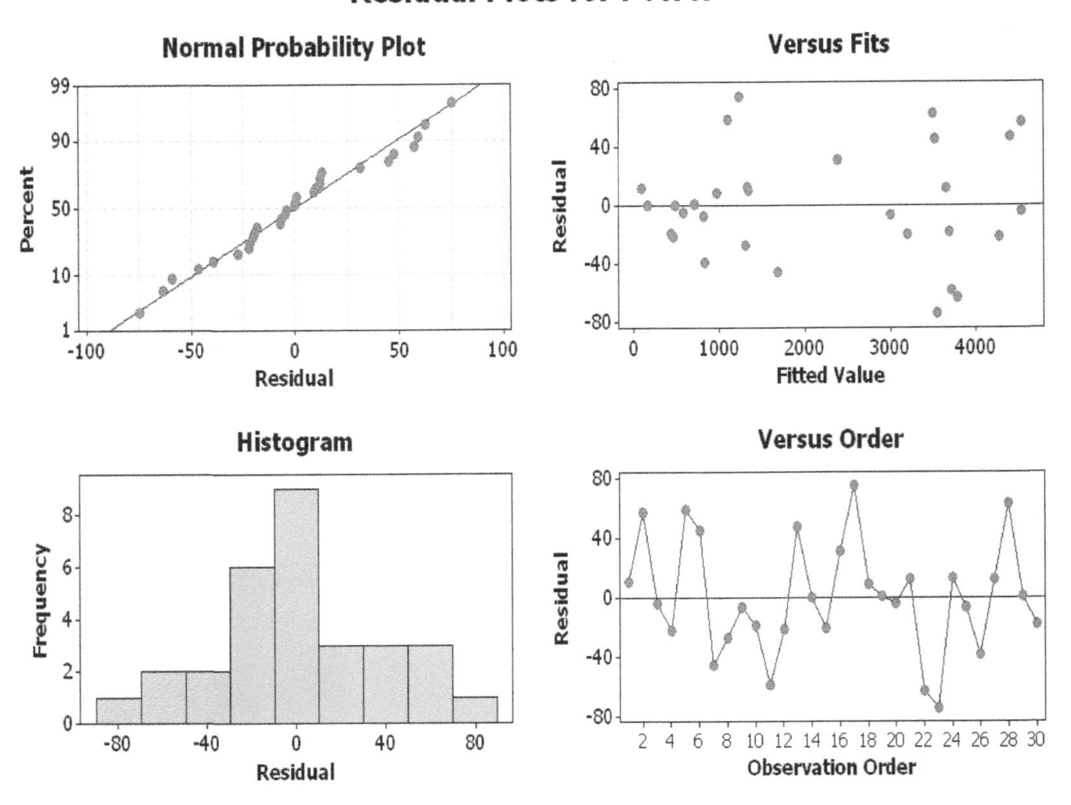

The pattern in the Versus Fits plot has gone. We have successfully produced a model with a quadratic term.

Example 7. Multi Colinearity

Miff also wants to buy a car and is interested in the power output of the engine. Miff believes that engine power is most affected by a different set of parameters to Prawn.

Analyse Miff's data to establish which of the predictors are significant.

Form the regression equation.

Worksheet: Engine Power2.

1. Open the Excel file and transfer the data into Minitab. Note that Miff has changed the predictors for this example.

	A	B	C	D	E	F
1	EngTemp	MPG	Air Density	Outside Temp		Power
2	74.6	41	3.825	13		66.63
3	47.3	32	3.825	8		52.70
4	93.6	55	3.55	12		75.20
5	87	48	3.9	11		74.80
6	68.6	60	1.975	34		67.05

2. Click Graph <<Matrix plot.

Graph

Matrix Plot...

3. Select Simple and then click OK.

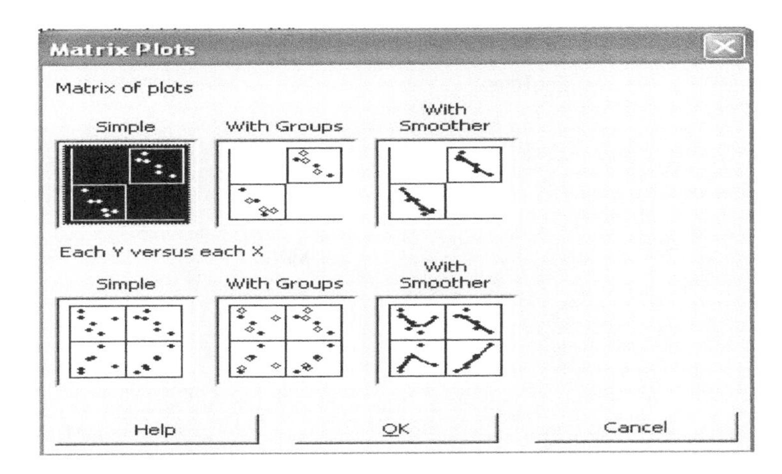

4. Select all the data columns as the Graph variables.

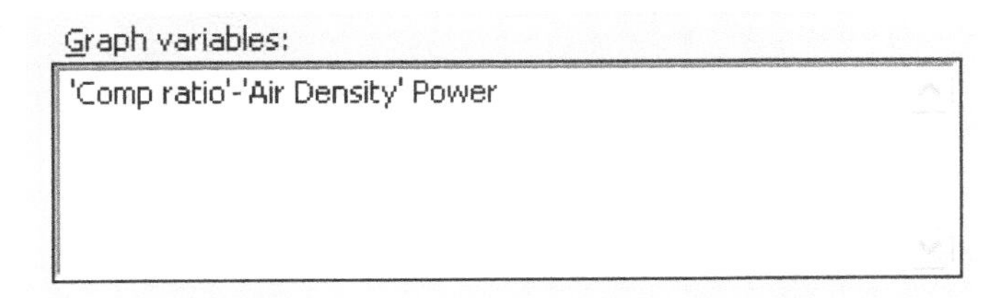

5. Click on Matrix Options.

6. Under Matrix Display select the radio button for Lower left (personal preference).
7. Click OK and OK again.

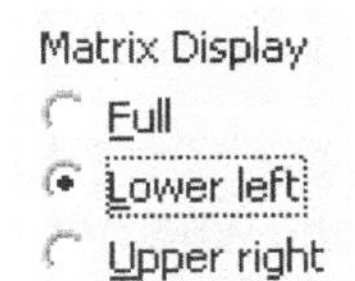

All the predictors appear to be independent from each other apart from Air Density and Outside Temp. MPG is about the only predictor that does not appear to be correlated with Power.

We will also check the correlation coefficients.

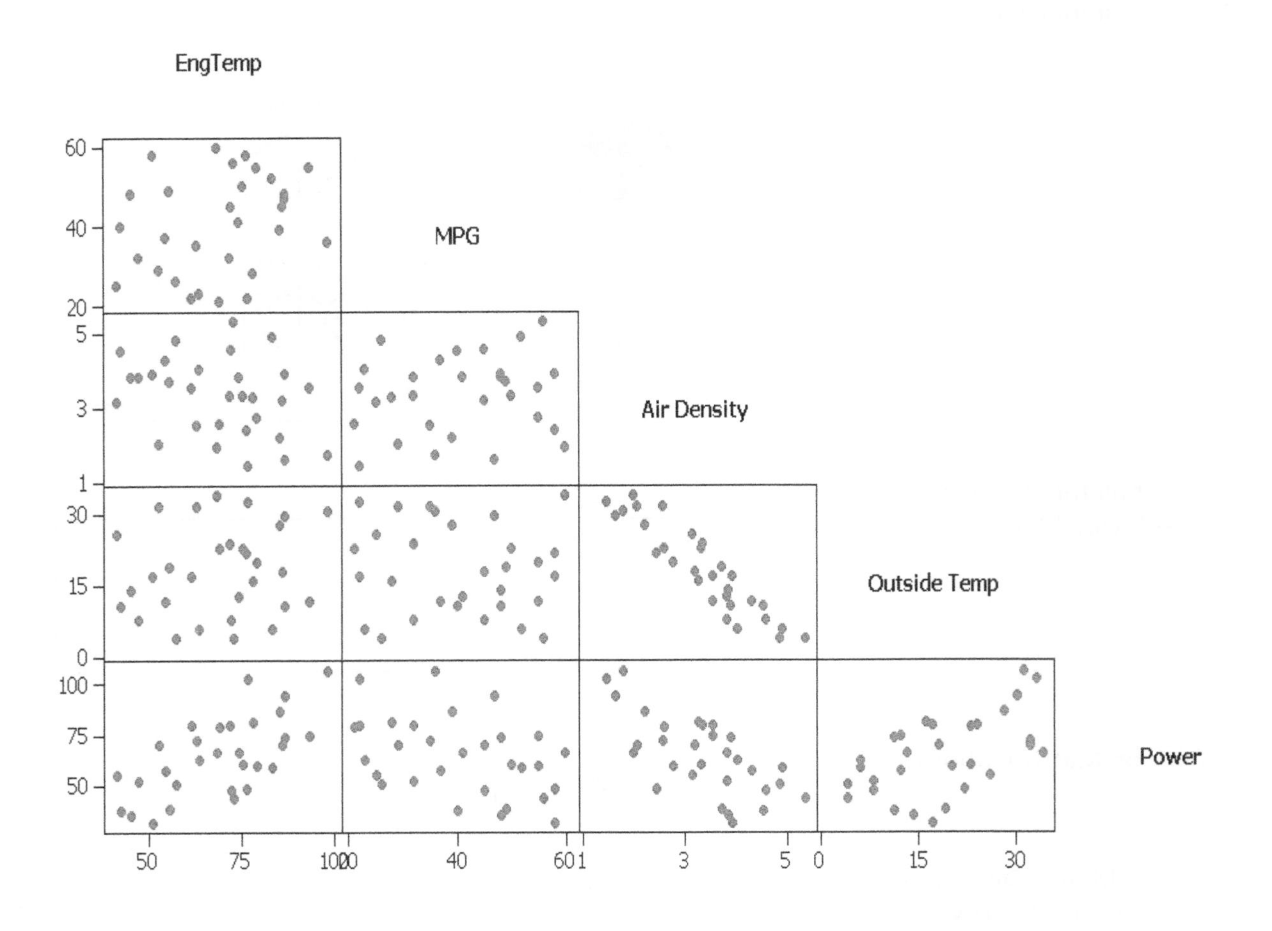

8. To check the strength of the correlation and whether it is statistically significant click Stat <<Basic Statistics <<Correlation.

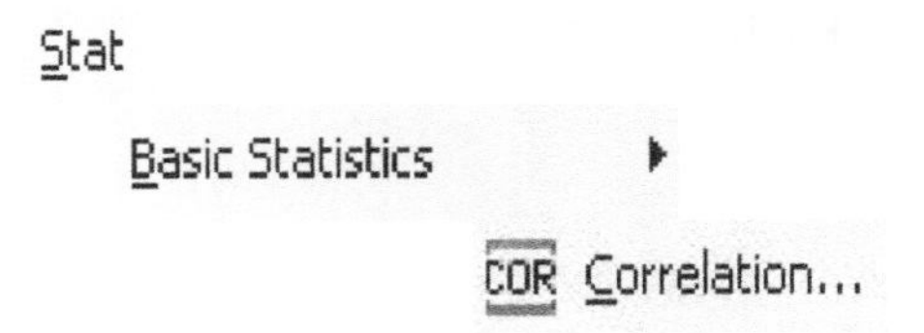

9. Select all the Variables and then click OK.
10. Select the session window icon to see the results.

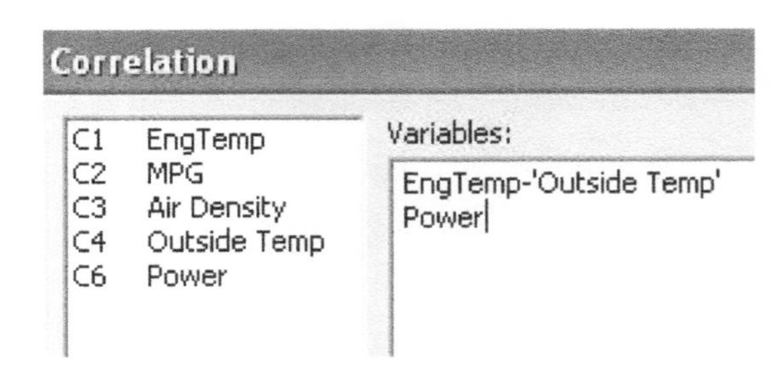

As suspected, out of the predictors, Air Density and Outside Temp are strongly correlated.

We will now carry out the regression analysis.

Correlations: EngTemp, MPG, Air Density, Outside Temp, Power

	EngTemp	MPG	Air Density	Outside Temp
MPG	0.270			
	0.150			
Air Density	-0.319	0.140		
	0.085	0.461		
Outside Temp	0.161	-0.101	-0.924	
	0.394	0.597	0.000	
Power	0.647	-0.408	-0.682	0.561
	0.000	0.025	0.000	0.001

Cell Contents: Pearson correlation
P-Value

11. Click on Stat <<Regression <<General Regression.

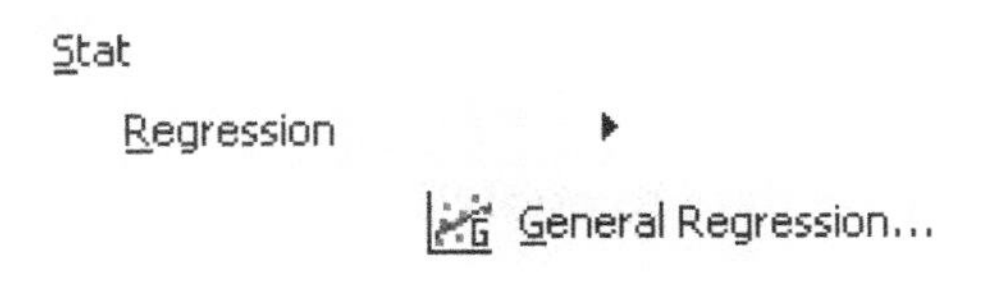

12. Select Power as the response and all the other variables as predictors. Click OK.

Response: Power
Predictors: EngTemp-'Outside Temp'

13. Click on the session window icon. The regression table shows that Outside Temp and Air Density are not significant. We should be able to remove them both from the model.

Predictor	Coef	SE Coef	T	P
Constant	35.23	18.73	1.88	0.072
EngTemp	0.88432	0.08335	10.61	0.000
MPG	-0.86582	0.09368	-9.24	0.000
Air Density	-2.063	3.129	-0.66	0.516
Outside Temp	0.5926	0.3246	1.83	0.080

S = 5.68767   R-Sq = 92.5%   R-Sq(adj) = 91.3%

14. Click on the Edit Last icon.

15. Change the selection of the predictors to EngTemp, MPG and Air Density. Then click OK.
16. Click on the session window icon.

Response: Power

Predictors: EngTemp MPG 'Air Density'

After removing Outside Temp notice that Air Density has become significant. This is because the two parameters were correlated and they were causing instability within the model. This is known as multi colinearity and the best way to find the best combination of predictors is to use a procedure like Best Subsets or Stepwise Regression. We will look at Predictor Selection Procedures after the exercise.

Predictor	Coef	SE Coef	T	P
Constant	66.857	7.441	8.98	0.000
EngTemp	0.82068	0.07904	10.38	0.000
MPG	-0.82707	0.09526	-8.68	0.000
Air Density	-7.403	1.163	-6.37	0.000

S = 5.93745   R-Sq = 91.5%   R-Sq(adj) = 90.6%

Exercise 4. Multiple Predictor Regression.

The drying rate of plasterboard within an industrial dryer is affected by a number of factors. Hawk wants to understand which factors are significant and then wants to form a model which can be used to optimise the drying rate.

Analyse the data to establish which of the predictors are significant to the response.

Form the regression equation and then check the residuals to establish if the model is valid.

Worksheet: Drying Rate.

All the predictors appear to be independent of each other.

Humidity and Drying Rate appear to have a strong negative correlation.

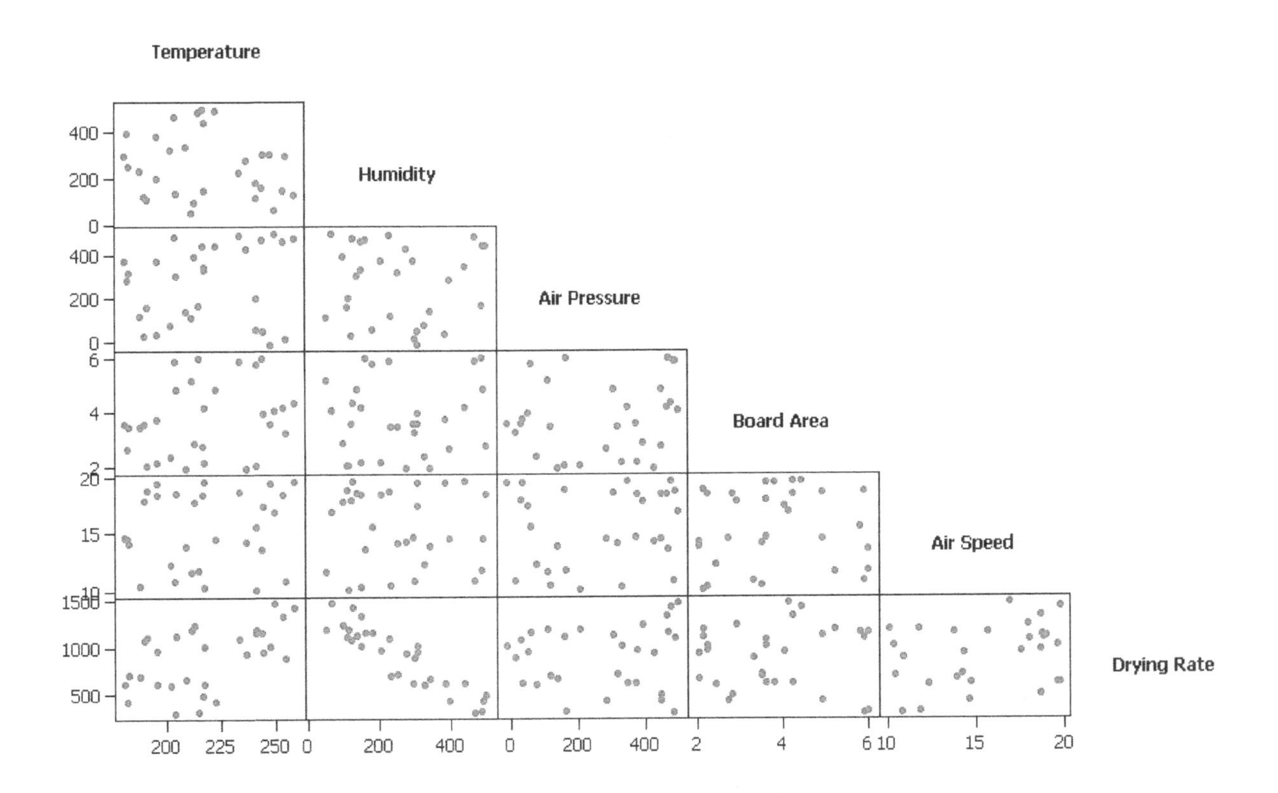

None of the predictors are correlated so when it comes to reducing our model we can remove more than one predictor at a time.

	Temperature	Humidity	Air Pressure	Board Area	Air Speed
Humidity	-0.198				
	0.294				
Air Pressure	0.141	-0.006			
	0.458	0.976			
Board Area	0.252	0.091	0.174		
	0.179	0.632	0.358		
Air Speed	0.099	-0.077	0.150	0.029	
	0.602	0.686	0.428	0.879	
Drying Rate	0.537	-0.908	0.093	0.003	0.303
	0.002	0.000	0.624	0.986	0.104

Cell Contents: Pearson correlation
P-Value

On the first run Air Pressure is the only predictor that is not significant.

Predictor	Coef	SE Coef	T	P
Constant	38.00	19.43	1.96	0.062
Temperature	4.88085	0.08065	60.52	0.000
Humidity	-2.01583	0.01403	-143.66	0.000
Air Pressure	0.02122	0.01087	1.95	0.063
Board Area	-5.061	1.481	-3.42	0.002
Air Speed	20.3942	0.5642	36.15	0.000

S = 9.98006   R-Sq = 99.9%   R-Sq(adj) = 99.9%

On the second run all the predictors are significant. The regression equation for this model is straightforward.

We need to check the residuals to ensure the model is valid.

Predictor	Coef	SE Coef	T	P
Constant	36.63	20.48	1.79	0.086
Temperature	4.89483	0.08473	57.77	0.000
Humidity	-2.01557	0.01480	-136.20	0.000
Board Area	-4.649	1.547	-3.01	0.006
Air Speed	20.5479	0.5892	34.87	0.000

S = 10.5257   R-Sq = 99.9%   R-Sq(adj) = 99.9%

The regression equation is
Drying Rate = 36.6 + 4.89 Temperature - 2.02 Humidity - 4.65 Board Area + 20.5 Air Speed

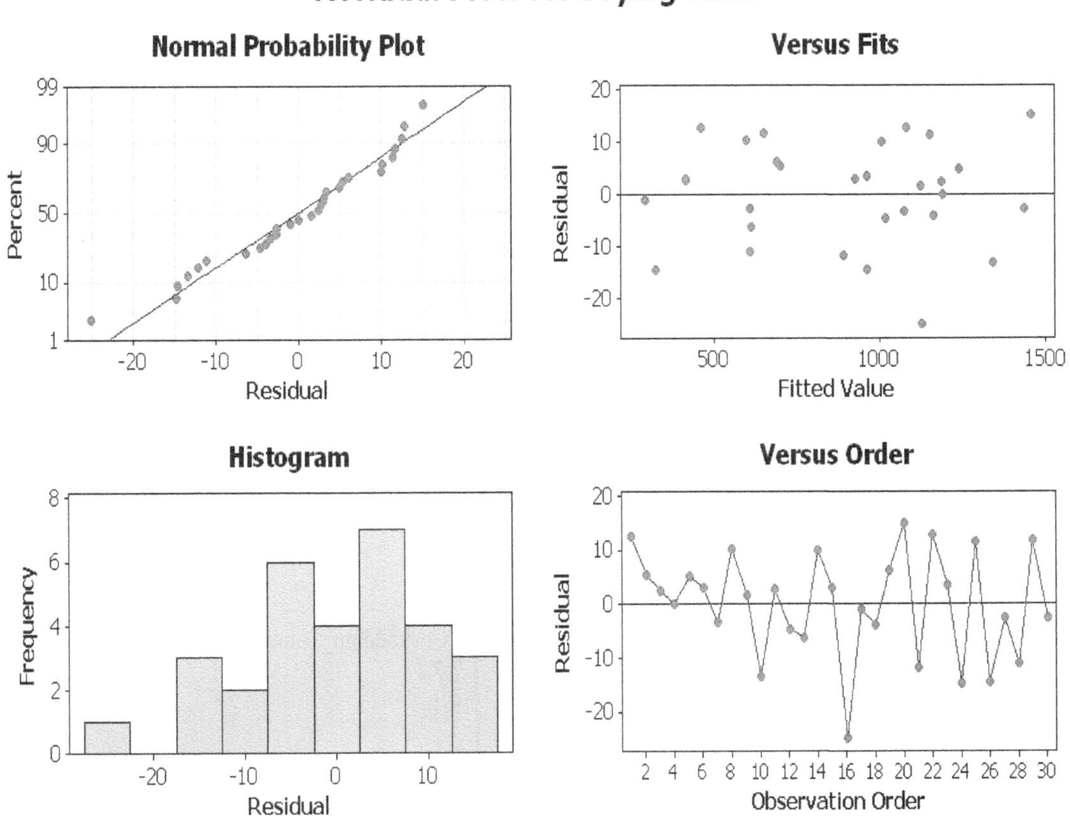

There do not appear to be any issues with the residuals.

Hawk now understands that the parameters affect drying rate are temperature, Humidity, Board Area and Air Speed. Air Pressure does not affect the drying rate.

9.8 Predictor Selection Procedure

Test	• Best Subsets and Step Wise Regression
What's it used for?	• They are used to select the best predictor variables when fitting a model. They do not carry out the regression.
Assumptions and Limitations	• Both procedures usually produce consistent results. • They become particularly useful when multi colinearity is causing problems in reducing the model. • A combination of metrics is used to show the best predictors for the model. • Neither procedure will attempt to find models with interaction or higher order terms..

Example 8. Predictor Selection Procedure.

Using the cannon data set from Example 5, use Best Subsets and Stepwise selection to evaluate the best combination of predictors.

Worksheet: Cannon.

1. Open the Excel file and transfer the data into Minitab.

Ball_Wt	Charge_W	Charge_Gr	Wadding_	Annulus_R	Barrel_Len	Distance Fired
5	1177	7.7	9	2.1	112	274.2
9.2	1132	19.1	8	2.1	116	329.3

2. Click Stat <<Regression <<Best Subsets

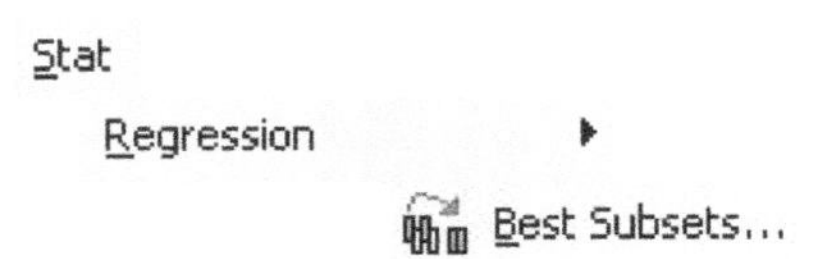

3. Select Distance Fired as the Response and C1 to C6 as the Free predictors. Then click OK.
4. Select the session window icon to see the results.

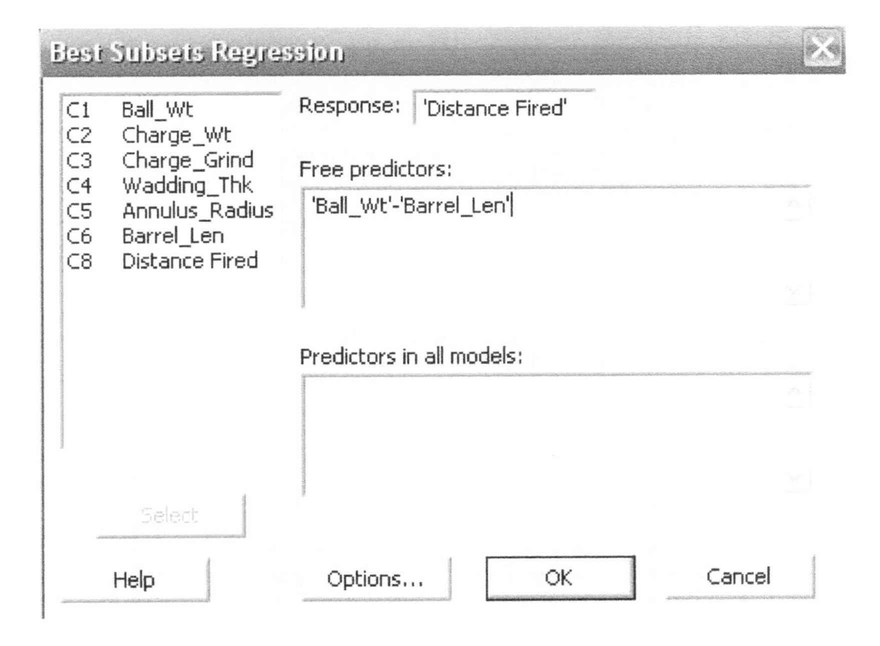

Minitab works through a number of models using different combinations of predictors. We select the best combination of predictors by looking at the results table.

Generally, we are looking for the highest R-Sq and R-Sq(adj) values. We use R-Sq when comparing models with the same number of predictors and R-Sq(adj) for differing numbers. We also want the lowest Mallows C_p and the lowest S value(S can be thought of as the standard deviation of residuals).

For the example we would select the model with Mallows $C_p = 3.5$ as this row gives us the best results for R-Sq, R-Sq (adj) and S. Incidentally, this gives the same combination of predictors as the result in Example 5.

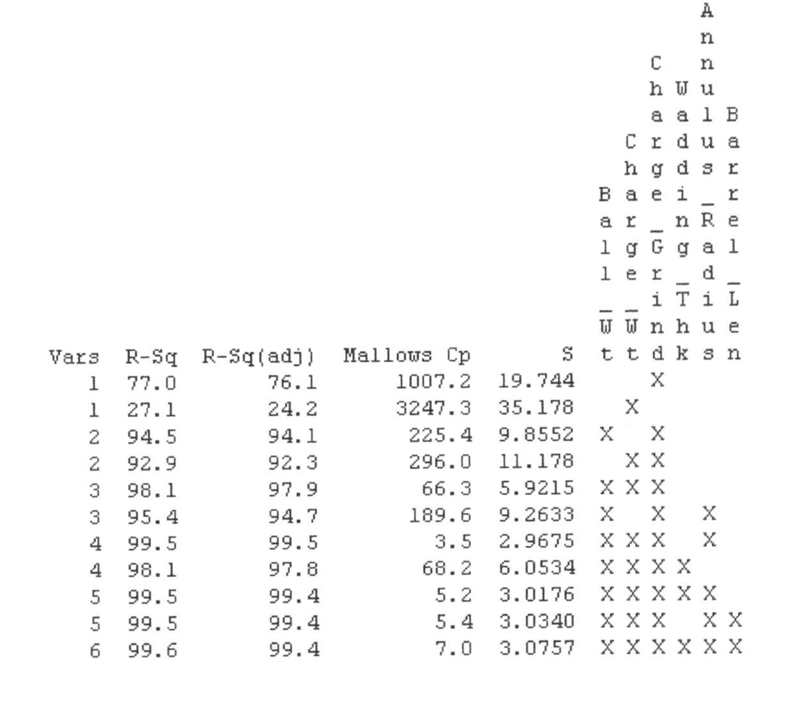

Response is Distance Fired

Vars	R-Sq	R-Sq(adj)	Mallows Cp	S	Ball_Wt	Charge_Wt	Charge_Grind	Wadding_Thk	Annulus_Radius	Barrel_Len
1	77.0	76.1	1007.2	19.744			X			
1	27.1	24.2	3247.3	35.178		X				
2	94.5	94.1	225.4	9.8552	X		X			
2	92.9	92.3	296.0	11.178		X	X			
3	98.1	97.9	66.3	5.9215	X	X	X			
3	95.4	94.7	189.6	9.2633	X		X		X	
4	99.5	99.5	3.5	2.9675	X	X	X		X	
4	98.1	97.8	68.2	6.0534	X	X	X	X		
5	99.5	99.4	5.2	3.0176	X	X	X	X	X	
5	99.5	99.4	5.4	3.0340	X	X	X		X	X
6	99.6	99.4	7.0	3.0757	X	X	X	X	X	X

Now we will carry out the same procedure but this time using Stepwise selection.

5. Click Stat <<Regression <<Stepwise.

6. Select Distance Fired as the Response and C1 to C6 as the Predictors. Then click OK.
7. Select the session window icon to see the results.

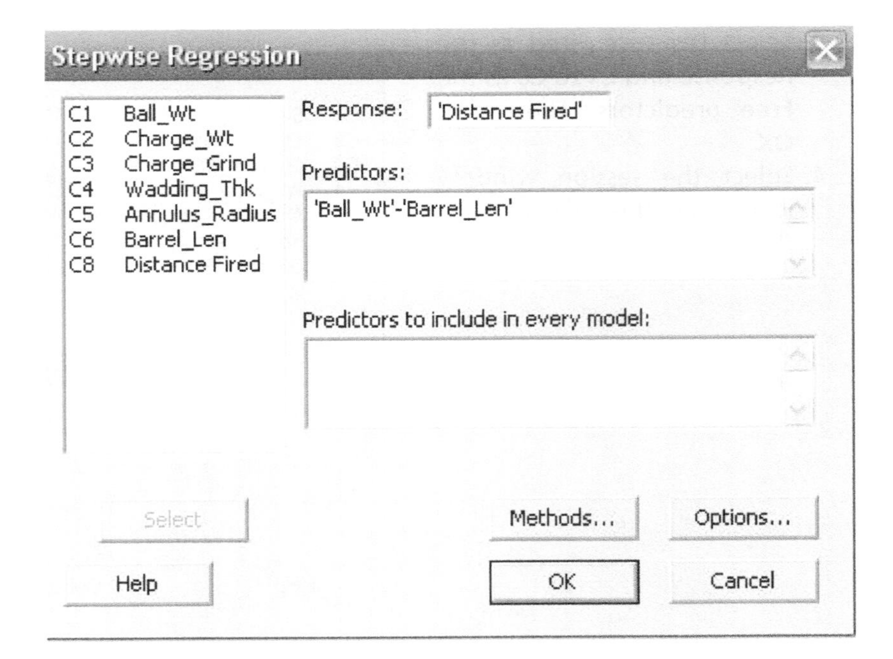

Again, the procedure works through a different combination of predictors until it finds the best model.

The Stepwise procedure can be modified in the initial set-up. It can be set to add predictors or subtract them. In the default mode it will add and then subtract. It can also be asked to stop after a number of iterations.

The result is the same as that for Best Subsets and the combination of predictors we selected in Example 5.

```
Step                 1        2       3       4
Constant        184.46   235.65   39.56   40.07

Charge_Grind      7.31     8.27    7.67    8.00
T-Value           9.16    20.02   29.01   57.89
P-Value          0.000    0.000   0.000   0.000

Ball_Wt                   -7.70   -5.17   -5.42
T-Value                   -8.74   -7.91  -16.48
P-Value                   0.000   0.000   0.000

Charge_Wt                         0.166   0.181
T-Value                            6.59   14.22
P-Value                           0.000   0.000

Annulus_Radius                             -8.7
T-Value                                   -8.34
P-Value                                   0.000

S                 19.7     9.86    5.92    2.97
R-Sq             77.04    94.51   98.10   99.54
R-Sq(adj)        76.12    94.05   97.85   99.46
Mallows Cp      1007.2    225.4    66.3     3.5
```

Exercise 5. Predictor Selection Procedure.

After we discovered the multi colinearity in Example 7 we did not complete the exercise. Using the data set from Example 7 use the Best Subsets and Stepwise selection procedures to evaluate the best combination of predictors for the regression model.

Worksheet: Engine Power2.

Using Best Subsets, the best model appears to be one with EngTemp, MPG and Outside Temp as predictors. It does not select Air Density as a predictor.

Response is Power

Vars	R-Sq	R-Sq(adj)	Mallows Cp	S	EngTemp	MPG	Air Density	Outside Temp
1	46.5	44.6	153.1	14.384			X	
1	41.8	39.8	168.7	14.999	X			
2	78.3	76.7	48.5	9.3207	X	X		
2	67.0	64.6	86.5	11.506	X		X	
3	92.4	91.5	3.4	5.6255	X	X		X
3	91.5	90.6	6.3	5.9375	X	X	X	
4	92.5	91.3	5.0	5.6877	X	X	X	X

The Stepwise procedure selects the same predictors as the Best Subsets procedure.

Step	1	2	3	4	5
Constant	108.01	57.63	66.86	35.23	23.43
Air Density	-12.8	-9.9	-7.4	-2.1	
T-Value	-4.93	-4.54	-6.37	-0.66	
P-Value	0.000	0.000	0.000	0.516	
EngTemp		0.591	0.821	0.884	0.912
T-Value		4.09	10.38	10.61	12.88
P-Value		0.000	0.000	0.000	0.000
MPG			-0.827	-0.866	-0.884
T-Value			-8.68	-9.24	-10.00
P-Value			0.000	0.000	0.000
Outside Temp				0.59	0.79
T-Value				1.83	6.94
P-Value				0.080	0.000
S	14.4	11.5	5.94	5.69	5.63
R-Sq	46.50	67.00	91.54	92.53	92.40
R-Sq(adj)	44.59	64.55	90.56	91.34	91.53
Mallows Cp	153.1	86.5	6.3	5.0	3.4

9.9 Nonlinear Regression

Test	• Nonlinear Regression
What's it used for?	• It is used when you cannot model the relationship with linear parameters. Linear models have only additive terms with only one parameter in each term.
Assumptions and Limitations	• Response variables must be continuous, predictors must be continuous or indicator variables (i.e. 0, 1). • The procedure uses an iterative algorithm to fit the best model and therefore requires starting values for the iteration. • Prior to running the procedure the expectation function needs to be specified. Due to the almost infinite number of nonlinear functions prior knowledge of the model is almost mandatory. • Nonlinear Regression is M16 only.

Example 9. Nonlinear Regression.

Zaf, the chemical engineering consultant, is conducting evaluation trials on a new granular active carbon (GAC) filter. He is checking its ability to remove sulfur dioxide (SO_2) from a gas stream. A gas stream with a constant 200 ppm SO_2 content is passed through the filter at varying flowrates. The exhaust concentration (Conc) of SO_2 is recorded against the flowrate.

Zaf has conducted a literature search and has found that a Power (concave) function will be the most appropriate model to fit.

Help zaf by forming a model for the relationship between Flow and Conc. Can a flow of 30 kg/min give a concentration of 20 ppm.

Worksheet: GAC Filter.

1. Open the file GAC filter and transfer the data into a new Minitab project window.

Flow	Conc
4.03	93.097
20.52	12.325
41.42	5.242
[illegible]	[illegible]

As a comparison we will display the results of using the Regression <<Fitted Line Plot procedure on this data. We have used a cubic model and from the session window we can see that the cubic term is significant.

The Fitted Line Plot shows a fairly good fit with an R-Sq(adj) of 92.9% which also indicates a good fit.

Sequential Analysis of Variance

Source	DF	SS	F	P
Linear	1	16768.4	86.22	0.000
Quadratic	1	7191.4	90.14	0.000
Cubic	1	2928.8	90.69	0.000

Conc = 120.3 - 11.59 Flow
+ 0.3860 Flow**2 - 0.004138 Flow**3

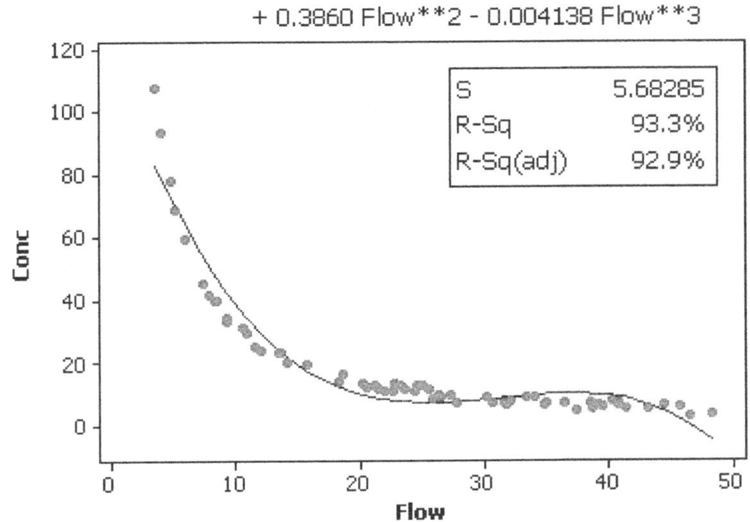

The Four in One residual plot indicates problems with the model. The histogram is extremely skewed and the versus fit plot shows that the residuals are not evenly distributed around the zero line. In fact we are seeing strong curvature indicating the data is not well explained by the model. Clearly, we need to attack this problem in a different way so we turn to Nonlinear Regression.

Residual Plots for Conc

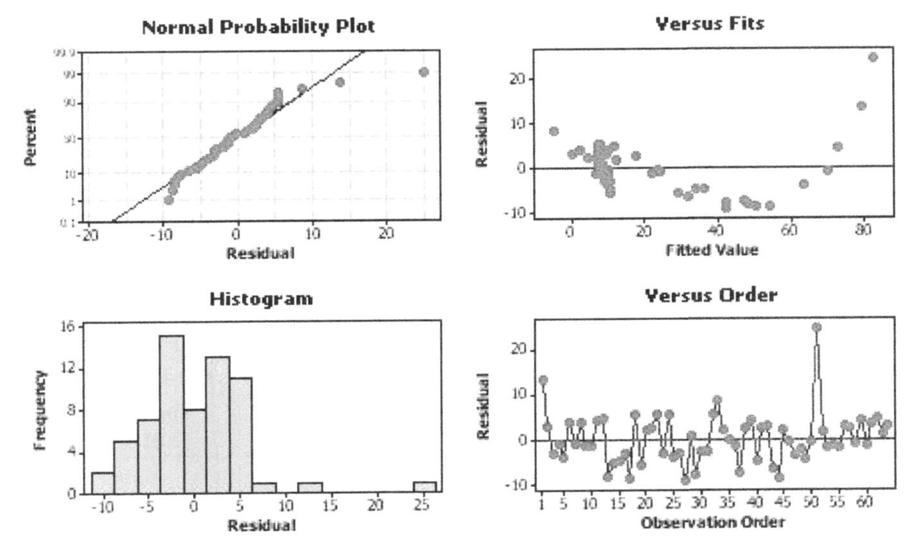

2. Click on Stat <<Regression <<Nonlinear Regression.

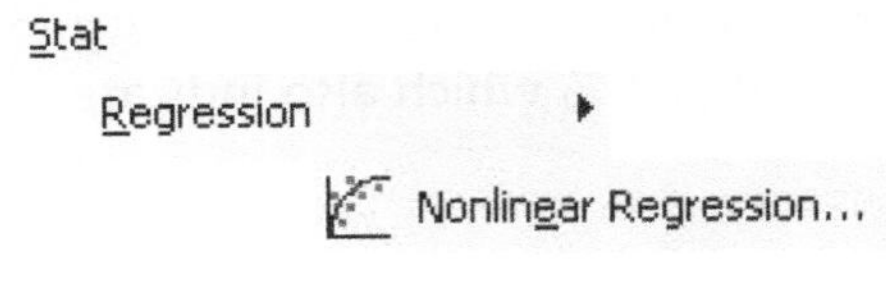

3. Enter Conc as the Response variable.
4. We have the option of using the calculator to form our own function or we can use a predefined function by clicking on Use Catalog. Click on Use Catalog.
5. Select the Power (concave) function from the list. We see that it uses one predictor and two parameters.
6. Click on OK.

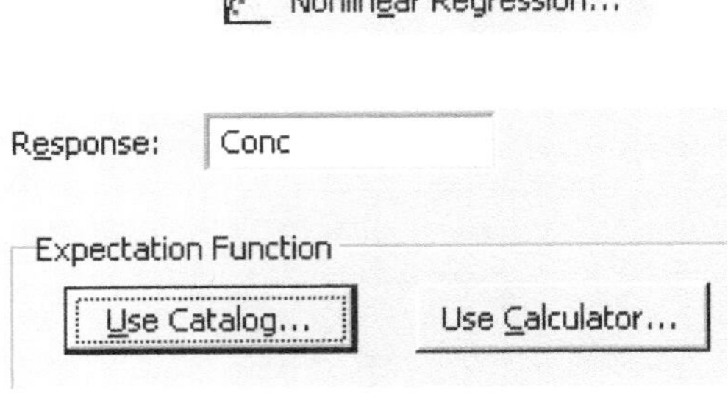

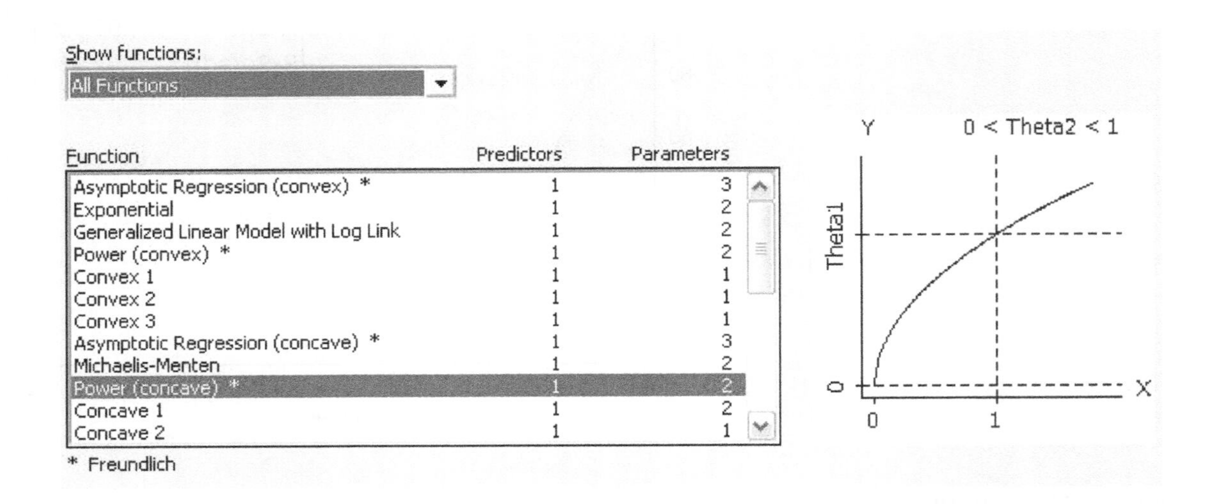

7. We can see the algebraic form of the equation we have selected. We must now let the procedure know which column represents X in the equation. Select Flow and then click OK.

Theta1 * X^Theta2

Choose an actual predictor for each placeholder:

Placeholder	Actual Predictor
X	Flow

8. As the Nonlinear Regression procedure uses an iterative process to produce the model we need to estimate starting values. Click on the Parameters button to enter the starting values for the iteration.
9. Looking at the proposed equation and the data that has been generated I guessed the starting values. (Actually, on the first occasion I used 0.01 but the iteration stopped after four loops.) Enter 0.1 as the starting values for Theta1 and Theta2 and then click OK.

Required starting values:

Parameter	Values	Locked
Theta1	0.1	
Theta2	0.1	

10. Click on the Graphs Button. Click on the tick box for Display prediction intervals as this will help us to answer a part of the question.
11. Select the Four in one residuals plot.
12. Select OK and OK again.

☑ Plot of fitted curve with data
☐ Display confidence interval
☑ Display prediction interval

Residuals
○ Individual plots
☐ Histogram of residuals
☐ Normal plot of residuals
☐ Residuals versus fits
☐ Residuals versus order
◉ Four in one

13. Within the Session Window we see the Lack of Fit test. We only get the results of this test if there are replicates within our data. If the *P* value was below our α level then there would be evidence to show that model did not fit the data. However, this is not the case.

```
Lack of Fit

Source          DF       SS        MS       F       P
Error           62  85.1460   1.37332
  Lack of Fit   61  84.3395   1.38262    1.71   0.552
  Pure Error     1   0.8064   0.80645
```

14. Go to the Fitted Line Plot. We can see the equation that has been developed and the Prediction Interval for new points has been plotted.

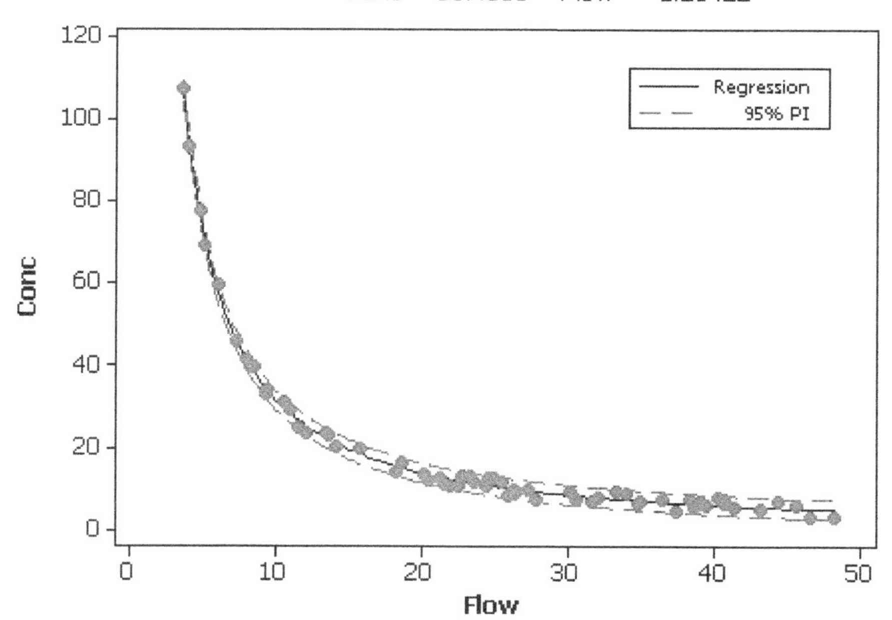

15. Right Click on the graph area and then select Add <<Reference Lines. Complete the menu box as shown with reference lines at X = 30 and Y = 20.

Show reference lines at Y values:
20

Show reference lines at X values:
30

As the intersection point of the two lines lies outside of the prediction interval we can be confident that a flow of 30 kg/min will not give a concentration of 20 ppm.

We need to check the residuals to validate the model.

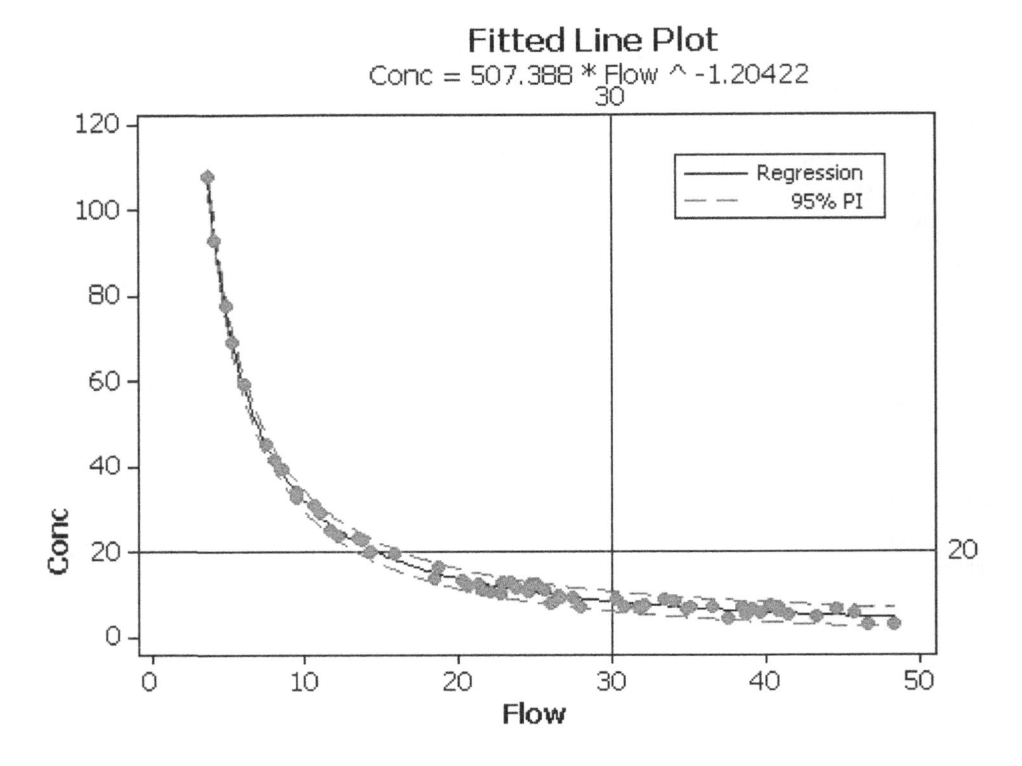

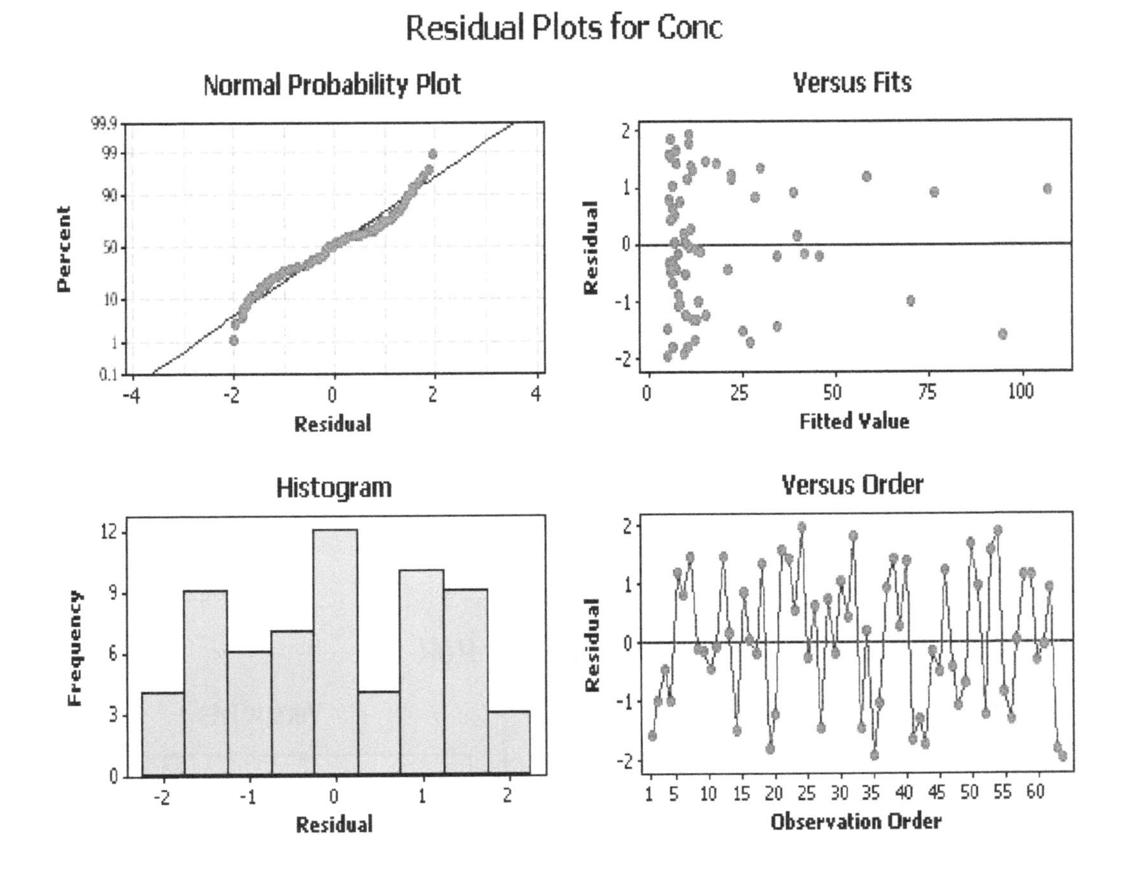

There do not appear to be any issues with the residuals.

Exercise 6. Nonlinear Regression.

The reaction rate of a new enzyme is being evaluated. The concentration of the substrate (Conc) is varied and the resulting rate of reaction (RoR) is measured.

For this type of study it is known that the Michaelis–Menten equation is usually appropriate to model enzyme kinetics.

Form a model for the relationship between Conc and RoR.

Worksheet: Enzyme.

Within the session window we see the Lack of Fit test. We see that the *P* value is above 0.05 which indicates that we cannot say that our model does not fit the data.

```
Lack of Fit

Source           DF         SS         MS     F      P
Error            98  0.0086888  0.0000887
  Lack of Fit    93  0.0084743  0.0000911  2.12  0.203
  Pure Error      5  0.0002145  0.0000429
```

The Fitted Line Plot shows the parameters within the Michaelis–Menton equation that have been developed for this study.

Starting values of 0.1 were again used for theta1 and theta2.

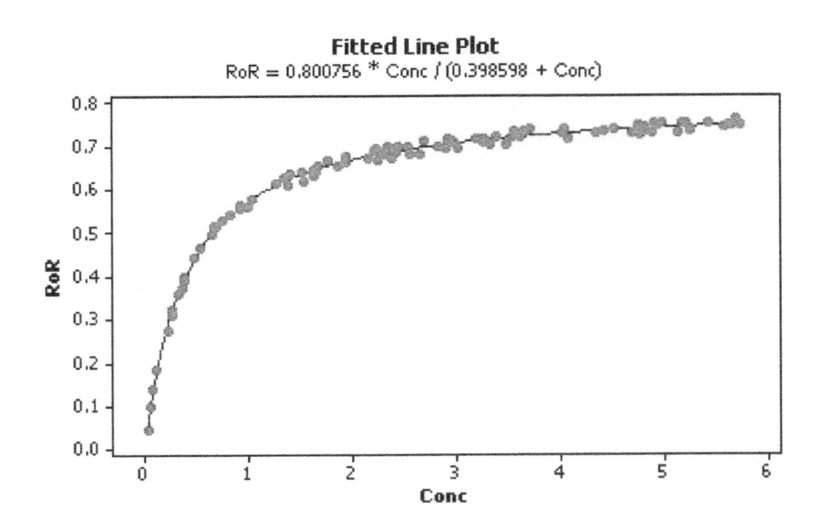

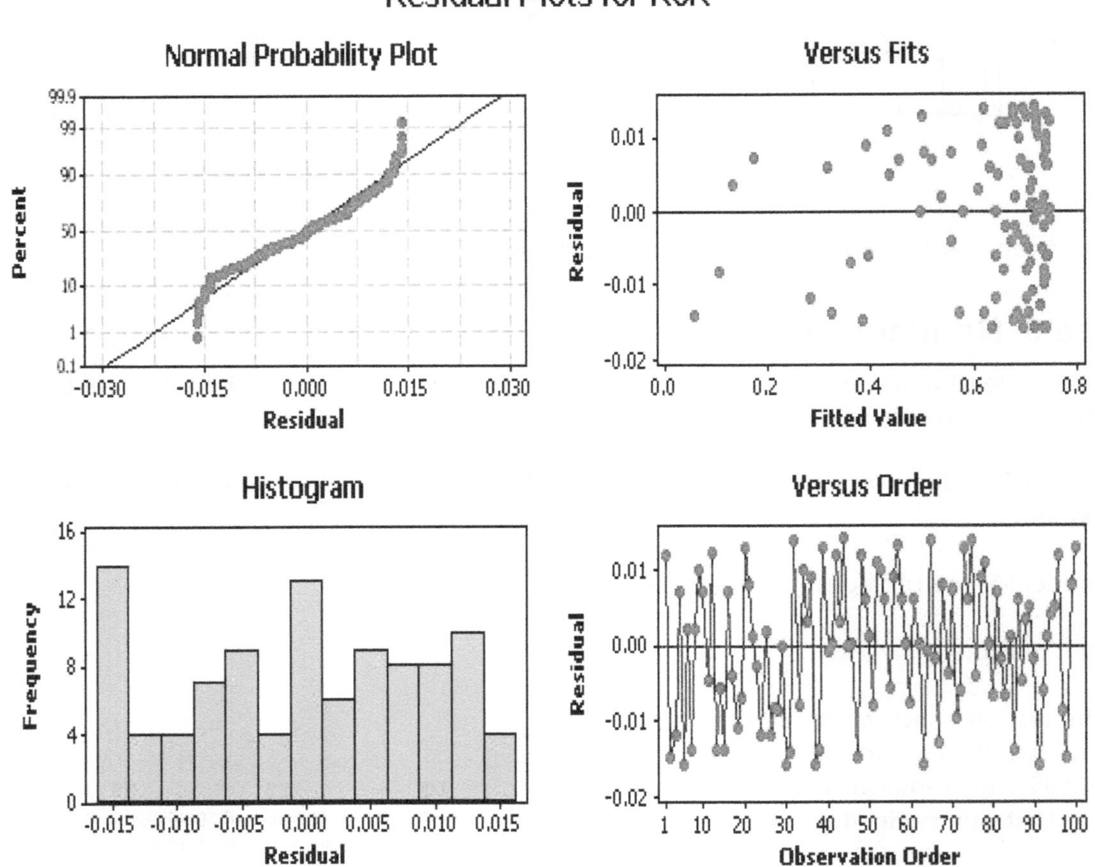

Although the residuals could be better we don't have enough abnormality to reject the model.

CHAPTER 10
Design of Experiment

10.1 Why Use Design of Experiment?

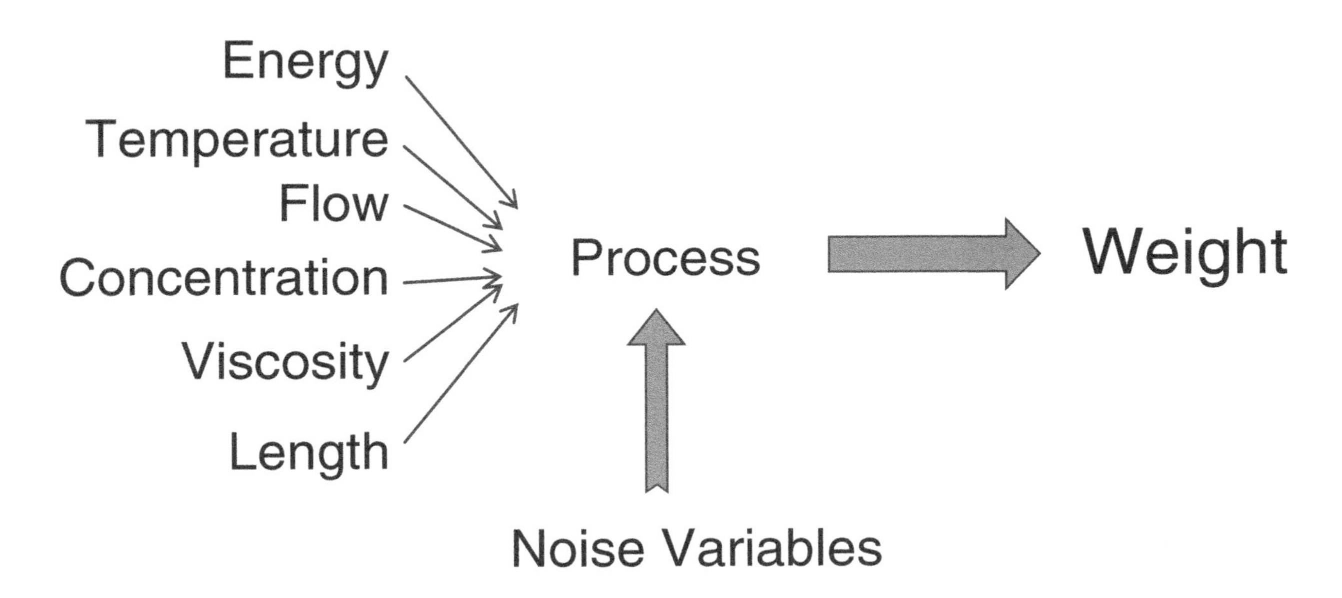

Design of experiment (DOE) is used to assess the input factors and interactions that are important to our response. In a series of designed experiments, where the number of experimental runs are optimised, the important factors and interactions are identified. We can then look at optimising the significant input settings in order to achieve our desired output.

10.2 Types of DOE

The type of DOE and the number of runs it uses depends on our objectives.

Screening DOE	Characterisation DOE	Response Surface Designs	Taguchi Designs
Used to reduce a large number of factors to the few significant factors that are relevant.	Used to evaluate main factors and their interactions. Leading us to a prediction equation.	Used to model curvature within the design	Used to reduce variation and reduce sensitivity to noise factors.

Problem Solving and Data Analysis using Minitab: A clear and easy guide to Six Sigma methodology, First Edition. Rehman M. Khan.

10.3 DOE Terminology

In order to explain some of the DOE terminology we are going to look at an example of a Two Level Full Factorial DOE.

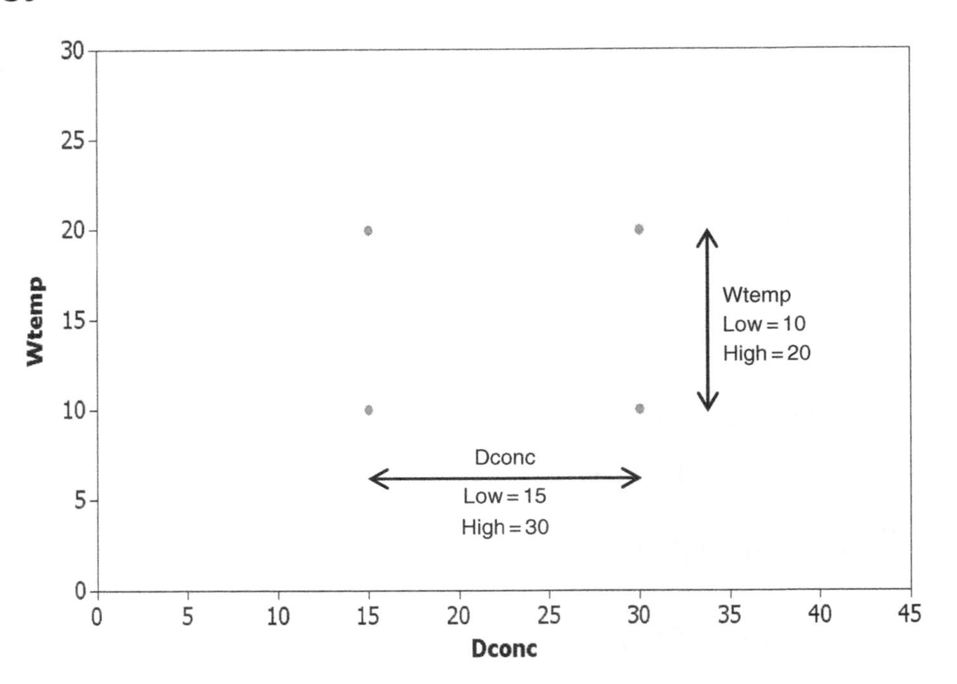

We are going to look at a mopping process. We are interested in how the detergent concentration and water temperature affect the grime concentration after mopping. Our factors are Dconc and Wtemp and the response is Grime.

We carry out four experiments using low and high levels for each factor as shown. As each combination of factors is run only once we say there are no *Replicates*. It is preferable to have Replicates within a DOE as it allows Minitab to calculate the variability within the experiment. (In reality our example DOE would not work without replicates.)

One of the main benefits of using DOE is that it optimises the number of experimental runs that are needed.

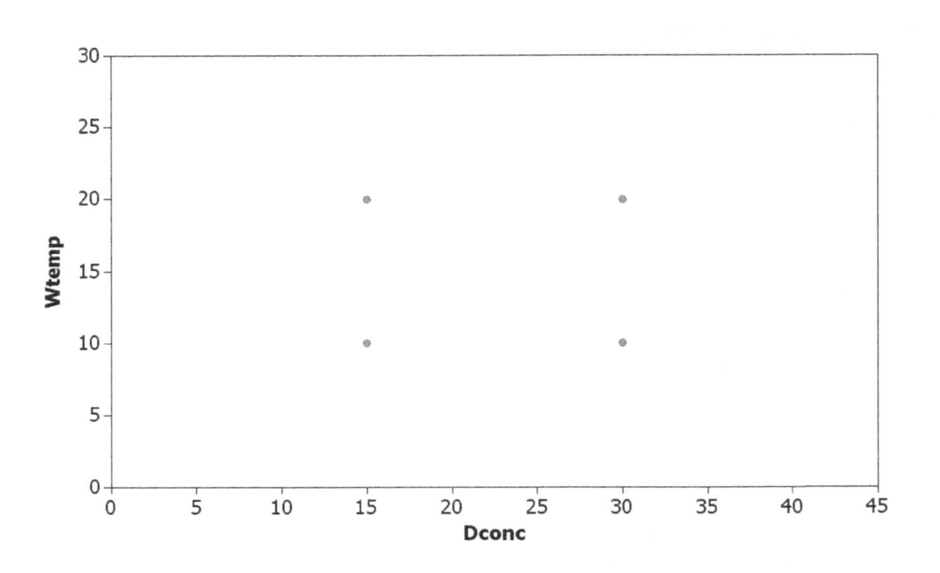

Let's just look at how many runs would be required if we had not used a Two Level Factorial approach but decided to use six levels for each factor.

From needing four experimental runs we have now gone to needing 36.

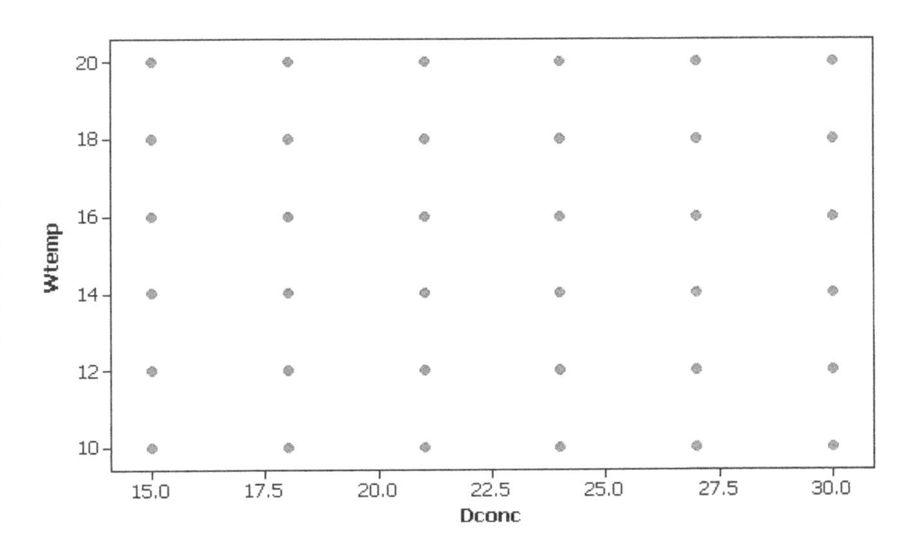

The experimenter needs to decide if the cost and time required to do the extra runs is beneficial in terms of the information that is likely to be gained and how much value that information will have.

As we are carrying out a Two Level Factorial DOE we look at each factor at a high and low levels only. We call these our *Corner Points* and we assume a linear response between the corner points. Other types of DOE have additional points to look for curvature.

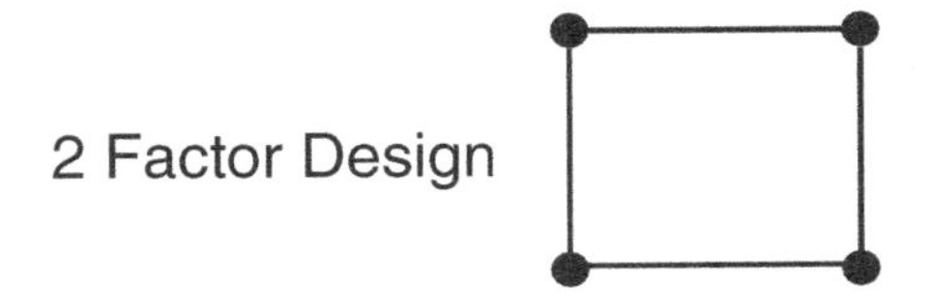

With this type of DOE:

- For two factors we need four runs.
- For three factors we need eight runs.
- For four factors we need 16 runs.

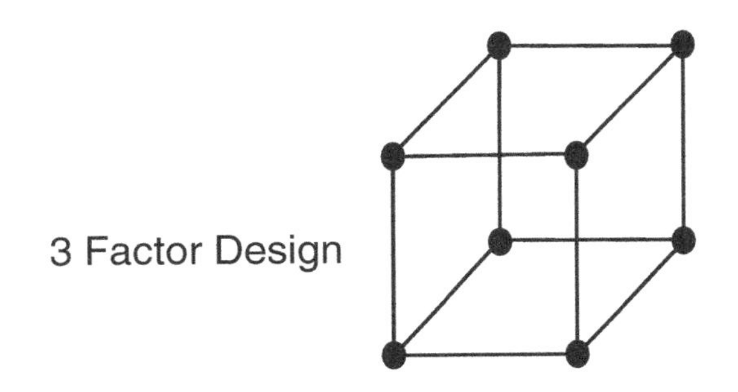

We call this a 2^k factorial, where k is the number of factors.

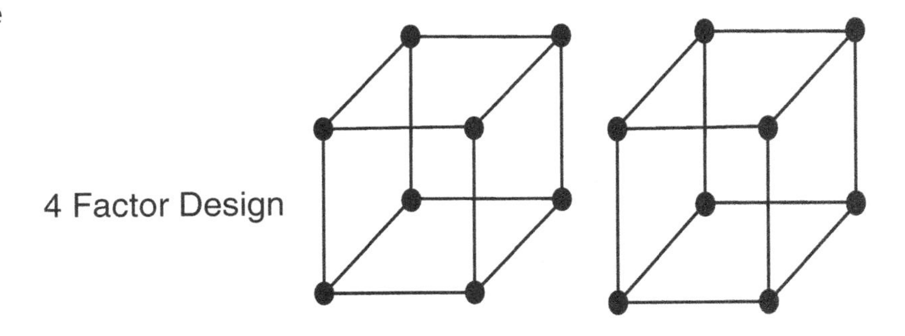

Let's have a look at the Design Matrix for our experiment.

Wtemp	Dconc
10	15
10	30
20	15
20	30

We can change the values into Coded units where the high values are represented by +1 and the low values by −1. We can also add the interaction term. Notice that the interaction term is the product of the coded units of the factors.

Wtemp	Dconc	Wtemp*Dconc
−1	−1	1
−1	1	−1
1	−1	−1
1	1	1
		Sum of Column = 0

Our design is *Balanced* because it contains the same number or runs at each high value and low value for each factor, that is two.

The design is *Completely Orthogonal* because the sum of the products of the coded factors is zero. Also, the Pearson correlation coefficient for the coded factors and interaction term is zero, that is No Correlation.

Wtemp	Dconc	Grime
10	15	212
10	30	135
20	15	187
20	30	90

Let's say we have carried out the experiment. The results are shown in the table. The higher the grime value the dirtier the sample was after mopping.

The Cube plot shows the results in a very graphical and straight forward manner.

Let's have a look at the Main Effects plot and Interaction plot and see how the results were calculated.

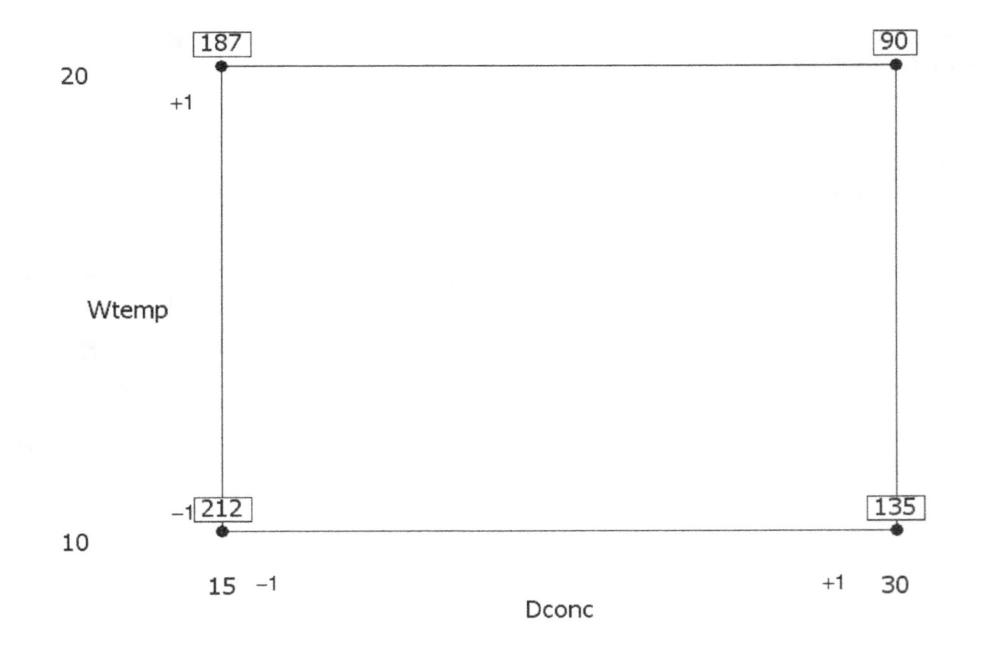

Let's start with the Main Effects plot and see how the Main Effects are calculated.

$$\text{Factor Effect} = \begin{matrix}\text{Average of}\\ \text{responses at}\\ \text{high levels}\end{matrix} - \begin{matrix}\text{Average of}\\ \text{responses at}\\ \text{low levels}\end{matrix}$$

$$\textit{Dconc Effect} = \frac{135 + 90}{2} - \frac{212 + 187}{2} = -87$$

Equation (10.1)

$$\textit{Wtemp Effect} = \frac{187 + 90}{2} - \frac{212 + 135}{2} = -35$$

Equation (10.2)

Let's have a look at the Interaction plot and see how the Interaction is calculated.

$$\text{Interaction} = \frac{212 - 135}{2} - \frac{187 - 90}{2} = -10$$

Equation (10.3)

Now that we understand more about the terminology let's work through the full example.

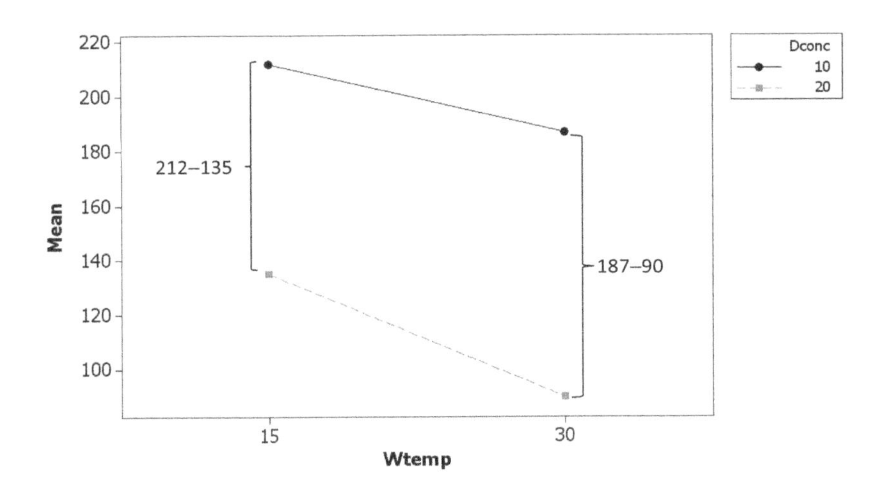

10.4 Two Level Factorial Designs

Procedure	• Two Level Factorial Designs
What's it used for?	• The design of experiment process is used to understand the effect of process inputs and their interactions on the outputs of the process.
Assumptions and Limitations	• A two level factorial DOE evaluates the changes in response when combinations of high and low levels of each factor are tested. • This type of DOE cannot detect curvature within a single run. • This type of DOE can be used for characterisation or screening but of the two it is more likely to be used for characterisation.

Example 1. Setting up the experimental design.

We are going to expand on the mopping example. We are interested in how the detergent concentration, water temperature and material of the sponge on the mop affect the grime concentration after mopping. Our factors are Dconc, Wtemp and Material and the response is Grime. Initially eight runs were carried out on day1 and then the eight runs were replicated on day2.

If the historical standard deviation is 10 and the effect of the factors is at least 20 what is the minimum power for this experiment?

Also, set up the experimental worksheet.

No File.

1. Click Stat <<Power and Sample Size <<2 Level Factorial Design.

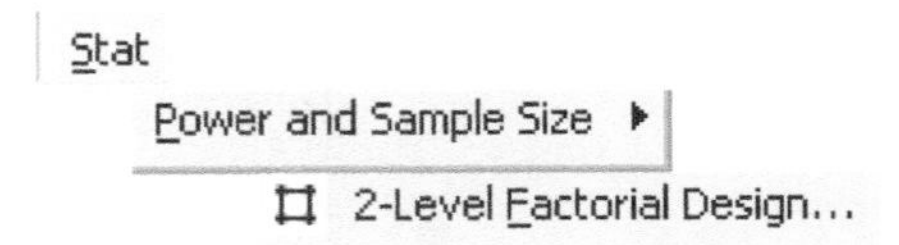

2. Complete the menu box as shown. Standard deviation and Effects are estimated from historical knowledge of the process.
3. Click on Designs.

Power and Sample Size for 2-Level Factorial Design

Number of factors: 3

Number of corner points: 8

Specify values for any three of the following:

Replicates: 2

Effects: 20

Power values:

Number of center points per block: 0

Standard deviation: 10

Design... Options... Graph... Help OK Cancel

4. In order to highlight how to use Blocks we said that the experiment was carried out over two days. We need to know if carrying out half the runs on one day and half on the next affects the results. In order to check if this has changed the experiment we click on Include blocks in model and we specify 2 as the number of blocks. In this way using Blocks allows us to assess whether a noise variable is affecting the experiment.
5. Click OK and OK again.

Power and Sample Size for 2-Level Factorial - De...

Number of blocks: 2

Number of terms omitted from model: 0

Include term for center points in model

Include blocks in model

Help OK Cancel

6. The graph that pops open shows the Power Curve against Effect. With the data that we specified we will have a 92.7% chance of recognising a difference if one exists.

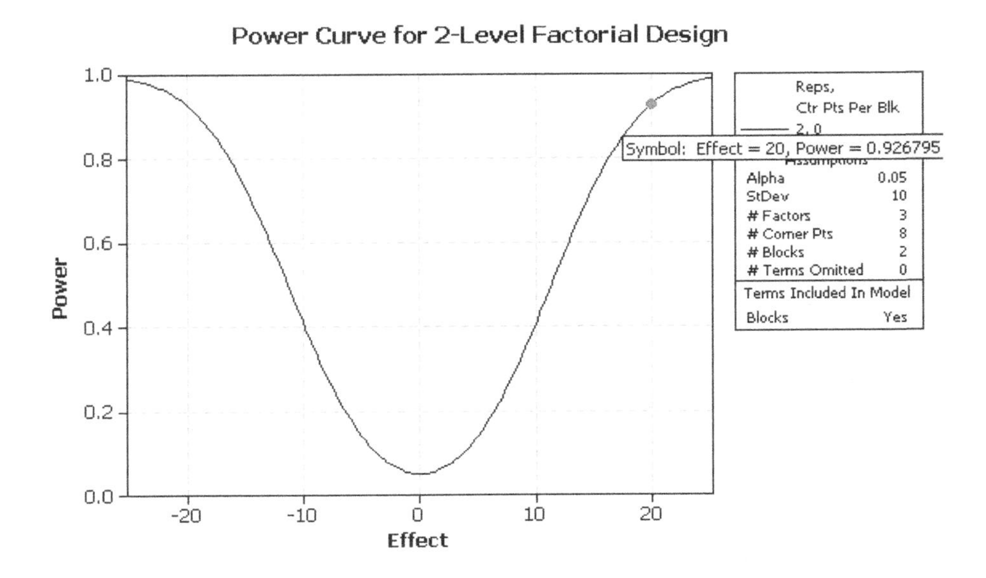

2. Go to the session window to see the summary of the data. As we are happy with the power we can now set up the worksheet for the experiment.

```
Power and Sample Size

2-Level Factorial Design

Alpha = 0.05  Assumed standard deviation = 10

Factors: 3   Base Design: 3, 8
Blocks:  2

Including blocks in model.


Center
Points
   Per                      Total
 Block   Effect   Reps   Runs      Power
     0       20      2     16   0.926795
```

3. Click Stat <<DOE <<Factorial <<Create Factorial Design...

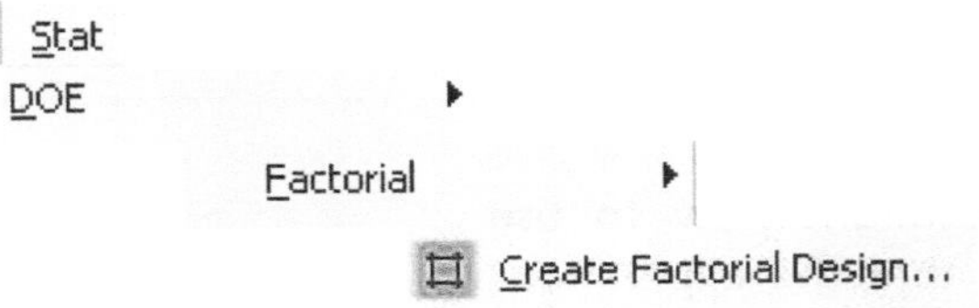

4. We are going to use a '2-level factorial' design which should be selected by default. Change the Number of factors to 3 and click on Display Available Designs.

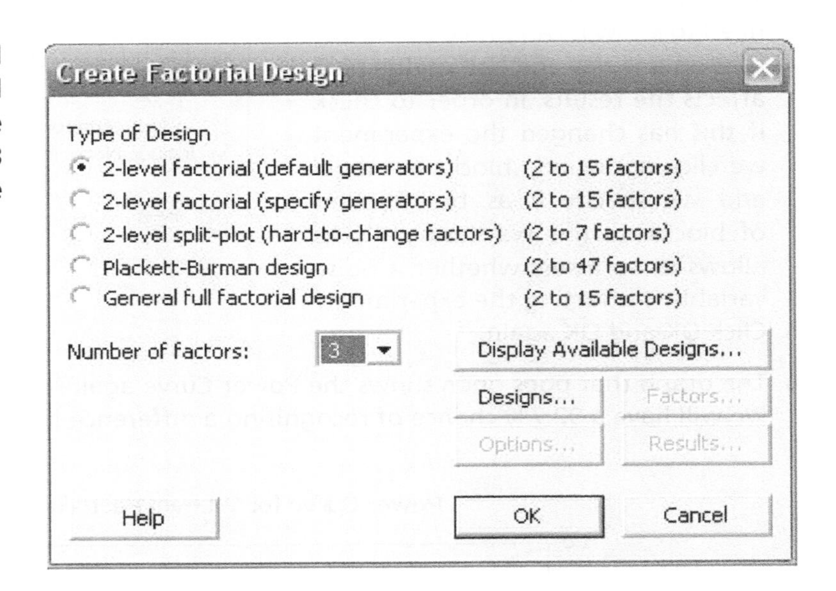

A matrix of available two level factorial designs is shown here. The roman numerals are the resolution of fractional designs, which we will cover later. We do not need to select anything from this menu. Click OK and then Designs.

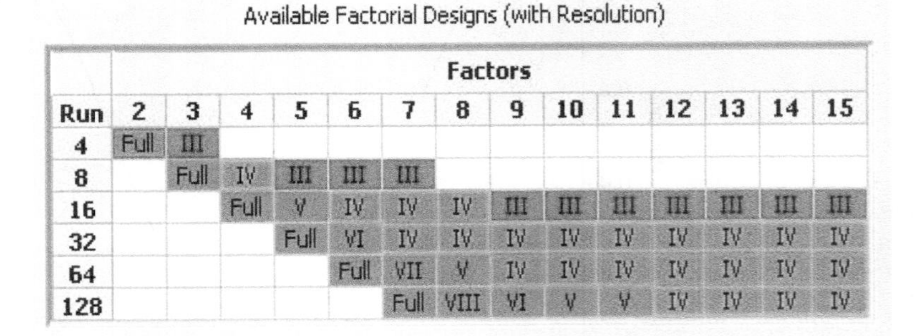

Available Factorial Designs (with Resolution)

	Factors													
Run	2	3	4	5	6	7	8	9	10	11	12	13	14	15
4	Full	III												
8		Full	IV	III	III	III								
16			Full	V	IV	IV	IV	III	III	III	III	III	III	III
32				Full	VI	IV	IV	IV	IV	IV	IV	IV	IV	IV
64					Full	VII	V	IV	IV	IV	IV	IV	IV	IV
128						Full	VIII	VI	V	V	IV	IV	IV	IV

5. Ensure Full Factorial is highlighted. Complete the remainder of the menu box as shown. Click OK.

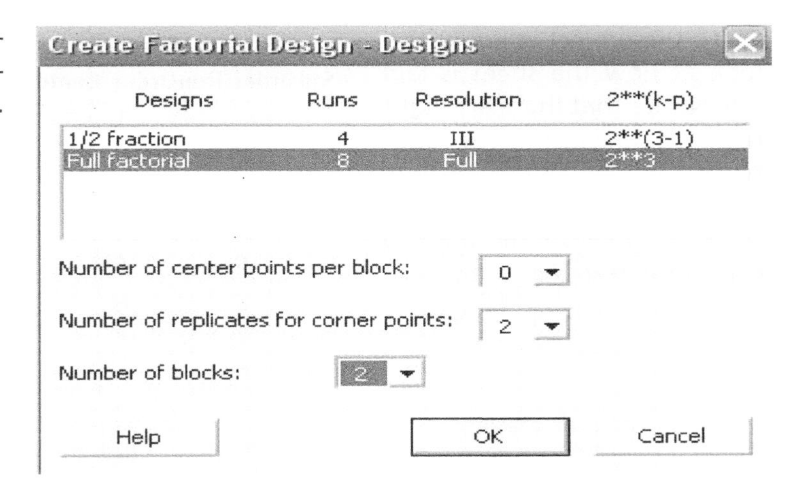

6. Notice how the remaining buttons become selectable. Click on the Factors button.

7. This is where we enter the name, type and values of high and low settings for our factors. Notice that we are using a categorical value for Material. Copy the menu box and click OK.

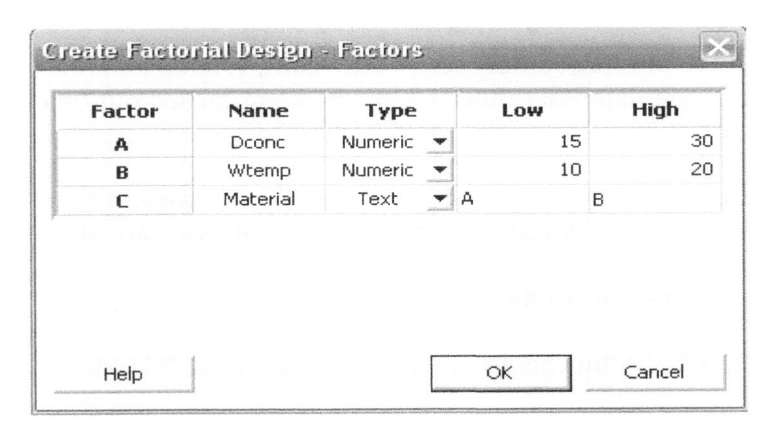

8. Click on the Options button. Folding a design is concerned with Fractional designs. For the example we are going to deselect Randomise runs so we can see the structure of the experimental runs in the project window. Normally we would randomise the runs to reduce human influence on the experiment. Click on OK and OK again to produce the worksheet.

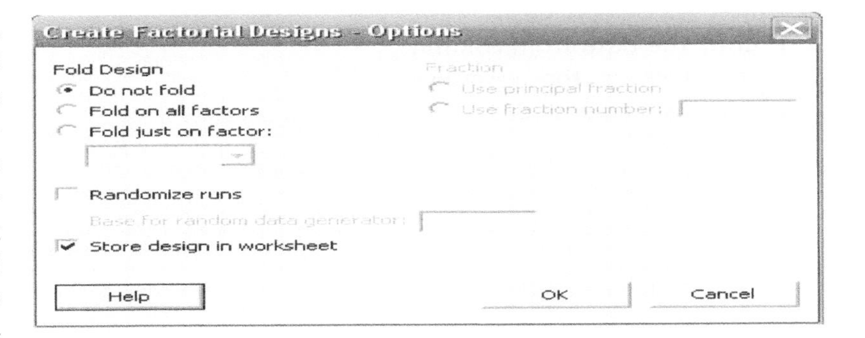

9. Go to the project window and look at how the sheet is laid out. Notice that there are eight runs within each block. Each block contains all the possible combinations of factors. This is the sheet that will be used to collect the response data.

C1	C2	C3	C4	C5	C6	C7-T
StdOrder	RunOrder	CenterPt	Blocks	Dconc	Wtemp	Material
1	1	1	1	15	10	A
2	2	1	1	30	10	A
3	3	1	1	15	20	A
4	4	1	1	30	20	A
5	5	1	1	15	10	B
6	6	1	1	30	10	B
7	7	1	1	15	20	B
8	8	1	1	30	20	B
9	9	1	2	15	10	A
10	10	1	2	30	10	A
11	11	1	2	15	20	A
12	12	1	2	30	20	A
13	13	1	2	15	10	B
14	14	1	2	30	10	B
15	15	1	2	15	20	B
16	16	1	2	30	20	B

Example 2. Analysing the experiment.

Just in case you forgot, our factors are Dconc, Wtemp and Material and the response is Grime. We now have the experimental results. Eight runs were carried out on day1 and then the eight runs were replicated on day2.

The researchers want to know which factors have affected the response and whether carrying out the replicates the next day affected the experiment. Also, are any of the interactions significant?

Use the worksheet window from example 1.

All data sets for this chapter are within Excel file 10 DOE.xls.

Worksheet: Mopping.

1. Copy the data from column H, which is the response data, into column C8.

H
Grime
218
169.6
186
143
204
164.2
175.7

2. Click Stat <<DOE <<Factorial <<Analyze Factorial Design
3. For Responses, select Grime.
4. Click on the Terms button.

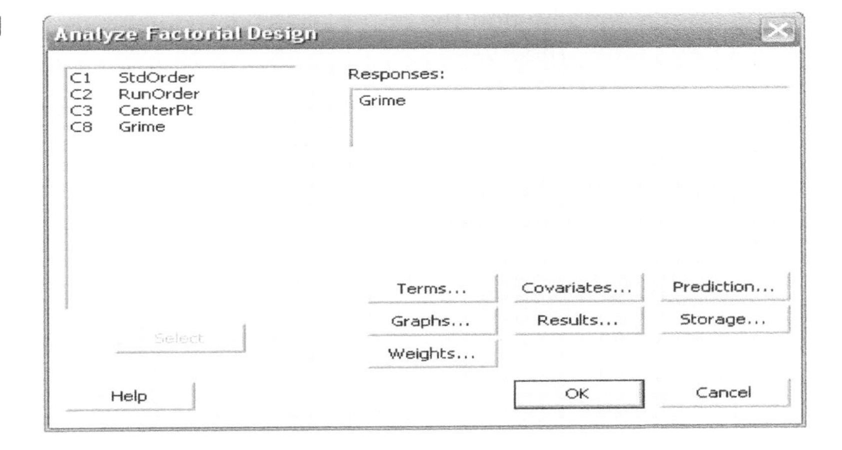

5. Ensure that blocks are included in the model and the selected terms appear as shown.
6. Click OK then click on the Graph button.

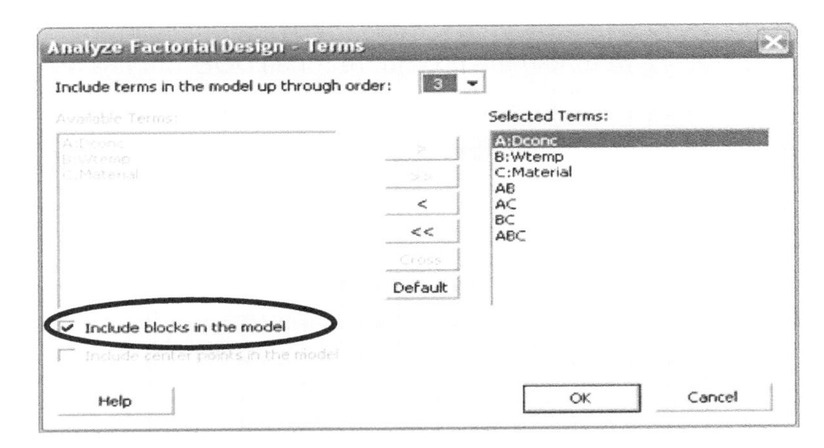

7. Under Effects Plots select Pareto. We won't select the residual plots until the model has been reduced.
8. Click OK and OK again.

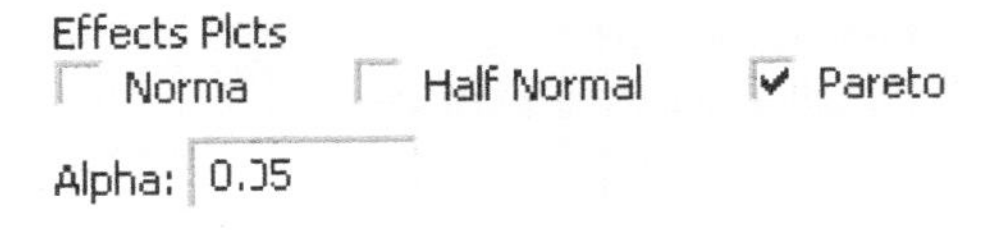

The Pareto Chart that is generated shows the significance of the terms. If any of the bars representing the terms go beyond the red line they are significant.

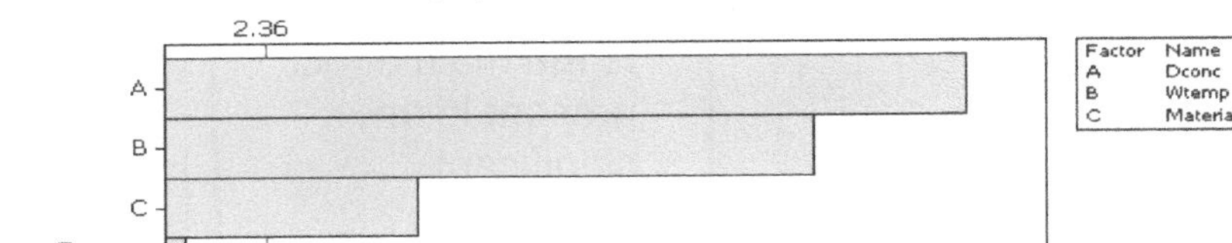

Within the session window find the Estimated Effects table. The *P* values listed will tell us which of the factors and interactions are significant. We also see that our Block was not significant, so carrying out the experiment over two days had no effect.

We will now discuss reducing the model.

Factorial Fit: Grime versus Block, Dconc, Wtemp, Material

```
Estimated Effects and Coefficients for Grime (coded units)

Term                    Effect     Coef  SE Coef       T      P
Constant                         173.90    1.114  156.11  0.000
Block                             -0.63    1.114   -0.56  0.592
Dconc                   -41.22   -20.61    1.114  -18.50  0.000
Wtemp                   -33.45   -16.72    1.114  -15.01  0.000
Material                -13.00    -6.50    1.114   -5.84  0.001
Dconc*Wtemp               0.12     0.06    1.114    0.06  0.957
Dconc*Material           -0.53    -0.26    1.114   -0.24  0.820
Wtemp*Material           -1.05    -0.52    1.114   -0.47  0.652
Dconc*Wtemp*Material     -0.98    -0.49    1.114   -0.44  0.675


S = 4.45581      PRESS = 726.100
R-Sq = 98.85%    R-Sq(pred) = 94.00%    R-Sq(adj) = 97.54%
```

This flow sheet shows the procedure that should be followed when removing insignificant terms from DOE models.

It is good practice to follow the procedure for all DOEs but it is not required for orthogonal designs with replicates.

Although not highlighted within the flow sheet we can remove all nonsignificant terms of the same order at the same time.

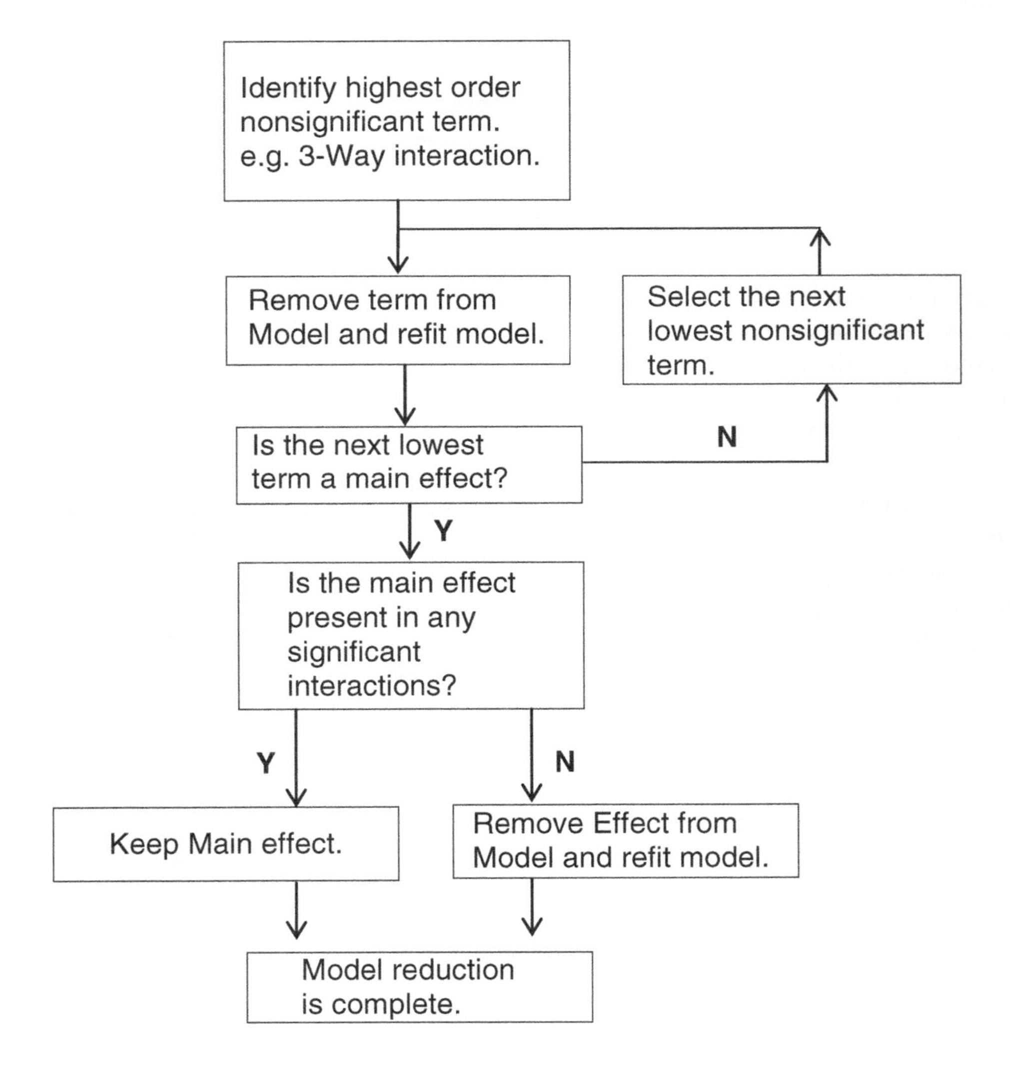

9. We are now going to reduce the model as per the flow sheet shown on the previous slide. Click on the Edit Last button.

10. Click on the Terms button.
11. Deselect Blocks and the 3-Way interaction from the model.
12. Click OK and OK again.

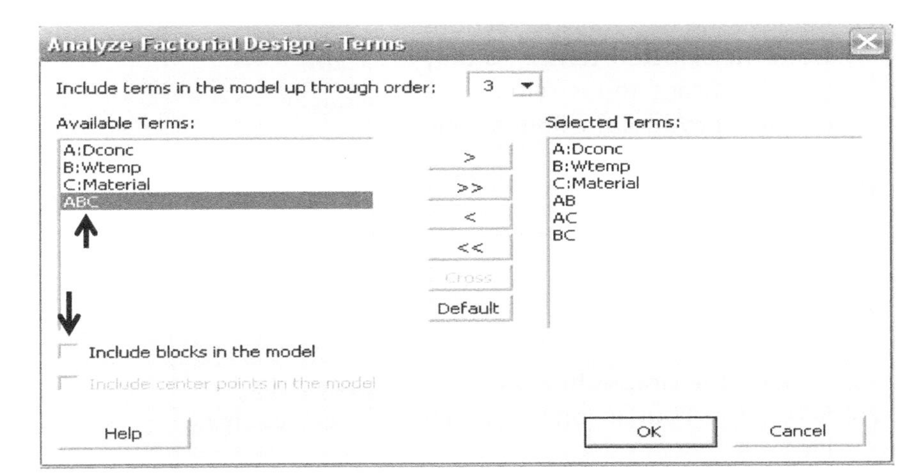

13. The three way interaction has been removed from the Pareto Chart and we can see that all of the two way interactions are also not significant.

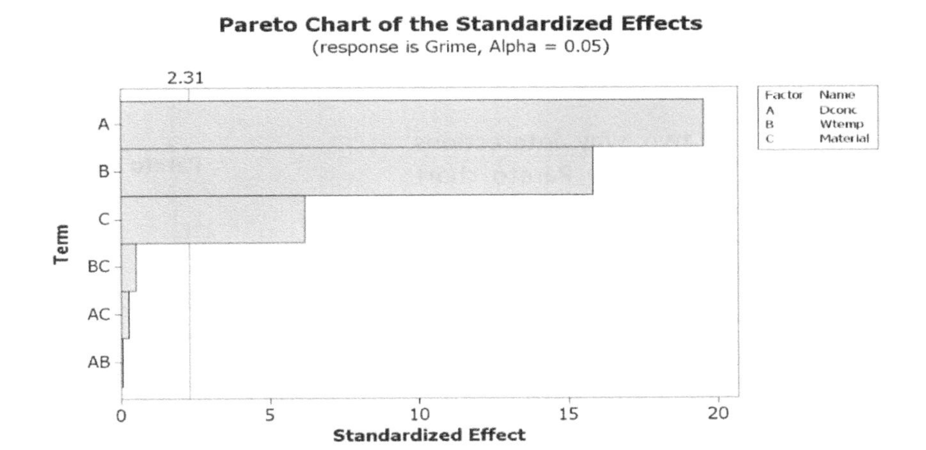

14. Within the session window the effects table also shows that the two way interactions are not significant.

Estimated Effects and Coefficients for Grime (coded units)

Term	Effect	Coef	SE Coef	T	P
Constant		173.90	1.017	170.94	0.000
Dconc	-41.22	-20.61	1.017	-20.26	0.000
Wtemp	-33.45	-16.72	1.017	-16.44	0.000
Material	-13.00	-6.50	1.017	-6.39	0.000
Dconc*Wtemp	0.13	0.06	1.017	0.06	0.952
Dconc*Material	-0.53	-0.26	1.017	-0.26	0.802
Wtemp*Material	-1.05	-0.53	1.017	-0.52	0.618

S = 4.06930 PRESS = 471.016
R-Sq = 98.77% R-Sq(pred) = 96.11% R-Sq(adj) = 97.95%

15. Click on the Edit Last button.

16. Click on the Terms button.

17. Remove the two way interactions from the selected terms. As only the significant terms remain we can now produce the Four in One Residual Plot.
18. Click OK.

Selected Terms:
A:Dconc
B:Wtemp
C:Material

19. Click on the Graphs button.
20. Select the Four in one Residual Plot.

21. Click on OK and OK again.

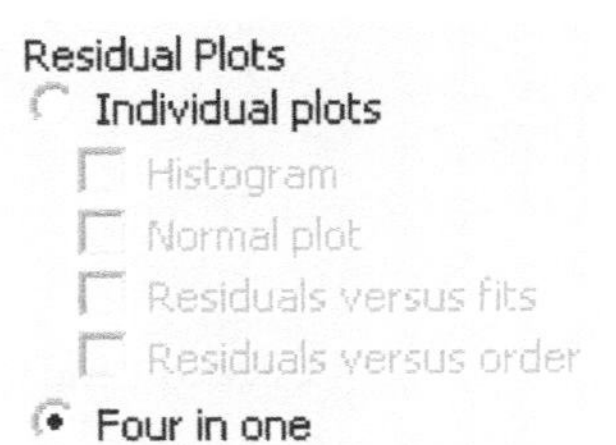

22. With the two way interactions removed from the Pareto chart only the main factors are showing as significant.

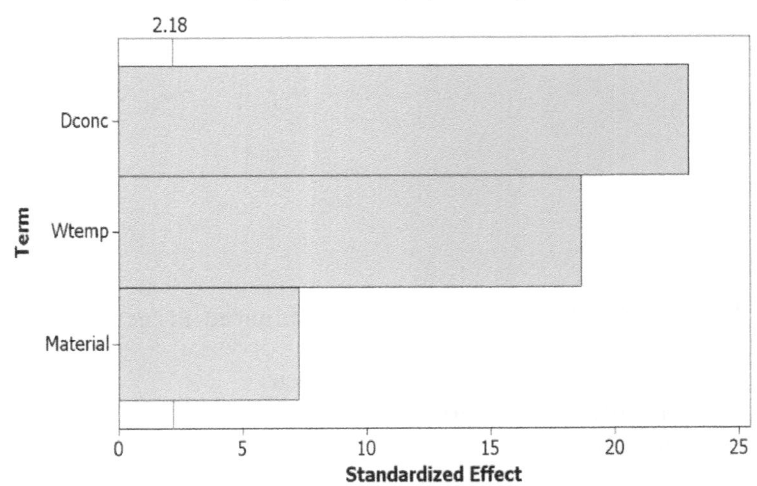

23. In the session window we see confirmation that the main factors are significant in the Estimated Effects table. R-Sq(adj) is 98.4. This means that 98.4% of the variation seen within Grime can be explained by the changes within our factors.

Estimated Effects and Coefficients for Grime (coded units)

Term	Effect	Coef	SE Coef	T	P
Constant		173.90	0.8974	193.79	0.000
Dconc	-41.22	-20.61	0.8974	-22.97	0.000
Wtemp	-33.45	-16.72	0.8974	-18.64	0.000
Material	-13.00	-6.50	0.8974	-7.24	0.000

S = 3.58942 PRESS = 274.858
R-Sq = 98.72% R-Sq(pred) = 97.73% R-Sq(adj) = 98.40%

Also listed within the session window we have a group of numbers listed under the line 'Estimated Coefficients for Grime using data in uncoded units'.

Term	Coef
Constant	285.913
Dconc	-2.74833
Wtemp	-3.34500
Material	-6.50000

These are the coefficients that can used to build a prediction equation for the Response. The full prediction equation for Grime is shown here for your convenience.

We also need to confirm the validity of the model by checking the residuals.

$$\text{Grime} = 285.9 - 2.74*\text{Dconc} - 3.35*\text{Wtemp} - 6.5*\text{Material}$$

where Material would have values of 1 or –1

Equation (10.4)

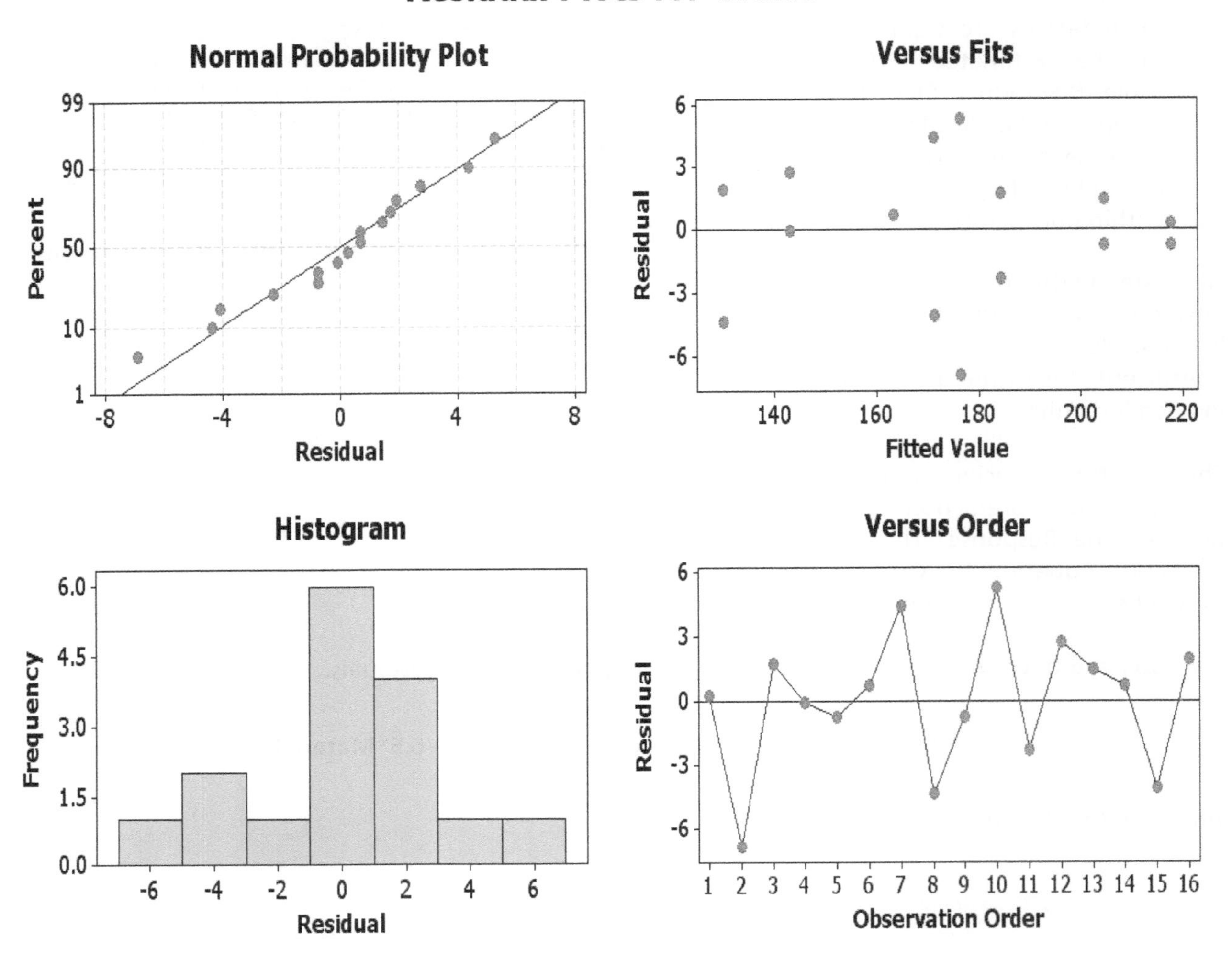

- Residuals are normally distributed on the normal probability plot.
- Residuals are equally distributed about the centre line on the versus fits plot.
- Histogram of residuals does not show extreme skewness.
- If the original data was collected in time order then the residuals should not show any ordered behaviour on the versus order plot.

As there are no issues with the Residuals we will now produce the Main Effects and Cube Plots.

24. Click Stat <<DOE <<Factorial << Factorial Plots

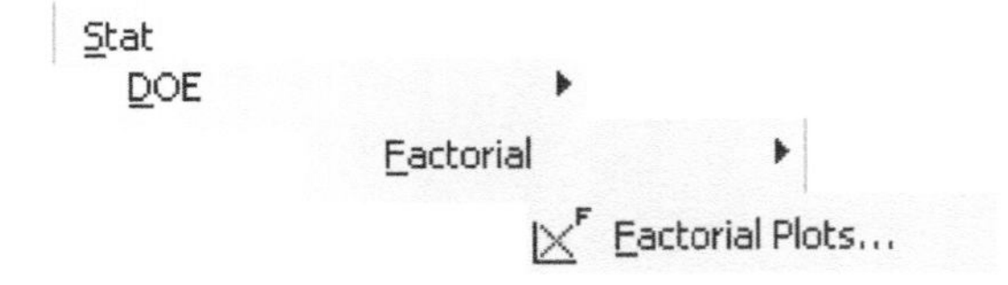

25. As there were no significant interactions we will only look at the Main Effect and Cube Plots. Go into Setup for Main Effects.

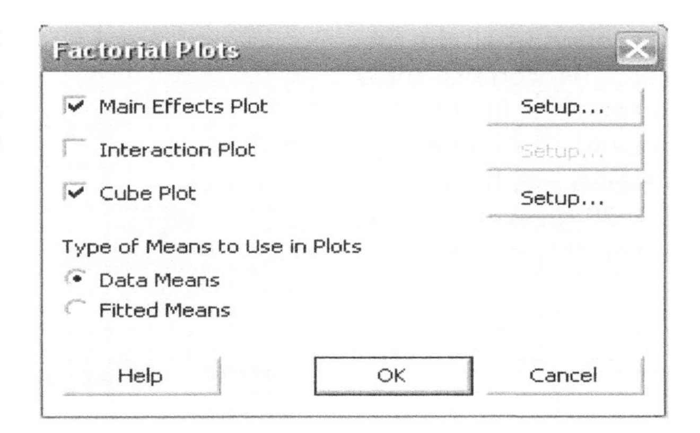

26. Select Grime in the Responses and then select the three main effects as the factors within the plots. Then click OK. After setting up the Cube Plot in exactly the same way click OK and OK again.

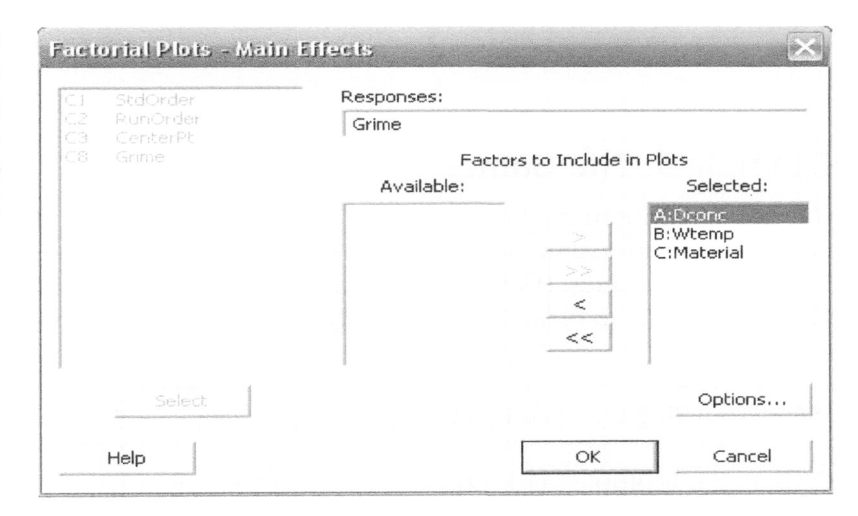

27. We want to minimise the Grime response. The Main Effect Plots show the levels we need to use in our factors to minimise Grime. As the y axis units are the same we can see the Main Effect that has the steepest line has the greatest effect on the response.

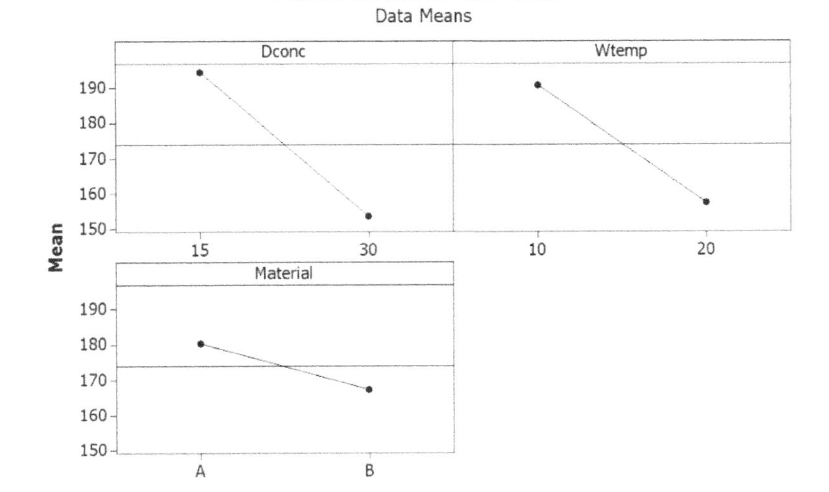

28. The cube plot shows us all the factors together. The corners show the average response for the corresponding factors. The corner representing Material = B, Dconc = 30 and Wtemp = 20 gives the lowest Grime value and this is marked on the cube.

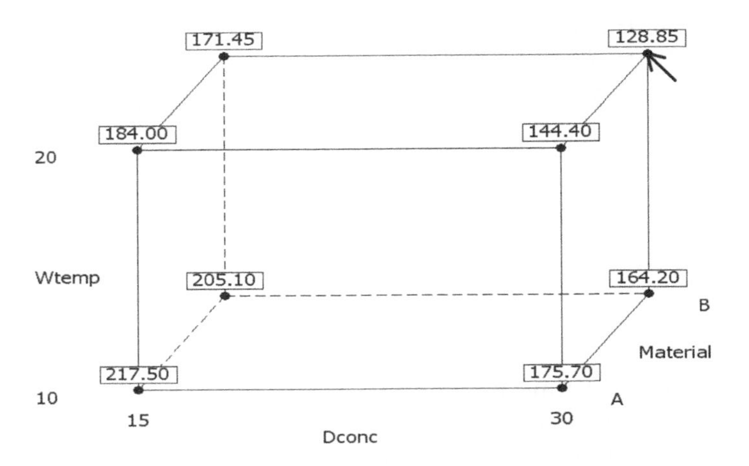

Example 3. The Gamer.

Humzah has bought the latest Cawl of Duty game for his PC. Unfortunately, the game does not run fast enough so he buys some new hardware for his PC. He runs a two level factorial DOE in order to optimise his hardware settings. His response is frames per second (FPS), however he realises that on occasions his CPU overheats. This means that he needs CPUtemp as a response as well. The factors that he is trying to optimise are:

FSB = Front Side Bus speed (200, 220)

CPUX = CPU Multiplier (10, 12)

Memspeed = Memory speed (333, 400)

GPUspeed = Graphics Processing Unit speed (500, 600).

Establish which of the factors and interactions are significant. Use the Results Optimizer to establish the settings to use to minimise CPUTEMP and maximise FPS. CPUTEMP is more important and the target for that is 65 with an upper limit of 70. He wants a target of 200 for FPS with a lower limit of 180.

Worksheet: Gamer.

As there are two responses we must carry out the DOE analysis twice to decide what the significant factors are for each response before we use the Results Optimizer. Also, as we are going to be importing the complete results table into Minitab we must first initialise the design. This is done by using the Define Custom Factorial Design procedure. Basically, we need to tell Minitab that the results we have imported in are from a DOE procedure.

1. Copy the data from the Excel file into a new Minitab project window.

2. Click Stat <<DOE <<Factorial <<Define Custom Factorial Design.

FSB	CPUX	Mem Speed	GPU Speed	FPS	CPU TEMP
200	10	333	500	153.41	58.72
220	10	333	500	153.41	67.47
200	12	333	500	155.91	73.55
220	12	333	500	158.01	84.01
200	10	400	500	182.10	72.80

3. Select the four factors. Two level factorial should be selected by default.

4. Click on the Low/High button.

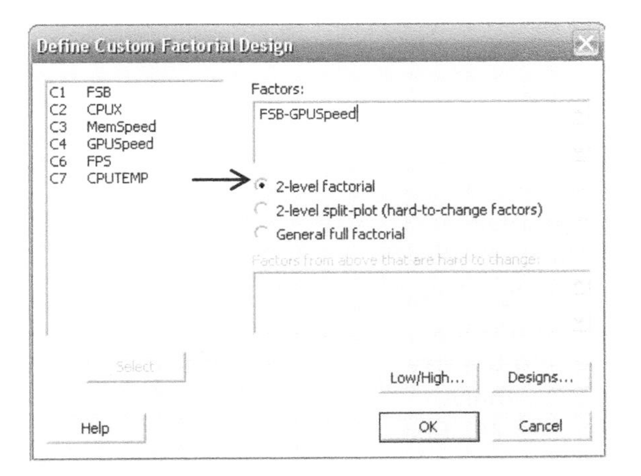

5. Minitab will automatically pick up the high and low values of all of the factors. Note that uncoded units is selected by default.

6. Click OK and then click on the Designs button.

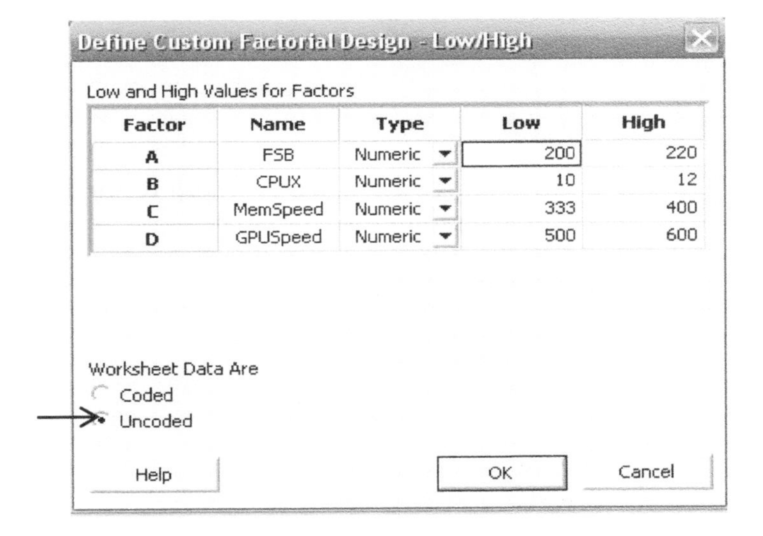

7. We don't need to change anything but we could specify the data order, center points and blocks within this menu.
8. Click OK and OK again. Now that we have defined our factorial design Minitab will be happy to analyse it for us.

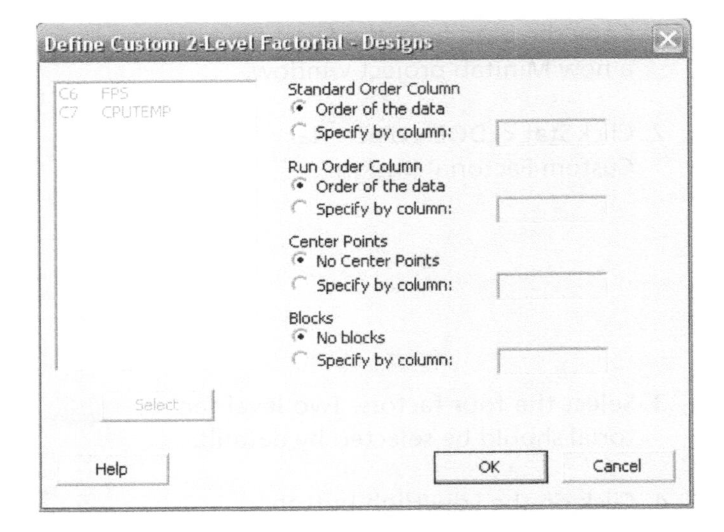

9. Click Stat <<DOE <<Factorial <<Analyse Factorial Design. Select FPS as the response. We are going to carry out the DOE analysis procedure for FPS first and then CPUTEMP.

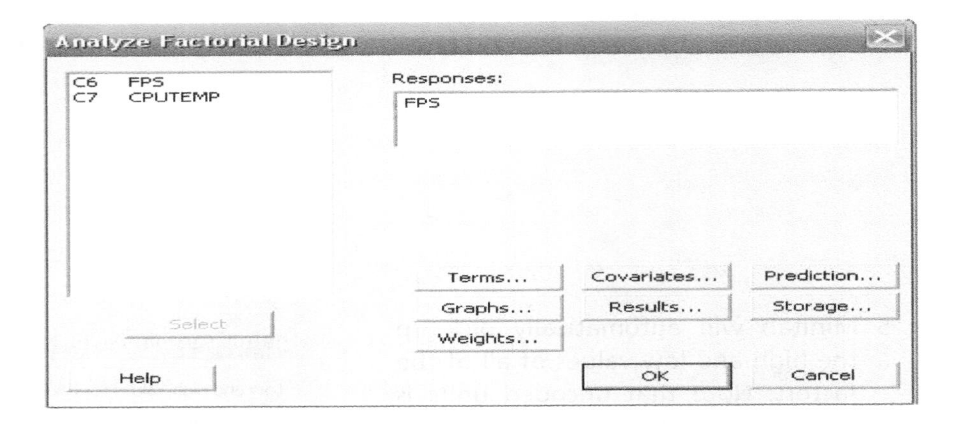

10. Click on the Terms button. All terms up to the fourth order should be preselected. Click OK.

11. Click on Graphs.

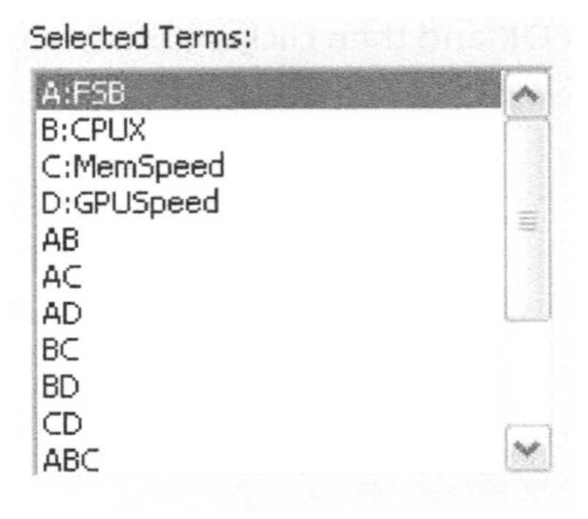

12. Under Effects Plots select Pareto.
13. Click OK and OK again.

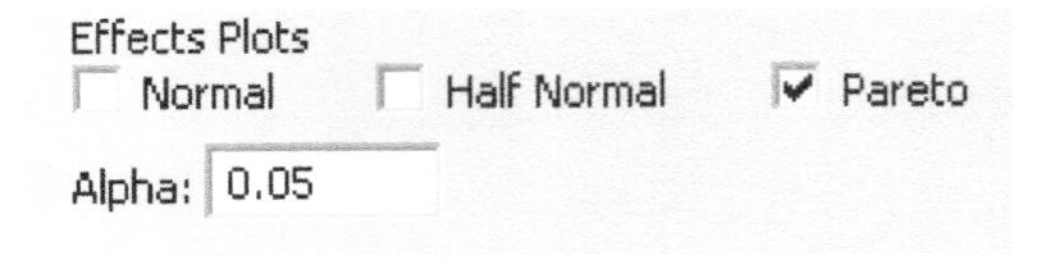

14. On the Pareto Chart all the bars that go beyond the red line are significant.

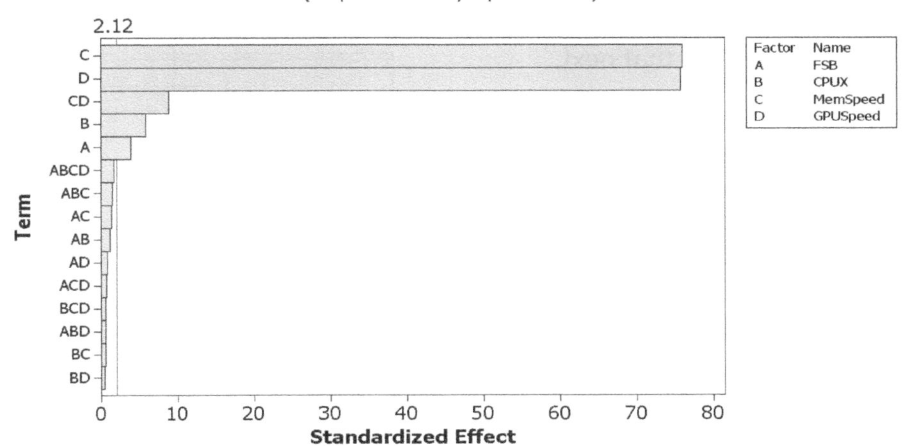

15. In the session window the Analysis of Variance table shows the breakdown of the terms that are significant. It can be seen that the three way and four way interactions are not significant. However, we will use the recommended stepwise procedure to reduce the model.

Analysis of Variance for FPS (coded units)

Source	DF	Seq SS	Adj SS	Adj MS	F	P
Main Effects	4	15269.7	15269.7	3817.43	2895.42	0.000
FSB	1	20.3	20.3	20.32	15.41	0.001
CPUX	1	45.8	45.8	45.84	34.77	0.000
MemSpeed	1	7624.3	7624.3	7624.33	5782.86	0.000
GPUSpeed	1	7579.2	7579.2	7579.21	5748.63	0.000
2-Way Interactions	6	108.7	108.7	18.11	13.74	0.000
FSB*CPUX	1	1.8	1.8	1.76	1.33	0.265
FSB*MemSpeed	1	2.7	2.7	2.70	2.05	0.171
FSB*GPUSpeed	1	0.8	0.8	0.81	0.62	0.444
CPUX*MemSpeed	1	0.4	0.4	0.43	0.32	0.577
CPUX*GPUSpeed	1	0.3	0.3	0.30	0.23	0.640
MemSpeed*GPUSpeed	1	102.7	102.7	102.68	77.88	0.000
3-Way Interactions	4	4.6	4.6	1.15	0.87	0.502
FSB*CPUX*MemSpeed	1	3.1	3.1	3.06	2.32	0.147
FSB*CPUX*GPUSpeed	1	0.4	0.4	0.43	0.32	0.577
FSB*MemSpeed*GPUSpeed	1	0.6	0.6	0.63	0.48	0.498
CPUX*MemSpeed*GPUSpeed	1	0.5	0.5	0.48	0.36	0.557
4-Way Interactions	1	4.0	4.0	3.99	3.03	0.101
FSB*CPUX*MemSpeed*GPUSpeed	1	4.0	4.0	3.99	3.03	0.101

16. Click on the Edit Last icon.

17. Then click on the Terms button.

18. Deselect the ABCD term or change the selection so that only terms up to order three are considered.

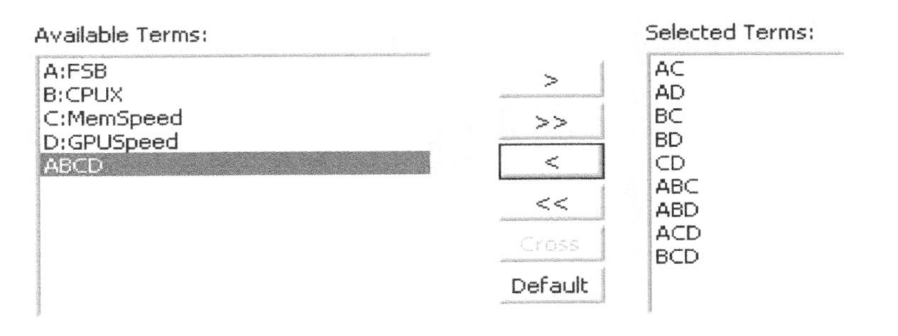

19. Click OK and OK again.

20. The Pareto chart shows that we can also remove all three way interactions so we will do that next.

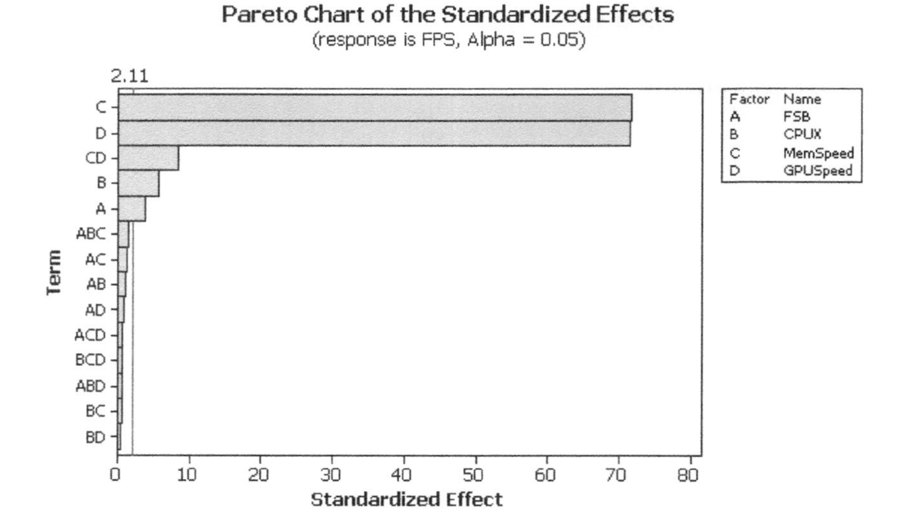

21. Click on Edit Last.
22. Then click on the Terms button.
23. Change the selection so that only terms up to order two are considered or manually remove all four of the three way interactions.
24. Click OK and OK again.

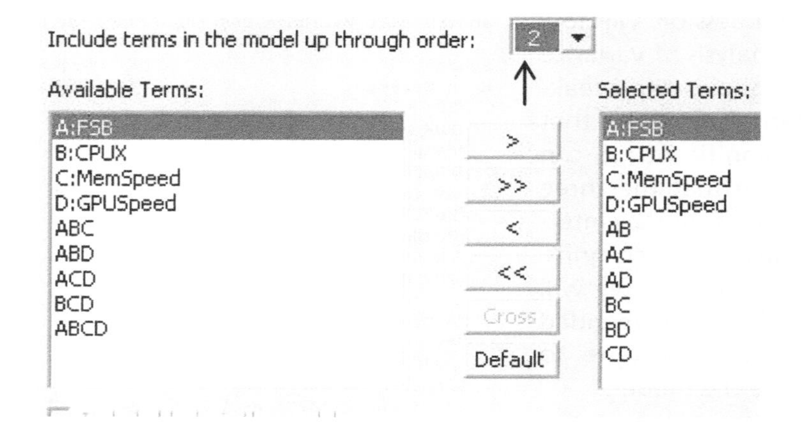

25. The Pareto chart shows that we can also remove all two way interactions with the exception of CD which is significant.

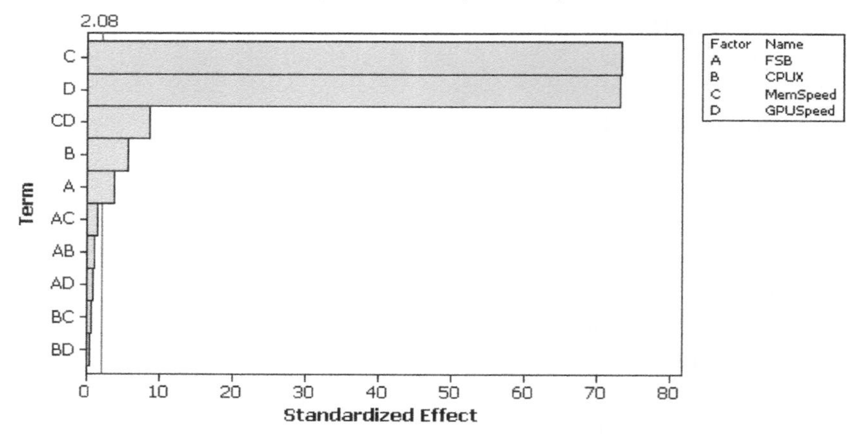

26. Click on the Edit Last icon.

27. Then click on Terms button.

28. Change the selection so that only CD and the main effects are selected. Click OK.

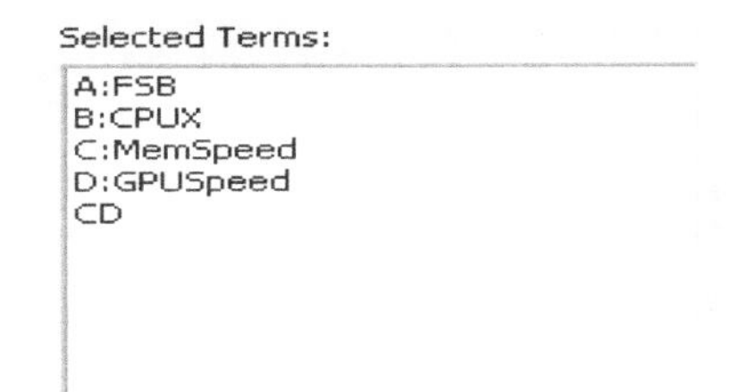

29. As this is the last iteration we will also produce the residuals plot. Click on the Graphs button and then select the Four in one residual plot. Click OK and OK again.

Residual Plots
Individual plots
Histogram
Normal plot
Residuals versus fits
Residuals versus order
Four in one

30. The Pareto chart shows only significant terms as expected.

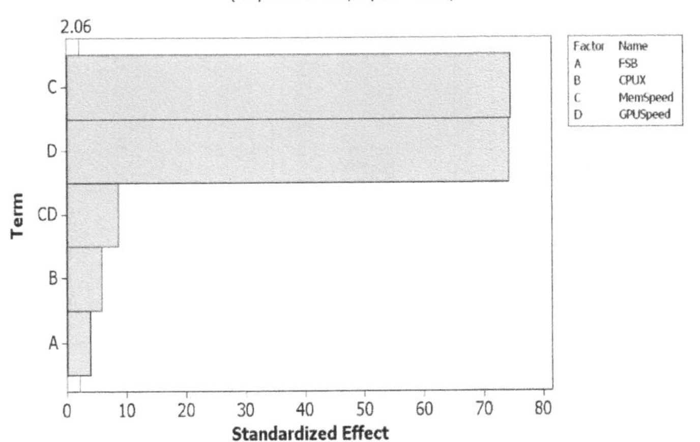

31. In the Session window the Estimated Effects table shows that the main effects are significant and the CD term which is the interaction between MemSpeed and GPUSpeed.

Term	Effect	Coef	SE Coef	T	P
Constant		184.183	0.2071	889.33	0.000
FSB	1.594	0.797	0.2071	3.85	0.001
CPUX	2.394	1.197	0.2071	5.78	0.000
MemSpeed	30.871	15.436	0.2071	74.53	0.000
GPUSpeed	30.780	15.390	0.2071	74.31	0.000
MemSpeed*GPUSpeed	3.583	1.791	0.2071	8.65	0.000

```
S = 1.17155      PRESS = 54.0563
R-Sq = 99.77%    R-Sq(pred) = 99.65%    R-Sq(adj) = 99.72%
```

We can confirm the validity of the model by looking at the residuals for FPS. After which we will look at the Main Effect and Cube plots.

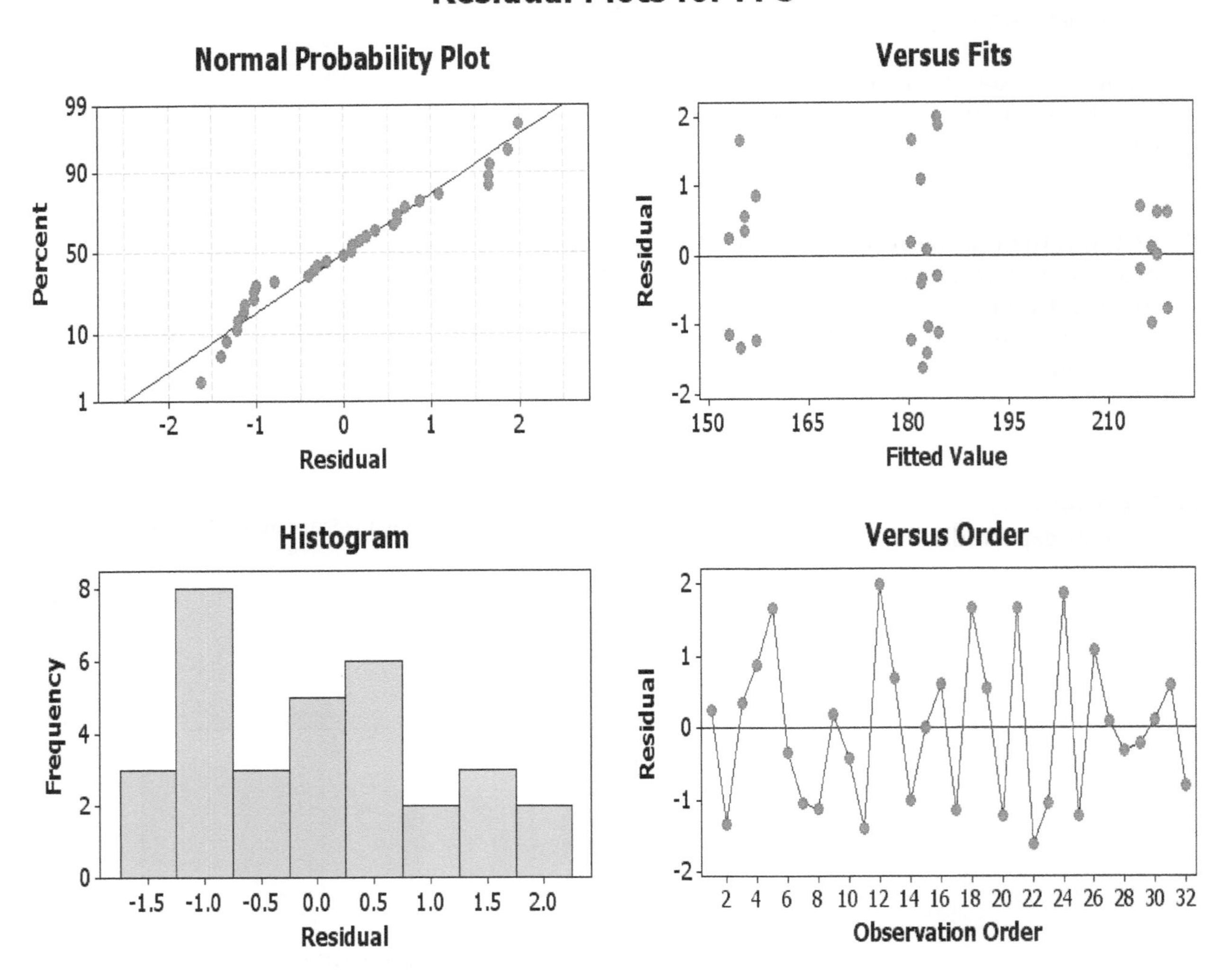

- Residuals are normally distributed on the normal probability plot.
- Residuals are equally distributed about the centre line on the versus fits plot.
- Histogram of residuals does not show extreme skewness.
- If the original data was collected in time order then the residuals should not show any ordered behaviour on the versus order plot.

As there are no issues with the Residuals we will now produce the Main Effects and Cube Plots.

32. Click Stat <<DOE <<Factorial <<Factorial Plots. Select Main Effects Plot and Cube Plot. We won't plot the Interaction Plot to see our single interaction term.

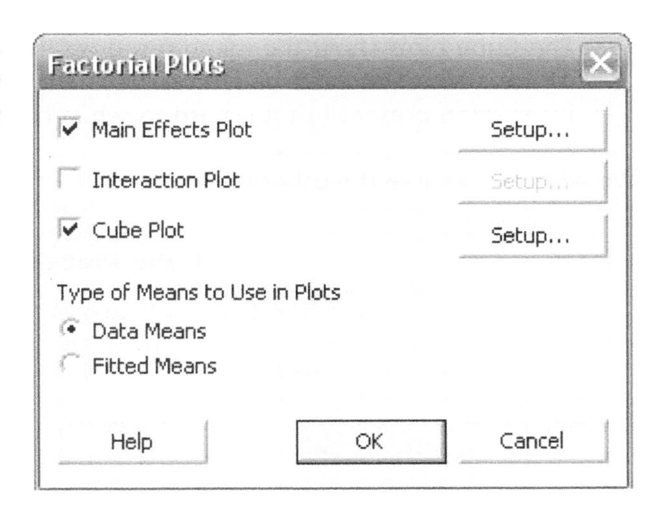

33. Both plots need to be set up in exactly the same way before they can be displayed. Click on Setup then select FPS as the Response and select all the main effects as factors to be plotted. Carry out the same setup for both plots and then click OK and OK again.

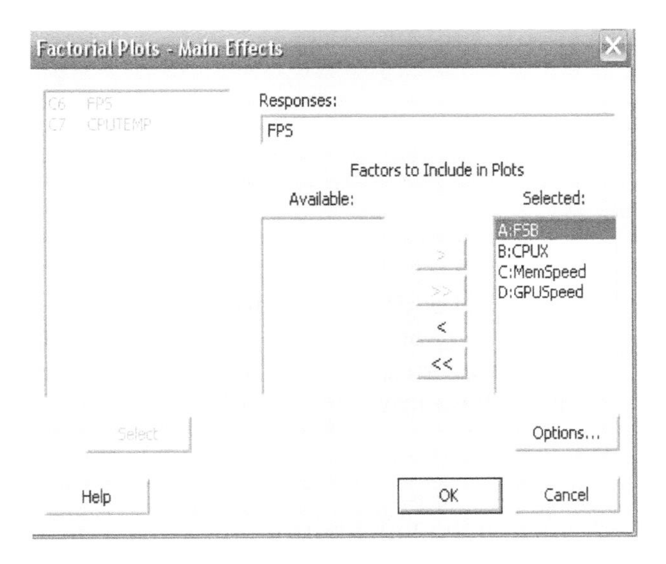

34. The Main effects plot shows that MemSpeed and GPUSpeed are having a relatively large effect on the response.

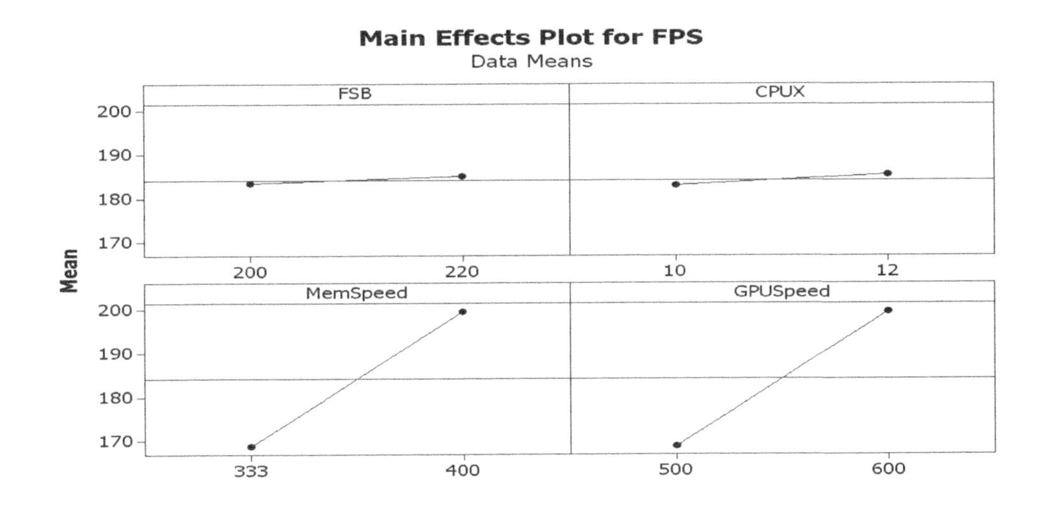

35. The Cube plots show the mean response that we got at the different levels. The changing levels of the main effects can be seen to be changing the response, bearing in mind that there is also an interaction present but it is hard to see in this cube plot.

We will now analyse the other response.

Cube Plot (data means) for FPS

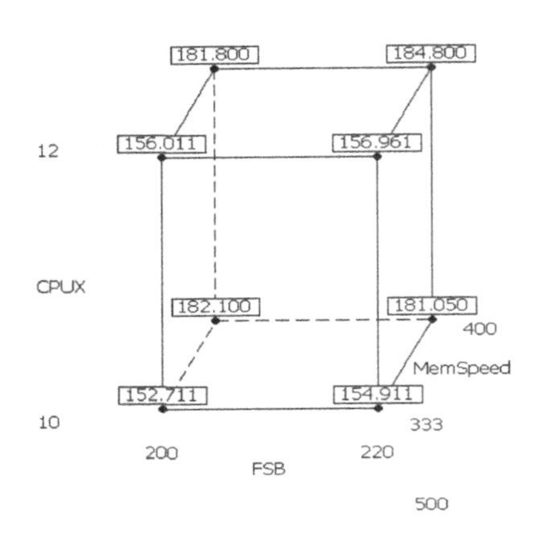

36. Click Stat <<DOE <<Factorial <<Analyse Factorial Design. We are going to carry out the DOE analysis procedure for CPUTEMP so that needs to be selected as the Response.

Responses:
CPUTEMP

37. Click on the Terms button. Ensure all terms up to the fourth order are preselected. Click OK.
38. Click on the Graphs button.

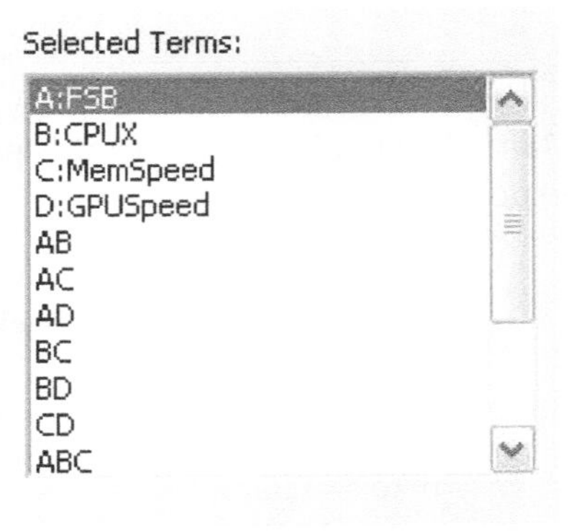

39. Deselect the Residual Plots. The Pareto plot remains selected.
40. Click OK and OK again.

Residual Plots
Individual plots
Histogram
Normal plot
Residuals versus fits
Residuals versus order
Four in one

41. Again, on the Pareto Chart all the bars that go beyond the red line are significant.

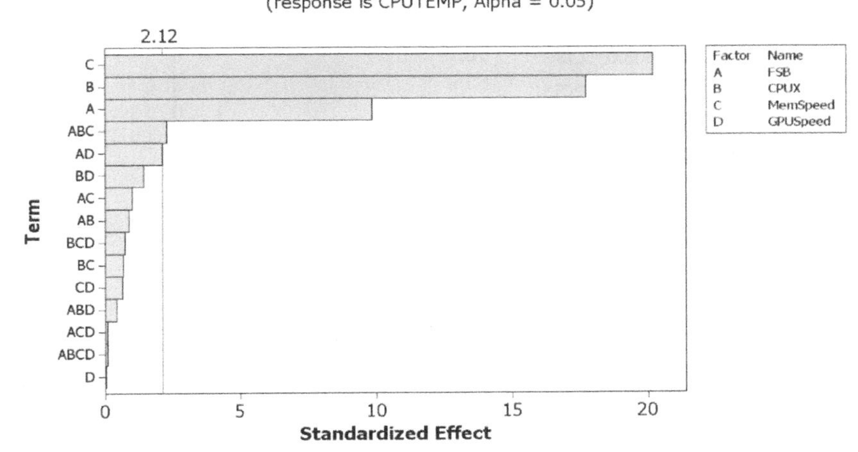

42. In the session window the Analysis of Variance table shows the breakdown of the terms that are significant. It can be seen that the four way interaction is not significant but one three way interaction is significant.
43. As a mini exercise I will leave it to you to reduce the model.

Estimated Effects and Coefficients for CPUTEMP (coded units)

Term	Effect	Coef	SE Coef	T	P
Constant		78.3992	0.3692	212.36	0.000
FSB	7.2567	3.6283	0.3692	9.83	0.000
CPUX	13.0733	6.5367	0.3692	17.71	0.000
MemSpeed	14.9017	7.4508	0.3692	20.18	0.000
GPUSpeed	0.0125	0.0063	0.3692	0.02	0.987
FSB*CPUX	0.6483	0.3242	0.3692	0.88	0.393
FSB*MemSpeed	0.7267	0.3633	0.3692	0.98	0.340
FSB*GPUSpeed	-1.5625	-0.7812	0.3692	-2.12	0.050
CPUX*MemSpeed	-0.4900	-0.2450	0.3692	-0.66	0.516
CPUX*GPUSpeed	-1.0375	-0.5187	0.3692	-1.41	0.179
MemSpeed*GPUSpeed	-0.4625	-0.2312	0.3692	-0.63	0.540
FSB*CPUX*MemSpeed	-1.6817	-0.8408	0.3692	-2.28	0.037
FSB*CPUX*GPUSpeed	-0.3125	-0.1563	0.3692	-0.42	0.678
FSB*MemSpeed*GPUSpeed	0.0625	0.0313	0.3692	0.08	0.934
CPUX*MemSpeed*GPUSpeed	0.5375	0.2687	0.3692	0.73	0.477
FSB*CPUX*MemSpeed*GPUSpeed	0.0625	0.0313	0.3692	0.08	0.934

44. The final Pareto chart shows that the main effects were all significant with the exception of D (GPUSpeed), however, this is kept in the model because there are significant interactions that contain GPUSpeed.

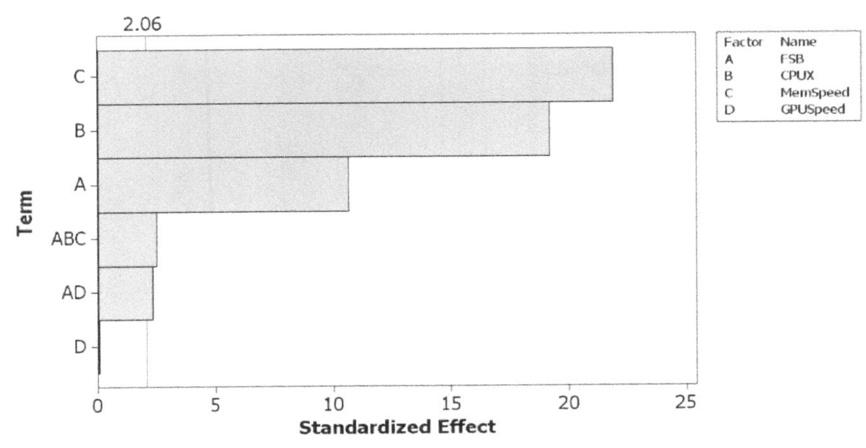

45. The Estimated Effects table also shows the same information.
46. As another mini exercise plot the Main Effects and Cube Plots. Do not put GPUSpeed into the cube plot.

Estimated Effects and Coefficients for CPUTEMP (coded units)

Term	Effect	Coef	SE Coef	T	P
Constant		78.3992	0.3405	230.23	0.000
FSB	7.2567	3.6283	0.3405	10.66	0.000
CPUX	13.0733	6.5367	0.3405	19.20	0.000
MemSpeed	14.9017	7.4508	0.3405	21.88	0.000
GPUSpeed	0.0125	0.0063	0.3405	0.02	0.986
FSB*GPUSpeed	-1.5625	-0.7813	0.3405	-2.29	0.030
FSB*CPUX*MemSpeed	-1.6817	-0.8408	0.3405	-2.47	0.021

S = 1.92630 PRESS = 151.987
R-Sq = 97.49% R-Sq(pred) = 95.89% R-Sq(adj) = 96.89%

47. The Main effects plot shows that CPUX and MemSpeed have a relatively large effect on the response, whereas FSB has slightly less effect and GPUSpeed has no effect at all.

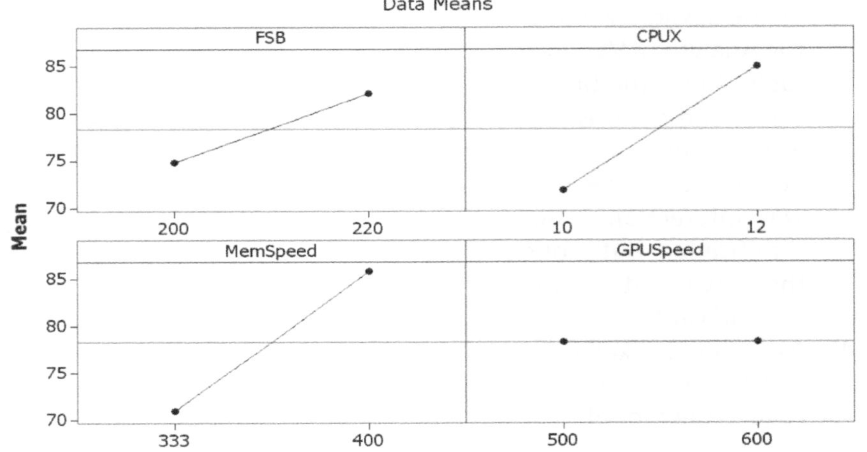

48. Part of the ABC interaction can be seen within the cube plot. As GPUSpeed did not have an effect we did not put it into the cube plot as an additional factor.

Cube Plot (data means) for CPUTEMP

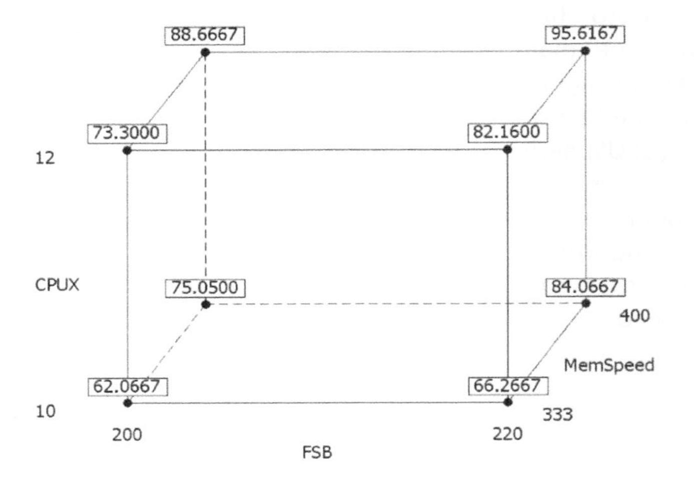

We now need to use the Results Optimizer to establish the levels of the factors that we want to use to minimise CPUTEMP and maximise FPS. It was said that CPUTEMP was more important and the target for that was 65 with an upper limit of 70. The target for FPS was 200 with a lower limit of 180.

49. Click Stat <<DOE <<Factorial <<Response Optimizer.
50. Select both of the responses. Then click on the Setup button.
51. Complete the header of the menu box as shown. Note how the menus are completed with our required response information.
52. Click OK and OK again.

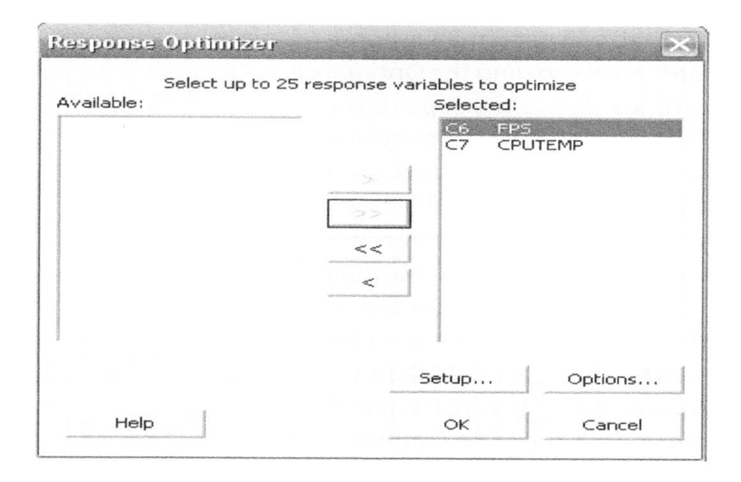

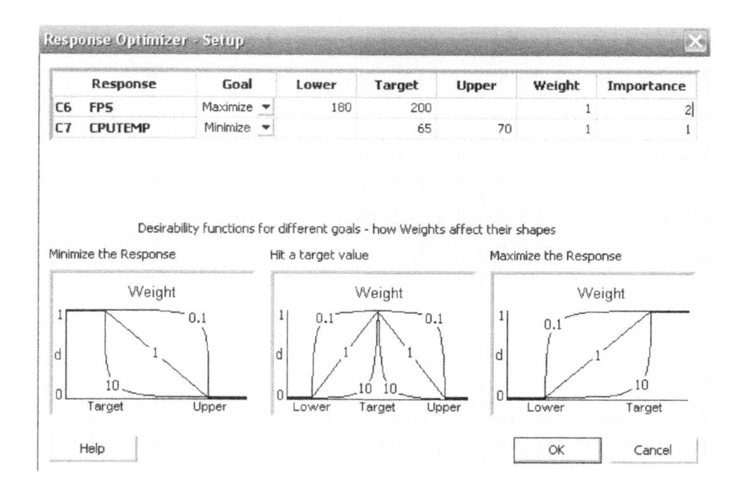

Response	Goal	Lower	Target	Upper	Weight	Importance
C6 FPS	Maximize	180	200		1	2
C7 CPUTEMP	Minimize		65	70	1	1

Minitab will open the Response Optimizer showing the Optimal settings. Along the top are the four main effects. The red value in square brackets shows their current settings and these values represent the best settings we can use to achieve the goals that have been set. However, if the user wishes to explore new values they can be changed by dragging the corresponding vertical line. The Optimal is given as a graphical matrix, on the left hand axis we have the Responses. An FPS of 193.6 and CPUTEMP of 67.5 are the optimal response values we can achieve. 'd' is the desirability factor for each response. The Composite Desirability is 0.614, as shown in the top left.

The lower plot shows the effects on the responses and desirability when the values of the Main Effects are changed by dragging the red lines. Have a go yourself and see how the values change.

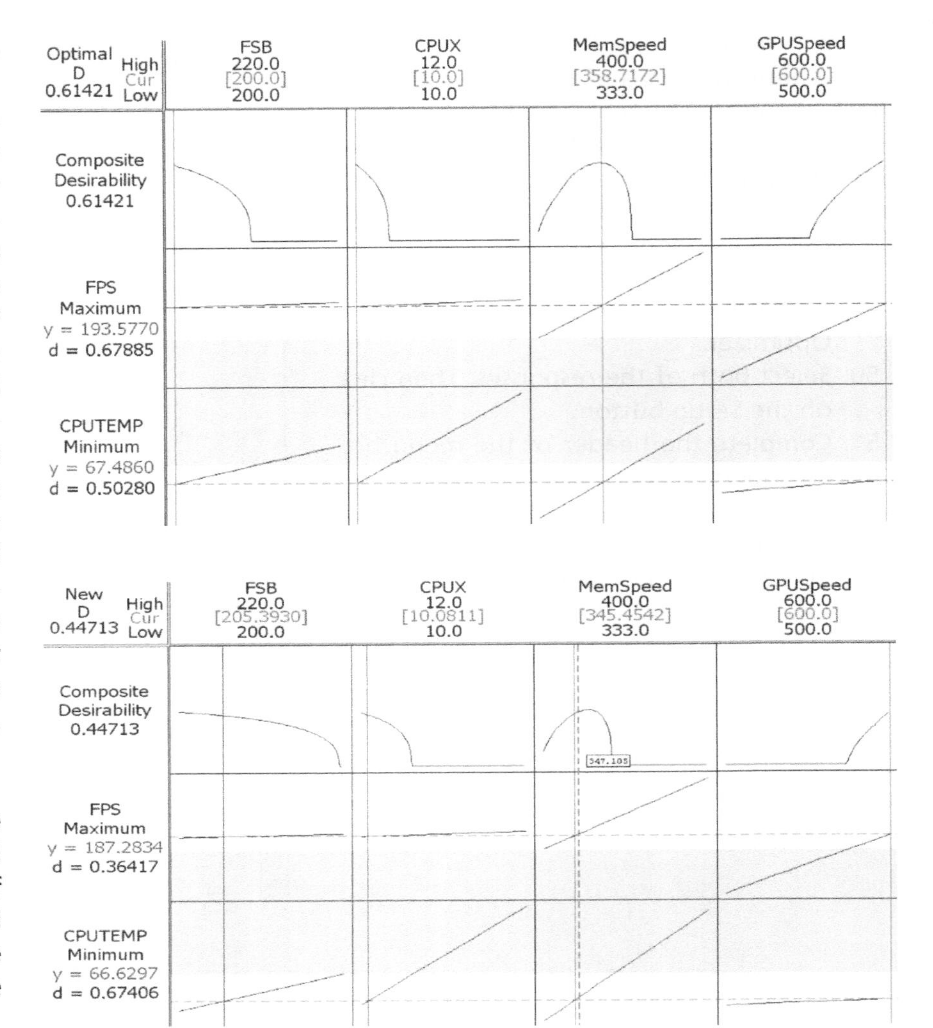

Exercise 1. Artificial Diamonds.

Princess Raeesa has developed a new press to make artificial diamonds. Pure carbon powder is placed into the press and then exposed to extreme pressure and temperature for a set amount of time to produce artificial diamonds. The settings of time, pressure, temperature and particle size of the carbon need to be optimised to give a yield of 50%.

If the yield is higher than 55% the diamonds lose clarity; and any lower than 45% and the process becomes uneconomic. A series of experiments were carried out at the following levels of each of the factors and the yield was measured.

Particle size = Partsize (10, 40)

Time = Time (400, 600)

Pressure = Press(30, 60)

Temperature = Temp(800, 900)

Establish the settings required to obtain a yield between 45 and 55%.

Worksheet: Diamond.

Using the Pareto Chart the model can be reduced to significant terms only.

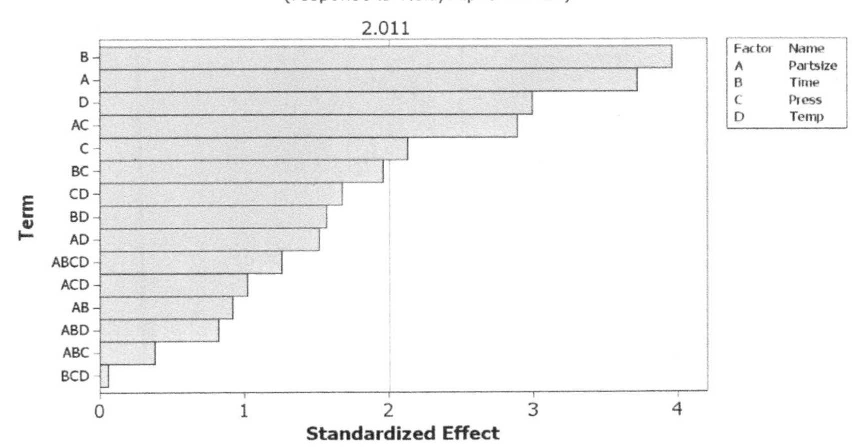

It can be seen that all the main terms are significant as is the Partsize*Press interaction. Although, not part of the question we will have a look at the main effects and interactions plot for the AC term.

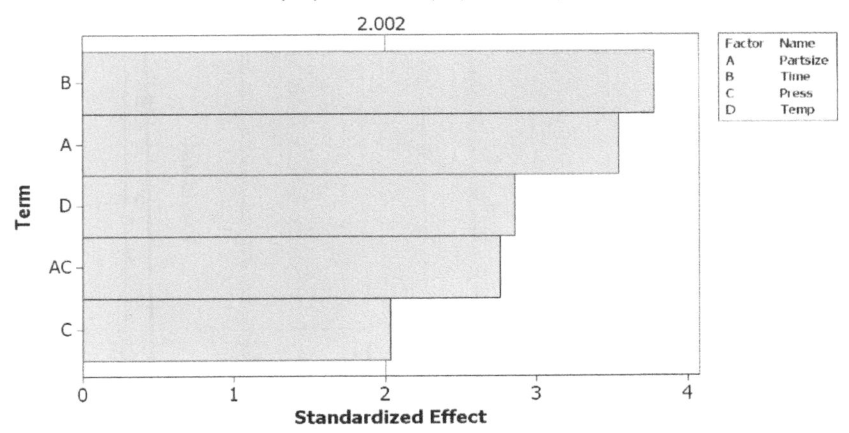

The main effects plots show that all the main effects have a positive effect on yield.

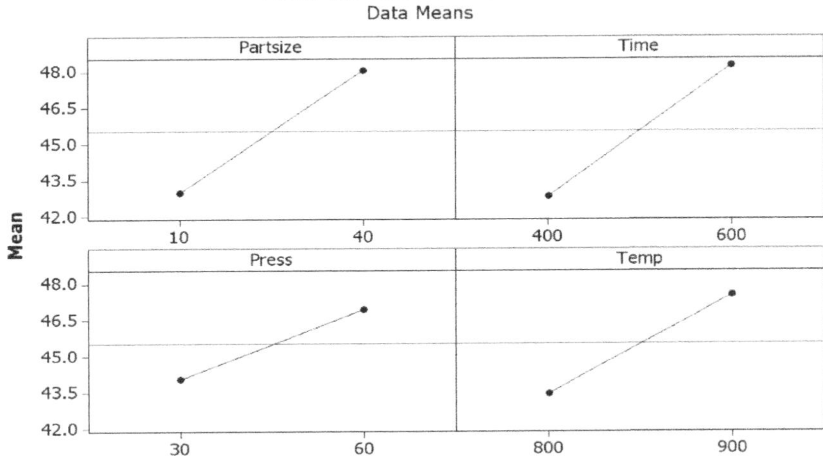

The interaction plot shows that there is a strong interaction between particle size and pressure.

We can now check the residuals to establish the validity of the model.

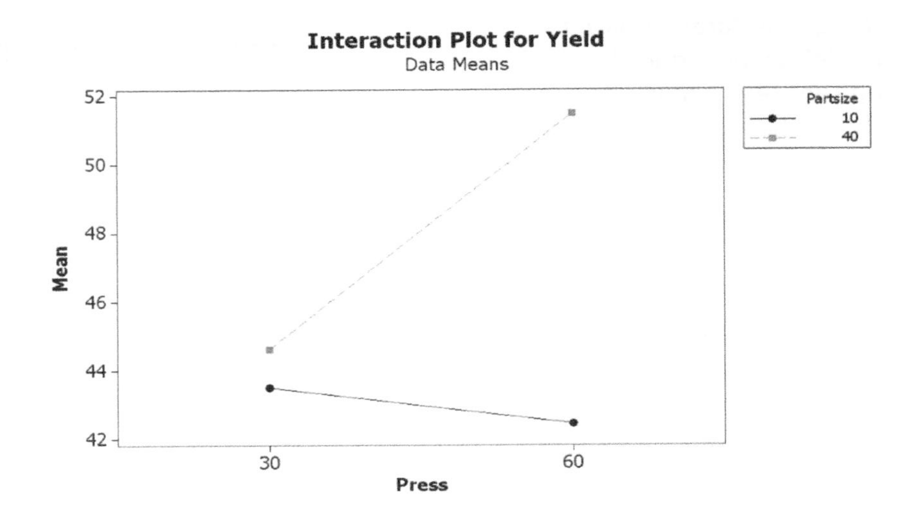

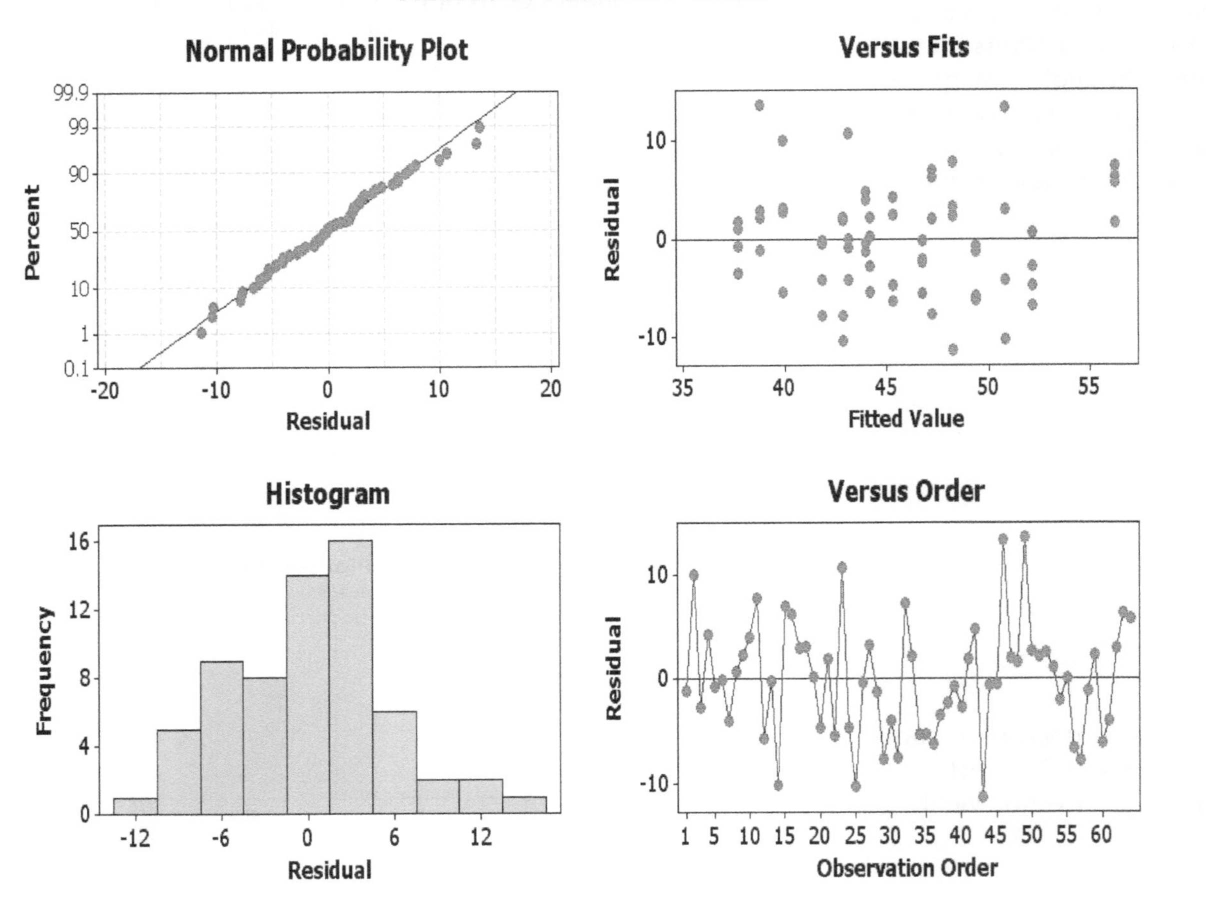

- Residuals are normally distributed on the normal probability plot.
- Residuals are equally distributed about the centre line on the versus fits plot.
- Histogram of residuals does not show extreme skewness.

- If the original data was collected in time order then the residuals should not show any ordered behaviour on the versus order plot.

We must now use the Response Optimizer to establish the settings we need to get a yield of 50%.

Minitab gives us a solution to get a yield of 50%.

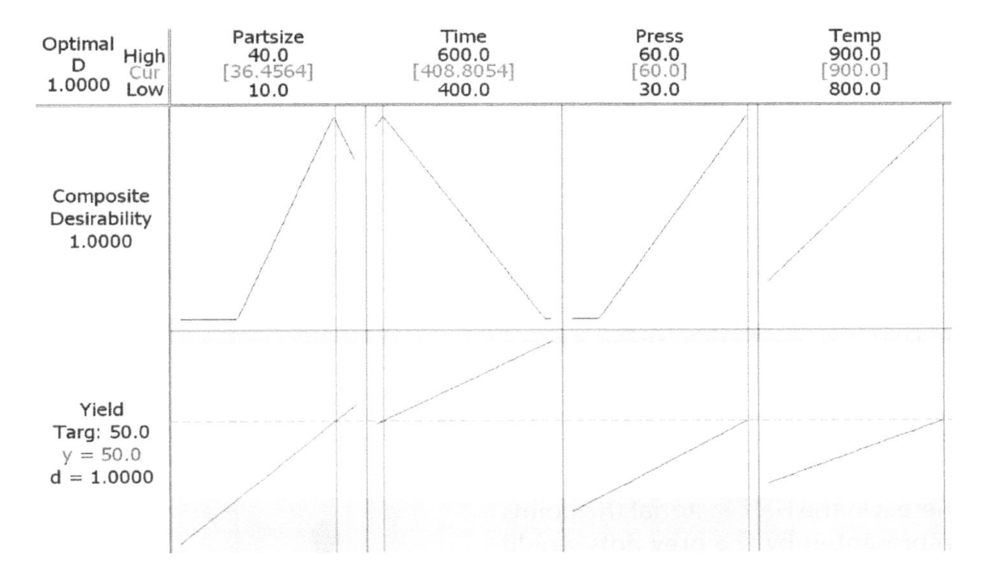

However, this plot shows a different solution. There is still freedom within the other parameters to change the settings and still obtain a Yield of 50%. (The accuracy of the slider stops us getting exactly 50%.)

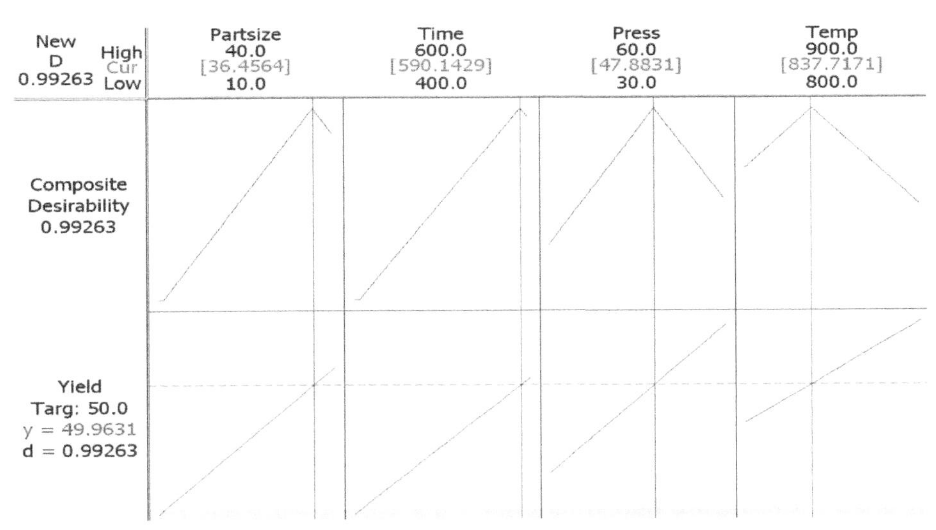

10.5 Fractional Factorial Designs

As can be seen from Minitab's table of available designs the experiment must have at least three factors for a fractional factorial design to be employed.

With Fractional designs we say the number of runs required is 2^{k-n}, where k is the number of factors and n indicates the fractional level of the design. Where Fractional Level = $\frac{1}{2^n}$ (e.g. when $n = 2$ we have a quarter fractional design).

A, 2^{6-3} would be an eight fractional design with six factors and it would require eight runs. Notice that we don't actually subtract one power from the other in order to show the type of experiment we are running. Locate the design in the table shown above. The 'III' indicates that the design has a resolution of three; this will be explained soon.

	Factors													
Run	2	3	4	5	6	7	8	9	10	11	12	13	14	15
4	Full	III												
8		Full	IV	III	III	III								
16			Full	V	IV	IV	IV	III	III	III	III	III	III	III
32				Full	VI	IV	IV	IV	IV	IV	IV	IV	IV	IV
64					Full	VII	V	IV	IV	IV	IV	IV	IV	IV
128						Full	VIII	VI	V	V	IV	IV	IV	IV

If we wanted to run a Full Factorial experiment with three factors it would require eight runs, whereas a Half Factorial with three factors would require four runs, ($2^{3-1} = 4$).

Both of these experiments can be represented by the cube plots shown here. On the Half Factorial the points represented by the grey dots would not be tested.

Let's say that the three factors were A, B and C. Let's have a look at the Coded levels we would be using in the experiment.

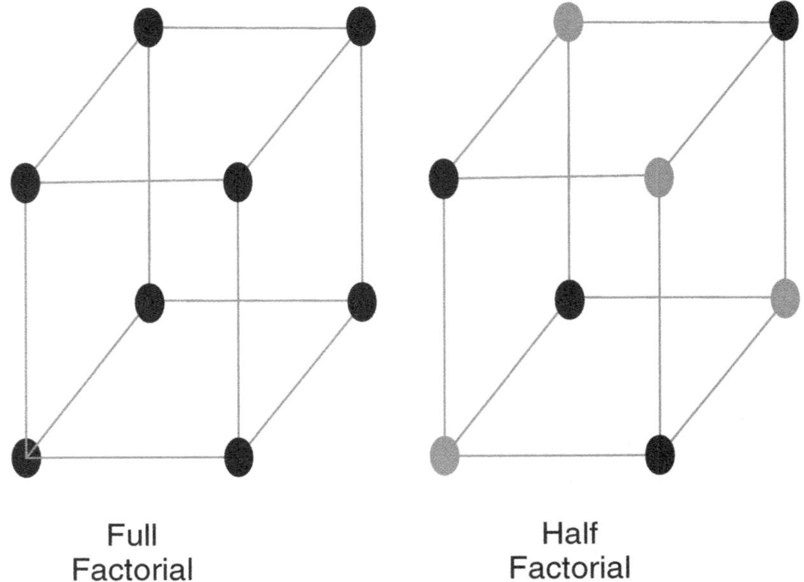

The table opposite shows the coded levels that would be used for the full factorial experiment. The table also contains interactions and the results of multiplying out the coded units.

A	B	C	A*B	A*C	B*C	A*B*C
1	1	1	1	1	1	1
–1	1	1	–1	–1	1	–1
1	–1	1	–1	1	–1	–1
–1	–1	1	1	–1	–1	1
1	1	–1	1	–1	–1	–1
–1	1	–1	–1	1	–1	1
1	–1	–1	–1	–1	1	1
–1	–1	–1	1	1	1	1

In order to form the half fractional experiment Minitab takes the lines where the A*B*C interaction is 1 and uses them to form the fractional experiment.

However, there is now a problem in that column A is now the same as B*C. This means that our estimation of the effect of A will be the same as the B*C interaction.

A	B	C	A*B	A*C	B*C	A*B*C
1	1	1	1	1	1	1
–1	–1	1	1	–1	–1	1
–1	1	–1	–1	1	–1	–1
1	–1	–1	–1	–1	1	1

We say that the main effects are 'confounded' or 'aliased' with the two way interactions as the same can be said for B & A*C and C & A*B.

It can be seen that the advantage of fractional designs is the reduced number of runs, which reduces time and no doubt cost. However, the assumption is made that lower order terms are responsible for the changes and not the higher order interactions.

We previously stated that this experiment had resolution III. This means that the first order terms (main effects) are confounded with the second order terms (two way interactions). As III = 1 + 2.

Similarly, for a resolution V design the following terms would be confounded:

First and fourth orders

Second and third orders.

To understand what the term Design Generator means it is better to look at things from the other direction. If we had a two level full factorial with two factors it would look like the table shown here.

A	B	A*B
1	1	1
-1	-1	1
-1	1	-1
1	-1	-1

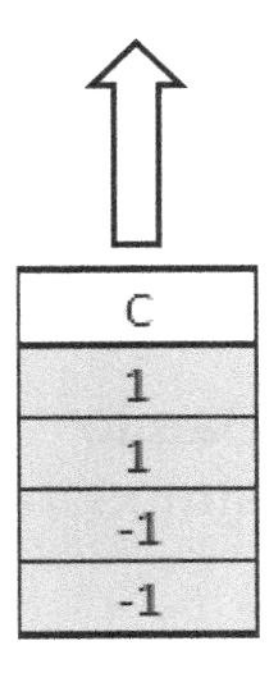

C
1
1
-1
-1

For a half fractional design we essentially squeeze another factor in which is identical to A*B. The combination of 1s and −1s are the same for C and A*B. This means that the estimate of the main effect will be the same. However, We assume that A*B will be negligible when compared to the effect of C.

We say that the Design Generator is C = AB because C is aliased with A*B. This term is the Design Generator even though other terms are confounded. In more complex designs the Design Generator can be manually selected.

A	B	C
1	1	1
-1	-1	1
-1	1	-1
1	-1	-1

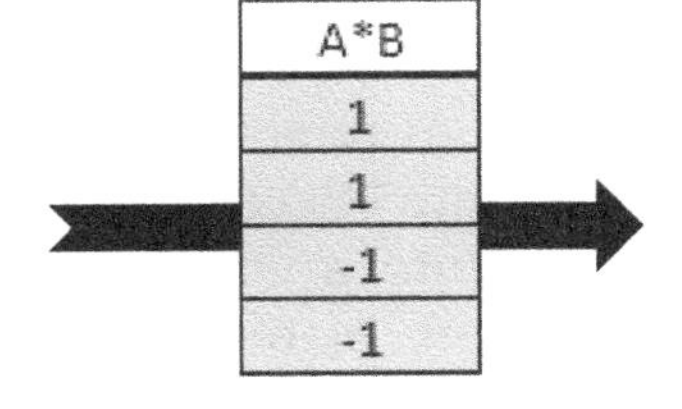

A*B
1
1
-1
-1

Negligible

Procedure	• Fractional Factorial Designs
What's it used for?	• The design of experiment process is used to understand the effect of process inputs and their interactions on the outputs of the process.
Assumptions and Limitations	• Fractional designs will tend to be used for screening important factors. • Fractional designs are used when it is thought that higher order terms are not having an impact on the response. • Fractional Designs will contain terms that are confounded. • This type of DOE cannot detect curvature within a single run. • This type of DOE is not orthogonal.

Example 4. The column.

Engineers want to understand the operation of their sulphur dioxide (SO_2) scrubbing column. They want to know what factors are important and affect the concentration of SO_2 (SO_2Conc) leaving the column in the gas stream.

The SO_2 is removed from the gas going into the column by spraying water into a column containing a packing material. The packing material provides a high surface area for contact between the gas stream and the water. The number of nozzles within the column can be adjusted. The engineers also monitor the pressure within the column and the pH of the water. The engineers can actually set the pH of the water using a separate control system.

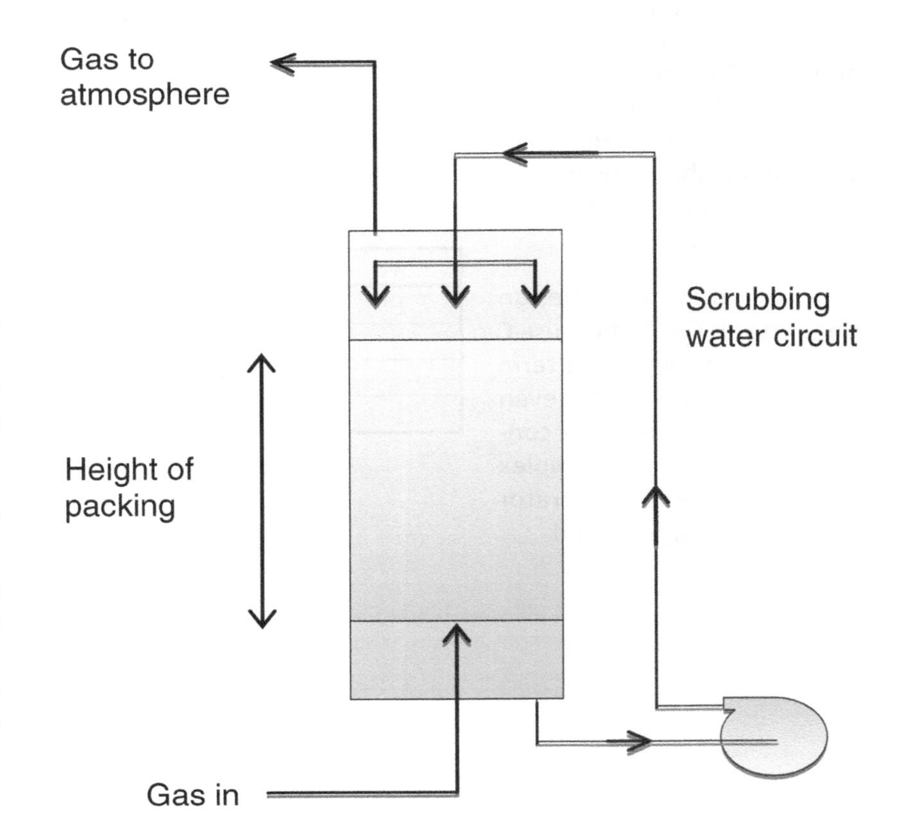

The factors that the engineers will be considering are:

GasFlow (20, 30)

GasPressIn (4, 8), gas inlet pressure

Nozzles (3, 4)

HofPacking (22, 30), height of packing

FRWater (100,150), flow rate of water

pH (3, 7)

ColPress (20, 50), column pressure.

A two level factorial would require 128 runs which would take far too long. The engineers are willing to consider a half fractional DOE. Set up the worksheet for them and note which terms are confounded.

No Data File.

1. Click Stat <<DOE <<Factorial <<Create Factorial Design…

2. In the menu box that appears change the Number of factors to 7.

Number of factors: 7

3. Click on the Display Available Designs button. We see that a half factorial design with seven factors has a resolution of VII. This means the following order terms will be confounded: first and sixth, second and fifth, third and fourth order terms. Click OK.

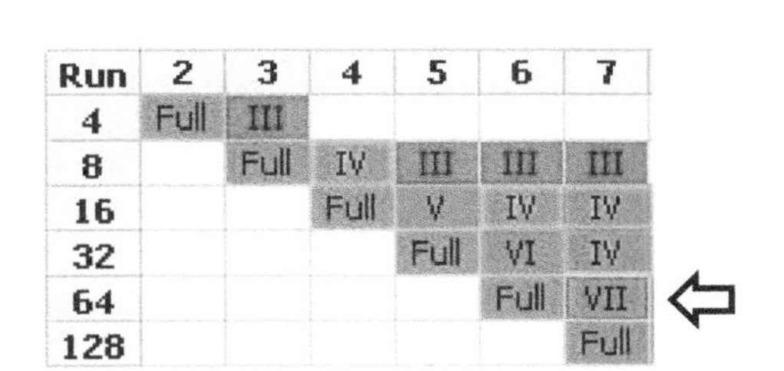

Run	2	3	4	5	6	7
4	Full	III				
8		Full	IV	III	III	III
16			Full	V	IV	IV
32				Full	VI	IV
64					Full	VII
128						Full

4. Click on the Designs button. Select the half fraction design then click on OK.

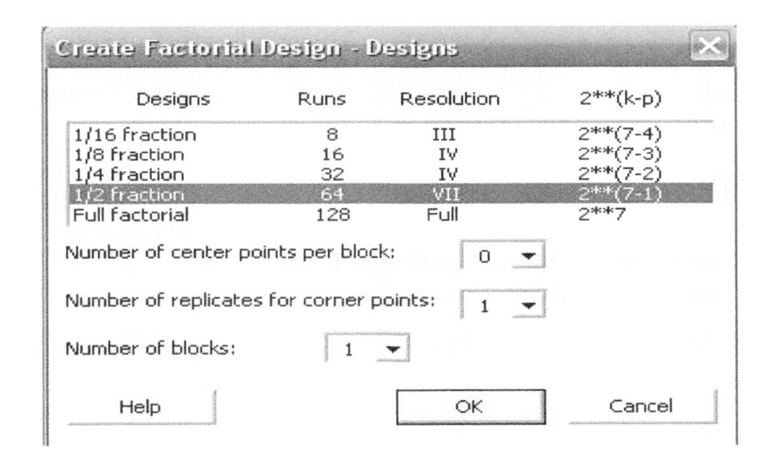

5. Click on the Factors button. Enter the details of all seven factors. Note that we are entering Nozzles as a text value as we can only have three or four nozzles physically within the column. Click OK when complete.

Factor	Name	Type	Low	High
A	GasFlow	Numeric	20	30
B	GasPressIn	Numeric	4	8
C	Nozzles	Text	3	4
D	HofPacking	Numeric	22	30
E	FRWater	Numeric	100	150

6. Click on the Options button. Ensure that Randomise runs is not selected. Normally, it would be better to randomise runs to reduce human influence but we want to be able to copy in the table of results later. Click OK to produce the experimental worksheet in the project window, as shown below.

☐ Randomize runs

C1	C2	C3	C4	C5	C6	C7-T	C8	C9	C10	C11
StdOrder	RunOrder	CenterPt	Blocks	GasFlow	GasPressIn	Nozzles	HofPacking	FRWater	pH	ColPress
1	1	1	1	20	4	3	22	100	3	50
2	2	1	1	30	4	3	22	100	3	20
3	3	1	1	20	8	3	22	100	3	20
4	4	1	1	30	8	3	22	100	3	50
5	5	1	1	20	4	4	22	100	3	20
6	6	1	1	30	4	4	22	100	3	50
7	7	1	1	20	8	4	22	100	3	50
8	8	1	1	30	8	4	22	100	3	20

The session window gives the design summary. The design generator is G = ABCDEF. The defining relation I = ABCDEFG allows the formation of the Alias Structure.

For example, if we want to know which terms are aliased with BC we multiply BC by I + ABCDEFG:

$BC (I + ABCDEFG) = BC + AB^2C^2DEFG.$

As $B^2 = 1$ and $C^2 = 1$, then

$BC (I + ABCDEFG) = BC + ADEFG.$

Therefore the BC interaction will be aliased with ADEFG interaction.

Fractional Factorial Design

```
Factors:   7   Base Design:          7, 64   Resolution:  VII
Runs:     64   Replicates:               1   Fraction:    1/2
Blocks:    1   Center pts (total):       0


Design Generators: G = ABCDEF


Defining Relation:  I = ABCDEFG


Alias Structure

I + ABCDEFG

A + BCDEFG
B + ACDEFG
```

Example 4. The column, continued.

With the trial sheet set up the engineers carry out all of the trial runs. Copy the results from the Excel file (column SO2Conc) and paste them into C12 of the Minitab sheet you have just created.

Find out which of the factors are significant and establish if any of the interactions are significant.

Produce the Main Effects plots and interaction plots if applicable.

Worksheet: Column.

The procedure to analyse fractional designs is the same as that for two level designs so we will not fully repeat the repetitive elements of the procedure.

1. Click Stat <<DOE <<Factorial <<Analyse Factorial Design. Then select SO2Conc as the Response.

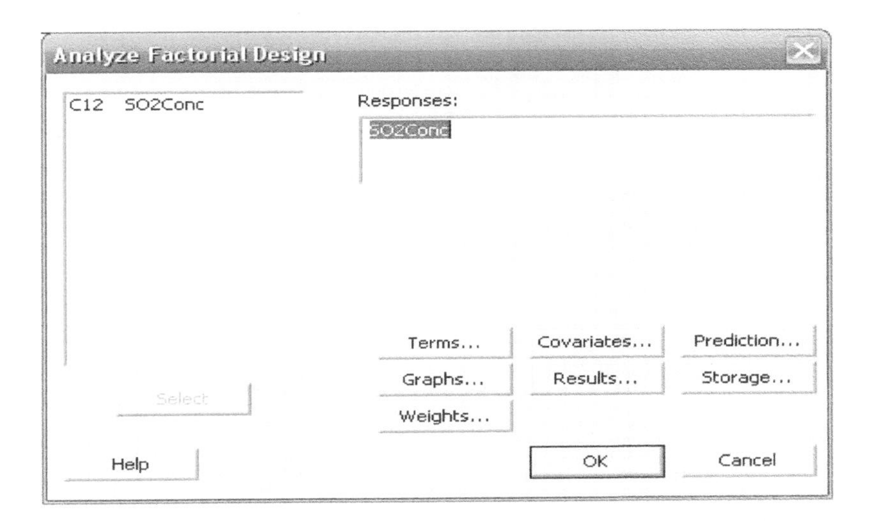

2. Click on the Terms button. Note that Minitab has automatically set the Selected Terms to only include up to third order terms. This is so that pairs of confounded terms are not included in the model by default. Click OK.

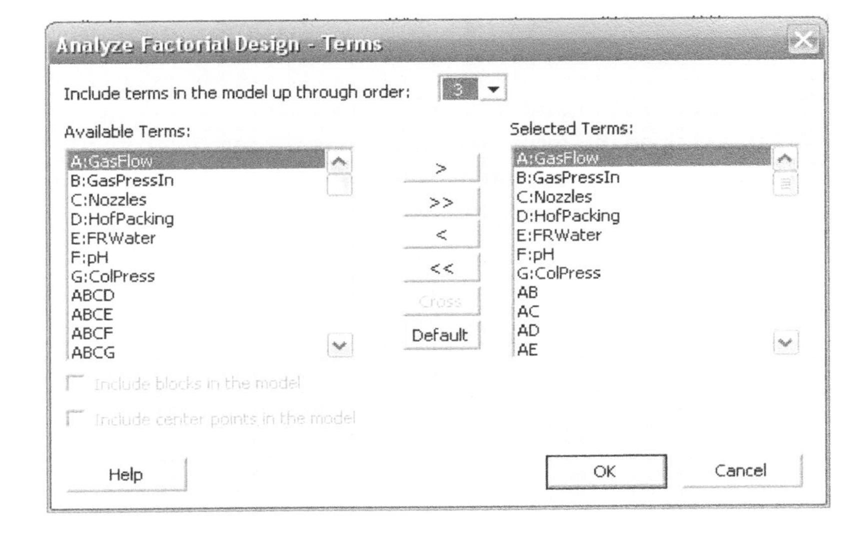

3. Click on the Graphs button.
4. Under Effects Plots select Pareto. We don't want residual plots until we have fully reduced the model.
5. Click OK and OK again.
6. The Pareto plot shows us that none of the third order terms are significant so we can eliminate them from the model. As the design did not contain any replicates we cannot use the ANOVA table in the session window to reduce the model. We will only use the Pareto chart.

Effects Plots
Normal Half Normal Pareto
Alpha: 0.05
Residuals for Plots:
Regular Standardized Deleted
Residual Plots
Individual plots
Histogram
Normal plot
Residuals versus fits
Residuals versus order
Four in one

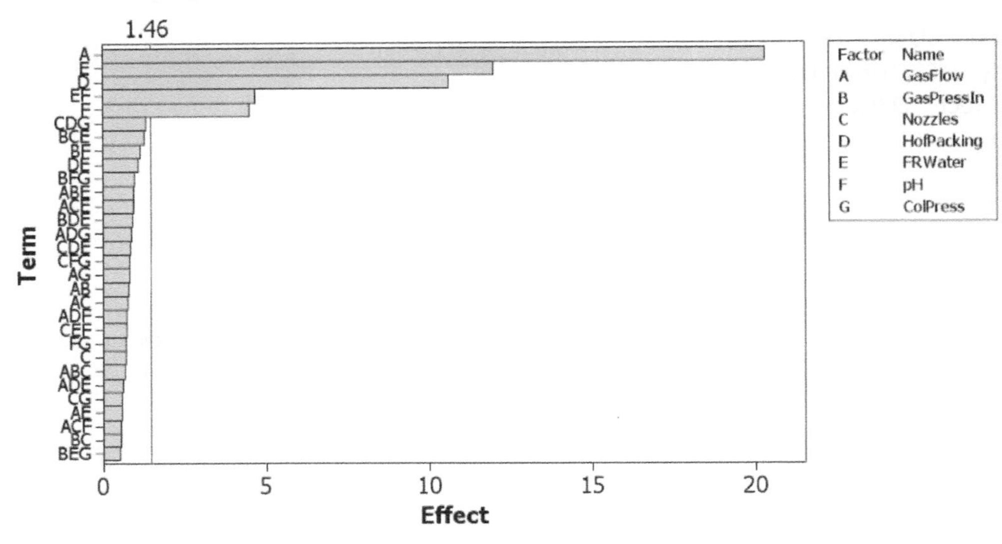

7. To reduce the model click on Edit Last icon.

8. Then click on the Terms button. Change the selection so only second order terms are included. Click on OK and OK again.

Include terms in the model up through order: 

9. It can be seen that EF is the only second order term that is significant.
10. Reduce the model again. The easiest way to do this is to change the selector so that only first order terms are included and then manually select EF so that it is included.

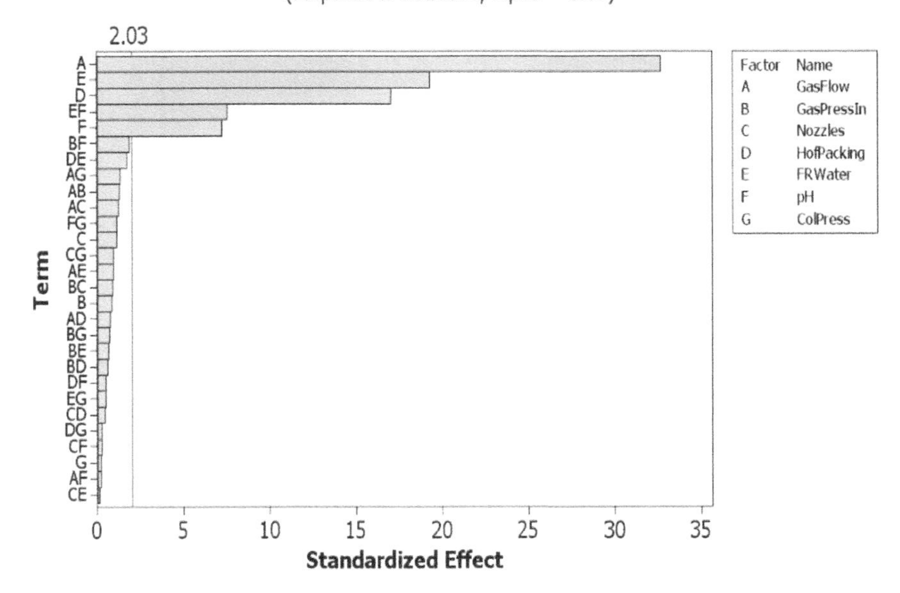

The Pareto plot shows that three of the main effects are not significant. As these are not included in any of the interactions they can also be removed from the model.

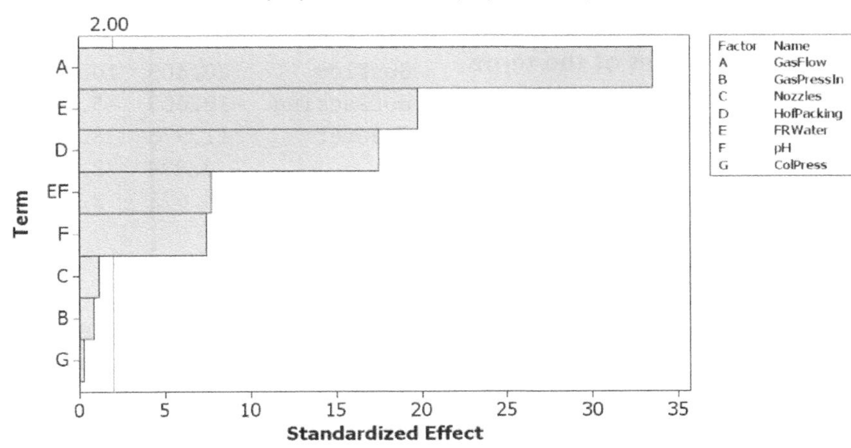

11. Click on the Edit Last icon and then the Terms button.
12. Change the final selection of terms so it is the same as that shown. Click OK.

Selected Terms:

A:GasFlow
D:HofPacking
E:FRWater
F:pH
EF

13. Click on the Graphs button and select the Four in one Residual Plots.
14. Click OK and OK again.

Residual Plots
- ( ) Individual plots
  - [ ] Histogram
  - [ ] Normal plot
  - [ ] Residuals versus fits
  - [ ] Residuals versus order
- (•) Four in one

15. The final Pareto plot shows all the terms that are significant.

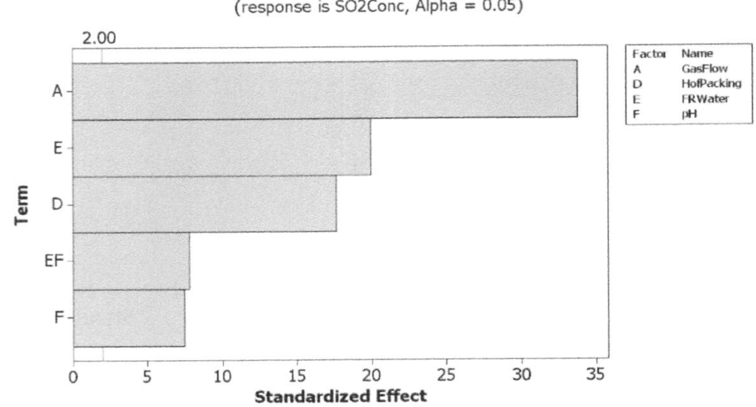

16. Within the session window the Effects table shows the T and *P* values of the terms that are significant. We also see that 96.89% of the variation within the response can be explained by the changes within our factors. Finally, we see the coefficients that can be used to form a prediction equation.

```
Term          Effect    Coef  SE Coef       T      P
Constant              25.314   0.3004   84.26  0.000
GasFlow       20.303  10.152   0.3004   33.79  0.000
HofPacking   -10.603  -5.302   0.3004  -17.65  0.000
FRWater       11.978   5.989   0.3004   19.93  0.000
pH             4.484   2.242   0.3004    7.46  0.000
FRWater*pH     4.672   2.336   0.3004    7.78  0.000


S = 2.40353      PRESS = 407.972
R-Sq = 97.14%    R-Sq(pred) = 96.51%    R-Sq(adj) = 96.89%

Term              Coef
Constant       2.66484
GasFlow        2.03031
HofPacking    -1.32539
FRWater      0.0059688
pH            -4.71875
FRWater*pH   0.0467187
```

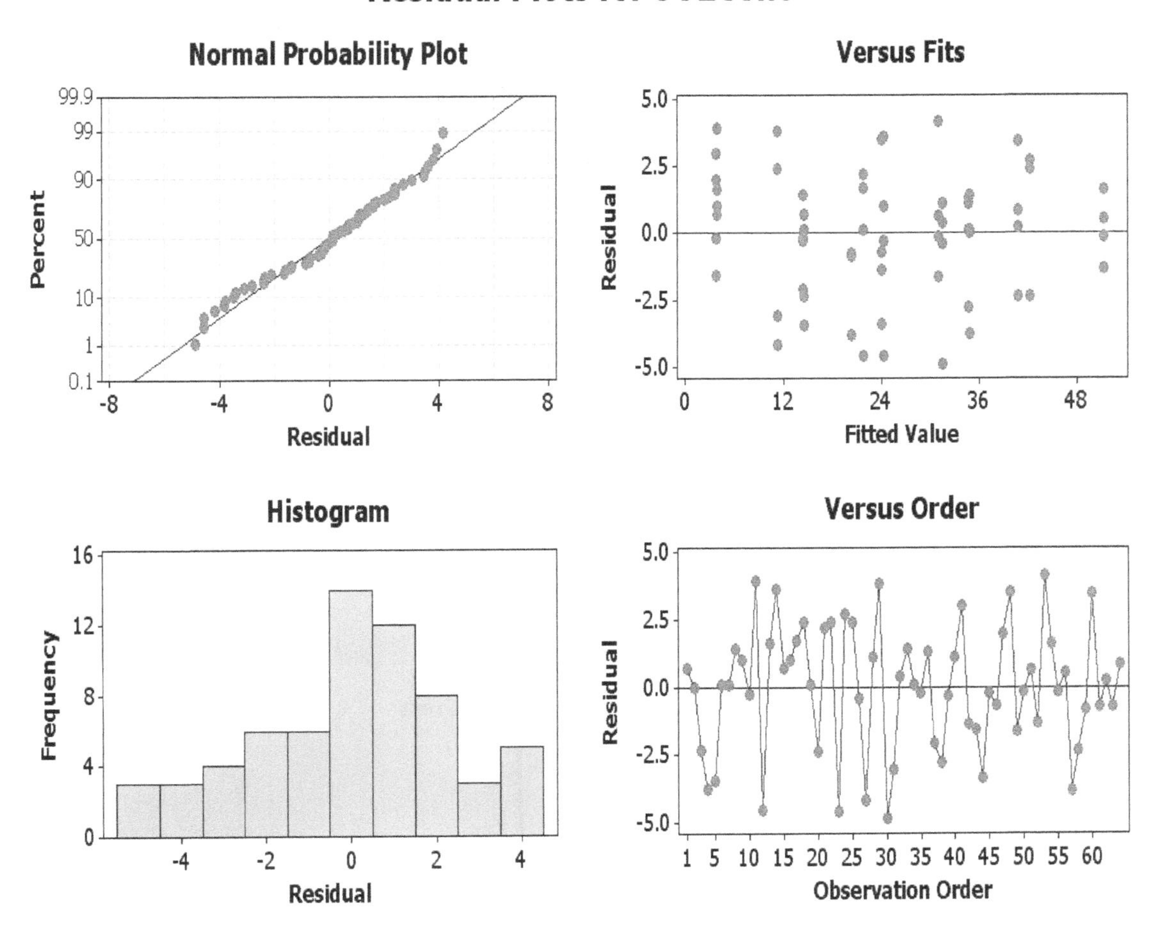

- Residuals are normally distributed on the normal probability plot.
- Residuals are equally distributed about the centre line on the versus fits plot.
- Histogram of residuals does not show extreme skewness.
- If the original data was collected in time order then the residuals should not show any ordered behaviour on the versus order plot.

As there are no issues with the Residuals we will now produce the Main Effects and Interaction Plots.

17. Click Stat <<DOE <<Factorial <<Factorial Plots. Select Main Effects Plot and Interaction Plot. We don't want a cube plot as the number of terms is fairly high.

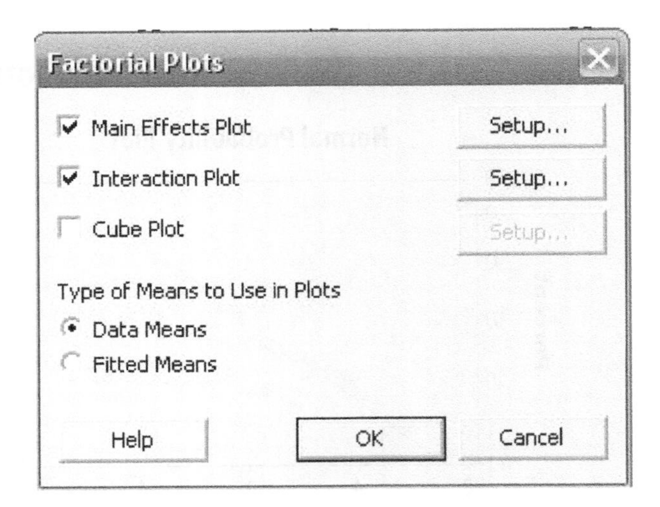

18. Click on Setup for the Main Effects Plot. Select SO2Conc as the Response. Select the main effects as shown. Click OK.

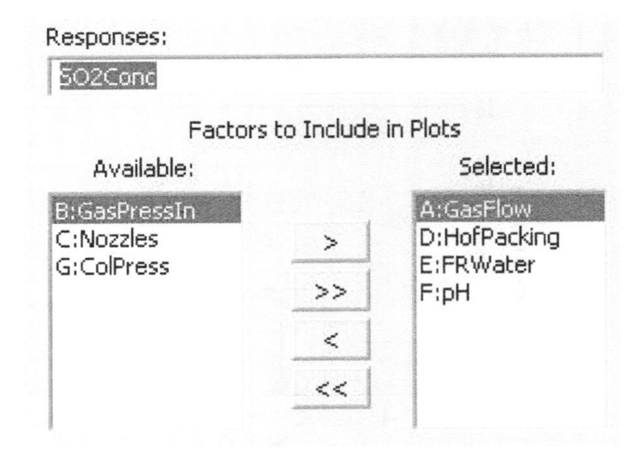

19. Then click on Setup for the Interactions Plot. Select SO2Conc as the Response. Select the E:FRWater and F:pH as the variables to be plotted. EF was the only significant interaction. Click OK & OK again.

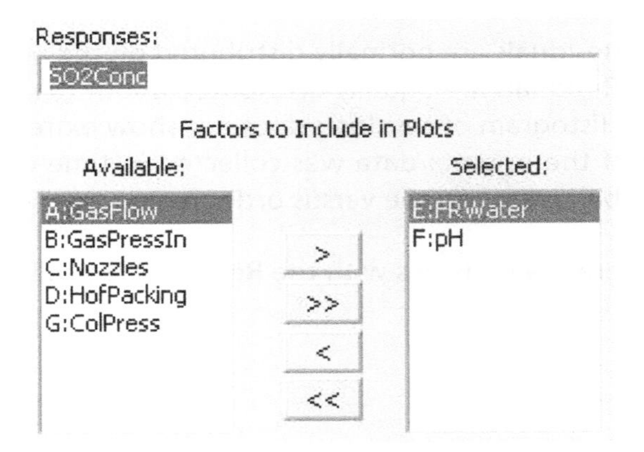

The interaction plot shows that when FRWater is low there is not much impact in changing the pH. However, when FRWater is high there is a much greater impact in changing the pH.

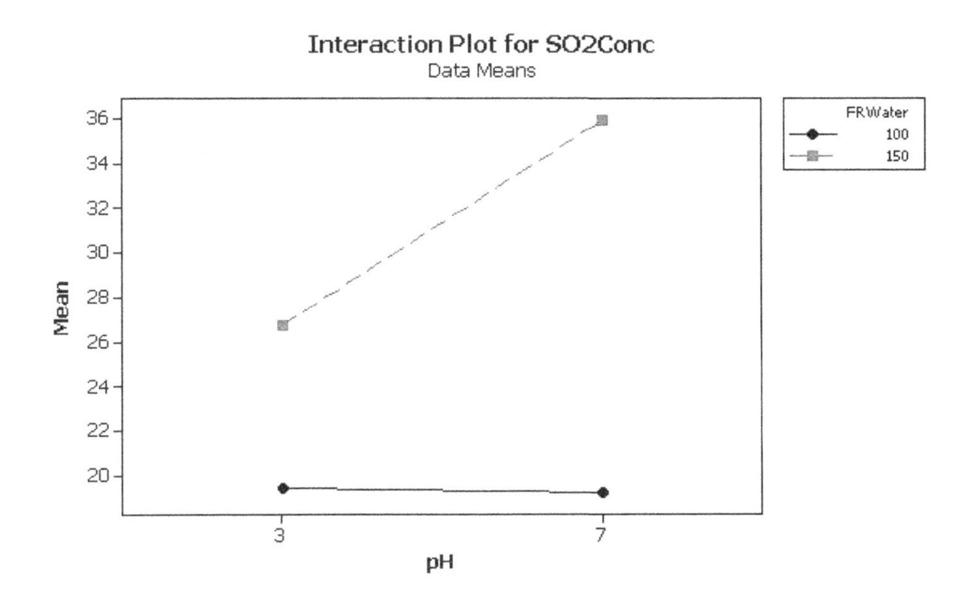

The Main Effect plots show the relative strength of the change caused by each of the factors. Note that HofPacking has a negative effect.

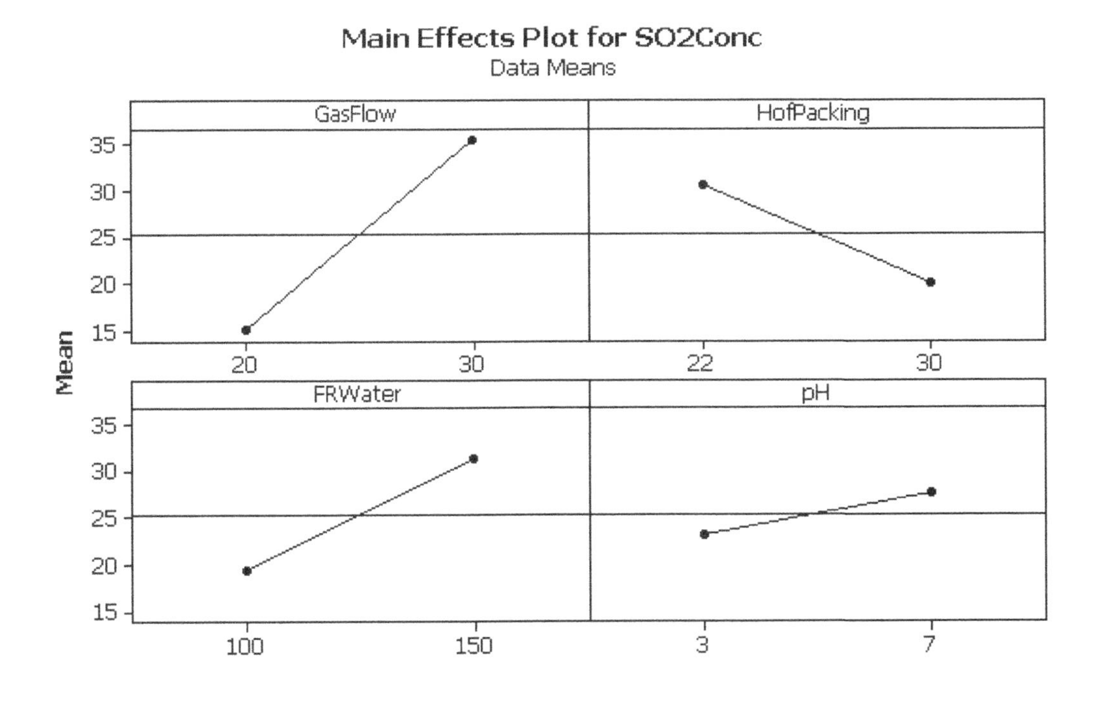

Exercise 2. The Reactor.

Engineers want to understand the operation of their reactor which they are using to convert hexane into perfluorohexane. They want to know what factors are important and how they affect the yield of the product, perfluorohexane, leaving the reactor.

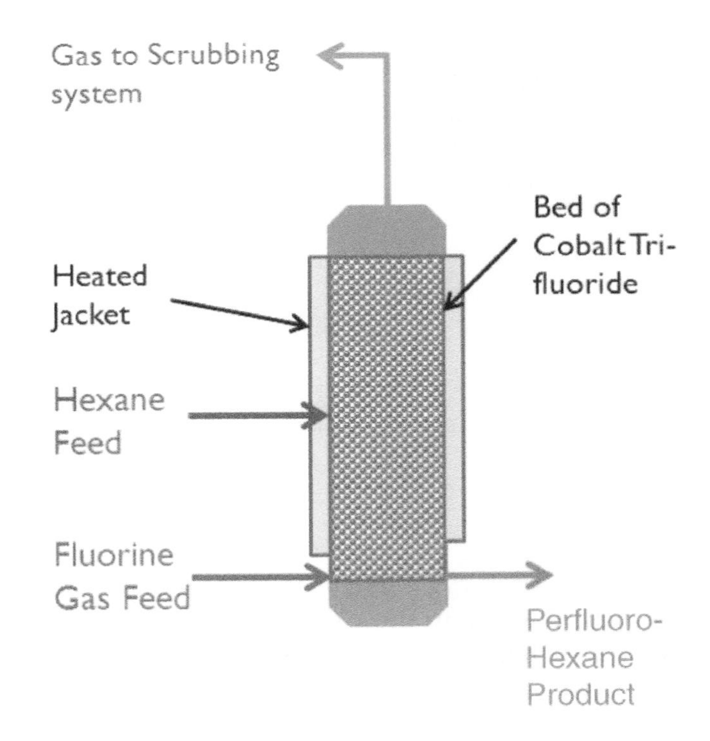

Fluorine gas is fed into a heated reactor. The reactor contains a bed of cobalt tri-fluoride (CoF_3). The CoF_3 captures the fluorine gas until it can react with the heated hexane feed. The hexane is converted to perfluorohexane only if all of the hydrogen molecules within hexane are replaced with fluorine atoms. The percentage of fully fluorinated perfluorohexane in the product stream specifies the yield.
The reactor is kept at a very slight negative pressure so that unreacted gases can be drawn into the gas scrubbing system.

The factors that the engineers considered to be important are:

Particle Size of CoF_3, Partsize (80, 120)

Reactor Temperature, RTemp (300, 400)

Reactor Pressure, RPress (–10, –2)

Hexane Feed Temperature, HexTemp (30, 60)

Feedrate of Fluorine, FRF2 (30, 45)

Feedrate of Hexane, FRHex(12,18)

The engineers carried out a half fractional factorial DOE. Analyse the experiment and decide which of the terms are significant.

Produce the Main Effects plots and Interaction plots if applicable.

Worksheet: Reactor.

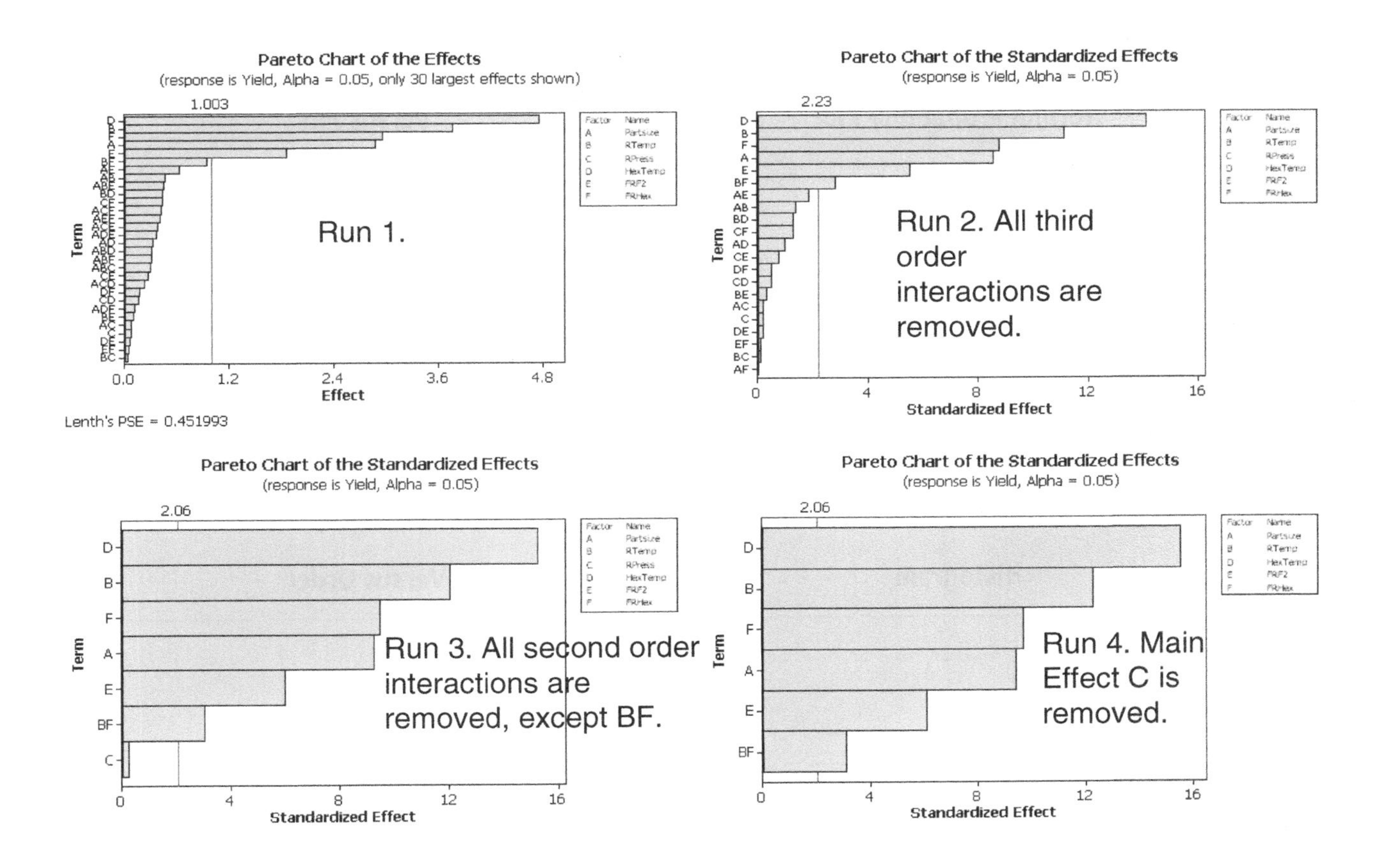

Pareto Chart of the Effects
(response is Yield, Alpha = 0.05, only 30 largest effects shown)
1.003
Run 1.
Term
Effect
Lenth's PSE = 0.451993
Pareto Chart of the Standardized Effects
(response is Yield, Alpha = 0.05)
2.23
Run 2. All third order interactions are removed.
Standardized Effect
Pareto Chart of the Standardized Effects
(response is Yield, Alpha = 0.05)
2.06
Run 3. All second order interactions are removed, except BF.
Standardized Effect
Pareto Chart of the Standardized Effects
(response is Yield, Alpha = 0.05)
2.06
Run 4. Main Effect C is removed.
Standardized Effect

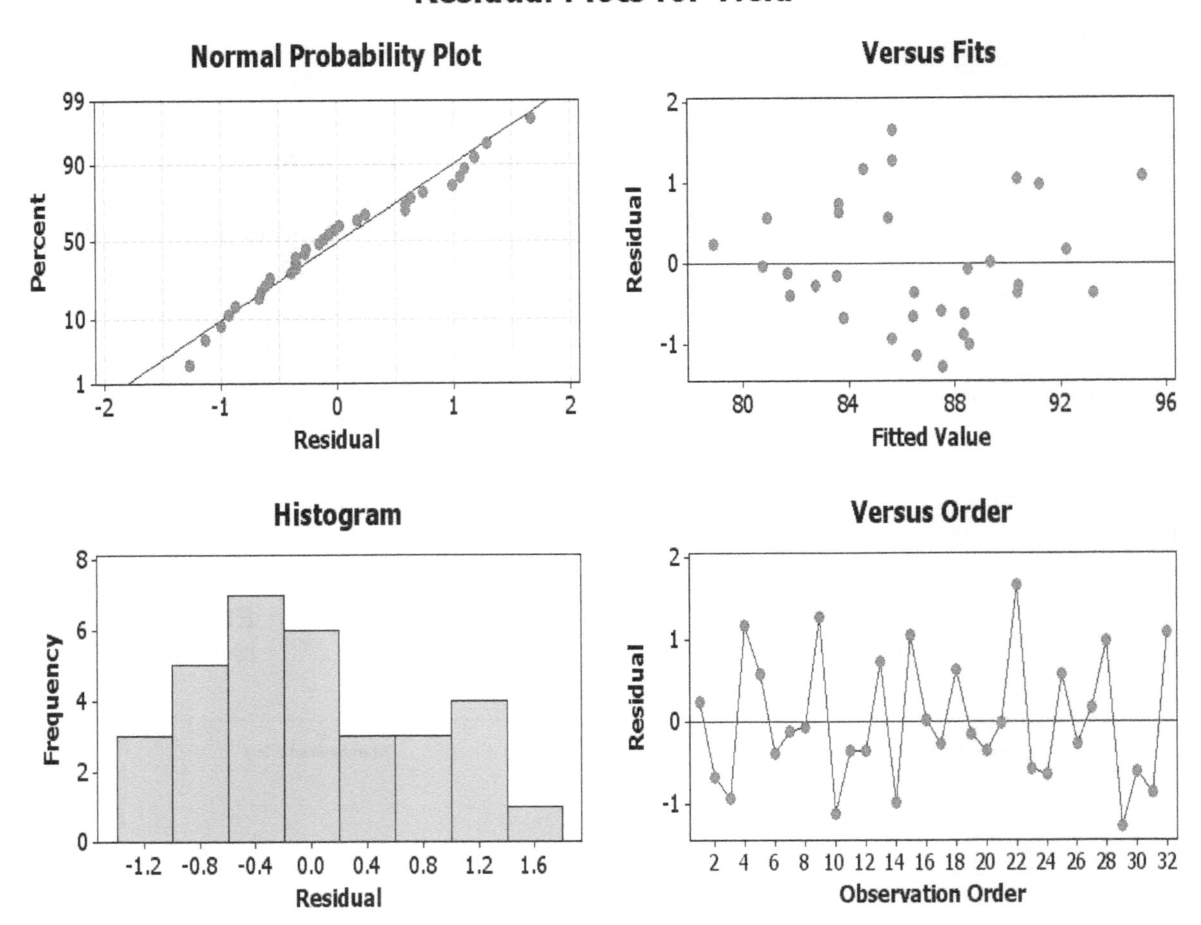

- Residuals are normally distributed on the normal probability plot.
- Residuals are equally distributed about the centre line on the versus fits plot.
- Histogram of residuals does not show extreme skewness.
- If the original data was collected in time order then the residuals should not show any ordered behaviour on the versus order plot.

There are no issues with the Residuals.

The Main Effects plot shows the relative strength of the change caused by each of the factors.

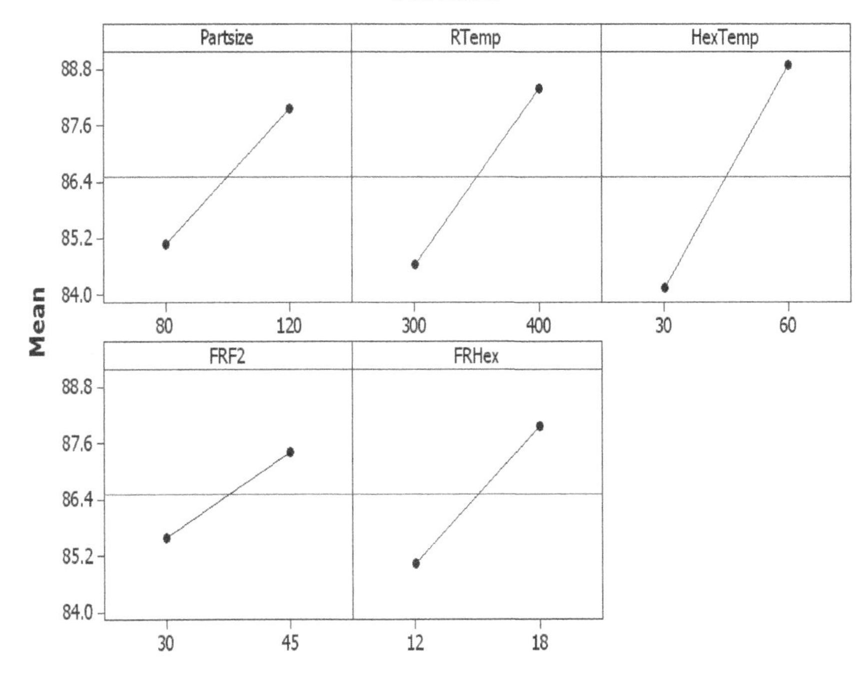

Although the interaction is significant it can be seen that the lines appear to be being close to parallel.

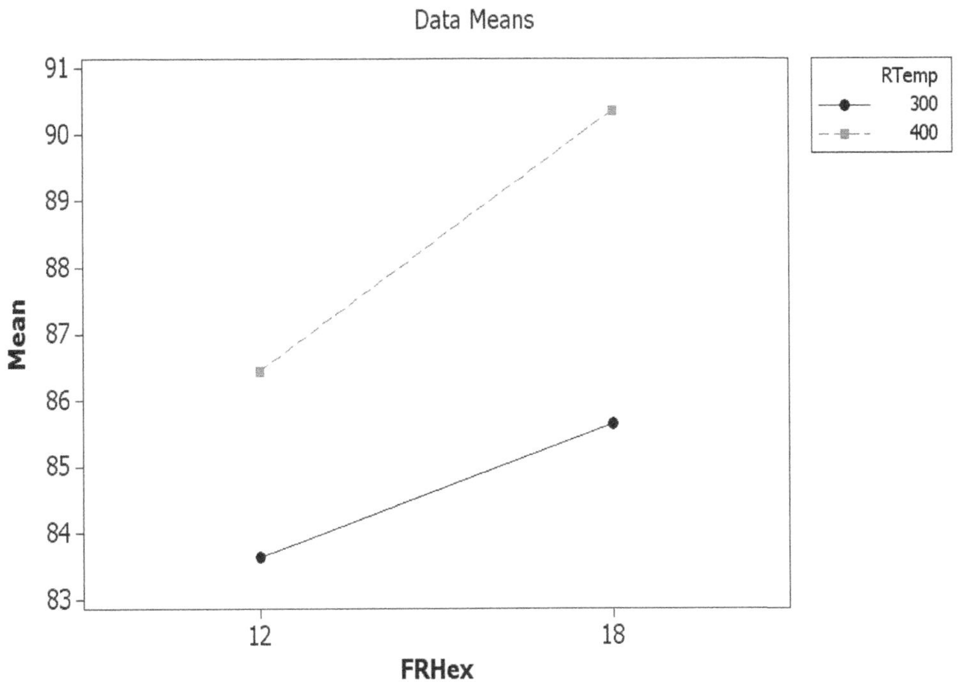

CHAPTER 11
Help

11.1 Help Overview

The final task of this course is to leave you with a method of helping yourself should you get stuck and also a method for you to increase your knowledge and understanding of Minitab. We are going to do that by showing you the Help system.

The Help system within Minitab offers a comprehensive library of information on Minitab. It includes information on the data you need to run a procedure, how to execute the procedure and how to interpret the results. It includes a glossary and a section to show you how calculations are carried out. In fact you will probably be hard pushed to ask a question that is not covered within the Help system. Therefore, we need to learn how to get to the answer.

Initially, navigation within the Help system may seem a bit confusing as there are multiple ways of getting to the same point. In fact, there are ways of going round in an endless loop as well. I will show you what I think are the straightforward routes.

I made the decision to use an independent data set in this course. This does mean that you have to download the data but the advantage of having an independent data set within Minitab is enormous. This allows the reader to run the examples again whilst using the Minitab help system as the guide. This will allow the reader to see the procedures from another perspective and effectively give them two ways of learning.

Clicking on the Help within the Toolbar will open the Help Drop Down menu as shown below.

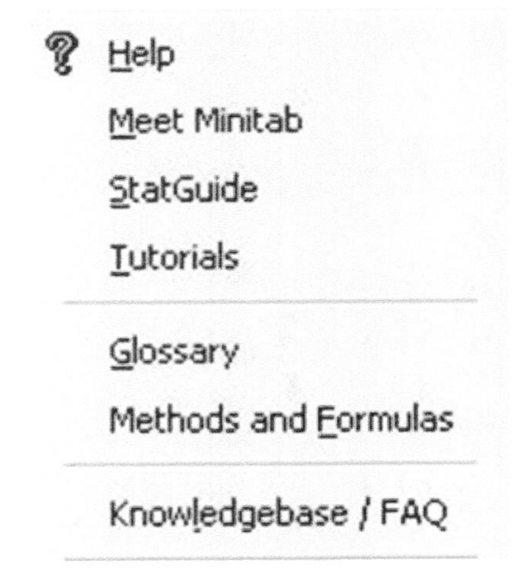

Help: The main menu can be used to navigate to most of the other help systems but this is also the main route to get help into all things Minitab.

Meet Minitab: This option takes you to the website and allows you to download the Meet Minitab document, or you may have received one with your copy of Minitab.

StatGuide: This option is another help system that focuses on interpretation of test results.

Tutorials: Think of these tutorials as the front end of the procedures that are used within the StatGuide.

Problem Solving and Data Analysis using Minitab: A clear and easy guide to Six Sigma methodology, First Edition. Rehman M. Khan.

Glossary: Explanation of terms used within Minitab.

Methods and Formulas: This option explains the maths that Minitab uses behind the scenes.

Knowledgebase/FAQ: This option takes the user onto the Minitab website for web based help.

11.2 Help! Help!

In order to get to the main help menu click Help <<Help. Or press F1 or click on the Help icon in the tool bar.

In the middle section of the Help menu you can see some of the previous options that we discussed as being within the drop down menu.

In order to show how to navigate using the help menu we will pretend that we wish to know more about One Way ANOVA .

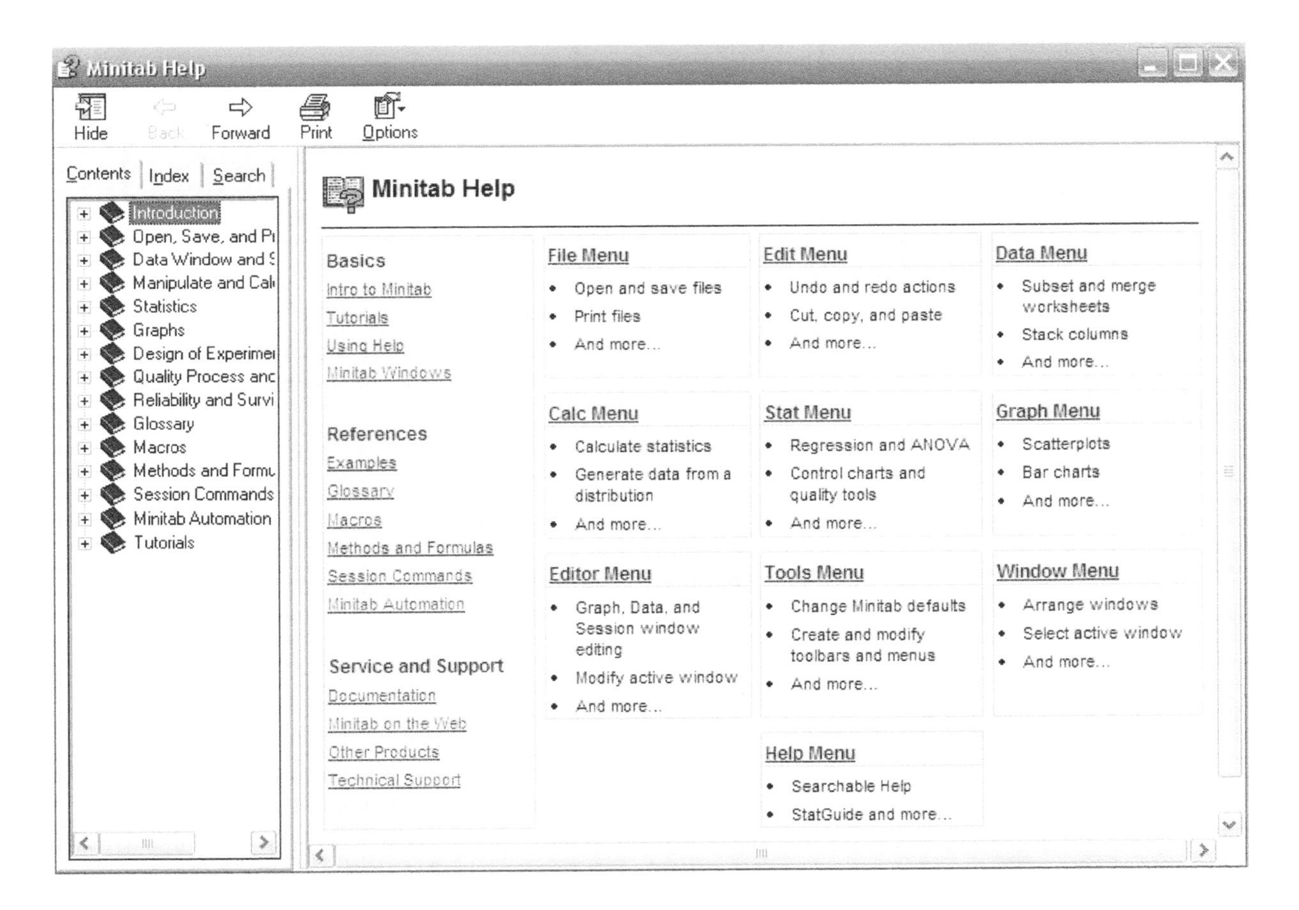

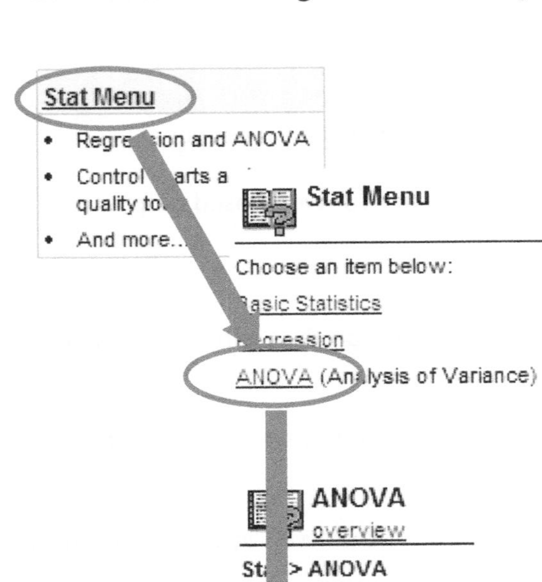

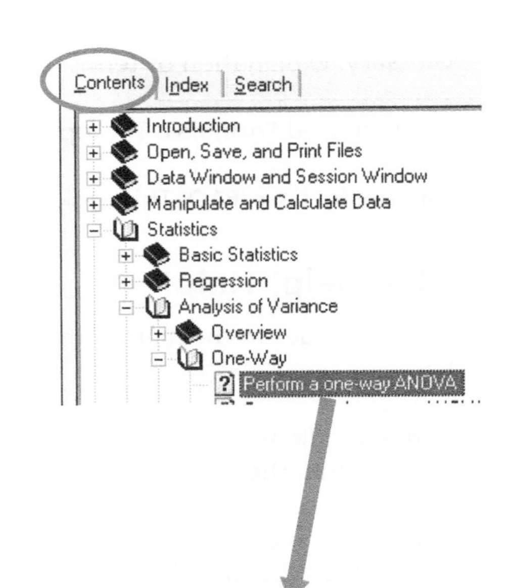

ANOVA
overview

St > ANOVA

Al s you to perform a

Select one of the follow

One-Way –

performs multiple comp

One-Way Analysis of Variance

overview how to example data see also

Stat > ANOVA > One-way

Performs a one-way analysis of variance, with the response variable in one column, factor levels in another. If each group is entered in its own column, use Stat > ANOVA > One-Way (Unstacked).

You can also perform multiple comparisons and display graphs of your data.

Dialog box items

Response: Enter the column containing the response.

Factor: Enter the column containing the factor levels.

Store Residuals: Check to store residuals in the next available column.

Store fits: Check to store the fitted values (level means) in the next available column.

Confidence level: Enter the confidence level. For example, enter 90 for 90%. The default is 95%.

<Comparisons>

<Graphs>

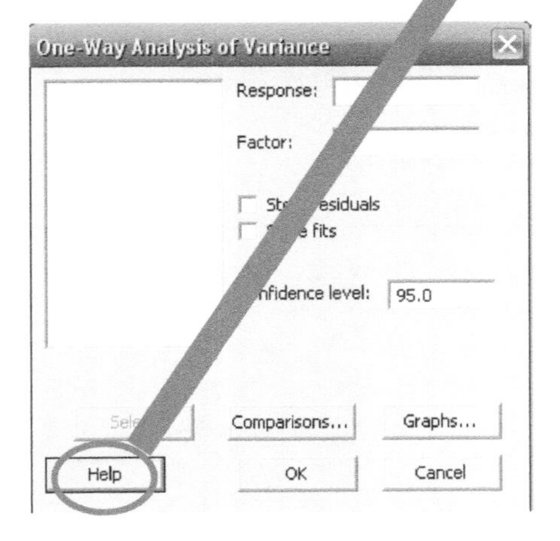

The diagram shows three ways to navigate to the help section on One Way ANOVA. Two of the ways are from the main help menu. You can use the menu boxes on the right hand side of the help screen or either of the three search systems on the left hand side: Contents, Index or Search. The third way of getting to the help page is via clicking the help button on the menu box for One-Way ANOVA (all menu boxes have the help button in the bottom left corner). Let's have a look at the help page in greater detail.

The subject specific help sections are usually further broken down into five subcategories. We will go through these.

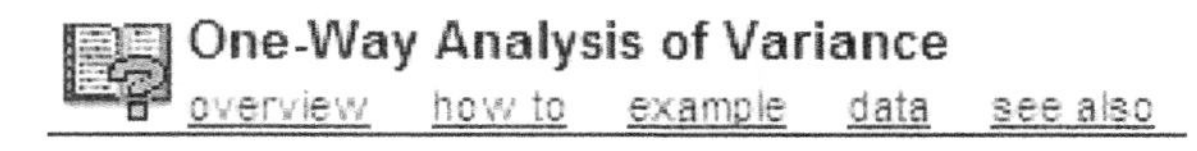

The Navigation bar in the top left corner is always available to move back and forward through the different levels of help.

 Hide  Back  Forward  Print  Options

Overview gives a summary of the procedure. Key terms are explained in dialog boxes if the users clicks on the link or they are taken to different help topics.

 **Analysis of Variance Overview**

see also

Analysis of variance (ANOVA) is similar to regression in that it is used to investigate and model the relationship between a response variable and one or more predictor variables. However, analysis of variance differs from regression in two ways: the predictor variables are qualitative (categorical), and no assumption is made about the nature of the relationship (that is, the model does not include coefficients for variables). In effect, analysis of variance extends the two-sample t-test for testing the equality of two population means to a more general null hypothesis of comparing the equality of more than two means, versus them not all being equal. Several of Minitab's ANOVA procedures, however, allow models with both qualitative and quantitative variables.

Minitab's ANOVA capabilities include procedures for fitting ANOVA models to data collected from a number of different designs, for fitting MANOVA models to designs with multiple response, for fitting ANOM (analysis of means) models, and graphs for testing equal variances, for confidence interval plots, and graphs of main effects and interactions.

How to is a very brief memory jogger on how to carry out the analysis.

 **To perform a one-way analysis of variance with stacked data**

main topic see also

1. Choose **Stat > ANOVA > One-Way**.
2. In Response, enter the column containing the response.
3. In Factor, enter the column containing the factor levels.
4. If you like, use any dialog box options, then click OK.

Example is a fully worked example with a problem scenario, brief explanation of the procedure and then the results with full interpretation. It is important to note that the data set is given in EH_AOV.MTW for this example. We will look at how to retrieve these later in the module.

Example of a one-way analysis of variance with multiple comparisons

main topic   interpreting results   session command   see also

You design an experiment to assess the durability of four experimental carpet products. You place a sample of each of the carpet products in four homes and you measure durability after 60 days. Because you wish to test the equality of means and to assess the differences in means, you use the one-way ANOVA procedure (data in stacked form) with multiple comparisons. Generally, you would choose one multiple comparison method as appropriate for your data. However, two methods are selected here to demonstrate Minitab's capabilities.

1 Open the worksheet EXH_AOV.MTW.
2 Choose **Stat > ANOVA > One-Way**.
3 In Response, enter *Durability*. In Factor, enter *Carpet*.
4 Click Comparisons. Check Tukey's, family error rate. Check Hsu's MCB, family error rate and enter *10*.
5 Click OK in each dialog box.

Data tells us the data requirements for the test. How data should be laid out and the type of data required.

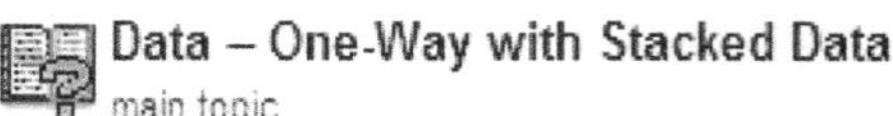

The response variable must be numeric. Stack the response data in one column with another column of factor level values identifying the population (stacked case). The factor level (group) column can be numeric, text, or date/time. If you wish to change the order in which text categories are processed from their default alphabetical order, you can define your own order. See Ordering Text Categories. You do not need to have the same number of observations in each level. You can use Make Patterned Data to enter repeated factor levels.

Note If your response data are entered in separate worksheet columns, use Stat > ANOVA > One-Way (Unstacked).

See Also is a collection of links to related topics.

see also

Related topics

- Methods and formulas
- ANOVA examples
- Discussion of multiple comparisons
- More complex ANOVA models
- Graph editing overview
- References - ANOVA

Related commands

- Stat > ANOVA > Main Effects Plot
- Stat > ANOVA > One-Way (Unstacked)
- Stat > Power and sample size > One-Way ANOVA

11.3 Tutorials

Tutorials and StatGuide work together to give another set of worked examples. This set is separate from those given within the Help topic specific menus. The Tutorial section deals with the front end of the procedure up to execution. However, StatGuide deals only with interpretation. The tutorials section in M15 looks different to that for M16. The M16 section is shown below.

To access the tutorial click <<Help <<Tutorials. Alternatively, click on Tutorials from the Main Help screen.

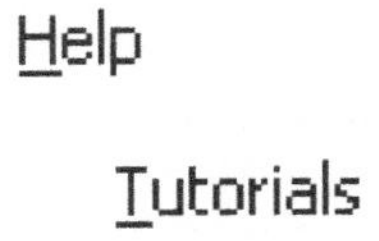

Help

Tutorials

or

Minitab Help

Basics

Intro to Minitab

Tutorials

Using Help

Minitab Windows

Navigate to the One Way ANOVA Tutorial by clicking on the contents tab and then select ANOVA <<One Way ANOVA.

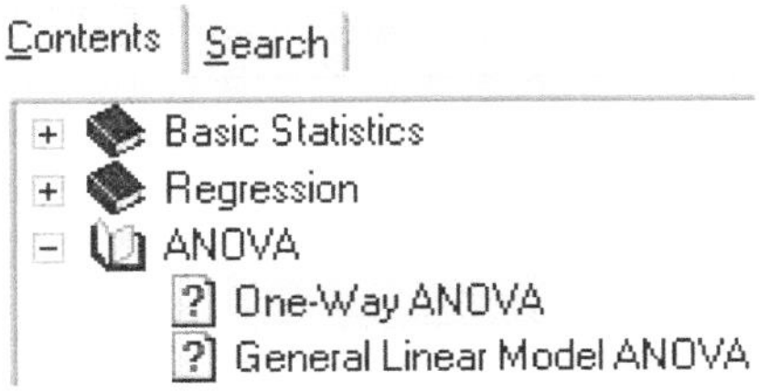

Contents | Search

- Basic Statistics
- Regression
- ANOVA
  - One-Way ANOVA
  - General Linear Model ANOVA

The Tutorials have three main sections which are accessed via three discrete looking tabs. The discrete tabs are Uses, Data and How To. The Back and Forward navigation buttons are also available.

Uses tells us how the procedure can be applied and then gives an example scenario.

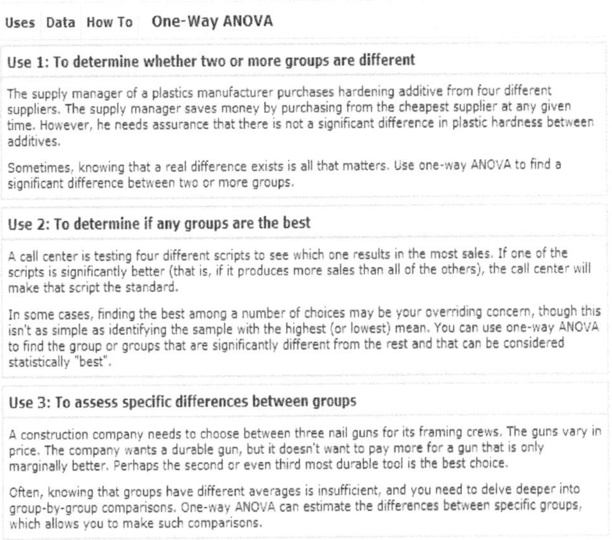

Uses Data How To **One-Way ANOVA**

Use 1: To determine whether two or more groups are different

The supply manager of a plastics manufacturer purchases hardening additive from four different suppliers. The supply manager saves money by purchasing from the cheapest supplier at any given time. However, he needs assurance that there is not a significant difference in plastic hardness between additives.

Sometimes, knowing that a real difference exists is all that matters. Use one-way ANOVA to find a significant difference between two or more groups.

Use 2: To determine if any groups are the best

A call center is testing four different scripts to see which one results in the most sales. If one of the scripts is significantly better (that is, if it produces more sales than all of the others), the call center will make that script the standard.

In some cases, finding the best among a number of choices may be your overriding concern, though this isn't as simple as identifying the sample with the highest (or lowest) mean. You can use one-way ANOVA to find the group or groups that are significantly different from the rest and that can be considered statistically "best".

Use 3: To assess specific differences between groups

A construction company needs to choose between three nail guns for its framing crews. The guns vary in price. The company wants a durable gun, but it doesn't want to pay more for a gun that is only marginally better. Perhaps the second or even third most durable tool is the best choice.

Often, knowing that groups have different averages is insufficient, and you need to delve deeper into group-by-group comparisons. One-way ANOVA can estimate the differences between specific groups, which allows you to make such comparisons.

Data gives us the data type requirements for the procedure and it also gives us a graphical display of how the data should be laid out.

Uses | Data | How To **Data for One-Way ANOVA**

What kind of data is required?

You must have continuous data that are divided into two or more groups. For example:

- Strength measurements for samples of four different brands of wire
- Customer satisfaction ratings for six different locations of a hotel chain
- Length of parts produced on three different manufacturing lines

This data can be organized as either of the following:

- A single stacked column of response data, with a corresponding column identifying the groups
- Unstacked response data where each group is in a separate worksheet column

Your samples must be independent—that is, the observed values in one sample do not depend on the observed values of another sample. Unlike a t-test, ANOVA can compare more than two groups.

What should my worksheet look like?

Stacked data

The continuous response data for all of the groups are in a single column (in this example, Strength) with a column of numeric, text, or date/time data that identifies the group of each observation (in this example, Additive).

	C1	C2
	Strength	Additive
1	9.5	1
2	8.4	1
3	9.7	1
4	9.3	2
5	8.7	2
6	9.1	2
7	9.6	3

How To is a detailed explanation for setting up the test procedure. However, as stated earlier it only goes up to the point of execution. The interpretation of results is given within the StatGuide and there is a link to the StatGuide at the bottom of the How To section.

Again, to allow you to work through the example Minitab tells you that the data set is PAINT.MTW. This is a different file to that given within the Help menus. As stated earlier, we will look at getting the example data sets later in the module. We will now have a look at the StatGuide.

Uses | Data | How To **Performing a One-Way ANOVA in Minitab**

Scenario

Investigators compared the hardness of four different blends of paint. Six samples of each paint blend were applied to a small piece of metal, cured, and measured for hardness.

Choose the appropriate analysis

A one-way ANOVA is appropriate for a comparison of the means because you have one factor (paint blend) and that factor has more than two levels (four paint blends). You also have a continuous response variable (paint hardness). The data are unstacked because each group's data are in separate columns. Consequently, you choose the unstacked version of the analysis. If the response data had been stacked in one column, you would have selected One-Way.

1. Open the worksheet PAINT.MTW.
2. Choose **Stat** > **ANOVA** > **One-Way (Unstacked)**.

Interpreting the output

Now what? For guidance on interpreting the results of this analysis, see StatGuide.

11.4 StatGuide

The main ways of getting to the StatGuide for One Way ANOVA are to click Help <<StatGuide or the StatGuide icon and then find the relevant topic from any of the tabs. The other is to get there via the relevant Tutorial.

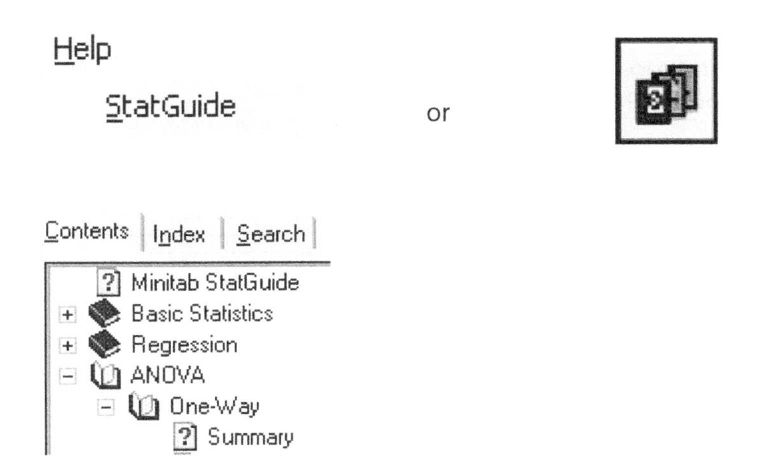

The StatGuide gives a very detailed explanation of the test results. Initially, it begins with a summary of the procedure but then it goes through the interpretation of the results for the example started within the Tutorial. For One Way ANOVA it continues with the PAINT.MTW example.

A one-way analysis of variance (ANOVA) tests the hypothesis that the means of several populations are equal. The method is an extension of the two-sample t-test, specifically for the case where the population variances are assumed to be equal. A one-way analysis of variance requires the following:

- a response, or measurement taken from the units sampled.
- a factor, or discrete variable that is altered systematically. The different values chosen for the factor variable are called levels of the factor. Each level of the factor in the analysis corresponds to a larger population with its own mean. The sample mean is an estimate of the level mean for the whole population.

A one-way ANOVA can be used to tell you if there are statistically significant differences among the level means. The null hypothesis for the test is that all population means (level means) are the same. The alternative hypothesis is that one or more population means differ from the others.

In addition to helping you evaluate whether all the level means are the same, Minitab also provides output to help you determine which level means are different when differences exist.

On the top right corner of the StatGuide window are the navigation arrows to move through different sections of the results. If you click on Topic a dialog box opens showing the breakdown of the procedure explanation. The window in the background is Graph Residuals versus Fits. Clicking the left arrow will make the section go back one stage onto Graphs Normal Probability Plot of Residuals.

It is also possible to use the Contents tab to open all the topics and select them by clicking on the section required.

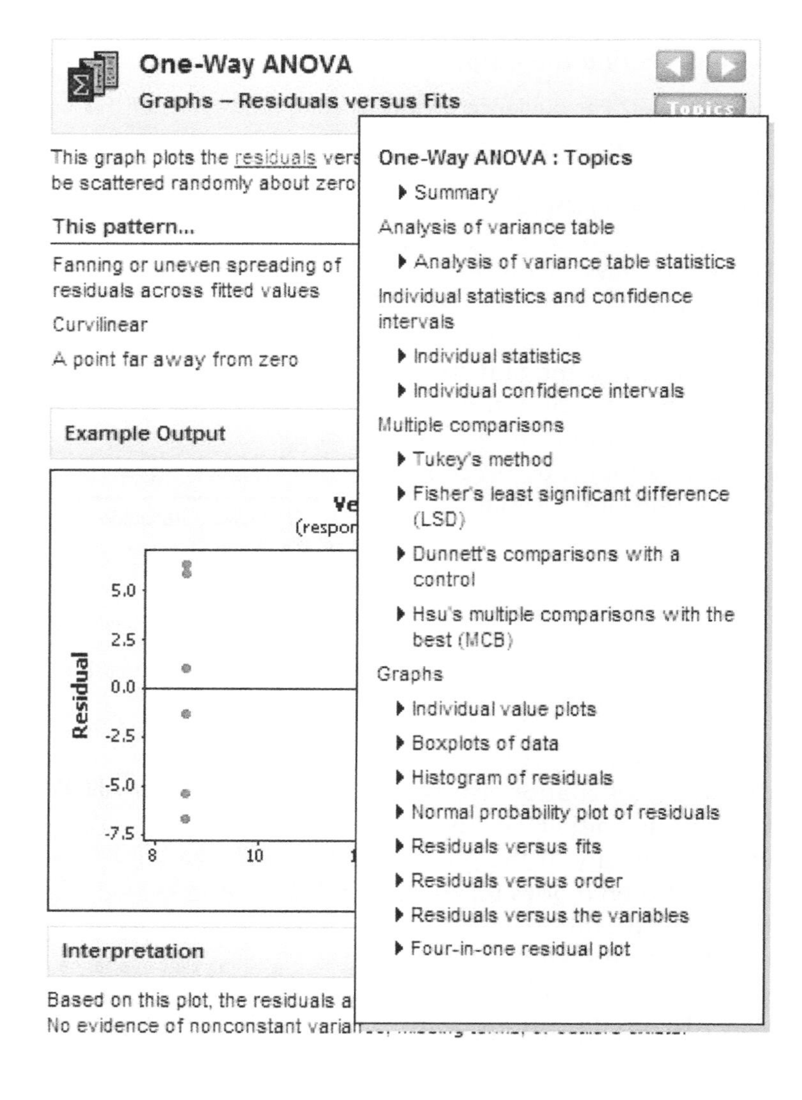

11.5 Methods and Formulas

To navigate to Methods and Formulas click on Help <<Method and Formulas or select it from the main help menu.

Help

Methods and Formulas

or

References

Examples

Glossary

Macros

Methods and Formulas

Session Commands

Minitab Automation

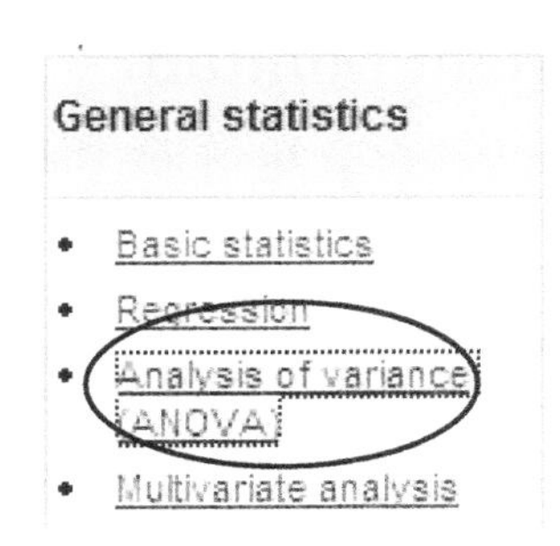

In order to then Navigate to the Methods and Formulas for One Way ANOVA click on the links as shown. Again, it is a very logical way of navigating to the required procedure.

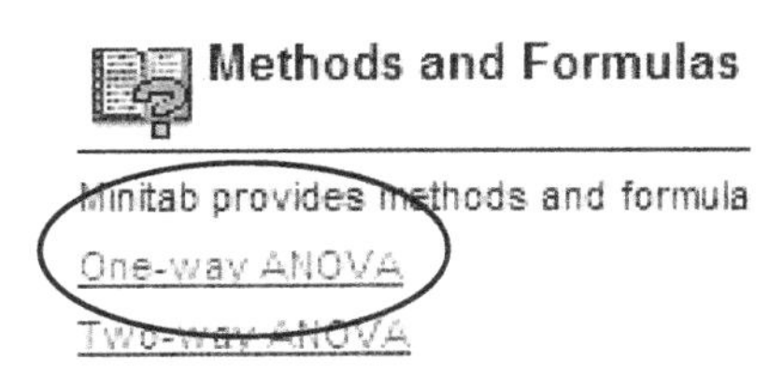

All of the Methods and Formulas for One Way ANOVA are shown under subheadings.

A single click on one of the links will open the requested information in the window below the headings. We have opened the formula for calculation R-Sq(adj).

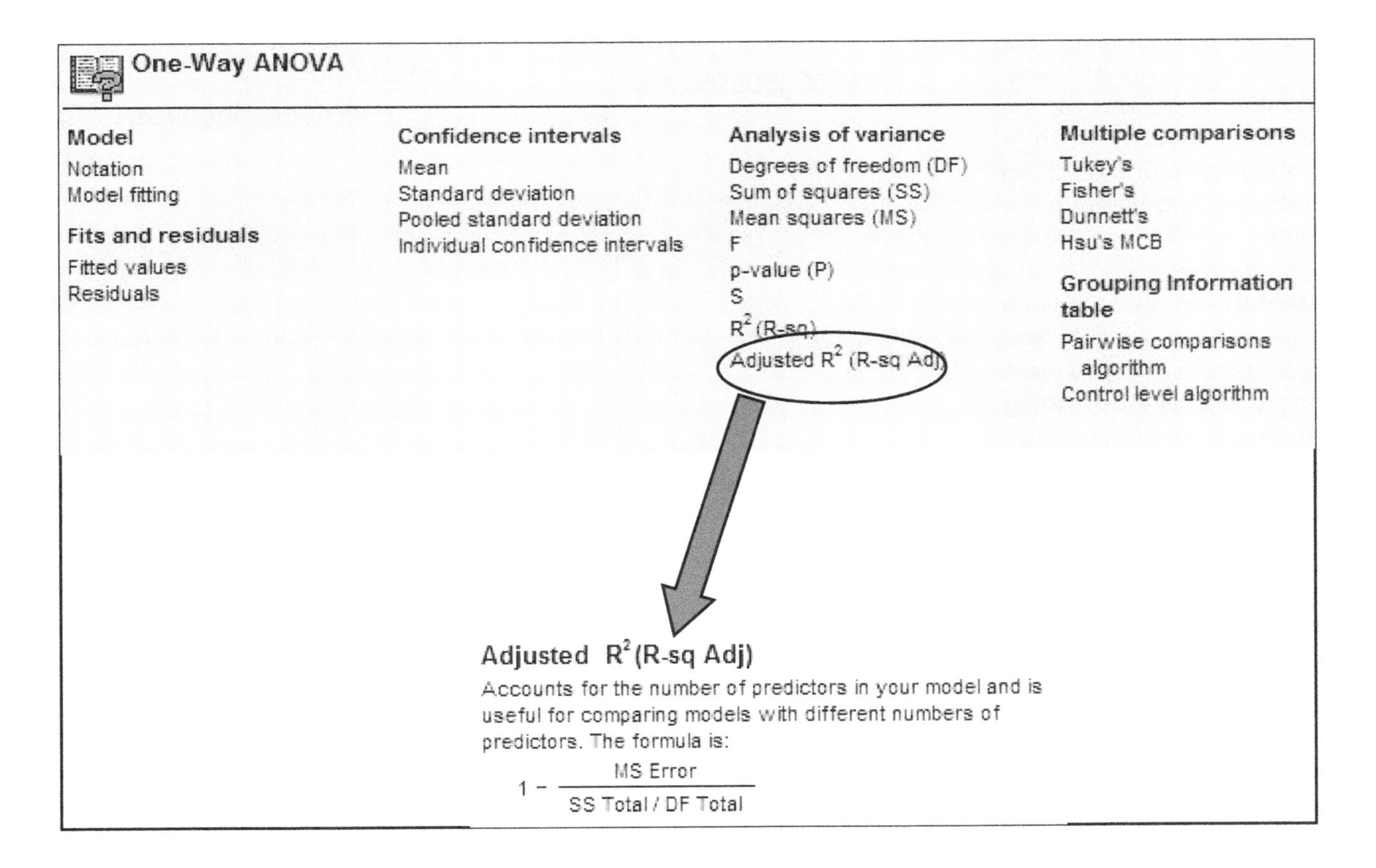

11.6 Meet Minitab

Meet Minitab is a booklet that helps the new user get started with Minitab. You may have got a copy with the Disc of Minitab or you can download the latest copy from the Minitab website by clicking Help <<Meet Minitab (on an internet enabled PC). This link will take you straight to the download page for Meet Minitab. Meet Minitab also contains fully worked examples and graphics. The folder of data sets for Meet Minitab will be with the Minitab sample data files.

Help

Meet Minitab

Minitab Documentation

Meet Minitab

Orient yourself to Minitab 16 and quickly begin using it to analyze your project data with this concise PDF guide.

11.7 Help on the Web

Minitab also offers extensive support on the internet. This is mainly available on their website at www.Minitab.com . This can be accessed directly with a web browser or by clicking Help << Knowledgebase/FAQ.

Minitab is also has a presence on Twitter and Facebook in case you were interested. We are going to stick with the help on their website for now.

Navigate to the Knowledgebase and enter One Way ANOVA into the Search Facility and click on Search.

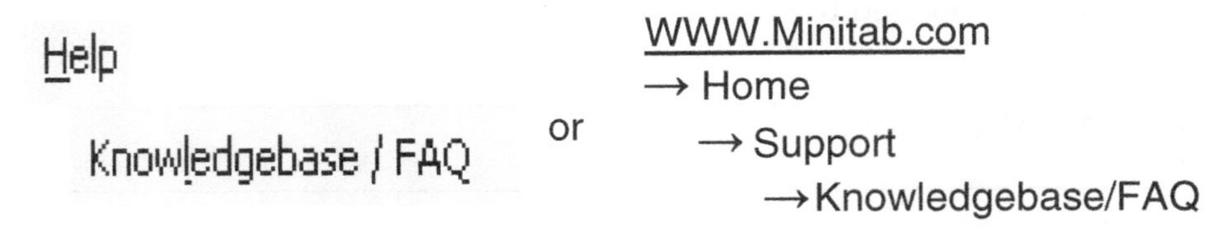

Knowledgebase / FAQ

Find the answers to many technical and customer support questions FAQ.

▸ Top Ten FAQ

Search All FAQ

One-Way ANOVA | Search

▸ Help with Search

Browse FAQ by Product

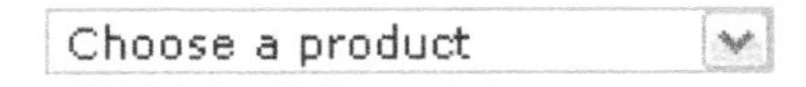

The Knowledgebase brings up a list of documents and FAQs relating to my query.

If I feel a document is relevant I can click on the link in the header to open it. ID 1267 has been opened to show you a part of the solution.

If you register with Minitab you can also send questions to the Help Team which they will answer promptly.

ID 1111: Calculating the confidence intervals for the means ...
... Calculating the confidence intervals for the means displayed with **one-way ANOVA** - ID 1111. ... 1. Click the link, **One-way ANOVA** Confidence Intervals Macro, below. ...

Calculating Levene's Test Using One -way ANOVA
... Page 1 Technical Support Document Calculating Levene's Test Using **One-way ANOVA** ... Technical Support Document Calculating Levene's Test Using **One-way ANOVA** ...

Calculating Power for One-Way ANOVA "By Hand"
... In Minitab Release 14 or later, Stat > Power and Sample Size > **One-Way ANOVA** gives the following output: MTB > Power; SUBC> **OneWay** 3; SUBC> Sample 10; SUBC ...

ID 1267: Formula for 95% confidence intervals in One-Way ANOVA ...
... Formula for 95% confidence intervals in **One-Way ANOVA** - ID 1267. Revised: 7/8/2008. ... How can I calculate the 95% confidence intervals for **One-Way ANOVA** "by hand?" ...

How can I calculate the 95% confidence intervals for One-Way ANOVA "by hand?"

Solution

The confidence interval for the mean of the ith factor level is calculated by:

$Mean_i \pm t_{[0.975,N-r]} * S_p / sqrt(n_i)$

where:
N is the total number of observations
r is the number of factor levels
n_i is the number of observations for the ith factor level
S_p is the pooled standard deviation

I want to show a couple of the more interesting FAQs. This one is ID-2613 and contains links to the Minitab White Papers for the Assistant. The White Papers contain the results of research conducted by Minitab into the Methods used by the Assistant and what conditions are required to validate the tests. They contain interesting background reading.

Description

What methods are used to perform the analyses and the Report Card checks in the Assistant in Minitab 16?

Solution

The Assistant in Minitab 16 was specially designed to help investigators in the field of quality to quickly select the most appropriate analysis for their data and to readily interpret the results. For this reason, we developed and implemented statistical methods that allow quality investigators to efficiently analyze their data, verify the assumptions for the analysis, and check the reliability of the results. The results are then presented in the Assistant's uniquely designed Reports and Report Cards.

The methods used by the Assistant are based on established statistical practice and theory, referenced guidelines in the literature, and extensive simulation studies performed by statisticians in Minitab's Research and Development department. You can learn more about the specific methodology and supporting research underlying each Assistant analysis by clicking the links to each white paper below.

Variables Control Charts (Xbar-R, Xbar-S, I-MR)

Attributes Control Charts (P chart, U chart)

Gage R&R Study (Crossed)

Attribute Agreement Analysis

Capability Analysis (Normal)

Binomial Capability Analysis and Poisson Capability Analysis

1-Sample t (also applies to Paired t)

2-Sample t

1-Sample % Defective

2-Sample % Defective

Chi Square Goodness of Fit, Chi-Square Test for Association, and Chi-Square % Defective

One-Way ANOVA

Regression

1-Sample Standard Deviation

2-Sample Standard Deviation and Standard Deviations Tests

11.8 Help on the Web and Datasets

This one is ID-2241. It contains useful information about the Minitab data sets. It also contains a file that is worth downloading. The file contains the names of the example data sets used within the StatGuide and Help system. A section of the spreadsheet is shown below. You can see the file names relating to One Way ANOVA.

Solution

Yes, sample data sets are included with Minitab. To view descriptions for each data set:

1. Choose **Help** > **Help**.
2. Click the **Contents** tab.
3. Browse to **Introduction** > **Data Sets** >**Sample Data Sets** > **Data set descriptions**.

For a spreadsheet of all sample data sets and the analyses to which they correspond, click the link, Sample Data Sets, below.

For instructions on opening a Minitab sample data set, see Knowledgebase ID 2108.

For instructions on accessing the Minitab sample data sets without installing Minitab, see Knowledgebase ID 808.

Related Documents

- **Sample Data Sets**
- **Knowledgebase ID 2108** (http://www.minitab.com/support/answers/answer.aspx?ID=2108)
- **Knowledgebase ID 808** (http://www.minitab.com/support/answers/answer.aspx?ID=808)

B	C	D
Minitab Tool	**StatGuide Data Set Name**	**Help Data Set Name**
ANOVA- Main Effects Plot	Alfalfa.MTW, Driving.MTW	Alfalfa.MTW
ANOVA- One-Way ANOVA	PaintHardness.MTW	Exh_AOV.MTW
ANOVA- One-Way ANOVA (Unstacked)	Paint.MTW	
ANOVA- Test for Equal Variances	Driving.MTW	Exh_AOV.MTW

11.9 Datasets

To open one of the example datasets click on File <<Open Worksheet. You can then either click on the Look in Minitab Sample Folder button or you can manually navigate to where the sample data files are stored.

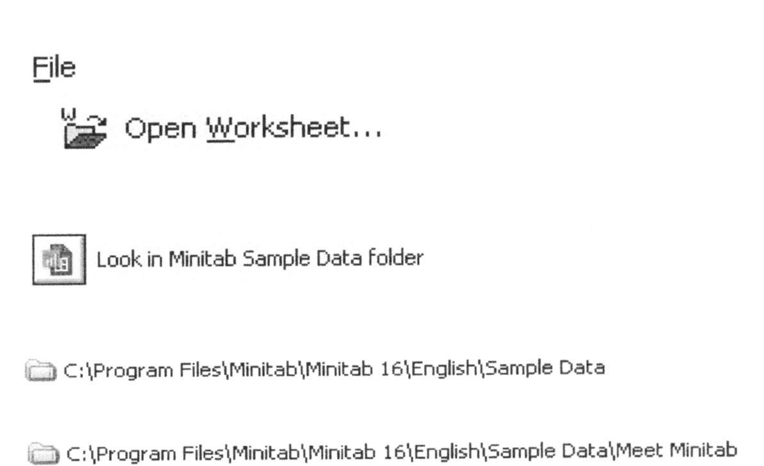

I have shown the location where my sample data files are stored and the location where the Meet Minitab data files are stored. However, depending on how you installed Minitab you may have a different location.

This ends the chapter and the book. I hope I have completed my task and opened the world of statistical problem solving using Minitab for you.

Index

Problem Solving and Data Analysis using Minitab: A clear and easy guide to Six Sigma methodology, First Edition. Rehman M. Khan.